ON THE PERSPECTIVE OF PAINTING

The MIT Press's publishing mission benefits from the generosity of our donors, including Samuel Jay Keyser.

On the Perspective of Painting

Piero della Francesca and Rocco Sinisgalli

Prologue by Martin Kemp

The MIT Press
Cambridge, Massachusetts
London, England

Rocco Sinisgalli and the MIT Press are honored to thank the Tavolozza Foundation for a generous grant that helped to support the production of this book.

The MIT Press
Massachusetts Institute of Technology
77 Massachusetts Avenue, Cambridge, MA 02139
mitpress.mit.edu

The MIT Press would like to thank the anonymous peer reviewers who provided comments on drafts of this book. The generous work of academic experts is essential for establishing the authority and quality of our publications. We acknowledge with gratitude the contributions of these otherwise uncredited readers.

This book was set in ITC Berkeley Oldstyle and Helvetica Neue LT by the MIT Press.
Printed and bound in China.

Library of Congress Cataloging-in-Publication Data

Names: Piero, della Francesca, 1416?–1492, author. | Sinisgalli, Rocco, 1947– translator, writer of added commentary. | Kemp, Martin, writer of prolog. | Piero, della Francesca, 1416?–1492. *De prospectiva pingendi*.
Title: On the perspective of painting / Piero della Francesca and Rocco Sinisgalli ; prologue by Martin Kemp.
Other titles: *De prospectiva pingendi*. English
Description: Cambridge, Massachusetts : The MIT Press, 2025. | Includes bibliographical references and index.
Identifiers: LCCN 2024061955 | ISBN 9780262049993 (hardcover)
Subjects: LCSH: Perspective—Early works to 1800.
Classification: LCC NC749 .P5413 2025 | DDC 750.1/8—dc23/eng/20250529
LC record available at https://lccn.loc.gov/2024061955

10 9 8 7 6 5 4 3 2 1

EU Authorised Representative: Easy Access System Europe, Mustamäe tee 50, 10621 Tallinn, Estonia
Email: gpsr.requests@easproject.com

Cover image: Piero della Francesca, *The Madonna and Child with Saints, Angels and Federico da Montefeltro*, © Pinacoteca di Brera, Milan

Contents

This book is dedicated to the memory of Henry Millon,
Dean Emeritus of the Center for Advanced Study in the Visual Arts (CASVA)
at the National Gallery of Art, Washington, DC.

Prologue

Martin Kemp

Precursors

Perspective was hugely important for painters in the *avant-garde* of the Renaissance. Piero's treatise was the first to be devoted specifically to this subject. The optical procedures of perspective not only provided painters with a systematic formula for the creation of the illusion of space on a flat surface but also allowed them to claim that painting is founded on a "theory" of a mathematical nature, to rival the ancient harmonic theories of music (a designated "liberal art," unlike painting). Indeed, the term *perspectiva* in Latin or *prospectiva* in Italian was in a broader sense the name for the science of optics. We translate the title of Piero's treatise as "The Perspective of Painting," when it could be understood in the fifteenth century as "The Optics of Painting."

The first demonstration of what we call perspective was not in a normal painting. It was accomplished by 1413 by the sculptor and architect Filippo Brunelleschi. He showed how the octagonal Baptistery in Florence could be rendered geometrically on the flat surface of a panel. It is likely that he used surveying techniques to project the building onto what we would now call the picture plane. He set up his resulting painting as a kind of peep show. He created an eyehole in the back of the panel through which we could look in a mirror at his painted rendering. The effect of the lost panel seems to have been very beguiling, but it was not of direct utility for painters making images of their normal subjects. Hence the delay in his invention being taken up. It was first comprehensively applied around 1426–1427 by Masaccio in his wall painting of *The Trinity*, in which the figures occupy an illusionistic space, probably plotted by techniques directly related to Brunelleschi's procedures. The result was a symmetrical view into a kind of chapel in which the receding horizontal lines of the architecture *within* the mural converged on what came to be categorized as a "vanishing point." God the Father, the Dove, and the crucified Christ inhabit this pictorial illusion, while the donors kneel at the threshold of the sacred space.

Perhaps the most sophisticated development of what Masaccio was doing was the *St. Lucy Altarpiece* by Domenico Veneziano in which a complicated tiled floor was rendered in meticulous perspective. Domenico provides a compelling space for the figures as well as working with the surface zones of the panel that are assigned to the protagonists. At the center is the Virgin with her Child, flanked by the Florentine saints, John the Baptist and Zenobius, together with St. Francis and Lucy. The importance of Domenico in our present context is that in 1439 he and Piero della Francesca were working together in Florence on lost frescoes.

The Painter-Mathematician

Piero's family were minor nobility in Borgo Sansepolcro, and he would have received a better education than most painters. He did not remain in Florence, later working in Sansepolcro and in a surprising number of locations, Rimini, Rome, Arezzo, Ancona, Pesaro, Bologna, and not least Urbino. It was in that lovely Renaissance hill town that he produced his most notable set of surviving works, the famous *Flagellation*, the *Senigallia Madonna*, the pendent portraits of *Federico da Montefeltro* and *Battista Sforza*, and the *Brera Altarpiece*, at which we will be looking further. Alongside his activities as a painter, Piero produced sophisticated manuscripts on mathematics: a treatise on applied mathematics (the *Trattato d'abaco*); an anthology of mainly geometrical essays by Archimedes; and two important works destined for the fine library of duke Federico da Montefeltro in Urbino, *De prospectiva pingendi* and *De quinque corporibus regularibus* (On the Five Regular Solids). These professional achievements in mathematics separate him decisively from other early authors who wrote about perspective, most notably Lorenzo Ghiberti, the sculptor, who compiled a selection of optical texts in his *Commentaries*, and even Leon Battista Alberti in his *Della pittura / De pictura* (On Painting), who devoted the first part of his treatise to pictorial space. Alberti provided a highly simplified account of optics inside and outside the eye, together with an oddly unrelated account of how to draw a basic perspectival projection of a tiled floor on which objects could be placed in due proportion.

For those who wished to endow painter's perspective with a scientific base, there were two main streams of optical science. The first arose in Greek antiquity with the treatise *On Optics* by Euclid, which centered on the geometry of light. The later *Optics* by Ptolemy ranged more widely over visual phenomena. The second tradition was created by Islamic scholars, above all Ibn al-Haytham (Alhazen), who looked in detail at the operation of the eye and the science of light in direct vision, reflection, and refraction. The latter was transmitted into Western medieval philosophy, not least in John Pecham's standard textbook *Perspectiva communis* in the thirteenth century. The tradition from al-Haytham to Pecham was much concerned with the observational and experimental dimensions of optics, including the "errors of sight." Of the various sciences in medieval natural philosophy, none was more sophisticated than optics, which was seen to manifest God's order in visible and detectable form.

Although it is likely that Piero was acquainted with medieval optical theory, the tone of his treatise was Euclidian, expounding with geometrical insistence the pure mathematics involved in the spatial plotting of geometrical figures onto foreshortened surfaces. Euclid is specifically cited six times in the first part or book of the treatise. Piero clearly admired the crystalline purity of Euclid's logic, not least his systematic building from first principles. Accordingly, Piero begins with simple flat configurations, culminating in the plan of a square building (Book 1, Proposition XXVII). In the second part of *De prospectiva pingendi*, Piero looks at solid bodies, beginning with a cube and extending into more complex subjects, like a house or temple (Book 2, Proposition IX). Book 3 constructs "as many bodies as possible." In reality, Piero looks at a few complex and difficult forms, including a torus (a faceted ring or *mazzocchio*), a twisted cube, bases and capitals of columns, human heads, and a half-dome (which proves to be difficult).

Complex shapes, which are not feasibly constructed as geometrical diagrams on foreshortened planes, are subject to a "second path." This involves the projection of designated points on a solid body onto a plane or intersection as seen from a fixed viewpoint. In the case of a human head, for example, "circuits" of points on horizontal bands around the head give rise to optical rays that converge on a designated point (the eye of the observer). The resulting pyramid of rays is cut by a flat intersection, on which the position of each ray is recorded. This cage-like array of circuits can be tilted at will to show a head foreshortened in any direction—as is required for the sleeping soldiers in Piero's *Resurrection* in the town hall of Borgo Sansepolcro. He tells us about the tools needed for such precise plotting of rays separated by very tiny distances. "Threads of equine silk" will represent light rays. Straight rulers of wood and paper are to be used. The paper ones are good for recording numbers and letters. Even with such precision, the tightly packed rays and points on the intersection hardly seem feasible, even for Piero. But for him precise accuracy seems to be an ethical matter whether or not the spectator can see the minutiae of the results. The artist must strive to be minutely accurate, even if only God could confirm that this is the case.

All of this serves to make us see Piero's art as dominated and even validated by geometrical optics of a linear kind. But, as he acknowledges at the start, what he likes to call "proportioning" is just one part of the optics of painting. He chooses this term, which means that he is concerned with the proper proportional representation of objects according to size and distance. "Proportioning" indicates that he is not dealing with all the wider questions of the optics of light. This is what he says:

> The whole theory of painting requires three parts: drawing, proportioning, and coloring. Drawing is the tracing of profiles and of outlines found in objects. Proportioning is the collocation of these profiles and outlines in their place through a process of proportioning. Coloring is the application of colors to objects which, whether light or dark, reveal themselves according to the varieties of lights. In this work, for the moment, we shall only talk about the proportioning, what is called perspective, mixing it also with a certain part of drawing. . . . We shall omit coloring, dealing only with that part which comes to describing objects through lines, angles, and proportions.

This is much in line with medieval optical texts, which typically deal with three major aspects of *perspectiva*: the passage and reception of direct light; the geometry of reflections from flat, concave, and convex surfaces; the behavior of refraction when light passes from one substance to another. We should not assume, because Piero writes on "proportioning," that he does not accord importance to all other parts of optics. Indeed, if we look intensely at how he uses light, we can see that he never neglects "coloring," that is to say, "the varieties of light." Let us try to do so by analyzing just one great picture, *The Madonna and Child with Saints, Angels and Federico da Montefeltro* (known as the *Brera Altarpiece*), in the Pinacoteca di Brera in Milan.

Perspective in Practice

We have prepared ourselves for the perspectival features, so we will begin in this more familiar territory. The very up-to-date architecture in the *Brera Altarpiece* was to find a suitable home in the church of San Bernardino in Urbino, designed by Francesco di Giorgio as the mausoleum for the duke. Francesco was a painter, sculptor, engineer, writer, and architect from Siena. Federico was concerned to attract the best Renaissance talents to his glittering court.

The perspective of the architecture is drawn with great precision, in keeping with Piero's elaborate procedures in his second book. The vanishing point lies within the Virgin's face. It may be that the subtly rotated heads of the angels were plotted with the intersection method of his third book. Piero was very meticulous about such things.

The painted building is set up as the east end and crossing of a compact church. In the upper corners of the panel, we can just see the cornices of the "nave," within which we stand. The sliver of a pilaster on the left signals that the

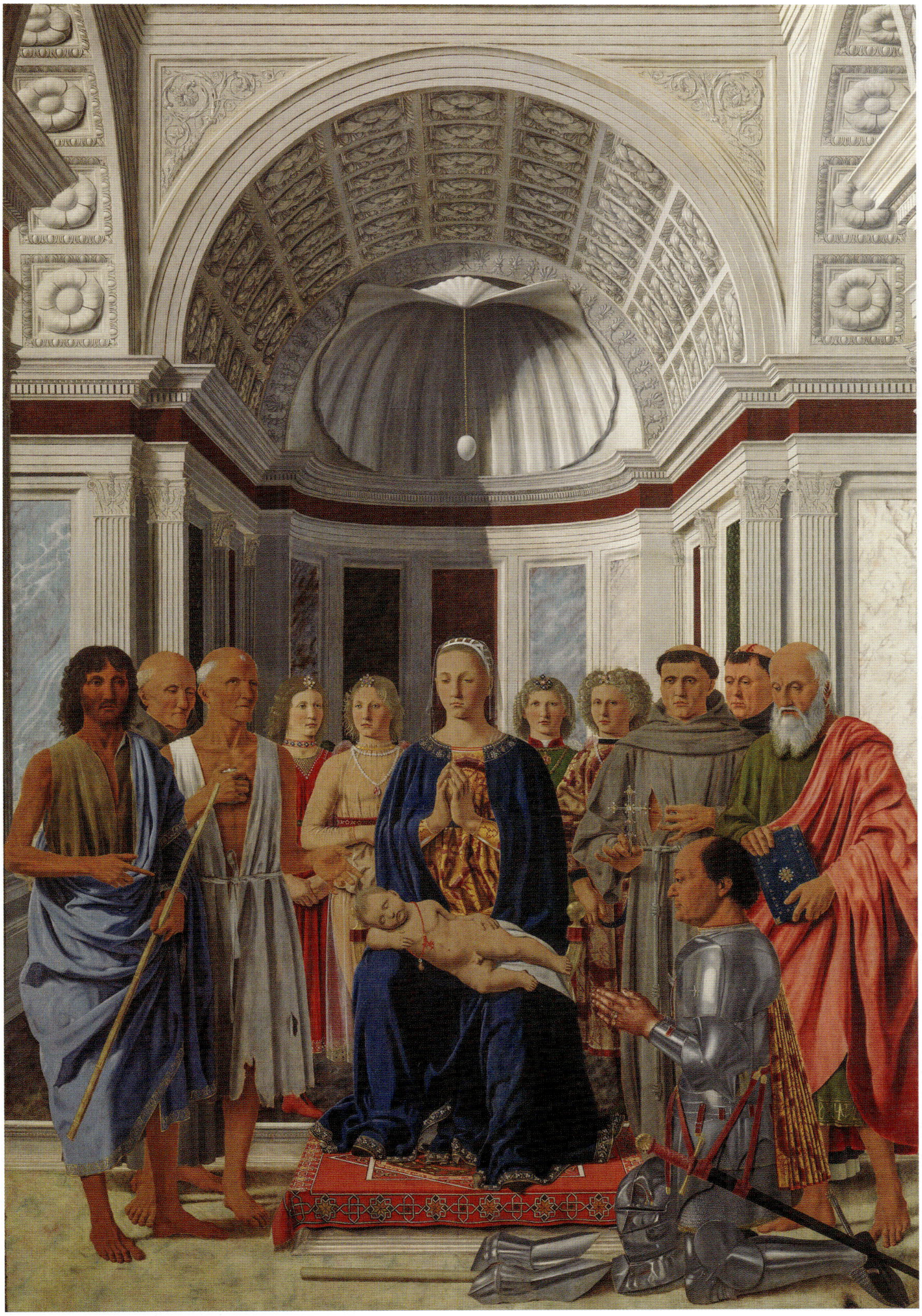

Figure 0.1 *The Madonna and Child with Saints, Angels and Federico da Montefeltro*, © Pinacoteca di Brera, Milan.

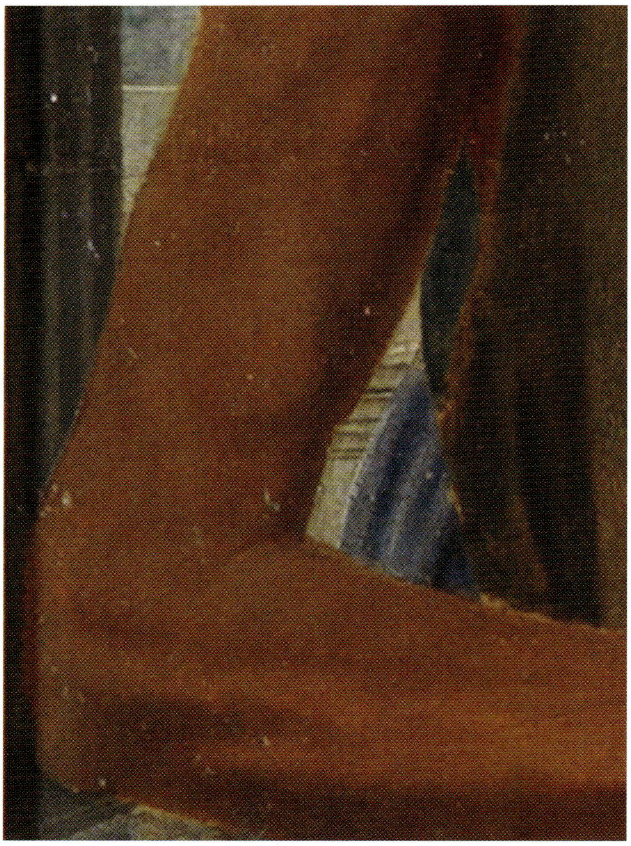

Figure 0.2 The elbow of St. John the Baptist and the pilaster of the crossing, *The Madonna and Child with Saints, Angels and Federico da Montefeltro*, © Pinacoteca di Brera, Milan.

Baptist, at the left end of the group, is on our side of the crossing, and that the curved group of figures stand on our side of the crossing rather than on the other side of it.

Indeed, our immediate impression is that the holy group occupies an area that extends at the back to the far side of the crossing. This play of logical depth and apparent closeness of the figures is similar to Domenico Veneziano's *St. Lucy Altarpiece*. Even more ambiguous is the egg hanging (apparently) vertically above the Virgin's head. We "see" it as directly above her, but it must be considerably deeper in the space of the niche at the back of the "choir." In fact, the egg must be very big. It is an ostrich egg, which was associated with Virgin birth and was not uncommon in Renaissance chapels.

What is Piero doing with this combination of spatial precision and spatial ambiguity? The answer is that his architecture asserts itself competitively with the real space of the actual building using his divine geometry. But the painting has to work above all as a picture in which the settings, real and illusionistic, make the best effect in presenting the duke kneeling before the Virgin and Child with her accompanying saints and angels. We are close, but not too close. The duke is closer but is not mingling. Decorum is preserved.

Surface Reflections

The most charming of all the reflections is that in the white egg that hangs from the beak at the center of the hinge of a scallop shell. Illuminated quite shallowly from the upper left, the left side of the egg is caressed by a strongly directional light. The curve of the shadow on the right of the egg is overwhelmed by reflected light from the niche and wall to the right. Such back reflections were not much liked by many artists, as Leonardo da Vinci testified. He tells us that the "theory" of reflected lights in shadows "has been put into practice by some, but there are many others who avoid it, and each side laughs at the other." The architecture is comprehensively infused with back reflections. A good example is in the underside of the cornice above St. Jerome at the right end of the group, which is notably bright, particularly if compared to that on the left.

The boldest of the reflective surfaces are the curving plates of Federico's armor. The most conspicuous is in the pauldron of his shoulder and upper left arm, which reflects an arched window somewhere in the nave of the chapel. Other less complete gleams shine in strips and curves on the plate armor. In the optical texts these brightest highlights were accorded the names of *splendore* and *lustro*. They behaved differently from other highlights. Their glare overcame the color of the surfaces and moved in response to the position of the observer.

The geometry of such mirrored reflections on convex surfaces played a conspicuous role in medieval optical texts. One intractable problem was specifically named after Alhazen: "find the point on a spherical convex mirror at which a ray of light coming from a given point must strike in order to be reflected to another given point."[1] The painting is richly endowed with little convex mirrors of crystal and pearl in the costumes of the Virgin and her angels. The costume of Piero's *St. Michael* in the National Galley, London, is adorned with translucent oval gems, which continue to gleam with refracted light even in the shadows. Piero clearly enjoyed painting such delicate delights.

Passing Through

The third of the major divisions in optical science was what we call refraction. That Piero was very alert to the particular phenomena of light passing through translucent materials is shown by the ointment flask held by Mary Magdalene in the fresco of the saint in the cathedral at Arezzo. On the left

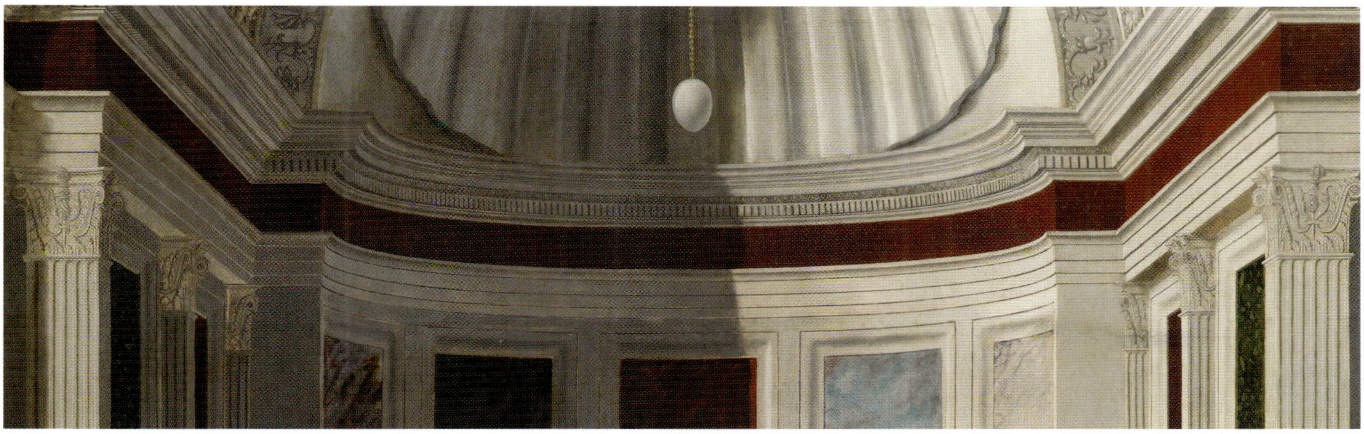

Figure 0.3 Back reflections and shadows in the lateral cornices, *The Madonna and Child with Saints, Angels and Federico da Montefeltro*, © Pinacoteca di Brera, Milan.

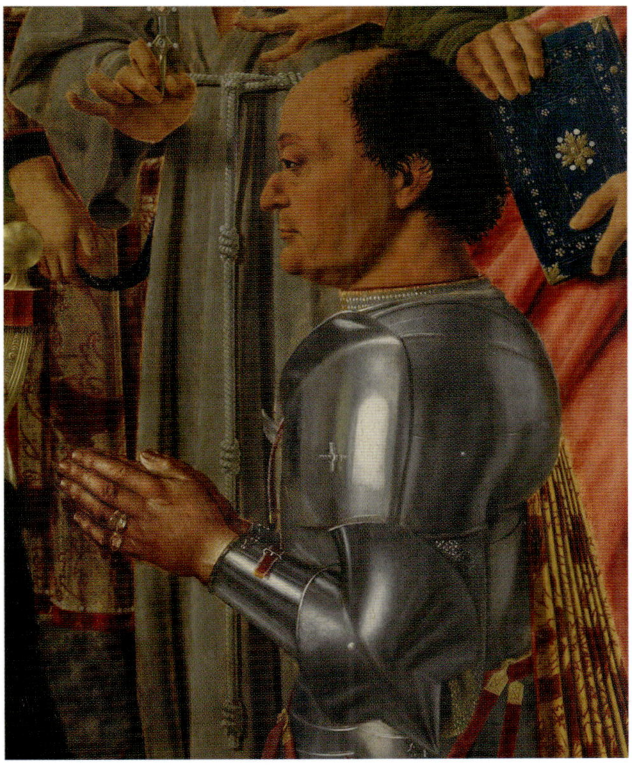

Figure 0.4 Federico's armor, *The Madonna and Child with Saints, Angels and Federico da Montefeltro*, © Pinacoteca di Brera, Milan.

of the flask, a series of vertical highlights denote the reflections of windows in the nave of the cathedral, while the light that has passed through the glass and transparent ointment results in a diffuse radiance across the right of the vessel. The difference between reflection and refraction is succinctly described. In the *Brera Altarpiece*, comparable reflections and refractions are present in the crystal cross held by St. Francis close to the stigmatum on his chest.

Larger-scale phenomena of reflection and refraction are very apparent in the watery narrative of Piero's *Baptism of Christ* in the National Gallery, London. Christ stands in shallow water, signaled by thin bands of white impasto above his ankles. Close to the picture, we see through the water to its stony bed. As our eye moves up and into more distant areas of the river, we see that colored objects are reflected on its surface. Piero has realized that we can see through water when we look at a steep angle, while reflection takes over at shallow angles. Roger Bacon in the thirteenth century noted that "there are certain bodies of moderate density, such as water, in which reflection and refraction occur simultaneously. For, by refraction, we see fish and stones in it, and by reflection the sun and moon, as experience teaches."[2] The transition between refraction and reflection is not as sharply defined as Piero indicates, but that has not deterred him from seeking to define an optical rule.

Some idea of how subtle were Piero's optical observations can be gleaned from the *Senigallia Madonna* in the Galleria Nazionale in Urbino. Behind the Virgin, a beam of light streams though the window. If we look very closely, we can just see that the beam is composed from a myriad of tiny spots of dancing light. What we are seeing are suspended particles of dust. This phenomenon had been noted by the Roman author Lucretius, and by Albertus Magnus and Dante. Piero is the first to paint it. For Dante, the particulate light shaped itself into a cross. Does it signal divine light in Piero's *Madonna*?

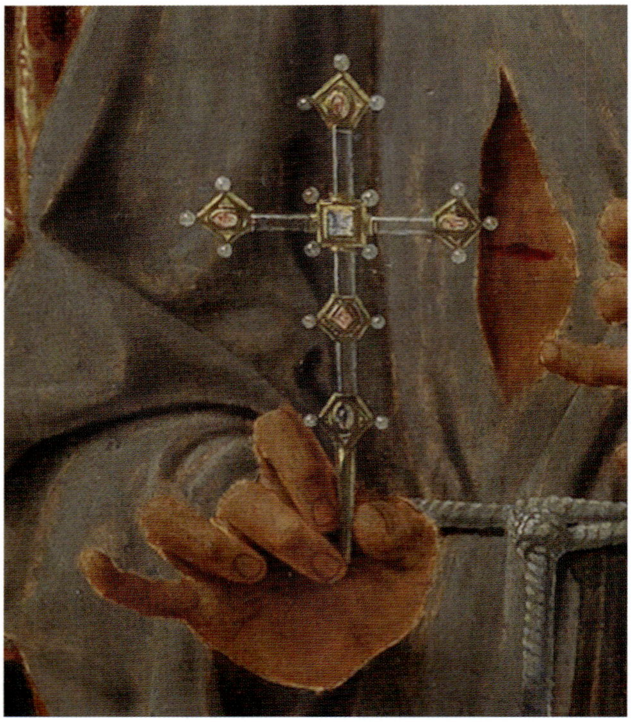

Figure 0.5 St. Francis's cross, *The Madonna and Child with Saints, Angels and Federico da Montefeltro*, © Pinacoteca di Brera, Milan.

In fact, Piero's meticulous science of light caused problems for the depiction of the divine light that often accompanied miraculous events in sacred art. If he portrayed the sublime light of optics in nature, what more was left for divine light? Piero's answer, like that of Dante, is that divine light did not follow the rules. In the *Flagellation* an irrational source of light arises to the right of Christ, flooding the coffers of the ceiling in the forecourt to Pilate's palace. Only Christ sees this light. When a diving angel is accompanied by an intense nighttime glare in the *Dream of Constantine* in the basilica of San Francesco, Arezzo, only Constantine sees the light, although his eyes are closed. Also in Arezzo, we see God the Father's hands casting diagonal rays of divine light toward the Virgin to signal her impregnation by the Holy Spirit.

Throughout his art, Piero achieves a special kind of visual rapture, compounded jointly from the divine geometry of light and its sensory manifestations. *De prospectiva pingendi* lies at the heart of this special achievement.

Acknowledgments

I had my first real immersion in Piero della Francesca's world in 2006 at the Clark Art Institute (Williamstown, Massachusetts), where I carefully read Kenneth Clark's *Piero della Francesca*. After that, I decided to deepen my interest in Piero's work by studying *De prospectiva pingendi*: this led me to write *Playing with Piero della Francesca: Geometry at the Service of the Arts*. This small book was published by Federighi Editori, Certaldo (Florence), in 2012; it is part of a series that I wrote and curated, called "Sciences and Arts in the Renaissance."

In 2012, I began to translate the vernacular and Latin versions of *De prospectiva pingendi* into modern Italian. I talked about my work to my dear friend Henry Millon, Dean Emeritus of the Center for Advanced Study in the Visual Arts at the National Gallery of Art in Washington, DC. I met with Henry several times, showing him the text I had translated to Italian and that my daughter, Paola, had turned to English. We talked about the drawings I had executed following Piero's words. Henry encouraged me like no other and prompted me to keep going. To him goes my most heartfelt gratitude. I dedicate this book to his memory. To Paola goes all my love for the great effort she put into the Italian-to-English translation.

During summer 2014, when the English translation had been completed, I was invited as a Guest Researcher at the Getty Research Institute in Los Angeles. I owe a debt of gratitude to Alexa Sekyra, Head of the Scholars Program, and Gail Feigenbaum, Associate Director, for offering me the opportunity to use the Getty's resources.

My stay at the Getty Center marked a turning point. I soon realized that I could not simply copy or photograph the drawings in *De prospectiva pingendi*. Being a professor of descriptive geometry at the Sapienza University of Rome, I felt the strong necessity not only to execute all the drawings present in Piero's work from scratch, but to take care of the construction process *ex novo*, especially with regard to the capital, the dome, and the two human heads. These latter topics required a long time and the collaboration of several other people.

Following Piero's design process from the very beginning made me proud of fulfilling my duties as a geometry scholar in several research centers: at the Getty Research Institute, at the Center for Advanced Study in the Visual Arts, at the Clark Art Institute, at the Centre Alexandre Koyré (Paris), and at the Royal Flemish Academy (Brussels). These centers of study allowed me to live with art and science at the same time.

Moreover, I remember with affection my former students Sergio Coppola and Davide Spagnoletto from the course in Descriptive Geometry which I taught at the Faculty of Architecture at La Sapienza. To them I gave some forty drawings regarding the theorems of the dome and some forty regarding the theorems of the capital and the text I had translated from Latin. Under my constant creative guidance and with the assistance of professor Luigi Tommasi—friend and collaborator in the course of Descriptive Geometry—these two exceptional students realized more than one hundred drawings for the dome and one hundred for the capital. From these students' computer processing, I was able to obtain all the drawings of the capital and the dome contained in Piero's text, with the exception of figure B.100 for the capital and figure B.200 for the dome. Sergio and Davide graduated from the Faculty of Architecture presenting theses on "The Dome by Piero della Francesca" and "The Capital by Piero della Francesca" respectively.

I have to mention one other former student, Selenia Izzo. Without her, this work would not appear in the way you see it in this book. Together we generated 600 drawings, which she produced using her computer under my constant creative guidance. I owe Selenia my sincere and wholehearted gratitude for having dedicated herself to this project with passion and competence. These 600 drawings concern the entire creative process of the two human heads, and several of them may be found in Appendix A: they are the ones numbered figures A.18, A.19, A.23 through A.51, A.53 through A.56, and A.61 through A.64. Some of them may be found in Propositions VIIIa and VIIIb of Book 3, with the exception of figures B.141 and B.173.

I would like to thank all those scholars with whom I have shared my research during lively conversations: Marc De Mey, Kim Veltman, Antoine Wilmering, Massimo Ciavolella, James Banker, Stefano Marconi, Antonio Severa, Ennio Polidori.

I would like to thank the Bibliothèque Municipale de Bordeaux for letting me publish the drawings in Ms Latin 616.

At last *but* not least, if I have been able to complete this work, I owe it to my wife, Raffaela. To her goes my utmost gratitude and affection for having supported me during these long, tiring years of study.

Rocco Sinisgalli

Introduction

Rocco Sinisgalli

Introduction and Methodology

As I embarked on the journey into Piero della Francesca's seminal work, the *De prospectiva pingendi*, I found myself immersed in a fascinating labyrinth of manuscripts, languages, and artistic insights. At the heart of my endeavor lay a meticulous effort to not only translate but reconstruct Piero's profound treatise on perspective, bridging the gap between his fifteenth-century vernacular and Latin expressions and the modern English language.

The manuscripts themselves offered a treasure trove of historical and artistic significance. Among the eight manuscripts, four were in the vernacular, providing a glimpse into Piero's original expressions in his native Sansepolcro dialect. Notably, the Parma and Reggio Emilia codices stood out as autograph manuscripts, attributed directly to Piero himself for both text and drawings, showcasing his intimate involvement in the crafting of his treatise. On the other hand, the Latin manuscripts, particularly the Bordeaux and Milan codices, offered a refined and structured rendition of Piero's ideas, presenting them to a broader, more cultured audience of the time.

Unraveling Piero's linguistic intricacies, I discovered a symbiotic relationship between him and his trusted collaborator, "Matteo," as they endeavored to translate the complex concepts and geometric principles embedded in the vernacular text into Latin. This partnership not only facilitated a broader dissemination of Piero's ideas but also underscored the evolving linguistic landscape of fifteenth-century Italy, where Latin served as a lingua franca of intellectual discourse.

My approach to translation went beyond mere linguistic conversion; it was a comparative endeavor aimed at uncovering nuances, reconciling discrepancies, and reconstructing missing elements of Piero's text. By juxtaposing the vernacular and Latin versions, I discerned subtle intricacies in Piero's thought process, shedding light on his artistic methodology and theoretical underpinnings. This meticulous comparative analysis not only enhanced the clarity and coherence of the translation but also provided invaluable insights into Piero's creative process.

One of the most intriguing aspects of my work was addressing discrepancies in the drawings, particularly in the third book of *De prospectiva pingendi*. Here I encountered instances of inaccuracies, incompleteness, and ambiguity, prompting me to undertake a task of reconstruction. Drawing from both the vernacular and Latin sources, I meticulously reconstructed the drawings, ensuring a harmonious and cohesive representation of Piero's vision across all three books of his treatise.

Beyond the realm of translation, I grappled with Piero's overarching contribution to the theory of perspective. *De prospectiva pingendi* stood as a beacon of scientific rigor, distinguishing itself from earlier and contemporary treatises by its systematic approach and methodological clarity. Piero's treatise heralded a new era of scientific inquiry in art, laying the foundation for future generations of artists and scholars to explore the depths of perspective theory.

Furthermore, my exploration delved into Piero's artistic oeuvre, uncovering thematic threads that intertwined seamlessly with the theoretical principles espoused in *De prospectiva pingendi*. From the architectural precision of his polyptychs to the statuary accuracy of his human figures, Piero's paintings served as tangible manifestations of his theoretical insights, bridging the gap between theory and practice in the realm of Renaissance art.

As I reflected on Piero's legacy, I was reminded of his interdisciplinary prowess, transcending the boundaries between art and science. His immersion in the arts and geometric sciences during his formative years in Florence laid the groundwork for his dual identity as both an artist and a mathematician. Through his treatise and his paintings, Piero emerged as a visionary figure, ushering in a new era of intellectual inquiry and artistic innovation.

In the following sections, I will delve deeper into the historical premises surrounding Piero della Francesca's contributions to the theory of perspective, exploring the development of perspective theory through Piero's works and its impact on Renaissance art and providing a comprehensive understanding of its evolution and significance.

Historical Context

Piero della Francesca (1412?–1492), author of *De prospectiva pingendi*, is one the giants of the Renaissance. He is seen as a pioneer of modern perspective, but at the same time he was also a great painter and the author of two minor mathematical treatises: the *Trattato d'abaco* which talks about commercial calculation and stereometric geometry, and the *Libellus de quinque corporibus regularibus* in which he analyzes the five regular bodies: tetrahedron, cube, octahedron, icosahedron, and dodecahedron. *De prospectiva pingendi* (written between 1475 and 1481) is the first functionally didactic and rigorously scientific treatise on the theory of perspective, differing from all other previous and contemporary works as well as those that followed at the beginning of the seventeenth century, when the *Perspectivae libri sex* (Pisauri, 1600) by Guidobaldo del Monte (1545–1607) appeared. I translated the latter from Latin to Italian for the first time under the title *I sei libri della prospettiva di Guidobaldo dei Marchesi del Monte* (Rome, 1984). Moreover, the phrase "rigorously scientific" cannot be utilized for the discursive thought used in Leon Battista Alberti's *On Painting* (1404–1472), although this work—as we will soon notice—has a scientific foundation at its base. The same phrase "rigorously scientific" cannot be assigned to the texts about painting and perspective by Leonardo da Vinci (1452–1519) because these have been collected without a real methodological engagement. Leonardo does use strong definitions about perspective, which he divides into linear perspective (artificial) and aerial perspective. Aerial perspective is then subdivided into aerial perspective of outlines (which become more shaded in relation to the distance of the observation point) and aerial perspective of colors (which change or become less sharp also in relation to the distance of the observation point). In any case, it is not my intention here to question the demonstrational skills of many authors who maintain the scientific level required by the treatment of perspective, which I have amply discussed in my book *Verso una storia organica della prospettiva* (Rome, 2000).

On Painting

Piero della Francesca puts an end to the protohistory of perspective in the Renaissance and kicks off a scientific period of the new discipline. For the mystic scientific ways of the medieval world, Piero thus substitutes the rational consciousness of the modern age. This change is documented by the state of knowledge about perspective during one of the most significant periods for Western artistic thought. The window of time in which Piero operates as a painter spans from 1440 to circa 1478, and it coincides with one of the most glorious and fascinating eras of Italian painting, a splendid creative season which teemed with many exceptional masters who do not need to be named because they are so well known. These masters' and Piero's pictorial works—from Masaccio (1401–1428) to Beato Angelico (1395–1455) to Filippo Lippi (1406–1469)—are by far most inspired by this new science of art. These painters' works confirm the acquisition of the rule of perspective that Filippo Brunelleschi (1377–1446) had experimented with well ahead of anyone else by means of "legitimate construction" and the experiment of mirrors, summarized below. We shall not go into discussions of that topic, nor of the origins of the theory of Renaissance perspective which include *perspectiva communis*, which is medieval optics.

On Science

In 1439, when Piero worked as an apprentice in the workshop of Domenico Veneziano (1410–1461) in Florence, he had the chance to immerse himself in the art of painting of well-known people mentioned above. It is very likely in Florence that he learned about Brunelleschi's experiment with mirrors and that he read *De pictura*, which Leon Battista Alberti (1404–1472) wrote and dedicated to Brunelleschi himself between 1435 and 1436. As we will see, this work is an excellent tool for drawing directly on the picture plane, in perspective, without the aid of a plan and elevation.

To quote from paragraph 53 of *De pictura*, in the translation I published with Cambridge University Press (2011):

> Furthermore, I wish that the painter be expert, as far as possible, in all liberal arts (we shall discuss these "arts" in the "Premise about Figurative Arts") but above all I desire in him the knowledge of geometry.
>
> Certainly, our rudiments, from which one extracts a whole, complete and precise technique of painting, are easily assimilable by a geometrician. I also believe that for those who ignore this science neither rudiments nor some procedures of painting can be sufficiently comprehensible. I, therefore, claim that geometry absolutely must not be neglected by painters.

Piero was certainly influenced by these thoughts, and he might have met Brunelleschi, as well as Alberti, for the first time in person in Florence in 1439. Alberti in fact was working there as the head of an ecclesiastical district as well as an "apostolic editor," attending the papal curia for the Ecumenical Council that took place in Florence between 1438 and 1439.

This first immersion in the arts and the geometrical science of perspective was certainly meaningful for Piero's development as a painter and as a treatise writer. These two sides of his persona become fundamental to him, joined and separate. As an artist, he focused on landscapes, colors, and lights: these were themes and techniques he first learned from Domenico Veneziano and later deepened during his stay in Ferrara in 1450, where he certainly met the Flemish painter Rogier van der Weyden (circa 1400–1464). Flemish art had charmed Italy to the point that it was said of its principal representative, Jan van Eyck (circa 1390–1441), "Johannes Gallicus nostri saeculi pictorum princeps judicatus est," literally "John the Gaul is considered to be the first of the painters of our times." It was Bartolomeo Facio (circa 1410–1457) who described him as such in the volume *De viris illustribus*, published in Naples in 1454. Facio concludes his work about Van Eyck with this sentence: "plenam notitiam habere non potui," or "I could not find complete information [about his work]." However, Facio mentions sunlight penetrating and illuminating the ambience even through small gaps; he mentions the small dog, the man and the woman being painted frontally and also from the back by means of a mirror—a clear reference to the *Arnolfini Portrait* (London, National Gallery of Art, 1434). Facio also mentions Van Eyck's descriptions of sites (hills, woods, villages, castles) and naturally the human figures—a clear reference to the *Ghent Altarpiece* (Ghent, St. Bavo's Cathedral, 1426–1432). These are works Facio had not seen, otherwise they would have been named correctly, but this demonstrates, a few years after Van Eyck's death, not only the fame of Flemish oil painting but also the popularity of these two specific works, showing landscapes and a hint of architecture, in Italy's cultural and artistic worlds.

It is important to mention several themes present in many of Piero's paintings. In the *Polyptych of Perugia* (Perugia, Galleria Nazionale dell'Umbria, 1450–1468), we see the *Annunciation* taking place in a classical architectural setting, in front of a portico, foreshortened in perspective with capitals and columns. In the lower part of the same polyptych, there is a representation of the *Madonna Enthroned with the Christ Child Blessing*; she is sitting on a throne, with the dome subdivided into lacunar ceilings. In the *Frescoes of St. Francis* (Arezzo, Coro di San Francesco, 1452–1460) there are two classical architectural settings with a detailed representation of the capitals and the columns: the first setting shows *The Meeting of Solomon and the Queen of Sheba*; and the second one the *Annunciation*. In the *Finding and Recognition of the True Cross*, an episode present in this fresco, we see a village, a temple, and a street with common buildings. The *Flagellation of Christ* (Urbino, Galleria Nazionale delle Marche, 1455–1460) represents a proper classical architectural setting with a detailed presentation of the capitals, the columns, and a foreshortened floor with perfect geometrical figures. Even the *Madonna and Child Enthroned with Four Angels* (Williamstown, Massachusetts, Clark Art Institute, 1460–1470) is part of a classical architecture with specific details of the capitals. We also have the *Holy Conversation* or the *Brera Madonna* (Milan, Pinacoteca di Brera, 1472–1474), placed in a classical architectural setting with a barrel vault with lacunars and a semicurved apse and a large shell. There is another work that is thought by several historians to have been Piero's doing: *The Perspective of the Ideal City* (Urbino, Galleria Nazionale delle Marche, 1470). In this painting, there is one well on each side; on the left, there is a building which shows a portico on the ground floor and on the last floor there is a loggia, recalling the architectural structures present in the Arezzo frescoes.

In his paintings, Piero represents buildings, temples, capitals, columns, vaults, domes with lacunars, and wells. These are themes that, in *De prospectiva pingendi*, are described, designed, and represented in perfect geometrical shapes. Moreover, for what concerns human figures, Piero represents the heads with unique statuesque accuracy, as may be noted when looking at his paintings. These explanations help us better understand the pictorial origins of the main perspective realizations present in *De prospectiva pingendi*.

On Scientific Applications

All the applications Piero executes in the first and second books of *De prospectiva pingendi* (briefly listed below) are logical, and they grow in difficulty. In Book 3, the applications become astonishing: in particular, I am referring to those of the capital, the dome, and the two human heads. In the latter case, the complex morphology has been subject to a close analysis led through horizontal sections and points that have been identified radially on each section: the operational solution is spectacular and it anticipates by many centuries the surprising possibilities offered by a modern computer. We may rightly think that, in its absolute chronological priority, *De prospectiva pingendi* is in fact the most influential precedent to numerous later works written by scholars who may or may not have known of its existence.

On the Figurative Arts

It is good to specify how scientific applications of perspective utilized in painting emphasized the role of figurative arts in Florence and more generally in Tuscany at the end of the fifteenth century. On the basis of remote customs, starting from the beginning of the early Middle Ages, painters were framed in the category of the *artes mechanicae* (mechanical arts): those arts that were practiced by using one's hands. By the end of the 1400s, it was widely accepted that painters should be considered part—with full attributes—of the large family of operators of the *artes liberales* (liberal arts): those arts that were practiced by using one's brain. The liberal arts were seven and were divided into two groups: the *artes sermocinales* and the *artes reales*. The *artes sermocinales* were three, called the trivium, and included grammar, rhetoric, and dialectics. The *artes reales* were four, called the arts of the quadrivium, and included arithmetic, geometry, astronomy, and music. Only the seven liberal arts were favored by the mighty; they also meant higher salaries for those who practiced them. Through *De prospectiva pingendi*, it was clearly demonstrated that painters' figurative arts should be accepted as an exact science. Luigi Vagnetti emphasizes this point in his *De naturali et artificiali perspectiva* in 1979; I dedicated my volume *I sei libri della prospettiva di Guidobaldo dei Marchesi del Monte* to this great teacher of mine.

On the Manuscripts and Editions of *De prospectiva pingendi*

There are eight manuscripts of *De prospectiva pingendi*. Four are in the vernacular, of which the Parma codex is an autograph manuscript by Piero for both the text and the drawings, and the Reggio Emilia codex is also an autograph manuscript by Piero for the drawings. There are four Latin manuscripts, of which those at Bordeaux and Milan are Piero's autograph manuscripts for the drawings. The Parma manuscript is represented by five modern editions: the German translation of 1899; the vernacular edition of 1942, reprinted in 1984; the American edition in photographic facsimile of 1992; the French translation of 1998; and the digital edition of 2016 (Venice). The Reggio Emilia manuscript is represented only by a photographic facsimile edition from 2008. In 2016, three volumes entitled *De prospectiva pingendi* were published: a critical edition of the vernacular text; a critical edition of the drawings; and a facsimile reproduction of the Parma codex. In 2017, three volumes entitled *De prospectiva pingendi* were published: a critical edition of the Latin text; a critical edition of the drawings; and a facsimile reproduction of the Bordeaux codex.

On the Vernacular and Latin Texts

Piero wrote *De prospectiva pingendi* in the language spoken in the fifteenth century in Sansepolcro, his natal town, where the local dialect was missing the lexical correctness, the scientific terminology, as well as the grammatical rules that are the essence of the written and spoken modern Italian language. As we will see, Piero then asked master "Matteo," whom Piero knew extremely well, to translate it into Latin. Naturally, the two worked together to translate the concepts and geometrical terms present in the vernacular treatise. Comparing the two texts, one can easily notice that the vernacular is hard to read; it creates doubts and perplexities. The Latin text instead has a precise syntax of cases and complements, as well as harmony of the verbs and terminology. This demonstrates that at the time, in Sansepolcro, the most appropriate language to present this work to a cultured and international audience was Latin.

On Operational Choices

I decided to work on a comparative translation, which means a contemporary translation of both Latin and vernacular texts of the treatise. This type of comparison gave me the chance to reconstruct the missing parts of the text. In fact, where necessary, the current translation uses the Latin when the vernacular is missing and vice versa. Through this comparative translation, I realized that several drawings in the third book presented obvious discrepancies. As we will see, some drawings are incorrect, incomplete, missing, unclear, or have an inadequate size. I am referring to the drawings of the capital, the two human heads, and the dome. In any case, these are discrepancies that do not damage the constructive process: that is, the methodological path connected to the construction process. I was obliged to reconstruct the drawings present in these three theorems according to the passages analyzed in both the vernacular and the Latin versions. Having taken into consideration the importance of this type of reconstruction, for the sake of uniformity I reconstructed almost all the drawings of all the other theorems present in all three books of *De prospectiva pingendi*.

On Working Criteria

To better understand the work carried out, it is good to observe the working criteria chosen to realize the comparative translation. Working criterion 1 concerns the capital and dome theorems, both present in the third book. Figure 0.6 shows translations of a passage into modern Italian and English in the center, with the texts of two vernacular manuscripts (Parma and Reggio Emilia) on the left and of two Latin manuscripts (Milan and Bordeaux) on the right. In working criterion 2 (figure 0.7) we have the text of the first human head, with texts arranged in four columns: on the left, we have the vernacular text of Parma and the Latin text of Milan, while on the right we have the modern Italian and English translations. For both these tables, naturally the notes (with all the necessary references concerning the other manuscripts) appear in the back matter of the present volume.

In working criterion 3 (figure 0.8) we have five columns concerning the second human head: on the left we have the vernacular text of Parma and the Latin text of Milan, in the center the modern Italian translation, and on the right the Latin text of Bordeaux, while in the last column on the right we have the English translation. Naturally the notes (with all the necessary references concerning the other manuscripts) appear in the back matter. Only by using this method, comparing these texts in such a way, was it possible for me to reconstruct everything with textual and graphic integrity.

This involved both the reconstruction, in plan and elevation, of the objects—which Piero teaches us to build with great geometric precision—and the reconstruction of the perspectives of the same objects. In order to avoid confusion and allow a clear and seamless understanding of the text (as specified in the Introduction), in some propositions of Book 3 (Proposition III to Proposition IX) I added precise "specific additions" to paragraphs with the aim of distinguishing the various increasingly complex passages.

On Several Differences between This Book and Other Editions of *De prospectiva pingendi*

What I have laid out in the previous paragraphs emphasizes the differences between this work and other existing editions. In the German edition from 1899 the drawings have been simply copied by hand; in the French edition from 1998 the drawings have been copied digitally without following the construction of scientific processes. The vernacular edition of Parma (1942) and its later reprint (1984) reproduce photographs of the original drawings. There are a photographic facsimile of the Parma manuscript (1992),

Figure 0.6 Working criterion 1: capital and dome.

Parma manuscript (folio 49 verso)	**Reggio Emilia manuscript** (folio 59 recto)			**Milan manuscript** (folio 69 recto)	**Bordeaux manuscript** (folio 69 verso)	**Notes**
Dal dato puncto, nel termine posto, il capitello descricto con proporzione degradare.	Dal dato puncto, nel termine posto il chapitello descripto com proportione degradare.	**Proposizione VII Da un punto dato, scorciare proporzionalmente, nel termine prestabilito, un capitello assegnato.**	**Proposition VII From a given point, shorten proportionally, in an established terminal, an assigned capital.**	A dato puncto, in constituto termino posito descriptum capitellum cum proportione degradare.	[A] dato [puncto] in constituto termino posito descriptum Capitellum cum proportione degradare.	The term "puncto" which means "point" is missing in the Latin manuscript of Bordeaux.
Adunque, per volere degradare il descricto capitello non lasciando il modo cominciato,	Adunque per volere degradare il descripto capitello non lasciando il modo cominciato	Ci apprestiamo, dunque, a scorciare un capitello assegnato; cosa che faremo senza allontanarci dal metodo intrapreso.	We will then shorten an assigned capital, which we will do without diverting from the method undertaken.	Degradaturis itaque nobis descriptum capitellum: nec ab incepto modo desistentibus:	[D]egradaturis itaque nobis descriptum capitellum nec ab incepto modo 1 desistentibus	
farai in propia forma il capitello, il quale tu voli fare, che se vegga da una faccia,	farai in propia forma il capitello il quale tu vuoli fare che se vegga da una faccia	Bisogna determinarlo in forma appropriata, in modo che sia visto da un lato.	One must determine it in appropriate shape, so that it is seen from one side.	quod facturi sumus, propria in forma constituendum est: ita ut ab latere uno perspiciatur.	quod facturi sumus propria in forma constituendum est[,] ita ut ab latere uno perspiciatur.	We prefer to translate "propria in forma" as "in appropriate shape," which corresponds to the one in plan and in elevation.
		Dimensionando il Capitello in Alzato	**Dimensioning the Capital in Elevation**			
del quale fa che sia la linea de la sua larghezza da piè, dove se ferma su la colonna .4.,	del quale fa che sia la linea de la sua largheza da pie dove se ferma su la colonna .4.	La linea della sua larghezza, dove si attacca alla colonna, sia uguale a 4.	The line of its width, where it sticks to the column, is equal to 4.	Sitque linea sue latitudinis, ubi in columna firmatur .4.	Sitque linea sue latitudinis ubi in columna firmatur .4.	It is clear that the measurement on which Piero bases the compositional characteristics of the capital is the unitary model, a standard quantity that serves as a basis for the measurement of other quantities of the same nature or even to carry out, in this case, the design of a geometrically "proportioned" capital.
la quale devidi per equali in puncto .K.,	La quale devidi per equali in puncto .K.	Venga questa linea divisa in due parti uguali nel punto K.	Divide this line into two equal parts at point K.	eaque per equalia dividatur in puncto K:	eaque per aequalia dividatur in puncto K.	
et sopra .K. mena la linea perpendicolare .AK., che sia .5.	et sopra .K. mena la linea perpendicolare .AK. che sia .5.	Si conduca, sopra K, la perpendicolare AK, la quale sia uguale a 5.	Trace perpendicular line AK equal to 5 above K.	Et super K ducatur perpendicularis linea AK que sit .5.	Et super K ducatur perpendicularis linea AK qu[a]e sit .5.	
poi tira una linea passante per .A., equidistante la linea .K., che sia .7.;	poi tira una linea passante per .A. equidistante la linea .K. che sia .7.	Si conduca poi la linea che passa per A ed è parallela a quella per K: sia questa linea uguale a 7.	Trace the line that goes through A and is parallel to the one that goes through K: this line will be equal to 7.	Trahatur deinde linea transitum faciens per .A. et equidistans à linea K que sit .7.	Trahatur linea deinde transitum faciens per .A. et [a]equidistans á linea K qu[a]e sit .7.	
poi devidi .AK. in secte parti equali, de le quali ne poni una de le secte socto .A., che sia .AC.;	poi devidi .AK. in secte parti equali delle quali ne poni una delle secte socto .A. che sia .AC.	Dopo ciò si divida AK in sette parti uguali, una delle quali venga posta al di sotto di A: sia essa AC.	After, divide AK into seven equal parts, and place one part below A: call this AC.	Post hec dividatur AK septem in partes equales: quarum una ponatur sub .A. que sit AC.	Post hec dividatur AK settem in partes equales quarum una ponatur sub .A. qu[a]e sit AC.	

Figure 0.7 Working criterion 2: first head.

Parma manuscript (folio 59 recto)	**Milan manuscript** (folio 79 recto)		
Nel termine asignato, col dato puncto, proportionalmente la testa degradare.	In termino constituto cum dato puncto capud proportionaliter degradare.	**Proposizione VIII(a) Nel termine prestabilito, assegnato con il punto dell'occhio, scorciare una testa in modo proporzionale.**	**Proposition VIIIa In the preestablished terminal, assigned with the point of the eye, shorten a head in a proportional way.**
		Due Profili di una Testa	**Two Profiles of a Head**
Sia commo nel principio di questo dixi che era de bisogno sapere desegnare quelle cose che l'omo vol fare in propia forma,	Quoniam in principio huius Libri diximus, opus esse ut quis designandi peritiam teneat eius rei, quam propria in forma facturus sit.	Poichè all'inizio di questo libro abbiamo detto che si deve rappresentare ciò che si farà in propria forma,	Since we have stated, at the beginning of this book, that one must represent what will be drawn in real shape,
pero desegna una testa in un occhio,	Igitur designetur capud quoddam, ita constitutum ut oculi unius demonstrationem habeat,	si disegnerà allora una testa, disposta in modo che mostri un solo occhio,	we will then draw a head so that it is only showing one eye,
cioe in canto con quello profilo che tu intendi degradare,	cum eo profilo cuius degradationem faciendam intelligis.	con il profilo che tu pensi di scorciare.	with the profile you believe has to be shortened.
et con quella ne fa poi un'altra in facia con doi occhi,	Cumque eo aliud quod faciem ostentet duobus oculis integratam.	Insieme con questo profilo costruisci un secondo che mostri la faccia integrata con due occhi.	Together with this profile, build another one that shows the face integrated with two eyes.
da quella medessima grandeza et tucte le parti correspondenti.	Sitque eadem altitudine ac reliquis conrespondentibus partibus.	Questo [secondo profilo] abbia la stessa altezza del primo e tutte le parti in modo che si corrispondano [tra loro].	This [second profile] has to have the same height of the first one and all the parts must correspond [to one another].
Prima tira la linea recta da la sommità della testa in uno occhio,	Trahatur primo linea recta a vertice capitis alterum oculum habentis	Si tiri dapprima una linea retta dall'estremità della testa, che ha un solo occhio.	First, trace a straight line from the end of the head with one eye.
contingente la sommità de la testa in doi occhi;	contingens verticem capitis quod duos oculos tenet,	Tale linea tocchi l'estremità della testa che ha due occhi.	This line must touch the end of the head that has two eyes.
et poi ne tira un'altra infra quella e la sommità de la fronte, che sia linea .A.;	ac insuper alia ducatur linea que medium locum teneat inter dictam supremam lineam et frontis partem superiorem, que sit linea .A.	Si conduca inoltre una seconda linea, che stia nel mezzo tra la detta linea che sta più in alto e la parte superiore della fronte: sia questa la linea A.	Moreover, trace a second line, which is in the middle between the line that is above and the part above the forehead: let this be line A.
poi ne mena un'altra a sommo la fronte, che sia linea .B;	Deinde alia ducatur linea supra frontem et sit linea B.	Si conduca poi un'altra linea sulla fronte: sia essa la linea B.	Trace another line on the forehead: let this be line B.
et una ne tira fra la fronta e il naso passante per l'occhio, che sia linea .C;	Alia preterea extendatur inter frontem et nasum, que per oculum vadat et signetur linea C.	Si conduca inoltre un'altra linea fra la fronte e il naso, la quale passi per l'occhio: si segni come linea C.	Trace another line between the forehead and nose which goes through the eye: mark it line C.
et linea l'altra passante per la gobba del naso, che sia .D.,	Aliaque ducatur transiens per gibbum nasi et signetur linea D.	E si conduca un'altra linea passante per la gobba del naso: si segni come linea D.	Trace another line that goes through the bump on the nose: mark it line D.
tira l'altra a pie del naso, che sia linea .E.	Trahatur alia per inferiorem terminum nasi que dicatur linea E.	Si conduca un'altra linea per l'estremità inferiore del naso: si indichi come linea E.	Trace another line that goes through the lower end of the nose: mark it line E.

Figure 0.8 Working criterion 3: second head.

Parma manuscript (folio 68 verso)	**Milan manuscript** (folio 89 verso)		**Bordeaux manuscript** (folio 83 recto)	
		Proposizione VIII(b) Seconda Testa		**Proposition VIIIb Second Head**
Et per che la testa che s'è facta he senza alcuno pende[re], intendo che se ne facci un'altra che abbia doi movimenti,	Et quoniam quod adhuc fecisti nullam deflexionem habet est intentio ut illud fiat quod duabus motibus sit:	Poichè [la testa] che hai fatta sinora non ha alcuna inclinazione, è [mia] intenzione che quella [la testa] sia [ora] prodotta mediante due movimenti.	Et quoniam quod adhuc fecisti nullam deflexionem habet est intentio ut illud fiat, quod duobus motibus sit:	Since the head you have built till now has no inclination, it is my intention now to build that head through two movements.
che l'alzi dinanze et penda in un canto, et che ella non sia equidistante al termine,	silicet ut a fronte elevetur, in latus perpendeat, nec a termino equidistantiam habeat:	Sarebbe a dire che [la testa] si innalzi frontalmente, che penda da un lato e che non sia equidistante al piano del termine.	silicet ut a fronte elevetur, [et] in latus perpendeat, nec a termino equidistantiam habeat:	This means that the head is lifted up frontally, leaning on one side, and it is not equidistant from the plane of the terminal.
per la quale porrai comprendere tucti gli altri movimenti de la testa,	quo facile tenere poteris ceteras omnes capitis motiones.	Tramite questo [modo di procedere] potrai comprocedere facilmente tutti gli altri movimenti della testa.	quo facile tenere poteris ceteras omnes capitis motiones.	Through this procedure you will easily understand all the other head movements.
		Prima Rotazione		**First Rotation**
Tu ai la testa in un occhio, la quale fu facta per la precedente, che ha .8. linee atraverso che sono .A. .B. .C. .D. .E. .F. .G. .H.	Habes caput oculi unius quod constat per id quod precessit, idque lineas octo ex tranuerso tenet[,] que sunt ABCDEFGH	Da quanto precede, hai la testa che risulta avere un solo occhio; la quale ha otto linee trasversali, le quali sono A, B, C, D, E, F, G, H.	Habes caput oculi unius quod constat, per id quod precessit, idque lineas octo ex tranverso tenet[,] que sunt ABCDEFGH	From what was done earlier, you have the head with one eye only, which has eight transversal lines, which are A, B, C, D, E, F, G, H.
ciascuna coli suoi segni,	quarum quelibet cum suis signis extat	Ciascuna di queste [linee] appaia con i propri segni.	quarum quelibet cum suis signis extat	Each of these lines appears with its own marks.
a la quale tira una linea che continga la puncta del naso e il mento,	ad quod duc lineam, que tangat acumen nasi, atque mentum	Conduci su questa [testa] una linea che tocchi la punta del naso e il mento.	ad quod duc lineam, que tangat acumen nasi atque mentum	Trace on this [head] a line that touches the point of the nose and the chin.
che se dica linea .IP. che facci pendare la testa iderieto quanto te piaci che penda la testa,	que sit linea IP que faciem unum oculum habentem in occipitium tantum inclinet quantum vis quod propensionis habeat.	Sia IP questa la linea; la quale renda la testa avente un solo occhio tanto inclinata dalla parte della nuca quanto tu vuoi che sia l'inclinazione.	que sit linea IP que faciem unum oculum habentem in occipitium tantum inclinet quantum vis quod propensionis habeat.	Let this line be IP, which makes the head with one eye only lean on the side of the back of the head as much as you want it to lean.
		Alzato del Circuito A		**Elevation of Circuit A**
et sopra la linea .IP. mena una linea perpendiculare contingente la sommità de la testa che sia linea .z. senza termine;	Ducas et super lineam quae perpendicularis sit, ac huius capitis verticem tangat et sine termino linea z.	Conduci anche perpendicolarmente alla linea IP una linea indeterminata z che tocchi la sommità della testa.	Ducas et super lineam quae perpendicularis sit, ac huius capitis verticem tangat et sine termino linea z [dicatur].	Also trace an unlimited line z perpendicular to line IP which touches the top of the head.
poi tira .1. .2. .3. .4. .5. .6. .7. .8. .9. de la linea .A. de la testa in un occhio, tucte equidistante la linea .z. senza termine,	Extendas preterea .1. 2. 3. 4. 5. 6. 7. 8. 9. lineae .a. capitis unoculi, quae omnes equidistent a linea z et sine termino sint.	Tira poi dai punti 1, 2, 3, 4, 5, 6, 7, 8, 9 della linea A della testa ad un solo occhio, delle linee, senza termine, tutte parallele alla linea z.	Extendas preterea 1. 2. 3. 4. 5. 6. 7. 8. 9. lineae .a. [unoculi capitis], quae omnes equidistent a linea z et sine termino sint.	From points 1, 2, 3, 4, 5, 6, 7, 8, 9 of line A of the head with one eye only, trace lines with no end parallel to line z.

a photographic facsimile of the Reggio Emilia codex (2008), and a digital edition of the vernacular of Parma (Venice, June 2016) which contain reproductions of the drawings in facsimile.

Finally, there are *Edizione critica del testo volgare, Tomo I* (Rome, December 2016) and *Edizione critica del testo latino, Tomo I* (Rome, 2017). These two editions offer a critical separate analysis of the two texts that does not compare or verify the drawings at all.

But to this *Edizione critica del testo volgare* (Tomo I) is added, in a separate volume, an *Edizione critica dei disegni* (Tomo II): here the drawings in Reggio Emilia's vernacular manuscript are analyzed. And again, to the *Edizione critica del testo latino* (Tomo I) is added, in a separate volume, an *Edizione critica dei disegni* (Tomo II): here the drawings in Bordeaux's Latin manuscript are analyzed. In addition, two separate volumes reproduce the anastatic prints of two manuscripts: the vernacular of Parma and the Latin of Bordeaux.

On Several Features That Differentiate the *Edizioni critiche*

What are the features that differentiate the two *Edizioni critiche*? They do not offer the translation of the two texts into a modern language. They offer a separate critical analysis of the vernacular and Latin texts which consists of comparing and specifying single terms: it is an analysis which includes two separate and distinct glossaries. These editions report missing letters, syllables, and words of the manuscripts that are being studied: this means that whether or not the text's accuracy coincides with the graphics of the drawings has not been analyzed.

What are the features that differentiate the two *Edizioni critiche dei disegni*? They offer an analysis that develops the drawings of the vernacular manuscript of Parma and the drawings of the Latin manuscript of Bordeaux separately. There is no text-drawing comparison in order to actually produce the drawings in each manuscript. In fact, it is not possible to follow the making of the proposed drawings present in these two volumes by reading the text. We can only follow the above-mentioned process, which we will now analyze. In each of the two volumes, all the drawings have two different editions: a diplomatic edition that is as compliant as possible with the original drawings; and a critical edition that integrates (or, better, corrects) the original drawings.

On the Ecdotics Adopted in the *Edizioni critiche dei disegni*

The new analysis of the drawings present in Tomo II (*disegni*) for the vernacular manuscript of Parma essentially bases itself on several specific researches that concern the sophistication of the marks present in the original drawings. The same method is adopted for the Latin manuscript of Bordeaux. For now, we will only take into consideration Tomo II (*disegni*) for the manuscript of Parma. The original drawings have been investigated through computer scans which allow us to recognize the various graphic techniques used for the ink marks, the compass marks, the marks that do not make it through the paper's thickness, the engraved marks, the holes, and the scratches. Specific graphic procedures are established through this research. These procedures are then adopted for the drawing represented in the diplomatic edition and for the drawing represented in the critical edition. The comparison between these marks and the geometrical structure of the drawings would allow us to understand the procedures used to execute the drawings, to check the dialogue between text and drawing, and naturally to check the space represented by the drawing. This is certainly a new and ambitious procedure, which is being applied for the first time to the drawings of a very important text of the fifteenth century. This process was named "ecdotics" (that is, "philology") of the drawings, and it investigates the reading procedure and the identification of a mark and consequently the graphic and historic process of the tracing of a drawing.

This methodology has led to conclusions that are different from the ones I propose in this volume. I believe this process is somewhat appropriate for these studies; but when applied to this specific case, the analysis lacked a comparative translation of Latin and vernacular and the simultaneous application of descriptive geometry to the drawings in question in the two texts.

On Several Differences That Distinguish a Fundamental Drawing of the *Edizioni critiche*

For now, we will only analyze a fundamental drawing present in Tomo II (*disegni*) (Rome, 2016). In working criterion 4 (figure 0.9), I have copied the final *disegno critico* (that is, "critical drawing") found on page 226 of Tomo II (*disegni*). This is an extremely important drawing because it represents the conclusion reached after a very long creative process of the tridimensional model of the human head built in Tomo II (*disegni*). This model is achieved by means of twelve tables of preparatory drawings.

Successively, the perspective of this model is realized by means of two series of other preparatory drawings: twelve "diplomatic drawings" and twelve "critical drawings."

If we compare working criterion 4 with working criterion 5 in figure 0.10 (which shows the drawing of the perspective I realized of the same first head), we immediately see the differences that distinguish the two perspectives.

The first thing that strikes us is the lack of the midsection of the head in the first perspective.

Moreover, this midsection needs to be subdivided into 12 parts, as instructed in the text. Please observe that this midsection is fully represented in the manuscript of Parma, while in the Reggio Emilia manuscript it lacks some parts: please compare the perspective of the head I drew in working criterion 5 and the two perspectives present in working criterion 7 relative to the Parma and Reggio Emilia manuscripts.

The second thing that strikes us is the lack of the upper extremity of the second ear (in working criterion 4), which is present in the Parma and Reggio Emilia manuscripts. Please compare working criterion 7.

The third thing that strikes us is the lack of a truly significant and important point when observing the perspective of the head present in working criterion 4. As may be seen in working criterion 5, this is point S which delimits the apparent contour of the head at the top. This point must be found above circuit A. In the perspective of the head in working criterion 4, the upper apparent contour of the head clearly goes through points 16, 1, 2, 3, 4 of circuit A. These points must be found undoubtedly below the upper apparent contour of the head, as may be noticed in the drawing of working criterion 5. This difference is ascertained in the analysis I conducted of folio 65 recto of the manuscript of Parma: please compare working criterion 6 concerning this (figure 0.11); in the first drawing at the top, the upper lateral elevation of the head has been drawn (that is, the upper part of the midsection of the head) with visual rays converging toward the eye.

Please observe that there is a minimal distance highlighted by the two arrows' extremities between the visual ray (which goes through point 1) and the upper curve of the midsection of the head. This distance must not be underestimated; in fact, its extreme importance is confirmed by the presence of six points of circuit A which lie below the upper apparent contour of the head in the perspective of Parma, and the presence of five points of circuit A which lie below the upper apparent contour of the head in the perspective of Reggio Emilia (please compare the upper drawings of working criterion 7, figure 0.12, regarding this).

For this reason, in the lower drawing of working criterion 6, above line A, I had to modify the upper part of the forehead by making it bigger so as to identify the visual rays 1-1 and SS. Please observe that the distance between visual ray 1-1 and the new exterior curve of the midsection of the head is not minimal anymore. This may be noticed by looking at the new distance between the points of the two arrows. In fact, visual ray SS identifies distance S-1 on vertical line K (the picture plane), which corresponds to the vertical separation that must happen, in perspective, between the upper apparent contour of the head and point 1 of circuit A. This distance corresponds to the distance between the two points of the vertical arrows in working criterion 5.

Please note that by enlarging the upper part of the original head in the Parma manuscript (compare the lower figure in working criterion 7) we find a less dark curve, which we may define as "deleted," which the amanuensis (or Piero himself) may have erroneously bent toward the vertical dotted line, just below the apparent contour of the head. This type of analysis is not present in Tomo II (*disegni*).

On Other Differences That Distinguish the *Edizioni critiche*

There are other observations that also make a difference between this edition and the *Edizioni critiche*.

Once more, in Tomo II (*disegni*), on page LXXI it says, "There is a misleading indication: adding some remarkable points to circuits C, D, E, and F only. In effect, these points have already been identified during the tracing of the circuits described in paragraphs III.8.8." The editors do not accept that Piero introduces additional subdivisions that are necessary to identify other points; these are the points of the nose, eyes, and mouth. I have identified these points and subdivisions separately by using dashed lines in figure B.112 in Appendix B. Again, as may be seen in figure A.26 in Appendix A, additional tangent points have been added to the visible images of each circuit; these were highlighted, in each case, on the right and on the left of the apparent contour of each circuit by using an asterisk. These are important points because (as we specify in detail in the third part of the Introduction, in the discussion of the remade circuit A) Piero specifically mentions them. In Tomo II (*disegni*), there is no mention of this issue, and this is confirmed in the figure present in working criterion 4, where the additional tangent points have not been highlighted on the apparent contour of the head. In fact, these points should have been added (where necessary) on the right and on the left of the image of each circuit. Although we do not know how things really went, or how events unfolded, I doubt that Piero himself could have realized the drawings present in Propositions VII, VIII, and IX of the third book. These are theorems about the capital, the two human heads, and the dome. This issue is never questioned in the *Edizioni critiche* of the texts and drawings.

We have discussed the perspective of the vernacular head present in the *Edizione critica dei disegni* (Rome, 2016), and we should make many other observations about the perspective of the head in the *Edizione critica dei disegni* (Rome, 2017).

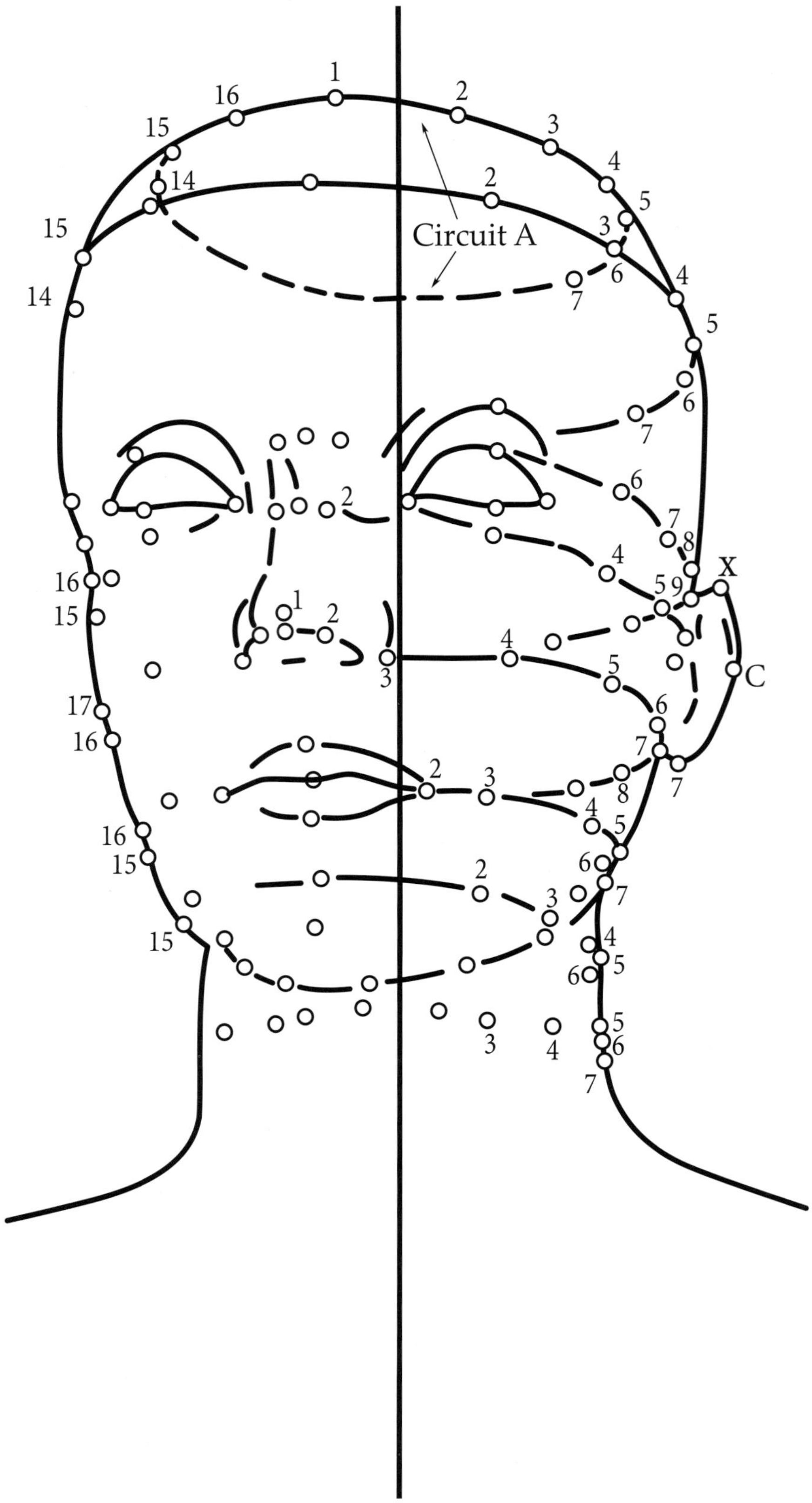

Figure 0.9 Working criterion 4: facsimile of critical drawing from *Edizione critica del testo volgare*, Tomo II, *Edizione critica dei disegni* (Rome, 2016), page 226.

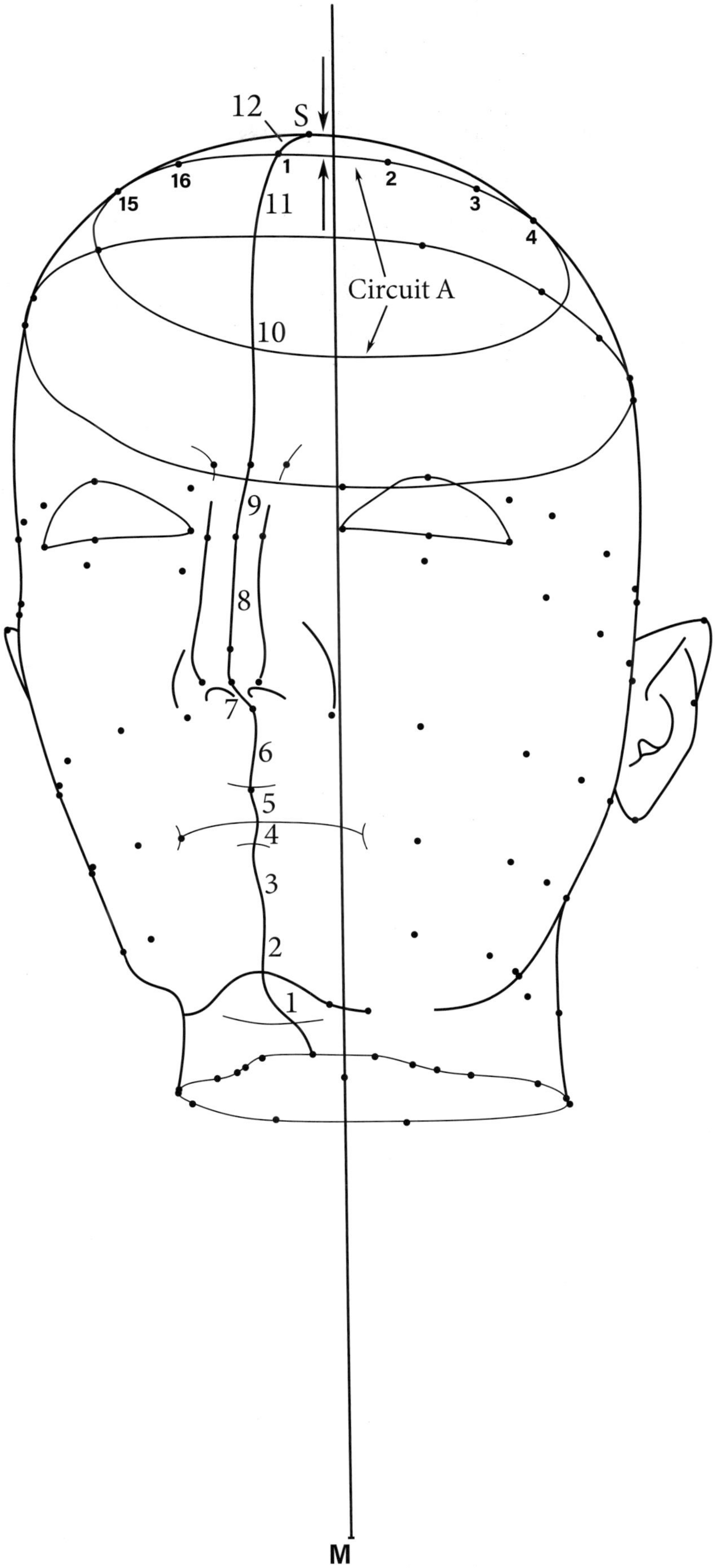

Figure 0.10 Working criterion 5: the outline of circuit A compliant with the original.

Parma manuscript,
folio 65 recto

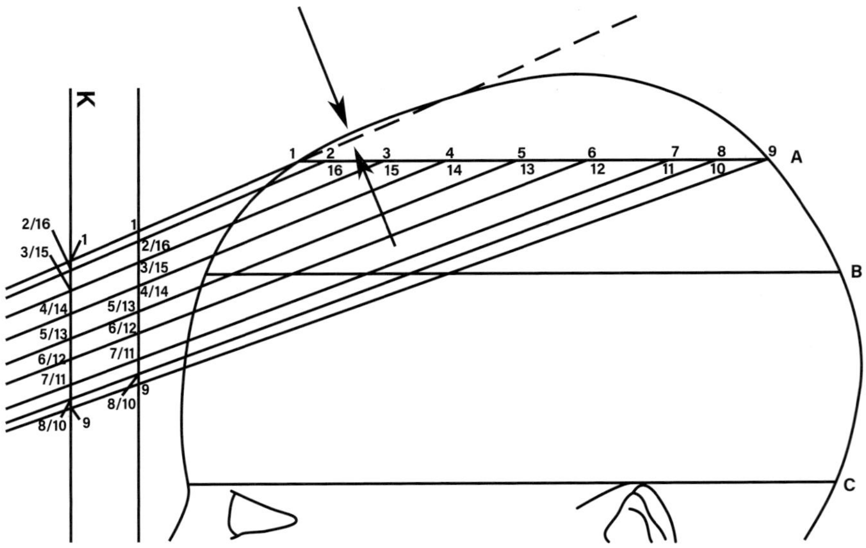

Reconstruction of the lateral elevation
of the head above the A circuit

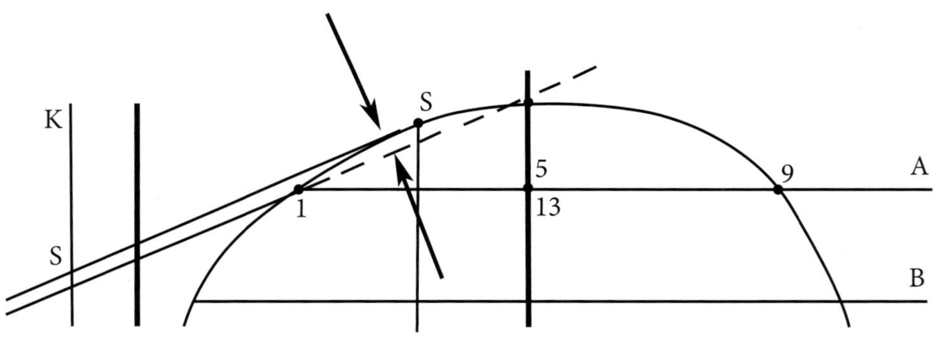

Figure 0.11 Working criterion 6: distances between arrows.

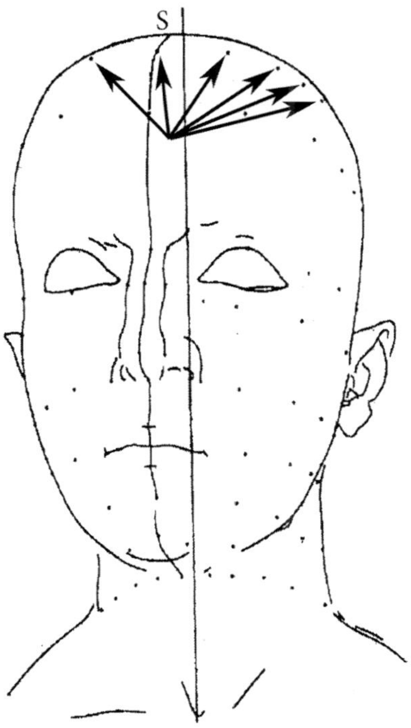

Parma manuscript, folio 66 verso

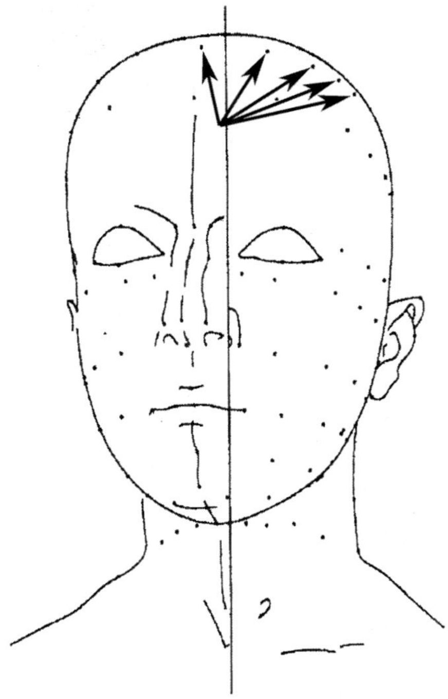

Reggio Emilia manuscript, folio 76 recto

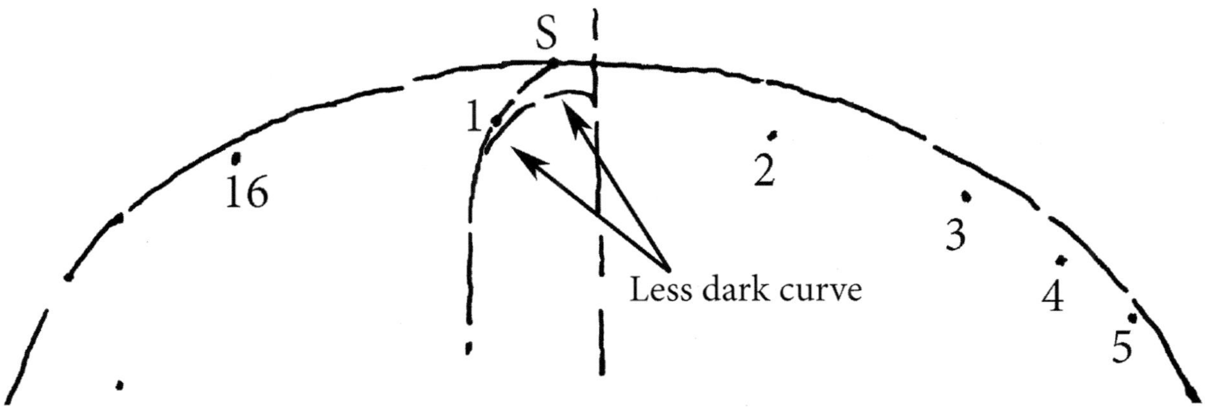

Parma manuscript, folio 66 verso: detail

Figure 0.12 Working criterion 7: the visible points of circuit A and the less dark curve.

History and Contents

Legitimate Construction and the Experiment of Mirrors

Filippo Brunelleschi is recognized worldwide for having rediscovered perspective at the beginning of 1425, which would become the science of art. In fact, Brunelleschi usually takes the credit for having realized perspective by means of plan and elevation, which are architectural tools, as well as the cut of the visual pyramid.[1] This method was called "legitimate construction" (see figure A.1 in Appendix A).

We do not know anything else, that is, we do not have any other information regarding Brunelleschi's perspective, apart from an experiment recorded by biographer Antonio di Tuccio Manetti (1423–1497).[2] This experiment was carried out by Brunelleschi utilizing mirrors and observing the Baptistery of Florence (figures A.2, A.3, A.4).[3] As I show in my book *Perspective in the Visual Culture of Classical Antiquity*, the science of mirrors was at the base of the science of perspective of the ancients, as it was at the base of the modern rediscovery of perspective.[4]

Abbreviated Construction

In 1435 and 1436, Leon Battista Alberti wrote his famous *On Painting*,[5] which he dedicated to Brunelleschi in his vernacular version. Without reproducing or referring to any drawing, Alberti teaches how to draw the perspective of a square directly on the picture plane. He does this through the aid of an auxiliary construction and a diagonal which has the sole aim of obtaining the correct shortening of the checkerboard. Figure A.5 shows the auxiliary construction that Alberti claimed to have prepared separately from the picture plane.[6] Figure A.6 shows the shortening on the picture plane of the basic square using the diagonal.[7] This shortening is realized directly on the picture plane with the aid of the auxiliary construction, which is shown below in the scale one wishes to use with respect to the picture plane. In figure A.7 you may see how to limit the plan of the buildings on the basic square using Alberti's method;[8] figure A.8 shows how to represent the heights of the buildings directly on the picture plane.[9]

The Veil

To represent small surfaces, like those of the face of a human being for example, Alberti suggested the use of "a veil woven of very thin threads and loosely intertwined, dyed with any color, subdivided with thicker threads according to parallel partitions, in as many squares as you like, and held stretched by a frame; which [veil] I place, indeed, between the object to be represented and the eye, so that the visual pyramid penetrates through the thinness of the veil."[10] In figure A.9 you may see an interpretation of Alberti's veil.[11] Figure A.10 illustrates two extracts from *On Painting*,[12] by means of three figures: the first by Masaccio, the second by Abraham Bosse (1604–1676), and the third by Albrecht Dürer (1471–1528). What I have tried to show by using few simple words until now, what I have tried to summarize using graphs, should make us understand the status of knowledge about geometrical perspective during the first part of the Renaissance (until circa 1470).

De prospectiva pingendi

Piero della Francesca wrote *De prospectiva pingendi* between 1470 and 1482. The title is in Latin for both the vernacular and Latin versions of this work, and it literally means "The perspective of what is to be painted." Less literal translations go like this: "On perspective for painting," "Perspective in painting," "Perspective of painting," and "On the perspective of painting." *De prospectiva pingendi* consists of three books that contain fifty-four propositions: thirty in the first book, twelve each in the second and third. Here follows a summary of all the contents in the propositions.

First Book

In the first book, the theme discussed concerns how to draw flat figures in perspective. From the first introductory lines of the text, we understand that Piero proceeds with practical applications and with great precision. These are the definitions through which Piero makes a premise about the point and the line, just as Alberti had: "A point is that which does not have parts; we understand it by imagining it. A line is length without width; it reflects the nature of a point because it also is perceived by the brain." The first eleven propositions analyze optical and geometrical problems that concern angles, lines, and square surfaces and their diagonals.

Proposition XII says: "From a given eye, in an established terminal, shorten the assigned plane." The perspective constructions begin here.

Proposition XIII says: "Reduce the shortened surface to a squared shape according to the previous demonstration." In fact, Piero shows the correspondence between the real square and the shortened square, operating simultaneously on the square's perspective and on the reconstruction of the square in plan. It is good to specify that the shortening of the square is at the base of all perspectives that will be executed in all the propositions of the first and second books.

Proposition XIV says: "Divide a squared figure shortened into equal parts."

Proposition XV says: "Produce the divisions in squares of a shortened squared surface divided into many equal parts." Piero teaches how to subdivide the same squared surface into many unequal parts. In reality, he places the square in real shape below the shortened square (figure A.11), which we can consider to be the rotated correspondent of the observed square if we place it on the other side of the terminal, as is obtained in the theory of mirrors.[13] This is the most important proposition, in which Piero demonstrates that the subdivisions of the square in perspective are proportionate to those that the diagonal produces in the square in real shape. Once the correspondence between the real and the shortened square is shown, Piero deals with the direct method, that is, the direct method of drawing any flat figure in perspective without the aid of plan and elevation, as long as it is placed on the square in real shape. This is a method that is still used today in many schools of art, architecture, and engineering.

Proposition XVI: "Turn a shortened quadrilateral surface into an octagonal figure."

Proposition XVII: "Turn an eight-angled shortened figure into a sixteen-angled one."

In the three following propositions Piero teaches us to draw a triangle of equal sides on a squared (shortened) surface; an equilateral hexagon on a shortened quadrilateral surface; and an equilateral pentagon on a shortened surface.

Proposition XXI: "Remove a given quantity of a shortened squared surface's perimeter."

Proposition XXII: "Add proportionally a given quantity around the perimeter of a shortened squared surface."

Proposition XXIII: "From a non-squared surface, as big as you wish, cut a squared surface." In this proposition the distance point appears for the first time.

Proposition XXIV: "Add other similar quadrilaterals, through diagonals, to a given shortened quadrilateral."

In the five following propositions Piero teaches us to draw an assigned square on a shortened surface; a regular assigned octagon on a shortened surface; assigned squares on a shortened surface; the perimeter of a base of a square building on a shortened surface; and an octagonal base of an assigned building on a shortened surface.

In Proposition XXX, Piero explains in what proportion the eye can expand, with its rays, the visual angle.

Second Book

Piero goes on to represent several solid bodies that are to be placed directly on the shortened square or on a figure similar to the square.

Proposition I: Piero teaches how to place, first, a shortened cubic body on a shortened squared surface according to the distance and the assigned terms of the mentioned shortened surface; and second, the same body so that the sides are not parallel to the ones of the plane at the base.

Proposition II: "On a shortened surface with a shortened base of eight sides, place a similar shortened body, which has eight faces without including the bases."

Proposition III: "On a shortened pentagonal area, shorten a body that has as many faces."

Proposition IV: "On a shortened hexagonal surface elevate a pillar that has as many faces."

Proposition V: "On a shortened plane elevate a column that has sixteen faces; let the column be appropriate to the proportion of the given surface."

Proposition VI: "Elevate a well with six equal and shortened faces on a shortened plane and surround it with steps that follow the conformation of the faces."

In Proposition VII, Piero teaches how to raise a shortened cube on a plane and surround it with a base and a cymatium; and how to raise the same shortened cube on the same plane in a different position.

Proposition VIII: "Place a shortened column on a shortened plane so that it is not parallel to the straight line of the shortened plane."

Proposition IX: "Place a quadrilateral house on a shortened plane."

Proposition X: "Proportionally trace an eight-faced temple on a shortened plane."

Proposition XI: "On a shortened plane trace a cross vault, that is, an arced vault which, drawn in squared form, is sustained by a four-walled construction."

In Proposition XII, Piero explains why the columns that are further away from the eye become bigger than the ones that are closer, especially when they are placed on equal bases.

Third Book

The theme developed in the third book is the direct method of putting into perspective flat and solid figures, that is, the method of plan and elevation, which is shown and applied in a clear and precise manner.

The following are the topics Piero discusses in the first four propositions: a square surface located on the floor; an assigned surface that has eight equal sides; four circles that have the same center and are subdivided into twelve equal parts; and an assigned torus divided into twelve equal parts which has eight circles that delimit its thickness.

In Proposition V, Piero examines an assigned cube which stands on one of its vertices with none of its sides parallel to the established terminal. He analyzes the problem of a solid's rotation, first in plan and then in elevation, teaching how to find the new plan and the new elevation scientifically: new plan and new elevation which are necessary to draw a cube, thus placed, in perspective.

In Proposition VI he examines the assigned base of a round column.

In Proposition VII he examines an assigned capital.

In the first part of Proposition VIII, which is called Proposition VIIIa, Piero teaches how to plan a human head geometrically, or scientifically, by finding its perspective from a frontal position with respect to the terminal. In the second part of Proposition VIII, called Proposition VIIIb, he takes the same head and places it in a random position with respect to the terminal. Thus, he teaches how to scientifically find the new plan and elevation necessary to represent the head in perspective. He does not ask himself about the "veil," which for Alberti was an extremely important experimental method for the "profiles of small surfaces." Piero speaks about the human head as any other solid object, specifying, though, that while he limits himself to executing eight sections of a head, an artist should execute "a multitude of circuits at one's own will." This is done with the aim of obtaining a more appropriate representation of the same head in perspective.

In Proposition IX Piero executes the perspective of a dome.

In Proposition X he executes the perspective of a spherical body.

In Proposition XI he obtains the perspective of a so-called fresh vase.

In Proposition XII he executes the perspective of a ring hanging from an elevated position.

First Method

The first method Piero presents for finding perspective (figure A.12) is inspired by the abbreviated construction of Alberti, with the addition of the overturning of the ground floor on the picture plane, or terminal, as you can see in figure A.11. The finding of perspective of a point in space is immediate, as is easy to verify in the three passages shown in figure A.12. Piero uses this first method starting from Proposition XV of the first book for the representation of flat figures which belong to the base plane, that is, the ground plane. He uses this first method again, throughout the whole second book, for the representation of solid figures that lean on these flat ones.

Second Method

The second method presented by Piero in Proposition I of Book 3 (figures A.13, A.14, A.15) is inspired by the legitimate construction of Brunelleschi. Piero begins with the operations in plan (figure A.13): he fixes an object, the square in plan; he fixes the terminal (that is, the picture plane) and the eye in plan; he joins the eye in plan with the points of the desired object; he finds the intersections of these lines (the visual rays in plan) on the ground line, that is, on the baseline of the terminal. He then marks these intersections on a wooden ruler, which he puts aside. He then goes on to the operations in elevation (figure A.14): he fixes the height of the eye in elevation; he reports the projections of the points, already considered in plan, in elevation. He then joins the eye in elevation with the projections of the points in elevation; he finds the intersections of these lines (the visual rays in elevation) on the terminal in elevation and marks these intersections on two paper rulers, which he puts aside. Having done this, Piero goes on to draw perspective through the use of the wooden and paper rulers. After placing the ground line on the terminal and the vertical axis MN at the center (figure A.15), he uses the wooden ruler as an abscissa, positive or negative on the ground line with respect to point M, which he takes on the horizontal axis of the ground line. Piero then uses two paper rulers at the same time to identify the heights of the points with respect to point M, which he considers to belong to vertical axis MN. Point M, in this case, is the origin of the ordinates of the points, starting from M, on axis MN. The certainty that Piero is using point M as the origin of the horizontal axis of the widths and as the origin of the vertical axis of the heights, in order to identify the points on the terminal, is shown by the fact that he fixes in each figure the axis for a point M to be drawn on the terminal. We find ourselves, in the history of perspective, facing a system of coordinates of points on the plane. The only exceptions are in Propositions X, XI, XII of the third book, where the terminal is not vertical but horizontal. These are two perspectives. One perspective goes from the top to the bottom, and the other one goes from bottom to top. Because the object of representation is found between the eye and the terminal here, Piero adopts the construction process of the shadows caused by the glim: in fact, he puts forward the process of the envelope curves which is associated with Christiaan Huygens (1629–1695) and Ehrenfried von Tschirnhaus (1651–1708).[14]

First Perspective Treatise

Having learned about Brunelleschi's legitimate construction and the experiment of mirrors, having learned about Alberti's abbreviated construction and the contributions to the two perspective methods, having carefully thought about the titles of all the propositions, we realize that *De prospectiva pingendi* is the first treatise of perspective of all time.

The First Drawing Manual and the First Manual of Architecture

When following the text in each single proposition, we realize that step after step, Piero explains all passages relative to the construction in plan and elevation of all the objects he analyzes. It could seem repetitive and redundant that he dwells on indicating, in plan and elevation, a great number of points and lines that delimit, for example, the capital, the dome, and the human head. It is obvious that if one wishes to find the perspective of an object, one must know to perfection how that objective is really made. Only if whoever reads Piero should also draw what he writes will they understand the importance of going through each passage. One will then easily understand that *De prospectiva pingendi* is also the first drawing manual and the first manual of architecture of modern times. All detailed passages that Piero follows to build numerous objects, in particular the capital, the dome, and the human head, demonstrate this.

Comparative Latin-Vernacular Translation

There are four *De prospectiva pingendi* manuscripts in Latin and four in the vernacular.[15] The present translation was carried out using the Latin manuscripts of Bordeaux and Milan, constantly compared with the vernacular versions of Parma and Reggio Emilia. Why is comparing the Latin editions to the vernacular editions the preferred approach? All manuscripts, including Parma and Reggio Emilia, present discrepancies and mistakes, not only in the text but in the drawings too. The most important errors are in the text and drawings of the capital, the dome, and the human head. As we will show, Piero cannot have personally transferred the drawings of the capital, the dome, and the human head in any of the four above-mentioned manuscripts. Since in all manuscripts there are page repetitions, missing or inverted sentences, lines, and letters, as I have repeatedly emphasized in the notes, I was compelled to translate from Latin to Italian and compare this to the vernacular. In the parts where the Latin translation was missing, the vernacular version showed correct sentences, and vice versa. I preferred the Latin version because this language is extremely accurate, especially when explaining complex scientific or geometric concepts. Moreover, Latin does not need interpretation, while the local vernacular dialect, spoken by Piero, was missing an adequate vocabulary. Take into consideration the comments of Luca Pacioli (circa 1445–1517), who was from Piero's hometown and was his admirer. He claims that the work was translated from the vernacular to Latin by a famous poet and orator, who was also Piero's friend. That the translator was Piero's friend is very important; this means that the collaboration between author and translator was efficient.[16] James R. Banker has identified the translator as "Maestro Matteo di ser Paolo d'Anghiari."[17]

Printed Editions

It was only in 1899 that *De prospectiva pingendi* was published for the first time in a translation from the Parma manuscript's vernacular to German, an edition curated by Costantin Winterberg.[18] In this first edition, the drawings were copied as they are seen in the Parma manuscript. In 1942, Parma's vernacular edition was published by Giusta Nicco Fasola.[19] This edition uses photos of the drawings in the Parma manuscript. In 1998, the French translation of Parma's vernacular was published.[20] In this edition, as in the edition of 1899, the drawings have been copied as they appear in the original Parma manuscript. We also have facsimile editions: the American edition of the Parma manuscript in 1992; the Aboca Edizioni edition which reproduced the Reggio Emilia manuscript in 2008; and finally the "critical editions" of the Parma and Bordeaux manuscripts published by the Istituto Poligrafico e Zecca dello Stato in 2017.[21]

All these editions do not take into account the constructive and geometrical process. This means that the editions do not compare scientifically or literally the texts and the drawings. Only by making these comparisons will one be able to find discrepancies and inconsistencies, which will be examined following this introduction and text by using detailed notes.

Operational Choices

Let us leave it to philologists to establish the time of transcription for Piero's and the amanuenses' manuscripts. Let's have handwriting experts decide who and how many were the amanuenses. We will concentrate on comparing texts and drawings utilizing the scientific method, in order to obtain a clear and truthful critical edition of a book that revolutionized the world of perspective. When necessary, in this work I have compared step by step the Latin manuscripts of Bordeaux and Milan with the vernacular manuscripts of Parma and Reggio Emilia. At times, I compared them with other manuscripts of the treatise as well, as specified in the notes. I have conducted a careful comparative analysis of the written and graphic operations in order to present this work, for the first time, in its literal and scientific integrity. Following this introduction, we will have a chance to understand the importance of having carried out these literal, graphic, and geometric checks. In the notes, I have reported the choices made in order to convey better scientific results. At times, I have used the present tense where the manuscripts used the future because it was necessary for the context of the sentences. When necessary, I have transcribed the drawings' letters in capital letters. Considering the multiple discrepancies between the Greek and Latin letters in the vernacular and Latin texts, I preferred to use Latin letters.

Specific Additions and Notes Subdivisions

From Proposition III to Proposition IX of Book 3, I have added subheadings to paragraphs with the aim of distinguishing the various increasingly complex passages. These subheadings will help in reading and following the text.

The following are the subheadings added to Proposition III of Book 3: Wooden Ruler A; Wooden Ruler B; Wooden Ruler C; Wooden Ruler D; Paper Ruler a; Paper Ruler b; Paper Ruler c; Paper Ruler d; The Shortening of the First Circle; The Shortening of the Second Circle; The Shortening of the Third Circle; The Shortening of the Fourth Circle.

I proceed in the same manner till Proposition IX of Book 3, about the dome, where the subheadings go as follows: The Barrel Vault or the Arch of the Dome; The Slice of the Dome; The Seven Circles; A Quarter of a Circle; A Second Circuit; The Subdivisions; The Semicircles in Plan; The Subdivisions of the Semicircles in Plan; Line A of the Figure of the Elevation; Line B and the Remaining Ones; The Terminal and the Eye Point in Plan; Wooden Ruler A; Wooden Ruler B; Wooden Rulers C, D, E, F, G, H, I, K, L, M, N, O, P, Q, R; The Terminal and the Eye Point in Elevation; Paper Ruler a; The Remaining Paper Rulers; The Shortening of the Dome; The Vertical Elongations of the Points; The Horizontal Elongations of the Points; Cornice.

De prospectiva pingendi in History

Certainly Luca Pacioli, who knew Piero intimately, informed Leonardo da Vinci of *De prospectiva pingendi*, who got himself a copy. Leonardo in fact reports in a note: "Borges will get you Archimedes from the bishop of Padua, and Vitellozzo [that book that comes] from Borgo Sansepolcro" (figure A.16a).[22] Moreover, Luca Pacioli could have been that unnamed perspective scholar whom Albrecht Dürer (1471–1528) frequented during his second trip to Italy while he stayed in Bologna. In 1886, the German scholar August Schmarsow recognized that Piero's work was well known to Dürer.[23] Schmarsow was impressed by the concurrence of the perspective order developed by Piero with the order developed by Dürer, as well as the concurrence of many drawings present in the treatises of Piero and Dürer.[24] It is important to add that Dürer was inspired by Piero's work even in the first volume he wrote about the symmetry of human bodies, in which the match between Piero's and Dürer's heads is blatant, although Dürer limited himself to considering only the vertical and horizontal rotations of the head, as one can see by observing the drawings in the *De symmetria partium humanorum corporum*, published posthumously in 1531 (figure A.16b).[25]

The first to confirm the fame of Piero della Francesca as an authority on perspective was Giorgio Vasari (1511–1574), who in 1550 stated: "He was employed by Guidobaldo Feltro the elder, Duke of Urbino, for whom he made many very beautiful pictures. Nevertheless, there were preserved some of his writings on geometry and perspective, in which sciences he was not inferior to any man of his own time, or perchance even to any man of any other time, as is demonstrated by all his works, which are full of perspectives."[26] It is important to specify that when Daniele Barbaro (1513–1570) published his work on perspective (1569) he used many theorems and drawings by Piero della Francesca, publishing them as if they were his own in a volume called *La pratica della perspettiva*. In this respect, what we have said about Dürer's human head we must say about Barbaro's, that it presents vertical and horizontal rotations. As can be seen in figure A.17, even the representations of Barbaro's head have been copied from Piero's.[27] Clearly both Dürer and Barbaro did not follow the complexity of the constructive process of the human head randomly placed in space, due to the discrepancies (which we will examine following this introduction) present in the texts and drawings of all manuscripts of the *De prospectiva*. The most striking incident was that Barbaro discredited Piero, stating: "Nobody that I know has left a text about perspective; unless we want to call rules and precepts some poorly explained limited practices that do not have a definite order or foundation. Amongst these, we can find some practices by Piero of Borgo Santo Stefano and others that could be needed by idiots."[28] Naturally this is without a doubt a failed attempt to push away any suspicion of plagiarism from his readers.[29] The first testimony of this type of plagiarism can be found in the book by Jacopo Barozzi da Vignola (1507–1573), published posthumously in 1582 by Ignazio Danti (1536–1586). In this book it says: "Amongst those contemporaries who have left some memory about the art [of perspective], the first who, chronologically speaking, presents the best form and method to write [about this art] was master Piero della Francesca of Borgo Sansepolcro. He left us three handwritten books with excellent drawings. Whoever wants to learn about the excellence of these books, can look at the way Daniele Barbaro copied a large part of them into his book on perspective."[30] This shows that Piero's work was known by the experts and scholars who studied this topic. In 1864, when the first history of modern and ancient perspective was published in Paris by Noël-Germinal Poudra (1794–1894), it spoke of Piero della Francesca as the most ancient author on perspective, confirming that he was remembered by many as the father of perspective.[31] Nevertheless the complete contents of *De prospectiva pingendi* were unknown, as they are today by historians of perspective, geometry, and mathematics who have not had the chance of analyzing the work in its completeness. The complete vision of Piero's work must include the perspective of the human head placed randomly in any position in space and realized scientifically for the first time.[32] This topic is still fundamental for representing any object scientifically in the computer era. This is what I announced in 2012: "The fact that Piero tried to follow a theoretical path—because he was certain of being useful to the rising problem between theory and practice in the science of perspective—is rightly emphasized by the constructive novelties he introduced and by the specific solutions he utilized. These were matters that concerned modern descriptive geometry, which he applied, in exemplary fashion, to represent human heads. The significance of these constructions has not escaped, so it seems, those who followed. However, no one—for what I know—has highlighted how these are cutting-edge even in contemporary computer representation."[33]

The Art of Perspective: Scientific and Geometric Approaches

First Head

Our focus now turns to the significant scientific processes that underlie the geometric planning and perspective representation of complex forms. These operations are introduced through an exploration of Piero della Francesca's innovative methods, where art and science intersect. This section highlights key concepts such as the use of lateral and frontal elevations, the precision of circuit planning, and the role of orthogonal axes in defining spatial relationships. Emphasis is placed on the meticulous attention to detail required to reconcile theoretical constructs with artistic execution. This analysis provides insight into Piero's ability to apply scientific rigor to artistic representation, offering a thorough exploration of his techniques for representing form in space.

Significant Scientific Operations It is good to focus our attention on significant scientific operations that are at the base of planning the first head and at the base of the perspective representation of the same head. When planning the head geometrically, that is, scientifically, for the first time in the history of art and science, the artist starts (figure A.18) by choosing a lateral elevation with one eye and a frontal elevation with two eyes.

Subsequently Piero plans eight circuits in plan, with the infallible precision of a scientist, assigning numbers on the plans. He then prompts to report the same numbers on the two elevations, lateral and frontal (figure A.19).

A Multitude of Circuits at One's Own Will While executing these operations, it is important that Piero has specified that "one may introduce a multitude of circuits at one's own will,"[34] where by "circuits" he means the paths of the sections with which we cut the head horizontally. This is a very important concept that Piero introduces to solve the issue of the apparent outline. From the four passages shown in figure A.20, we may understand how science is incapable still today of delimiting the head in its construction process, if not through the hand of the artist or through a computer, where the latter and the former will have to intervene time after time to unite the outlines of the various sections. In figure A.20a we are executing the head in elevation with one eye only. In figure A.20b we cut it horizontally in eight random sections. Figure A.20c shows the spatial view of the eight sections. In figure A.20d we can see the need to unite the circuits' representations with an apparent outline defined by the artist's hand or through a computer, just as we would do today through meshes.

Orthogonal Axes in Plan The identification of a point in plan and in space using a compass is another fundamental theme that catches the attention of whoever is executing Piero's creative process. Please observe figure A.21: Piero identifies a point in the circuit described in plan, for example point 2, by using two orthogonal axes (MI and MV) that originate from point M. Then observe figure A.22: in space, the third axis that coincides with vertical MP originates from P and not M: P is the starting point of the sections with which the head is cut; therefore point 2* (which is in space) belongs to the horizontal plane that cuts the head going through point A, which is on the MP axis.

The Remade Circuit A and the Circuit A Compliant with the Original

We must premise that at the end of the representation of the first head, Piero encourages his readers to find the apparent outline of a circuit of the head. In fact, he writes (figure A.23): "it may happen that when the thread is led above the circuits, at the beginning or at the end of the object being considered, there is no correspondence to any mark; the extreme lines, in fact, that originate from the eyes surround the entire object, touching it; it happens that their mark is not found above the indicated marks but that the first line, that is, the thread, touches the circuit between points 14 and 15. Mark point 14*."[35] After drawing the first dotted circuit A (figure A.23) in conformity with the original A,[36] I was obliged to identify point 14* in plan. This point is found on the point of contact of the tangent 14*-14* on circuit A which has been rebuilt; this is not drawn with a dotted curved line. I did the same thing for point 7* in plan. This point is found on the point of contact of the tangent 7*-7* on circuit A which has been rebuilt; this is not drawn with a dotted curved line. To draw these tangents, I was obliged to widen the circuit in conformity with the original, drawn using a dotted curved line, so as to allow: (1) the first tangent line, drawn from the eye in plan, to touch the circuit between points 14 and 15 at point 14*; (2) the second tangent line, drawn from the eye in plan, to touch the circuit between points 7 and 6 at point 7*. I could not miss the chance to take into consideration and to shorten both circuits: the one that has been remade and the one that is compliant with the original, with the objective of a useful comparison between perspectives. Therefore in figure A.23 we have found the paper ruler and the wooden ruler relative to the circuit that has been remade. In figure A.24 we have found the paper ruler and the wooden ruler relative to the circuit which is compliant with the original.

The Images of All Visible Points of the Circuits The first step Piero takes to realize the head's perspective is the one that makes me reproduce (figure A.25), one by one, on the terminal, the images of all the visible points of the various circuits by means of the wooden and paper rulers.

The Visible Lines of All Circuits The second step concerns the extension of the marks (figure A.26). This means uniting the visible points in each circuit, which means that the artist's hand will draw the visible lines of all circuits.

The Twelve Units and the Vertical Section of the Head Piero warns us: "Certainly there are twelve units."[37] To establish these units, it is necessary to draw the tangent from the eye to the vertical section of the head, as you can see in figure A.27.

Therefore, point S, which we have just found, allows us to identify in perspective (figure A.28), that is, on the terminal or the picture plane, the subdivision into twelve parts of the vertical section of the head.

The Protrusions of the Head The third step (figure A.29) concerns "the protrusions, as is evident, of the eyes, the nose, the mouth, and whatever remains so that you can draw them correctly."[38]

The Apparent Outline Piero warns us (figure A.30): "We must proceed this way for the remaining outlines."[39] This means drawing the ears' outlines and the outline that delimits the whole head. Naturally figure A.31 reproduces the head in its artistic completeness, without numbers and circuits, delimited in its most appropriate apparent outline.

Vernacular Manuscripts: Perspectives' Comparisons As announced above, in figure A.32 we are showing the perspective of circuit A that is compliant with the original, and in figure A.33 the new perspective of the head, without numbers and circuits, deriving from the representation of the apparent outline of the original circuit A. We cannot neglect a comparison of the two perspectives (figure A.34): the first relates to the remade circuit A and the second one to the circuit A that is compliant with the original, with the perspectives shown in the Parma and Reggio Emilia manuscripts. As may be noticed in these comparisons, Piero has made a note about the tangents, with the aim of teaching the necessity of adding more numbers on the circuits in order to plan a better apparent outline of the head. The perspective with circuit A that is compliant with the original seems to resemble even more the perspectives of the Parma and Reggio Emilia manuscripts.

Vernacular Manuscripts: Incongruences It is good to understand what the incongruences are relative to the human head and what effective graphic data we have available concerning Proposition VIIIa of Book 3. These incongruences have led me to develop more than 600 drawings to reproduce the entire creative process of the human head which Piero della Francesca realizes, using the scientific method for the first time in the history of art and science.

Parma Manuscript, Folio 59 Verso Premise: In the drawing on the left in figure A.35[40] I extracted the lateral elevation of the head (that is, "the head with one eye only," as Piero calls this elevation), emphasizing the four sections AA, BB, CC, DD. Below I have reproduced the four corresponding upper circuits with the letters that identify them in plan. I have reproduced the same head with one eye only in the drawing on the right, emphasizing the four sections EE, FF, GG, HH. Below I have reproduced the four corresponding lower circuits with the letters that identify them in plan. Please observe that the ears do not appear in plan in either the upper or the lower circuits.

Drawing on the left: Please notice the following incongruences (figure A.35) in the sections relative to the four upper circuits. Between sections BB and CC, at the back of the head, from the point I have marked with X, I have drawn a dotted tangent line (since it is only imaginary). This tangent ends in plan exactly at point X. It is obvious that the circuit which is marked in plan with an X has to have a section in elevation: this section cannot be the one in elevation that goes through the two X. This is emphasized by the fact that points B and C on the right side of the head in elevation are aligned on a vertical line. These points coincide in plan at point C/B. Considering vertical BC which ends at point C/B in plan, the circuits in plan should be three, but it is obvious that they must be four.

Drawing on the right: Please notice the following incongruences relative to the four lower circuits (figure A.35). Segment EE, which represents circuit E in elevation, is not tangent to the inferior extremity of the ear. If projected in plan, segment GG, which represents circuit G in elevation, ends at the right on no circuit. If projected in plan, segment HH, which represents circuit H in elevation, ends at the left on no circuit, and at the right it ends well beyond the last external circuit. The vertical line EE on the right does not go through the first internal circuit in plan. The right extremity of section EE does not correspond to this first internal circuit in elevation (identified by the vertical dotted line). On folio 68 verso of the Reggio Emilia manuscript, which corresponds to folio 59 verso of the Parma manuscript, we are missing the elevations of the four upper circuits and the four lower circuits.[41]

Parma Manuscript, Folio 61 Recto Premise: In the drawing on the left in figure A.36,[42] I have reproduced the lateral elevation of the head, that is, the head with one eye only, as Piero writes. I have emphasized the four sections AA, BB, CC, DD. Below I have reproduced the corresponding four upper circuits, with the letters that identify them, in plan. In the drawing on the right I have reproduced the same lateral elevation of the head, emphasizing the four sections EE, FF, GG, HH. Below I have reproduced the corresponding four lower circuits with the letters that identify them.

Drawing on the left: Concerning the sections relative to the four upper circuits (figure A.36), please observe the following incongruences. Between sections BB and CC, from the point I marked with X at the back of the head, I marked a tangent line, which I drew with a dotted line because it is imaginary. This tangent line ends in plan exactly at point X. It is obvious that the circuit, marked in plan with an X, does not have a corresponding section in elevation. Also, consider that points B and C on the right side of the head in elevation are aligned on a vertical line, and in plan they coincide at point C/B: therefore in plan the circuits should be reduced to three, when they have to be four. Please observe that on the left, section BB in plan does not have the extremity that matches on a circuit. Notice that two circuits go through point C, placed on the left in plan, but as can be seen in elevation, only circuit C should go through.

Drawing on the right: Concerning the four lower circuits (figure A.36), notice the following incongruences. On the left, in plan, circuit G is broken and does not connect to the appropriate circuit on the right. Moreover, notice point E on the right in plan: on the left of this point E there appears the first internal circuit which does not have a corresponding section in elevation, as is shown by the vertical dotted line. In plan again, on the right, between points E and G there lies the right extremity of section FF of elevation: this extremity does not have a corresponding circuit in plan. On folio 70 recto of the Reggio Emilia manuscript, which corresponds to folio 61 recto of Parma, we are missing the plans of the four lower circuits.[43]

Parma Manuscript, Folio 63 Verso Please observe the following incongruences (figure A.37). In the drawing above, where we can see the upper circuits, on line IP we have reproduced the same numbers and the same letter M that appear on line KL. Because line KL identifies the terminal as seen in plan, the numbers and the letter M should only be on line KL. The incongruence happens twice because the numbers and letter M have been reproduced on straight line IP even in the drawing below, in the plan of the lower circuits. Moreover, in the drawing below, on the right, there are only three circuits, but there should be four. Again, in the drawing below, on the left, there are two arcs that break without turning into circuits: one of the two arcs should have continued into a fourth circuit which does not appear on the right. On folio 71 verso of the Reggio Emilia manuscript, which corresponds to folio 63 verso of Parma, we are missing the points of view in plan: both the point of view relative to the four upper circuits and the point of view relative to the four lower circuits. Please also notice that on this folio 71 verso of Reggio Emilia, there are no indications of the ears, in the plans of either the four upper circuits or the four lower circuits.

Parma Manuscript, Folio 65 Recto Please observe the following incongruences (figure A.38). On line IP we have reproduced the same numbers that appear on line KL. Because line KL identifies the terminal in plan, that is, the picture plane, the numbers should be only on line KL. Another incongruence is represented by line IP, which should be tangent to the nose instead of cutting it. Folio 74 verso of the Reggio Emilia manuscript, which corresponds to folio 65 recto of Parma, is missing the point of view in elevation and does not present any numbers on line KL, which represents the terminal.[44]

Parma Manuscript, Folio 61 Recto and Folio 63 Verso; Reggio Emilia Manuscript, Folio 70 Recto In figure A.39, we are considering what we may define as the arch of the chin. In figure A.39a, we have reproduced the drawing and the numbers, relative to the arch of the chin, that appear on the head with one eye only on folio 61 recto of Parma. For these numbers we cannot find any corresponding numbers in the plan on folio 61 recto of the same manuscript. Figure A.39b reproduces the drawing and the numbers, relative to the arch of the chin, that appear in the head with one eye only on folio 70 recto of Reggio Emilia. Remember that this folio is missing the plans of the four lower circuits. In figure A.39c, taken from folio 63 verso of Parma, we have the only representation in plan of the arch of the chin. This representation does not have a correspondent in the representation in elevation and is identified with numbers 1′, 2′/16′, 3′/15′.[45]

Latin Manuscripts: Incongruences: Bordeaux Manuscript, Folio 74 Verso Premise: In the drawing on the left in figure A.40,[46] I have reproduced the lateral elevation of the head, that is, the head with one eye only, emphasizing the four sections AA, BB, CC, DD. Below, I have reproduced the corresponding four upper circuits with the letters that identify them in plan. In the drawing on the right I have reproduced the same lateral elevation of the head, emphasizing the four sections EE, FF, GG, HH. Below I have reproduced the corresponding four lower circuits with the letters that identify them in plan. Please observe that the ears do not appear in plan, in either the upper or the lower circuits. Take into consideration: (1) that in the original figure, the numbers appear on the plans of the four upper circuits and of the four lower circuits, in compliance with how they are shown in figure A.18, previously examined; and (2) that in the lateral elevation of the head the letters A, B, C, D, E, F, G, H appear, indicating the sections of the circuits.[47]

Drawing on the left: Concerning the four upper circuits, please observe the following incongruences. Between sections BB and CC, at the back of the head, I traced a tangent line from a point which I marked with an X; this is a dotted line because it is imaginary. This tangent ends exactly at point X in plan. It is obvious that the circuit marked in plan with an X does not have a corresponding section in elevation. Because points B and C placed on the right, in elevation, are aligned on a vertical line (which also happens in the Parma manuscript, folio 59 verso), in plan these points end at point C/B. Considering that the vertical line BC ends in plan at point C/B, the circuits in plan should be three, but it is obvious that for construction they have to be four. On the left, in plan, there are two circuits that go through point C; still on the left, in elevation, we have no correspondence of the circuits. Observe on the left, in plan, that the circuit for D has just been hinted: it does not continue above or below.

Drawing on the right: Concerning the four lower circuits, please observe the following incongruences. Segment EE, in elevation, which represents circuit E, cuts the ear instead of being tangent to the lower extremity of the ear. The vertical line HH on the left does not exactly go through the extremity of the circuit in plan. On the left, in plan, this circuit—which should have gone through H—breaks above horizontal line HM, without connecting to another circuit.

Bordeaux Manuscript, Folio 76 Recto Figure A.41 shows the lateral elevation of the head with one eye only, without the frontal elevation of the head with two eyes and without any drawing of the plans of the four upper circuits or the four lower circuits. This drawing does not make sense. To understand the incoherence, please imagine figure A.19 (previously examined) without the plans of the eight circuits and without the frontal elevation of the head with two eyes. Even the London and Paris manuscripts show lateral elevations of the head with one eye only without the frontal elevations of the head with two eyes and without any drawings for the plans of the four upper circuits and the four lower circuits. The Latin manuscript of Milan shows the frontal elevation and lateral profile side by side on the same folio, without any hint of the plans of the circuits. It is obvious that these are incoherent drawings, without any connection to constructive logic.

Bordeaux Manuscript, Folio 77 Verso Figure A.42 presents the frontal elevation of the head with two eyes, without the lateral elevation of the head with one eye only and without any drawing of the plans of the four upper and the four lower circuits. This drawing does not make sense. To understand the incoherence, please imagine figure A.19 (previously examined) without the plans of the eight circuits and without the lateral elevation of the head with one eye only. Even the London and Paris manuscripts present the frontal elevation of the heads drawn on separate pages from the lateral profiles, without any hint of the plans of the circuits. The Latin manuscript of Milan instead presents the two drawings in figures A.41 and A.42 side by side, although there is no hint of the plans of the circuits. Again, these are incoherent drawings that show no connection to constructive logic.

Bordeaux Manuscript, Folio 79 Recto In figure A.43 we are comparing the points of view on folio 79 recto of Bordeaux with the points of view on folio 63 verso of Parma. The points of view in Parma are indicated by the bottom ends of the dotted lines. The latter have been proportionally overlapped to the plans and to the points of view present in the Bordeaux manuscript.[48] As one may observe, the points of view of Bordeaux, as well as the points of view present in the other three Latin manuscripts, are placed further to the right in comparison to the points of view in Parma. For this reason, the perspective in the Latin manuscripts will be different from that in the vernacular manuscripts.[49]

Bordeaux Manuscript, Folio 79 Recto: Details Let's premise that the lower circuits are represented in plan in the top drawing in figure A.44, and that the upper circuits are represented in plan in the bottom drawing.

In the top drawing, the left missing arc of the last circuit H concerning the neck (see figure A.40) should go through the five unnumbered marked points that can be distinguished on the folio. The two points marked with the arrows do not lie on straight lines M-16 and M-15. As you may notice, point 2 appears twice on the first internal circuit drawn on the left, and again twice on the terminal in plan.

In the bottom drawing, the vertical axis for M, which does not go through the beginning of the ears in figure A.40, now passes through the beginning of the ears, which have been drawn in plan here. Two circuits go through point 1 which lies on the horizontal line which goes through M; this cannot happen for the simple reason that points D, C, and B have been placed on three vertical well-distinguished lines, as can be noticed in the elevation on the left of figure A.40. A circuit goes through a point identified by the arrow next to point 1 which lies on the horizontal line that goes through M. This circuit is only hinted; it does not continue above or below. Observe again how, on the left, there are six circuit projections while there should be four.

Bordeaux Manuscript, Folio 80 Verso, Compared to the Parma Manuscript, Folio 65 Recto In figure A.45 we are comparing the point of view on folio 65 recto of the Parma manuscript to the point of view on folio 80 verso of the Bordeaux manuscript. Please observe that in the Parma manuscript, the principal optical axis—marked by a dotted line—goes below the head (on the chest of the figure, which has not been drawn). In the Bordeaux manuscript and in all Latin manuscripts, the principal optical axis—also marked by a dotted line—goes between the chin and the lower lip. This difference, together with the different position of the eye in plan, analyzed in figure A.43, naturally leads to a representation in perspective of the head that is different from the one present in the vernacular manuscripts.

Bordeaux, Milan, London, Paris Manuscripts In figure A.46, please consider what we may define as the arch of the chin. In the elevations of this feature in all the Latin manuscripts there are only numbers and drawings, as one will notice, that are different from one another. The only representation in plan, as we have seen, is the one on folio 63 verso of the Parma manuscript which is examined in figure A.39c.

Latin Manuscripts: Comparison of Perspectives; Bordeaux Manuscript, Folio 82 Recto Considering the different point of view in plan and elevation, the perspective in the Latin manuscripts must be completely different from that in the vernacular manuscripts. As may be observed in figure A.47, the perspective in the Bordeaux manuscript (folio 82 recto) is completely different from that in the vernacular manuscripts of Parma and Reggio Emilia, as shown in figure A.34.[50]

The First Two Perspectives of the Latin Head First perspective, with the remade circuit A: The perspective is found from the new point of view, just as the point of view is presented in the Latin manuscripts. The first drawing in figure A.48 shows the perspective of the Latin head with the remade circuit A, just as this circuit appears in figure A.23.

Second perspective, with circuit A compliant with the original: The perspective is found from a new point of view, just as the point of view is presented in the Latin manuscripts. The second drawing in figure A.48 shows the perspective of the Latin head with circuit A compliant with the original, just as this circuit appears in figure A.24. As can be observed in the third drawing, the perspective that appears in the Bordeaux manuscript, folio 82 recto, is different both from the one with the remade circuit A and from the one with circuit A compliant with the original.

The Planning of the New Latin Head It was necessary to plan a new head so as to have a perspective that is more compliant with the one in the Latin manuscripts. Figure A.49 shows that this operation consisted mainly in shortening the distances between some circuits, in leaving circuits C, D, E, F unvaried, and in planning the remaining circuits. Figure A.50 shows the details relative to the elevation of the head with one eye only.

The Second Two Perspectives of the New Latin Head Figure A.51 presents the perspective of the head with the remade circuit A, along with the perspective of the head with circuit A compliant with the original. As may be verified by observing the third drawing in this figure, the perspective that appears in Bordeaux, folio 82 recto, is still very different from the one with remade circuit A, while it is slightly similar to the one with circuit A compliant with the original.

Second Head

Significant Scientific Operations It is good to focus our attention on several significant scientific operations that are at the base of the planning of the second head and of the representation in perspective of the same. In the second part of Proposition VIII in Book 3, Piero places the first head in a random position in space: he does this through three operations (figure A.52). The first operation consists of rotating the elevation of the head with one eye only around the lower extremity of the tip of the nose. The second operation consists of finding the new elevation of the head with two eyes. The third operation consists of rotating this new elevation of the head with two eyes around the lower extremity of the tip of the nose. Consequently, Piero proceeds to: (1) draw the new plans with the projections of all the circuits of the head thus placed in space (figures A.53, A.54); (2) choose on the plane of the new plans (figure A.55) both the terminal (that is, the picture plane) and the eye; and (3) find the wooden rulers. He then goes on to: (1) draw in elevation (figure A.56) on a vertical plane orthogonal to the terminal the heights of all visible points present on the circuits (the apparent outlines of the single circuits and the apparent outline that delimits the head in this new position have been added to these points; the tracing of these apparent outlines is not present in any manuscript, Latin or vernacular; in any case, these apparent outlines well summarize the conclusive steps to representing the new elevation of the head inclined in space); (2) report the height of the eye on the same elevation beyond the terminal, in compliance with the position of the eye, a position that was chosen in plan; and (3) find the paper rulers. These are fundamental steps to realize the second perspective. The understanding of the scientific process which Piero executes—with great accuracy of language—has not been fully comprehended till now because of the incongruences found in the vernacular and Latin texts, as well as and above all because of the inconsistencies present in the drawings of the vernacular and Latin manuscripts. Please take into consideration that in rotating the elevation of the head with one eye only (figures A.53 and A.54) Piero artistically touches the four lower circuits, curving them in the back.

Vernacular Manuscripts: Incongruences It is good to understand what some incongruences consist of, as they are seen in the manuscripts which are the effective graphic data we have at our disposal. The second drawings of the vernacular manuscripts of Parma and Reggio Emilia, of which a reproduction compliant with the originals is offered (figure A.57), are very similar to each other. These are the drawing on the Parma manuscript, folio 71 recto, and the drawing on the Reggio Emilia manuscript, folio 81 recto.

As you may understand from the question mark I have drawn, I am asking myself: Where are the new plans of the circuits, as they were seen in figures A.53 and A.54? They are missing. This is an unforgivable negligence, which does not allow us to read or comprehend the text. Perhaps it is this type of absence that prevented Dürer and Daniele Barbaro from understanding the process of representation of this second head.[51]

Latin Manuscripts: Incongruences Figure A.58, which is the second drawing of the manuscript of Bordeaux, presents the only elevation of the head with one eye only, with the numbers of several circuits in plan. These numbers are connected to their respective numbers on the head with one eye only through horizontal lines. It is obvious that the points in plan present in figures A.53 and A.54 cannot be reproduced in plan without the presence of the same points on corresponding inclined elevations of the head with two eyes. These inclined elevations appear in figure A.59. The manuscript of Bordeaux, on folio 86 verso, in fact presents a very important and significant drawing, without which one cannot understand the constructive process elaborated by Piero in its fullness. This drawing of folio 86 verso of Bordeaux, apart from not being present in the two vernacular manuscripts, is also missing in the Latin manuscript of Milan. The latter, in fact, on folio 93 verso presents a drawing (figure A.60)[52] without a representation of the plans of the circuits, apart from the horizontal lines that can be seen in the figure. The manuscript of London presents the same two drawings that appear in the manuscript of Bordeaux (figures A.58 and A.59); in the same manner, the manuscript of Paris presents the same two drawings again. Let us be careful, though! The drawing in figure A.58 is placed vertically and not horizontally in the manuscript of Paris.

Vernacular and Latin Manuscripts: Perspectives' Comparisons
Figure A.61 presents the perspective of all circuits, with circuit A compliant with the original, alongside the perspective of the same head without the indication of the circuits. Figure A.62 presents the perspective of all circuits, with the remade circuit A, alongside the perspective of the same head without the indication of the circuits. Careful, though! In the perspectives of figures A.61 and A.62 we can see both ears, while in the vernacular and Latin manuscripts we can only see the right ear when looking at the head. Therefore I have chosen to plan the head with less protruding ears, as can be seen in figure A.63. In this figure, in fact, we can observe the perspective of circuit A compliant with the original, as well as the perspective of the remade circuit A.

Comparisons with the Original Perspectives of Parma and Bordeaux
In the first two drawings of figure A.64, the perspective of the head inclined in any position on folio 76 verso of the Parma manuscript is compared to the perspective of the head inclined in any position on folio 96 recto of the Bordeaux manuscript. In the second two drawings of figure A.64, the perspective of the head with protruding ears and with circuit A compliant with the original is compared to the perspective of the head with ears that are less protruding and circuit A compliant with the original.[53] As can be observed, the perspective with the less protruding ears is very similar to the perspectives of Parma and Bordeaux that aren't substantially different from each other.

The Capital

Significant Scientific Operations In Proposition VII of Book 3, Piero starts by describing in detail the constructive process of the capital using these steps:[54] dimensioning the capital in elevation; dimensioning the volute in elevation; the cymatium; dimensioning circuits A, B, and C of the cymatium in plan; other points on circuits A, B, C of the cymatium in plan. Piero then considers the leaves and the addition of the flowers. He identifies all points present on the leaves' relative circles by considering successively circles K, I, H, G, and F of the leaves in elevation. Piero identifies in detail the points present on the volute using these steps: circuit E′ in plan and elevation; circuit E in plan and elevation; circuit D in plan and elevation. He then goes on to consider all the points present on the cymatium following these steps: circuit C in plan and elevation; circuit B in plan and elevation; circuit B′ in plan and elevation; circuit A in plan and elevation. At this point, Piero fixes "the terminal and the eye point in plan," and he finds "the wooden rulers A, B, C, D, E, F, G, H, I, K." He then fixes "the terminal and the eye point in elevation" and moves on to the "paper rulers a, b, b′, c, d, e, e′, f, g, h, i, k," that is, the "paper rulers or rulers of the height," and then follows to "the shortening of circuits A, B, B′, C, D, E, E′, F, G, H, I, K."

Latin and Vernacular Manuscripts: Incongruences To better understand the choices that have been made to realize this volume, we will now analyze the incongruences which are at the base of Proposition VII of Book 3. The manuscript of Parma, at folio 52 recto (figure A.65), presents both the elevation and the plan of the capital in white (that is, without any number specifications, except for the small letters that identify the horizontal sections of the capital). Also the manuscript of Reggio Emilia, at folio 63 recto, presents the elevation without any number specifications, while in plan there are numbers and small letters.

The Latin manuscript of Milan, like the manuscript of Parma, presents the two drawings of the plan and the elevation in white. All other Latin manuscripts partially show numbers and letters. In the drawings relative to the manuscript of Parma, folio 54 verso, which I redrew as close as possible to the original drawings (figure A.66), we can see (together with some numbers and letters) the plan, the elevation, the terminal, a paper ruler, a wooden ruler, and some radii of the first upper circuit which converge toward the respective points of view. The lack of adequate drawings generates confusion, and it is impossible to follow the planning of the capital. Please remember that on the elevation of the volute (where the radii converge toward the point of view) there are no numbers: they are too many and it was not possible to transcribe them. They may be identified in the tables in the text of Proposition VII.[55] The representation of the leaves is problematic too. Please look at the drawings in figure A.67. The representation of the leaves is missing on folio 69 verso of Milan, and it does not appear in the plans of all the drawings of this manuscript. The representation of the leaves is present in all other manuscripts, Latin and vernacular. In these manuscripts, as may be observed, the leaves are represented in plan and are very different from one another. There are seven circles in Parma and Reggio Emilia, there are five complete circles in Milan (which could be seven incomplete ones). In Bordeaux, London, and Paris there are five circles. It is obvious that no manuscript offers the possibility of following the planning of the leaves through the drawings present in these manuscripts. The incongruences are present in the perspectives, too; please observe the copies I extracted from folio 57 recto of Parma and from folio 72 verso of Bordeaux (figure A.68).

The tip of the flower on the left in Parma (figure A.68a) does not go on circuit C (i.e., on the base of the last kyma); rather it goes well above it. The tip of the flower in the center goes below circuit C. The two tips of the flowers present in Bordeaux (figure A.68b) both go on circuit C. The volute on the left in Parma and Bordeaux coincides with circuit C, while in the center the upper line of the strip that connects the two volutes goes above circuit C. Please observe again, in both Parma and Bordeaux, the curvature of the strip that unites the two volutes on the left (i.e., the Ionic part of the capital according to the type chosen here). Considering the accuracy with which Piero writes and plans the capital, since, in both the vernacular texts and the Latin texts, there is no trace of this curvature, we must consider it anomalous. This is certainly a modification that has been applied by whoever copied Piero's original drawing into the manuscripts. For this reason, this curvature has not been reproduced in the reconstruction of the capital (figure A.68c).

The Dome

Significant Scientific Operations Let's move on to an analysis of the drawings in Proposition IX of Book 3. Piero begins to describe in detail the constructive process of the dome through these steps: the barrel vault or the arch of the dome; the slice of the dome; the seven circles; a quarter of a circle; a second circuit; the subdivisions; the semicircles in plan; the subdivisions of the semicircles in plan; line A of the figure of the elevation; line B and the remaining ones. Only at this point, having identified all the points of the coffers in plan and elevation, does he state that he has found the elevation and the width of the dome in the appropriate form. He can then go on to fix "the terminal and the eye point in plan" and to find "the wooden rulers A, B, C, D, E, F, G, H, I, K, L, M, N, O, P, Q, R." He then goes on to fix and find the terminal and the eye point in elevation; paper ruler a; the remaining paper rulers; the shortening of the dome; the vertical elongations of the points; the horizontal elongations of the points; the cornice.

Vernacular and Latin Manuscripts: Incongruences In figure A.69 (which reproduces the drawing of the slice of the dome found on folio 77 verso of the Parma manuscript), the last two internal lines on the right do not converge on point T, but on a point placed before point T.[56] The representation of the dome in plan appears in figure A.70, which reproduces folio 78 verso of the Parma manuscript. Using this drawing, we cannot follow the constructive process of the dome, no matter how hard we try.[57] Nor can it be followed through the elevation drawings in figure A.71. To offer an idea of the complexity of the constructive process of the dome, I tried to reproduce, in compliance with the original, the last figure present on folio 82 recto of the Parma manuscript, as can be seen in figure A.72a. Observing the drawing carefully, we realize that all the coffers are missing the lines that join the vertices of the single squares. Keep in mind that all the drawings relative to the perspective of the dome, in all the manuscripts, Latin as well as vernacular, present incongruences which have now been described as relative to the edges of the coffers. In figure A.72b, in addition to the incongruences now described, we show how easy it would have been to draw circuit Q—which is missing in all the drawings of all the manuscripts, both Latin and vernacular—in order to identify the seven internal squares of the coffers placed at the top; and how—in all the drawings of all the manuscripts, both Latin and vernacular—the four side coffers placed on the right, and the four side coffers placed on the left, are missing the external vertical stripes, although these stripes narrow down until they disappear in the points marked, on the right and on the left, on circuit B.

Final Considerations

It is obvious that Piero certainly executed the drawings of the head at full scale, because the detailed development and the constructive arguments of these texts require it. He then had to reduce these drawings' size to a few centimeters (roughly to the size of a wristwatch) with the aim of adapting them (or, better, of having the amanuenses adapt them) on a folio or cartha of the manuscripts. Naturally when copying such small drawings, the amanuenses made mistakes, the incongruences which I have described for the first time in detail. It is obvious that Piero cannot have personally drawn the two heads in the manuscripts, and he cannot possibly have personally executed the drawings of the capital and the dome: someone must have copied them for him. The drawings examined here are those present in the vernacular manuscripts of Parma and Reggio Emilia and the Latin manuscripts of Milan and Bordeaux. It is good to remember that these manuscripts are all considered Piero's autograph for the drawings. The Parma manuscript is considered to be Piero's autograph also for the text. Following the analysis we carried out here, we cannot consider the two heads, the capital, and the dome in these manuscripts to be Piero's autograph drawings. Nor can we consider the Parma text (relating to these three theorems) to be Piero's autograph: this can be verified by carefully reading the incongruences relative to the text of Parma, which are highlighted step by step in the notes.[58] However, it must be specified that these incongruences in the two heads, the capital, and the dome do not damage the scientific validity of the work. As we know nothing about Piero's autograph drawings concerning the dome, the capital, and the two heads, it is natural to ask: Where are the original drawings that Piero executed of the capital and dome? Where are the full-scale drawings (1:1 scale) of the two heads that Piero had to execute?

Let's leave it to future scholars to deepen the research about the drawings, as well as the task of finding the most appropriate solution concerning the historical procedures in drafting and transcribing the vernacular and Latin manuscripts.

In the translation of *De prospectiva pingendi* that follows, we have chosen to reproduce Piero's so-called original drawings from a Latin manuscript, offering readers the opportunity to observe the drawings *de visu* (through direct view), while all the reconstructed drawings have been included in Appendix B. These two opportunities clearly highlight the objectives, characteristics, and exceptional value of Piero's work. Since the Latin version of *De prospectiva pingendi* was written after the vernacular, it should be considered a "better" version, not only because it was written in a clear and precise language like Latin, but also because scholars of many nationalities could have read the manuscript. For these reasons, we have chosen the drawings present in the Latin manuscript of Bordeaux, Bibliothèque Municipale, Ms Latin 616. It should be noted, however, that this manuscript is missing some lost folios; so that in the third book two drawings are missing in Proposition IV, four drawings in Proposition V, and three drawings in Proposition VI.

On the Perspective of Painting

Piero della Francesca

translated by Rocco Sinisgalli

Book 1

The whole theory of painting requires three parts: drawing, proportioning, and coloring. Drawing is the tracing of profiles and of outlines found in objects. Proportioning is the collocation of these profiles and outlines in their place through a process of proportioning. Coloring is the application of colors to objects which, whether light or dark, reveal themselves according to the varieties of lights. In this work, for the moment, we shall only talk about the proportioning, what is called perspective, mixing it also with a certain part of drawing. Without the possibility of drawing, in fact, perspective rightly does not appear in any painting. We shall omit coloring, dealing only with that part which comes to describing objects through lines, angles, and proportions. That part consists of five parts:[1] in sight, which is the eye; in the form of the seen object; in the distance between the eye and the seen object; in the lines that extend from the edge of the seen object to the eye; in the terminal[2] between the eye and the observed object where the work is expected to be placed. Of these five parts we have thus posed the eye as the first one. We shall not speak about it except when painting process demands it. Therefore, let us say that the eye occupies first place because it is the place toward which are sent all objects, seen from different angles. It happens that, when things that are seen are located the same distance away from the eyes, a bigger object is presented at a greater angle and a smaller one at a smaller angle: this difference makes us understand the shortening of objects.[3] We have said that the second part is the form of the object. Without form, in fact, the object could not certainly be perceived by the intellect nor seen by the eyes. The distance from the eye to the object constitutes the third part; without this [distance] the object to be observed would touch the eye.

And if the object were bigger with respect to the semicircular conformation of the eye, this would not be able to receive the object. The fourth part is made up of the lines, which, leaving from the object's edges, reach the eye: within these lines the eye receives and discerns the object. The fifth part is the terminal on which the object to be represented is placed and on which the measurements of the object have to be precise. If the terminal were missing, in fact, there could be no representation nor a definite description of the object. In addition, it is necessary that we understand under what conditions one draws an object by means of lines on a plane, which we will realize in real shape by means of lines.

Thus, having premised these things we shall proceed with our work on that part which is called perspective, and we shall divide it into three books. In the first one we will talk about points, lines, flat surfaces. In the second one, we shall go over cubical bodies which have a squared basis, solids, rounded as well as multisided columns. In the third one we will examine the proportions of heads, of capitals, of torchi, relative to numerous bases and to other bodies that have different positions.

A point is that which does not have parts; we understand it by imagining it. A line is length without width; it reflects the nature of a point because it also is perceived by the brain. However, since we are dealing with perspective which needs precise descriptions which can be caught by the eyes, it is necessary to present the point and the line with different definitions. Thus the point will be a thing that is very small that can be perceived by the eye. The line will be a length that extends continually from point to point: the eye will be able to perceive its width. The surface comprises length and width and is delimited by lines. There are numerous different kinds of surfaces. There are triangular, quadrangular, pentagonal, hexagonal, octagonal surfaces as well as ones with different faces, as is shown by examples in the figures [figure B.1].[4]

Proposition I[5]

Each dimension is shown to the eye by an angle.

This alone is obvious. A point, in fact, will not contain any dimension, although, in painting procedure, I shall trace a minimum quantity which it is necessary to give to them [the points], so that any other extension is considered larger than the dimension of a point. When the lines originate from the extremities of a dimension and arrive in a point, they create an angle, and that point is the eye.

This occurs as an angle. For example, let A [figure B.2] be a point and BC the dimension from whose extremities lines extend to point A, tracing lines BA, CA, and BC. In fact, A is any given point and so are B and C. Since the lines are traced from one point to the other, since the points are not placed on a straight line, they necessarily form a triangle. Without doubt we say that A is a point from which arises the faculty by means of which we see. The lines which originate from the extremities BC create in that point that angle. We shall conclude that quantity BC is seen by the eye at a predetermined angle.

Bordeaux, Bibliothèque Municipale, Ms Latin 616, folio 1 verso.

Proposition II

All dimensions observed at the same angle, even if positioned in different ways, are presented to the eye in the same way.

For example, let A be the eye, from which lines AB and AC extend [figure B.3]. Then, let there be more dimensions: BC, EF, GH. Each dimension is the same to the eye, that is, at angle A which we have indicated to be the eye. Lines extend from the eye, between which there are the intervening dimensions that touch those lines. No intervening base extends beyond or is detached from them. This being the situation, [bases] are perceived as equal in size. Thus, we shall say that they are delivered to the eye as the same. The ray AC is a straight line led through H and F; and none of those bases goes above the ray or detaches from it. And the ray AB goes through G and E as a straight line that touches the extremities of those [bases].[6] Therefore, let this be the conclusion: that the bases hit the eye according to the same dimension: which is what has been proposed.[7]

Bordeaux, Bibliothèque Municipale, Ms Latin 616, folio 2 recto.

Proposition III

If one places additional right-angle bases at their angle [in the eye], the one that reveals a larger angle is larger or closer.

Let BC and EF be two bases; base BC has angle A and base EF has angle D, and angle A is larger than angle D [figure B.4]. Then I say that base BC is larger than base EF, or that it [BC] is closer to angle A than EF is to angle D.

Proof. Let BAC and EDF be two angles. Above angle D draw angle A by drawing DG that exceeds EF. We have in fact decided that angle A is bigger than angle D. Extend EF. Let them meet at G. I say that base BC is larger then base EF by the quantity FG, as is demonstrated in Euclid's twenty-fourth proposition.[8] Do not extend base BC more than base EF. I say that BC is closer to angle A than base EF is to angle D. Imagine two triangles as we had earlier, that is, BAC and EDF, and let angle A be bigger than angle D and then widen angle D [till it is] equal to angle A and let it be limited by line DG. And from point F draw a parallel line to DE: this parallel line is FI. It intersects DG at point I. Draw from I a parallel line to EF. This parallel line divides line DE at point H. I say that HI is proportional to BC. BC is equal to EF and EF is equal to HI. Then they are equal. As a result, base BC is closer to angle A than EF is to angle D by the entire quantity HE. And this clarifies what has been proposed.

Bordeaux, Bibliothèque Municipale, Ms Latin 616, folio 2 verso.

Proposition IV

If lines were drawn from a point above two bases that have the same size and one was closer than the other one, the closer one will have a wider angle at the point [corresponding to the angle].

For example, let A [figure B.5] be the point and BC and DE are, naturally, the bases that have the same size, and let BC be closer to point A than DE is. I say that BC creates a larger angle than the one made by DE.

Proof. Draw line AF as you wish.[9] Draw two equal bases on this line, BC and DE. BC will be closer to point A [than DE]. Draw AB and AC from point A. From the same point, drawn AD and AE. ABD will be aligned, and AE will intersect line BC at point G. This being the situation, by the preceding proposition, angle A that has base BC is bigger than angle A whose base is DE by as much as the angle that generates base GC. Base BG's angle, in fact, is equal to the angle of base DE because angle A is the angle of both bases.

Thus, angle A of base BC is larger than angle A of base DE by the amount of the angle of base GC which is part of base BC.

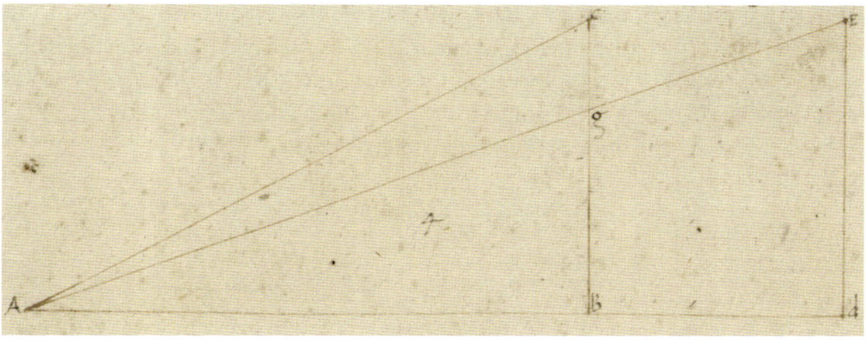

Bordeaux, Bibliothèque Municipale, Ms Latin 616, folio 2 verso.

Proposition V

If two straight lines are drawn from a given point above a base, and above [these] same lines one draws a second [base] parallel and equal in length to the first, the second one would be divided and would divide the preceding two lines in the same proportion. That part of the line which was traced above, the part that is limited between the two lines, will have with the [first] base the same proportion that the whole lines have with the parts delimited by the point.

Having established point A [figure B.6], draw lines AB and AC which have base BC; if an equal and parallel line is added to BC, and this is called DF, which cuts AC at point E and touches AB at point D, it will intersect the two initial lines in the same proportion. This happens because the product of FE and EA is the same as the one of DE and EC, and because the product of FE and AD is the same as the one of ED and DB, and because the product of DB and AE is the same as the one of EC and AD from the moment in which they form a proportion.

There is an alternative way to prove it with numbers [figure B.7]. As we have shown,[10] if AC is 21, AB 18, BC 6, let 6 also be line DF which we shall add; let this line divide AC at point E in 14 and let it touch AB at point D in 12. In the previous demonstration [figure B.6], we have said that DE is equal to 4, EF to 2, EC to 7, and DB to 6. I say that the proportion between 7 and 14 is the same as between 2 and 4, which is also found between 6 and 12. And the proportional ratio between 2 and 6 is the same as between 6 and 18 and between 7 and 21. In conclusion, we can then say that the two lines are divided by a unique proportion and that DE is to BC as AD is to AB and AE is to AC. And AE is to AC as DE is to BC, a ratio that is the same as the one of AD to AB. And with the same procedure the ratio between BC and DE is equal to the one between AC and AE: this is very clear. A proof with numbers may be carried out with AB [equal to] 18, AC to 21, BC to 6, AD to 12, BD to 6,[11] AE to 14, EC to 7, DE to 4. Let's say that the same proportion between DE [equal to] 4 and between BC [equal to] 6 exists also between AD which is equal to 12 and AB which is equal to 18 and the proportion between AE which is 14 and AC which is equal to 21. This is what had been proposed.

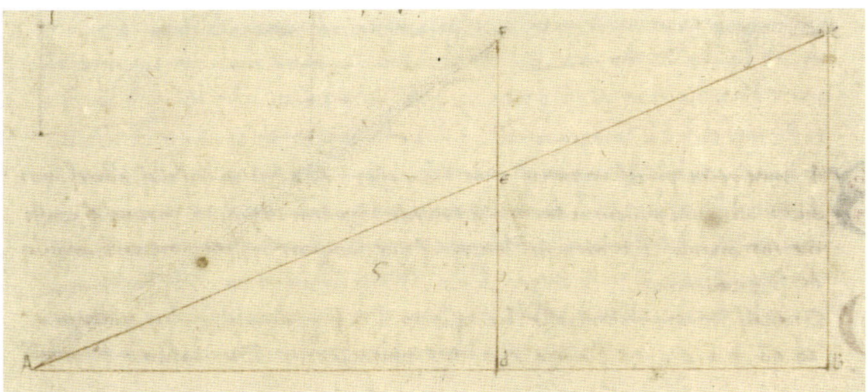

Bordeaux, Bibliothèque Municipale, Ms Latin 616, folio 3 recto.

Proposition VI

If there were two equal lines perpendicular to a single line and parallel to each other; and if from any point there were two lines drawn to the extremities of the closer line and two more lines were drawn to the extremities of the further one; I say that if the bottoms of the perpendiculars were placed lower than the [chosen] point, the one furthest away and lower than the [furthest] line presents itself to the eye higher than the lower line that is closer. And, if the other ends of the perpendiculars lie above the [chosen] point, the furthest extremity appears lower.

BC is a straight line [figure B.8]; at each end are two equal perpendicular equidistant lines BD[12] and CE. Draw a partially elevated point A. BD is closer to point A. Point C will be seen above B; point E lower than D.[13]

Proof. Draw two lines, AB and AD, and two more AC and AE which divide BD at F and G. I say that F will be higher than B by the length BF, and that E will be lower than D by the length DG, and that FG and CE narrow toward the [chosen] point [A] as the same dimension, as was proved in the second proposition of this book and in the tenth proposition of Euclid's *De aspectuum diversitate*.[14]

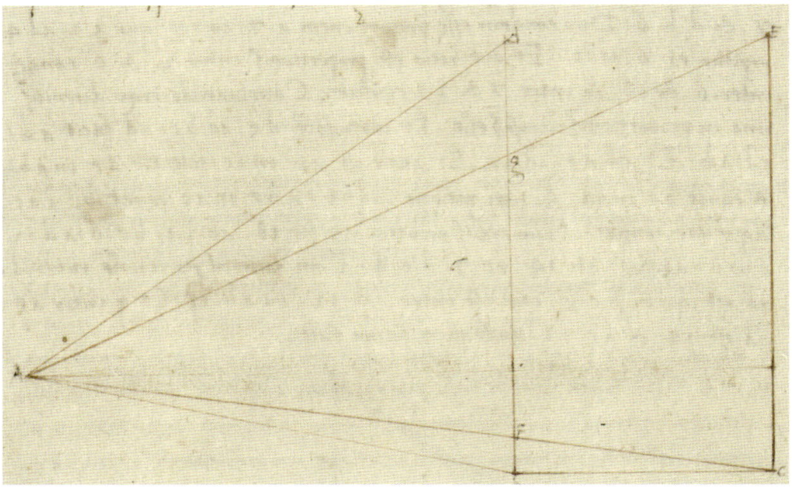

Bordeaux, Bibliothèque Municipale, Ms Latin 616, folio 3 verso.

Proposition VII

If a straight line is divided into equal parts and lines generated from those subdivisions ended at one point, in this same point these lines would not generate equal angles, and the shorter lines would generate wider angles than those generated by longer lines.

BC is a straight line [figure B.9] divided by points D, E, F, G; BA, DA, AE, FA, GA, CA are drawn to point A. I say that EF is below a larger angle than ED. The sides of base EF, in fact, are shorter than those of base ED. In the same way, the sides of base DE are shorter than those of base BD. For this reason, among angles that have the same bases, those with shorter sides relative to those with longer sides generate larger angles, as Euclid demonstrates in his twenty-fourth proposition.[15]

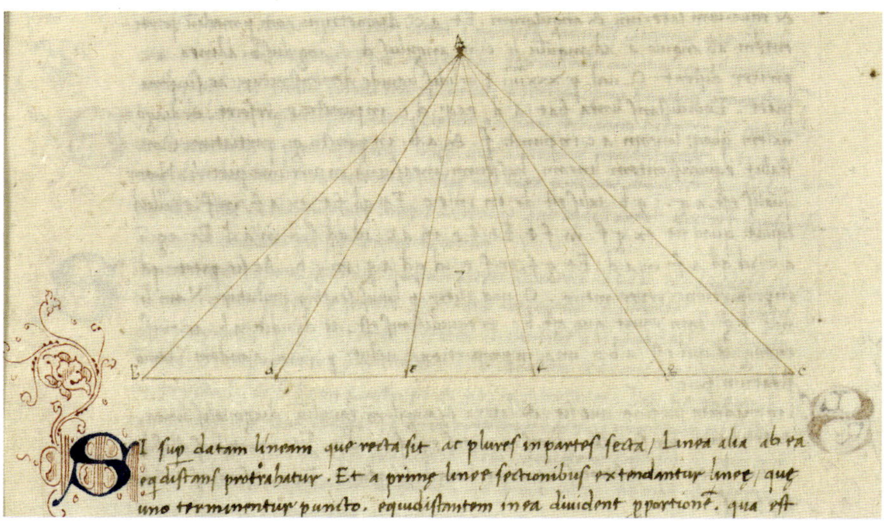

Bordeaux, Bibliothèque Municipale, Ms Latin 616, folio 4 recto.

Proposition VIII

If, above a given straight line, divided into equal parts, we draw another line which is equidistant to this one, and from the subdivisions of the first line we draw lines ending in one point, these lines will divide the equidistant line in the proportion of the given line.

Draw line BC [figure B.10], which is divided at [points] D, E, F, G. Draw another line equidistant from this one. Call it HI. Draw lines BA, DA, EA, FA, GA, CA to point A, dividing HI at K, L, M, N. I say that this [HI] was divided according to the proportion of the given line, because BD is to DE as HK is to KL and EF is to FG as LM is to MN and FG is to GC as MN is to NI.[16] Triangle ABD is similar to triangle AHK, as triangle ADE is to triangle AKL, and triangle AEF is to triangle ALM.

In this way, therefore, they are linked by proportion. And the proportion between AB and BC is also present between AH and HI. When the larger bases are in proportion, the smaller ones will be also. And the angles of triangle ABD will be identical to the angles of triangle AHK. Therefore [the figures] are proportional, as shown in Proposition XXI of Euclid's sixth book.[17] This is what was proposed.

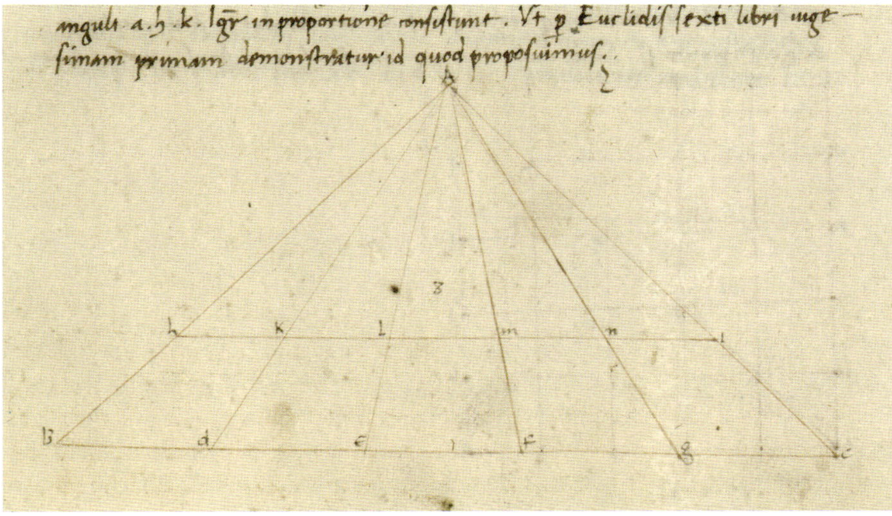

Bordeaux, Bibliothèque Municipale, Ms Latin 616, folio 4 recto.

Proposition IX

If we draw a diagonal line on a surface of four sides, that will cut the surface into two equal parts; and if we draw a line parallel to the sides, this line will divide the diagonal line into two parts and itself in the same proportion.

ABCD is a surface with four equal sides, and AC is a diagonal [figure B.11]. AC divides surface ABCD into two equal parts. We have established, in fact, that the surface is a quadrilateral, with equal sides and angles, and that the diagonal line AC divides it in half, from angle A to angle C,[18] since angle D and angle B are equally distant from line AC. This is proved by Euclid's Proposition XXXIV,[19] confirming the first part of our statement. Draw a line parallel to AD, intersecting DC at point E and the diagonal line AC at point F and line AB at point G.

The parallel line subdivides itself and each of these lines proportionally. In fact, AG is to GB as DE is to EC. And what we obtain from AF and FE is equal to what we obtain from FG and FC; and [what we obtain from] FE and AC is equal to [what we obtain from] FC and AD, and [what we obtain from] AG and AC is equal to [what we obtain from] AF and AD, and [what we obtain from] GF and FE is equal to [what we obtain from] AG and GB. Therefore they are in proportion. We can prove this in a different way through the sixth proposition of this book. Line EG, in fact, is equal and parallel to BC and divides [sides] AB, AC of triangle ABC in the same proportion, as was also demonstrated at the same time in the fifth [proposition of this book].

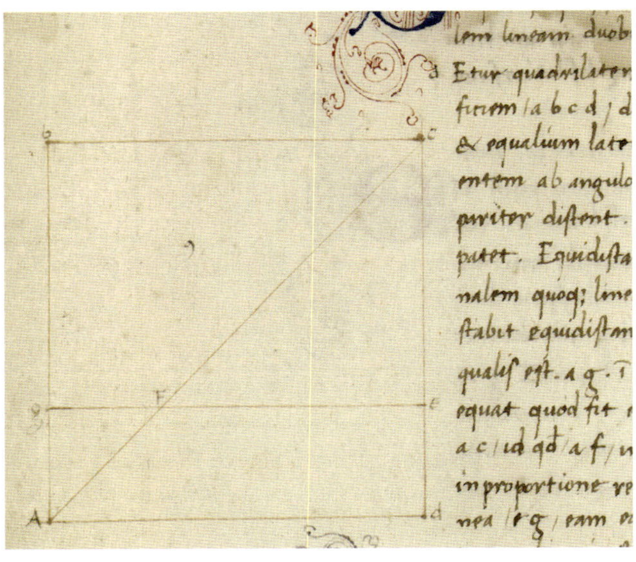

Bordeaux, Bibliothèque Municipale, Ms Latin 616, folio 4 verso.

Proposition X

If one draws a diagonal line, on a square surface with equal sides and angles, and that diagonal is divided into many equal parts and from those subdivisions one draws equidistant lines to the four sides of the surface, [these equidistant lines] will divide the surface in similar surfaces.

Let ABCD be the surface [figure B.12]; AC is the diagonal line, divided into equal parts at points E, F, G, H. I say that if we draw a line parallel to AB through E and another [vertical] parallel to AD; and if we draw the parallel line to AB through F and another [vertical] parallel to AD; and if we draw the parallel line to AB through G and another [vertical] parallel to AD; and if we draw the parallel line to AB through H and another [vertical] parallel to AD; these lines will form twenty-five similar surfaces to ABCD, with equal sides and angles. Draw, as previously mentioned, the surface and the diagonal line divided at points E, F, G, H, and draw the parallel line to AB through E dividing AD at I and BC at K. Draw the parallel line to AD through E and another [vertical] line that divides AB at L and DC at M. Draw the parallel line to AB through F, intersecting AD at N and BC at O. Draw another line through F parallel to AD that intersects AB at P and DC at Q. And extend through G the parallel line to AB, dividing AD at R and BC at S. Draw through G another [vertical] line parallel to AD that intersects AB at T and DC at V. Draw the parallel line through H to AB, dividing AD at X and BC at Y. In the same way, according to this procedure trace through H another [vertical] parallel line to AD that intersects AB at z and DC at Z. I say that all these squares that are part of surface ABCD are similar to it and to each other. In fact, they are formed by parallel lines and angles that are identical to each other. Therefore, their sides are proportional to the sides that form surface ABCD. For this reason, it is appropriate to conclude that those surfaces which we talked about are similar in shape, as demonstrated in the thirtieth [proposition] of Euclid's first book.[20]

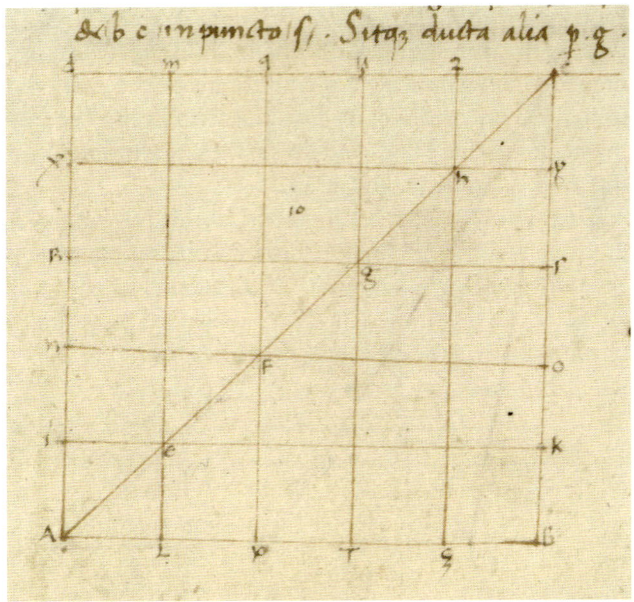

Bordeaux, Bibliothèque Municipale, Ms Latin 616, folio 4 verso.

Proposition XI

If we have a quadrilateral figure, with the sides divided into nonequal subdivisions, and from these we draw parallel lines to the sides, and from the angles is drawn a diagonal, [this] line would divide the [parallel lines] according to a determined proportion, and if from those subdivisions we draw equidistant crosswise lines, we would have surfaces that have the same proportion.

Let ABCD be the proposed figure, formed by four sides [figure B.13]. Divide side AB in nonequal parts through points E and F, drawing the equidistant lines to AD through E and F, intersecting DC at G and H. Draw diagonal AC which divides EG at I and FH at K. Extend the horizontal parallel line to AB through I, dividing AD at L, FH at P, and BC at M. Draw a parallel line to AB through K, dividing AD at N, EG at Q, and BC at O. Three of these squared figures are equilateral, six instead are made with equal sides of those three. Indeed, because of this, it is clear that [these six] are not equilateral; however, their surfaces are in proportion and are composed from nonequal lines, but lines that are parallel to each other.

The demonstration that follows shows that the six figures are proportional to each other. Thus, in fact, AE is to AL as EI is to EF, and LI is to LN as IP is to IQ, and NQ is to ND as QK is to QG, and as EI is to EF so FP is to FB, and as IQ is to IP so PM is to PK, and as KQ is to QG so KO is to KH. These things [that we have said] on the proportion of these [sides] are a proof. However, to make this procedure clearer, we shall demonstrate it using numbers. All sides of ABCD are 9 units. We have divided each of these [sides] into three nonequal parts, [for example] in E and in F. And let's say that AE is equal to 2, EF equal to 4, and FB equal to 3. Thus [it follows that] DG[21] is equal to 2, GH to 4, and HC to 3.

And we have traced the diagonal that goes through points I, K, subdividing it proportionally. And from those points of subdivision we have drawn [horizontal] parallel lines to AB that meet AD at points L and N and BC at points M and O, EG in Q and FH in P. This way AE, AL, LI, IE are equal [to each other], and so are IP, PK, KQ, QI, and KO, OC, CH, HK. The diagonal, in fact, which generates these [divisions], starts from the angles of the surface composed of equal sides

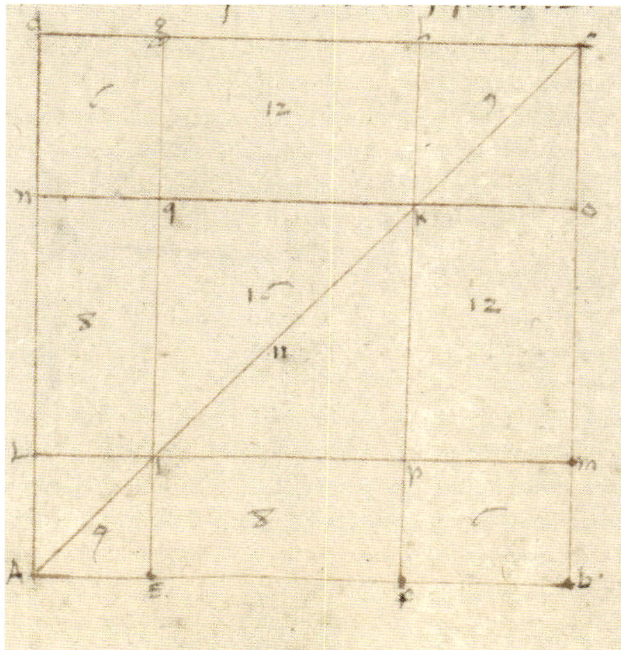

Bordeaux, Bibliothèque Municipale, Ms Latin 616, folio 5 recto.

and angles and bisects them. This way we have equal-sided squares, as is shown in Proposition XXIV of Euclid's sixth book.[22] Earlier, we posited that AE is [equal to] 2. Thus, AEIL will have all its sides equal to 2. We decided that EF is equal to 4, therefore IPKQ will have all its sides equal to 4. We decided that FB is [equal to] 3; thus KOCH will have all its sides equal to 3; and EFPI will have one side equal to 2 and the other one equal to 4; and FBMP will have one side equal to 2 and the other one equal to 3; and PMOK will have one side equal to 3 and the other one equal to 4; and LIQN will have one side equal to 2 and the other one to 4; and NQGD will have one side equal to 2 and the other equal to 3; and QKHG will have one side equal to 3 and the other one to 4. If the side of the first square is multiplied by itself we obtain 4, and 2 times 4 is 8 and 4 times itself is 16 and 2 times 3 is 6 and 3 times 4 is 12. All these numbers are in proportion. The surface, in fact, which is 4 is proportioned to the surface which is 8 as surface LQ which is 8 is proportioned to surface IK which is 16. And in the same way, surface NG which is 6 is proportioned to surface QH which is 12, because EP which is 8 is proportioned to surface FM which is 6.[23] Surface PO which is 12 is proportioned to surface QH which is 12 and this to KC which is 9. The surfaces are in proportion to each other, like the divisions of the square's sides. The proportion between 2 and 4 also exists between 4 and 8 and between 8 and 16 and between 6 and 12; and the proportion between 4 and 3 also exists between 8 and 6, between 16 and 12, between 12 and 9: numbers that are connected by a similar ratio. Having established these things, we have certainty about the proportioning of squared figures.

Up to this point, we have considered the proportion of nonshortened lines and surfaces and, moreover, in which way the diagonals divide quadrilateral surfaces into two equal parts, as well as how subdivisions created by parallel lines on the same surfaces are proportioned.

Now, however, because we have to speak about shortened lines and surfaces, we must give precise indications about proportion, in order to understand what is meant by "proportionally." Certainly proportions are unlimited, and the one discussed is neither double like 2, 4, 8 nor sesquialtera like 4, 6, 9, nor sequitertia like 9, 12, 16, nor triple or quadruple. I do not call a shortened proportion one like 4, 8, 12, 15, nor 4, 6, 7, 8, 9, but one that takes place according to the distance from the eye to the site where the shortened objects are placed, and the one that goes from this terminal to the shortened objects. Here is a precise indication of what has just been said. Let there be four parallel lines, let them be one braccio apart from each other,[24] let them be one braccio long, and let them be placed between parallel lines. And from the first line, that constitutes the established place, to the eye, let there be 4 braccia. Thus, from the second line to the first one there will be a "sexquiquarta" ratio; from the third line to the second one, at the determined site there will be a "sexquiquinta" ratio; from the fourth line to the third one at the determined site there will be a ratio of "sexquisexta."[25] To make this clearer, the proportion between these four lines is the same as between the numbers 105, 84, 70, 60.[26] But if the distance between the eye and the established place varies, so does the proportion. Those four lines we talked about, in fact, will change in relation to the proportion, when we will back off 6 braccia away [from the established site] for example, in comparison to the original view of things to observe. And they will be like these four numbers 84, 72, 63, 56, which are not in the same proportion as the first ones.[27] The distance of the eye from the reality of the first [established] place, in fact, is not in the same proportion with the distance [of the eye] from the second [established] site. Therefore, if the ends are changed, so is the proportion; and there is always a proportion from the first line to the second, and it coincides with the proportion from the eye to the first one and from the second one to the eye; that is, the proportion that we have from the line that begins from the eye and ends on the first line is the same as from the eye to the second line. Because the changes of these proportions cannot be easily demonstrated through numbers, I shall try to demonstrate the changes through the lines in the shortened surfaces.

Proposition XII

From a given eye, in an established terminal, shorten the assigned plane.

Let A be the given eye [figure B.14], above line DC which is perpendicular to DA,[28] and let DC be subdivided at point B which is the established terminal. FB is a perpendicular line above B; and BC is the assigned plane to be shortened. Draw a line from point A to point C at the end of the assigned plane. This line intersects BF at point E. Therefore, I say that BE naturally is the shortened assigned plane, because BE is seen by the eye just as BC appears in the established terminal.

Proof. Draw AB: we shall obtain triangle ABC. Bases BC and BE will be included in the same angle. They present themselves in the same way to the eye, as is shown in the second proposition of this book. We shall say that BE is the shortened assigned surface. To prove this, we shall adopt an alternative way so that when we have skillfully executed the first shortening, we shall easily comprehend what remains. Earlier, I said that given the eye, we must consider it in the following way: the place where the eye is placed is also the place where we have to be positioned to observe the assigned surface. The assigned plane corresponds to a quantity of indeterminate size, within the surface, until the established terminal where we have to execute the shortening of the [assigned] surface. This gap of the space is the distance that the eye has with respect to a wall or to a picture plane or any other thing where we are ready to place things we wish to shorten.[29] After the eye has been placed, to place it in the following way, high or low, far or close, according to our needs to realize the work, we make the assigned surface BC 20 braccia, the distance DB from the eye is 10 braccia, the eye, placed in A, has to be elevated 3 braccia. Draw AC, which will divide BF at point E, as I mentioned earlier. E, in its assigned boundary, is higher than B by length BE. In fact, A is elevated above BC, as demonstrated in the tenth proposition of Euclid's *De aspectuum diversitate*.[30] We shall then say that BE is 2, a number that corresponds to ⅔ of the height [of the eye]. The eye was placed, in fact, three braccia high and ⅔ corresponds to 2, because the line that starts from point A divides the parallel line [at the height of the eye] in proportion, so that the proportion of DC to BC is the same as between BE and DA; and DA is 3 and BE is 2. Also DC is 30 and BC is 20. 20 is to 30 as 2 is to 3. We shall say that BE is the shortening obtained.

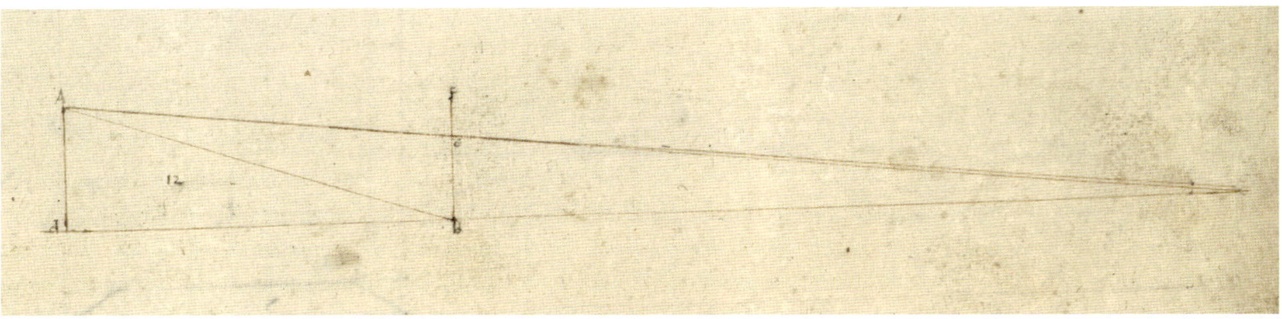

Bordeaux, Bibliothèque Municipale, Ms Latin 616, folio 6 recto.

Proposition XIII

Reduce the shortened surface to a squared shape according to the previous demonstration.[31]

Let line DC be divided [figure B.15] at point B and draw BF perpendicular to it. A is above D. Draw a perpendicular line which ends at C and is equal to BC: let it be GC. From point G extend a line parallel to BC: let it be GF. Let's say that this figure BC, CG, GF, FB is a square with equal angles.[32] Draw from A lines AC and AG which will divide BF at two points. AC will intersect it at E and AG at H. I say that E is presented to point A higher than B. A, in fact, is higher than B, and H appears lower than F, because A is lower than F, as demonstrated in the tenth and eleventh propositions of the *De aspectuum diversitate* by Euclid.[33] Therefore BE will appear equal to BC on the assigned terminal and EH will appear to be equal to GC and HF equal to FG. Draw AF and AB. We shall obtain three triangles, each with two bases: ABC, in effect, as clear as it is, has the two bases BC and BE, and triangle ACG has the two bases GC and HE. In the same way, triangle AGF has two bases which are FG and FH. By the second proposition of this book, it is obvious that base BE appears equal to base BC.

In fact, they are underneath angle A, and base EH seems to be equal to base CG because they are underneath the same angle. And base HF appears to be equal to FG because they are together under one angle. And the proportion between AB and BE is the one between DC and BC; and the one of FH and CG is also present for AH and HG. And when the distances of things are—with respect to the height of the eye—in the same proportion with the shortened thing, there clearly is a correct shortening. We shall then say that EH and CG are on a plane and that BE is part of a squared object. Now draw a line from point A parallel to DC and without limits. Divide line BC into two equal parts at point I. And draw a perpendicular line above I and in the place where this perpendicular line cuts the line which begins at point A, parallel to DC, draw point a. Extend the parallel to BC from point E to divide CG at K. Then draw from point a the line that will reach point B and that divides EK at point d. Successively, extend the line from a to point C; this line will divide EK at point e. At these conditions, I say that shortened plane which is BCde was reduced to a squred figure.

This is the demonstration. Examine whether de is equal to EH, that is, referred to the length CG. I say therefore that [de] is equal or similar to [EH]. Between AB and AD, in fact, we have the same proportion that exists for AC and AE. And the one between de and BC is the same as the one between EH and GC. Hence, being proportional they will be equal or similar. They are equal because we have established that BC is equal to the length of one and the other. This way, what we have proposed is clear. If someone asks: why is the eye in the middle [at point a]? Because that position is convenient to observe the work. And the eye itself can be placed where you want it, as long as it does not go beyond the limit of the figure, as will be shown in the last figure of this [first] book. And wherever you shall place the eye, you shall have the same proportion.

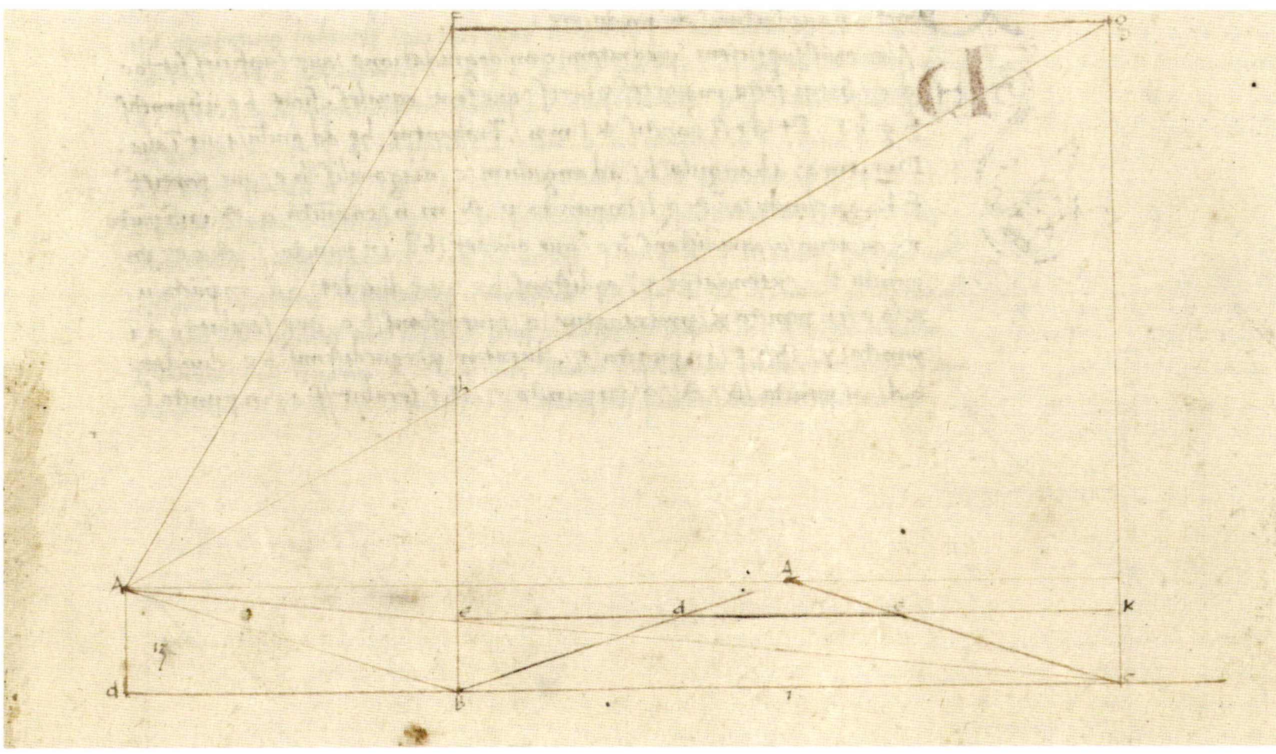

Bordeaux, Bibliothèque Municipale, Ms Latin 616, folio 7 recto.

Proposition XIV

Divide a squared figure shortened into equal parts.

BCDE is the shortened figure [figure B.16]. Let A be the eye, as we have said previously. The eye has been placed above the shortened surface of the square. The eye corresponds to the place established earlier, as demonstrated. We shall proceed indeed with these things because this method produces the same effect in a short time. Let BCDE be a [shortened] square. Let A be the eye. Divide BC into many parts at your own will. Divide it at F, G, H, I. Then draw [a line] from point F to point A, and [extend lines] from points G, H, I to point A. Divide DE at points K, L, M, N. I say that DE has been sectioned according to the same proportion as BC. In fact BF is to DK what BC is to DE,[34] and FG is to KL what GH is to LM, and HI to MN what BC is to ED, so that they constitute a proportion.

This can be demonstrated in a different way. Because BC and DE lie underneath the same angle, and in the same way do BF and DK, and so do FG with KL and so do GH with LM, and we have the same ratio with HI and MN, and IC with NE all lie underneath the same angle; as a consequence the bases are equidistant; as a result they share the same proportion, as demonstrated for the fifth proposition[35] of this book. This is what was proposed.

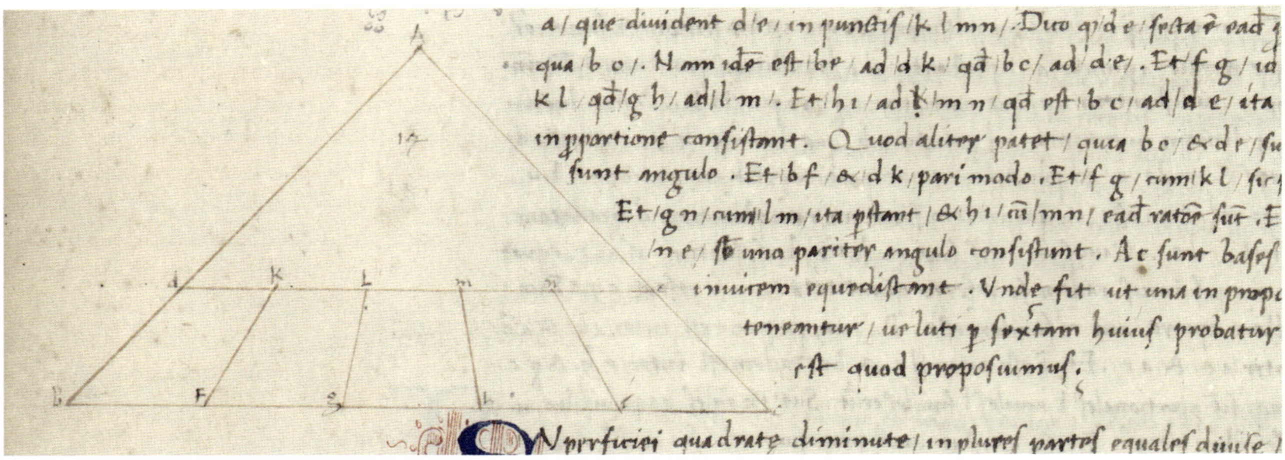

Bordeaux, Bibliothèque Municipale, Ms Latin 616, folio 7 verso.

Proposition XV

Produce the divisions in squares of a shortened squared surface divided into many equal parts.

Let one have a shortened squared surface, whose surface is BCDE [figure B.17] and is divided into many equal parts, as BC is subdivided at points F, G, H, I and DE at points K, L, M, N. Draw lines from these [subdivisions] to the eye, as in the preceding [figure]. Draw diagonal BE from B to E. It will intersect FK at O, GL at P, HM at Q, and IN at R. Draw through O the equidistant line [ST] to BC. It will divide BD at S and CE at T. Draw through P the equidistant line [VX] to BC. It will divide BD at V and CE at X. Draw through Q the equidistant line [Yz] to BC. It will cut BD at Y and CE at z. Draw through R the equidistant line [Z-9] to BC. It will divide BD at Z and CE at 9 and will intersect FA at 1, GA at 2, and HA at 3.

Then those divisions will be reduced to a squared shape, as I said before: this can be proved in the following way. Draw a square in its real shape below line BC; let it have size BCDE as we proposed before, and let it have the same subdivisions in parts of BC, so as to obtain squared shapes with equal sides. Draw diagonal BE. I say that it will divide those sections on the same points according to which the shortened square is divided by the diagonal. Therefore, if the diagonal of the square in real shape [BE] cuts the parts in proportion, the corresponding diagonal [BE] will divide the shortened squared surface in the proportion present in the shortened figure in the following way. As AD and AB are in proportion, so are DK and BF. And the proportion present between KL and FG is the same one as between LM and GH, and as MN is to HI, so NE is to IC; and the proportion between AZ and AB is the same one as between Z-1 and BF, between 1-2 and FG, between 2-3 and GH, between 3-R and HI, between R-9 and IC. And in the same way, we continue in the remaining ones, so that they are in proportion. And the executed divisions are formed by squares: as had to be demonstrated.

But if it were to happen that the square were not divided into equal parts, the diagonal would divide it in proportion, as a second figure demonstrates [figure B.18]. Let BCDE be the real square. Draw diagonal lines BE and CD, which divide the surface into four equal parts: any line that is traced parallel to the sides [of the square] will divide them

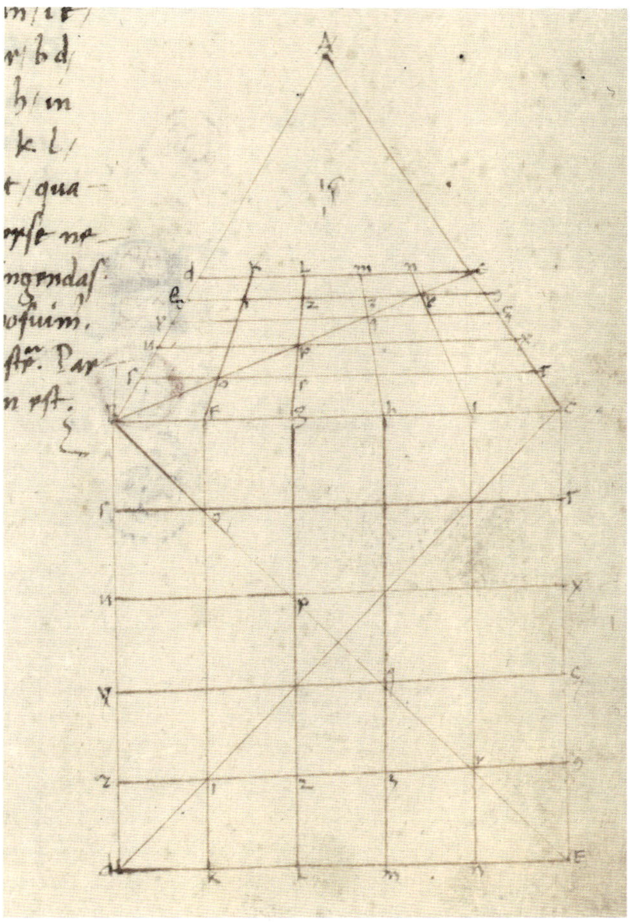

in proportion. For example, let BCDE be, as we have said, the figure in real shape. We have to enter for ¾ of the surface [in dividing this surface]. I shall then take BF which is ¾ of BC. I shall draw a parallel line to BD through F, which will divide the diagonals at two points. BE in fact will be divided at point H, DC at point G, and DE at point I; parts so divided are proportional to each other. In fact, BF is to BC as BH is to BE, and in the same way DG is to DC as BF is to FC and as DI is to IE. And if we draw another parallel line to DE that goes through H, BD will be divided at point K and CE at point L. This way, BK will be equal to BF because FH is to HI as BF is to FC and the diagonal line divides FI and KL in the same point H. And because the diagonal [BE] that is detached in a square also produces another square, it happens that BF, FH, HK, KB are connected to each other by a ratio of equality. And, previously, we have said that we should reach ¾ of the surface of that square that we claimed to be BCDE. We shall say therefore that line KL is placed below ¾ of the surface: it divides, in fact, BD and CE in parts equal to BF, which, as we have said, contains ¾.[36]

Bordeaux, Bibliothèque Municipale, Ms Latin 616, folio 8 recto.

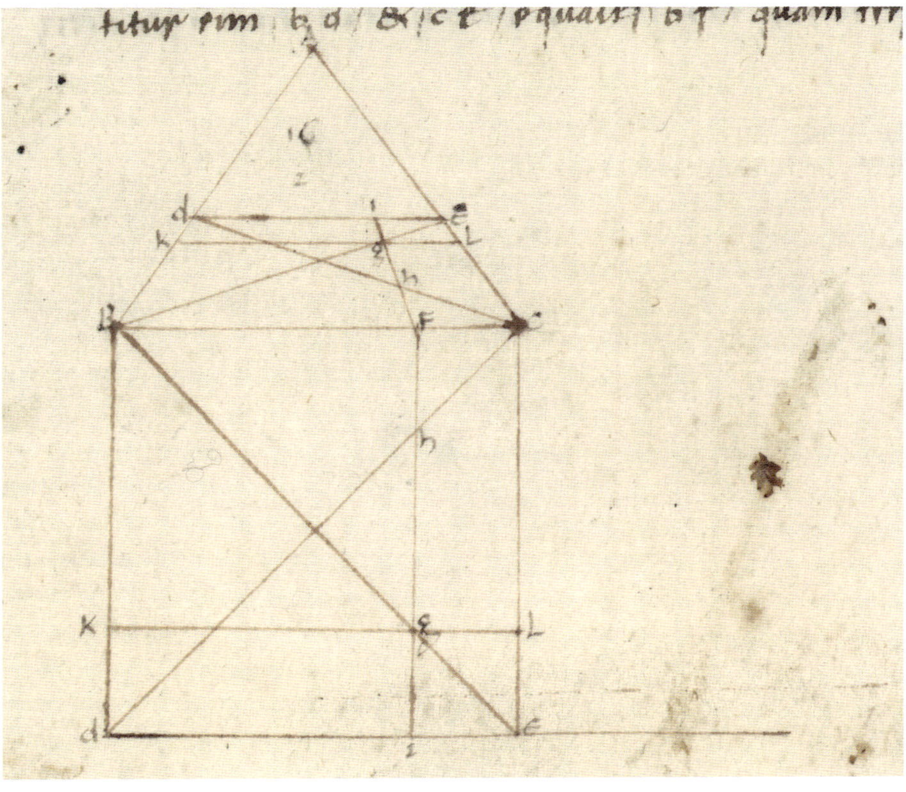

Bordeaux, Bibliothèque Municipale, Ms Latin 616, folio 8 recto.

Proposition XVI

Turn a shortened quadrilateral surface into an octagonal figure.

Let BCDE be a shortened quadrilateral surface and A the eye point [figure B.19].[37] Draw a real figure with four equal sides below line BC; one of its sides has to be as long as BC. In the real square BCDE, draw eight sides in real shape. And divide BC at F and G, and BD at M and N, and DE at K and L and CE at H and I, so that FG is equal to GH, and GH to HI and HI to IK and IK to KL and KL to ML and ML to MN and MN to NF, so that they are equal to each other.

Afterward, draw diagonals BE and DC which will intersect at point O. Then draw HN which will divide diagonal BE at point P and diagonal CD at point Q. Draw IM which will divide diagonal BE at point S and diagonal CD at point R. Also, draw diagonals of the shortened surface BECD. Then extend from F and G the lines to point A, which will meet at four points [the diagonals]. The line from F will cut, in fact, BE at P and DC at R and DE at L. And the [line] from G will cut BE at S and DC at Q and will cut DE at K. At this point, draw PQ equidistant to BC, which cuts BD at N and CE at H. RS is parallel to BC and divides BD at M and CE at I. And draw GH, IK, LM, NF. This way, in fact, we shall have completed the octagonal figure. This emerges from the fact that we proved in the thirteenth[38] proposition of this book: that shortened BCDE exactly coincides with BCDE in real shape. In addition to this, for the eleventh proposition of this book, we have demonstrated that in a square subdivided into unequal parts, the diagonal divides these parts proportionately. Or rather, for the twenty-fourth proposition of Euclid's sixth book[39] it is sufficiently demonstrated that all the surfaces of the quadrilateral figure [which are] around a diameter have a similar shape ratio. They are located, in fact, around the diagonals which correspond to diameters which divide the surfaces in the same way as they divide the quadrilateral figure and also the shortened shape. Therefore, we shall say that GFNMLKIH is the shortened squared shape which has been turned into a figure of eight sides.

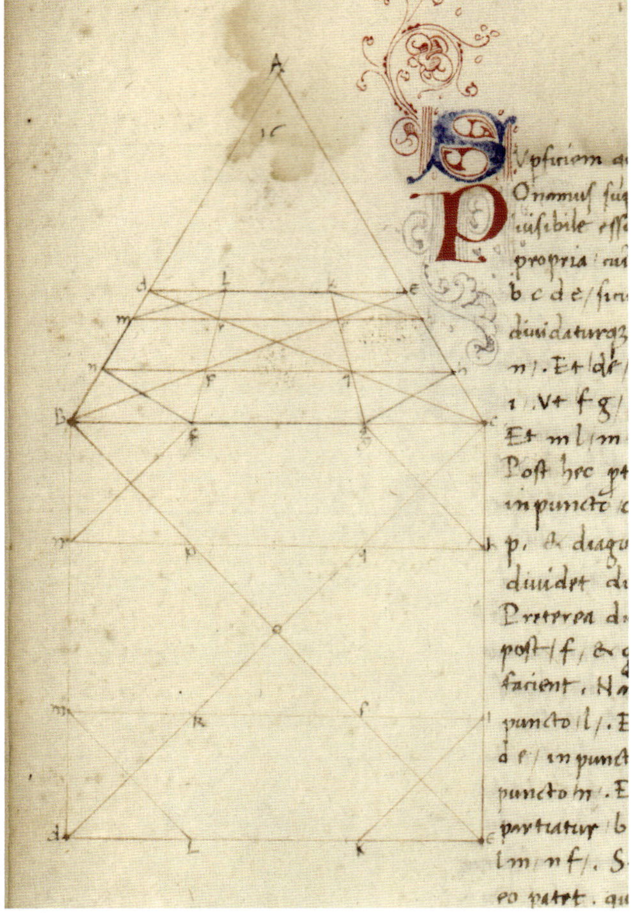

Bordeaux, Bibliothèque Municipale, Ms Latin 616, folio 8 verso.

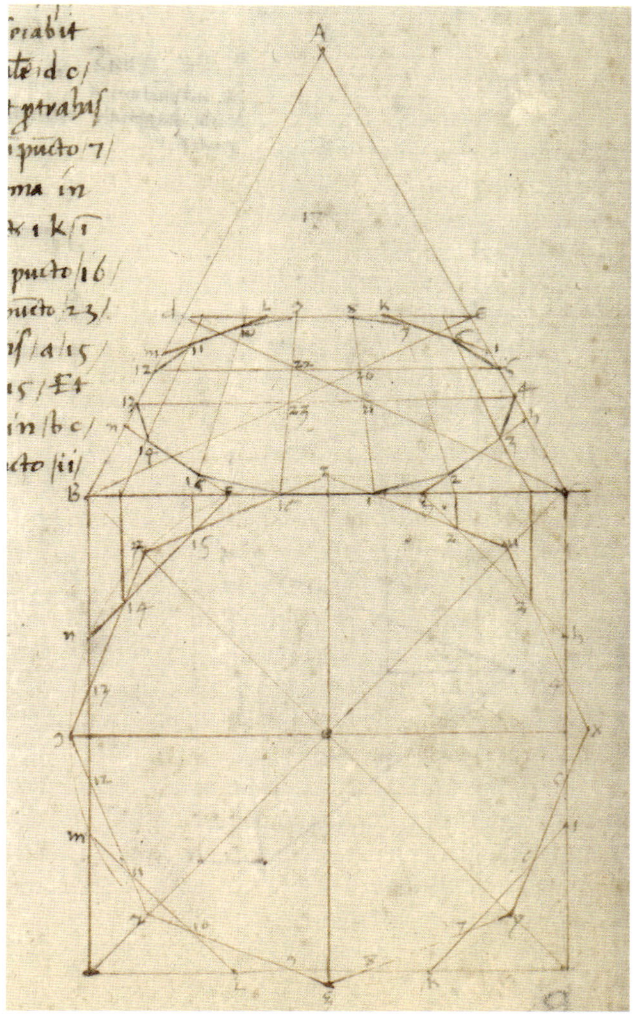

Bordeaux, Bibliothèque Municipale, Ms Latin 616, folio 9 recto.

Proposition XVII

Turn an eight-angled shortened figure into a sixteen-angled one.

We have drawn an eight-angled figure, which is [formed by] FG, FN, NM, ML, LK, KI, IH, HG; and which is located in BE according to a square [figure B.20]. We want to make it a sixteen-sided figure. We first have the octagon in real shape, which touches the shortened surface in the way we have demonstrated earlier.[40] Place the fixed point of the compass in the center of this surface, which is [point] O. Extend the other movable point of the compass to F and rotate according to that quantity, which will allow it to touch all eight of the angles of the octagon. In addition, divide each of the sides of the octagon into equal parts and separate them with a point. Now draw, from the center to the circumference,[41] a line that passes through those points and intersects that circle.

The line that passes between F and G will divide the circle at point T; the line traced between G and H will divide the circle at point V, the line traced between H and I will intersect it at point X, the line traced between I and K will go above the circle at point Y, the line traced between K and L will intersect it at point z, the line traced between L and M will intersect it at point Z, the line traced between M and N will intersect it at point S, and the line traced between N and F will intersect it at point R. Draw TV, which will cut FG at point 1 and GH at point 2; and draw VX, which will cut GH at point 3 and HI at point 4; and draw XY, which will divide HI at point 5 and IK at point 6; and draw Yz, which will intersect IK at point 7 and KL at point 8; and draw zZ, which will cut KL at point 9 and LM at point 10. And draw ZS, which will divide LM at point 11 and MN at point 12; and draw SR, which will intersect MN at point 13 and NF at point 14; and draw RT, which will divide NF at point 15 and FG at point 16. At this point, we have found sixteen sides in real shape. Because these sides must be shortened, you need to draw equidistant lines parallel to BD and CE through points 1, 2, 3, 16, 15, 14: these parallel lines meet line BC. And from the contact points on line BC, draw, toward point A, which rises above the eight sides, the shortened lines, which are FG [in real length], GH,[42] HI, IK, KL, LM, MN, NF shortened. And draw the line from point 1 on FG to point A, which cuts diagonal BE at point 20, diagonal DC at point 21, and line KL of the shortened octagon at point 8. And draw the line that goes from point 2 [to point A] and divides GH at point 2 and IK at point 7. Draw the line that starts from point 3 of the octagon in real shape until it touches BC and intersects GH at point 3 and IK at point 6 of the shortened figure. And extend the line from point 16 to point A, which will divide FG at point 16, diagonal DC at[43] point 22, diagonal BE at point 23, and KL at point 9. And draw to point A a line that starts from the contact of point 15: this line will intersect FN of the shortened octagon at point 15 and LM at point 10. And from the contact of the line through 14 with BC, draw a line to point A, which will divide FN at point 14 and LM at point 11. And draw, through 20 and 22, a line parallel to BC which will cut HI at point 5 and MN at point 12. And draw through 21 and 23, which are on the diagonals, a line parallel to BC,[44] cutting HI at 4 and MN at 13. All these characteristics take us to the shape of a sixteen-sided polygon. [In the shortened drawing,] draw lines 1-2, 3-4, 5-6, 7-8, 9-10, 11-12, 13-14, and 15-16. I say that the figure has a proportionate shortening. We have posed, in fact, a shortened octagon as we have seen in the preceding proposition. Later, when drawing an octagon with eight sides in real shape, we have increased the number of sides to sixteen, of the same length. We have then drawn the lines parallel to BD from the angles [of the octagon], making line BC a terminal, from which we have drawn the lines toward point A to form there the octagon and the two diagonals as they also divide the octagon in real shape. What we proposed has been achieved.

Proposition XVIII

Draw a triangle of equal sides on a squared [shortened] surface.

Let BCDE [figure B. 21] be the squared shortened surface on which we have to draw an equilateral triangle. I shall draw a squared surface in its real shape which touches BC according to an equilateral side. BCDE will certainly be this [square], in which I will trace triangle FGH with equal sides. I shall then draw a diagonal BE on the real-shape square. I shall then draw from point F a line parallel to BC which will intersect diagonal BE at point I. Successively, I will draw from point G a line parallel to BC which will intersect the diagonal at K. Moreover, I will draw from point H a line parallel to BC, which will intersect the diagonal at L. And I will draw a line parallel to BD from point I, which will cut BC at point M. And I will draw a line parallel to BD through K which will meet BC at point N. And I will draw a line parallel to BD through L which intersects BC at point O. This being the situation, I will draw, in the shortened figure, diagonal BE and I will draw a line from point M to point A. Where [this line] will intersect the diagonal,[45] I shall draw point I. And I will draw a line from point N to point A, and where it will intersect the diagonal BE I will insert point K. And I will draw a line from point O to point A, and where it will cut the diagonal BE I will insert point L.

And I will draw [from these points I, K, and L] parallel lines to BC until they touch BD and CE. After this, I shall go back to the figure in real shape. And I will draw a line from F parallel to BD, which meets BC at point P. And I will draw a line from G parallel to BD, which meets BC at point R.[46] And I will draw a line from H parallel to BD, which meets BC at point Q. I also will draw the line from P to A, and where it will intersect the horizontal line which passes through I, I will insert point F. And I will draw the line from Q to A, and where it will intersect the horizontal line which passes through L, I will insert point H. And I will draw the line from R to A, and where it will intersect the horizontal line which passes through K, I will insert point G. Finally, I will draw FG, GH, HF. In this way you will get the equilateral triangle FGH, reduced proportionally.

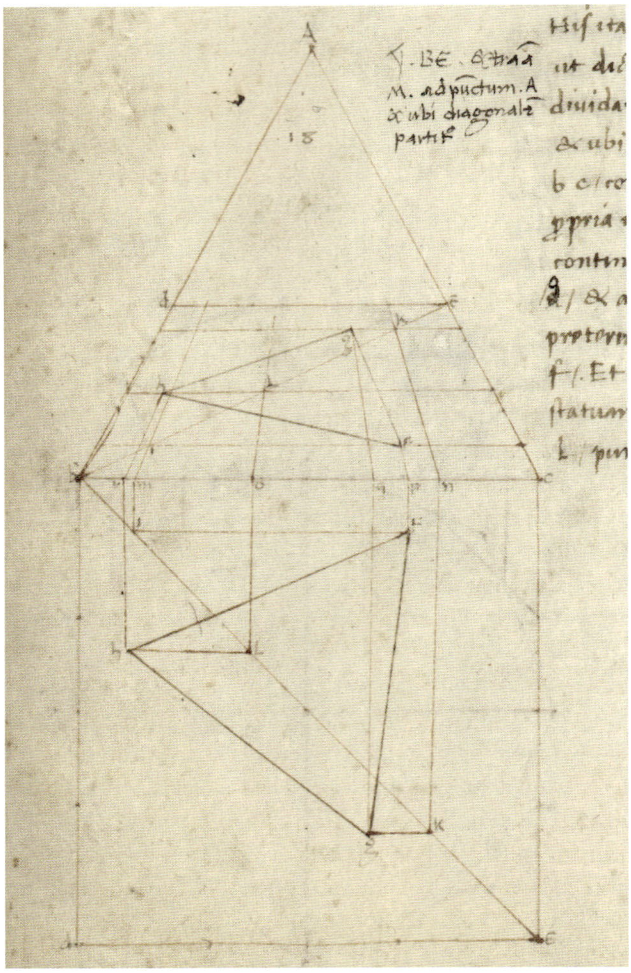

Bordeaux, Bibliothèque Municipale, Ms Latin 616, folio 9 verso.

Proposition XIX

Draw an equilateral hexagon on a shortened quadrilateral surface.[47]

Let BCDE be a shortened quadrilateral surface, where we have to draw a hexagon with equal sides [figure B.22]. For an orderly succession, we will set in real shape, below the shortened surface, the shared line BC and the surface in real shape BCDE. I will draw a circle in the surface[48] of the size of the desired hexagon. The semidiameter of this circle will be equal, therefore, to any side of the hexagon FGHIKL. And we will draw the diagonal BE on the square in real shape. And we will draw a line from F parallel to BC, to intersect the diagonal at point M. And we will draw a line from G parallel to BC, to intersect the diagonal at point N. And we will draw a line from H parallel to BC to intersect the diagonal at point O. And we will draw a line from I parallel to BC to intersect the diagonal at point P. And we will draw a line from K parallel to BC to intersect the diagonal at point Q. And we will draw a line from L parallel to BC to intersect the diagonal at point R. And from all these [intersections] we will draw lines equidistant to BD, till they meet BC. The line that passes through M will intersect BC at point 1; through N at point 2; through O at point 3; through P at point 4; through Q at point 5; through R at point 6. Then we will draw the diagonal BE on the shortened surface. From point 1 we will draw a line to point A. Where this line will intersect the diagonal [BE], insert point M. From point 2 we will draw a line to point A. Where this line will intersect the diagonal [BE], insert point N. From point 3 we will draw a line to point A. Where this line will intersect the diagonal [BE], insert point O. From point 4 we will draw a line to point A. Where this line will intersect the diagonal [BE], insert point P. From point 5 we will draw a line to point A. Where this line will intersect the diagonal [BE], insert point Q. From point 6 we will draw a line to point A. Where this line will intersect the diagonal [BE], insert point R. Then draw from all these points [M, N, O, P, Q, R] lines parallel to BC till they meet BD and CE. These things accomplished, go back to the six sides in real shape. Draw the parallel lines to BD from each angle. At first, draw a line that meets BC, from F at point 11, from G at point 12, from H at point 13, from I at point 14, from K at point 15, from L at point 16. From all these points [of intersection] there must be an extension toward point A.[49] Draw, then, a line from point 11 [to point A], and where this line will intersect [the line parallel] through M [to BC], insert point F. Draw a line from point 12 [to point A], and where this line will intersect the line parallel through N [to BC], insert point G. Draw a line from point 13 [to point A], and where this line will intersect the line parallel through O [to BC], insert point H. Draw a line from point 14 [to point A], and where this line will intersect the line parallel through P [to BC], insert point I. Draw a line from point 15 [to point A], and where this line will intersect the line parallel through Q [to BC], insert point K. Draw a line from point 16 [to point A], and where this line will intersect the line parallel through R [to BC], insert point L. Draw, then, lines FG, GH, HI, IK, KL, LF. And so the hexagon we have proposed to draw proportionally is solved. From the things that were said, in fact, it can be seen that the diagonals generate the same thing in the shortened forms as in the quadrilaterals in real shape.

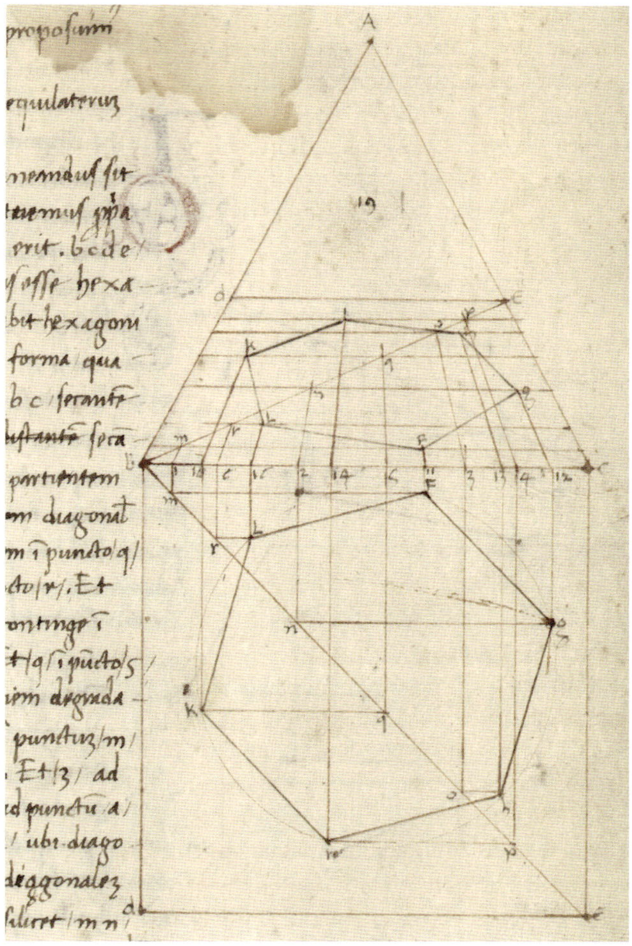

Bordeaux, Bibliothèque Municipale, Ms Latin 616, folio 10 recto.

Proposition XX

Draw an equilateral pentagon on a shortened surface.

Let us establish a shortened surface as was shown in Proposition XIV of this book. Let BCDE be the shortened surface [figure B.23]; the square in real shape with side BC is below it. Draw on it a pentagon FGHIK with equal sides and angles, and draw lines parallel to BC from these points till they touch diagonal BE. First draw a line from F which will intersect the diagonal at point L. Then draw a line from G which will intersect the diagonal at point M. Draw a line from H which will intersect the diagonal at point N. Draw a line from I which will intersect the diagonal at point O. Draw a line from K which will intersect the diagonal at point P. From points L, M, N, O, P draw all the lines parallel to BD till they meet BC. From point L draw a line that meets BC at point 1. From M, a line that intersects BC at point 2. From N, a line that intersects BC at point 3. From O, a line that intersects BC at point 4. From P, a line that intersects BC at point 5. From these points of contact we must draw the lines to point A.[50] This point is placed above the shortened plane and diagonal BE is drawn. From point 1, draw a line that will intersect the diagonal at point L. From point 2, draw a line that will intersect the diagonal at point M. From point 3, draw a line that will intersect the diagonal at point N. From point 4, draw a line that will intersect the diagonal at point O. From point 5, draw a line that will intersect the diagonal at point P. Then draw lines parallel to BC from points L, M, N, O, P until they intersect BD and CE in the shortened square. After you have done these things, return to the non-shortened pentagon and draw lines parallel to line BD from points F, G, H, I, K, till they intersect line BC. The line running from F touches line BC at 11. The line running from G touches line BC at 12. The line running from H touches line BC at 13. The line running from I touches line BC at 14. The line running from K touches line BC at 15. Then draw a line from point 11 to point A, and where this line will intersect the line parallel to BC through point L, insert point F. Draw a line from point 12 to point A, and where this line will intersect the line parallel to BC through point M, insert point G. Draw a line from point 13 to point A, and where this line will intersect the line parallel to BC through point N, insert point H. Draw a line from point 14 to point A, and where this line will intersect the line parallel to BC through point O, insert point I. Draw a line from point 15 to point A, and where this line will intersect the line parallel to BC through point P, insert point K. Finally, draw lines FG, GH, HI, IK, KF. Under these conditions, in the shortened plane the equilateral pentagon FGHIK is drawn proportionately. This is what we had said we would demonstrate.

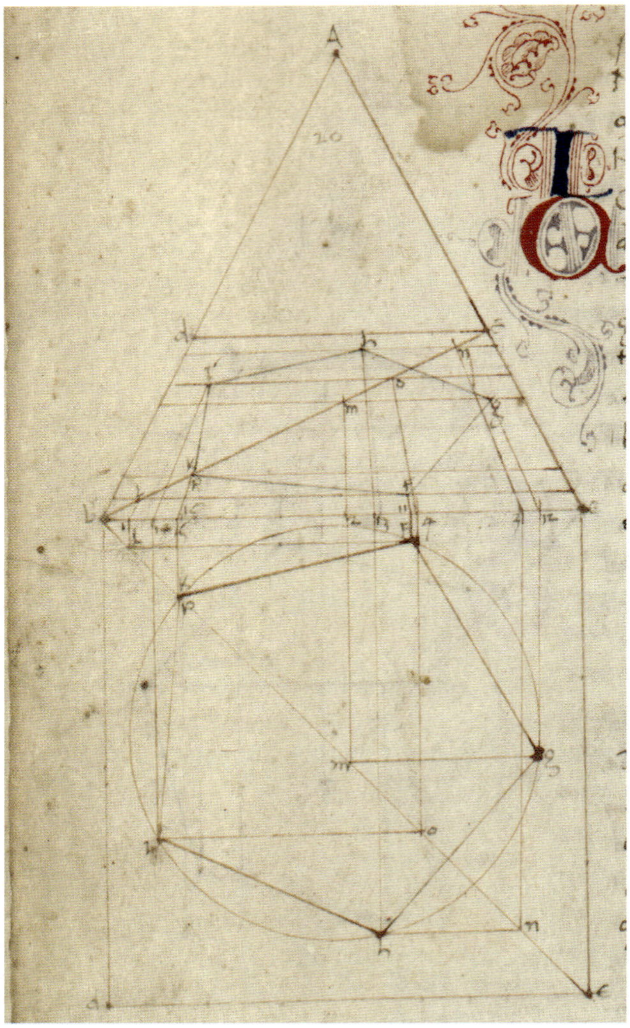

Bordeaux, Bibliothèque Municipale, Ms Latin 616, folio 10 verso.

Proposition XXI

Remove a given quantity of a shortened squared surface's perimeter.

Let BCDE be the shortened surface; and let the quantity that will be taken in the decrease be according to quantity BL [figure B.24]. I shall draw diagonals BE and CD. I will then draw a line from L to point A. This line will intersect BE at F, DC at H, and DE at K. From point F, I will draw a line parallel to BC which will intersect BD at O, CE at P, and DC at G. Moreover, I will draw from H a line parallel to BC intersecting BD at Q. It will divide diagonal BE at I and CE at R. And I will draw from point A a line that passes through I and G. This line will intersect DE at N and BC at M. According to this procedure, I shall say that the shortened surface has been diminished along its perimeter in proportion to quantity BL which had been defined earlier. Then let BCDE be a square that retains a real shape. And draw around the square a quantity equal to BL. This quantity will cut BC at L. Draw through L a line parallel to BD which will intersect DE at K. Draw a line parallel to CE from M, which—as we supposed—coincides with quantity BL, that is, MC. The parallel to CE will intersect DE at N. Then draw BO equal to BL. Draw a parallel to BC through O. This will intersect LK at F, MN at G, and CE at P. Let DQ be equal to DK and extend through Q a line parallel to DE which will intersect LK at H, MN at I, and CE at R. And draw diagonal BE in the real-shape square, passing through F and I, and the diagonal DC, passing through H and G. And from the divisions on BC, that is, BL and MC, equal to the given quantity, draw from point L the line to point A, which will intersect diagonal BE at point F and diagonal DC at point H, as in the quadrilateral figure in real shape. Then draw a line from point M to point A. This line will divide the diagonals at points G and I. Draw from these points [G and I] the parallel lines that pass through those intersections and touch BD at O and Q, and CE at P and R,[51] as may be seen in the surface in real shape. In this way, it is demonstrated that the shortened surface has been diminished by the quantity that was given, by BL, and that was removed from the perimeter of BCDE. The size of the diminished shape is equal to FGHI, which occurs also in the real-shape surface. In fact, the parallel lines are divided by the diagonals in both shapes and the two surfaces are divided proportionally, as was sufficiently demonstrated in Propositions XI and XVI of this book. Thus, the quantity assigned through BL has been subtracted from the shortened surface BCDE in the perimeter, as has been said numerous times. We wanted to demonstrate this.

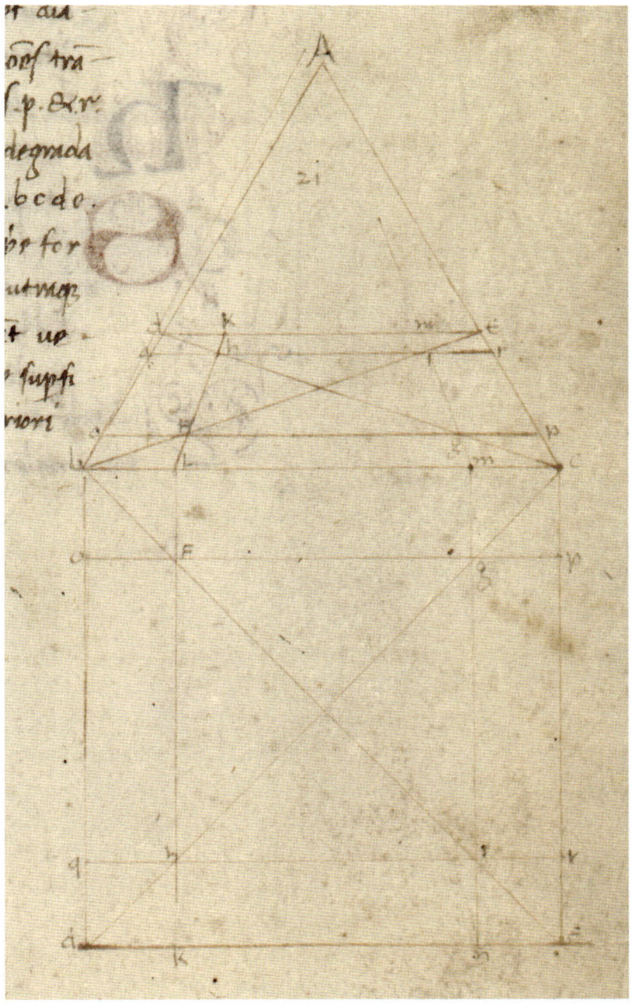

Bordeaux, Bibliothèque Municipale, Ms Latin 616, folio 11 recto.

Proposition XXII

Add proportionally a given quantity around the perimeter of a shortened squared surface.

As we have previously said, let BCDE [figure B.25] be a shortened squared surface, to be increased by the given quantity BL, which I will place externally to line BC. BL will be on the extension of BC. From point A I will draw a line that goes through L to F. I will draw diagonal BE so that it goes through F and E. I will draw the other diagonal that goes through CD, intersecting AF at point H. I will draw from F a line parallel to BC, intersecting the diagonal through C at point G. I will draw from H a line parallel to DE, intersecting the diagonal through E at point I. I will draw from A the line AG through I.[52] The quantity BL has been added to the surface BCDE. As soon as the square surface in real shape FGHI is made, with side equal to FG, draw diagonals FI and GH. Draw side DB according to the shortened surface so that [DB] meets FG at point P. Draw EC so that it meets FG at point M. Draw through P a line parallel to FH of the square in real shape, which will intersect diagonal FI at point B and diagonal HG at point D. Draw through M a line parallel to GI, which will intersect diagonal GH at point C and diagonal FI at point E. And extend line BC and DE to form a quadrilateral shape similar to surface FGHI. Side FG of the square in real shape is equal to side FG of the shortened figure. The diagonals divide the equidistant lines of the shortened figure according to how [the other diagonals] divide, in proportion, the equidistant lines of the real shape. From this it is clear that a quantity equal to BL was added to quadrilateral BCDE. As we have generated a reduction through diagonals in the preceding proposition, likewise here we have generated an addition through diagonals. With the aid of these [diagonals], we can realize reduction and addition proportionally, as was demonstrated in Propositions XI and XV [of this book].[53] With the diagonals we added BL to the shortened surface and to the square BCDE, as proposed.

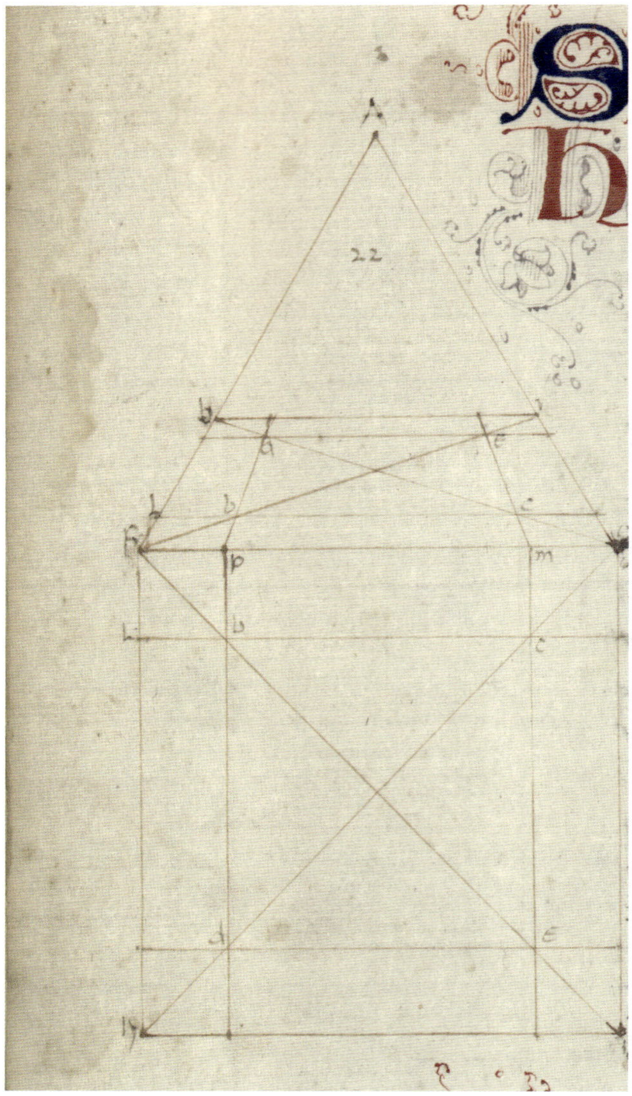

Bordeaux, Bibliothèque Municipale, Ms Latin 616, folio 11 verso.

Proposition XXIII

From a non-squared surface, as big as you wish, cut a squared surface.

Let a surface, from which we are going to cut a squared surface, be given, and let width and length be known [figure B.26]. We will cut a squared figure from the extension in length of that surface. With this aim, let the shortened surface BCDE be 50 braccia long and 10 wide. Before surface BCDE is shortened, let BC be 10 braccia and BD 50. I shall draw the width 10 braccia and diagonal BE. Because 10 braccia is one fifth of 50 braccia, we will divide BC into five equal parts, marked by points F, G, H, I. We will draw the line that will divide diagonal BE at K from F to A. And we will draw from K a parallel to BC, which will cut BD[54] at L and CE at M. After this, I say the quadrilateral figure BLCM is cut from the shortened plane.

Here is the demonstration. Let the quadrilateral in its real shape be drawn. The quadrilateral is 50 braccia long and 10 braccia wide. Let this quadrilateral be NOPQ. NP is 50 braccia and NO is 10 braccia. Draw diagonal NQ and subdivide NO into five equal parts at R, S, T, V. Let a line parallel to NP pass through R, which will intersect the diagonal at X and PQ at Y. Extend a line parallel to NO from X, which will intersect NP at z and OQ at Z. These lines will form a square NOzZ in real shape, which was cut from the plane NOPQ. This plane was generated by the diagonal which intersects line RY at X which corresponds to the fifth part of NO, as we mentioned earlier. Although figure BCDE is shortened, it corresponds to NOPQ. Line BC is equal to NO and is divided into five equal parts. One of the five parts has been selected, that is, BF. Draw a line [from F] to A; this line

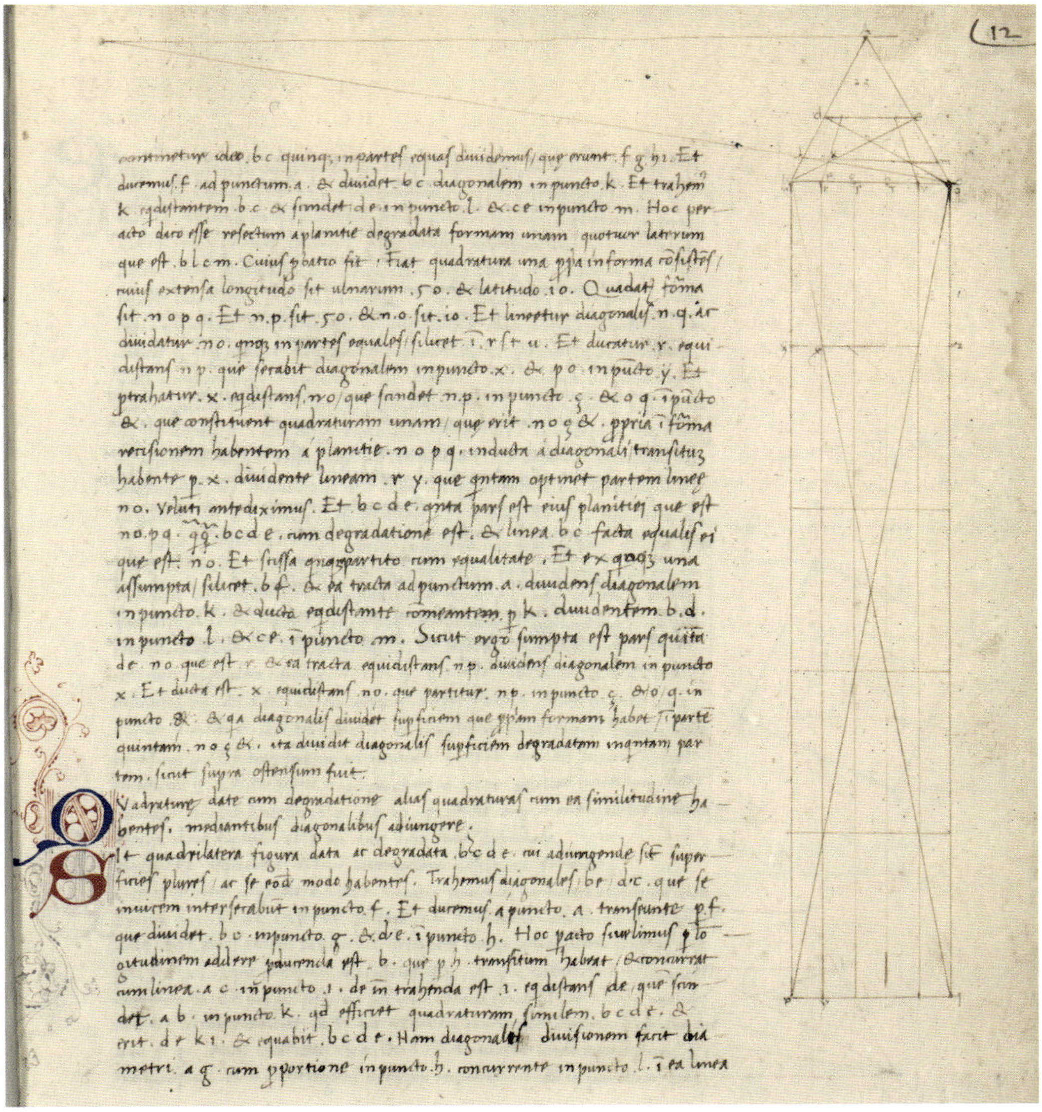

Bordeaux, Bibliothèque Municipale, Ms Latin 616, folio 12 recto.

divides the diagonal at K. Draw through K the parallel to BC, which divides BD at point L and CE at point M. This way the fifth part of DE and NO was selected, which is given by R. From R a line parallel to NP was drawn, and it divides the diagonal at X. From X a line parallel to NO was drawn, and it divides NP at z and OQ at Z. And because the diagonal divides the surface in real shape, in the fifth part NOzZ the diagonal divides the shortened surface in the fifth part in this same way, as was shown earlier.[55] However, if we did not know the dimension of the length and width of the plane [in real shape], I will draw a parallel to BC from point A; this will be as long as the distance from the eye to the terminal. I will mark the eye at point O there. From this point I will draw line OC which will divide line BD at L. I say that BL has cut from the shortened plane BCDE the quantity BC, equal to BL. Draw a line parallel to BC through L which will cut diagonal BE at K and CE at M. I say that BLCM corresponds to the quadrilateral that has to be cut from the shortened plane BCDE [not squared]. This happens because the line, which goes from the eye O and ends at C, intersects BD at L, such that C appears by BL higher to the eye when compared to B, as demonstrated in Proposition XII.[56]

Proposition XXIV

Add other similar quadrilaterals, through diagonals, to a given shortened quadrilateral.

Let BCDE be the given shortened quadrilateral, to which we must add several similar surfaces [figure B.27]. We will draw diagonals BE and DC which will intersect at point F. We will draw from point A a line that passes through F and that intersects BC at G and DE at H. After, if we want to add, in length, a quadrilateral we must draw [from B] a line that passes through H and meets AC at I. Then we must draw [from I] a line parallel to DE, which will intersect AB at K. This will produce quadrilateral DEKI, which will be similar to quadrilateral BCDE. The diagonal divides, in fact, the diameter line AG proportionally at point H and runs toward point I which is on line AC. The line parallel to DE was drawn from I; it will touch AB at point K. The proportion between AK and AB is equal to that between KI and BC. And the proportion between AK and AD is equal to the one of KI and DE. Therefore, we have added in depth to the quadrilateral BCDE the quadrilateral DEKI which is similar to the first one [BCDE]. However, if we need to add a quantity in width, we will draw from F a parallel to BC, which will intersect BD at L and CE at M. Then I shall extend DE to P so that[57] EP is equal to DE. We will extend from B the line that passes through M till it meets P. And we will draw from A the line that passes through P till it meets O. CO is then equal to BC, and EP is equal to DE. The proportion between DE and BC is equal to that between EP and CO; and this proportion coincides with that between DP and BO. Thereby, we have added in width the similar quadrilateral figure COEP to quadrilateral BCDE. However, if we need to extend in length and in width, so as to make the figure a quadrilateral made of four quadrilaterals equal to BCDE, I will use CO which is equal to BC. And I will draw from point A line AO. Then I will draw the diagonal line from B, which goes through E and divides AO at Q. Then I will draw a parallel to BC through Q, which will divide AB at K. These lines constitute the quadrilateral figure BOKQ which is similar to BCDE. This way, three equal and similar equilateral figures [to the first quadrilateral] have been added to quadrilateral BCDE by using the definitions of the diagonals and the equidistant lines, as was shown earlier.

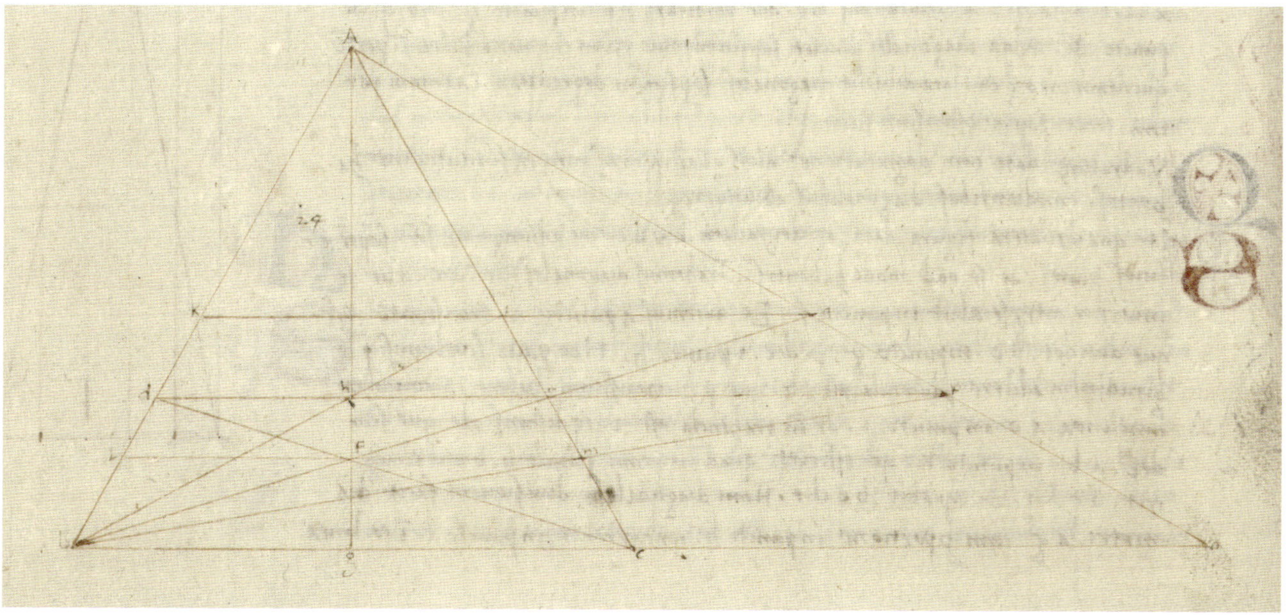

Bordeaux, Bibliothèque Municipale, Ms Latin 616, folio 12 verso.

Proposition XXV

Draw an assigned square on a shortened surface.

Let BCDE be a shortened surface with point A [figure B.28]. Let FGHI be the assigned square [in real shape]. I will draw BCDE, the surface on which I will draw the assigned square FGHI, as I have realized in the [following] demonstration. From the points in this figure [in real shape] I will draw lines equidistant to BC. Firstly, there will be a parallel for G to BC, which will cut diagonal BE at 2. Then there will be a parallel for F, which will cut the diagonal at 1. And I will draw a parallel for H, which will divide the diagonal at 3. And I will draw a parallel for I, which will cut the diagonal at 4. And I will draw from 1 a parallel to BD, which will intersect BC at 5. And I will draw from 2 a parallel line to BD, which will intersect BC at 6. And I will draw from 3 a parallel line to BD, which will intersect BC at 7. And I will draw from 4 a parallel line to BD, which will intersect BC at

8. And I will draw from G a parallel line to BD, which will intersect BD at L. And I will draw from F a parallel line to BD, which will intersect BC at K. And I will draw from H a parallel line to BD, which will intersect BC at M. And I will draw from I a parallel line to BD, which will intersect BC at N. All these lines will also be drawn on the shortened surface. Draw diagonal BE first. Then, from 5, draw the line to A which intersects the diagonal at 1. From 6, draw the line to A which intersects the diagonal at 2. From 7, draw the line to A which intersects the diagonal at 3. From 8, draw the line to A which intersects the diagonal at 4. Then draw the parallel lines to BC and DE from 1, 2, 3, 4. Draw the line from K to A. Where this line [KA] intersects the line through 1 which is parallel to BC, insert point F. Draw the line from L to A. Where this line intersects the line through 2 which is parallel to BC, insert point G. Draw the line from M to A. Where this line intersects the line through 3 which is parallel to BC, insert point H. Draw the line from N to A. Where this line intersects the line through 4 which is parallel to BC, insert point I. Draw lines FG, GH, HI, IF. This way the assigned quadrilateral figure has been achieved, as confirmed through rules that have been stated.

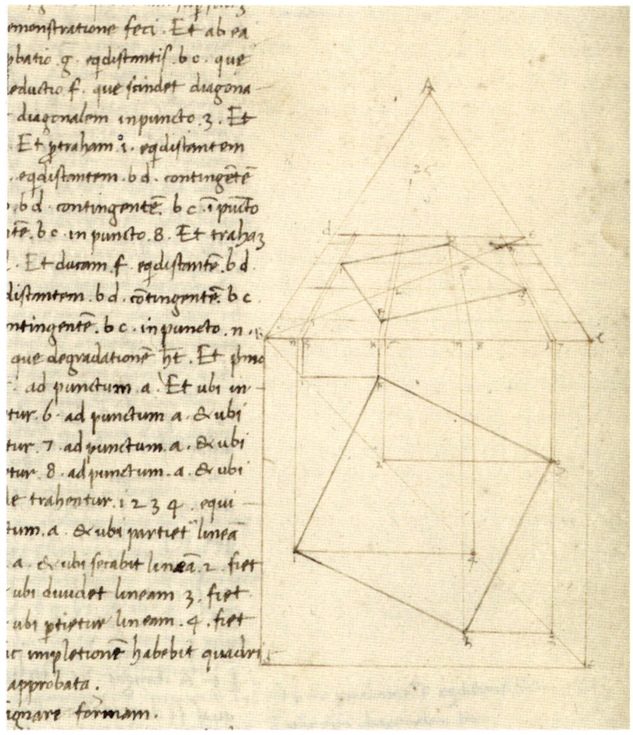

Bordeaux, Bibliothèque Municipale, Ms Latin 616, folio 13 recto.

Proposition XXVI

Draw a regular assigned octagon on a shortened surface.

Let BCDE be the shortened surface where we have to draw the assigned octagon [figure B.29]. Let BCDE be in real shape, where we will draw the assigned octagon after drawing the quadrilateral figure FGHI where we will draw the diagonals FH and GI that intersect at point K. With a compass's point at K, draw a circle with radius KF. The circumference goes through FGHI. Divide FG into two parts at L. Draw the line that passes through K and L parallel to FI, which meets the circle at M and N. Divide FI into two equal parts at O. Draw a line parallel to FG through O, which goes through K and touches [the circle] at P and Q. Draw MQ, which will intersect FG at R and GH at S. Draw QN, which will intersect GH at T and HI at V. Draw NP, which will divide HI at point X and IF at point Y. Draw PM, which will divide IF at point z and FG at point Z. Draw RS, TV, XY, zZ. This being the situation, we have realized the octagon in real shape. Now we must draw lines parallel to BC from all the points, to intersect diagonal BE of the surface in real shape. That is, the line that starts at R will intersect diagonal BE at 1. The line that starts at S will intersect diagonal BE at 2. The line that starts at T will intersect diagonal BE at 3. The line that starts at V will intersect diagonal BE at 4. The line that starts at X will intersect diagonal BE at 5. The line that starts at Y will intersect diagonal BE at 6. The line that starts at z will intersect diagonal BE at 7. The line that starts at Z will intersect diagonal BE at 8. From all these divisions executed on diagonal BE, draw lines equidistant to BD until they touch BC. Firstly, the line that starts from 1 will touch BC at 11; the line that starts from 2 will touch [BC] at 12; that from 3 will touch [BC] at 13; that from 4 will touch [BC] at 14; that from 5 will touch [BC] at 15; that from 6 will touch [BC] at 16; that from 7 will touch [BC] at 17; and that from 8 will touch [BC] at 18. Moreover, draw lines parallel to BD from the remaining angles until they touch BC. Draw through R the line that touches BC at 21; through S the line that touches BC at 22; through T the line that touches BC at 23; through V the line that touches BC at 24; through X the line that touches BC at 25; through Y the line that touches BC at 26; through z the line that touches BC at 27, through Z the line that touches BC at 28. All these figures are drawn in real shape. To shorten them, we must draw diagonal BE in the shortened surface. Then draw a line from 11 to A, and where it intersects diagonal BE insert 1. Draw a line from 12 to A; where it intersects the diagonal insert 2. Draw a line from 13 to A; where it intersects the diagonal insert 3. Draw a line from 14 to A; where it intersects the diagonal BE insert 4. Draw a line from 15 to A; where it intersects the diagonal insert 5. Draw a line from 16 to A; where it intersects the diagonal insert 6. Draw a line from 17 to A; where it intersects the diagonal insert 7. Draw a line from 18 to A; where it intersects the diagonal insert 8. Then draw from all the points 1, 2, 3, 4, 5, 6, 7, 8 lines parallel to BC, until they reach BD and CE. Draw a line from 21 to point A; where it intersects the line through 1 insert R. Draw a line from 22 to point A; where it intersects the line through 2 insert S. Draw a line from 23 to point A; where it intersects the line through 3 insert T. Draw a line from 24 to point A; where it intersects the line through 4 insert V. Draw a line from 25 to point A; where it intersects the line through 5 insert X. Draw a line from 26 to point A; where it intersects the line through 6 insert Y. Draw a line from 27 to point A; where it intersects the line through 7 insert z. Draw a line from 28 to point A; where it intersects the line through 8 insert Z. Then we must draw these lines: RS, ST, TV, VX, XY, Yz, zZ, ZR. This way the assigned octagon, which was shortened proportionally, is completed. In fact, the lines that divide the diagonal in the plane in real shape also divide the diagonal in the shortened plane. Moreover, the lines which go from the sides of the octagon to BC, when they pass through the lines which divide the diagonal of the plane in real shape, go from BC to point A through the diagonal of the shortened plane. This way we have demonstrated what we had proposed.

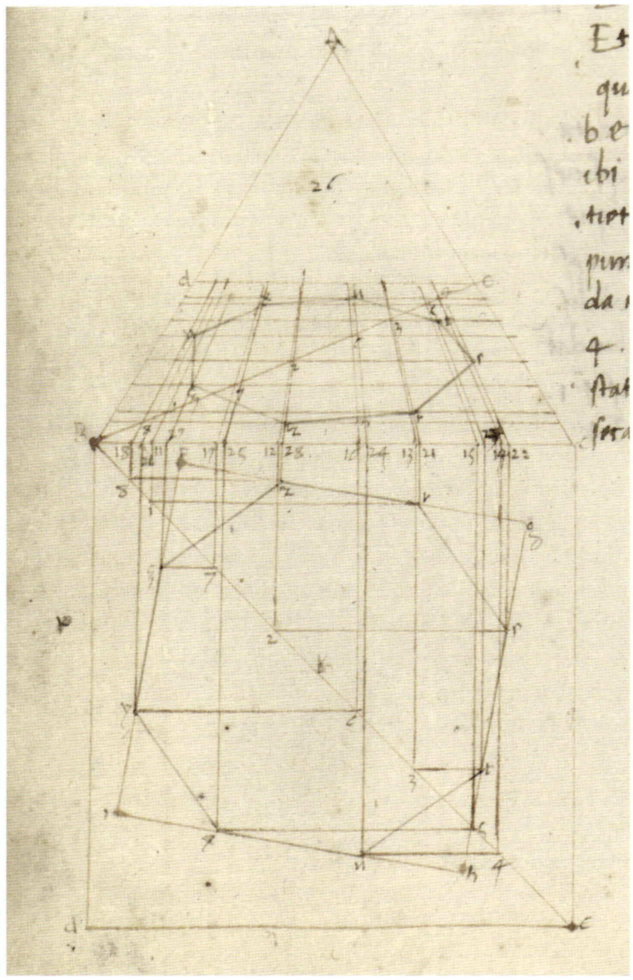

Bordeaux, Bibliothèque Municipale, Ms Latin 616, folio 13 verso.

Proposition XXVII

Draw assigned squares on a shortened surface.

Let BCDE be the shortened surface and with point A [figure B.30]. Draw two assigned squares on it that are not placed in the same way. As in the previous propositions, let BCDE be the plane in real shape in which we will draw two squares in real shape: FGHI and KLMN. Let BE and CD be the diagonals of surface BCDE, a square in real shape. From the sides of the assigned surface FGHI draw lines equidistant to BC which divide diagonal BE. First, draw a parallel to BC from F, which will intersect the diagonal BE[58] at 1. Then draw a parallel from G so that it intersects BE at 2. Draw a parallel from H so that it intersects BE at 3. Draw a parallel from I so that it intersects BE at 4. Then draw from 1 the parallel to BD so that it intersects BC at 5. Draw from 2 the parallel to BD so that it intersects BC at 6. Draw from 3 the parallel to BD so that it intersects BC at 7. Draw from 4 the parallel to BD so that it intersects BC at 8. Then draw from F the parallel to BD so that it intersects BC at 11. Draw from G the parallel to BD so that it intersects BC at 12. Draw from H the parallel to BD so that it intersects BC at 13. Draw from I the parallel to BD so that it intersects BC at 14. Up to this point, the first square in real shape has been prepared, drawing lines. The second square KLMN remains to be processed by lines to be represented. First, we will draw the line parallel to BC from K which will intersect diagonal DC at 21. From L we will draw a parallel to BC which will intersect diagonal DC at 22. From M we will draw a parallel to BC which will intersect diagonal DC at 23. From N we will draw a parallel to BC which will intersect diagonal DC at 24. From these points, we will draw lines parallel to CE until they meet BC. From 21 draw a line that touches BC at point 25. From 22 draw a line that touches BC at point 26. From 23 draw a line that touches BC at point 27. From 24 draw a line that touches BC at point 28. Now we must take care of the sides, from which we will draw the parallel lines to CE until they meet BC. Draw a line parallel to CE from K till it touches BC at point 31. Draw a line parallel to CE from L till it touches BC at point 32. Draw a line parallel to CE from M till it touches BC at point 33. Draw a line parallel to CE from N till it touches BC at point 34. Because by tracing these lines you have completed the squares in real form, you are now ready to draw the same lines on the shortened surface. Draw diagonals BE and DC in the shortened surface and divide BC into two equal parts at P. Having done this, consider the first square. Draw a line from 5 to A; where it intersects diagonal BE insert 1. Draw a line from 6 to A; where it intersects diagonal BE insert 2. Draw a line from 7 to A; where it intersects diagonal BE insert 3. Draw a line from 8 to A; where it intersects diagonal BE insert 4. Draw line AP. Through points 1, 2, 3, 4 draw lines parallel to BC till they intercept BD and AP. Then draw a line from point 11 to point A, and where it intersects a line parallel to BC at 1, insert F. Draw a line from 12 to A, and where it intersects the line at 2 insert G. Draw a line from 13 to A, and where it intersects the line at 3 insert H. Draw a line from 14 to A, and where it intersects the line at 4 insert I. Then we will draw FG, GH, HI, IF. This way you have the first square. Draw the second square. Draw a line from 25 to A, and where it intersects diagonal DC insert 21. Draw a line from 26 to A, and where it intersects diagonal DC insert 22. Draw a line from 27 to A, and where it intersects diagonal DC insert 23. Draw a line from 28 to A, and where it intersects diagonal DC insert 24. From all these points, draw lines parallel to BC till lines CE and AP are intercepted. Then draw the line from 31 to A, and where it intersects the line at 21 insert K. Draw the line from 32 to A, and where it intersects the line at 22 insert L. Draw the line from 33 to A, and where it intersects the line at 23 insert M. Draw the line from 34 to A, and where it intersects the line at 24 insert N. Having drawn lines KL, LM, MN, NK, you will have completed the second square. This is what we had proposed. However, if the same surfaces were to be turned into octagons, we would subdivide the sides and realize the relative perimeters as we did in the preceding examples.

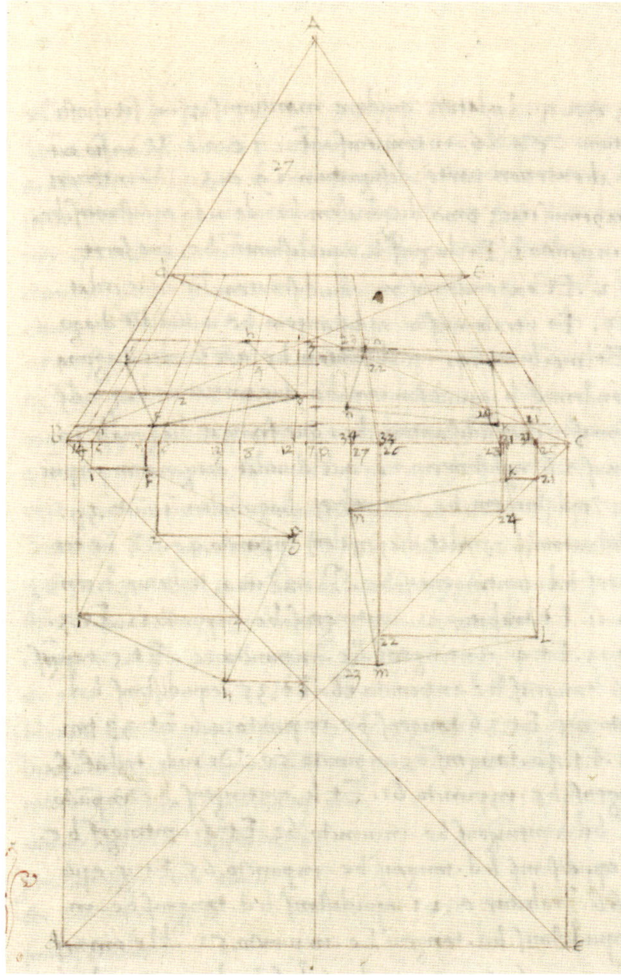

Bordeaux, Bibliothèque Municipale, Ms Latin 616, folio 15 recto.

Proposition XXVIII

Draw the perimeter of a base of a square building on a shortened surface.

The aim is to place the shape of an assigned building and its subdivisions in the shortened surface BCDE [figures B.31, B.32].[59] To avoid not following the undertaken order, we will draw the plane of the surface in real shape, that is, BCDE below the [shortened] BCDE plane. Then we will draw the surface with perimeter FGHI on this surface. Let each side be 16 braccia long, or of any size you wish. Let the first room be KLMN, the second one OPQR, and the third one STVX. Let the first room have a door in the middle whose external width is identified by 21 and 22 and whose internal width is identified by 23 and 24. The [external] width of the side door is 25 and 26 and its internal width is 27 and 28. Room STVX has a door whose external width is identified by 29 and 30 and whose internal width is identified by 31 and 32.

Firstly, we will draw diagonal BE [figure B.32], then from point F a parallel to BC which intersects the diagonal at 1. We will draw from K a parallel to BC which intersects the diagonal at 2. We will draw from M a parallel to BC which intersects the diagonal at 3. We will draw from O a parallel to BC which intersects the diagonal at 4. We will draw from Q a parallel to BC which intersects the diagonal at 5. We will draw from H a parallel to BC which intersects the diagonal at 6. We will draw from 25 a parallel to BC which intersects the diagonal at 35. We will draw from 26 a parallel to BC which intersects the diagonal at 36. We will draw from 29 a parallel to BC which intersects the diagonal at 39. We will draw from 30 a parallel to BC which intersects the diagonal at 40. For all these points we now need to draw lines parallel to BD until they meet BC. First, draw from 1 a line that meets BC at 11. From 2 draw a line that touches BC at 12. From 3 draw a line that touches BC at 13. From 4 draw a line that touches BC at 14. From 5 draw a line that touches BC at 15. From 6 draw a line that touches BC at 16. And from 35 draw a line that touches BD at 45. From 36 draw a line that touches BC at 46.

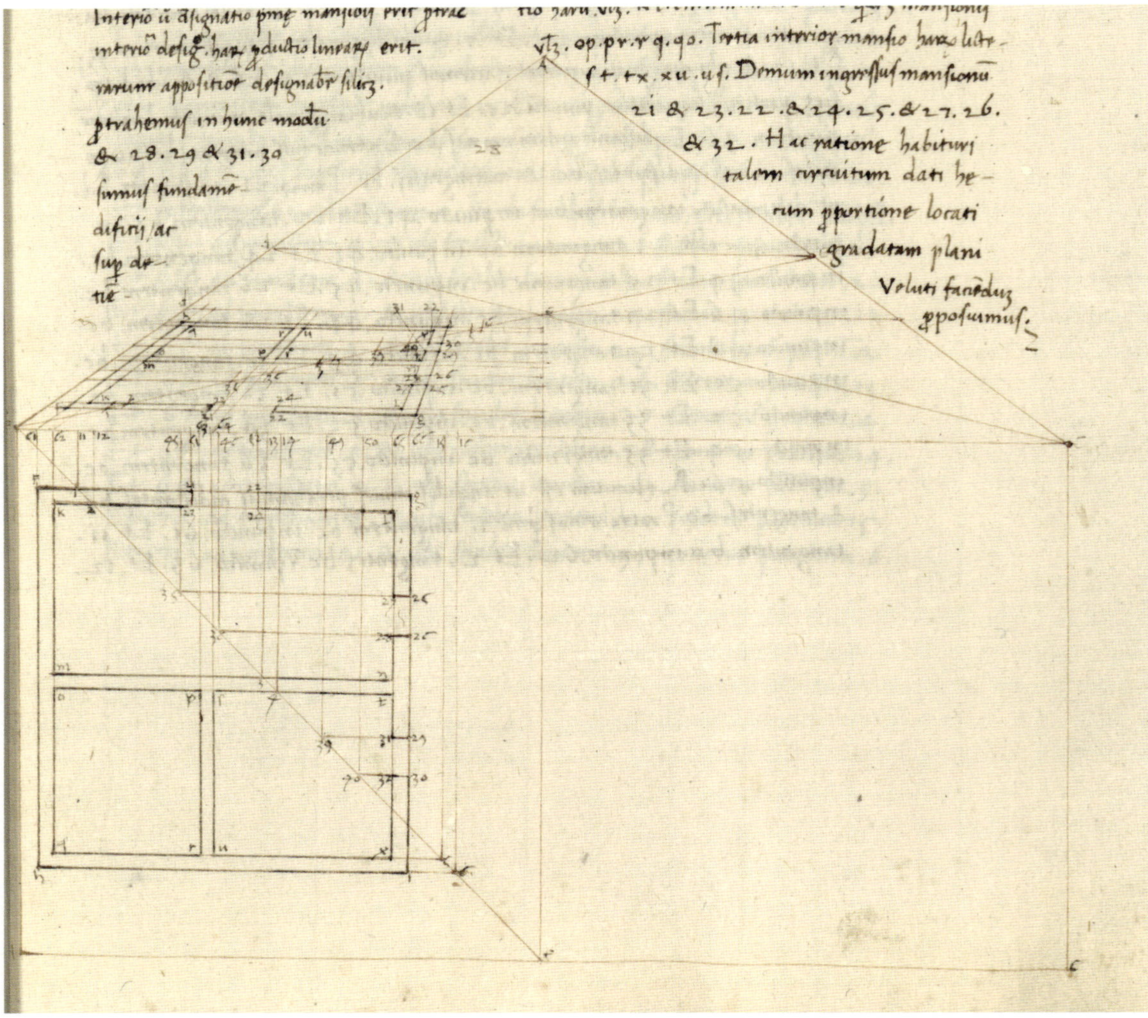

Bordeaux, Bibliothèque Municipale, Ms Latin 616, folio 16 recto.

From 39 draw a line that touches BC at 49. From 40 draw a line that touches BC at 50. Then from F draw a parallel to BD that touches BC at 61. From K draw a parallel to BD that touches BC at 62. From P draw a parallel to BD that touches BC at 63. From S draw a parallel to BD that touches BC at 64. From L draw a parallel to BD that touches BC at 65. From G draw a parallel to BD that touches BC at 66. From 21 also draw a parallel to BD that touches BC at 51. From 22 draw a parallel to BD that touches BC at 52. For all these points 11, 12, 13, 14, 15, 16, 45, 46, 49, 50, draw lines to point A of the shortened plane; the lines intersect the diagonal BE. Draw a line to A from 11, which intersects the diagonal at 1. Draw a line to A from 12, which intersects the diagonal at 2. Draw a line to A from 13, which intersects the diagonal at 3. Draw a line to A from 14, which intersects the diagonal at 4. Draw a line to A from 15, which intersects the diagonal at 5. Draw a line to A from 16, which intersects the diagonal at 6. Draw a line to A from 45, which intersects the diagonal at 35. Draw a line to A from 46, which intersects the diagonal at 36. Draw a line to A from 49, which intersects the diagonal at 39. Draw a line to A from 50, which intersects the diagonal at 40. From all these points on the shortened surface 1, 2, 3, 4, 5, 6, 35, 36, 39, 40 draw lines parallel to BC till they meet BD and CE. Then we will draw a line to A from 61: where it will intersect the line [that goes through] point 1, there will be point F; where it will intersect the line [that goes through] point 6, there will be point H. We will draw a line to point A from 62: where it will intersect the line [that goes through point] 2, there will be K; where it will intersect the line [that goes through point] 3, there will be M; where it will intersect the line [that goes through point] 4, there will be O; where it will intersect the line [that goes through point] 5, there will be Q. We will draw a line to point A from 63: where it will intersect the line [that goes through point] 4, there will be P; where it will intersect the line [that goes through point] 5, there will be R. We will draw the line to point A from point 64: where it will intersect the line [that goes through point] 4, there will be S; where it will intersect the line [that goes through point] 5, there will be V. We will draw the line to point A from point 65: where it will intersect the line [that goes through point] 2, there will be L; where it will intersect the line [that goes through point] 35, there will be 27; where it will intersect the line [that goes through point] 36, there will be 28; where it will intersect the line [that goes through point] 3, there will be N; where it will intersect the line [that goes through point] 4, there will be T; where it will intersect the line [that goes through point] 39, there will be 31; where it will intersect the line [that goes through point] 40,[60] there will be 32; where it will intersect the line [that goes through point] 5, there will be X. After this, draw a line from 66 to A: indicate G where it intersects the line [that goes through point] 1; indicate 25 where it intersects the line [that goes through point] 35; indicate 26 where it intersects the line [that goes through point] 36; indicate 29 where it intersects the line [that goes through point] 39; indicate 30 where it intersects the line [that goes through point] 40; indicate I where it intersects the line [that goes through point] 6. Draw the line from 51 to A: indicate 21 where it intersects the line [that goes through point] 1; indicate 23 where it intersects the line [that goes through point] 2. Draw the line from 52 to A: indicate 22 where it intersects the line [that goes through point] 1; indicate 24 where it intersects the line [that goes through point] 2. Finally, draw lines FG, GI, IH, HF, which constitute the external border of the figure [in the reduced surface]. The internal border of the first room consists of lines KL, LN, NM, MK. The internal border of the second room consists of lines OP, PR, RQ, QO. The internal border of the third room consists of lines ST, TX, XV, VS. We will indicate entrances of the rooms by drawing lines from 21 to 23, from 22 to 24, from 25 to 27, from 26 to 28, from 29 to 31, from 30 to 32. According to this procedure, we will be able to obtain the layout of the base of the assigned building and replicate it proportionally on the shortened surface, as we proposed to do.

Proposition XXIX

Draw an octagonal base of an assigned building on a shortened surface.

Having decided to draw a regular octagon on a shortened surface, we will first draw, below line BC, the plane BCDE in real shape [figure B.33]. We will draw the octagon on this plane as shown in Proposition XXVI[61] of this book. Let 1, 2, 3, 4, 5, 6, 7, 8 be the points of the external perimeter and let K be the center. We will draw lines K-1, K-2, K-3, K-4, K-5, K-6, K-7, K-8, and we will draw line 11-12 parallel to 1-2, line 12-13 parallel to 2-3, line 13-14 parallel to 3-4, line 14-15 parallel to 4-5, line 15-16 parallel to 5-6, line 16-17 parallel to 6-7, line 17-18 parallel to 7-8, line 18-11 parallel to 8-1. We will draw diagonal BE, and from 1 a line parallel to BC. Insert 21 where this parallel intersects diagonal BE. We will draw a line parallel to BC from 11. Insert 22 where this parallel intersects diagonal BE. We will draw a line parallel to BC from 2. Insert 23 where this parallel intersects diagonal BE. We will draw a line parallel to BC from 12. Insert 24 where this parallel intersects diagonal BE. We will draw a line parallel to BC from 3. Insert 25 where this parallel intersects diagonal BE. We will draw a line parallel to BC from 13. Insert 26 where this parallel intersects diagonal BE. We will draw a line parallel to BC from 4. Insert 27 where this parallel intersects diagonal BE. We will draw a line parallel to BC from 14. Insert 28 where this parallel intersects diagonal BE. We will draw a line parallel to BC from 5. Insert 29 where this parallel intersects diagonal BE. Do the same for 15. Insert 30 where this parallel meets the diagonal. Do the same for 6. Insert 31 where this parallel meets the diagonal. Do the same for 16. Insert 32 where this parallel meets the diagonal. Do the same for 7. Insert 33 where this parallel meets the diagonal. Do the same for 17. Insert 34 where this parallel meets the diagonal. Do the same for 8. Insert 35 where this parallel meets the diagonal. Do the same for 18. Insert 36 where this parallel meets the diagonal. All these lines are parallel to BC. We will draw lines parallel to BD from all the points on this diagonal, till they meet BC. Firstly, we will draw a parallel to BD from 21, which will meet BC at 41. From 22 I will draw a parallel to BD until it meets BC at 42. From 23 I will draw a parallel to BD until it meets BC at 43. From 24 I will draw a parallel to BD until it meets BC at 44. From 25 I will draw a parallel to BD until it meets BC at 45. From 26 I will draw a parallel to BD until it meets BC at 46. From 27 I will draw a parallel to BD until it meets BC at 47. From 28 a parallel to BD until it meets BC at point 48. From 29 a parallel to BD until it meets BC at point 49. From 30 a parallel to BD until it meets BC at point 50. From 31 a parallel to BD until it meets BC at point 51. From 32 a parallel to BD until it meets BC at point 52. From 33 a parallel to BD until it meets BC at point 53. From 34 a parallel to BD until it meets BC at point 54. From 35 a parallel to BD until it meets BC at point 55. From 36 a parallel to BD until it meets BC at point 56. From all the corners, it remains to draw lines parallel to BD till they meet BC. First, we will draw from 1 a line that touches BC at 61. Then we will draw from 11 a line that touches BC at point 62. From point 2 a line that touches BC at point 63. From point 12 a line that touches BC at point 64. From point 3 a line that touches BC at point 65. From point 13 a line that touches BC at point 66. From point 4 a line that touches BC at point 67. From point 14 a line that touches BC at point 68. From point 5 a line that touches BC at point 69. From point 15 a line that touches BC at point 70. From point 6 a line that touches BC at point 71. From point 16 a line that

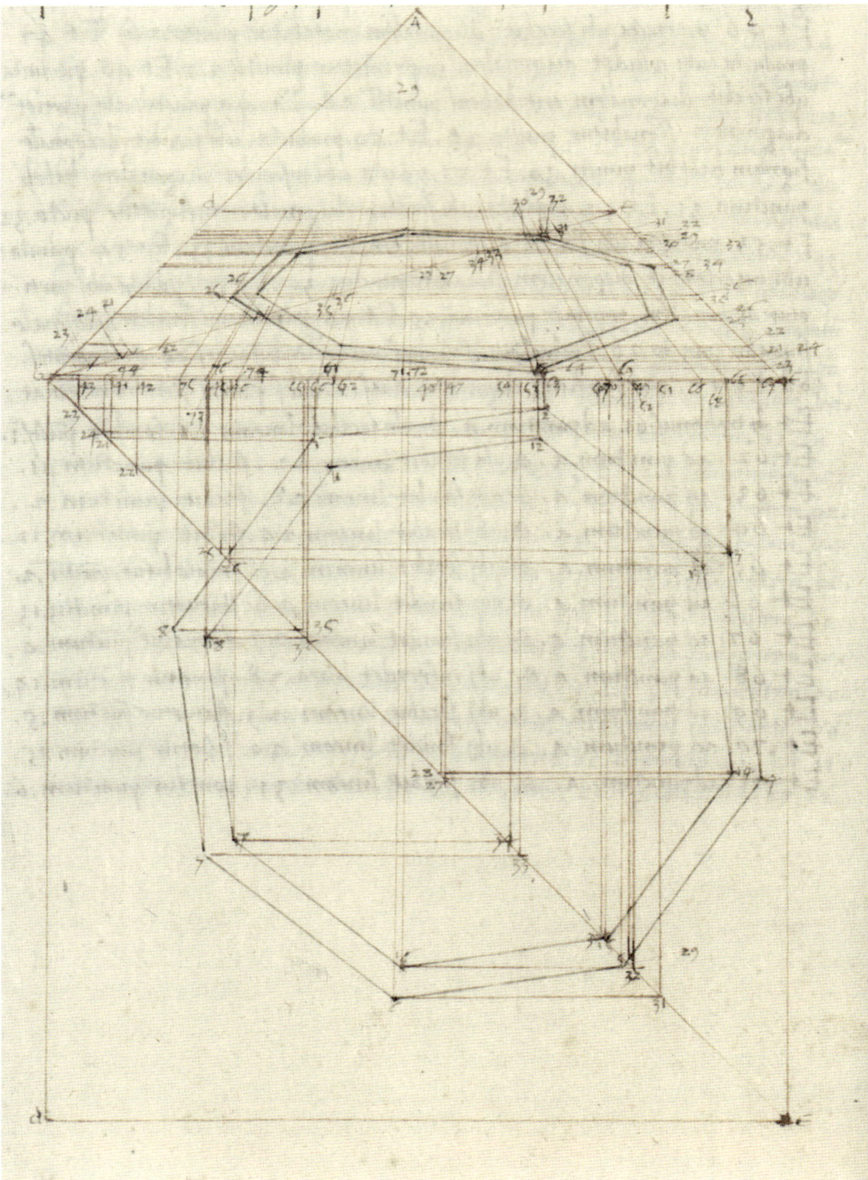

Bordeaux, Bibliothèque Municipale, Ms Latin 616, folio 17 verso.

touches BC at point 72. From point 7 a line that touches BC at point 73. From point 17 a line that touches BC at point 74. From point 8 a line that touches BC at point 75. From point 18 a line that touches BC at point 76.

Up to this point we have drawn the lines in real shape. Now we will draw them on the shortened surface toward point A, starting at point 41. Where the line traced from this point [41] intersects diagonal BE insert 21. Where the line traced from 42 intersects the diagonal BE insert 22. Where the line traced from 43 intersects the diagonal BE insert 23. Where the line traced from 44 intersects the diagonal BE insert 24. Where the line traced from 45 intersects the diagonal BE insert 25. Where the line traced from 46 intersects the diagonal BE insert 26. Where the line traced from 47 intersects the diagonal BE insert 27. Where the line traced from 48 intersects the diagonal BE insert 28. Where the line traced from 49 intersects the diagonal BE insert 29. Where the line traced from 50 intersects the diagonal BE insert 30. Where the line traced from 51 intersects the diagonal BE insert 31. Where the line traced from 52 intersects the diagonal BE insert 32. Where the line traced from 53 intersects the diagonal BE insert 33. Where the line traced from 54 intersects the diagonal BE insert 34. Where the line traced from 55 intersects the diagonal BE insert 35. Where the line traced from 56 intersects the diagonal BE insert 36. And we will draw from all these points parallel lines to BC until they meet BD and EC of the shortened surface, which is divided into two equal parts.[62] And we will draw from 61 the line to A, and where this line intersects the line that goes through 21, insert 1. We will draw from 62 the line to A, and where this line intersects the line that goes through 22, insert 11. We will draw from 63 the line to A, and where this line intersects the line that goes through 23, insert 2. We will draw from 64 the line to A, and where this line intersects the line that goes through 24, insert 12. We will draw from 65 the line to A, and where this line intersects the line that goes through 25, insert 3. And we will draw from 66 the line to A, and where this line intersects the line that goes through 26, insert 13. We will draw from 67 the line to A, and where this line intersects the line that goes through 27, insert 4. We will draw from 68 the line to A, and where this line intersects the line that goes through 28, insert 14. We will draw from 69 the line to A, and where this line intersects the line that goes through 29, insert 5. We will draw from 70 the line to A, and where this line intersects the line that goes through 30, insert 15. We will draw from 71 the line to A, and where this line intersects the line that goes through 31, insert 6. We will draw from 72 the line to A, and where this line intersects the line that goes through 32, insert 16. And we will draw from 73 the line to A, and where this line intersects the line that goes through 33, insert 7. We will draw from 74 the line to A, and where this line intersects the line that goes through 34, insert 17. We will draw from 75 the line to A, and where this line intersects the line that goes through 35, insert 8. We will draw from 76 the line to A, and where this line intersects the line that goes through 36, insert 18. Then we will draw lines 1-2, 2-3, 3-4, 4-5, 5-6, 6-7, 7-8, 8-1 and the corresponding internal lines, that is, 11-12, 12-13, 13-14, 14-15, 15-16, 16-17, 17-18, 18-11. Having drawn these lines, the base of the octagonal building is completed. I believe we have said enough about surfaces. If one dedicates himself to these things, he will easily obtain, via what we have said, the proportioning and the method for all other surfaces. He will also have easy access to all that follows.[63]

Proposition XXX

But since it is necessary that most of those who are not expert in this art desist from error, when they say that in dividing the shortened surface by braccia, this [the surface] becomes larger than that in real form, we must specify why it happens.

The reason is that they neglect the distance between the eye and the preestablished terminal, where we must draw the objects. They also ignore in what proportion the eye can expand, with its rays, the visual angle. To not keep anyone in doubt whether or not perspective is based on scientific truth, there must be a demonstration of the distance and how much an angle can expand in the eye, in order to remove any doubt.[64] Quadrilateral BCDE [on a real surface] is formed by equal parallel lines [figure B.34]. Its internal part is identified by figure FGHI having line FG parallel to BC, line FH [parallel] to BD, line GI [parallel] to CE, and line HI [parallel] to DE. Draw diagonals BE and CD. Let BE go through F and I, CD through G and H. Let the two diagonals meet at point A where the eye is located. Divide the surface between the two figures into equal parts. Firstly, divide BC at points 1, 2, 3, 4, 5, 6, 7, 8. Then divide FG at points 12, 13, 14, 15, 16, 17. Divide BD at points 21, 22, 23, 24, 25, 26, 27, 28. Divide FH at points 32, 33, 34, 35, 36, 37. Through these parts [of the perimeter], one shall understand all the remaining ones. Draw lines F-1, 12-2, 13-3, 14-4, 15-5, 16-6, 17-7, G-8. This will be the appearance of the first side. The shape of the second will appear after drawing lines 21-F, 22-32, 23-33, 24-34, 25-35, 26-36, 27-37, 28-H.

We realize that all these lines are parallel and equal to each other and relate to point A which we have called the eye. The latter is divided by diagonals BE and DC into four equal parts, which form one eye. In the head the eye has a circular shape and only shows the fourth part. Therefore, we must say that point A exists as though it were made by four eyes: the first one for the part opposite line FG; the second for the part opposite line FH; the third one for the part opposite line GI; the fourth one for the part opposite line HI. In fact, if there were four men, each one of them looking at the figure on their own wall, they would do the same thing as the eye [A] does. We can say that the eye is circular and that, from the intersection of the two nerves in the center of the crystal humor, there lies the ability to see. From this point there are rays that leave the eye in straight lines; these regulate the fourth part of the eye. We have assumed, for example, that the rays form a right angle at the center and the lines coming from this angle end at points F and G. We shall say then that line FG is a higher quantity than the eye opposite it is able to see. If this line goes beyond the diagonal, the other eye would be below the fourth part of the circuit; but this cannot be true because the diagonals of the perfect square divide the circle into four equal parts. Line FG lies within that limit so that the eye cannot see a greater one. Because of this, [when one exceeds that limit] the shortened quantity is greater than the nonshortened because it ends, with gaze, in the part of the other eye.

Proof. Draw points B, 1, 2, 3, 4, 5, 6, 7, 8, and C and the lines to point A. The line from point B will be the diagonal that goes through point F of line FG. Add quantity 1 to B for an extension of line BC: let this be called BK. Add quantity F to 21 for an extension of line 21-G: let this be called 21-L. And draw KL, which closes square BKL21. If the line is drawn from K to point A, this line will divide 21-F at point M. I can claim that the shortening KL is greater than 21-L [which is not shortened] by quantity 21-M. KL in fact always appears to be equal to LM, which is greater than L-21. As said, the shortened part is greater than the nonshortened one. The eye, in fact, cannot observe, in that plane, point K, which is that part of the eye opposite line FH. And although the eye can make out FG, the intellect cannot comprehend, or understand, the differences, unless as though they were a stain seen from afar, not being able to

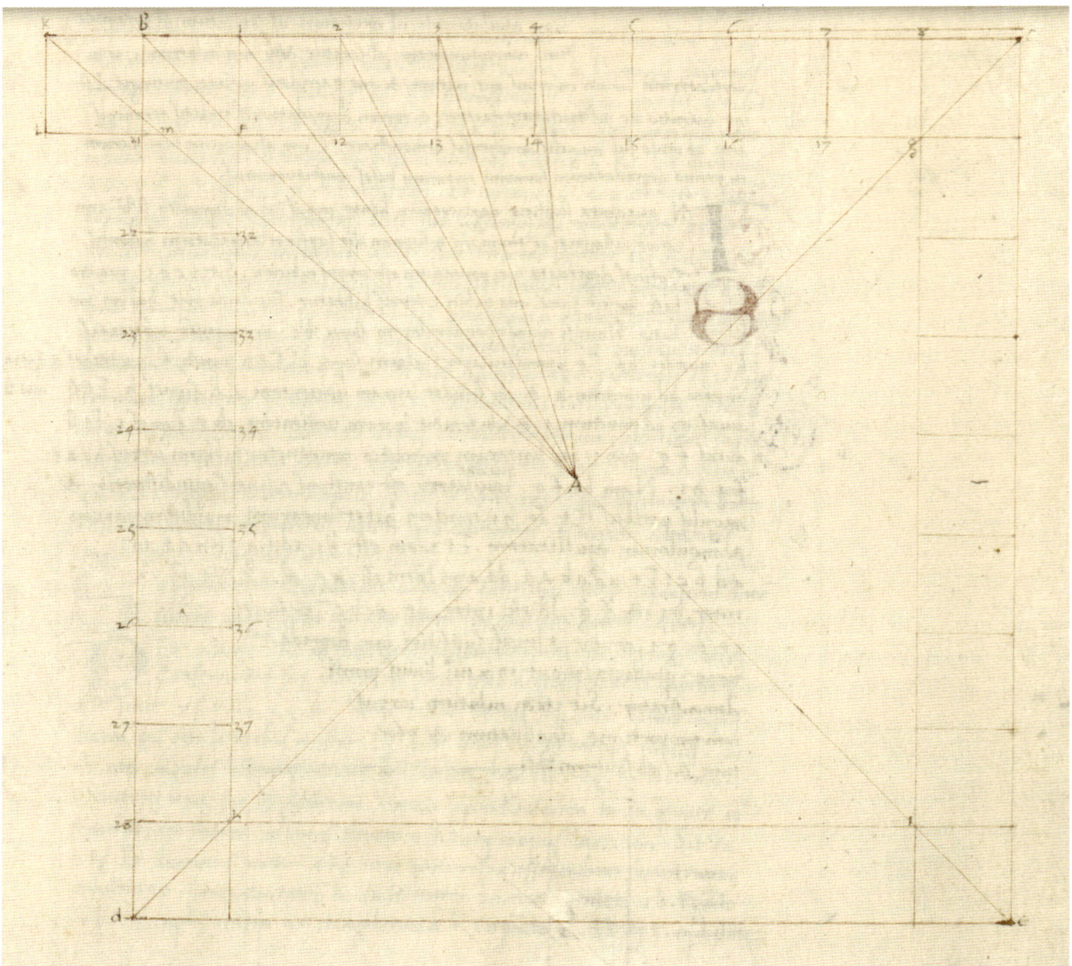

Bordeaux, Bibliothèque Municipale, Ms Latin 616, folio 19 recto.

ascertain whether it is a man or an animal. This is the way F and G exist with respect to point A. Because of those things of which we do not perceive differences, unless as though they were spots, it is necessary to take into consideration a smaller quantity of line FG and a smaller angle than a right angle.[65] To allow the eye to make out objects opposite it, these have to be presented to the eye at an angle smaller than a right angle, which will be two-thirds of a right angle. Three angles, in fact, make an equilateral triangle when they are the same length. And because this line [FG] is present in elements, we shall express it with numbers so as to understand this work plan more clearly. Let's say that if your work will occupy a width of 7 braccia, then you will have to see it from a 6-braccia distance. If your work were bigger, you should move away from it proportionally. If it were smaller, it does not matter, as long as you stand at a distance mentioned above, because you can move away as much as you want from that plane of work. The eye, in fact, has the strength of observing the entire [preestablished] plane without moving. If you will observe the reasons we have claimed, it will be clear to you that it is they [the detractors] who are wrong and not the perspective procedure. To them, the shortened object becomes bigger than the nonshortened one.[66]

Book 2

Each body has three dimensions: length, width,[67] height. The extremities of a body are its surfaces. The shapes of a body vary and are multiple. We may find a cubic body, a nonequilateral tetragon, a round body, one with sides, one with the shape of a round pyramid,[68] one with the shape of a pyramid with sides, and one with many and diverse sides, as we may see in natural and accidental things. I would like to discuss these [bodies] and show, in the assigned terms, their shortenings according to the eye, through visual angles, once we have fixed, as the bases of these bodies, the shape of several surfaces shortened in the first book.

Proposition I

Place a shortened cubic body on a shortened squared surface according to the distance and the assigned terms of the mentioned shortened surface.

Let BCDE be the shortened surface on which one may build a cubic body; let the surface be the base, that is, a face of the cube [figure B.35]. This is a very easy elevation. In fact, if line BF perpendicular to BC is raised on B and is as long as BC; if another perpendicular line CG, as long as BC, is raised on C, if you draw another nonlimited perpendicular line on D and another one on E; if from point F you draw the line to point A and where this one intersects the line elevated from D, you mark point H; if from point G you draw the line to point A and where this one intersects the line that elevates from E, you mark point I; and if you draw FG and HI: this line tracing will yield cubic body BCDE-FGHI. Figure BCFG, in fact, has equal sides and is formed by parallel lines and equal angles. Figure DEHI has the same equality in parallel lines and angles. And the proportion of HI and FG is the same as DE and BC. And the proportion of AI and AG is the same AE and AC. And the proportion of HI and FG is the same as between DE and BC.[69] And the sides CE and EI are composed by lines of the squared shortened surface, as shown in [Proposition] XIV of the first book. In this way, the cube shortened proportionally has been completed and perfected.

However, if the same body were to be placed in the shortened plane so that the sides are not parallel to the ones of the plane [at the base] [figure B.36], we shall draw on the plane the shortened surface FGHI, as has been demonstrated

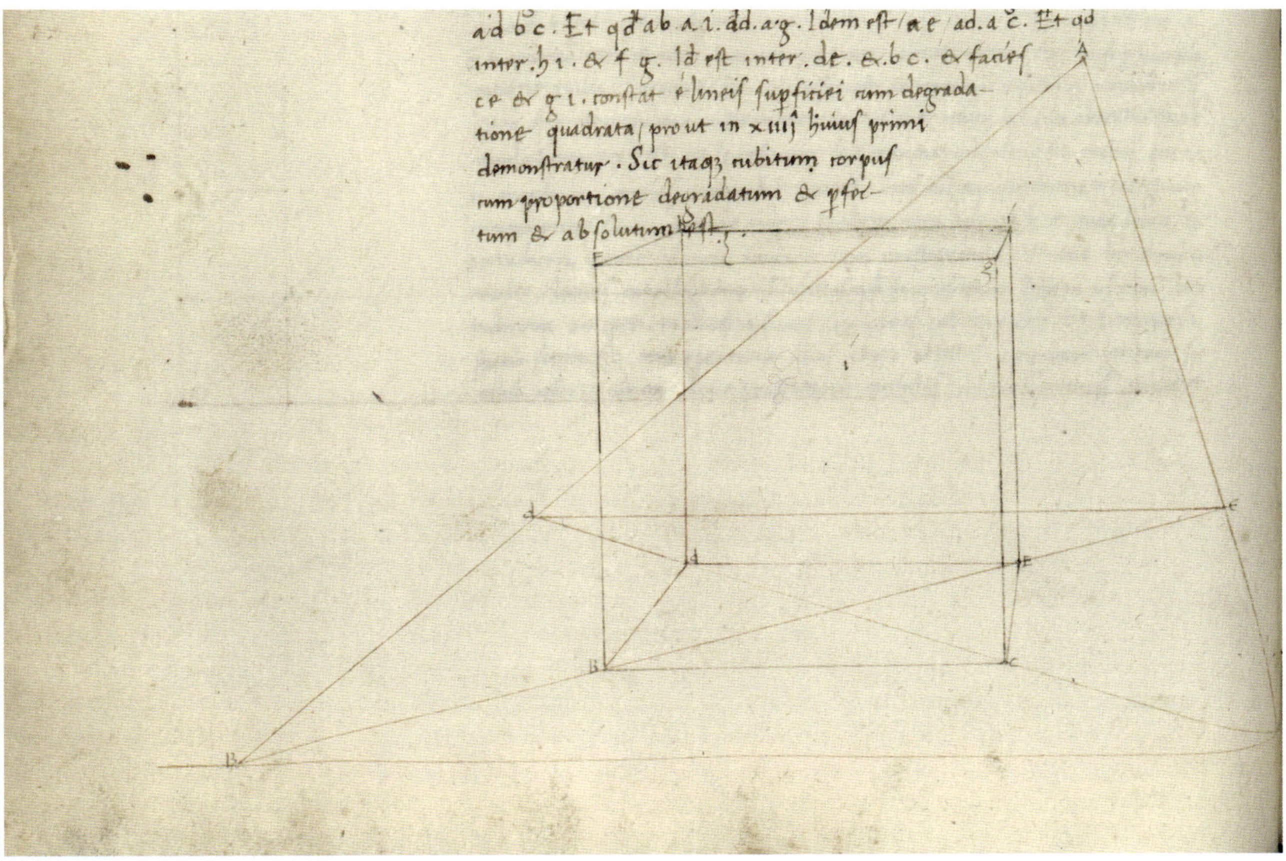

Bordeaux, Bibliothèque Municipale, Ms Latin 616, folio 19 verso.

for Proposition XXV of the first book. And we shall conduct from F a line parallel to BC: where [this parallel] will intersect line BA, we shall have point O. And we shall conduct [a line] parallel to BC from G: where this [parallel] intersects line BA, we shall have point P. And we shall conduct from H [a line parallel to BC]: where this parallel intersects line BA, we shall have point Q. And we shall conduct from I [a line parallel to BC]: where this parallel intersects line BA, we shall have point R. We shall also draw the vertical line OS on O, the vertical line PT on P, the vertical line QV on Q, the vertical line RX on R, the vertical line BZ on B equal to length FG in real form. We shall draw line AZ passing through S, T, V, and X. We shall draw lines, without a limit, from F, G, H, and I parallel to OS. Successively we shall draw from S a line parallel to BC, and where it will intersect the line that rises from F without a limit, we shall mark point K. We shall draw from T a line parallel to BC, and where it will intersect the line that rises from G without a limit, we shall mark point L. We shall draw from V a line parallel to BC, and where it will intersect the line that rises from H without a limit, we shall mark point M. We shall draw from X a line parallel to BC, and where it will intersect the line that rises from I without a limit, we shall mark point N. We shall then draw lines KL, LN, NM, MK. By tracing these lines we have completed the cubic body proportionally on the shortened surface given, which was our intention. Bases OS, PT, QV, RX, which have their [own] position in the respective shortened spaces, present themselves in the same angle, as shown by the parallel lines, which extend from the angles in the shortened quadrilateral.

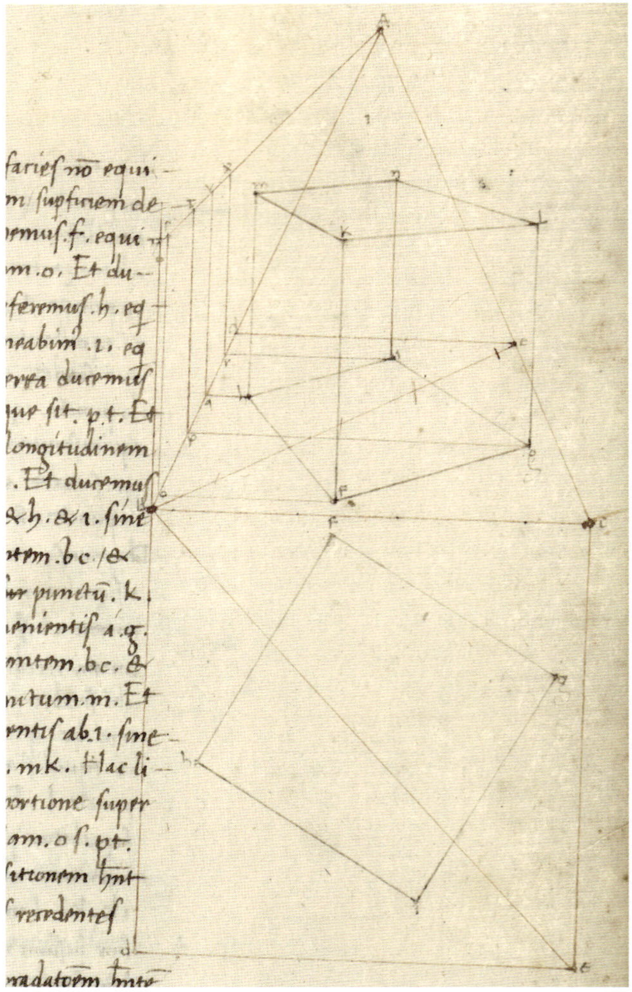

Bordeaux, Bibliothèque Municipale, Ms Latin 616, folio 20 recto.

Proposition II

On a shortened surface with a shortened base of eight sides, place a similar shortened body, which has eight faces without including the bases.

By means of the twenty-sixth [proposition of the first book],[70] as was shown in preceding drawings, let us place figure RSTVXYzZ with eight shortened sides on a shortened surface [figure B.37]. We are about to build, on this base, a body which will have eight faces and two similar bases which have eight angles each. Similar to the first one, let FGHIKLMN be the second base. Thus, we shall draw from point B of the plane [at the base] the vertical line BP, as long as we want the eight-sided body to be. And we shall draw line AP, because we have established that A is the eye appearing in its place. We shall draw a line parallel to BC from R till it meets AB at point 1; we shall draw a parallel to BC from S till it meets AB at point 2; we shall draw a parallel to BC from T till it meets AB at point 3; we shall draw a parallel to BC from V till it meets AB at point 4; we shall draw a parallel to BC from X till it meets AB at point 5; we shall draw a parallel to BC from Y till it meets AB at point 6; we shall draw a parallel to BC from z till it meets AB at point 7; we shall draw a parallel to BC from Z till it meets AB at point 8. We shall draw a line parallel to BP from 1 till it meets AP at point 11; we shall draw a parallel to BP from 2 till it meets AP at point 12; we shall draw a parallel to BP from 3 till it meets AP at point 13; we shall draw a parallel to BP from 4 till it meets AP at point 14; we shall draw a parallel to BP from 5 till it meets AP at point 15; we shall draw a parallel to BP from 6 till it meets AP at point 16; we shall draw a parallel to BP from 7 till it meets AP at point 17; we shall draw a parallel to BP from 8 till it meets AP at point 18. I will then draw unlimited lines, from points RSTVXYzZ, which are parallel to BP. We shall draw a line parallel to BC from point 11: where this will intersect the line that comes from R, we shall mark F. We shall draw a line parallel to BC from point 12: where this will intersect the line that comes from S, we shall mark G. We shall draw a line parallel to BC from point 13: where this will intersect the line that comes from T, we shall mark H. We shall draw a line parallel to BC from point 14: where this will intersect the line that comes from V, we shall mark I. We shall draw a line parallel to BC from point 15: where this will intersect the line that comes from X, we shall mark K. We shall draw a line parallel to BC from point 16: where this will intersect the line that comes from Y, we shall mark L. We shall draw a line parallel to BC from point 17: where this will intersect the line that comes from z, we shall mark M. We shall draw a line parallel to BC from point 18: where this will intersect the line that comes from Z, we shall mark N. We shall trace in the end lines FG, GH, HI, IK, KL, LM, MN, NF. In this way is completed the proportionally shortened eight-faced body, without including the bases. It was, in fact, built on the octagonal figure, through the twenty-sixth proposition of the first book.[71] From the angles of this figure we traced the lines parallel to BC till they intersected line AB. Then lines parallel to BP were traced till they intersected line AP. These parallels were shortened proportionally, because they are included under the same angle in their shortened extremities.

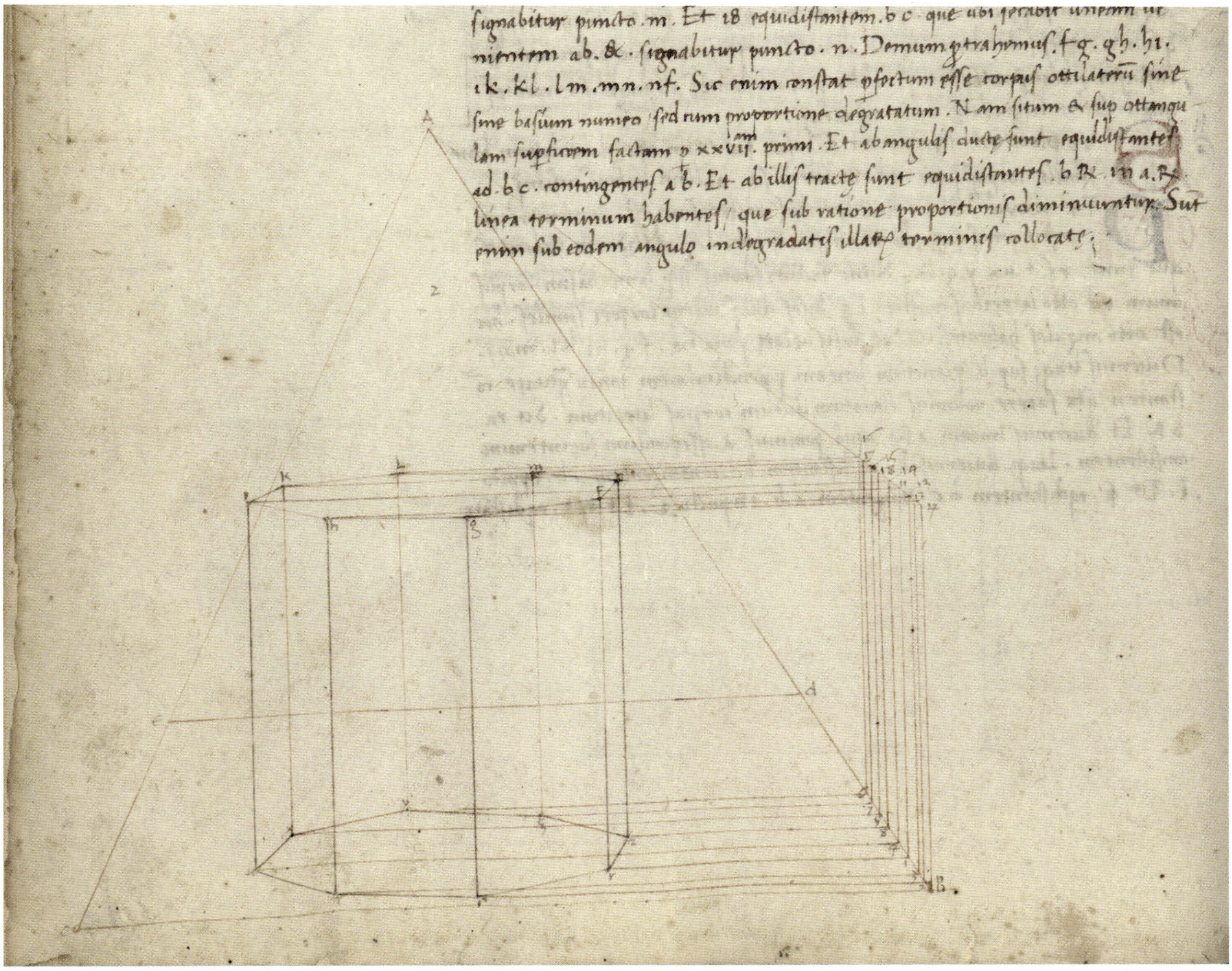

Bordeaux, Bibliothèque Municipale, Ms Latin 616, folio 20 verso.

Proposition III

On a shortened pentagonal area, shorten a body that has as many faces.

We have the pentagonal surface FGHIK performed through the twentieth proposition of the first book. On this surface we are about to place a pentagonal body, which follows the sides of surface FGHIK[72] and has figure LMNOP as a second base above [figure B.38]. Since our shortened plane is BCDE, we shall draw on B the perpendicular line Bp, as high as we want, which is the height of the said solid. And we shall trace from F a line parallel to BC till it meets BD at point Q; and we shall trace from G a line parallel to BC till it meets BD at point R; and we shall trace from H a line parallel to BC till it meets BD at point S; and we shall trace from I a line parallel to BC till it meets BD at point T; and we shall trace from K a line parallel to BD till it meets BD at point V. Then we shall trace from the eye A line Ap; and from Q a line parallel to Bp, which shall intersect Ap at point 1. And from R a line parallel to Bp, which shall intersect Bp at point 2; and from S a line parallel to Bp, which shall intersect Ap at point 3; and from T a line parallel to Bp, which shall intersect Ap at point 4; and from V a line parallel to Bp, which shall intersect Ap at point 5.

We will then trace from points F, G, H, I, K lines parallel to Bp that have no end. Moreover, we shall trace from point 1 a line parallel to BC, and where it will intersect the line that rises from F, we shall have point L. We shall trace from 2 a line parallel to BC, and where it will intersect the line that rises from G, we shall have point M. We shall trace from 3 a line parallel to BC, and where it will intersect the line that rises from H, we shall have point N. We shall trace from 4 a line parallel to BC, and where it will intersect the line that rises from I, we shall have point O. We shall trace from 5 a line parallel to BC, and where it will intersect the line that rises from K, we shall have point P. In the end we shall trace lines LM, MN, NO, OP, PL. By tracing these lines we have placed the body with its faces on the surface of the pentagon in a proportionate manner. Because there exist a lot of ways and methods to shorten, and these all tend toward the same aim, this method seemed to be more easy to demonstrate and we shall follow it throughout this second book.

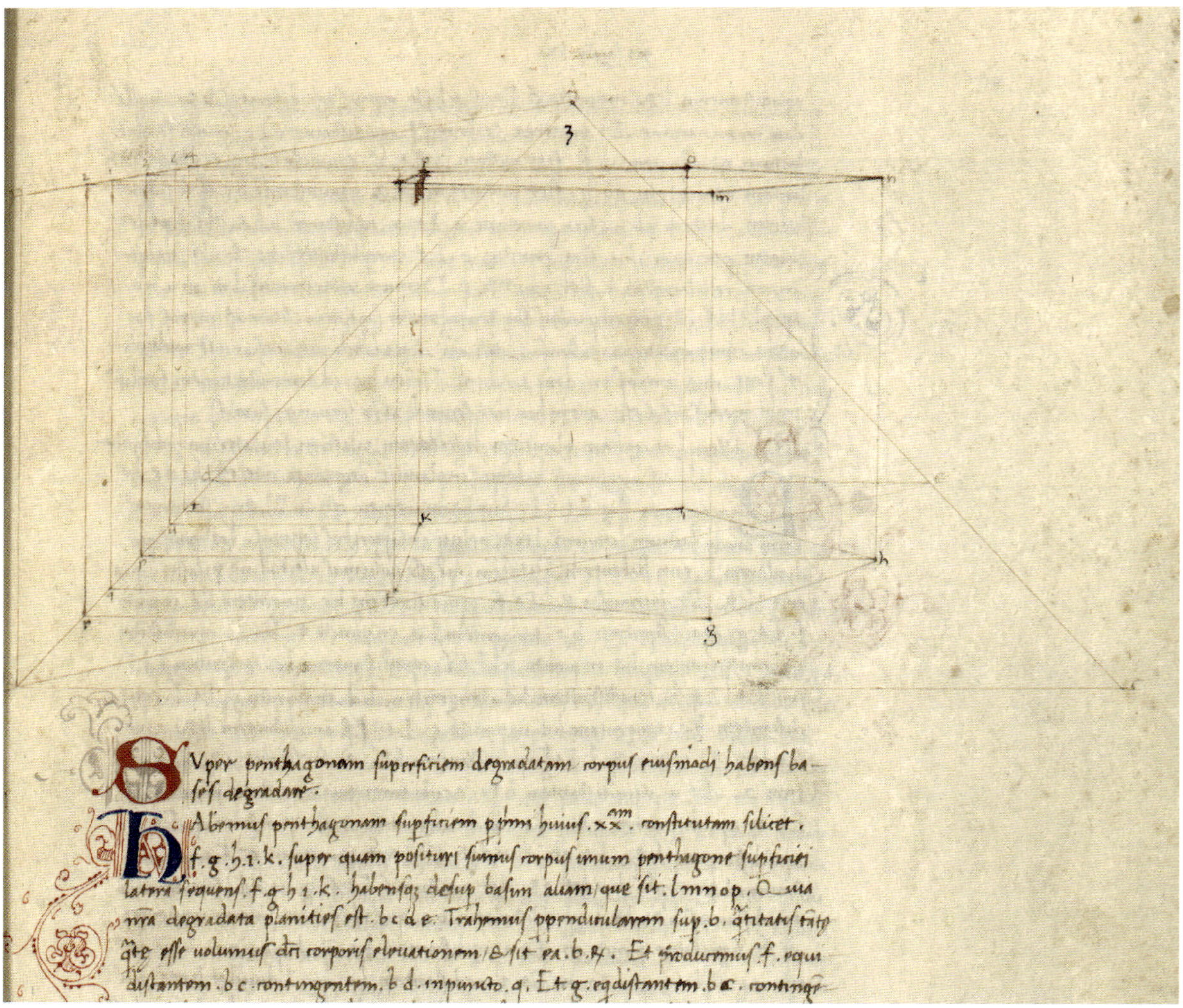

Bordeaux, Bibliothèque Municipale, Ms Latin 616, folio 21 recto.

Proposition IV

On a shortened hexagonal surface elevate a pillar that has as many faces.

By means of the twentieth proposition of the first book, we have, in the shortened plane BCDE, hexagon FGHIKL, on which we intend to place a pillar through the measurement of its sides [figure B.39]. As we did, then, in the previous cases, we shall trace perpendicular Bp on B; this will be as long as we want the pillar to be. And we shall trace line Ap and from F a line parallel to BC, till it meets BD at point S; and from G a line parallel to BC, till it meets BD at point T; and from H a line parallel to BC, till it meets BD at point V; and from I a line parallel to BC, till it meets BD at point X; and from K a line parallel to BC, till it meets BD at point Y; and from L a line parallel to BC, till it meets BD at point Z; and from S a line parallel to Bp, till it meets Ap at point 1; and from T a line parallel to Bp, till it meets Ap at point 2; and from V a line parallel to Bp, till it meets Ap at point 3; and from X a line parallel to Bp, till it meets Ap at point 4; and from Y a line parallel to Bp, till it meets Ap at point 5; and from Z a line parallel to Bp, till it meets Ap at point 6. And for points F, G, H, I, K, L we shall trace lines with no end which are parallel to Bp. We shall trace a line parallel to BC from 1, and where it will intersect the line that rises from F, we shall mark point M; a line parallel to BC from 2, and where it will intersect the line that rises from G, we shall mark point N; a line parallel to BC from 3, and where it will intersect the line that rises from H, we shall mark point O; a line parallel to BC from 4, and where it will intersect the line that rises from I, we shall mark point P; a line parallel to BC from 5, and where it will intersect the line that rises from K, we shall mark point Q; a line parallel to BC from 6, and where it will intersect the line that rises from L, we shall mark point R. In the end, we shall trace lines MN, NO, OP, PQ, QR, RM. Having completed this, we have executed the six-faced pillar which we had announced.

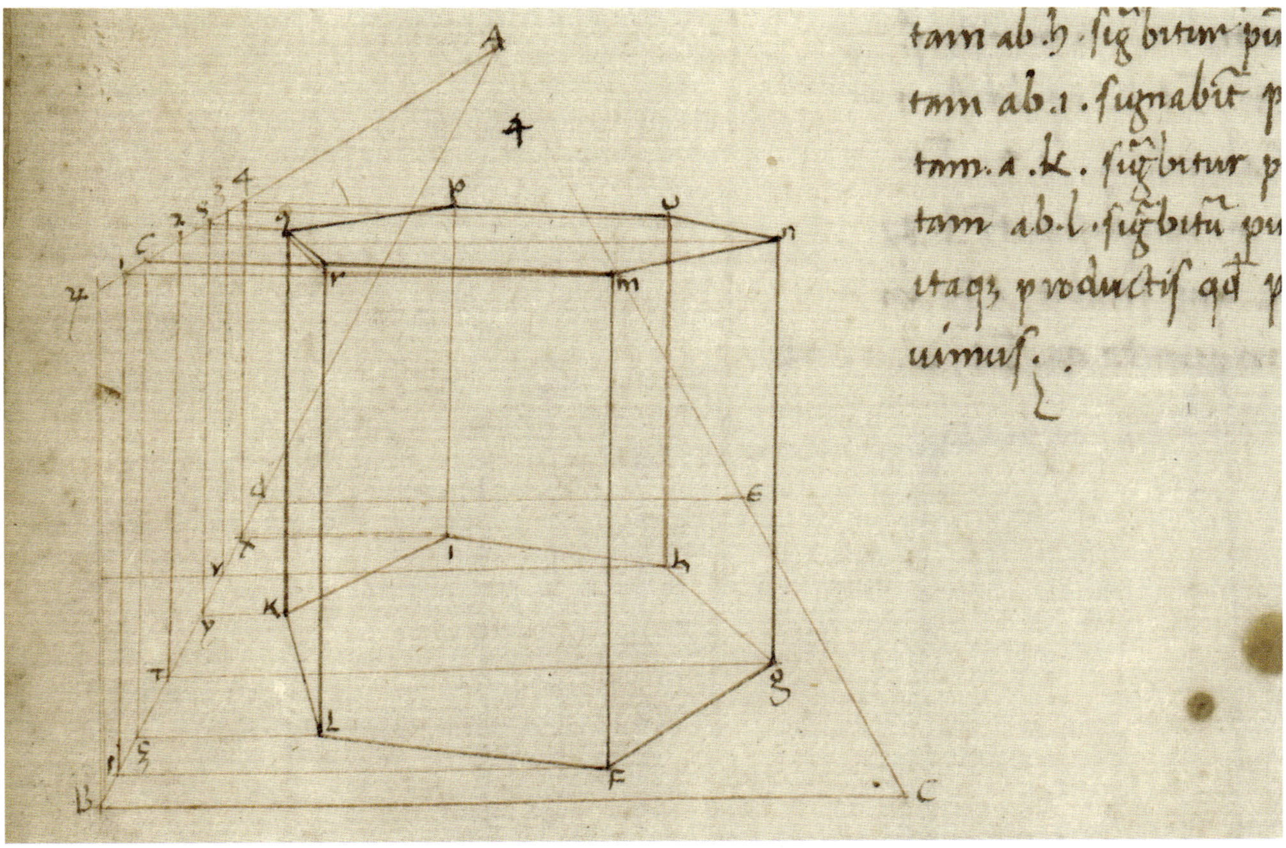

Bordeaux, Bibliothèque Municipale, Ms Latin 616, folio 21 verso.

Proposition V

On a shortened plane elevate a column that has sixteen faces; let the column be appropriate to the proportion of the given surface.

The aim of this demonstration is to place, above the shortened plane, our column that has sixteen equal faces [figure B.40]. These [faces] will be shortened in proportion to the shortened plane BCDE,[73] which has been realized through the fourteenth [proposition] of the first book. We shall therefore place on the mentioned plane the surface with sixteen faces, as shown in the seventeenth proposition of the first book. The given figure is 1-2-3-4-5-6-7-8-9-10-11-12-13-14-15-16. Our shortened plane is BCDE. The point of the eye is A. First of all, I shall trace from point 1 a line parallel to BC till it touches BD at point 21; from point 2 a line parallel to BC till it touches BD at point 22; from point 3 a line parallel to BC till it touches BD at point 23; from point 4 a line parallel to BC till it touches BD at point 24; from point 5 a line parallel to BC till it touches BD at point 25; from point 6 a line parallel to BC till it touches BD at point 26; from point 7 a line parallel to BC till it touches BD at point 27; from point 8 a line parallel to BC till it touches BD at point 28; from point 9 a line parallel to BC till it touches BD at point 29; from point 10 a line parallel to BC till it touches BD at point 30; from point 11 a line parallel to BC till it touches BD at point 31; from point 12 a line parallel to BC till it touches BD at point 32; from point 13 a line parallel to BC till it touches BD at point 33; from point 14 a line parallel to BC till it touches BD at point 34; from point 15 a line parallel to BC till it touches BD at point 35; from point 16 a line parallel to BC till it touches BD at point 36. Moreover, I shall trace vertical line Bp on B, making it as long as the column to be drawn, and I shall mark the line from p to A. I shall then trace from point 21 a line parallel to Bp, and where it will intersect line Ap, I shall mark point 41. I shall trace from point 22 a line parallel to Bp, and where it will intersect line Ap, I shall mark point 42. I shall trace from point 23 a line parallel to Bp, and where it will intersect line Ap, I shall mark point 43. I shall trace from point 24 a line parallel to Bp, and where it intersects line Ap, I shall mark point 44. I shall trace from point 25 a line parallel to Bp, and where it will intersect line Ap, I shall mark point 45. I shall trace from point 26 a line parallel to Bp, and where it will intersect line Ap, I shall mark point 46. I shall trace from point 27 a line parallel to Bp, and where it will intersect line Ap, I shall mark point 47. I shall trace from point 28 a line parallel to Bp, and where it will intersect line Ap, I shall mark point 48. I shall trace from point 29 a line parallel to Bp, and where it will intersect line Ap, I shall mark point 49. I shall trace from point 30 a line parallel to Bp, and where it will intersect line Ap, I shall mark point 50. I shall trace from point 31 a line parallel to Bp, and where it will intersect line Ap, I shall mark point 51. I shall trace from point 32 a line parallel to Bp, and where it will intersect line Ap, I shall mark point 52. I shall trace from point 33 a line parallel to Bp, and where it will intersect line Ap, I shall mark point 53. I shall trace from point 34 a line parallel to Bp, and where it will intersect line Ap, I shall mark point 54. I shall trace from point 35 a line parallel to Bp, and where it will intersect line Ap, I shall mark point 55. I shall trace from point 36 a line parallel to Bp, and where it will intersect line Ap, I shall mark point 56. Moreover I will trace from points 1, 2, 3, 4, 5, 6, 7, 8, 9, 10, 11, 12, 13, 14, 15, 16 lines parallel to Bp and indeterminately long. I will then trace a line parallel to BC from point 41, and where this line will intersect the line that rises from 1, I shall mark point 61. From point 42 a line parallel to BC, and where it will intersect the line that rises from 2, I will mark point 62. From point 43 a line parallel to BC, and where it will intersect the line that rises from 3, I will mark point 63. From point 44 a line parallel to BC, and where it will intersect the line that rises from 4, I will mark point 64. From point 45 a line parallel to BC, and where it will intersect the line that rises from 5, I will mark point 65. From point 46 a line parallel to BC, and where it will intersect the line that rises from 6, I will mark point 66. From point 47 a line parallel to BC, and where it will intersect the line that rises from 7, I will mark point 67. From point 48 a line parallel to BC, and where it will intersect the line that rises from 8, I will mark point 68. From point 49 a line parallel to BC, and where it will intersect the line that rises from 9, I will mark point 69. From point 50 a line parallel to BC, and where it will intersect the line that rises from 10, I will mark point 70. From point 51 a line parallel to BC, and where it will intersect the line that rises from 11, I will mark point 71. From point 52 a line parallel to BC, and where it will intersect the line that rises from 12, I will mark point 72. From point 53 a line parallel to BC, and where it will intersect the line that rises from 13, I will mark point 73. From point 54 a line parallel to BC, and where it will intersect the line that rises from 14, I will mark point 74. From point 55 a line parallel to BC, and where it will intersect the line that rises from 15, I will mark point 75. From point 56 a line parallel to BC, and where it will intersect the line that rises from 16, I will mark point 76. With these lines we may complete the sixteen faces by joining 61 with 62, 62 with 63, 63 with 64, 64 with 65, 65 with 66, 66 with 67, 67 with 68, 68 with 69, 69 with 70, 70 with 71, 71 with 72, 72 with 73, 73 with 74, 74 with 75, 75 with 76, 76 with 61. This is the surface placed above. The surface below is identified through these numbers: 1, 2, 3, 4, 5, 6, 7, 8, 9, 10, 11, 12, 13, 14, 15, 16. These are the corners that determine the column with sixteen sides, in conformity with the method of the shortened surface. Thus, in this way, according to the rules of the procedures, many or more faces may be represented, according to buildings' conformation. Faces that fall into place according to the same shortening relationship.

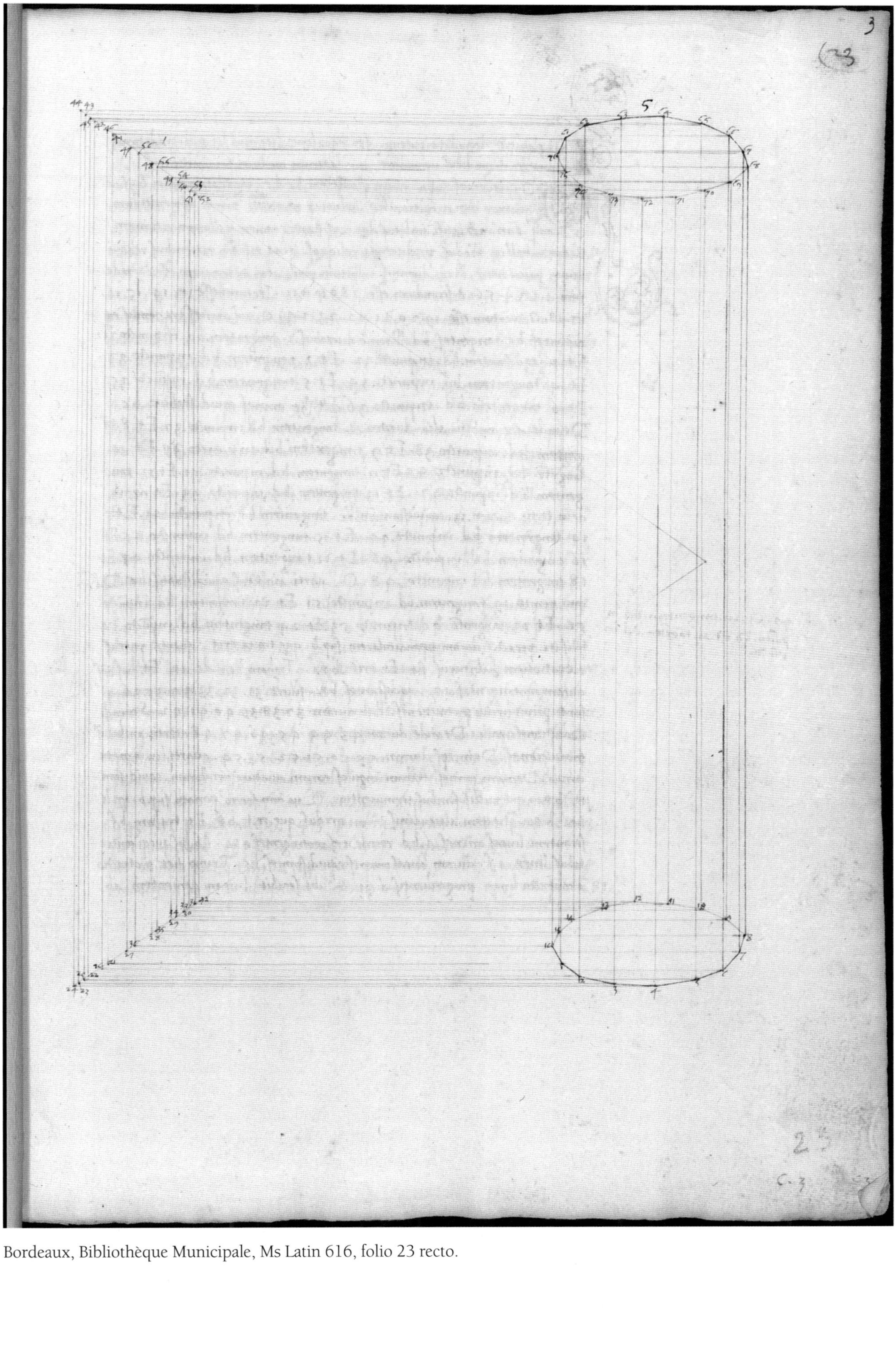

Bordeaux, Bibliothèque Municipale, Ms Latin 616, folio 23 recto.

Proposition VI

Elevate a well with six equal and shortened faces on a shortened plane and surround it with steps that follow the conformation of the faces.

We shall establish on our shortened plane BCDE[74] the surface of a well with four perimeters [figure B.41], as we have been taught in the twenty-ninth proposition of the first book. In fact, in that [surface of the twenty-ninth proposition] there are only two perimeters, but through the procedure of those [two] we understand how to reproduce the remaining ones, since the remaining ones may be created through the same procedures as well. We shall say, then, that the first shortened external perimeter is 1, 2, 3, 4, 5, 6; the second one is 7, 8, 9, 10, 11, 12; the third one is 13, 14, 15, 16, 17, 18; the fourth one is 19, 20, 21, 22, 23, 24. From all these points we shall draw lines parallel to BC till they touch BD. Firstly, we shall draw from point 1 the line that touches BD at point 31; then from point 2 the line that touches BD at point 32; then from point 3 the line that touches BD at point 33; then from point 4 the line that touches BD at point 34; then from point 5 the line that touches BD at point 35; then from point 6 the line that touches BD at point 36. All these lines are parallel to BC. Successively, considering the second perimeter,[75] I shall draw from point 7 the line that touches BD at point 37; from point 8 the line that touches BD at point 38; from point 9 the line that touches BD at point 39; from point 10 the line that touches BD at point 40; from point 11 the line that touches BD at point 41; from point 12 the line that touches BD at point 42. Successively, considering the third perimeter, I shall draw from point 13 the line that touches BD at point 43; from point 14 the line that touches BD at point 44; from point 15 the line that touches BD at point 45; from point 16 the line that touches BD at point 46; from point 17 the line that touches BD at point 47; from point 18 the line that touches BD at point 48. Firstly, I shall draw from point 19 of the fourth perimeter a line parallel to BC, which touches BD at point 49. From point 20 the line that touches BD at point 50. From point 21 the line that touches BD at point 51. From point 22 the line that touches BD at point 52. From point 23 the line that touches BD at point 53. From point 24 the line that touches BD at point 54. Having done these things, I shall draw on B vertical line Bp, which will be as tall as the height of the well. I shall then trace line Ap. And from the extensions of the points of the first perimeter, that is, from points 31, 32, 33, 34, 35, 36, I shall trace lines parallel to Bp till they touch Ap. And I will trace [the parallel lines] from the extensions of points 37, 38, 39, 40, 41, 42 of the second perimeter. And I will trace [the parallel lines] from the extensions of points 43, 44, 45, 46, 47, 48 of the third perimeter. And I will trace [the parallel lines] from the extensions of points 49, 50, 51, 52, 53, 54 of the fourth perimeter. Finally I will draw, from all the angles of those four perimeters, the parallel lines, with no end. Having traced them, I will place on line Bp a line of size BF which is equal to the height of the first step [of the well], and I will trace line AF which cuts all the lines that leave from BD and touch Ap. Also from the contact points with line AF I shall trace lines parallel to BC. Firstly, I shall generate the extension from the contact of the line that begins at point 31: where it will intersect the line that comes from angle 1, I shall mark point 61. I shall generate the extension from the contact of the line that begins at point 32: where it will intersect the line that comes from angle 2, I shall mark point 62. I shall generate the extension from the contact of the line that begins at point 33: where it will intersect the line that comes from angle 3, I shall mark point 63. I shall generate the extension from the contact of the line that begins at point 34: where it will intersect the line that comes from angle 4, I shall mark point 64. I shall generate the extension from the contact of the line that begins at point 35: where it will intersect the line that comes from angle 5, I shall mark point 65. I shall generate the extension from the contact of the line that begins at point 36: where it will intersect the line that comes from angle 6, I shall mark point 66. These are the elevations of the first step. Regarding the contact points of the second step with line AF, I shall execute an extension from the contact of the line that originates at point 37: where it will intersect the line that rises from 7, I shall trace point 67. I shall execute an extension from the contact of the line that originates at point 38: where it will intersect the line that rises from 8, I shall trace point 68. I shall execute an extension from the contact of the line that originates at point 39: where it will intersect the line that rises from 9, I shall trace point 69. I shall execute an extension from the contact of the line that originates at point 40: where it will intersect the line that rises from 10, I shall trace point 70. I shall execute an extension from the contact of the line that originates at point 41: where it will intersect the line that rises from 11, I shall trace point 71. I shall execute an extension from the contact of the line that originates at point 42: where it will intersect the line that rises from 12, I shall trace point 72. Till now we have completed the surface of the first step. Now, with the aim of establishing the procedure of the second step, I shall draw, starting from F, on line Bp, segment FG, as long as BF, and I shall draw line GA which will intersect all the lines parallel to Bp, and from the contact points that will be generated I will draw lines that are equidistant to BC. First of all, I shall draw from the contact of the line that originates at point 37 [a line equidistant to BC]: where it will intersect the line that rises from 7, I shall mark point 73. And I shall draw from the contact of the line that originates at point 38 [a line equidistant to BC]: where it will intersect the line that rises from 8, I shall mark point 74. And I shall draw from the contact of the line that originates at point 39 [a line equidistant to BC]: where it will intersect the line that rises from 9, I shall mark point 75. And I shall draw from the contact of the line that originates at point 40 [a line equidistant to BC]: where it will intersect the line that rises from 10, I shall mark point 76. And I shall draw from the contact of the line that originates at point 41 [a line equidistant to BC]: where it will intersect the line that rises from 11, I shall mark point 77. And I shall draw from the contact of the line that originates at point 42 [a line equidistant to BC]: where it will intersect the line that rises from 12, I shall mark point 78. These extensions are part of the second step, without the floor; to draw now the floor of the same step, I shall draw from the contact of the line that originates at point 43 [a line equidistant to BC]: where it will intersect the line that rises from 13, I shall mark point 79. And I shall draw from the contact of the line that originates at point 44 [a line equidistant to BC]: where it will intersect the line that rises from 14, I shall mark point 80. I shall draw from the contact of the line that originates at point 45 [a line equidistant to BC]: where it will intersect the line that rises from 15, I shall mark point 81. And I shall draw from the contact of the line that originates at point 46 [a line equidistant to BC]: where it will intersect the line that rises from 16, I shall mark point 82. And I shall draw from the contact of the line that originates at point 47 [a line equidistant to BC]: where it will intersect the line that rises from 17, I shall mark point 83. And I shall draw from the contact of the line that originates at point 48 [a line equidistant to BC]: where it will intersect the line that rises from 18, I shall mark point 84. Through these extensions, the floor of the second step will be perfected. I shall now draw lines from the contact points that produce lines parallel to Bp on line Ap. First, I shall draw, from the contact [with Ap] of the [vertical] line that goes from point 43, [a line equidistant to BC]; and where it will meet the [vertical] line that leaves point 13, I shall draw point 85. And I shall draw, from the contact of the line that goes from point 44, [a line equidistant to BC]; and where it will meet the line that leaves point 14, I shall mark point 86. And I shall draw, from the contact of the line that goes from point 45, [a line equidistant to BC]; and where it will meet the line that leaves point 15, I shall mark point 87. And I shall draw, from the contact of the line that goes from point 46, [a line equidistant to BC]; and where it will meet the line that leaves point 16, I shall mark point 88. And I shall draw, from the contact of the line that goes from point 47, [a line equidistant to BC]; and where it will meet the line that leaves point 17, I shall mark point 89. And I shall draw, from the contact of the line that goes from point 48, [a line equidistant to BC]; and where it will meet the line that leaves point 18,

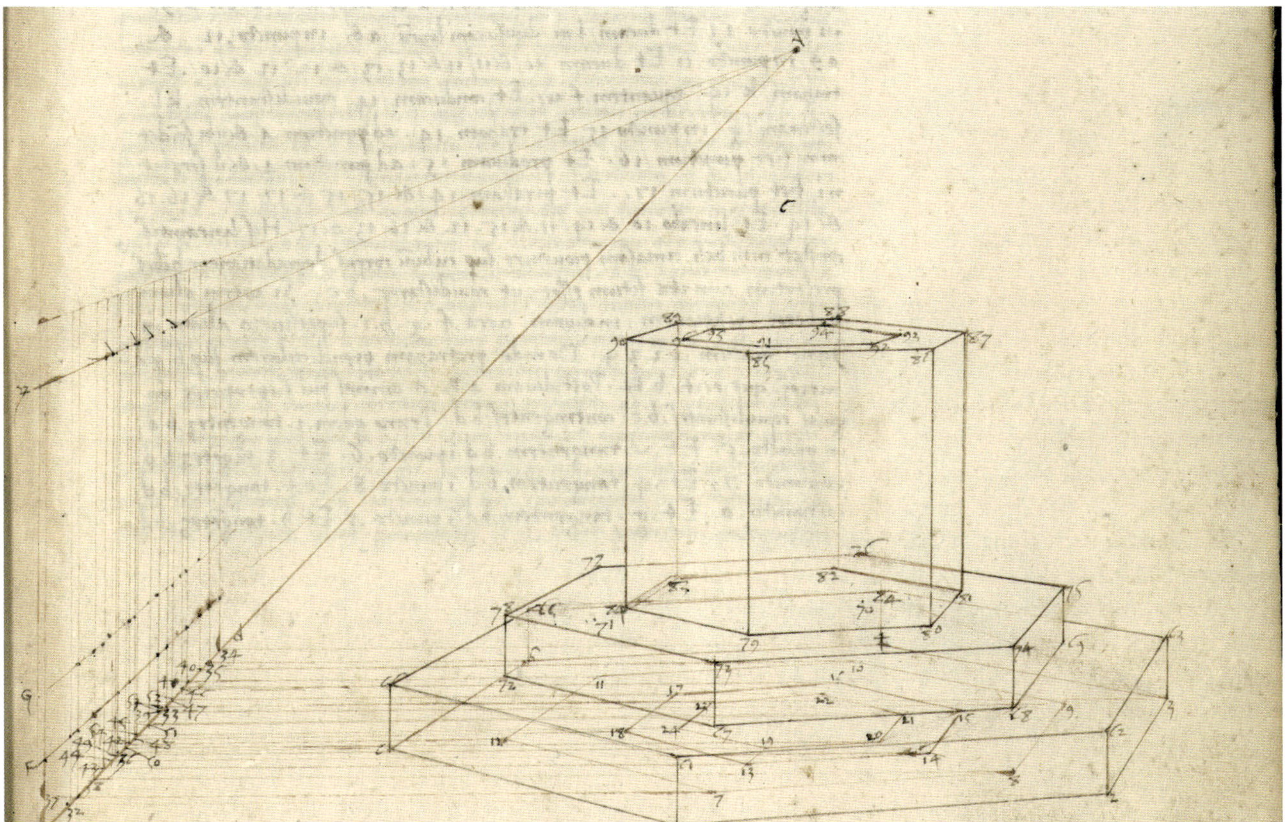

Bordeaux, Bibliothèque Municipale, Ms Latin 616, folio 25 recto.

I shall mark point 90. In this way, we will have completed the third perimeter. For the fourth perimeter I shall draw from the contact of the line that goes from point 49 [a line equidistant to BC], and where it will meet the [vertical] line that leaves point 19, I shall mark point 91. And I shall draw from the contact of the line that goes from point 50 [a line equidistant to BC], and where it will meet the line that leaves point 20, I shall mark point 92. And I shall draw from the contact of the line that goes from point 51 [a line equidistant to BC], and where it will meet the line that leaves point 21, I shall mark point 93. And I shall draw from the contact of the line that goes from point 52 [a line equidistant to BC], and where it will meet the line that leaves point 22, I shall mark point 94. And I shall draw from the contact of the line that goes from point 53 [a line equidistant to BC], and where it will meet the line that leaves point 23, I shall mark point 95. And I shall draw from the contact of the line that goes from point 54 [a line equidistant to BC], and where it will meet the line that leaves point 24, I shall mark point 96. Through these extensions we have therefore completed the fourth perimeter. Successively I will trace lines 61-62, 62-63, 63-64, 64-65, 65-66, 66-61, which mark the first perimeter. For the second one I will trace lines 67-68, 68-69, 69-70, 70-71, 71-72, 72-67. For the third one I will draw lines 73-74, 74-75, 75-76, 76-77, 77-78, 78-73. For the fourth one we will have these segments: 79-80, 80-81, 81-82, 82-83, 83-84, 84-79. For the fifth one I will draw these lines: 85-86, 86-87, 87-88, 88-89, 89-90, 90-85. The sixth one will have the following conformation: 91-92, 92-93, 93-94, 94-95, 95-96, 96-91. Using these tracings we have represented the well with two steps on a surface of six sides,[76] as we said we would do. Moreover, this same procedure is needed for bodies of 5, 8, 16 faces, or with any other odd or even number of faces you want them to be.[77]

Proposition VII

Raise a shortened cube on a plane and surround it with a base and a cymatium.

We shall have our plane BCDE[78] and above it, as has been shown in the first proposition of this book, the cubic body which has as shortened base FGHI and as upper face the figure KLMN [figure B.42]. To follow the proposed disposition, I shall lead, at my own will, line FO out of FG, extending the base. I will draw line GP which is equal to FO. I will draw AO and AP, extending both of them, and I shall draw diagonals FI, GH. I will extend diagonal FI till I intersect the line that goes through A and O, at point Q, and the line that goes through A and through P, at point T. And I will extend diagonal GH till I intersect the line that goes through A and P, at point R, and the line that goes through A and through O, at point S. And I shall draw lines QR, RT, TS, SQ. Then I shall draw above F, on line FK, segment FV of length FO, and I shall draw a line parallel to FG from V which intersects GL at point X. And I shall draw the line to point A from point V: where it will intersect HM I shall draw point Y. And I shall draw the line from point A to point X: where it will intersect IN I shall draw point Z. I shall draw VX, XZ, ZY, YV and I shall draw VQ, XR, YS, ZT. Then I shall draw from point K, on KL's extension, a line that is equal to FO which will be called Kz, and I shall draw Ls equal to Kz. I shall draw from point A a line that goes through z and a line through s. Then I shall draw the diagonals. Firstly, diagonal KN will intersect Az at point 10 and As at point 13. And I shall draw diagonal LM which will intersect line Az at point 12 and As at point 11. And I shall draw the lines from 10 to 11, from 11 to 13, from 13 to 12, from 12 to 10. And I shall draw the line from K to 14 equal to FV. And I shall draw from 14 a line parallel to KL, which will cut LG at point 15. And I shall draw from 14 the line to point A, and where it will intersect HM there we will have point 16. And I shall draw the line from point 15 to point A, and where it will intersect NI there we will have point 17. And I shall trace lines 14-15, 15-17, 17-16, 16-14. And I shall trace lines 10-14, 11-15, 12-16, 13-17. From these tracings, the cymatium appears, together with the base, above the shortened cubic body, precisely when the position [of that cubic body] is parallel to BC.

However, if one was to have a different position,[79] I would draw around surface FGHI another surface 1-2-3-4 [figure B.43]. Then I would draw above point B of the plane the vertical Bp, and I would draw Ap.

And from all the angles of this surface I would trace lines parallel to BC, till they meet BD. First of all I shall draw from 1 the line that intersects BD at point 5. And from 2 the line that intersects BD at point 6. And from 3 the line that intersects BD at point 7. And from 4 the line that intersects BD at point 8. Then I would draw from F the line that intersects BD at point O. From G the line that intersects BD at point P. From H the line that intersects BD at point Q. From I the line that intersects BD at point R. From all these points we shall draw lines parallel to Bp till they touch Ap. Moreover I shall draw from points F, G, H, I lines parallel to Bp with no end. Then I shall split Bp at point S and BS will be as tall as the height of the base, the extension as big as is intercepted by line FH on line 1-3.[80] Successively, I will draw SA, which will intersect all the lines that leave BD and are parallel to Bp. And I shall cut Bp at point T so that Tp is equal to BS; and I shall draw TA which will intersect the lines parallel to Bp; then I shall draw from the contact point of the line that leaves O, on line AS, the parallel to BC, and where this one will intersect the vertical line that comes from F, I shall mark point 11; and I shall draw from the contact point of the line that leaves P, on line AS, the parallel to BC, and where this one will intersect the vertical line that comes from G, I shall mark point 12; and I shall draw from the contact point of the

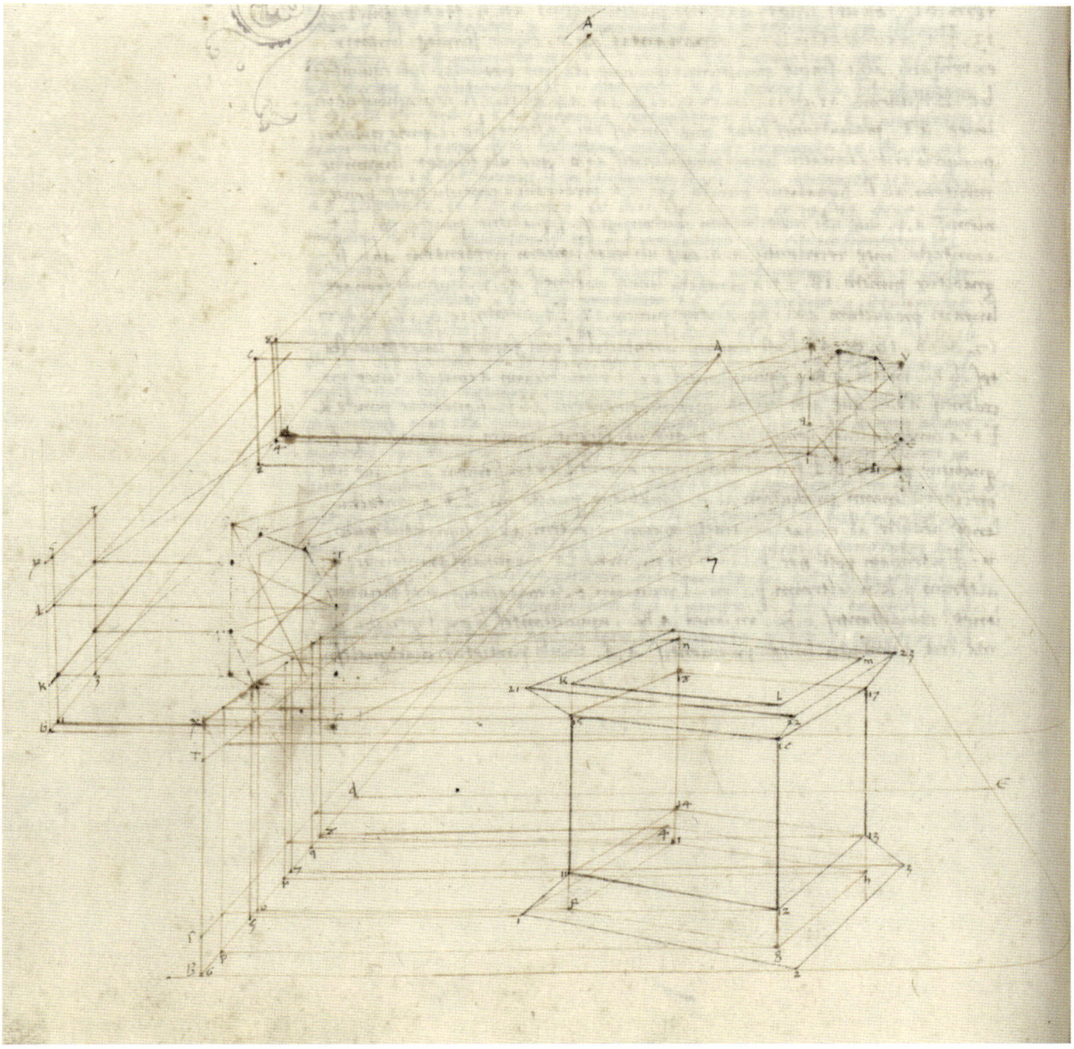

Bordeaux, Bibliothèque Municipale, Ms Latin 616, folio 26 verso.

line that leaves R, on line AS, the parallel to BC, and where this one will intersect the vertical line that comes from H, I shall mark point 13; and I shall draw from the contact point of the line that leaves Q, on line AS, the parallel to BC, and where this one will intersect the vertical line that comes from I, I shall mark point 14; all the lines [the horizontal ones] which have been drawn so to be parallel to BC. And I shall draw lines 11-12, 12-13, 13-14, 14-11. Then, from the points of contact of line AT, I shall draw all the lines parallel to BC. The first tracing will certainly originate from the contact of the line that leaves O: where [a line parallel to BC] will intersect the vertical line that comes from F, I shall draw point 15. And I shall draw from the contact of the line that leaves P [a line parallel to BC]: where this parallel will intersect the vertical line that comes from G, I shall draw point 16. And I shall draw from the contact of the line that leaves Q [a line parallel to BC]: where this parallel will intersect the vertical line that comes from H, I shall draw point 17. And I shall draw from the contact of the line that leaves R [a line parallel to BC]: where this parallel will intersect the vertical line that comes from I, I shall draw point 18.[81] And I shall draw lines 15-16, 16-17, 17-18, 18-15. Then, from the contacts of the lines parallel to Bp on line Ap, I shall draw lines parallel to BC. First of all, from the contact of the line that leaves O, I shall draw [a line parallel to BC]: where this parallel will intersect the vertical line that comes from F, I shall draw point K. And from the contact of the line that leaves P, I shall draw [a line parallel to BC]: where this parallel will intersect the vertical line that comes from G, I shall draw point L. And from the contact of the line that leaves Q, I shall draw [a line parallel to BC]: where this parallel will intersect the vertical line that comes from H, I shall draw point M. And from the contact of the line that leaves R, I shall draw [a line parallel to BC]: where this parallel will intersect the vertical line that comes from I, I shall draw point N. Successively, I shall trace lines KL, LM, MN, NK. And I shall draw the diagonals: one passing through KM and the other one through LN. Then I shall trace, from the contacts of the lines parallel to Bp on line Ap, lines equidistant to line BC. The first tracing will originate from the contact of the line coming from point 5: where the line parallel to BC will intersect the diagonal passing through K, there I shall mark point 21. From the contact of the line coming from point 6: where the line parallel to BC will intersect the diagonal passing through L, there I shall mark point 22. From the contact of the line coming from point 7: where the line parallel to BC will intersect the diagonal passing through M, there I shall mark point 23. From the contact of the line coming from point 8: where the line parallel to BC will intersect the diagonal passing through N, there I shall mark point 24. I shall then trace lines 1-11, 2-12, 3-13, 4-14, which are the corners of the base. To obtain the corners of the cymatium I shall trace lines 15-21, 16-22, 17-23, 18-24. In this way has been completed what we had promised: that we would have completed, using linear procedures, a cubic body with its base and the cymatium.[82]

Proposition VIII

Place a shortened column on a shortened plane so that it is not parallel to the straight line of the shortened plane.

Let BCDE be the shortened plane where we can place a column with eight faces, with the base and the shaft not parallel to BC [figure B.44]. Firstly, we must draw the surface with the length and width required by the length of the columns, according to what the procedure teaches us in the twenty-seventh proposition of the first book.[83] This surface will be FGHI. Trace from points F, G, H, I lines parallel to BC till they touch BD. And from F trace the parallel line that touches BD at point 1; from G trace the parallel line that touches BD at point 2; from H trace the parallel line that touches BD at point 3; from I trace the parallel line that touches BD at point 4. Then I shall trace above B the vertical line Bp; which will be as long as the thickness of the column;[84] and I will join point p to point A. Then I shall trace from points 1, 2, 3, 4 lines parallel to Bp till they intersect Ap. I shall first trace, through 1, the parallel line that touches Ap at point 5; then I shall trace, through 2, the parallel line that touches Ap at point 6; then I shall trace, through 3, the parallel line that touches Ap at point 7; then I shall trace, through 4, the parallel line that touches Ap at point 8. To complete the eight faces instead, we will operate as we are reminded by the twenty-seventh proposition of the first book.[85] We shall draw in real shape a squared figure, which is as big as we want the size of the column. Let fghi be the square, in which we shall draw the equilateral octagon rstvxyz′z. I shall place [a] size [equal to] fr above B: it will correspond to BK. I shall place size rs above K: it will correspond to KL. I shall place size sg above L: it will correspond to LM. These quantities enclosed together by the letters BM are equal to fg. Now trace the lines from points B, K, L, M to point A; these lines will meet all the lines parallel to Bp. After this, trace on F and on G two vertical lines with no limit and two other lines parallel to Bp on H and on I. Trace from the points of contact of line Ap all the lines parallel to line BC. First of all, from the contact of the parallel [horizontal] line originating from 5, which intersects the [vertical] line which comes from F, I shall trace point T. From the contact of the parallel line originating from 6, which intersects the line which comes from G, I shall trace point V. From the contact of the parallel line originating from 7, which intersects the line which comes from H, I shall trace point s′. From the contact of the parallel line originating from 8, which intersects the line which comes from I, I shall trace point Z. Successively, trace from the points of contact of line LA all the parallel lines to BC.[86] First of all, from the contact of the parallel [horizontal] line originating from 5, which intersects the [vertical] line which comes from F, I shall trace point R. From the contact of the parallel line originating from 6, which intersects the line which comes from G, I shall trace point S. From the contact of the parallel line originating from 7, which intersects the line which comes from H, I shall trace point X. From the contact of the parallel line originating from 8, which intersects the line which comes from I, I shall trace point Y. Successively, trace from the points of contact of line KA all the parallel lines to BC.[87] First of all, from the contact of the parallel [horizontal] line originating from 5, which intersects the [vertical] line which comes from F, I shall trace point N. From the contact of the parallel line originating from 6, which intersects the line which comes from G, I shall trace point O. From the contact of the parallel line originating from 7, which intersects the line which comes from H, I shall trace point P. From the contact of the parallel line originating from 8, which intersects the line which comes from I, I shall trace point Q. Then we will trace lines NO, PQ, and the diagonals Fs′ and GZ. And trace RX, which will divide diagonal Fs′ at point k; I will trace NP, which will cut Fs′ at point l; I will trace SY, which will cut GZ at point m; I will trace OQ, which will cut GZ at point 9. I shall then trace through l, parallel to FT, the line that touches FH at point 11 and Ts′ at point 13. I shall trace through k, parallel to FT, the line that touches FH at point 12 and Ts′ at point 14. I shall trace through 9, parallel to GV, the line that touches GI at point 21 and VZ at point 23. I shall trace through m, parallel to IZ, the line that touches GI at point 22 and VZ at point 24. Then trace lines 11-N, R-13, 14-X, 12-P, 21-O, S-23, 24-Y, Q-22. Finally draw these lines: 14-24, XY, PQ, 12-22, 11-21, NO, RS, 13-23. At these conditions we have the column, composed of faces, which we had said we would show.

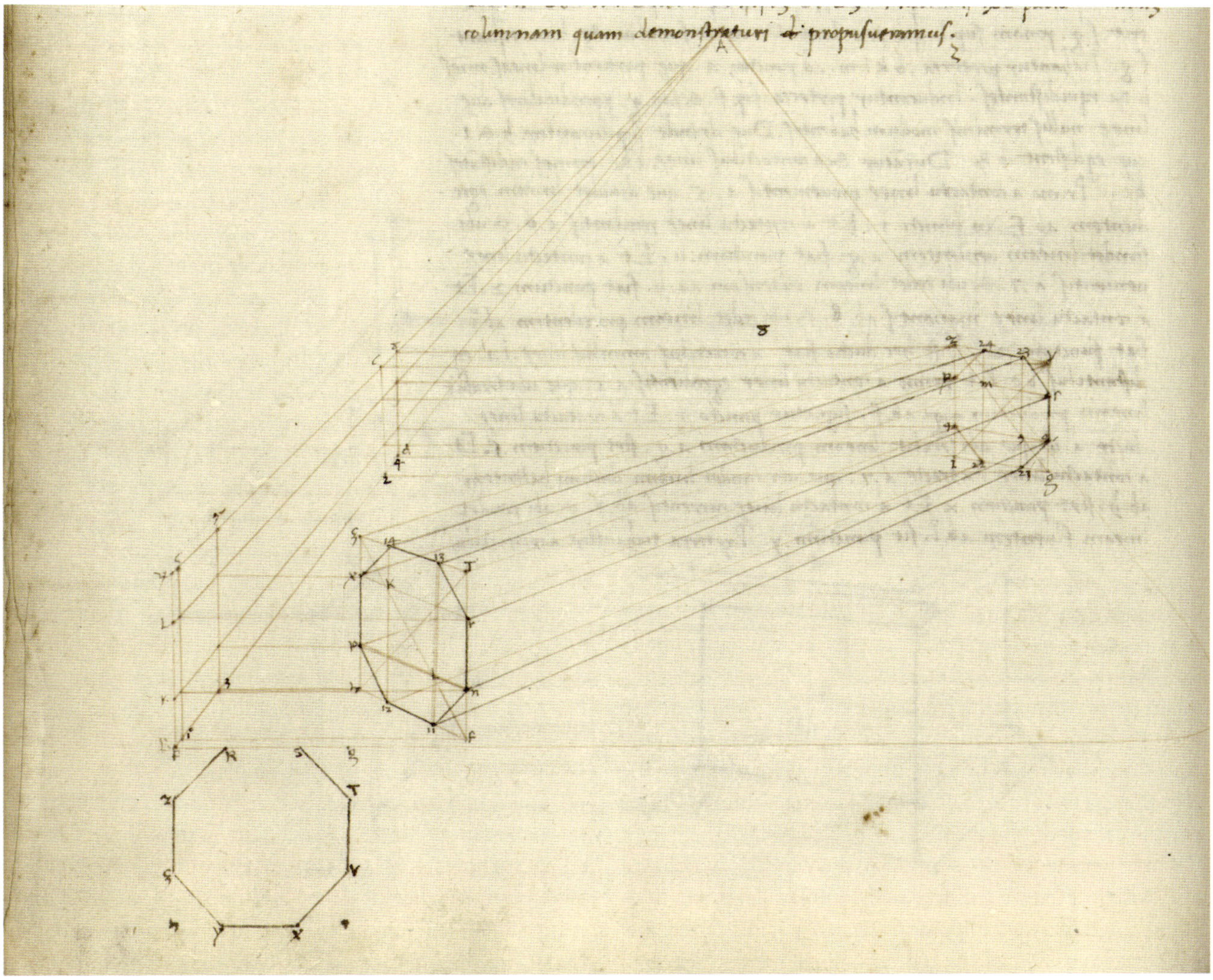

Bordeaux, Bibliothèque Municipale, Ms Latin 616, folio 27 verso.

Proposition IX

Place a quadrilateral house on a shortened plane.

Because we have a shortened plane BCDE[88] on which we shall place a house with four faces, we shall draw on this plane a surface as we have done in the twenty-eighth proposition of the first book [figure B.45].[89] FGHI will be the surface of the external part and KLMN the one of the internal part. Let the size of the wall be F-9, G-10. Thus, we shall draw perpendicular lines on the ends of FGHI. We shall draw the vertical line FO on F, the vertical line GP on G which is as high as FO. And we shall draw the vertical line HQ on H and the vertical line IR on I which is equal to HQ. We shall draw from A a line to point P which will go through R. We shall draw from O a line to point A which will pass through Q. We shall draw OP parallel to FG and QR parallel to OP. These lines represent the heights of the house on which we will place the roof. Firstly, we shall add to line OP, from each external part of the house, the quantity that I want to come out from the roof, which will be as long as OS and PT. Successively we shall draw diagonals OR, PQ which will be longer than their end and which intersect in V. And from point A we shall draw the line going through S which will intersect diagonal OR at point X, and where it intersects the diagonal going through Q, we shall mark point z. And from point A we shall draw the line going through T: where this line will intersect diagonal through P we shall mark Y, and where the same line will intersect the diagonal going through R, we shall mark point Z.

Successively, we shall divide OP into two equal parts at point s, and above s we shall draw the perpendicular line of size sp which is equal to the sixth part of XY. We shall draw on V a perpendicular line without end, and from p a line to point A: when this line meets the perpendicular line coming from V we shall mark point a, from which we will trace the lines that go toward X, Y, Z, and z. Having obtained in this way the calculation of the roof, it remains to take the height to which we will put the support of the window, which will be F-5 high. Firstly I will trace the line parallel to FG from 5 which will meet GP at point 6. Then I will indicate below 5 the thickness 5-1 of the windowsill. Moreover, we shall trace from 1 the line parallel to 5-6 which will meet G-6 at point 2. We shall trace from 5 to point A the line which will divide HQ at point 8. We shall trace from 6 to point A the line which will divide IR at point 7. We shall trace from 1 to point A the line which will divide H-8 at point 4. We shall trace from 2 to point A the line which will divide I-7 at point 3. We shall trace beyond 5 [on line 5-6] a line that is as long as the protrusion of the windowsill: let this be 5-300 [on one side] and 6-301 [on the other side]. We shall draw from A a line that goes through 300, with no end, and we shall trace diagonal 7-5 which will cut the line that goes through 5 at point 11. We shall trace the other diagonal line, which goes through 6-8, with no end. We shall trace from 11 the parallel line to 5-6 which will cut the diagonal that goes through 6-8 at point 12. We shall trace from 12 the line to point A which will meet the diagonal that goes out from 5-7 at point 13. We shall trace from 13 the parallel line to line 1-2 which will meet the diagonal that goes out from 8 at point 14. Finally, we shall trace lines 1-11, 2-12, 3-13, 4-14. In this way we will have completed the windowsill, on which we will divide line 5-6 into eleven equal parts because we have to arrange five windows. The first one of these [eleven parts] is 5-31, the second one 31-32, the third one 32-33, the fourth one 33-34, the fifth one 34-35, the sixth one 35-36, the seventh one 36-37, the eighth one 37-38, the ninth one 38-39, the tenth one 39-40, the eleventh one 40-6.[90] We will then establish on point 5 the height of the windows, which I will choose at my own will to be as tall as line 5-15. We will then draw from 15 a line parallel to 5-6, which will intersect line 6-P at point 16. Above points 31, 32, 33, 34, 35, 36, 37, 38, 39, 40, we will trace lines parallel to 5-15 till they meet 15-16. Let the parallel line originating from 31 meet it [15-16] at point 41, the one originating from point 32 meet it at point 42, the one originating from point 33 meet it at point 43, the one originating from point 34 meet it at point 44, the one originating from point 35 meet it at point 45, the one originating from point 36 meet it at point 46, the one originating from point 37 meet it at point 47, the one originating from point 38 meet it at point 48, the one originating from point 39 meet it at point 49, the one originating from point 40 meet it at point 50. To now build the width of the walls which is F-9, we shall extend from 9 the parallel line to F-15 which will meet line 15-16 at point 17. And we shall trace from point 15 the line to point A which intersects 9-17 at point 18. We shall trace from point 18 a line parallel to 15-16 which intersects 6-16 at point 19. We will join 41 to point A: where it intersects line 18-19, we shall mark point 51. We will join 43 to point A: where it intersects line 18-19, we shall mark point 53. We will join 45 to point A: where it intersects line 18-19, we shall mark point 55. We will join 47 to point A: where it intersects line 18-19, we shall mark point 57. We will join 49 to point A: where it intersects line 18-19, we shall mark point 59. Then I shall trace from points 51, 53, 55, 57, 59 lines parallel to 31-41 till they touch line 5-6.

After we will have done this we shall complete the windows, refined with each procedure, on the nonshortened face. We shall now demonstrate on what condition we may obtain, on the shortened face [figures B.45, B.46], five windows of the same size. We shall obtain the following:[91] if we will join point 16 to point A, this line will intersect line 7-R at point 10*.[92] We will trace diagonal 16-7, and will divide 6-16 into equal parts as there are on line 5-6: these parts will be indicated by points 61, 62, 63, 64, 65, 66, 67, 68, 69, 70.

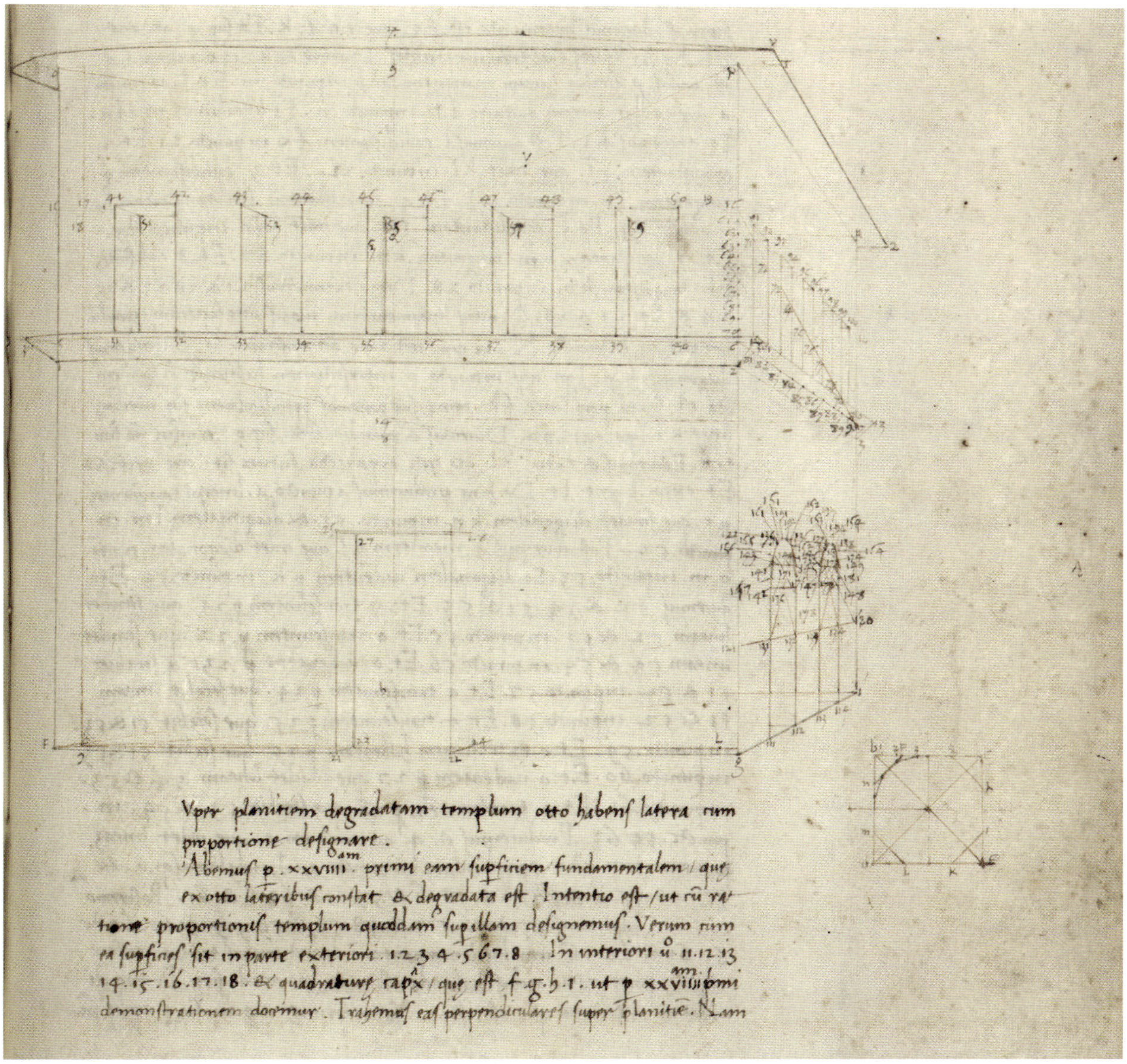

Bordeaux, Bibliothèque Municipale, Ms Latin 616, folio 31 recto.

We will trace first a line from point 61 to point A which cuts diagonal 16-7 at point 71. We will trace a line from point 62 to point A which will cut diagonal 16-7 at point 72. We will trace a line from point 63 to point A which will cut diagonal 16-7 at point 73. We will trace a line from point 64 to point A which will cut diagonal 16-7 at point 74. We will trace a line from point 65 to point A which will cut diagonal 16-7 at point 75. We will trace a line from point 66 to point A which will cut diagonal 16-7 at point 76. We will trace a line from point 67 to point A which will cut diagonal 16-7 at point 77. We will trace a line from point 68 to point A which will cut diagonal 16-7 at point 78. We will trace a line from point 69 to point A which will cut diagonal 16-7 at point 79. We will trace a line from point 70 to point A which will cut diagonal 16-7 at point 80. We will also trace from all these points all the lines parallel to 6-16. Firstly I will draw from point 71 the line that touches 6-7 at point 81 and line 16-10* at point 91. I will trace from point 72 the line that touches 6-7 at point 82 and line 16-10* at point 92. I will trace from point 73 the line that touches 6-7 at point 83 and line 16-10* at point 93. I will trace from point 74 the line that touches 6-7 at point 84 and line 16-10* at point 94. I will trace from point 75 the line that touches 6-7 at point 85 and line 16-10* at point 95. I will trace from point 76 the line that touches 6-7 at point 86 and line 16-10* at point 96. I will trace from point 77 the line that touches 6-7 at point 87 and line 16-10* at point 97. I will trace from point 78 the line that touches 6-7 at point 88 and line 16-10* at point 98. I will trace from point 79 the line that touches 6-7 at point 89 and line 16-10* at point 99. I will trace from point 80 the line that touches 6-7 at point 90 and line 16-10* at point 100. With the help of these transformations, we will have the repartition of the windows according to the shortening method. To obtain the width of these [windows], we will place on the nonshortened face, on line 15-16, the length that goes from 15 to 17*[93] which will be equal to 16-20. I will trace the line from point 20 to point A, and I shall draw all the parallel lines to 15-16 from points 92, 94, 96, 98, 100. Firstly I will certainly draw the one from point 92 which will cut 20-A at point 102;[94] I will trace the one from point 94, which will cut 20-A at point 104; I will trace the one from point 96, which will cut 20-A at point 106; I will trace the one from point 98, which will cut 20-A at point 108; I will trace the one from point 100, which will cut 20-A at point 110. Having done this we shall trace through points 102, 104, 106, 108, 110 the lines parallel to 6-16 till they touch 6-7.

Up till now we have talked about drawing the windows. Now we will discuss how to realize the doors, that is, the procedure to execute them.[95] Firstly, I shall draw the one that is not shortened and that is externally limited by number 21 and 22 [figure B.45], above which we shall trace vertical lines with no limit. Then we will take the distance between 21 and 22 and we shall double it and place it on the line erected above 21: this will be 21-25. We shall draw from point 25 the parallel line to FG which [the parallel] will intersect the line that is above 22 at point 26. We shall trace from point 21 the line to point A which shall intersect KL at point 23, and we shall trace from point 23 a line parallel to 21-25 till it touches 25-26. We shall trace a line from 25 to point A; this line will intersect the line that originates from 23 at point 27. We shall trace a line parallel to 25-27 from 27 till it touches line 22-26. This will be what we called a door, which will be in symmetry with the windows before mentioned in accordance with the specified ways of representation. To demonstrate what methods may be used to realize the windows and the doors with a half-round arch, we will delimit two curved doors; let the first one be indicated by the external part with 111-112 and the other one with 113-114, always from the same part. We shall trace lines parallel to line GP from all these points without establishing a limit. Since in the twenty-eighth proposition of the first book it appears clearly how to proceed with the height in real form, we will take it and place it on G: this will be G-121. And because we want the doors to be twice as tall as their width, we shall place the height once more on 121: we shall have 122. This [the height of the door] will correspond to G-122. Successively, we shall trace a line from 121 to point A: this will cut the line originating from 111 at point 131; the one originating from 112 at point 132; the one originating from 113 at point 133; the one originating from 114 at point 134. Then we shall trace a line from 122 to point A: this will cut the line originating from 111 at point 135; the one originating from 112 at point 136; the one originating from 113 at point 137; the one originating from 114 at point 138. Then we shall trace the diagonals 131-136, 132-135 and the ones of the other door 133-138 and 134-137. We shall take the length above, between 131 and 135 [figures B.45, B.47]. With this length [to be intended as a side] we shall draw the square in real shape bcde, and in it we shall draw figure fghiklmn with eight sides in real shape, according to the procedure in the sixteenth proposition of the first book. We shall take length bf and place it below point 135: this will yield segment 135-141. We shall take length fg and place it below point 141: this will yield segment 141-142. Then we shall trace a line from 141 to point A: this will cut the diagonal through 135 at point 143 and the diagonal through 136 at point 144; and it will cut line 132-136 at point 145, and it will cut line 133-137 of the second door at point 146, and it will cut the diagonal originating in 137 at point 147 and it will cut the diagonal originating in 138 at point 148. And it will cut line 134-138 at point 149. For these points I shall trace the lines parallel to 131-135. We shall trace a line from point 143 that touches line 135-136 at point 151. We shall trace a line from point 144 that touches line 135-136 at point 152. We shall trace a line from point

147 that touches line 137-138 at point 153 of the second door. We shall trace a line from point 148 that touches line 137-138 at point 154. Finally we shall draw lines 141-151, 152-145, 146-153, 154-149. Moreover, we need to change the octagon to a sixteen-sided figure in real shape according to the procedure shown in the seventeenth proposition of the first book; therefore we will divide bf at point 1, 1-f at point 2, and fg at point 3. We shall take length b-1 and place it below point 135: this will yield segment 135-155. We shall take length f-2 and place it above point 141: this will yield segment 141-156. We shall take length f-3 and place it below point 141: this will yield segment 141-157. Successively, we shall trace the line from 155 to point A which will cut 141-151 at point 161, line 151-145 at point 162, and will cut line 146-153 in the second door at point 163 and line 154-149 at point 164. We shall trace a line from point 156 to point A which will cut line 141-151 at point 171, line 152-145 at point 172, line 146-152 of the second door at point 173, line 154-149 at point 174. We shall also trace a line from point 158[96] to point A which will cut the diagonal through 135 and 132 at point 175. This line will cut diagonal 131-136 at point 176, line 132-136 at point 177, line 133-137 at point 178, diagonal 137-134 at point 179, diagonal 133-138 at point 180, and it will divide line 134-138 at point 181. Then we shall trace from point 175 a line parallel to 131-135 which will touch line 135-136 at point 191. And we shall trace from point 176 a line parallel to 131-135 which will touch line 135-136 at point 192. And we shall trace from point 179 of the second door a line parallel to 131-135 which will touch line 137-138 at point 193. And we shall trace from point 180 a line parallel to 137-133 which will touch line 137-138 at point 194. And we shall trace the lines of the first door: 157-171, 161-191, 192-162, 172-177, and for the second door lines 178-173, 163-193, 194-164, 174-181.

These are the details concerning the doors. We still need to show how big the width of the wall is and where the leaves, needed to close the doors, meet. To do this, we shall trace from point 30, which is indicated on line FG, a line to point A [figure B.45]. Then we shall trace from point 112 [figure B.47] a line parallel to FG. Where this line will cut line 30-A, we shall mark point 200. We shall trace a line parallel to FG from point 114; this line will cut 30-A at point 201. And we shall trace from 200 a line parallel to line 112-136.[97] And we shall trace from 201 a line parallel to line 114-138.[98] We shall take length G-30 and we will attach it to 122: there will be line 122-300 which is parallel to FG. We will trace the line from point 300 to point A. We will trace from 152 a line parallel to FG which will cut line A-300 at point 252. We will trace from 145 a line parallel to FG which will cut the line coming from point 200 at point 253.[99] We will trace from 154 a line parallel to FG which will cut line 300-A at point 254. We will trace from 149 a line parallel to FG which will cut the line coming from 201 at point 249. And we will draw lines 252-253, 254-249. We will trace from point 192 a line parallel to FG which will cut line 300-A at point 195. We will trace from point 162 a line parallel to FG which will cut line 252-253 at point 197. We will trace from point 172 a line parallel to FG which will cut line 252-253 at point 198. We will trace from point 177 a line parallel to FG which will cut the line coming from 200 at point 203.[100] Successively, we shall draw from point 194 of the second door a line parallel to FG which will cut line 300-A at point 196. We shall draw from point 164 of the second door a line parallel to FG which will cut line 254-249 at point 204. We shall draw from point 174 of the second door a line parallel to FG which will cut line 254-249 at point 205. We shall draw from point 181 of the second door a line parallel to FG which will cut the line coming from 201 at point 206. Successively, we shall draw lines 195-197, 197-198, 198-199 of the first door. These will be lines 196-204, 204-205, 205-206 of the second door. The house with doors we have proposed has been completed if we are aware of the lines that, placed in the right place, complete the whole work.[101]

Proposition X

Proportionally trace an eight-faced temple on a shortened plane.

By the twenty-ninth [proposition] of the first book, we already have in foreshortening that fundamental surface that is made of eight faces. We intend to trace proportionally, above this, a temple [figure B.48]. This surface, in reality, corresponds externally [to the outline] 1, 2, 3, 4, 5, 6, 7, 8 and internally [to outline] 11, 12, 13, 14, 15, 16, 17, 18, and the square that holds it [the surface] is FGHI, as we have learnt from Proposition XXVI in Book 1.[102] We will trace the perpendicular lines [from these points] to the surface. In fact, we will trace above F line FK which is as long as FG, above G the line equal to GL, and above H and I two lines without end. We will trace from K the line to point A which is the eye: this line will intersect the one originating from H at point M. We will trace the line from point L to point A which will intersect the line produced from point I at N. We will draw lines MN and KL. Then we will trace from 1 a line parallel to FK which will intersect KL at point 21. Then we will trace from 2 a line parallel to GL which will intersect KL at point 22. We will trace from 3 a line parallel to GL which will intersect LN at point 23. We will trace from 4 a line parallel to IN which will intersect LN at point 24. We will trace from 5 a line parallel to IN which will intersect NM at point 25. We will trace from 6 a line parallel to GL which will intersect MN at point 26.[103] We will trace from 7 a line parallel to HM which will intersect KM at point 27. We will trace from 8 a line parallel to FK which will intersect KM at point 28. Finally we will draw lines 21-22, 22-23, 23-24, 24-25, 25-26, 26-27, 27-28, 28-21. Through these line extensions we have delimited the eight-faced temple. Now, to demonstrate the roof procedure we will trace diagonals KN, LM which will meet at point O. Then I will take the sixth part of FK and I shall retrace this quantity vertically, upon the half of KL in PQ. Then I will trace an unlimited vertical line upon O. I will trace, externally to KL, line KS, and make it as long as I'd like the roof to lean out, and on the other side LT. Then, from point A, I will trace the line passing through T which will cut diagonal KN at point 51 and diagonal LM at point 52. Then we will trace a line to point A passing through S; this line will cut the diagonal passing through M at point 53 and the diagonal going through K at point 54. We will trace line 52-54 and line 51-53. We will trace from point O a line going through point 21 which will cut line 52-54 at point 55. We will trace from point O a line going through point 22 which will cut line 52-54 at point 56.

We will trace from point O a line going through point 23 which will cut line 51-52 at point 57. We will trace from point O a line going through point 24 which will cut line 51-52 at point 58. We will trace from point O a line going through point 25 which will cut line 51-53 at point 59. We will trace from point O a line going through point 26 which will cut line 51-53 at point 60. We will trace from point O a line going through point 27 which will cut line 53-54 at point 61. We will trace from point O a line going through point 28 which will cut line 53-54 at point 62. We will trace the line going from point Q to point A which will cut the perpendicular line originating from O at point V. We will trace lines V-55, V-56, V-57, V-58, V-59, V-60, V-61, V-62.[104] Then we will trace lines 55-56, 56-57, 57-58, 58-59, 59-60, 60-61, 61-62, 62-55. Having traced these lines we have completed the roof. Now we must execute the door procedure: which will happen in the following way. Firstly, divide line 1-2 of the first face into two equal parts at point 29. Trace vertical line 29-30 on point 29 at your own will, depending on how tall we'd like the door to be. Take the fourth part of line 29-30, which I will reproduce on the two sides of 29 on line 1-2: this will generate line 29-31 on one side and 29-32 on the other side. Trace upon 31 the vertical line 31-33, make it as long as 29-30, and upon 32 vertical line 32-34 of the same length. Then, above, I will trace line 33-34. We have said that internally line 11-12 corresponds to face 1-2. Trace a line, from point 31 to point A, which will cut line 11-12 at point 35.

Then I will trace a line from point 32 which will cut line 11-12 at point 36. Then I will trace the line from point 35 which is parallel to line 31-33 till it touches 33-34. I will trace a line, from point 33 to point A, which will cut the line originating from point 35 at point 37. I will trace a line from point 37 parallel to 33-34. I will trace a line from point 36 parallel to 32-34 which will cut the line coming from 37 at point 38. We have said enough about the door we said we would show.

Now, because the temple loses its grace if it is not illuminated, it must be said what the circular opening will be through which it takes the light. We will have to draw it as tall as the height of the door on the face, which offers us the first view.[105] Therefore, we shall draw from point 33 a line parallel to 1-21 till it touches line 21-22 at point 40. We shall draw from point 34 a line parallel to 2-22 till it touches line 21-22 at point 41. We shall trace [a line parallel to] 21-22 till it touches 33-40 at point 42, which will cut 34-41 at point 43.

Then we shall trace from point 35 the parallel line to 1-21 till it touches line 42-43. Then we will take the length of line 42-43 and we will reproduce it above 42 and 43: this will yield lines 42-44 and 43-45. Then we shall trace line 44-45 and a line that goes from point 42 to point A which will cut the line originating from 35 at point 46. We shall trace a parallel to 42-43 from point 46. We shall trace a parallel to 2-22 from point 38. This parallel will intersect the line originating from 46 at point 48. We shall trace a line from point 44 to point A which will cut the line originating from 37 at point 47. We shall trace a line from point 45 to point A which will cut line 38-48 at point 49. Moreover, we will trace diagonals

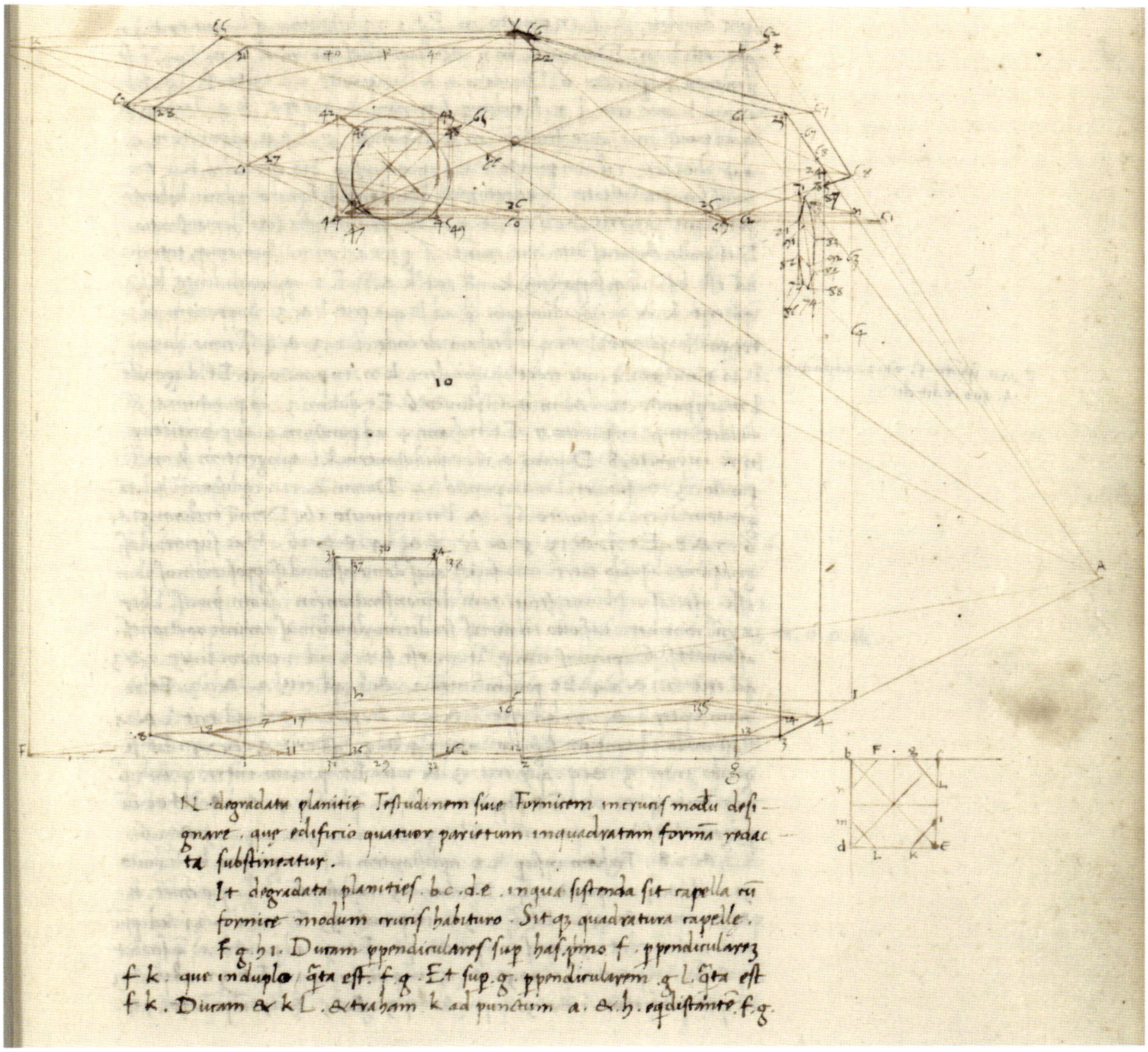

Bordeaux, Bibliothèque Municipale, Ms Latin 616, folio 33 recto.

42-45, 43-44 which will meet at point X. We will trace the other diagonals 46-49, 47-48 which will meet at point Y. We will then place the compass fixed point on X and twist the movable end till it touches line 42-43. We will then place the compass fixed point on Y and twist the movable end till it touches lines 46-48 or 46-47. Having traced these lines, we will be able to easily obtain the circular opening procedure which illuminates the temple.[106] However, because the temple is made of eight faces, the intention is that the circular opening acts as an ornament four times as it does in the first face.[107] The second opening will be in the third face, the third one in the fifth one and the fourth one in the seventh one, while the remaining ones will be missing. Thus, for the second opening we will draw line 42-43 till it touches line GL at point 61. Then we will trace the extension of line 44-45 which will touch GL at point 62. Then I will trace a line from point 61 to point A which will cut IN at point 63. Then I will trace a line from point 62 to point A which will cut IN at point 64. Then I will trace a line from point 40 to point A which will cut the diagonal KN at point 65 and the diagonal ML at point 66. We will trace a line parallel to KL from point 65 till it touches LN at point 67. We will trace a line parallel to KL from point 66 till it touches LN at point 68. We will trace a line parallel to GL from point 67 which will cut line 61-63 at point 71 and will divide line 62-64 at point 72. We will trace a line parallel to GL from point 68 which will cut line 61-63 at point 73 and will divide line 62-64 at point 74. We will then take the length of line 71-72 and I will make this length the side of a square in real shape; this side will be bcde. Inside this square we will draw the octagon in proper form, fghiklmn. From this figure we shall take length bf and we will put it under 71; this will yield line 71-81. We shall take length fg and put it under 81; this will yield line 81-82. We will then trace a line from point 81 to point A which will divide line 73-74 at point 83. We will trace a line from point 82 to point A which will divide line 73-74 at point 84. We will then trace diagonal 71-74 which will cut line 81-83 at point 91 and line 82-84 at point 92. We will then trace from point 91 a line parallel to 71-72 till it touches 71-73 at point 85 and line 72-74 at point 86. We will then trace from point 92 a line parallel to 73-74 till it touches 71-73 at point 87 and line 72-74 at point 88. Finally, we will draw lines 81-85, 87-83, 84-88, 82-86.

This way it is clear that the eight faces have their own way of being represented. However, if we want the temple to have 16 sides, follow the procedure that was explained earlier about the arches, the doors, and thickness. In fact, one procedure draws on another if someone looks carefully.[108]

Proposition XI

On a shortened plane trace a cross vault, that is, an arced vault which, drawn in squared form, is sustained by a four-walled construction.

Let BCDE be the shortened plane on which we want to place a chapel with cross vault [figure B.49]. Let FGHI be the base square. I will trace the perpendicular lines above these points. Firstly, I will trace perpendicular FK above F which will be twice as long as line FG; then I will trace above G the perpendicular GL, as long as FK. I will trace KL and the line from point K to point A. And I will trace from point H a line parallel to FK which will cut AK at point M. And I will trace from point I a line parallel to GL, and IN will be equal to HM. I will draw MN. I will draw diagonals KN, LM which will meet at point O. I will divide KL at point X and I will trace length XL below L: this will yield LP. And I will reproduce the same length below K. This will yield KQ. Then I will draw the line from point Q to point A: this will cut HM at point S. I will trace the line from P to point A which will divide IN at point R. Then I will take the length of line KL, and with this length [as a side] I will draw square bcde in proper form [figure B.50], as was shown in the sixteenth proposition of the first book.[109] I will draw an octagon whose angles will be fghiklmn. I will take length bf and I will reproduce it below point K [figure B.49], this will yield K-1; and I will reproduce it on line KL, this will yield K-2. And I will reproduce this length on the horizontal line through L, this will yield L-3; and below L it will generate L-4. Then I will draw lines 1-2 and 3-4. Firstly, I will draw the line from point 1 to point A, which will cut MS at point 5. I will draw the line from point 2 to point A, which will cut diagonal KN at point 11 and diagonal LM at point 12 and MN at point 6. Then I will draw the line from point 3 to point A, which will cut MN at point 7. I will draw the line from point 4 to point A, which will cut NR at point 8. I will draw from point 11 a line parallel to KL till it touches KM at point 13 and LN at point 14. I will draw from point 12 a line parallel to KL till it touches KM at point 15 and LN at point 16. Then I will draw lines 5-6, 7-8, 1-13, 5-15, 4-14, 8-16. Tracing the above will clearly show that the eight faces that we had set out to draw have been realized.[110] Now, according to the twenty-seventh proposition of the first book, we will divide the octagon in real shape into sixteen parts by adding the portions concerning these parts. We will in fact take length f-1 [figure B.50] and reproduce it on line 2-3 [figures B.49, B.51], which will yield 2-24. We will trace it on line 2-1 which will yield 2-23. We will trace it on line 1-23 which will yield 1-22. We will trace it below 1 which will yield 1-21. We will trace it on line 24-3 which will yield 3-25. I will trace it between 3 and 4 which will yield 3-26. I will trace it between 4 and 26 which will yield 4-27. I will trace it below 4 which will yield 4-28. Having completed this, I will execute the following adjustments: 21-22, 23-24, 25-26, 27-28. I will draw a line parallel to KL from point 22 till it touches K-1 at point 31 and L-4 at point 34. I will

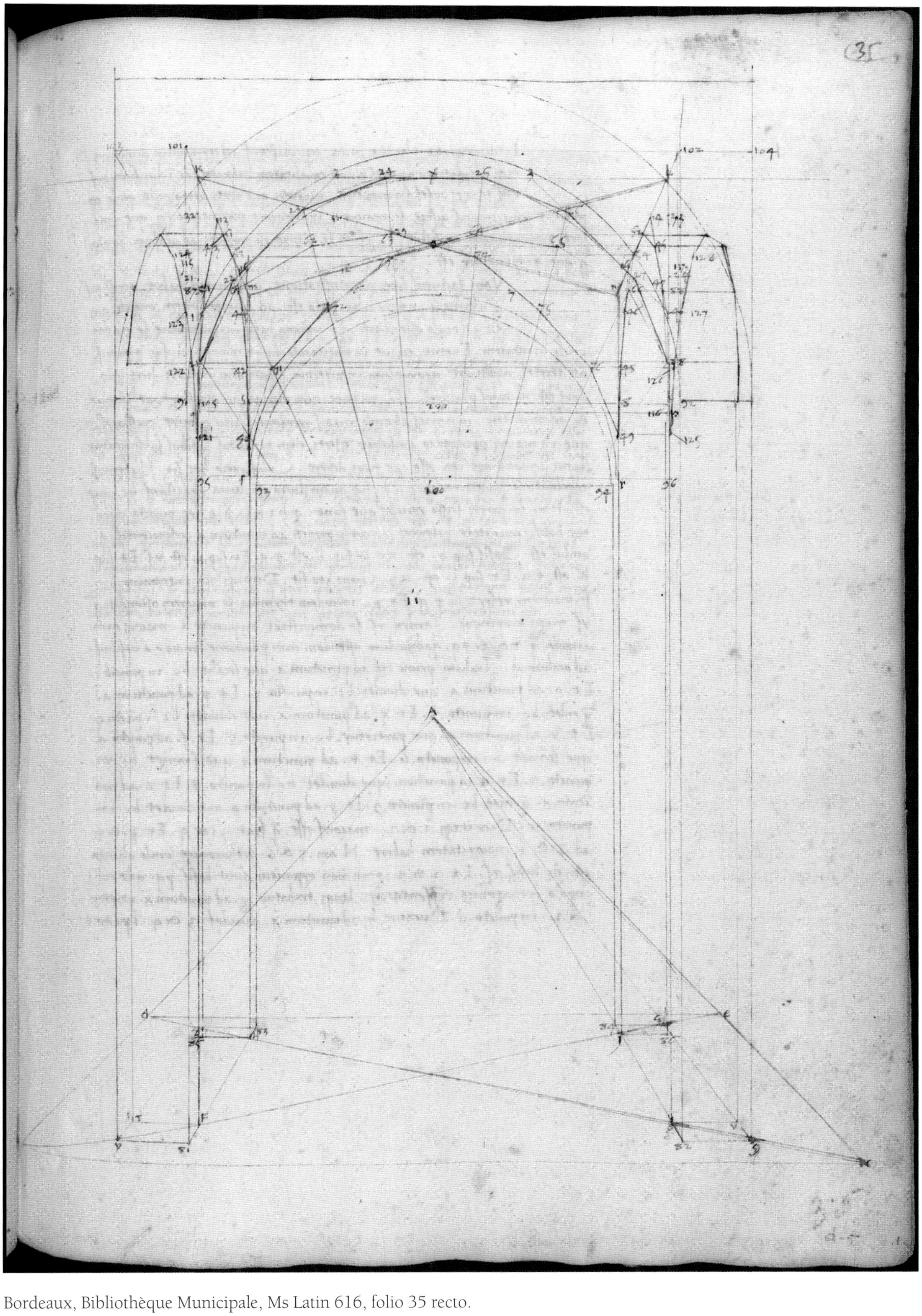

Bordeaux, Bibliothèque Municipale, Ms Latin 616, folio 35 recto.

draw a line parallel to KL from point 23 till it touches K-31 at point 32 and L-34 at point 33. I will trace a line from point 21 to point A which will cut 5-S at point 40. I will trace a line from point 31 to point A which will cut 1-13 at point 41 and 5-15 at point 42. I will trace a line from point 32 to point A which will cut 1-13 at point 43 and 5-15 at point 44.[111] On the other side, I will trace a line from point 33 to point A which will cut 4-14 at point 45 and 8-16 at point 46. I will trace a line from point 34 to point A which will cut 4-14 at point 47 and 8-16 at point 48. I will trace a line from point 28 to point A which will cut R-8 at point 49. I will trace a line from point 24 to point A which will cut diagonal KN at point 29 and diagonal LM at point 30. Then I will draw a line parallel to LK from point 29 till it touches 13-15 at point 51 and 14-16 at point 52. Then I will draw a line parallel to LK from point 30 till it touches 13-15 at point 53 and 14-16 at point 54. Finally trace these lines: 21-41, 43-51, 53-44, 42-40. This constitutes the first perimeter of the face that has been shortened. The second one is formed by these lines: 28-47, 45-52, 54-46, 48-49. This way we will have the shortened arcs; from their points I will trace the lines from an arc to another one following an order. Firstly, I will trace 40-49, 42-48, 44-46, 53-54, 51-52, 43-45, 41-47, 21-28. I will then trace the line from point 22 to point A which will divide 41-47 at point 61 and 42-48 at point 71. I will then trace the line from point 23 to point A which will divide 43-45 at point 62 and 44-46 at point 72. I will then trace the line from point 24 to point A which will divide 51-52 at point 63 and 53-54 at point 73. I will then trace the line from point 25 to point A which will divide 51-52 at point 64 and 53-54 at point 74. I will then trace the line from point 26 to point A which will divide 43-45 at point 65 and 44-46 at point 75. I will then trace the line from point 27 to point A which will divide 41-47 at point 66 and 42-48 at point 76. I will trace lines 21-61, 61-62, 62-63, 63-O, O-74, 74-75, 75-76, 76-49. This is the representation of a cross vault. For the other one I will trace lines 28-66, 66-65, 65-64, 64-O, O-73, 73-72, 72-71, 71-40. We have produced the cross vault which we had proposed, without pillars.

However, if someone wishes to position it on some pillars, they will have to place, out of line FG, a length at their own will, as the side of the pillar: on one side we will have FT and on the other one GV. He will draw from point A a line that goes through T and another one that goes through V, with no end. He will trace diagonal FI till it touches AT at point Y and AV at point z. He will draw diagonal GH till it touches AT at point Z and AV at point 9. Then he will draw lines Y-9 and Zz and then he will trace a line from point A which goes through F and will cut Y-9 at point 81 and zZ at point 83. Then he will draw a line from point A which goes through I and G and which cuts Y-9 at point 82 and Zz at point 84. Then he will trace HI which touches AY at point 85 and A-9 at point 86. Moreover, trace the unlimited parallel lines to FK from points Y and 81. Trace the parallel to HS from point 85 which will be as long as HS. Trace the unlimited parallel to HS from 83. Trace from point A the line passing through S and Q which will cut the line originating from 83 at point 93 and the line originating from 81 at 91. Trace the unlimited parallel lines to GL from 82 and 9. Trace from 84 and 86 other lines with no end. Trace from point A the line that goes through R and P and which divides the line that originates from 84 at point 94 and divides the line originating in 82 at 92. Trace MN which will cut the line originating in 85 at 87 and the line originating in 86 at point 88. Trace SR which will cut the line originating in 85 at 95 and the line originating in 86 at point 96. Trace from A the line passing through K which cuts the line originating in 81 at point 101. Trace from point A the line that goes through L and cuts the line originating in 82 at point 102. Trace line 101-102 which cuts the line originating in Y at point 103 and the line originating in 9 at 104. Draw lines from point A to points 103 and 104.

Then trace from point 5 a line parallel to KL which touches line 95-87 at point 105. Trace from point 15 a line parallel to KL which divides line 103-A at point 115. Draw line 105-115. Trace from point 40 a line parallel to KL which touches line 87-95 at point 121. Trace from point 42 a line parallel to KL which touches line 105-115 at point 122. Trace from point 44 a line parallel to KL which touches line 105-115 at point 123. Trace from point 53 a line parallel to KL which touches line 103-A at point 124. Having done this we will have obtained the representation of an arch. In reality, to obtain the procedure of the second arch I will trace from point 8 a line parallel to KL which will cut the line originating in 96 and in 88 at point 116. I will trace from point 16 a line parallel to KL which will cut line 104-A at point 117. Then I will trace line 116-117. I will trace from point 49 a line parallel to KL which will cut line 96-88 at point 125. I will trace from point 48 a line parallel to KL which will cut line 116-117 at point 126. I will trace from point 46 a line parallel to KL which will cut line 116-117 at point 127. I will trace from point 54 a line parallel to KL which will cut line A-104 at point 128. Having done this I will trace the lines of the first arch; these are: 121, 122, 123, 124. These will be the lines of the second arch: 125, 126, 127, 128. Since these are arches, we must draw in the shape of a circle those that are located frontally. At first I will draw line 93-94. I will then divide it into two equal parts at point 100, and above point 100 I will place the fixed end of the compass. With the movable end I will draw a circle with radius 100-93, moving it [the movable end] from 93 to 94. I will then draw line 91-92 and I will divide it into two equal parts at point 200 and, placing the fixed end of my compass on 200, I will move the movable end until it touches 91 and 92. Having done this we will have completed the vault we had set out to draw.

Proposition XII

If a line is traced parallel to the terminal, on a plane to be shortened,[112] and if it [the line] is divided into many equal parts and if on the points of division we place equal bases, each positioned orthogonally to the eye, the farther one will be represented in the terminal bigger than the closer one, when it [the farther] is presented to the eye with a smaller angle than the closer one.

The demonstration I am about to execute is not less important than the one in the last proposition of the first book to determine the width of the angle in the eye as the right size of the base that same eye is opposite. It occurs, in fact, that in building, there are columns of this kind: round or with a polygonal base, at the same time, as one can perceive in the structures of the arcades, where numerous columns are needed. In this case, many—while establishing their work with care—are surprised to discover that the columns that are further away from the eye become bigger than the ones that are closer, especially when they are placed on equal bases. I have decided to demonstrate that it is thus and that it has to be done this way. Let this be the procedure.

We have the plane to be shortened [figure B.52],[113] on which line FM parallel to BC has been divided into equal parts marked by points G, H, I, K, L. And in these points trace equal bases, orthogonally to point A which is the eye. Let NO be the base on G, PQ the one on H, RS the one on I, TV the one on K, XY the one on L. At these conditions, I say that NO is represented in the terminal BC as bigger than PQ is, and that PQ is represented in the determined place BC as bigger than RS which is closer, and that nonetheless RS is represented at point A with an angle which is wider than NO's or PQ's, as I shall demonstrate when tracing bases' lines up until eye A. Firstly, I shall draw a line from point N to point A which will divide BC at point 1. I shall draw a line from point O to point A which will divide BC at point 2. I shall draw a line from point P to point A which will divide BC at point 3. I shall draw a line from point Q to point A which will divide BC at point 4. I shall draw a line from point R to point A which will divide BC at point 5. I shall draw a line from point S to point A which will divide BC at point 6. I shall draw a line from point T to point A which will divide BC at point 7. I shall draw a line from point V to point A which will divide BC at point 8. I shall draw a line from point X to point A which will divide BC at point 9. I shall draw a line from point Y to point A which will divide BC at point 10. I say that line 1-2 is longer than line 3-4 and that 3-4 is longer than 5-6, because 5-6 is orthogonally opposed to eye A, like base RS, and because 3-4 is not opposed to the eye like base PQ, which comes under the same angle orthogonally. Draw, therefore, the line that goes from G to point A which will divide line 1-2 at point D. Draw the line that goes from H to point A which will divide line 3-4 at point E. Draw the line that goes from I to point A which will divide line 5-6 at point f. These subdivisions will be proportionate. In fact the proportion between line 5-6 with respect to line 3-4 is the same as between line fI and line EH. The proportion of line 3-4 with respect to line 1-2 is equal to the one between line EH and line DG. The proportion of line 5-6 with respect to line 1-2 is equal to the one between line fI and line DG. The product of line 1-2 and line fI is equal to the product of line 5-6 and line DG.[114] And the product of line 5-6 and line 1-2 is equal to the product of line 3-4 and itself. Similarly, the proportion between fI and fA is equal to the one between EH and EA. And the proportion between EH and EA is equal to the one between DG and DA. So it happens that they are in proportion. Since AG is greater than AH and AH is greater than AI, it follows that DG is greater than EH because DG is [proportional] with respect to the greatest [line] AG, as EH is with respect to the smallest AH. For the same reason, EH is greater than fI. Thus, I shall conclude by saying that line 1-2 is greater than line 3-4 because 1-2 is related to DG more than line 3-4. Line 1-2, in fact, which is proportionate to DG is greater than line 3-4 which has the same proportion with EH which is smaller [than DG]. I claim that line 3-4 is greater than line 5-6 for the same reason, because line 5-6 is in the same proportional relationship with fI which is really small. For this reason, it is necessary that in the terminal, what is further seems bigger, which is what we had said we would prove. Nonetheless the thing that is closest to point A, which is the eye, presents itself at an angle which is greater than the furthest base: the bases in fact are all equal, as has been widely shown in the fourth proposition of the first book.

Bordeaux, Bibliothèque Municipale, Ms Latin 616, folio 36 verso.

Book 3

Since a lot of people who dedicate themselves to painting criticize the method of perspective—without grasping the force of lines and angles that are produced by perspective, through which each representation of borders and lines occurs proportionally—it seemed appropriate to demonstrate how this procedure is beneficial to painters in executing a work correctly. Certainly perspective, which appears to have acquired its name from the term "to see at a distance," corresponds to nothing more than the appearance of things that exist at a certain distance from the eye and that, respecting specific conditions, stand in a proportional relationship conforming to [their] distance. If this procedure is not followed, shortening of objects will not be executed correctly. Because painting is nothing more than showing surfaces and shortened[115] bodies placed in the terminal [the picture plane] in conformity with how the real objects appear to the eye, and because there is always one part that is closer to the eye than another, and because the closest one appears under a wider angle than the furthest one, and because the power of the intellect cannot judge by itself how large they [the foreshortenings of objects] are, that is to say, how large is the nearest [part of a surface] and how large is the farthest, it is necessary, for those who wish to do so, to follow the efficacy of perspective. This possibility offers a description of all quantities proportionally and their shortening through the mediation of lines. Thus, after acquiring this type of painting, numerous ancient painters such as Aristomenes, Tasius, Polides, Apelles, Andramides, Nitheus, Zeuxis became very well known.[116] Moreover, many who flowered in this kind of art managed to devote their hands carefully to the work they had undertaken. If this thing were taken care of during our times by those experts of art who aspire to an acknowledgment by those who do not know perspective, they might obtain fame that not will darken with age, since perspective shines every day more illuminated by the testimony of experts.[117] With an intention to study these things for the glory of art during our time, I deemed it important to present a number of issues, thus undertaking the task of writing a work[118] for those who—let the future be my witness!—will find described, in this work, the relative part of perspective which is missing to painting. For what concerns perspective there are already two books. The first book contains procedures by which it is shown why on a flat surface the shortenings occur in multiple ways. The second book shows the way to shorten squared objects with more facets, positioned perpendicularly above horizontal surfaces. In the third one, I sought myself to draw as many bodies as possible. Since we must discuss the shortening of bodies placed between various surfaces and positioned in different ways, and we shall also discuss more difficult bodies, I will then follow a second path together with a second shortening procedure which are different from the ones discussed in previous demonstrations, although they are very similar and have the same effect. Therefore now, in conformity with the procedure followed earlier, I will expound another procedure, remembering two things. The first is the simplicity of demonstrating and comprehending; the second is the multitude of lines that need bodies to be drawn if we renounce the first procedure.[119] The eye and the spirit may be confused by the multitude of lines. Without the multitude of those lines, in fact, the bodies cannot be shortened to perfection, nor without difficulty. Thus I will follow a second procedure which will show the shortening of objects piece by piece. In this [procedure] it is necessary to understand what one wishes to do and be able to position it, aware of the goal, in real shape on the plane, because the power of the lines, together with the art [of perspective] as a guide, will produce the shortened object as it appears in the terminal [plane] of the visual lines. Therefore, it is necessary to know how to make proportionately all the contours of what you want to do and place them above the plane in the appropriate places in real shape.[120] For this reason, I will proceed with maximum diligence in the demonstrations that follow.

Proposition I

Shorten a square surface [located] on the floor, giving the explanation.

To show the path one needs to follow, I will carry out two or three flat-surface demonstrations, so that, through these [demonstrations], one may easily obtain the shortening of bodies. Therefore, I shall establish in real shape a square surface BCDE [figure B.53], and I will place the eye at point A which will be as far as the surface needs to be from those who will observe it. In this point A I will place a needle[121] or a nail that has a thread of equine silk [horsehair]. I will then trace line FG, parallel to BC, which will constitute the terminal [plane] between the eye and the surface [of the square]. I will trace point M which must be drawn in each surface or body.[122] It does not matter where it is located. It [M] is an element, in fact, that will be obvious when executing the construction. When these things are completed, one must have some wooden rulers without deformations. Take one ruler and make it coincide with FG so that it does not move. Then take the other end of the silk thread and bring it to point B [in the corner] of the surface: mark b where the thread touches the ruler. Take the thread to point C [in the corner] of the surface: mark c where the thread touches the ruler. Take the thread to point D [in the corner] of the surface: mark d where the thread touches the ruler. Take the thread to point E [in the corner] of the surface: mark e where the thread touches the ruler. Take the thread to point M: mark m where the thread touches the ruler. Having completed these things, mark point A on the ruler and call this ruler A, which having been called A can be taken out and put aside. This ruler, in fact, contains the way and the procedure [to obtain] the width.

Because one must examine how line DE of the surface BCDE is more elevated than BC, lift point A [figure B.54] above line CE[123] to a point high enough to see the surface without getting closer to FG, which constitutes the terminal, and without getting away from it. Having positioned eye A [in elevation], with the thread of silk or of equine silk, as I mentioned earlier, make a paper ruler so that it coincides with FG and draw EC which will divide the paper ruler at point a: let this be the [paper] ruler a. Lead the thread above E, and where it hits the paper ruler mark e-d. Lead the thread above C and where it hits the same ruler mark c-b in the same point. Put this ruler aside.

Make another ruler in the same way: let it have the same dimensions and marks as the one that was drawn earlier: let it have, like the first one, mark a. After, one needs to draw a straight line where you want to locate the shortened surface: let this be FG, divided into equal parts at point M [figure B.55]. One must trace perpendicular line MN through M and perpendicular lines FH and GI through F and G. One must then take the two paper rulers with mark a. Then place one of the two rulers so that it coincides with FH and the other one so that it coincides with GI, with the letter a of each ruler on line FG. Take the wooden ruler which takes letter A and is called the width ruler. Place it on the two paper rulers so that it touches points e-d of the two rulers.

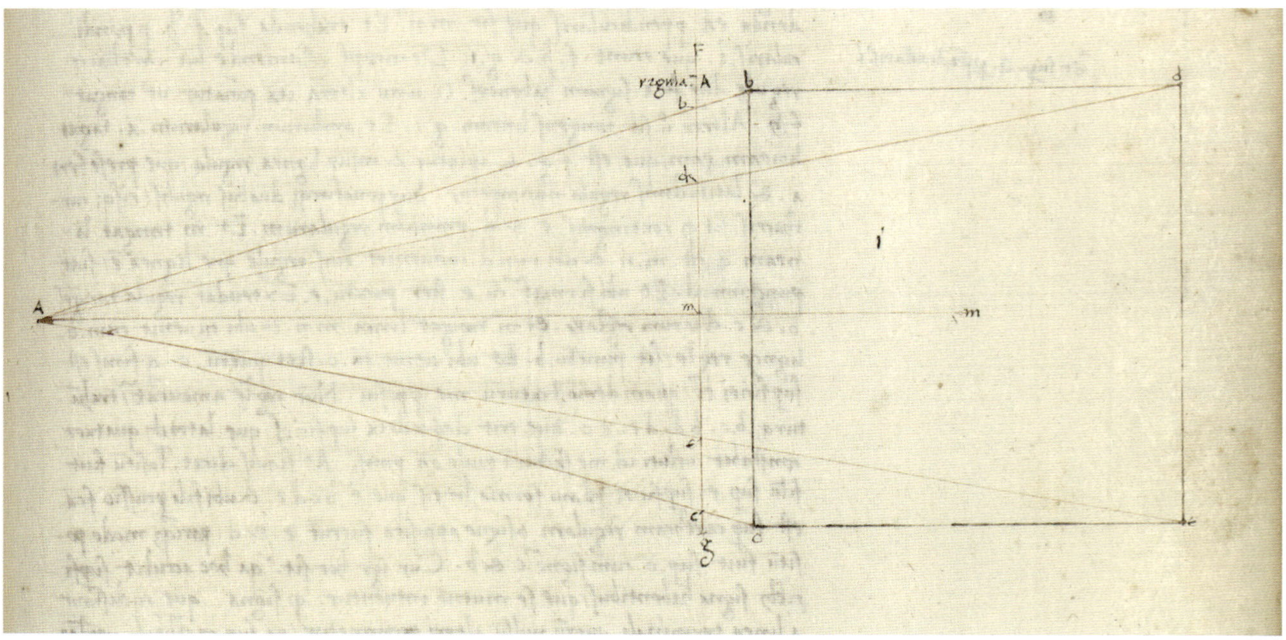

Bordeaux, Bibliothèque Municipale, Ms Latin 616, folio 38 recto.

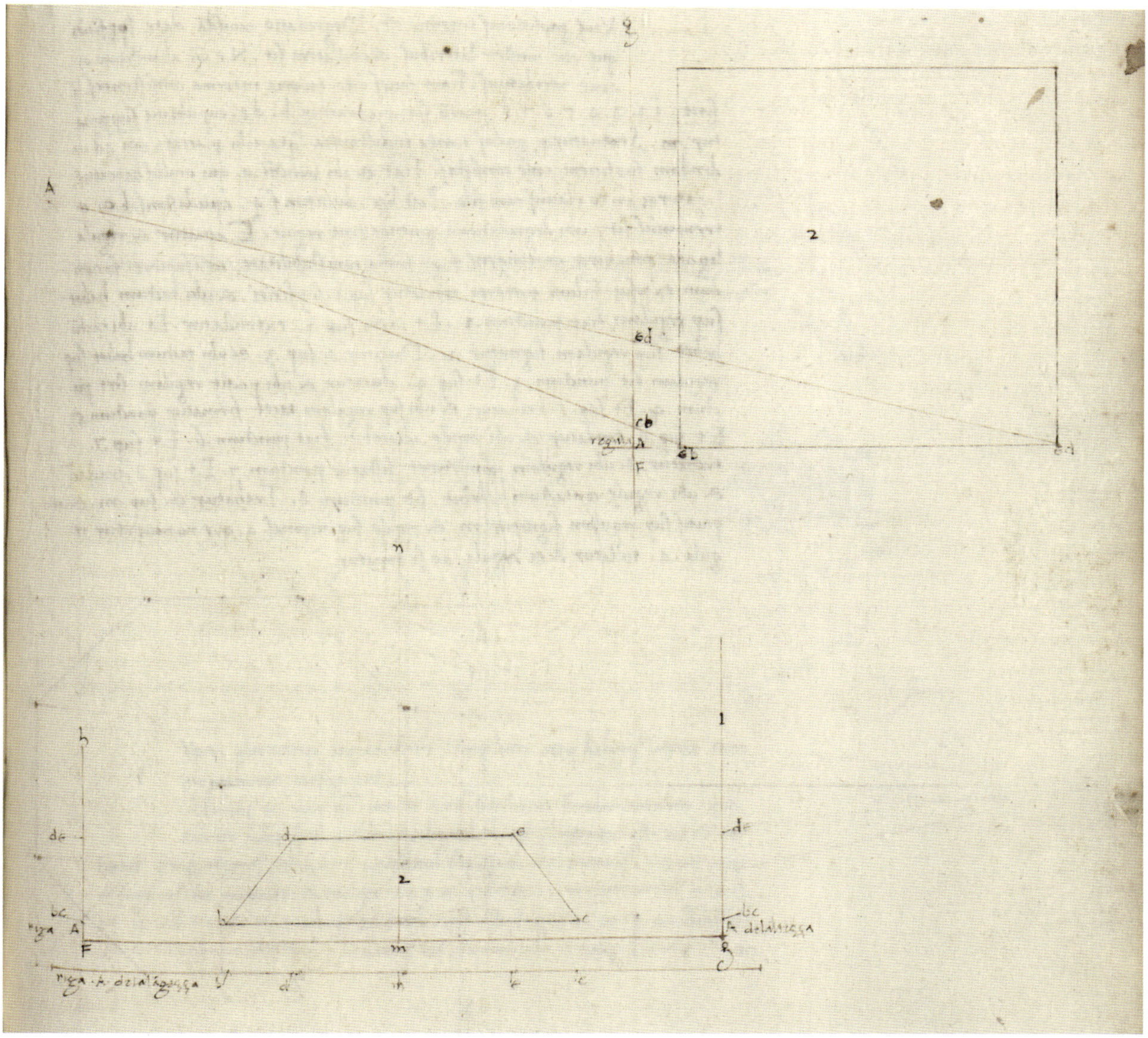

Bordeaux, Bibliothèque Municipale, Ms Latin 616, folio 39 recto.

Point m is on line MN. Where you see point d of the wooden ruler mark d, where you find point e mark e. Move the ruler till it joins c-b of the two rulers and let m be on MN. Where you see point b of the wooden ruler mark b, where you find point c mark c, points that constitute the end of the surface which I wanted to demonstrate. Now remove the rulers and draw lines bc, bd, de, ec: this will be the shortened surface, formed by four sides which I promised I would realize. However, if someone says: The thread has been placed on E of the surface in real shape BCDE and where the thread hits the paper ruler, point e and point d have been drawn and in the same way on point c, point b has been placed [figures B.53, B.54]. Why does this happen? This happens on the surfaces that have corresponding points, because the points that are equidistant from the terminal line so that none is above another, on the paper ruler which contains the height module, are drawn at the same point. Therefore, when the corners of a figure are placed so that none is above the other, as point c is with point b, and point e with point d, on the same elevation, and when line FG, which is the terminal, is equidistant from BC and DE, and by paper ruler I mean the height; when the thread is moved on point C of the surface, we see that there are point C and B. They are found at the same height. For this reason you will mark a point c-b on the paper ruler [figure B.54]. The same happens for E, for which point e-d will be marked.

However, the wooden ruler, which contains the width measurement, does not follow the same rule. On that ruler, every mark is reproduced as it is found, although it can happen that two marks will meet in a single point.

Proposition II

Proportionally shorten an assigned surface that has eight equal sides.

The aim of this proposition is to shorten an assigned equilateral surface of eight sides. To avoid distorting the order we have instituted in the figures, let the eight sides be given in real shape through points 1, 2, 3, 4, 5, 6, 7, 8 [figure B.56]. Let BCDE be the square in which to place M where you prefer. Place the eye at the distance you wish, where you stand to look at the surface. Let O be the place of the eye and put a nail with a thread on it. After, trace line FG parallel to BC, which is the terminal where the rulers of the shortenings must be placed. Take a wooden ruler and place it on FG so that it is very stable, as I proposed in the preceding surface.[124] Move the thread to point 1 of the surface: where it touches the ruler mark point 1. Move the thread to point 2 of the surface: where it touches the ruler mark point 2. Move the thread to point 3 of the surface: where it touches the ruler mark point 3. Move the thread to point 4 of the surface: where it touches the ruler mark point 4. Move the thread to point 5 of the surface: where it touches the ruler mark point 5. Move the thread to point 6 of the surface: where it touches the ruler mark point 6. Move the thread to point 7 of the surface: where it touches the ruler mark point 7. Move the thread to point 8 of the surface: where it touches the ruler mark point 8. Move the thread to point M of the surface: where it touches the ruler mark point m. Mark A on the ruler and call it ruler A: the ruler of width.[125] Take out this ruler and put it aside.

Now one needs to find [on the terminal] the elevation, that is, how deep the surface is [figure B.57]. In fact, we have said that BC is the width, thus CE[126] is the depth, that is, how deep it goes: in fact C is closer to the eye than E. Because you have line FG, the terminal, located between the eye and the [octagonal] surface, and because you have point O which is the eye, consider how much the eye must be elevated with the same point O on line CE. Place this point there and stick the nail with the thread on it. Take a paper ruler and make it coincide with the terminal FG so that it is stable. After, extend CE which will divide it at point a: because of a, this ruler will be called ruler a. One needs to place the thread on C, and where the thread touches the ruler we must mark points 1–8. Points 1–8 are distant from line FG just as point C; they lie on the same plane and none is higher than the other. For this reason C is needed for 1 and for 8, the same way 2 is needed for 2 and 7, 3 for 3 and 6, and E for 4 and 5. Place the thread on 7: where it touches the ruler mark points 2 and 7. Place the thread on 6: where it touches the ruler mark point 3 and 6. Place the thread on E: where it touches the ruler mark point 4 and 5. Take out the ruler and make a similar one that contains the marks with marked quantities and let it have as a mark the letter a. After doing this, trace a straight line in the place

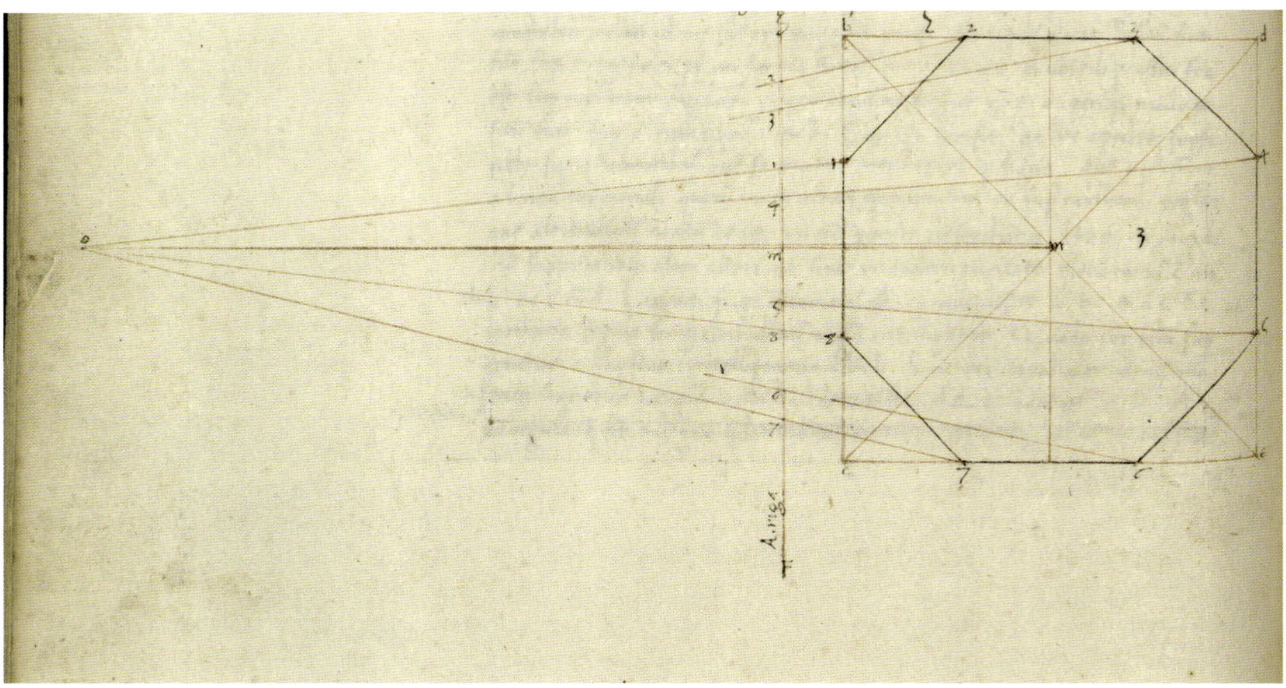

Bordeaux, Bibliothèque Municipale, Ms Latin 616, folio 39 verso.

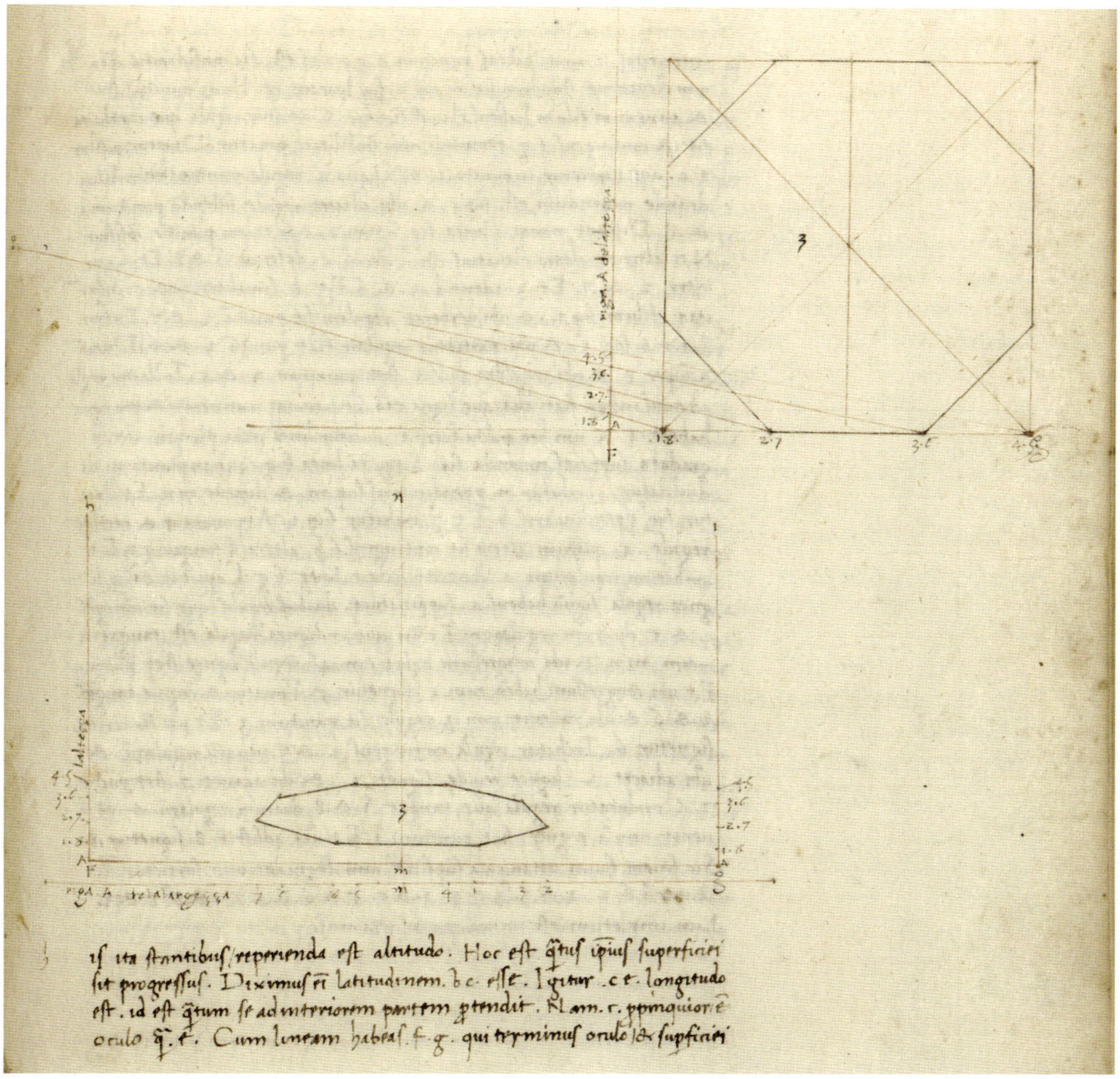

Bordeaux, Bibliothèque Municipale, Ms Latin 616, folio 40 recto.

where one will place the shortened surface [figure B.58]. Let this straight line be FG, which is divided at point M. Draw perpendicular line MN with N above M. Draw a perpendicular line to H above F, and above G a perpendicular line to I. Place paper rulers marked by a: make one coincide with FH and the other one with GI; let the ends a of each ruler touch line FG. Take the wooden ruler marked A and place it on the two paper rulers so that it touches points 4, 5 of the two rulers. Let point m of the wooden ruler touch line MN. Where you see point 4 of the wooden ruler mark 4. Where you see point 5 of the wooden ruler mark 5. Take the wooden ruler and place it so that it touches points 3, 6 of the two paper rulers. Where you see point 3 of the wooden ruler mark 3. Where you see point 6 of the wooden ruler mark 6. Take the wooden ruler and place it so that it touches points 2, 7 of the two paper rulers. Where you see point 2 of the wooden ruler mark 2. Where you see point 7 of the wooden ruler mark 7. Take the wooden ruler and place it so that it touches points 1, 8 of the two paper rulers. Where you see point 1 of the wooden ruler mark 1. Where you see point 8 of the wooden ruler mark 8. This way you have completed the shortening of the octagonal surface. Trace lines 1-2, 2-3, 3-4, 4-5, 5-6, 6-7, 7-8, 8-1. Now we have achieved what we had proposed.

Proposition III

Shorten four circles that have the same center and are subdivided into twelve [equal] parts.[127]

Because in previous demonstrations we only needed one wooden ruler and two paper rulers, considering that each ruler which contains the measurement of the width requires two paper rulers, [and] because we now need to realize the shortening of four circles, it is necessary to have four wooden rulers on which we will mark the width of four circles. We need eight paper rulers which correspond to the four wooden ones. On the paper rulers we will have the height measurement, as one may understand from demonstrations. On some rulers we have multiple points, on others fewer, in conformity with their relative outlines. First draw four [concentric] circles; the first one is on the outside and contains the others [figure B.59]. It will be marked by the letter A, the second by the letter B, the third by C, and the fourth by D. Let M be the center of these circles. Divide circle A into twelve equal parts at points 1, 2, 3, 4, 5, 6, 7, 8, 9, 10, 11, 12.

Then take a ruler and place it on 1 and M. Mark where the ruler divides circle B, circle C, and circle D, each with a point 1. Place the ruler also above 2 and M. Mark where the ruler divides circle B, circle C, and circle D, each with a point 2. Do the same for all points until point 12. Then trace straight line KL, the terminal, as near as one wishes to the circles [figure B.60]. Then get further away from the terminal KL, as one wishes to look at those circles, and in that place locate point O for the eye by placing a nail and a thread.

Wooden Ruler A Take the wooden ruler marked A and make it coincide with KL so that it does not move. Take a thread and extend it to point 1 of circle A, and where it touches the ruler A mark point 1. Extend the thread to point 2, and where it touches the ruler mark 2. Extend the thread to point 3, and where it touches the ruler mark 3. Extend the thread to point 4, and where it touches the ruler mark 4. Extend the thread to point 5, and where it touches the ruler mark 5. Extend the thread to point 6, and where it touches the ruler mark 6. Extend the thread to point 7, and where it touches the ruler mark 7. Extend the thread to point 8, and where it touches the ruler mark 8. Extend the thread to point 9, and where it touches the ruler mark 9. Extend the thread to point 10, and where it touches the ruler mark 10. Extend the thread to point 11, and where it touches the ruler mark 11. Extend the thread to point 12, and where it touches the ruler mark 12. Extend the thread to point M, and where it touches the ruler mark m. When you have completed these things, you have the designation of the first circle on ruler A: now we need to take off this ruler and put it aside.

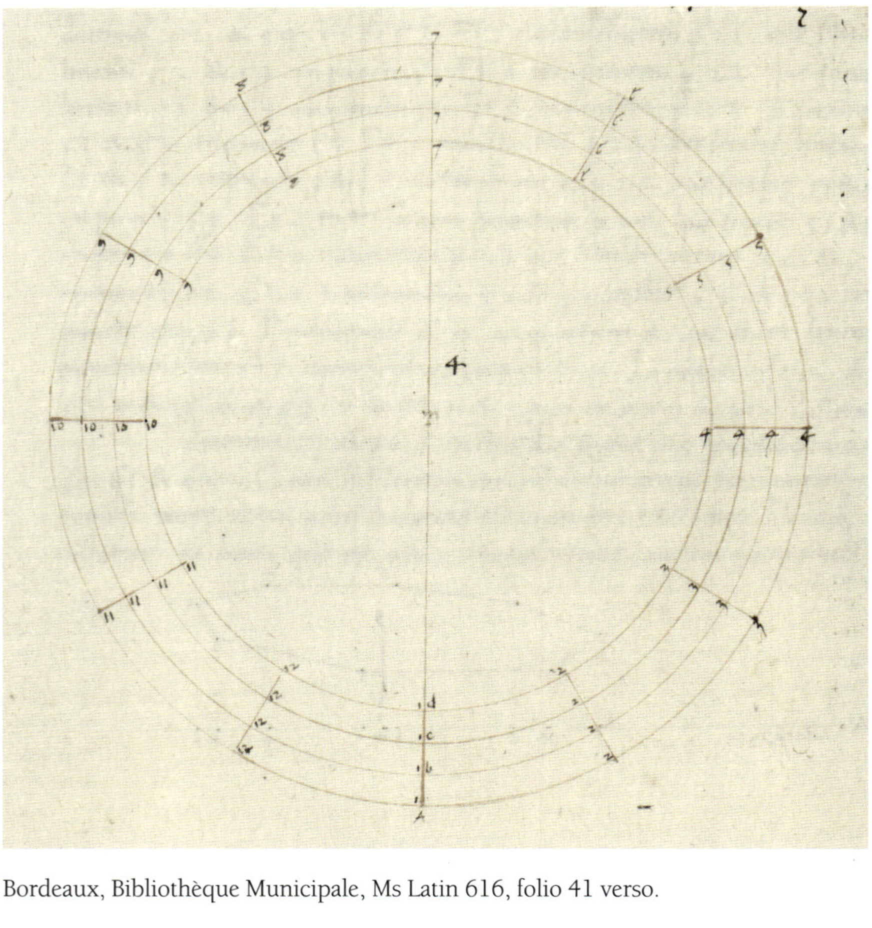

Bordeaux, Bibliothèque Municipale, Ms Latin 616, folio 41 verso.

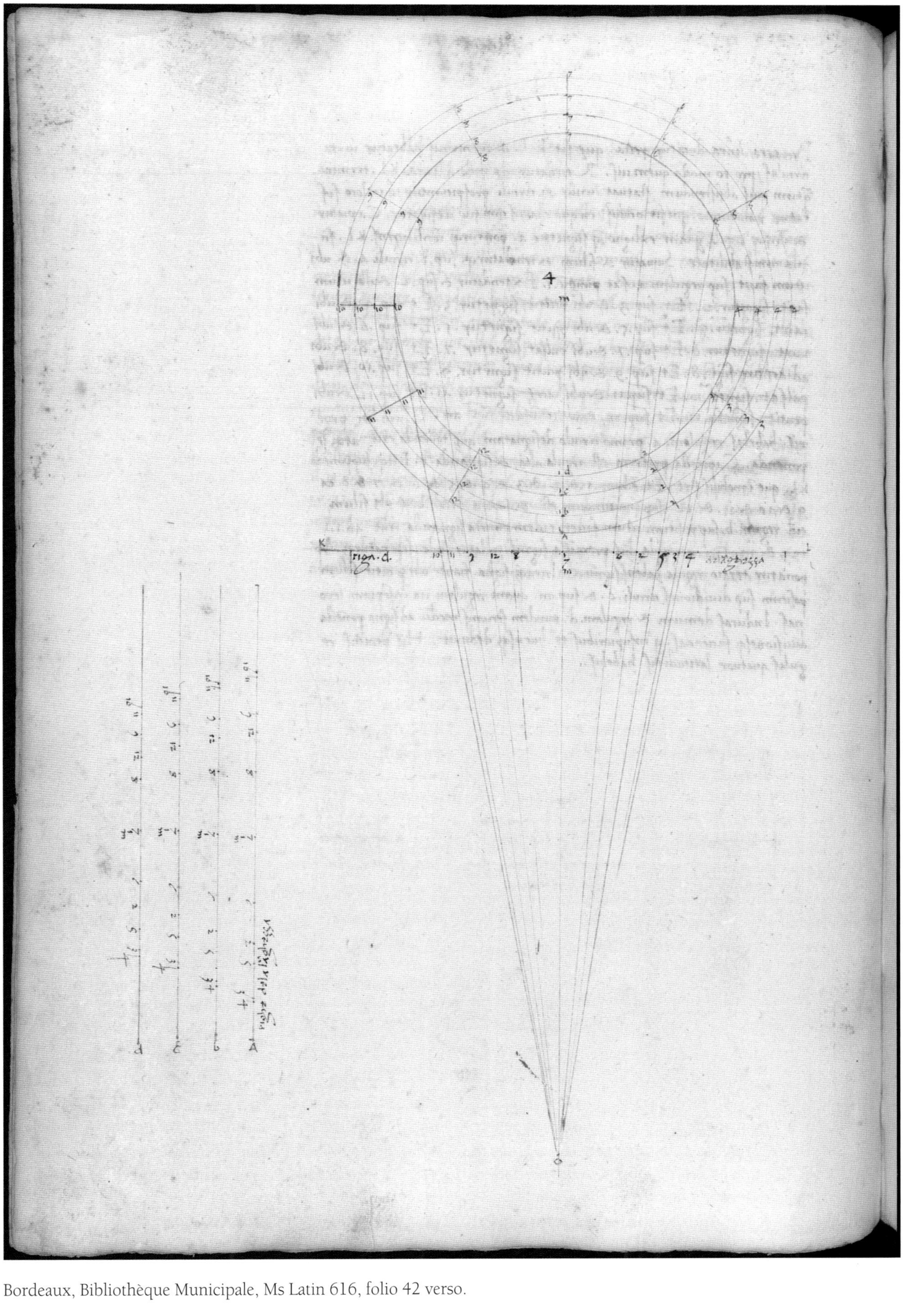

Bordeaux, Bibliothèque Municipale, Ms Latin 616, folio 42 verso.

Wooden Ruler B Now we need to take a second ruler which we will mark B, and we will place it on KL so that it does not move. We need to take the thread and place it on points 1, 2, 3, 4, 5, 6, 7, 8, 9, 10, 11, 12, and M of the circle marked B: mark points 1 to 12 and m where the thread intersects ruler B. Take off this ruler marked with these signs and put it aside.

Wooden Ruler C Take a third ruler marked by C. Mark on this ruler the points where the thread intersects the ruler when we place it on the subdivisions of circle C and on M. Take this marked ruler off and put it aside.

Wooden Ruler D Finally take the ruler D, which is also needed to mark similar points and divisions as indicated in the preceding steps. Having done this, you will have four rulers of the width. We need to shorten the length. This is the procedure to follow [figure B.61]. First, trace from point A of circle A a line that goes through M and divides the circle at point F, and [divides] circle B at point G, and circle C at point H, and circle D at point I, and line KL which is the terminal perpendicularly at point P. We have said that A coincides with 1 and F with 7.

Trace line 2-12 which cuts line AF at point 2-12. Trace line 3-11 which cuts line AF at point 3-11. Trace line 4-10 which cuts line AF at point 4-10. Trace line 5-9 which cuts line AF at point 5-9. Trace line 6-8 which cuts line AF at point 6-8. Trace a line from point P which is as long as the one that is going from KL to Q.[128] Let PQ be on the extension of line FA and trace the perpendicular line OQ on Q of the height you desire from which you will observe the circles. At this point O, place a nail with a thread.

Paper Ruler a Take a paper ruler and place it on KL so that it does not move [figure B.61]. Then trace AQ which cuts the ruler at point a. Take the thread and extend it to point 1 of circle A: mark 1 where the thread cuts the ruler. Place the thread on 2-12: mark 2-12 where the thread cuts the ruler. Place the thread on 3-11: mark 3-11 where the thread cuts the ruler. Place the thread on 4-10: mark 4-10 where the thread cuts the ruler. Place the thread on 5-9: mark 5-9 where the thread cuts the ruler. Place the thread on 6-8: mark 6-8 where the thread cuts the ruler. Place the thread on 7: mark 7 where the thread cuts the ruler. After, take this ruler off and make a similar one; mark it with a and put it aside.

Paper Ruler b Draw line 2-12 of circle B which intersects BG at point 2-12. Draw line 3-11 of circle B which intersects BG at point 3-11. Draw line 4-10 of circle B which intersects BG at point 4-10. Draw line 5-9 of circle B which intersects BG at point 5-9. Draw line 6-8 of circle B which intersects BG at point 6-8. Now take a paper ruler and place it on KL so that it does not move. Trace AQ which cuts the line at point b: this will be ruler b. Then take a thread and place it on point 1 with an apex [1′].[129] All these measures are with an apex.[130] Mark 1 where the thread that goes through 1′[131] intersects the ruler. Mark 2-12 where the thread that goes through 2′-12′ intersects the ruler. Mark 3-11 where the thread that goes through 3′-11′ intersects the ruler. Mark 4-10 where the thread that goes through 4′-10′ intersects the ruler. Mark 5-9 where the thread that goes through 5′-9′ intersects the ruler. Mark 6-8 where the thread that goes through 6′-8′ intersects the ruler. Mark 7 where the thread that goes through 7′ intersects the ruler. After, take this ruler off and make a similar ruler marked with b and put it aside.

Paper Ruler c Trace line 2-12 in circle C, which cuts CH at point 2″-12″: we must identify points [intersecting] CH with two apexes. Trace line 3-11 in circle C, which cuts CH at point 3″-11″. Trace line 4-10 in circle C, which cuts CH at point 4″-10″. Trace line 5-9 in circle C, which cuts CH at point 5″-9″. Trace line 6-8 in circle C, which cuts CH at point 6″-8″. Having done this, we now need to place another paper ruler which does not move on KL. We need to trace QA which cuts the ruler at point c: we will call this ruler c. Moreover, we need to take the thread and place it on point 1″. Point 1 is found where the thread touches the ruler. We then need to take the thread and place it on point 2″-12″, on point 3″-11″, on point 4″-10″, on point 5″-9″, on point 6″-8″, and on point 7″. As mentioned earlier, all these points will be reported on the ruler. When you have taken it from that location, marking it with c, trace another similar [marked also with a c], and put them both aside.

Paper Ruler d Successively, trace line 2-12 of circle D, which cuts DI at point 2‴-12‴, with three apexes, which we will have to place in the remaining ones [intersection points]. Trace line 3-11, which cuts DI at point 3‴-11‴. Trace line 4-10 in circle D, which cuts DI at point 4‴-10‴. Trace line 5-9 in circle D, which cuts DI at point 5‴-9‴. Trace line 6-8 in circle D, which cuts DI at point 6‴-8‴. Then take a paper ruler; place it on KL so that it does not move. Then take a thread and place it on point 1‴. Let point 1 be where the thread touches paper ruler d. Proceed in the same manner for points 2-12, 3-11, 4-10, 5-9, 6-8, 7, marking them where the thread touches ruler d. Trace QA which divides the ruler at point d. Let there be a ruler marked as d. Make a ruler similar to the one called d and put them both aside.

The Shortening of the First Circle Having done these things, trace straight line FG which you will divide at point M [figure B.62]. Trace the perpendicular line above M till it touches N: you will obtain line MN; and trace two lines perpendicular to FG, above F till K, above G till L. Then take the two paper rulers marked with a. Make sure that one coincides with FK and the other one with GL. Make sure that both points a of the two rulers are well fixed on line FG. Successively, take wooden ruler A; place this ruler [A] on the two paper rulers a so that it touches point 7 of the [two] rulers. Point m is always on line MN. Where on this line it [point m] coincides with 7 of the wooden ruler, mark point 7. When I say it coincides, the reference is always to the wooden ruler.[132] Put the ruler so that it goes through point 6-8 of the two rulers. Mark point 6, where it coincides[133] with point 6 of the same ruler. Mark point 8, where it coincides with point 8 of the same ruler. Put the ruler so that it goes through point 5-9 of the two rulers. Mark point 5, where it coincides with point 5 of the same ruler. Mark point 9, where it coincides with point 9 of the same ruler. Put the wooden ruler so that it goes through point 4-10 of the two rulers. Mark point 4, where it coincides with point 4 of the same ruler. Mark point 10, where it coincides with point 10 of the same ruler. Put the ruler so that it goes through point 3-11 of the two rulers. Mark point 3, where it coincides with point 3 of the ruler. Mark point 11, where it coincides with point 11 of the ruler. Put the ruler so that it goes through point 2-12 of the two rulers. Mark point 2, where it coincides with point 2 of the same ruler. Mark point 12, where it coincides with point 12 of the ruler. Put the ruler so that it goes through point 1 of the two rulers. Mark point 1, where it coincides with point 1 of the same ruler. We must always understand that point m touches line MN.[134] In this way you will have circle A. Draw lines 1-2, 2-3, 3-4, 4-5, 5-6, 6-7, 7-8, 8-9, 9-10, 10-11, 11-12, 12-1. After this, remove the rulers and put them aside.

The Shortening of the Second Circle For what concerns the second circle, take the two paper rulers marked with b and make sure that one coincides with FK and the other one with GL and that point b of both rulers touches line FG. Take wooden ruler B; place this ruler on the two paper rulers so that it touches point 7 of the [two] rulers. Point m is always on line MN, even if it's not said. On the two points indicated with 7, you have placed the ruler. Mark point 7, where it coincides with point 7 of the wooden ruler. Put the ruler so that it goes through point 6-8 of the two rulers. Mark point 6, where it coincides with point 6 of the same ruler. Mark point 8, where it coincides with point 8 of the ruler. Put the ruler so that it goes through point 5-9 of the two rulers. Mark point 5, where it coincides with point 5 of the wooden ruler. Mark point 9, where it coincides with point 9 of the ruler. Put the ruler so that it goes through point 4-10 of the two rulers. Mark point 4, where it coincides with point 4 of the ruler. Mark point 10, where it coincides with point 10 of the ruler. Put the ruler so that it goes through point 3-11 of the two rulers. Mark point 3, where it coincides with point 3 of the ruler. Mark point 11, where it coincides with point 11 of the ruler. Put the ruler so that it goes through point 2-12 of the two rulers. Mark point 2, where it coincides with point 2 of the ruler. Mark point 12, where it coincides with point 12 of the ruler. Put the ruler so that it goes through point 1 of the two rulers. Mark point 1, where it coincides with point 1 of the ruler. Take the rulers off now that circle B has been completed. Draw lines 1-2, 2-3, 3-4, 4-5, 5-6, 6-7, 7-8, 8-9, 9-10, 10-11, 11-12, 12-1.

The Shortening of the Third Circle For what concerns the third circle, take the two paper rulers marked with c and make sure that one is on FK and the other one on GL and that point c of both rulers touches line FG. Then take wooden ruler C; place this ruler on the two paper rulers so that it touches point 7 of both paper rulers. Mark point 7, where it coincides with point 7 of the wooden ruler. Put the ruler so that it goes through point 6-8 of the two rulers. Mark point 6, where it coincides with point 6 of the same ruler. Mark point 8, where it coincides with point 8 of the ruler. Put the ruler so that it goes through point 5-9 of the two rulers. Mark point 5, where it coincides with point 5 of the wooden ruler. Mark point 9, where it coincides with point 9 of the ruler. Put the ruler so that it goes through point 4-10 of the two rulers. Mark point 4, where it coincides with point 4 of the ruler. Mark point 10, where it coincides with point 10 of the ruler. Put the ruler so that it goes through point 3-11 of the two rulers. Mark point 3, where it coincides with point 3 of the ruler. Mark point 11, where it coincides with point 11 of the ruler. Put the ruler so that it goes through point 2-12 of the two rulers. Mark point 2, where it coincides with point 2 of the ruler. Mark point 12, where it coincides with point 12 of the ruler. Put the ruler so that it goes through point 1 of the two rulers. Mark point 1, where it coincides with point 1 of the ruler. Take the lines off now that circle C has been completed. Draw lines 1-2, 2-3, 3-4, 4-5, 5-6, 6-7, 7-8, 8-9, 9-10, 10-11, 11-12, 12-1.

The Shortening of the Fourth Circle After this, take the two paper rulers so that one coincides with FK and the other one with GL and point d of both rulers touches line FG. Then take wooden ruler D; place this ruler on the two paper rulers so that it touches point 7 of the [two] paper rulers. We must always understand that point m touches line MN.[135] Mark point 7, where it coincides with point 7 of the wooden ruler. Put the ruler so that it goes through point 6-8 of the two rulers. Mark point 6, where it coincides with point 6 of the same ruler. Mark point 8, where it coincides with point 8 of the ruler. Put the ruler so that it goes through point 5-9 of the two rulers. Mark point 5, where it coincides with point 5 of the ruler. Mark point 9, where it coincides with point 9 of the ruler. Put the ruler so that it goes through point 4-10 of the two rulers. Mark point 4, where it coincides with point 4 of the ruler. Mark point 10, where it coincides with point 10 of the ruler. Put the ruler so that it goes through point 3-11 of the two rulers. Mark point 3, where it coincides with point 3 of the ruler. Mark point 11, where it coincides with point 11 of the ruler. Put the ruler so that it goes through point 2-12 of the two rulers. Mark point 2, where it coincides with point 2 of the ruler. Mark point 12, where it coincides with point 12 of the ruler. Put the ruler so that it goes through point 1 of the two rulers. Mark point 1, where it coincides with point 1 of the ruler. Following these conditions you will have completed circle D. Take the rulers off and put them aside. Finally draw lines 1-2, 2-3, 3-4, 4-5, 5-6, 6-7, 7-8, 8-9, 9-10, 10-11, 11-12, 12-1. We have the four circles which are divided into twelve equal parts with the shortenings that respect what I had promised to demonstrate.

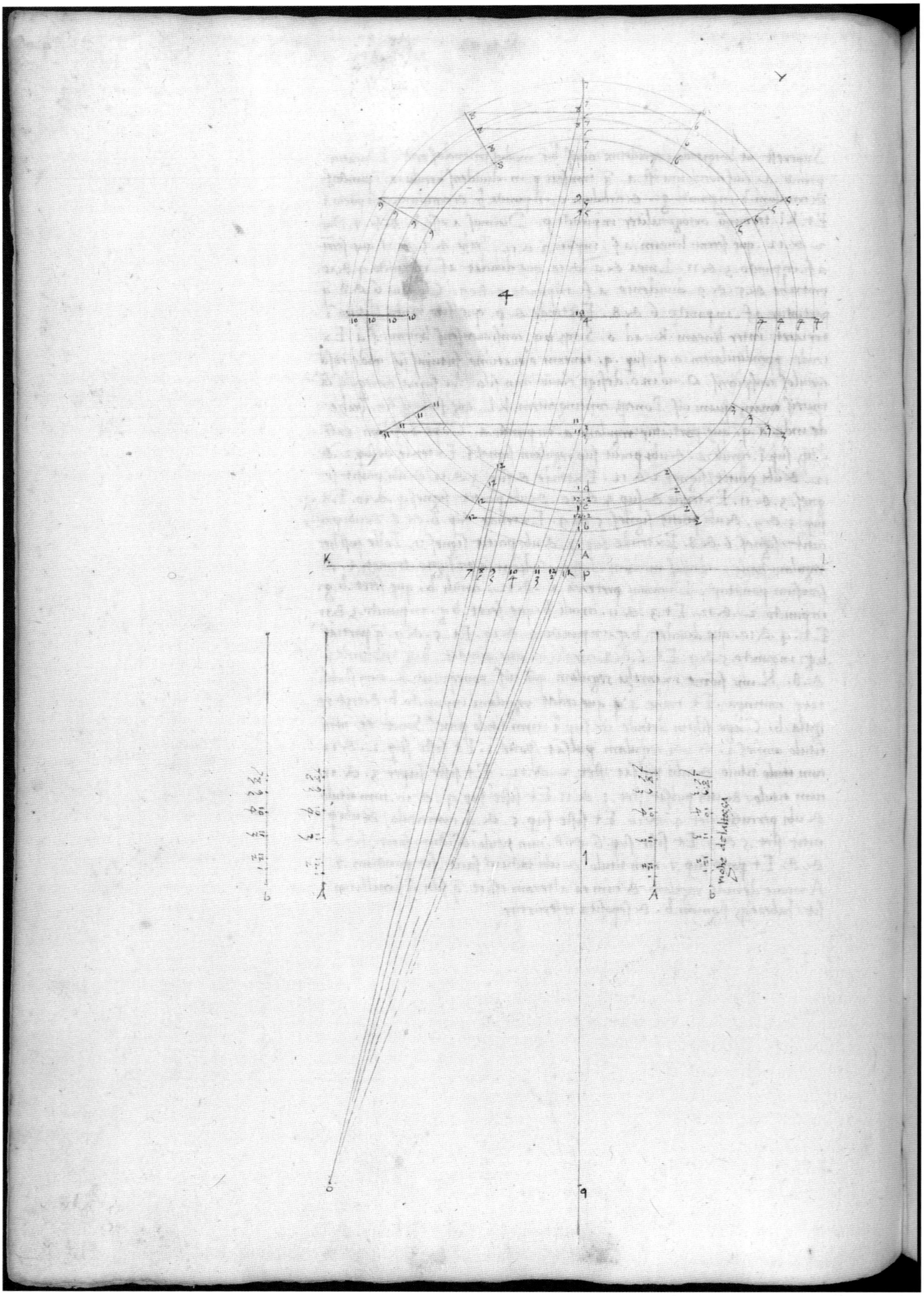

Bordeaux, Bibliothèque Municipale, Ms Latin 616, folio 43 verso.

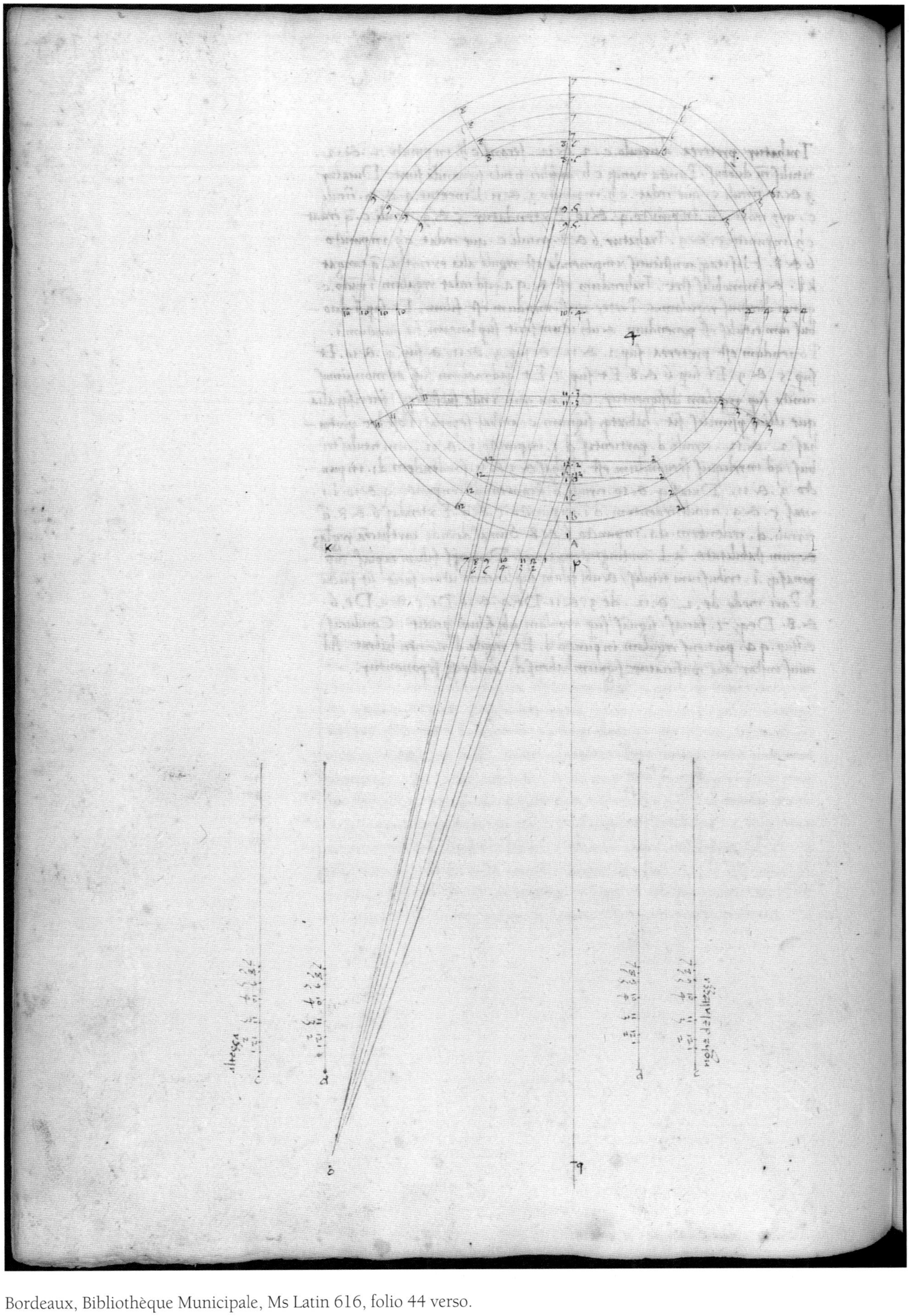

Bordeaux, Bibliothèque Municipale, Ms Latin 616, folio 44 verso.

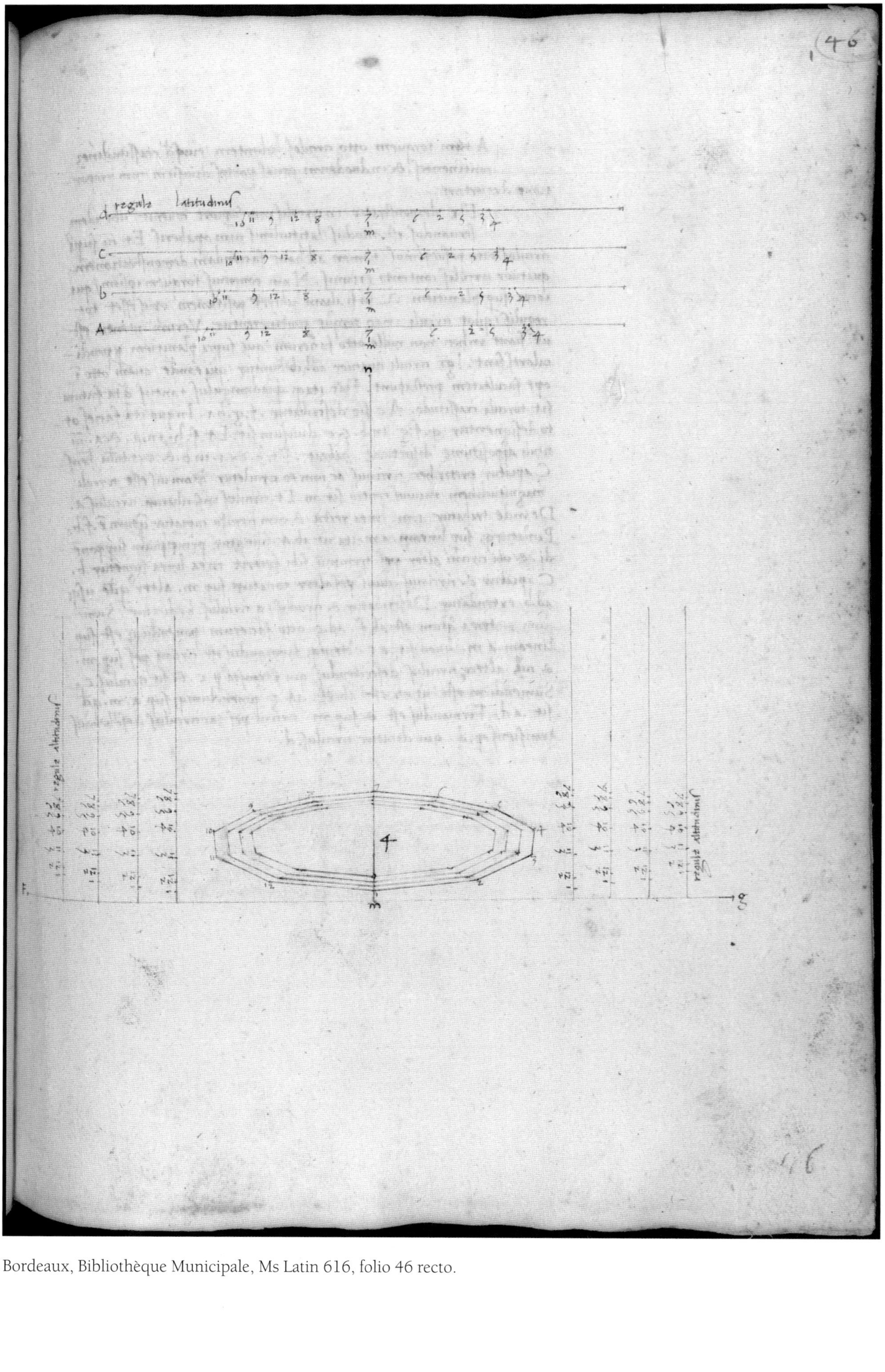

Bordeaux, Bibliothèque Municipale, Ms Latin 616, folio 46 recto.

Proposition IV

Proportionally shorten an assigned torus divided into twelve equal parts which has eight circles that delimit the thickness.

This demonstration is similar to the previous one in the circles, where one must maintain the procedure of the width when we operate. Although we will operate with eight circles later, for now we will use four circles in this demonstration. We will in fact place the torus on a [horizontal] plane. If we had a different position, we would need as many rulers as there are circles in the torus. In reality, my intention is to place some of the eight faces orthogonally to the horizontal plane. We will thus use four circles, which will offer us the same possibility of operating with eight.[136] Therefore, give the square the same thickness that you would like the torus to have. Represent it with FGHI [figure B.63]. Draw eight equal sides in this figure so that FG is divided at B and at C, FH is divided at A and at A′, HI is divided at B′ and at C′, GI is divided at D and at D′. Take a compass and trace a circle in proportion to the size one wishes to give the torus. Let M be the center of the circle. Let this circle be identified by the letter A. Trace straight line AM and with the compass take length FB [in the square] and place it on line MA starting from A. Apply the principle of superimposition.[137] When the other point of the compass stops, mark B on that line [MA]. Take the compass. Place one point on M and move the other point till it reaches B: draw the circle that is called circle B. Then, with the compass, take length FC of the octagon in the square and take it back to line

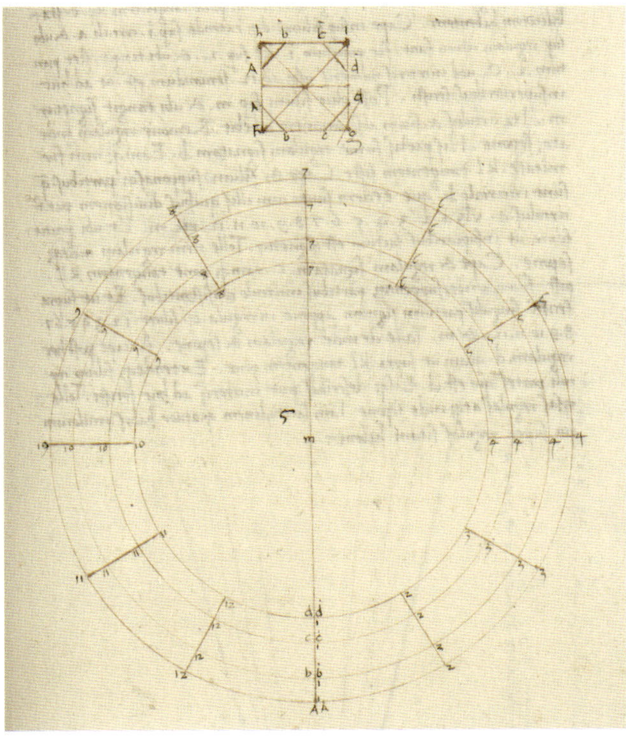

Bordeaux, Bibliothèque Municipale, Ms Latin 616, folio 47 recto.

MA: AC is obtained. In the same way we must fix one point of the compass at M and with the other one we must draw a circle that goes through C. You get the circle called C. Take length FG [in the square] and place it on line AM. We obtain AD. We must also fix the point of the compass in M and with the other one we must draw a circle that goes through D. You get the circle called D. Divide the circles that we have drawn into twelve equal parts, as has been done in the following way in the previous demonstration: 1, 2, 3, 4, 5, 6, 7, 8, 9, 10, 11, 12. Proceed this way for any circle, as shown in the figure.

Wooden Ruler A Place line KL as the terminal. Having done this, place the eye which will be as far as one wishes the person who looks at the circle to be. Let point O be in this place. Place a needle and a thread here as was done in the previous propositions. Take a wooden ruler A, make it coincide with KL so that it does not move. Then take the thread and place it on point 1 of circle A: when it touches the ruler, mark point 1. Then take the thread and place it on point 2: when it touches the ruler, mark point 2. Repeat this procedure till point 12, as you have done in the preceding propositions.[138] Then trace the thread on M, and where it touches the ruler mark point m. In this way we will have completed circle A. Take the ruler away and put it aside.

Wooden Ruler B Having done this, take a ruler marked B, make it match on KL so that it does not move. Take the thread too. Carry it to the parts that are on circle B which correspond to the subdivisions that are on circle A, that is, points 1, 2, 3, 4, 5, 6, 7, 8, 9, 10, 11, 12, and M. Where the thread touches the ruler, mark it as was done for the previous [numbers] of ruler A.[139] Take the ruler away and put it aside.

Wooden Ruler C Take a ruler marked C, place it so that it coincides with KL. Take the thread and carry it to the parts that are on circle C, and, as you have done earlier, put marks on the single subdivisions of the ruler C: that is, 1, 2, 3, 4, 5, 6, 7, 8, 9, 10, 11, 12, and M. Take the ruler away and put it aside.

Wooden Ruler D After, take a ruler marked D, place it so that it coincides with KL. Extend the thread to the subdivisions[140] of circle D and mark these subdivisions as you have done till now with all the other circles. Take the ruler away and put it aside. You now have the width of four circles which have been placed on the rulers.

The Heights of the Circles For what concerns the procedure of the height, consider the eight sides in real form [of the little square], with the aim of having the heights of the circles. These heights correspond to the heights of the torus which, in the upper part, appears to be delimited by letters ABCD and in the lower part by letters A′B′C′D′ [figure B.63]. Place this octagon on a straight line so that side B′C′ matches with it [figure B.64].

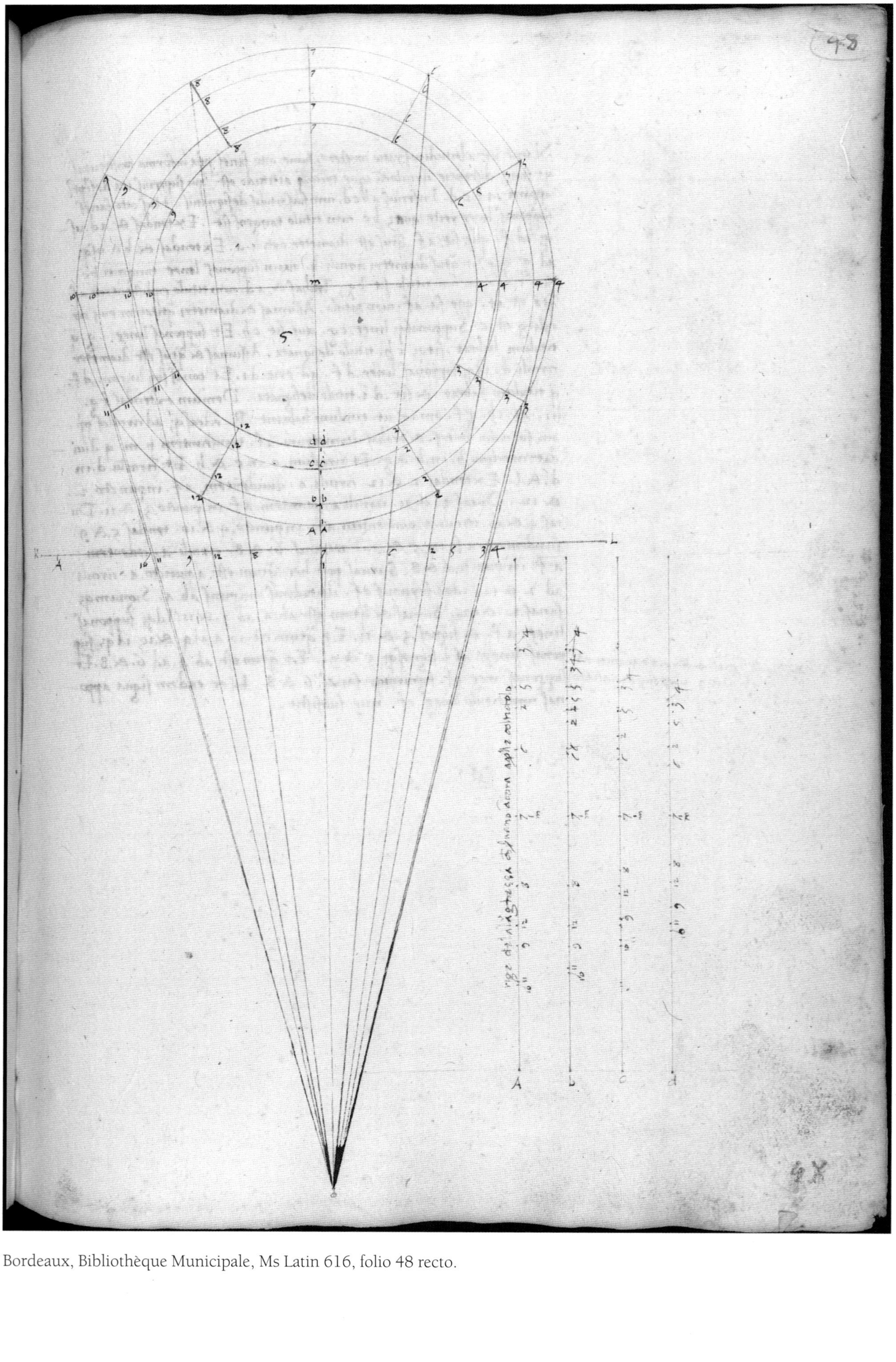

Bordeaux, Bibliothèque Municipale, Ms Latin 616, folio 48 recto.

Extend AD to F and obtain AF, which is as long as the diameter of circle A. Also extend BC to G so that the distance coincides with the diameter of circle B. Place this distance on the line that coincides with B′C′: let this line be B′G′. Trace A′D′ parallel to AF, which [parallel] is as long as A′F′. Take the length of the diameter of circle C and place it on line CG: obtain CH. Then take this length on line C′G′ and obtain line C′H′.

Take the length of the diameter of circle D and place it on line DF and you will have obtained DI. Place [the same length] on line D′F′[141] and obtain D′I′. Finally trace lines FG, HI and H′I′, G′F′. Go back to the circles in real shape and trace diameter AF, which passes through M and which divides circle B at points B and G, circle C at points C and H, and circle D at points D and I. Trace line 2-12 of circle A, which divides line AF at point 2-12. Trace line 3-11 of circle A, which divides line AF at point 3-11. Trace line 4-10 of circle A, which divides line AF at point 4-10. Trace line 5-9 of circle A, which divides line AF at point 5-9. Trace line 6-8 of circle A, which divides line AF at point 6-8.

Then take the distance from point A of the circle to point 2-12 and place it on line AF starting from point A in height: this point will mark 2-12.[142] Take the distance between point A and 3-11; place it on line AF and mark point 3-11. Take the distance between point A and 4-10; place it on line AF and mark point 4-10. Take the distance between point A and 5-9; place it on line AF and mark point 5-9. Take the distance between point A and 6-8; place it on line AF and mark point 6-8. Trace these same marks on line A′F′ which lies below. After this, mark the second circle B. Trace line 2-12: where it intersects the diameter mark point 2-12. Trace line 3-11 of circle B: where it intersects the diameter mark point 3-11. Trace line 4-10: where it intersects BG mark point 4-10. Trace line 5-9: where it intersects BG mark point 5-9. Trace line 6-8: where it intersects BG mark point 6-8. Take the quantity from B to 2-12 of circle B and place it on line BG of the height. Mark point 2-12; and place it on line B′G′ and mark 2-12. Take the quantity from B to 3-11 of line B′G′ in circle B and place it on the line of the height and mark point 3-11.

Do the same thing on line B′G′. Take the quantity from B to 4-10 of line BG in circle B and place it on line BG of the height and mark point 4-10. Do the same thing on line

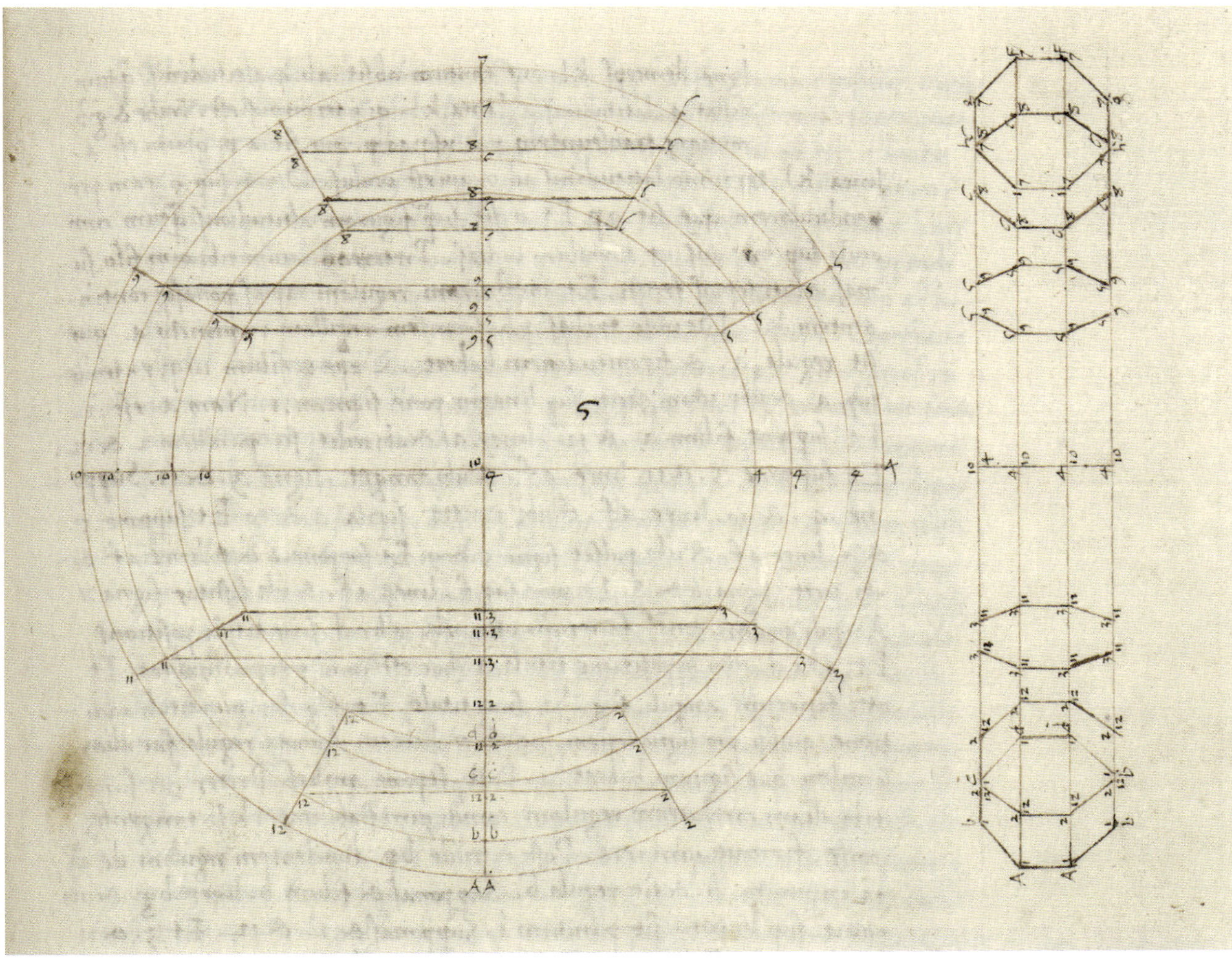

Bordeaux, Bibliothèque Municipale, Ms Latin 616, folio 50 recto.

B′G′. Take the quantity from B to 5-9 of line BG in circle B and place it on line BG of the height and mark point 9. Do the same thing on line B′G′. Take the quantity from B to 6-8 and place it on line BG of the height and mark point 6-8. Do the same thing on line B′G′. After this, mark above CH all the parts of circle C which are placed on line CH of the height and on the line of height C′H′.

Then place on diameter DI all the parts of circle D and place the same ones on the line of height D′I′ which is also a line of the height, which is what you have done in circle A and in circle B. Do the same for points 2-12, 3-11, 4-10, 5-9, 6-8, as you see in the figure.

Paper Rulers Trace line KL [figure B.64], which is distant from line a of the height as much as point A of the width is distant from line KL which is the terminal. Trace also G′B′ so that it goes through K till P. Let this be KP, which is the distance of the eye O from the line of the terminal KL in width. Also trace the perpendicular OP above P, and let O be on the figure of the height as tall as you want to be with the eye to observe the torus.

Paper Ruler a Moreover, take the thread and the needle as you have often already done. Take a paper ruler and place it so it coincides with KL. Then trace B′P which intersects the ruler at point a on which ruler a stays still. Take the thread and place it on A. Place point 1 where the thread intercepts [the ruler]: in fact A corresponds to 1. Place the thread on 2-12 of line AF: place point 2-12 where it hits the ruler. Place the thread on 3-11 of line AF: place point 3-11 where it hits the ruler. Place the thread on 4-10 of line AF: place point 4-10 where it hits the ruler. Place the thread on 5-9 of line AF: place point 5-9 where it hits the ruler. Place the thread on 6-8 of line AF: place point 6-8 where it hits the ruler. Place the thread on F of line AF: place point 7 where it hits the ruler. Certainly all eight angles of the faces, that is, ABCD and A′B′C′D′, are identified by number 1. And the eight angles of the sides, that is, FGHI and F′G′H′I′, are identified by number 7. Having taken out the ruler, make a similar one identified by a and put both to the side.

Paper Ruler b Then take another paper ruler and place it so that it coincides with KL and block it with wax. Then extend B′P which divides the paper ruler at point b: you will have ruler b. Then take the thread and bring it to point B of line BG: mark point 1 where it touches the ruler. Place the thread on 2-12, on 3-11, on 4-10, on 5-9, on 6-8 and mark all these points where the thread touches the ruler.[143] Take the thread and bring it to point G of line BG: mark point 7 where it touches the ruler. Take the ruler off and make a similar one identifying it with b, and then put both the rulers aside.

Paper Ruler c Take another paper ruler and place it on KL so that it stays still. Then extend B′P so that it cuts the ruler at point c: call this ruler c. Take the thread and bring it to point C of line CH: mark point 1 where it touches the ruler. Place the thread on 2-12, on 3-11, on 4-10, on 5-9, on 6-8. Mark all these points where the thread touches ruler c.[144] Take the thread and bring it to point H of line CH: mark point 7 where it touches the ruler. Take the ruler off and make a similar one identifying it with c, and then put both the rulers aside.

Paper Ruler d Take another paper ruler and place it on KL so that it stays still. Then extend B′P which cuts the ruler at point d: call this ruler d. Take the thread and bring it to point D of line DI: mark point 1 where it touches the ruler. Place the thread on 2-12, on 3-11, on 4-10, on 5-9, on 6-8 of line DI. Mark all these points where the thread touches ruler d. Take the thread and bring it to point I of line DI: mark point 7 where it touches the ruler. Take the ruler off and make a similar one identifying it with d, and then put both the rulers aside.

Paper Ruler a′ Take another paper ruler and place it on KL so that it stays still. Then extend B′P which cuts the ruler at point a: call this ruler a′. Take the thread and bring it to A′ of line A′F′: mark point 1 where it touches the ruler. Place the thread on 2-12, on 3-11, on 4-10, on 5-9, on 6-8 of line A′F′. Mark all these points where the thread touches the ruler. Moreover lead the thread to point F′ [of line A′F′]: mark point 7 where it hits the ruler. Take the ruler off and make a similar one identifying it with a′, and then put both the rulers aside.

Paper Ruler b′ Take another paper ruler and place it on KL as described earlier. Then extend B′P which cuts the ruler at point b′: call this ruler b′. Take the thread and bring it to B′ of line B′G′:[145] mark point 1 where it touches the ruler. Place the thread on 2-12, on 3-11, on 4-10, on 5-9, on 6-8 of line B′G′. Mark all these points on ruler b′ where the thread touches the ruler. Moreover lead the thread to point G′: mark point 7 where it hits the ruler. Take the ruler off and make a similar one identifying it with b′.

Paper Ruler c′ After this, make another similar paper ruler and place it on KL. Then extend B′P which cuts the ruler at point c′: call this ruler c′. Take the thread and bring it to C′ of line C′H′: mark point 1 where it touches the ruler. Place the thread on 2-12, on 3-11, on 4-10, on 5-9, on 6-8 of line C′H′. Mark all these points on ruler c′. Lead the thread to point H′: mark point 7 where it hits the ruler. Take ruler c′ off and make a similar one identifying it with c′. Place both of them aside.

Paper Ruler d′ After this, make another similar paper ruler and place it on KL so that it stays still. Then extend B′P which cuts the ruler at point d′: call this ruler d′. Take the thread and bring it to D′ of line D′I′: mark point 1 where it touches the ruler. Place the thread on 2-12, on 3-11, on 4-10, on 5-9, on 6-8 of the line of the height D′I′ and mark all these points where the thread hits the ruler. Lead the thread to I′ of the line of the height D′I′: mark point 7 where it hits the ruler. Take this ruler off and make a similar one with d′ and put both aside. You now have the heights of the torus on the paper rulers.

The Shortening of the Torus We now need to choose the space in which we will realize the shortened torus [figure B.65], a place that can either be made of paper or of any other material on which the straight line KL can be drawn wider than the torus that we are about to make. Divide it into two equal parts at point M. Trace above M the perpendicular line MN from N, above K the perpendicular line PK from P, and above L the perpendicular line LQ from Q.

The Shortening of Circle C Having done this, take the two paper rulers with c and place one on PK and the other on LQ, so that points c of the two rulers touch line KL and the rulers are still. Then take wooden ruler C: make sure it touches point 7 on both paper rulers. And point m of the wooden ruler touches line MN: this should always happen. Mark point 7 where point 7 of the wooden ruler is. When I say it coincides, the reference is always to the wooden ruler.[146] Place the ruler so that it touches point 6-8 of the two [paper] rulers: let point 6 be where point 6 of the same ruler is, let point 8 be where point 8 is. Place the ruler so that it touches point 5-9 of the two rulers: let point 5 be where point 5 of the ruler is, let point 9 be where point 9 is. Place the ruler so that it touches point 4-10: let point 4 be where point 4 of the ruler is, let point 10 be where point 10 of the ruler is. Place the ruler so that it touches point 3-11 of the two rulers: let point 3 be where point 3 of the ruler is, let point 11 be where point 11 is. Place the line so that it touches point 2-12 of the two rulers: let point 2 be where point 2 of the same ruler is, let point 12 be where point 12 is. Place the line so that it touches point 1 of the two rulers: let point 1 be where point 1 is. Take all the rulers off now since you have the first circle, C. Draw lines 1-2, 2-3, 3-4, 4-5, 5-6, 6-7, 7-8, 8-9, 9-10, 10-11, 11-12, 12-1.

The Shortening of Circle B For the second circle, that is, B, proceed in the following way. Take the two paper rulers identified with b and place one on KP and the other one on LQ so that points b of the two rulers touch line KL and the rulers are still. Then take wooden ruler B: make sure it touches point 7 of both paper rulers: mark point 7 where point 7 of the two rulers is. Place the ruler so that it touches point 6-8 of the two rulers: let point 6 be where point 6 of the same ruler is, let point 8 be where point 8 is. Place the ruler so that it touches point 5-9 of the two rulers: let point 5 be where point 5 of the wooden ruler is, let point 9 be where point 9 is. Place the ruler so that it touches point 4-10 of the two rulers: let point 4 be where point 4 of the ruler is, let point 10 be where point 10 is. Place the ruler so that it touches point 3-11 of the two rulers: let point 3 be where point 3 is, let point 11 be where point 11 is. Place the ruler so that it touches point 2-12 of the two rulers: let point 2 be where point 2 of the ruler is, let point 12 be where point 12 is. Place the ruler so that it touches point 1 of the two rulers: let point 1 be where point 1 is. Take the rulers off and draw lines 1-2, 2-3, 3-4, 4-5, 5-6, 6-7, 7-8, 8-9, 9-10, 10-11, 11-12, 12-1. These lines form circle B.[147]

The Shortening of Circle A To obtain circle A, take the two paper rulers identified with a and place one on KP and the other one on LQ so that points b of the two rulers touch line KL and the rulers are still. Then take wooden ruler A: make sure it touches point 7 of both paper rulers. And always make sure that point m of the wooden ruler touches line MN. Mark point 7 where point 7 of the two rulers is. Place the ruler so that it touches point 6-8 of the two rulers: let point 6 be where point 6 of the ruler is, let point 8 be where point 8 is. Place the ruler so that it touches point 5-9 of the two rulers: let point 5 be where point 5 of the wooden ruler is, let point 9 be where point 9 is. Place the ruler so that it touches point 4-10 of the two rulers: let point 4 be where point 4 is, let point 10 be where point 10 is. Place the ruler so that it touches point 3-11 of the two rulers: let point 3 be where point 3 is, let point 11 be where point 11 is. Place the ruler so that it touches point 2-12 of the two rulers: let point 2 be where point 2 of the [wooden] ruler is, let point 12 be where point 12 is. Place the ruler so that it touches point 1 of the two rulers: let point 1 be where point 1 is. Take the rulers off and draw lines 1-2, 2-3, 3-4, 4-5, 5-6, 6-7, 7-8, 8-9, 9-10, 10-11, 11-12, 12-1. Having done this we will have obtained circle A.

The Shortening of Circle A′ Now we will look at circle A′, which is built this way.[148] Take the two paper rulers identified with letter A′. Place one on KP and the other one on LQ so that points A′ of the two rulers touch line KL. Then take wooden ruler A′: place it on the two paper rulers; make sure it touches point 7 of both paper rulers. And point m of the wooden ruler—as we said—will touch line MN. Mark point 7 where point 7 of the two rulers is. We will continue this way, that is, with the application of the marks on line A′ whenever there is a contact.[149] Now take the rulers off and draw lines 1-2, 2-3, 3-4, 4-5, 5-6, 6-7, 7-8, 8-9, 9-10, 10-11, 11-12, 12-1. We have obtained circle A′.

The Shortening of Circle B′ Take the two paper rulers identified with letter B′: this mark will indicate the ruler. Place one on KP and the other on LQ so that points B′ of the two rulers touch line KL. Also take wooden ruler B′ so that it touches points 7, 6-8, 5-9, 4-10, 3-11, 2-12, 1 of the two rulers; and point m of the wooden ruler will touch line MN: where the marks match the marks on the wooden ruler, mark them with points 7, 6, 8, 5, 9, 4, 10, 3, 11, 2, 12, 1. Take the rulers off now and trace lines 1-2, 2-3, 3-4, 4-5, 5-6, 6-7, 7-8, 8-9, 9-10, 10-11, 11-12, 12-1. Through these tracings we have completed circle B′.

The Shortening of Circle D After this, take the two paper rulers identified with letter D. Place one ruler on KP and the other on LQ so that points D of the two rulers touch line KL. Then take the wooden ruler indicated with D; place it so that it touches points 7, 6-8, 5-9, 4-10, 3-11, 2-12, 1 of the two rulers. Where the marks match the marks on the wooden ruler, mark them as earlier with points 7, 6, 8, 5, 9, 4, 10, 3, 11, 2, 12, 1. Through these tracings we have completed circle D.[150] Take the rulers off now and put them aside. Trace lines 1-2, 2-3, 3-4, 4-5, 5-6, 6-7, 7-8, 8-9, 9-10, 10-11, 11-12, 12-1.

The Shortening of Circle D′ Take the two paper rulers identified with letter D′. Place one ruler on KP and the other on LQ so that points D′ of the two rulers touch line KL. Also take the wooden ruler indicated with D′ so that it touches points 7, 6-8, 5-9, 4-10, 3-11, 2-12, 1 of the two rulers. Where the marks match the marks on the wooden ruler, mark them with points 7, 6, 8, 5, 9, 4, 10, 3, 11, 2, 12, 1. Take the rulers off now. Trace lines 1-2, 2-3, 3-4, 4-5, 5-6, 6-7, 7-8, 8-9, 9-10, 10-11, 11-12, 12-1. Through these transformations, we have completed circle D′.

The Shortening of Circle C′ Take the two paper rulers identified with letter C′. Place one ruler on KP and the other on LQ so that points C′ of the two rulers touch line KL. Take the wooden ruler indicated with C′. Place it so that it touches points 7, 6-8, 5-9, 4-10, 3-11, 2-12, 1 of the two rulers. Mark point 7 where point 7 of the ruler is. Mark point 6 where point 6 is. Mark point 8 where point 8 is. Mark point 5 where point 5 is. Mark point 9 where point 9 is. Mark point 4 where point 4 is. Mark point 10 where point 10 is. Mark point 3 where point 3 is. Mark point 11 where point 11 is. Mark point 2 where point 2 is. Mark point 12 where point 12 is. Mark point 1 where point 1 is. Take off the rulers now and draw lines 1-2, 2-3, 3-4, 4-5, 5-6, 6-7, 7-8, 8-9, 9-10, 10-11, 11-12, 12-1. Circle C′ has been completed. In all these circles, one must follow the method that was utilized in the representation of the first and second circle.[151]

Proposition V

Proportionally shorten an assigned cube which stands on one of its vertices with none of its sides parallel to the established terminal.[152]

Cube in Plan and in Elevation: First Rotation We will first place cube ABCD and FGHI in real shape, in the way one can understand from the figure [figure B.66].[153]

We will trace straight line RS where we are about to represent it. We will then take the surface in real shape BCHG of one of the bases of the cube and we will place corner G on line RS.[154] Let corner B[G] be tilted with respect to this line as much as we want the cube to be tilted.[155] We will place corner C higher than the other corners [of the inferior surface],[156] and we will draw it so that it fits on a line parallel to RS. We will trace from B and H unlimited parallel lines to RS. We will also trace the line that is perpendicular to RS and goes through H and point I, and we will trace the line that goes from corner C to point D, and we will trace the line that goes from B to point A. And we will trace the line that goes [from D] to point F of line RS up to point V.[157] We will also take the length of side BC and we will extend it from D to DC.[158] We will place it by extending it from F to FG. We will also trace CG which will intersect the line coming out of H at point H and the line coming out of B at point B. We will then mark points G, C, and B of the squared surface so that they are equidistant from DV.

Second Rotation Let's now consider the surface generated by the first surface which corresponds to DC, IH, AB, FG [figure B.67]. Let's place it on line DV. That is, let's place point F so that it touches line DV and point G above DV as much as we'd like the side to be tilted.[159] Trace FG which is equal to FG of the surface generated by the first surface which is DC, IH, AB, FG. Trace line FD which should make a right angle at point F [with FG] and which is equal to FD of the surface generated by the first. Also trace line GC as long as FD and trace DC. After this, we will take the length from F to A of the surface generated from the first. We will place it in a similar way above FD and we will mark point A. We will take the length from F to I of the same surface and we will place it on FD in a similar way and we will mark I. We will also trace the line parallel to DC from I which will touch GC at point H. And we will trace the line parallel to FG from A which will touch GC at point B.

The Third Figure: Plan of the Tilted Cube We will trace a line parallel to RS from D [figure B.68]. We will mark point D where this parallel meets the line coming from C of the first surface. We will trace a line parallel to RS from I. We will mark point I where this parallel meets the line coming from H of the first surface. We will trace a line parallel to RS from A. We will mark point A where this parallel meets the line coming from B of the first surface. We will trace a line parallel to RS from F. We will mark point F where this parallel meets the line coming from G of the first surface. We will trace a line parallel to RS from C. We will mark point C where this parallel meets the line coming from C of the first surface. We will trace a line parallel to RS from H. We will mark point H where this parallel meets the line coming from H of the first surface. We will trace a line parallel to RS from B. We will mark point B where this parallel meets the line coming from B of the first surface. We will trace a line parallel to RS from G. We will mark point G where this parallel meets the line coming from G of the first surface. Let's now trace lines AD, DI, AF, FI and BC, CH, BG, GH.

The Choice of the Terminal and the Point of View Since this is the third figure [figure B.69], that is, the cube drawn on the plan which has a special shape,[160] on which one must take the shortening of the width, mark point M on this cube.

Then draw line KL, which moves away, as you wish, from the corner B of the cube in a special shape, so that KL is not parallel to side AB of the cube but is further from A than from B. Then step away to point O from line KL which is the terminal, standing at a distance from which you will look at the cube. Place point O at this point with thread and needle as you have already done before.

Wooden Ruler E Successively, take a wooden ruler and place it on KL so that it stays still [figure B.69]. Then take the thread and place it on point A of the cube in a special shape, which is the third figure.[161] Mark point A where the thread hits the ruler. Place the thread on point B, mark point B where the thread hits the ruler. Place the thread on point C, mark point C where the thread hits the ruler. Place the thread on point D, mark point D where the thread hits the ruler. Place the thread on point F, mark point F where the thread hits the ruler. Place the thread on point G, mark point G where the thread hits the ruler. Place the thread on point H, mark point H where the thread hits the ruler. Place the thread on point I, mark point I where the thread hits the ruler. Place the thread on point M, mark point M where the thread hits the ruler. Take the ruler off now, because through what has already been said we have obtained the dimensioning of the width. Put it aside: call it ruler E.

Elevation of the Tilted Cube To obtain the procedure of the heights we will first trace from corner B of the cube drawn in a special shape a line parallel to line KL which is the terminal [figure B.70]. We will then trace from corner A and corners C, D, F, G, H, I, the indeterminate parallel lines to line KL. We will trace on these the perpendicular line PQ which intersects the line that starts from B at point B, the one that starts from A at point A; the one that starts from C at point C; the one that starts from D at point D; the one that starts from F at point F; the one that starts from G at point G; the one that starts from H at point H; the one that starts from I at point I. Moreover, we will take all the lengths that are found starting from line DV in the second figure DC, IH, AB, FG. This figure is inclined by angle F on line DV. We will then take the length between line DV and G—which I will do for the remaining points—and we will place it on the line that starts from G, that is to say on the line that starts from the corner of the cube drawn on the plan in a special shape. We will mark point G there. We will take the length between line DV and point A and we will place it on the line that starts from A, and we will mark point A.[162] We will take the length between line DV and point B and we will place it on the line that starts from B, and we will mark point B. We will take the length between line DV and point C and we will place it on the line that starts from C, and we will mark point C. We will take the length between line DV and point D and we will place it on the line that starts from D, and we will mark point D. We will take the length between line DV and point F and we will place it on the line that starts from F, and we will mark point F.[163] We will take the length between line DV and point H and we will place it on the line that starts from H, and we will mark point H. We will take the length between line DV and point I and we will place it on the line that starts from I, and we will mark point I. All these points are considered realized on a figure of the elevation, and they are placed on lines which we have traced from the corners of the cube drawn on the plan in a special shape, starting from a line that intercepts them all perpendicularly. Let's say that this is the figure of the elevation in a special shape:[164] a figure that now needs to be shortened in a space between the rulers.

The Terminal and the Point of View in Elevation Extend line PQ beyond line KL, which is the terminal, by the quantity that lies between line KL and the eye O [figure B.71]. Let this quantity be pQ. Trace a line parallel [to pQ] above p and make it as high as you want the view of the cube to be: let this be point O.

Paper Ruler E Place a thread and a needle in it. Then take a paper ruler which is placed on line KL so that it stays still. Trace pQ which divides the ruler at point E, and indicate the ruler with point E. Place the thread on point C of the figure in height: mark C where the thread hits the ruler. Place the thread on point D: mark D where the thread hits the ruler. Place the thread on point H: mark H where the thread hits the ruler. Place the thread on point I: mark I where the thread hits the ruler. Place the thread on point B: mark B where the thread hits the ruler. Place the thread on point A: mark A where the thread hits the ruler. Place the thread on point G: mark G where the thread hits the ruler. Place the thread on point F: mark F where the thread hits the ruler. You have completed the procedure of the height. Take the ruler off and make a similar one with the same marks.

The Shortening of the Cube Draw straight line KL, which occupies the place where one must draw the shortened cube [figure B.72]. Divide KL into two equal parts at point M. Trace the perpendicular lines: that is, one from point M to point N, one from point K to point X, and one from point L to point Y. Place one of the two paper rulers to coincide with KX and the other one with LY. Let the rulers stay still and let points E of both rulers lie on KL. Place the wooden ruler so that it touches points C of the two paper rulers and so that M touches line MN. Mark point C where you find point C of the wooden ruler. Place the wooden ruler so that it touches points D of the two rulers. Mark point D where you find point D of the wooden ruler. Place the wooden ruler so that it touches points H of the two rulers. Mark point H where you find point H of the wooden ruler. Place the wooden ruler so that it touches points I of the two rulers and point M of the wooden ruler touches line MN: which should always happen even if it's not always said.[165] Mark point I where you find point I of the wooden ruler. Place the wooden ruler so that it touches points B of the two rulers. Mark point B where you find point B of the wooden ruler. Place the wooden ruler so that it touches points A of the two rulers. Mark point A where you find point A of the wooden ruler. Place the wooden ruler so that it touches points G of the two rulers. Mark point G where you find point G of the wooden ruler. Place the wooden ruler so that it touches points F of the two rulers. Mark point F where you find point F of the wooden ruler. Trace lines AB, BC, CD, DA, CH, HI, ID, AF, FG, GB. With these additions we will have the shortening of the cube which we had proposed.

Proposition VI

Proportionally shorten the assigned base of a round column.[166]

To follow the order we have undertaken,[167] we will first trace the base in real shape of which we know, once and for all, one face, as we can observe in the figure of its elevation [figure B.73].

The Elevation Let the base be as tall as half the width of the column and at the bottom as wide as the column plus two-fifths of the width of the column. Therefore we will put that the width be equal to GH. We will then divide the height into twelve equal parts. We will place four of these parts above G which will correspond to the height of point A, in conformity with the flat tile that is called the plinth. We will place three parts above A which will correspond to point C, in conformity with the round part which is called the torus. We will place the half of a part above point C: we will mark this position with point C′. We will place two parts above point C′: we will mark this position with point E which is called the belt. We will place the half of a part above point E as the listel fillet which is on the belt, and we will indicate [this position] with point E′. The remaining parts regard the upper roundness: we will mark this position with point E″.[168] We will now draw these parts with an appropriate order. We will trace a [horizontal] line F in the middle of the upper roundness. We will trace a second [horizontal] line D in the middle of the belt. We will trace another [horizontal] line B in the middle of the lower roundness.[169]

The Plan To realize the figure in real shape in width, take quantity GH of the flat tile, that is, the base of the plinth, and draw a quadrilateral surface [figure B.73]. Let any of the sides of this surface be as long as GH. Let the angles be equal like the sides. Let GHIK be the surface. Draw the diagonals of this surface and let their intersection be point M. Take the half of line A of the base and draw a circle with center M which is also the center of the surface where one realizes the measurement of the width.[170] Take the half of line B which is in the middle of the first roundness. Place the compass end at point M of the squared surface and with the other end trace the circle till you reach the starting point of the circle. Take the half of line C. Place the end of the compass at point M and with the other end trace the circle with that length. You will have obtained circle C. Take the half of line D. Place the end of the compass at point M and with the other end draw circle D. Take the half of line E. Place the end of the compass at point M and with the other end trace the circle with that length. You will have obtained circle E. Take the half of line F. Place the end of the compass at point M and with the other end trace the circle with that length. You will have obtained circle F. This way you will have obtained the circles.

The Subdivisions of the Circles in Plan After this, divide the sides of the squared surface GHIK into equal parts [figure B.73]. First divide GI into two equal parts. And from the point of this subdivision let come out the line that passes through M and that cuts at point 1 all these circles, that is, A, B, C, D, E, F. Let the circles be intersected at point 9 on the side of HK. Let the diagonal on the side of G intersect each of these circles at point 3 and the diagonal on the side of K intersect each of these circles at point 11. Then divide GH into two equal parts, and from the point of this subdivision let come out the line that passes through M and that cuts all these circles each at point 5. Let the circles be intersected at point 13 on the side of IK. Let diagonal IH intersect each of these circles at point 7 and on the side of I at point 15.[171] Now divide 1-3 into two equal parts at point 2 and trace through 2 the line that goes through M and that divides all these circles at point 2: on the opposite side it divides them all at point 10. Now divide 3-5 into two equal parts at point 4 and trace through 4 the line that goes through M and that divides all these circles at point 4: on the opposite side it divides them all at point 12. Now divide 5-7 into two equal parts at point 6 and trace through 6 the line that goes through M and that divides all these circles at point 6: on the opposite side it divides them all at point 14. Now divide 7-9 into two equal parts at point 8 and trace through 8 the line that goes through M and that divides all these circles at point 8: on the opposite side it divides them all at point 16.

The Numbers in Elevation of Circle A Divide line GH, which is located in the plinth of the figure in elevation,[172] into two equal parts at point M [figure B.73]. For this subdivision trace the line perpendicular to GH that intersects the parallel lines through points A, B, C, D, E, F: let the intersection take place at points 5. Take the distance from line 5-13 to point 2 of circle A and place it on one side and on the other of line A in elevation. Mark 2-16 on the right side and mark 8-10 on the left side.[173] Take the distance from line 5-13 to point 1 of circle A. Place the end of the compass on 5 of line A in elevation. With the other end of the compass mark point 1 on the right side and point 9 on the left side. Take the distance from line 5-13 to point 3 of circle A and place it on line A in elevation. Mark 3-15 on the right side and mark 7-11 on the left side.[174] Take the distance from line 5-13 to point 4 of circle A and place it on line A in elevation. Mark 4-14 on the right side of 5 and mark 6-12 on the left side.

The Numbers in Elevation of Circle B Take the distance from line 5-13 to point 1 of circle B [figure B.73]. Place this distance on line B of the elevation: mark 1 on the right side and mark 9 on the left.[175] Take the distance from line 5-13 to point 2 of circle B and place it on line B of the elevation. Mark 2-16 on the right side of 5-13 and mark 8-10 on the left side. Take the distance from line 5-13 to point 3 of circle B and place it on line B of the elevation. Mark 3-15 on the right side of 5 and mark 7-11 on the left side.[176] Take the distance from line 5-13 to point 4 of circle B and place it on line B of the elevation. Mark 4-14 on the right side of 5 and mark 6-12 on the left side.

The Numbers in Elevation of Circle C Take the distance from line 5-13 to point 2 of circle C and place it on line C of the elevation [figure B.73]. Mark point 2-16 on the right side and mark 8-10 on the left side. Take the distance from line 5-13 to point 1 of circle C and place it on line C of the elevation. Mark point 1 on the right side and mark 9 on the left side. Take the distance from line 5-13 to point 3 of circle C and place it on line C of the elevation. Mark point 3-15 on the right side of 5-13 and mark 7-11 on the left side. Take the distance from line 5-13 to point 4 of circle C. Place it on line C of the elevation. Mark point 4-14 on the right side of 5 and mark point 6-12 on the left side.

The Numbers in Elevation of Circle D Take the distance from line 5-13 to point 1 of circle D [figure B.73]. Place it on line D of the elevation. Mark point 1 on the right side of 5-13 and mark point 9 on the left side.[177] Take the distance from line 5-13 to point 2 of circle D. Place it on line D of the elevation. Mark point 2-16 on the right side of 5 and mark point 8-10 on the left side. Take the distance from line 5-13 to point 3 of circle D. Place it on line D of the elevation. Mark point 3-15 on the right side of 5 and mark point 7-11 on the left side. Take the distance from line 5-13 to point 4 of circle D and place it on line D of the elevation. Mark point 4-14 on the right side of 5 and mark point 6-12 on the left side.

The Numbers in Elevation of Circle E Take the distance from the line to point 1 of circle E[178] and place it on line E of the elevation [figure B.73]. Mark point 1 on the right side of 5 and mark point 9 on the left side. Take the distance from line 5-13 to point 2 of circle E and place it on line E of the elevation. Mark point 2-16 on the right side of 5 and mark point 8-10 on the left side. Take the distance from line 5-13 to point 3 of circle E and place it on line E of the elevation. Mark point 3-15 on the right side of 5 and mark point 7-11 on the left side. Take the distance from line 5-13 to point 4 of circle E and place it on line E of the elevation. Mark point 4-14 on the right side of 5 and mark point 6-12 on the left side.

The Numbers in Elevation of Circle F Take the distance from line 5-13 to point 1 of circle F[179] and place it on line F of the elevation [figure B.73]; mark point 1 on the right side of 5 and mark point 9 on the left side. Take the distance from line 5-13 to point 2 of circle F and place it on line F of the elevation; mark point 2-16 on the right side of 5 and mark point 8-10 on the left side. Take the distance from line 5-13 to point 3 of circle F and place it on line F of the elevation; mark point 1 on the right side of 5-13 and mark point 7-11 on the left side. Take the distance from line 5-13 to point 4 of circle F and place it on line F of the elevation; mark point 4-14 on the right side of 5 and mark point 6-12 on the left side. And you have the representation of the figure in appropriate shape in elevation.

The Terminal and the Point of View in Plan To draw the width of the circles present in the squared surface GHIK we must follow these steps. We will draw circles A, B, C, D, E, F [figure B.74] which we will shorten: this shortening will be indicated on the surface of the rulers.[180] We will draw line PQ, parallel to GI which is a side of the base: let this line be the terminal where the rulers are placed. We will then go away from line PQ to the distance from which we want to observe that prospect. We will mark point O in which we will fix the needle with the thread.

Wooden Ruler A We will then take a wooden ruler and place it along PQ, and we will name it ruler A [figure B.74]. We will place the thread on point 1 of circle A. We will mark point 1 where it hits the ruler. We will place the thread on 2 [of circle A]. We will mark point 2 where it hits the ruler. We will place the thread on 3. We will mark point 3 where it hits the ruler. We will place the thread on 4. We will mark point 4 where it hits the ruler. We will place the thread on 5. We will mark point 5 where it hits the ruler. We will place the thread on 6. We will mark point 6 where it hits the ruler. We will place the thread on 7; we will mark point 7 where it hits the ruler.[181] We will place the thread on 8; we will mark point 8 where it hits the ruler.[182] Do the same thing till point 16, marking all the points where the thread hits the ruler.[183] We will place the thread on M. We will mark point M where it hits the ruler.

The Other Wooden Rulers B, C, D, E, F These will be circle A's subdivisions, which we will repeat in the remaining circles: that is, circles B, C, D, E, and F [figure B.76].[184] We will report the subdivisions we obtain in circle B on ruler B with point M. We will report the ones we obtain from circle C on ruler C with point M. We will report the ones we obtain from circle D on ruler D with point M. We will report the ones we obtain from circle E on ruler E with point M. We will report the ones we obtain from circle F on ruler F with point M.[185]

Wooden Ruler G When we have completed these things, we need to take another wooden ruler and make it coincide with PQ: name it ruler G [figure B.76].[186] We will take the thread and place it on K. We will mark K where it hits the ruler. We will place the thread on I: we will mark point I where it hits the ruler. We will place the thread on H: we will mark point H where it hits the ruler. We will place the thread on G: we will mark point G where it hits the ruler. We will place the thread on M: we will mark point M where it hits the ruler. These will be the marks of the squared surface of the base. Take the ruler and place it aside.

The Terminal and the Point of View in Elevation You have the rulers of the width. To obtain the height,[187] draw unlimited line HG [figure B.75]; that is, in the figure of the elevation, the line of the plinth's ground floor which constitutes the base of the plinth. Draw perpendicular line PQ on this base, which is as far distant from G as the distance between line PQ [figure B.74]—which is the terminal—and line GI of the surface in width. After this, take the distance that there is in width between terminal PQ and eye O, place it on the line that starts at H and goes through G and R [figure B.75]: call it HR. Trace the perpendicular line OR above R; make it as high as you wish to look at the base from. Place the needle with the thread on this point O as you have often done till now.

Paper Ruler G′ Then take a paper ruler and place it on PQ[188] so that it stays still [figure B.75]. Do the same thing for all the other rulers. Draw line HG of the base, which constitutes the end of the plinth, so that it intersects the ruler at point G: this is ruler G′. Let the thread go through H: mark H-K where it hits the ruler. Let the thread go through G: mark G-I where it hit the ruler. Take the ruler off and draw another one with mark G′. Put both aside.

Paper Ruler G Moreover, place another paper ruler so that it coincides with PQ [figure B.75]. Trace HG which divides the paper ruler at point G. Let the thread go through H which gives the length of the height: mark H-K where it hits the ruler. Let the thread go through G: mark G-I where it hits the ruler. Take this ruler off and make a similar one with mark G. Place both aside.

Paper Ruler A Place another paper ruler so that it coincides with PQ [figure B.75]. Trace HG which divides the paper ruler at point A: this is ruler A. Let the thread go through point 1 of line A in elevation: mark 1 where it hits the ruler. Let the thread go through point 2 of line A: mark 2-16 where it hits the ruler.[189] Let the thread go through point 3: mark 3-15 where it hits the ruler.[190] Let the thread go through point 4: mark 4-14 where it hits the ruler.[191] Let the thread go through point 5: mark 5-13 where it hits the ruler.[192] Let the thread go through point 6: mark 6-12 where it hits the ruler.[193] Let the thread go through point 7: mark 7-11 where it hits the ruler. Let the thread go through point 8: mark 8-10 where it hits the ruler. Let the thread go through point 9: mark 9 where it hits the ruler. And you have reported the marks of line A on ruler A. Take this ruler off and make a similar one with mark A. Put both aside.

Paper Ruler B After this, place another paper ruler so that it coincides with PQ [figure B.76].[194] Trace HG which divides the paper ruler at point B. Let the thread go through point 1 of ruler B in height: mark 1 where it hits the ruler. Always, when I say "put," "place," "trace," or "lead," one must understand "on the line" and when I say "hits" one must understand "on the ruler."[195] Let the thread go through point 2: mark 2-16 where it hits the ruler.[196] Let the thread go through point 3: mark 3-15 where it hits the ruler.[197] Let the thread go through point 4: mark 4-14 where it hits the ruler.[198] Let the thread go through point 5: mark 5-13 where it hits the ruler.[199] Let the thread go through point 6: mark 6-12 where it hits the ruler.[200] Let the thread go through point 7: mark 7-11 where it hits the ruler. Let the thread go through point 8: mark 8-10 where it hits the ruler. Let the thread go through point 9: mark 9 where it hits the ruler. These are the points of line B on ruler B. Take this ruler off and make a similar one with mark B. Put both aside.

Paper Ruler C′ After this, place another paper ruler so that it coincides with PQ [figure B.76]. Trace HG which divides the paper ruler at point C′. Let the thread go through point 1, through point 2, through point 3, through point 4, through point 5, through point 6, through point 7, through point 8, through point 9[201] of line C in elevation. Mark 1 where it hits the ruler at point 1. Mark 2-16 where it hits the ruler at point 2.[202] Mark 3-15 where it hits the ruler at point 3.[203] Mark 4-14 where it hits the ruler at point 4.[204] Mark 5-13 where it hits the ruler at point 5.[205] Mark 6-12 where it hits the ruler at point 6.[206] Mark 7-11 where it hits the ruler at point 7. Mark 8-10 where it hits the ruler at point 8. Mark 9 where it hits the ruler at point 9. Take this ruler off and make a similar one with mark C′. Put both aside.

Paper Ruler C After this, place another paper ruler so that it coincides with PQ [figure B.76]. Trace HG which will intersect the paper ruler at point C. Let the thread go through point 1, through point 2, through point 3, through point 4, through point 5, through point 6, through point 7, through point 8, through point 9[207] of line C in height. Mark the ruler where the thread hits it as you have done earlier. You will take this ruler off and make a similar one with mark C. Put both aside.

Paper Ruler D After this, place another paper ruler so that it coincides with PQ [figure B.76]. Trace HG which will intersect the paper ruler at point D. Let the thread go through point 1 of ruler D in elevation. Let the thread go through point 2, through point 3, through point 4, through point 5, through point 6, through point 7, through point 8, through point 9 of line D.[208] Mark the ruler where the thread hits it as you have done earlier. Take this ruler off and make a similar one with mark D. Put both aside.

Paper Ruler E″ Place another paper ruler so that it coincides with PQ [figure B.76]. Trace HG which will intersect the paper ruler at point E″. Let the thread go through point 1, through point 2, through point 3, through point 4, through point 5, through point 6, through point 7, through point 8, through point 9 of ruler E″.[209] Mark the ruler where the thread hits the ruler.[210] Take this ruler off and make a similar one with mark E″.[211] Put both aside.

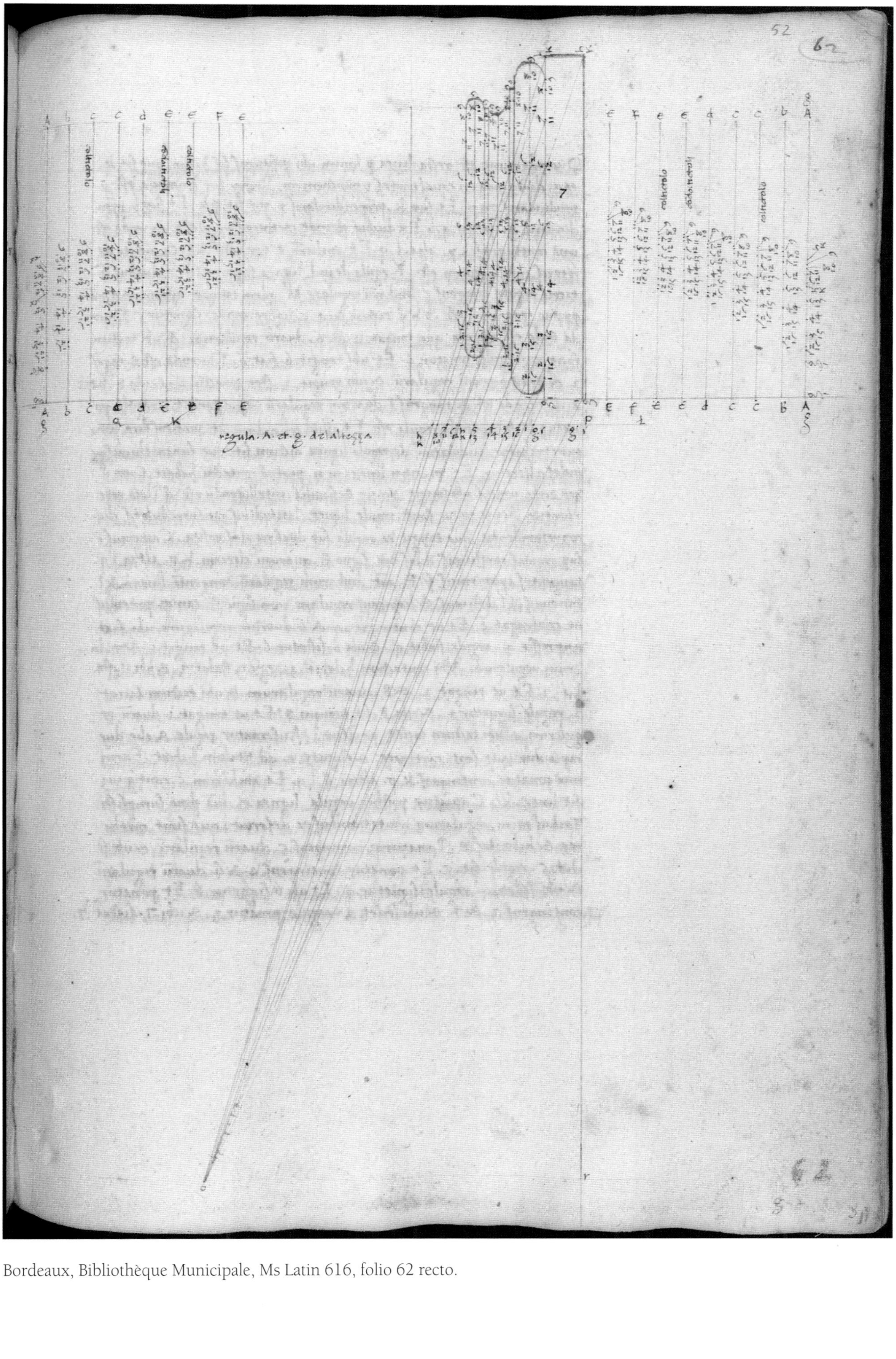

Bordeaux, Bibliothèque Municipale, Ms Latin 616, folio 62 recto.

Paper Ruler E′ You will designate another paper ruler with the same aim [figure B.76]. Place another paper ruler so that it coincides with PQ. Trace HG which will intersect the paper ruler at point E′.[212] Take the thread and observe what you have done till now in the representation. After removing this ruler, draw a similar one and put both aside.

Paper Ruler F Place another paper ruler so that it coincides with PQ [figure B.76]. Trace HG which will intersect the paper ruler at point F: you have ruler F.[213] After this, take the thread and place it on the ruler as you have done till now. You will mark the points where it touches the ruler. In addition, you will also draw another ruler identified with F, and you will place both aside.

Paper Ruler E Then place another paper ruler so that it coincides with PQ [figure B.76]. Trace HG which will intersect the paper ruler at point E. Let the thread go through all the marks of ruler E in elevation; mark the ruler where the thread hits it.[214] Take this ruler off and make a similar one for the appropriate use.

The Shortening of the Base Up until this point we have reported the extension of the widths and the heights on the rulers. Now we need to draw a straight line KL in the place where we want to place the base: a line to be divided into two equal parts[215] at point M [figure B.77]. Moreover, we need to trace the perpendicular line MN and on K a perpendicular line till P which is PK and on L a perpendicular line till Q which is QL.

Rulers E Then take the two paper rulers marked with E [figure B.77]. We have to place one on KP and the other on LQ. Let point E of both rulers touch KL: do this for all the remaining rulers. We will then place wooden ruler E above the two paper rulers so that it touches point 5 of the two rulers. Let point M touch line MN: this must always take place. Mark point 5 where point 5 of the wooden ruler touches it. We must then place the ruler so that it goes through 4-6 of the two [paper] rulers. Mark point 4 where you find point 4. Mark point 6 where you find point 6. We must then place the ruler so that it goes through 2-8 of the two rulers. Mark point 2 where you find point 2. Mark point 8 where you find point 8. We must then place the ruler so that it goes through 1 of the two rulers. Mark point 1 where you find point 1.[216] We must take the ruler off and remember that every time I say "place," "put," "trace"[217] I am referring to the wooden ruler which will be placed above the paper rulers. Point M must always be in contact with line MN. When I say these words: "where it touches," "where it hits," or similar things, I mean the place where the marks of the wooden ruler are, which is relative to the width. These marks correspond to those placed on the two paper rulers where the wooden ruler touches the two paper rulers.

Rulers F Take the two paper rulers marked with F [figure B.77]. Place one on line KP and the other on line LQ, with point F of both rulers [placed] so that it touches line KL. Take the wooden ruler with mark F and place it on point 5 of the two rulers. Where it corresponds to point 5 of the ruler, mark 5. Where it corresponds to point 4-6 of the two rulers: where it corresponds to 4, mark 4; where it corresponds to 6, mark 6. Where it corresponds to point 3-7 of the two rulers: where it corresponds to 3, mark 3; where it corresponds to 7, mark 7. Where it corresponds to point 2-8 of the two rulers: where it corresponds to 2, mark 2; where it corresponds to 8, mark 8. Mark 1 where it touches point 1.[218]

Rulers E′ Take the rulers off and make two more paper rulers marked with point E′ [figure B.77]. Place one of the two rulers on line KP and the other on line LQ, with point E of both rulers [placed] so that it touches line KL. Take the wooden ruler with mark E which you have taken at first [which is needed for three paper changes that have letter E] and place it on point 5 of the two rulers. Mark 5 where you find point 5 of the ruler. Where there is point 4-6 of the two rulers: where it corresponds to 4, mark 4; where it corresponds to 6, mark 6. Where you find point 3-7 of the two rulers: where it corresponds to 3, mark 3; where it corresponds to 7, mark 7. Where there is point 2-8 of the two rulers: where it corresponds to 2, mark 2; where it corresponds to 8, mark 8. Place the ruler that passes through point 1 of the two rulers. Mark 1 where it touches point 1 of the two rulers.[219]

Rulers E″ Take these rulers off and draw two more paper rulers marked with point E″ [figure B.77]. Place one of the two rulers on line KP and the other on line LQ with point E of the two rulers so that it touches line KL. Take the wooden ruler with mark E and place it so that it touches the parts of the two rulers while point M touches line MN. We will mark the points where there is the pressure made by signs of the wooden ruler as was already done in the other cases.[220]

Rulers D Take the rulers off and draw two more paper rulers marked with point D [figure B.77]. Place one of the two rulers on line KP and the other on line LQ with point D of both rulers so that it touches line KL. Take the wooden ruler with mark D as you have done earlier with the other ones.

Rulers C Having taken these rulers off, draw two more paper rulers marked with point C [figure B.77]. Place one of the two rulers on line KP and the other on line LQ. Make sure that point C of both rulers touches line KL. Take the wooden ruler with mark C and place it on the marks of the two rulers as you have often done.

Rulers C′ Having taken these rulers off, draw two more paper rulers marked with point C′ [figure B.77]. Place one of the two rulers on line KP and the other on line LQ. Point C′ of the rulers will be on line KL. After taking line C′, we will operate as was previously done.

Rulers B Having taken these rulers off, draw two more paper rulers marked with point B [figure B.77]. Place one of the two rulers on line KP and the other on line LQ. Point B of the two rulers will be on line KL. After taking line B, we will operate as was previously done.

Rulers A Having taken these rulers off, draw two more paper rulers marked with point A [figure B.77]. Place one of the two rulers on line KP and the other on line LQ. Point A of the two rulers will be on line KL. After taking a wooden ruler, we will follow the procedure as was previously done, marking signs that need to be placed. We will take these rulers off too.

Rulers G Having taken two more paper rulers marked with point G, we will place them so that one of the two rulers will be on line KP and the other on line LQ and so that point B of the two rulers will touch line KL [figure B.77].[221] Take wooden ruler G and place it so that it touches line HK of the two rulers while point M touches line MN. Mark H where we have H on the ruler. Mark K where we have K on the ruler. Draw the ruler that touches GI of both rulers. Mark G where we have G, mark I where we have I.

Rulers G′ Having taken these two rulers off, place two paper rulers marked with point G′ [figure B.77]: place one of the two rulers on line KP and the other on line LQ with point G of the two rulers so that it touches line KL. We will take the wooden ruler marked with G and we will place it on HK of the two rulers: mark H where we have H of the ruler, mark K where we have K of the ruler. Place the ruler also on GI of the two rulers: mark G where we have G of the ruler, mark I where we have I. Take these rulers off and put them aside.

The Representation of the Base Draw lines GH, HI, IK, KG [figure B.77].[222] These [lines] are part of the plinth.[223] In the same manner, you will trace the lines that form the same plinth, that is, GH, HI, IK, KG.[224] Draw the nine circles with an appropriate outline, first by using lines 1-2, 2-3, 3-4, 4-5, 5-6, 6-7, 7-8, 8-9, 9-10, 10-11, 11-12, 12-13, 13-14, 14-15, 15-16, 16-1.[225] This layout is the representation of one circle. We must operate this way to draw one after the other all the remaining circles. Having done this you will have completed the base with its shortening.

Final Recommendations However, if the transcriptions of the numbers are an issue for someone who thinks of them as an obstruction to the positioning of the base, because they occupy a lot of space, we could use really very small numbers.[226] when drawing the circles through the points, you will make sure that when you have identified a circle, you will complete it by drawing the outline. Without drawing the outlines, the multitude of circles could lead to errors. Unless someone—through a lot of practice—becomes expert enough to draw the outline of the circles one outline after the other. Since the circles of the width and of the base in proper form are divided into eight equal parts; since from point O, which is the eye, the thread was traced above these parts, that is, 1, 2, 3, 4, 5, 6, 7, 8, till 16,[227] although we could have drawn more parts for convenience; since, often, it can happen that the line that leaves the eye, that is, the thread from point O, touches the circles in other different points which we have drawn, that is, 1, 2, 3, 4, 5, 6, 7, 8, till 16;[228] when this happens in the place where there is the contact we will mark the circles with the additions of signs. Report these marks on the lines that act as the measurement of the height of the base, as was observed earlier.[229]

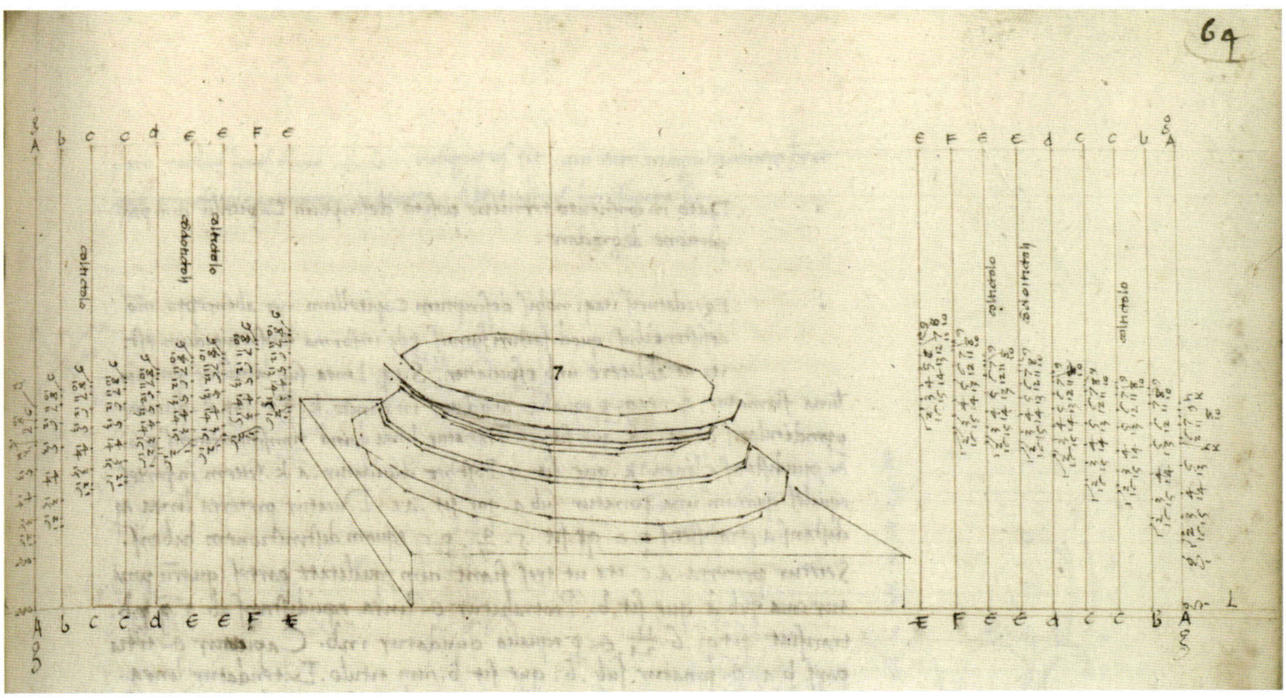

Bordeaux, Bibliothèque Municipale, Ms Latin 616, folio 64 recto.

Proposition VII

From a given point,[230] shorten proportionally, in an established terminal, an assigned capital.

We will then shorten an assigned capital, which we will do without diverting from the method undertaken. One must determine it in appropriate shape,[231] so that it is seen from one side.

Dimensioning the Capital in Elevation The line of its width, where it sticks to the column, is equal to 4 [figure B.78].[232] Divide this line into two equal parts at point K. Trace perpendicular line AK equal to 5 above K. Trace the line that goes through A and is parallel to the one that goes through K: this line will be equal to 7. After, divide AK into seven equal parts, and place one part below A: call this AC. Trace the line going through C and parallel to the one that goes through A: this line should be equal to $5\frac{4}{7}$, and let it be divided into equal parts at point C. Divide AC so as to have three equal parts. Place one part below A: mark it with point B. Trace a line going through B and parallel to the line that goes through A: let this be equal to $6\frac{11}{21}$ and let it be divided into equal parts at point B. Take the third part of BA and place it below B: mark it with point B′. Trace the line that goes through B′ and is parallel to that one through B: this is the stick of a fillet.[233] These [two] lines are on the cymatium. Then divide CK into three equal parts at point F and H. Let the line that goes through F be $4\frac{1}{2}$ long; the one through H $4\frac{1}{4}$ long: let them be traced parallel to the line that goes through K. Take the fifth part of CK and place it below C: this is E. Through E trace the parallel to the line through C: let it be equal to $4\frac{1}{2}$. Divide CE into two equal parts at point D and trace the line that goes through D parallel to the line through C: let this be equal to 5. Take the third part of DE and place it below E, mark line E′: this [third part] will be the stick of a fillet.[234] Take the seventh part of FK and place it below F: this is FG. Trace the line that goes through G and is parallel to that through F: let this be equal to $5\frac{4}{7}$. Take the fourth part of HK and place it below H: this is HI. Trace the line that goes through I and is parallel to that one through H: let this be equal to $4\frac{1}{2}$.

Dimensioning the Volute in Elevation Successively, one must go on to elaborate the tendrils, which you will call, if you want, either twigs or supports.[235] Let the volute be curved [starting] from the beginning of the circuit that starts from the fourth part of line C [figure B.78], with a constant [curved] path so as to be [agreeably] far from the end of line D.[236] Touch line F with this circuit; go through the point of the fourth part of line E and touch line D realizing a good outline, as you see in the figure. Proceed in the same manner on the other side. Moreover, there are the thicknesses and the parts of the remaining tendrils which can be seen. We must dispose them as the procedure of the following figure will show us.

The Cymatium The median part of the cymatium [figure B.79] is enriched with a flower which is as high as the cymatium.[237] This is the figure of the elevation. The width procedure will be shown this way. Let there be a square in real shape.[238] Let any side be as long as line A of the capital. Let the square be PQRS. Trace diagonals PS, QR which intersect at point M which will be the center. Take the quantity that goes from

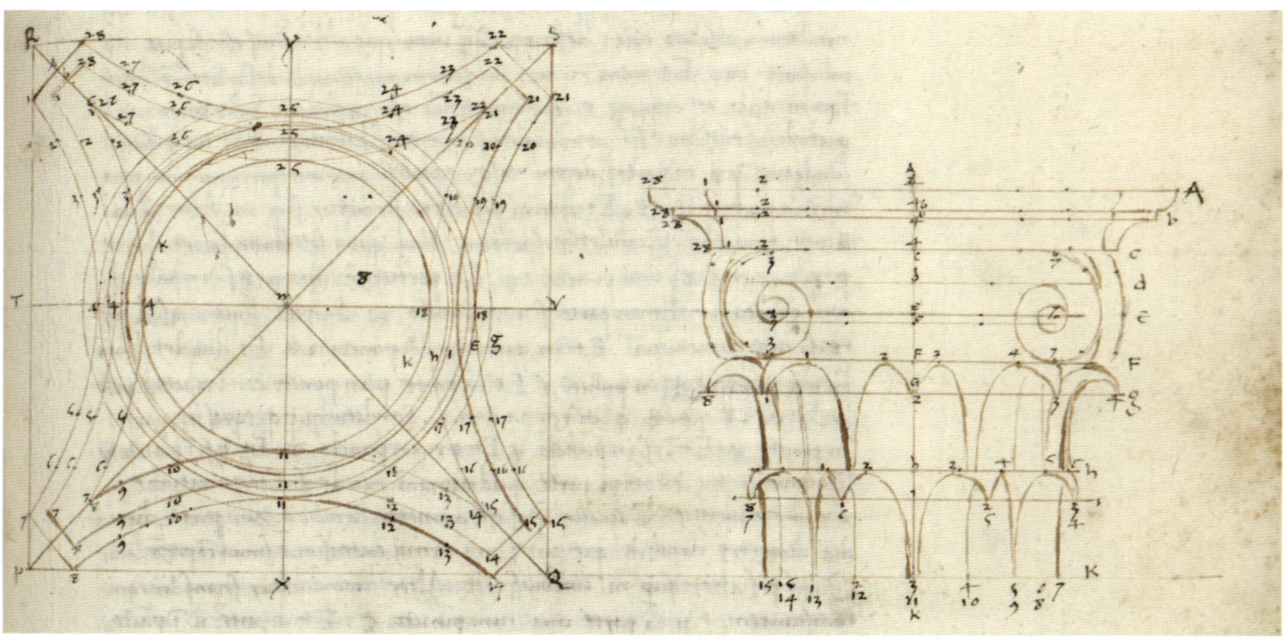

Bordeaux, Bibliothèque Municipale, Ms Latin 616, folio 65 recto.

K to the end of the relative line which is in the figure of the elevation. With this length as the radius, place one end of the compass on M and draw circle K with the other end. Take half of line I and place one end of the compass on M and with the other end draw around M circle I. Take half of line H and place one end of the compass on M and with the other end draw around M circle H. Take half of line G and place one end of the compass on M and with the other end draw circle G. Take half of line F and place one end of the compass on M and with the other end draw circle F. Take half of line E and place one end of the compass on M and with the other end draw circle E. Take half of line D and place one end of the compass on M and with the other end draw circle D. This way you will have completed the representation of the circles. Although in the elevated figure the circles are eight, and you have drawn five circles around M, circle E also serves as circle E′, as circle F, as circle I: they are in fact of equal size. This lets the number be complete.[239]

Dimensioning Circuit A of the Cymatium in Plan Having done this, divide PQ [figure B.79] into ten equal parts. Place one of these parts on P: we have P-8. Place one end of the compass on M and extend the other end until it touches 8. Trace this end around with this length: it will section line PQ on the side of Q at point 14. It will section line QS on the side of Q at point 15 and on the side of S at point 21. It will section line RS on the side of S at point 22 and on the side of R at point 28. It will section line PR on the side of R at point 1 and on the side of P at point 7. After, mark lines 1-28, 7-8, 14-15, 21-22. Divide PQ into equal parts at point T, RS at point V, PR at point X, QS at point Y. Trace line TV so that it goes beyond the two sides. Do the same for XY. Take the diameter of circle D and the semidiameter of circle K. Let these two lengths form the opening of the compass. Place one compass's end still on M and move the other end around. This second end cuts line TV on T's side at point z and on V's side at point Z. It cuts line XY on the side of X at point 9 and on the side of Y at point 4. Place one end of the compass on z, extend the other end till point 8, and trace the arc of the circle tangent to circle D: the arc will end at point 14.[240] Place one end of the compass on Z, extend the other end till point 22, and trace the arc of the circle tangent to circle D: the arc will end at point 28. Place one end of the compass on 9, extend the other end till point 1, and trace the arc of the circle tangent to circle D: the arc will end at point 7. Place one end of the compass on 4, extend the other end till point 15, and trace the arc of the circle tangent to circle D: the arc will end at point 21. This shall be circuit A.

Dimensioning Circuit C of the Cymatium in Plan Place one end of the compass on z [figure B.79], extend the other end till it reaches circle K, and trace the arc of the circle till it meets the diagonals. Place one end of the compass on Z and trace the arc up to circle K and the diagonals. Place one end of the compass on 9 and trace the arc up to circle K and the diagonals. Place one end of the compass on 4 and trace the arc up to circle K and the diagonals. Take the distance between circle K and circle D and place it on the diagonal beginning at line 8-7 and mark a point. After, place one end of the compass on M and extend the other end to the point that has been placed on the diagonal. Trace a circle that divides the two arcs tangent to circle K, on the side where 8 is on point 8 and on the side where 7 is on point 7; on the side where 1 is on point 1 and on the side where 28 is on point 28; on the side where 22 is on point 22 and on the side where 21 is on point 21; on the side where 15 is on point 15 and on the side where 14 is on point 14. This will be circuit C.

Dimensioning Circuit B of the Cymatium in Plan Take the third part of the distance between circuit A and circuit C [figure B.79] and place it on line TV so that it touches circuit A proceeding toward circuit C. Place the end of the compass on z and extend the other end till the point of the third part marked between circuit A and circuit C. Draw the arc until you reach [lines] 8-7 and 14-15. With this length [as a radius] trace in a similar way the arcs of circles that have centers at points Z, 9, and 4: you shall have circuit B.

Other Points on Circuits A, B, C of the Cymatium in Plan Trace 1-1 which will cut circle B at point 1 [figure B.79]. Trace 7-7 which will cut circle B at point 7. Trace 8-8 which will cut circle B at point 8. Trace 14-14 which will cut circle B at point 14. Trace 15-15 which will cut circle B at point 15. Trace 21-21 which will cut circle B at point 21. Trace 22-22 which will cut circle B at point 22. Trace 28-28 which will cut circle B at point 28. In the end trace 7-8, 14-15, 21-22, 1-28, which are identified with the same marks in the three circuits A, B, C. Divide 1-X into two equal parts and place a point from which you can trace a line parallel to RS. This parallel line will divide circuit A at point 3 and at point 19. That is, 19 is on the side of QS. Proceed in the same way for the remaining ones. It will divide circuit B at point 3 and point 19, circuit C at point 3 and point 19. Divide 1-3 into two equal parts at point 2. Trace the line parallel to RS for 2: this parallel line will divide circuit A at point 2,[241] circuit B at point 2, and circuit C at point 2. On the other side it will divide circuit C at point 20, circuit B at point 20, and circuit A at point 20. Divide X-7 into two equal parts and place a point from which you can trace a line parallel to PQ: this parallel line will divide circuit A at point 5, circuit B at point 5, and circuit C at point 5. It will divide on the other side circuit C at point 17, circuit B at point 17, and circuit A at point 17. Divide 5-7 into two equal parts and place a point from which you can trace the line parallel to PQ: this parallel line will divide circuit A at point 6, circuit B at point 6, and circuit C at point 6. On the other side it will divide circuit C at point 16, circuit B at point 16, and circuit A at point 16. For what concerns side PQ, divide 8-T into two equal parts. Trace from this point of subdivision a line parallel to PR: this parallel line will divide circuit A at point 10, circuit B at point 10, and circuit C at point 10. On the other side it will divide circuit C at point 26, circuit B at point 26, and circuit A at point 26. Divide 8-10 into two equal parts and trace [from this point of subdivision] a line parallel to PR: this parallel line will divide circuit A at point 9, circuit B at point 9, and circuit C at point 9. On the other side it will divide circuit C at point 27, circuit B at point 27, and circuit A at point 27.

Divide T-14 into two equal parts and trace from this point of subdivision a line parallel to QS: this parallel line will divide circuit A at point 12, circuit B at point 12, and circuit C at point 12. On the other side it will divide circuit C at point 24, circuit B at point 24, and circuit A at point 24. Divide 12-14 into two equal parts at point 13 and trace from this point of subdivision a line parallel to QS. This parallel line will divide circuit A at point 13, circuit B at point 13, and circuit C at point 13. On the other side it will divide circuit C at point 23, circuit B at point 23, and circuit A at point 23. These are the extensions of the cymatium.

The Leaves Now one must divide the circles, which are divided into eight equal parts by the diameters and the diagonals [figure B.80]. We will divide each of these parts into two equal parts so that they become 16: that is, points 1, 2, 3, 4, 5, 6, 7, 8, 9, 10, 11, 12, 13, 14, 15, 16. These subdivisions are on circle K, where one can find the origin of the leaves below and above. The first leaf below is between 1 and 3. The others are between 3 and 5, 5 and 7, 7 and 9, 9 and 11, 11 and 13, 13 and 15, and 15 and 1.[242] This is the origin of the leaves: these origins are found on circle K. The apex or the point of the first leaf is 1 and is on circle l; the apex of the second leaf that is between 3 and 5 is 2; of the third it is 3, and so on up until leaf number 8: all these [apexes] are on circle l. In circle H, of the leaves below, the first leaf that is between 1 and 3 is marked with 16 and 1, the second one is marked with 2 and 3, the third one with 4 and 5. One should proceed in the same manner for any leaf till number eight; the marks will be 16. Also the leaves above will have their origin in circle K. The first one is marked with 2, 16; the remaining ones with 2, 4, with 4, 6, with 6, 8, with 8, 10, with 10, 12, with 12, 14, with 14, 16. The vertices of these leaves, or their ends, are on circle G. The vertex or the point of the first leaf [marked] with 2 and 16 will be identified with 1. One will proceed in the same way for the other vertices 2, 3, 4, 5, 6, 7, 8, which are all on circle G. On circle F, the first leaf that [on K] is marked 2, 16, corresponds to 1 and 16 here. The remaining [points of the] leaves are 2 and 3, 4 and 5, 6 and 7, 8 and 9, 10 and 11, 12 and 13, 14 and 15.

The Addition of the Flowers We must also indicate the flowers, which are four. We must arrange them so that each one occupies the center of each face [figure B.79]. These must be as tall as the cymatium which touches these four lines: PQ, QS, SR, RP. The flowers will be marked this way in four places. The flower which will be on side PR is marked with 3′; let 4′ be in the middle, opposite [to the first one] we have 5′. The one on face PQ has 10′, then there are 11′ and 12′. The one on face QS has 17′, then there are 18′ and 19′. The one on face RS has 24′, then there are 25′ and 26′.

The Leaves in Elevation: Circle K Moreover, one will have to represent the tendrils, or similar supports or, if you'd like, the twigs in accordance with what we will see in the figure of the height [figure B.80].[243] One will have to proceed to make distinctions with numbers 1, 2, 3, 4, 5, 6, 7 up until the last one, as one can observe in the figure. Up until now we have made the figure of the width. We need to set all these parts in the figure of the height. Therefore we shall place on line K of the height in the extreme part on the right 15, and in the left part 7.[244] We will take the distance between line XY[245] and 4 of circle K of the width and we will place it on the right side of K of the elevation: we will make 2-12, and 4-10 on the left.[246] We will take the distance between line XY and 5 of circle K of the width [this must always be intended], and we will place it on line K of the elevation [this must always be done]: we will make 5-9 on the left and 1-13 on the right. We will take the distance between line XY and 6 of circle K of the width and we will place it on the right side of K on point 16-14, on the left on point 6-8; then finally place 3-11 on K.[247]

The Leaves in Elevation: Circle I For what concerns circle I [figure B.80] we will take the distance between line XY and 2 of circle I. We will place it on line l of the height; we will make 2-5 on the left of line I and 1-6 on the right. We will take the distance between line XY and 3 and we will place it on line l; we will make 8-7 on the right and 3-4 on the left.

The Leaves in Elevation: Circle H We will take the distance between line XY and 2 of circle H [figure B.80] and we will place it on line H of the height; we will have 1-10 on the right and 2-9 on the left.[248] We will take the distance between line XY and 3 of circle H and we will place it on line H of the height; we will have 11-16 on the right and 8-3 on the left. We will take the distance between line XY and 4 of circle H and we will place it on line H of the height; we will have 4-7 on the left and 12-15 on the right. We will take the distance between line XY and 5 of circle H and we will place it on line H of the height; we will have 13-14 on the left and 6-5 on the right.[249]

The Leaves in Elevation: Circle G For what concerns circle G [figure B.80],[250] we will take the distance between line XY and 3: we will place this distance on line G of the height. We will have 5-3 on the left and 7-1 on the right. We will take the distance between line XY and 4 and we will place it on line G; we will have 8 on the right and 4 on the left. Successively, place 2-6 on line G.

The Leaves in Elevation: Circle F For what concerns circle F [figure B.80], we will take the distance between line XY and 3 of circle F and we will place it on line F of the height. We will have 11-2 on the right and 10-3 on the left.[251] We will take the distance between line XY and 4 of circle F and we will place it on line F of the height. We will have 4-9 on the left and 12-1 on the right. We will take the distance between line XY and 5 of circle F and we will place it on line F of the height. We will have 16-13 on the right and 5-8 on the left. We will take the distance between line XY and 6 of circle F and we will place it on line F. We will have 14-15 on the right and 6-7 on the left.

Circuit E′ in Plan and Elevation For what concerns circuit E′ [figure B.81],[252] take the distance between line XY and 5 of circuit E′ and we will place it on line E′ of the height. We will have 5-31 on the right and 29-7 on the left. Take the distance between line XY and 8 and we will place it on line E′. We will have 8-28 on the left and 4-32 on the right. Where you have E′ mark 6 and 30.[253]

Circuit E in Plan and Elevation For what concerns circuit E [figure B.82], take the distance between line XY and 12 of the tendrils, or if you would like, supports, and we will place it on line E. We will have point 48-36 on the right and 12-24 on the left. Take the distance between line XY and 13 and we will place it on line E. We will have 13-23 on the left[254] and 47-37 on the right. Take the distance between line XY and 11 and we will place it on line E. We will have 1-35 on the right and 11-25 on the left. Take the distance between line XY and 10 and we will place it on line E of the height. We will have 10-26 on the left and 2-34 on the right. Take the distance between line XY and 9 of circuit E and we will place it on line E. We will have 9-27 on the left and 3-33 on the right. Take the distance between line XY and 8 and we will place it on line E. We will have 4-32 on the right and 8-28 on the left. Take the distance between line XY and 7 and we will place it on line E. We will have 5-31 on the right and 7-29 on the left.[255] Mark 6-30 where E is.[256] Successively, take the distance between line XY and 14 of the tendrils or of the twigs and place it on line E. We will have 46-38 on the right and 14-22 on the left. Take the distance between line XY and 15 and we will place it on line E. Mark line E. We will have 45-39 on the right and 15-21 on the left. Take the distance between line XY and 16 and we will place it on line E. We will have 44-40 on the right and 16-20 on the left. Take the distance between line XY and 17 of circuit E and place it on line E; mark line E. We will have 43-41 on the right and 17-19 on the left.[257] Take the distance between line XY and 18 and we will place it on line E. We will have 42 on the right and 18 on the left. Mark 6-30 where you have E. These are the points of circuit E, which is part of the tendrils.

Circuit D in Plan and Elevation For what concerns circuit D [figure B.83], take the distance between line XY and 10 and place it on line D of the height. We will have 2-34 on the right and 10-26 on the left. Take the distance between line XY and 11 and place it on line D of the height. We will have point 11-25 on the left and 1-35 on the right.[258] Take the distance between line XY and 9 and place it on line D. We will have point 9-27 on the left and 3-33 on the right.[259] Take the distance between line XY and 8 and place it on line D of the height. We will have point 4-32 on the right and 8-28 on the left. Take the distance between line XY and 7 and place it on line D. We will have point 5-31 on the right and 7-29 on the left.[260] Take the distance between line XY and 6 and place it on line D; mark point 6-30. Take the distance between line XY and 12 and place it on line D. We will have point 12-24 on the left and 48-36 on the right.[261] Take the distance between line XY and 13 and place it on line D. We will have point 13-23 on the left and 47-37 on the right.[262] Take the distance between line XY and 14 and place it on line D. We will have point 14-22 on the left and 46-38 on the right. Take the distance between line XY and 15 and place it on line D. We will have point 15-21 on the left and 45-39 on the right.[263] Take the distance between line XY and 16 and place it on line D.[264] We will have point 44-40 on the right and 16-20 on the left. Take the distance between line XY and 17 and place it on line D. We will have point 43-41 on the right and 17-19 on the left. Take the distance between line XY and 18 and place it on line D. We will have point 42 on the right and 18 on the left. These are the points of circuit D, and they are part of the tendrils.

Circuit C in Plan and Elevation For what concerns circuit C [figure B.84], take the distance between line XY and 7 and place it on line C of the height. We will have point 7-15 on the left and 1-21 on the right. Take the distance between line XY and 6 and place it on line C. We will have point 6-16 on the left and 2-20 on the right. Take the distance between line XY and 5 and place it on line C. We will have point 5-17 on the left and 3-19 on the right. Take the distance between line XY and 4 and place it on line C. Mark 4-18.[265] Take the distance between line XY and 8 and place it on line C. We will have point 8-14 on the left and 28-22 on the right. Take the distance between line XY and 9 and place it on line C. We will have point 27-23 on the right and 9-13 on the left. Take the distance between line XY and 10 and place it on line C. We will have point 10-12 on the left and 26-24 on the right. Mark 11 and 25 on line C.[266]

Circuit B in Plan and Elevation For what concerns circuit B [figure B.85], take the distance between line XY and 7 and place it on both lines B.[267] We will have point 7-15 on the left and 1-21 on the right so that each [part] contains its signs.[268] Take the distance between line XY and 6 and place it on line B. We will have point 6-16 on the left and 2-20 on the right. Take the distance between line XY and 5 and place it on line B. We will have point 5-17 on the left and 3-19 on the right.[269]

Circuit B′ in Plan and Elevation [For what concerns circuit B′] [figure B.86], take the distance between line XY and 5′ and place it on line B′. We will have point 3′-19′ on the right and 5′-17′ on the left.[270] Let each number have the appropriate index.[271] Take the distance between line XY and 11 and place

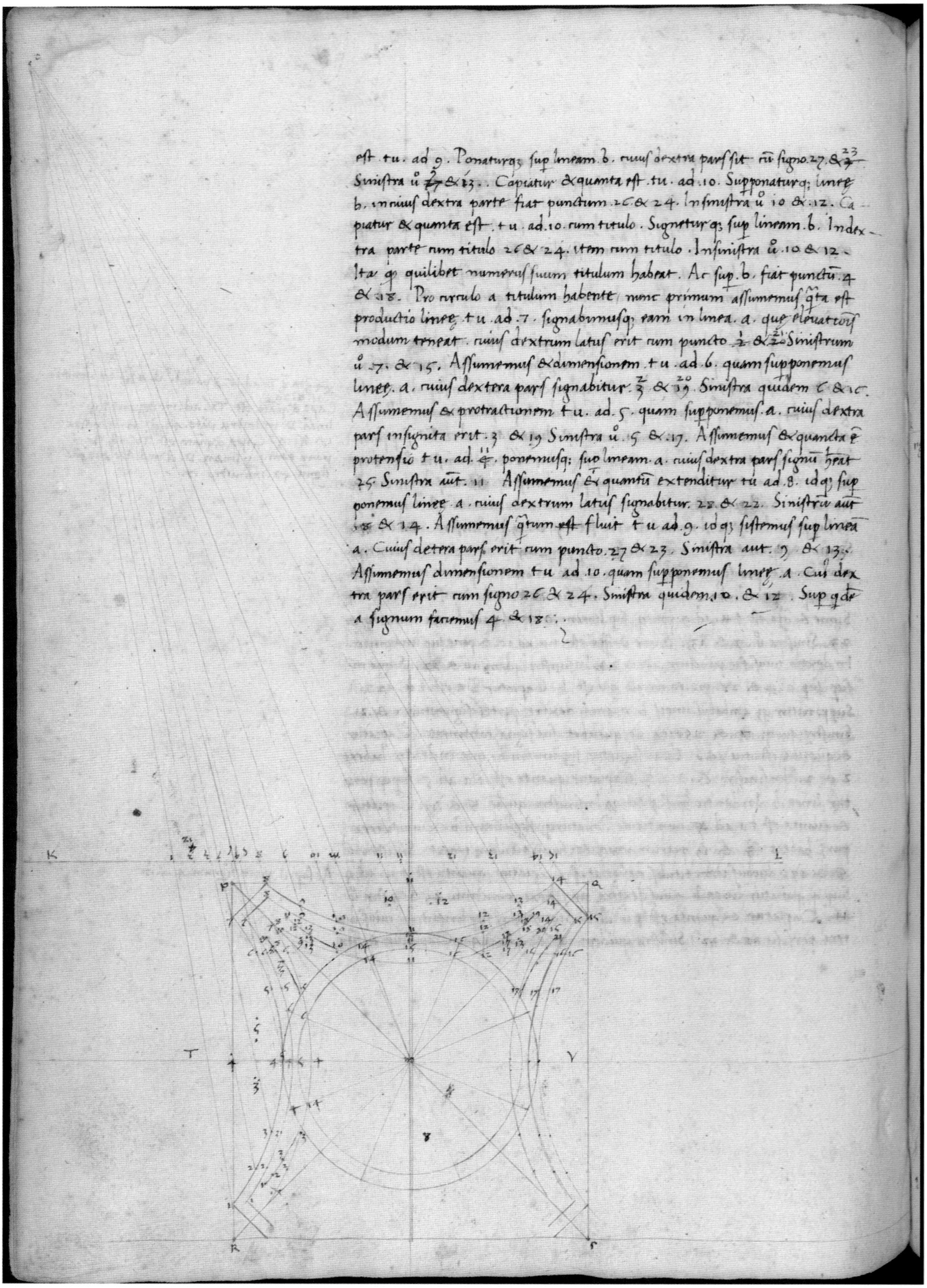

est tu. ad 9. Ponaturq; sup lineam .b. cuius dextra pars sit cu signo 27 & 23
Sinistra u 7 & 13. Capiatur & quanta est tu. ad 10. Supponaturq; linee
b. in cuius dextra parte fiat punctum 26 & 24. In sinistra u 10 & 12. Ca
piatur & quanta est tu ad 10 cum titulo. Signetur q; sup lineam .b. In dex
tra parte cum titulo 26 & 24. Item cum titulo. In sinistra u 10 & 12.
Ita q quilibet numerus suum titulum habeat. Ac sup .b. fiat punctu 4
& 18. Pro circulo a titulum habente nunc primum assumemus qta est
productio linee tu ad 7. signabimusq; eam in linea .a. que elevatiois
modum teneat. cuius dextrum latus erit cum puncto 21 & 20 Sinistrum
u 7 & 15. Assumemus & dimensionem tu ad 6. quam supponemus
linee .a. cuius dextra pars signabitur 2 & 20 Sinistra quidem 6 & 16.
Assumemus & protractionem tu ad 5. quam supponemus .a. cuius dextra
pars insignita erit 3 & 19 Sinistra u 5 & 17. Assumemus & quanta e
protensio tu ad 4. ponemusq; sup lineam .a. cuius dextra pars signu habeat
25 Sinistra aut 11 Assumemus & quantu extenditur tu ad 8. idq; sup
ponemus linee .a. cuius dextrum latus signabitur 28 & 22. Sinistru aut
8 & 14. Assumemus qtum est fluit tu ad 9. idq; sistemus sup linea
a. Cuius detera pars erit cum puncto 27 & 23. Sinistra aut 9 & 13.
Assumemus dimensionem tu ad 10. quam supponemus linee .a. Cui dex
tra pars erit cum signo 26 & 24. Sinistra quidem 10 & 12. Sup q de
a signum faciemus 4 & 18.

Bordeaux, Bibliothèque Municipale, Ms Latin 616, folio 68 verso.

it on line B′. We will have point 11 on the left and 25 on the right. Take the distance between line XY and 8 and place it on line B′. We will have point 8-14 on the left and 28-22 on the right. Take the distance between line XY and 9 and place it on line B′. We will have point 9-13 on the left and 27-23 on the right. Take the distance between line XY and 10 and place it on line B′. We will have point 10-12 on the left and 26-24 on the right. Take the distance between line XY and 10′ and place it on line B′. We will have point 10′-12′ on the left and 26′-24′ on the right. Let any number have the appropriate apex.[272] Mark 4 and 18 on line B.

Circuit A in Plan and Elevation For what concerns circuit A [figure B.87], take the distance between line XY and 7 and place it on line A, which is in elevation.[273] Mark point 7-15 on the left and 1-21 on the right. Take the distance between line XY and 6 and place it on line A. Mark point 6-16 on the left and 2-20 on the right. Take the distance between line XY and 5 and place it on line A. Mark point 5-17 on the left and 3-19 on the right. Take the distance between line XY and 4 and place it on line A. Mark point 4-18. Take the distance between line XY and 8 and place it on line A. Mark point 8-14 on the left and 28-20 on the right. Take the distance between line XY and 9 and place it on line A. Mark point 9-13 on the left and 27-23 on the right. Take the distance between line XY and 10 and place it on line A. Mark point 10-12 on the left and 26-24 on the right. We will place on line A points 25 on the right and 11 on the left.[274]

The Terminal and the Eye Point in Plan Up to here the position of all the points was indicated on the two figures in the appropriate shape.[275] We still need to execute the shortening of the figures through rulers. So first trace the line through Q and P with no end [figure B.88]. Next to this line, trace line KL parallel to PQ, as far as you would like the capital to be distant from it. Let this perpendicular line[276] be on this side of line PQ. Get as far from KL as you would like to be looking at the capital and mark point O there. Place the needle with the thread at this point, as has often been said.[277]

Wooden Ruler A Take the wooden ruler marked A and place it on KL so that it stays still [figure B.89].[278] Place the thread on 28 of circuit A: where it intercepts the ruler, mark point 28. Place the thread on 1: mark point 1 where it intercepts the ruler. Place the thread on 2: mark point 2 where it intercepts the ruler. Place the thread on 3: mark point 3 where it intercepts the ruler. Place the thread on 4: mark point 4 where it intercepts the ruler. Place the thread on 4′: mark point 4′ where it intercepts the ruler.[279] Place the thread on 5: mark point 5 where it intercepts the ruler. Place the thread on 6: mark point 6 where it intercepts the ruler. Place the thread on 7: mark point 7 where it intercepts the ruler. Place the thread on 8: mark point 8 where it intercepts the ruler. Place the thread on 9: mark point 9 where it intercepts the ruler. Place the thread on 10: mark point 10 where it intercepts the ruler. Place the thread on 11: mark point 11 where it intercepts the ruler. Place the thread on 11′: mark point 11′ where it intercepts the ruler.[280] Place the thread on 12: mark point 12 where it intercepts the ruler. Place the thread on 13: mark point 13 where it intercepts the ruler. Place the thread on 14: mark point 14 where it intercepts the ruler. Place the thread on 15: mark point 15 where it intercepts the ruler.[281] Place the thread on M: mark point M where it intercepts the ruler. Having done this, take the ruler away.

Wooden Ruler B Place ruler B[282] so that it coincides with KL [figure B.88]. Mark on it all the points that are on circuit B [figure B.90].

Wooden Ruler C Following the same procedure, do the same for circuit C by placing the points [figures B.88, B.90]. Having done this, mark M on all the rulers. These are borders relative to the cymatium. Take this ruler off too.

Wooden Ruler D Place ruler D so that it coincides with KL [figure B.88]. Lead the thread above the marks of circuit D and of the tendrils. Make a mark where the thread hits ruler D and on point M. Take the ruler off and put it aside [figure B.90].

Wooden Ruler E Take ruler E and place it so that it coincides with KL [figure B.88]. Place the thread on the marks of the tendrils and of circuit E: that is, from the side where the marks are visible [to the eye],[283] and make a mark where the thread hits the ruler and point M [figure B.90].

Wooden Rulers F, G, H, I, K Having taken this ruler off and put it aside, do the same thing for circle F on ruler F [figures B.88, B.90]. Take ruler G as well and report all the marks of circle G. Take this ruler off and in that same place, dispose ruler H so that it coincides with KL. Report on this ruler all the marks of circle H, following the method mentioned before to mark all the points where the thread touches the ruler. After ruler H you will place, through the same procedure, ruler I, as has been done up to this point for the other rulers. This will be executed also for ruler K, by reporting the marks of circle K. Up until now we have the width on the wooden rulers.

The Terminal and the Eye Point in Elevation We now must execute the procedure for the height. Trace the perpendicular line KL [figure B.91],[284] which is as far from line AK of the figure of the height as line XY of the width is from line KL which is the terminal. After, take the distance from line KL and O [in plan] and place it starting from KL in elevation. Mark point O [in elevation] below the capital as you would like it to be. Place the needle with the thread on point O.

Paper Ruler a Trace a straight line underneath the capital and let it pass through point K: let it be Kp [figure B.91].[285] Take a paper ruler so that it coincides with KL and let it start a little below Kp [of the elevation]. Do the same thing for the remaining rulers. This is ruler a [figure B.92].[286] Place the thread on 7-15 [of line A of the elevation]; where it hits the ruler, mark 7/15. Place it also on 6-16 of line [A] of the height, which must always be kept in mind. Mark point 6-16 where the thread hits the ruler. Also place it on 5-17 of line A: mark point 5-17 where it hits the ruler. Also place it on 4: mark point 4 where it hits the ruler. Also place it on 18 of line A: mark point 18 where it hits the ruler.[287] Also place it on 8-14 of line A: mark point 8-14 where it hits the ruler. Also place it on 9-13: mark point 9-13 where it hits the ruler. Also place it on 10-12 of line A: mark point 10-12 where it hits the ruler. Also place it on 11 of line A: mark point 11 where it hits the ruler. Also place it on 3 of line A: mark point 3 where it hits the ruler. Also place it on 2 of line A: mark point 2 where it hits the ruler. Also place it on 1 of line A: mark point 1 where it hits the ruler. Also place it on 4′ of line A: mark point 4′ where it hits the ruler.[288] After, take ruler a off [figure B.93] and make one similar to it. Place both aside.

Paper Rulers b, b′, c, d, e, e′, f, g, h, i, k Having done this, take another paper ruler and put it so that it coincides with KL [figure B.91]. Then trace pK, which becomes a ruler at point b. Take the thread and place it on the partitions of line B which lies on the dimension of the height. Where the thread hits the ruler, mark the points as on ruler a. Take the paper ruler off [figure B.93]. Make a similar one and put it aside. Take another paper ruler and put it on KL so that it stays still [figure B.91]. Extend pK which cuts it at point b′. Take the thread and extend it on each side of line B′ and report all points on the aforesaid ruler [b′]. After taking it off [figure B.93], make a similar one and put both aside. Take another paper ruler and place it on KL [figure B.91]. Trace pK which cuts it at point c. Take the thread and extend it on line C which lies on the representation of the height. When it singularly hits the ruler, report all the points on the ruler as has been done up until now.

After taking this ruler off [figure B.93], make a similar one and put both aside. Take another paper ruler and place it on KL [figure B.91]. Extend pK which will cut it at point d. Call this ruler d. Take the thread and extend it on all points of line D of the height. Report any marks on the ruler where the thread hits the ruler. After taking this ruler off [figure B.93], make a similar one and put both aside. Take another paper ruler and place it on KL [figure B.91]. Extend pK which cuts it at point e. Call this ruler e. Take the thread and extend it on line E which lies on the dimension of the height. Report any marks on ruler e where the thread hits the ruler. Having taken this ruler e off [figure B.93], make a similar one and put both aside. Take another paper ruler and place it on KL [figure B.91]. Extend pK which cuts it at point e′. Call this ruler e′. Take the thread and extend it on line E′ [of the height]. Report any marks on ruler e′ where the thread hits the ruler. After taking this ruler off [figure B.93], make a similar one and put both aside. Take another paper ruler and place it on KL [figure B.91]. Extend pK which cuts it at point f. Stretch the thread on all points of line F [of the height]. Make marks where the thread hits the ruler [f]. After taking this ruler off [figure B.93], make a similar one and put both aside. Take another paper ruler and place it on KL [figure B.91]. Extend pK which cuts it at point g. Call this ruler g. Take the thread to all the points of line G of the elevation and make marks where the thread hits the ruler. After taking this ruler off [figure B.93], make a similar one and put both aside.

Take another paper ruler and place it on KL [figure B.91]. Extend pK which cuts the ruler at point i. Take the thread to all the points of line I [of the elevation] and make marks on the ruler. Having taken this ruler off [figure B.93], make a similar one and put both aside. Take another paper ruler and place it on KL [figure B.91]. Extend pK which cuts the ruler at point k. Take the thread to all the points of line K. Make marks on the ruler, one by one, where the thread hits it. Take this ruler off [figure B.93] and make a similar one. In this way, through these points, we have obtained the shortenings of the heights and widths placed on the rulers [figure B.91]. Extend pK which cuts the ruler at point h. Take the thread to all the points of line H of the elevation and make marks where the thread hits the ruler. After taking this ruler off [figure B.93], make a similar one and put both aside.

Wooden Rulers or Rulers of the Width These are the rulers of the width: A, B, B′, C, D, E, E′, F, G, H, I, K [figure B.90].[289] Ruler A is marked by numbers 1, 2, 3, 4, 4′, 5, 6, 7, 8, 9, 10, 11, 11′, 12, 13, 14, 15. Ruler B is marked by the same numbers that B′ has, and these are 28, 12, 3, 3′, 5′, 5, 6, 7, 8, 9, 10, 10′, 11′, 12, 13, 14, 15. Ruler C is marked by 1, 2, 3, 4, 5, 6, 7, 8, 9, 10, 11, 12, 13, 14, 15. On ruler C there are the same marks that there are on ruler A.[290] Ruler D is marked by 2, 3, 4, 5, 6, 7, 8, 9, 10, 11, 12, 13, 14, 15, 16, 17, 18, 19, 20, 21, 22, 26.[291] Ruler E is marked by 1, 2, 3, 4, 5, 6, 7, 8, 9, 10, 11, 12, 13, 14, 15, 16, 17, 18, 19, 20, 21, 22, 23, 24, 25. Ruler E′ is marked by 1, 2, 3, 4, 5, 6, 7, 8, 9, 10, 11, 12, 13, 14, 15, 16, 17, 18, 19, 20.[292] Ruler F is marked by 2, 3, 4, 5, 6, 7, 8, 9. Ruler G is marked by 1, 2, 3, 4, 5, 6. Ruler H is marked by 1, 2, 3, 4, 5, 6, 7, 8. Ruler I is marked by 1, 2, 3, 4, 5.[293] Ruler K is marked by 2, 3, 4, 5, 6, 7, 8, 9, 10.[294] We have the wooden rulers which contain the extension of the width.

Paper Rulers or Rulers of the Height The paper rulers are the following [figure B.93]. Two are marked by letter a, two by letter b, two by letter b′, two by letter c, two by letter d, two by letter e, two by letter e′,[295] two by letter f, two by letter g, two by letter h, two by letter i, two by letter k. Ruler a is marked by 7, 1, 6, 2, 5, 3, 4, 4′, 8, 28, 9, 10, 11, 11′, 12, 13, 14, 15.[296] Paper ruler B is marked like this: 8, 14, 9, 13, 7, 15, 10, 12, 11, 6, 5, 4, 3, 2, 1.[297] Paper ruler b′ is marked like this: 7, 1, 6, 2, 5, 3, 3′, 4, 5′, 8, 28, 9, 10, 10′, 11′, 12′,

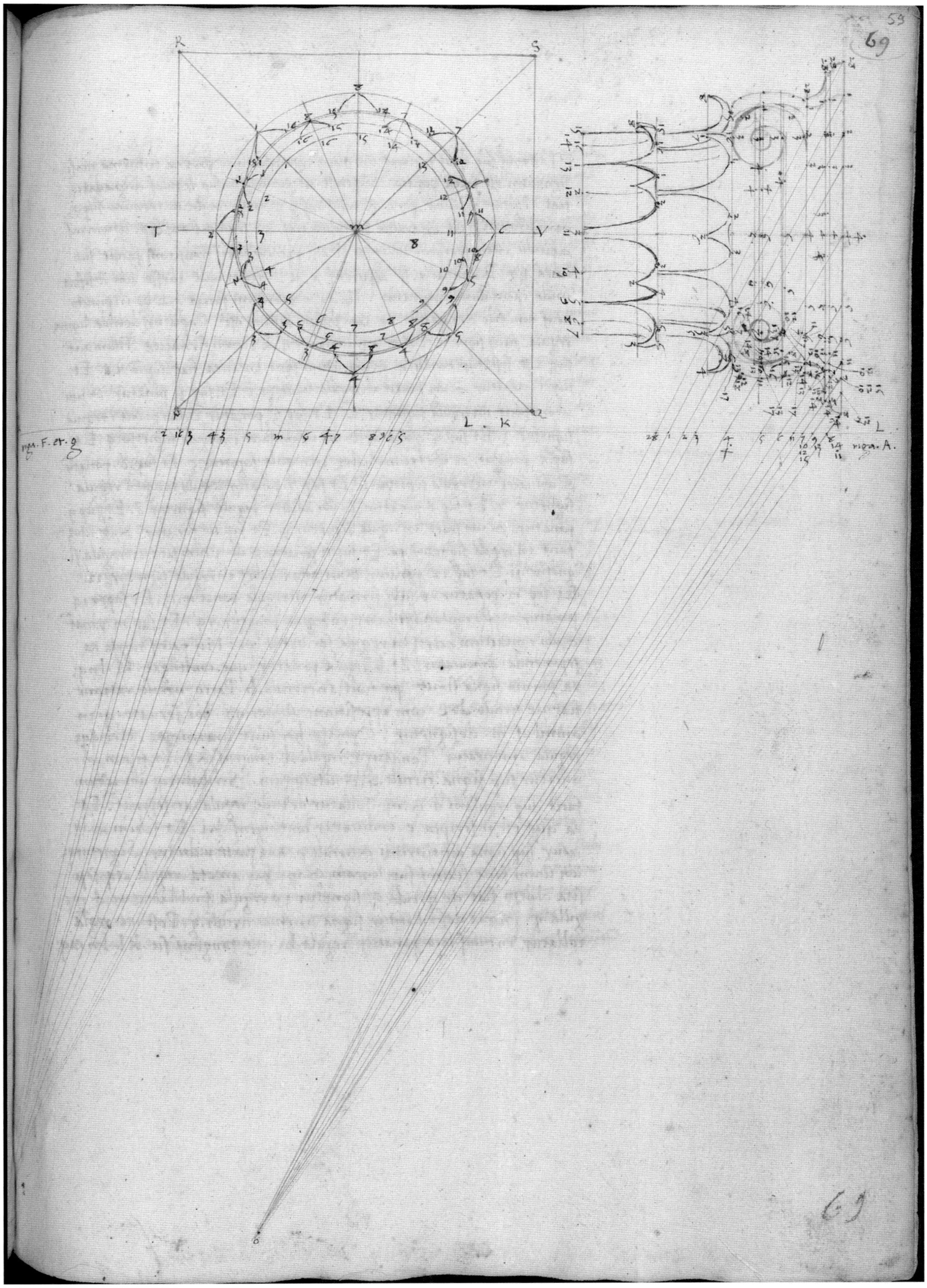

Bordeaux, Bibliothèque Municipale, Ms Latin 616, folio 69 recto.

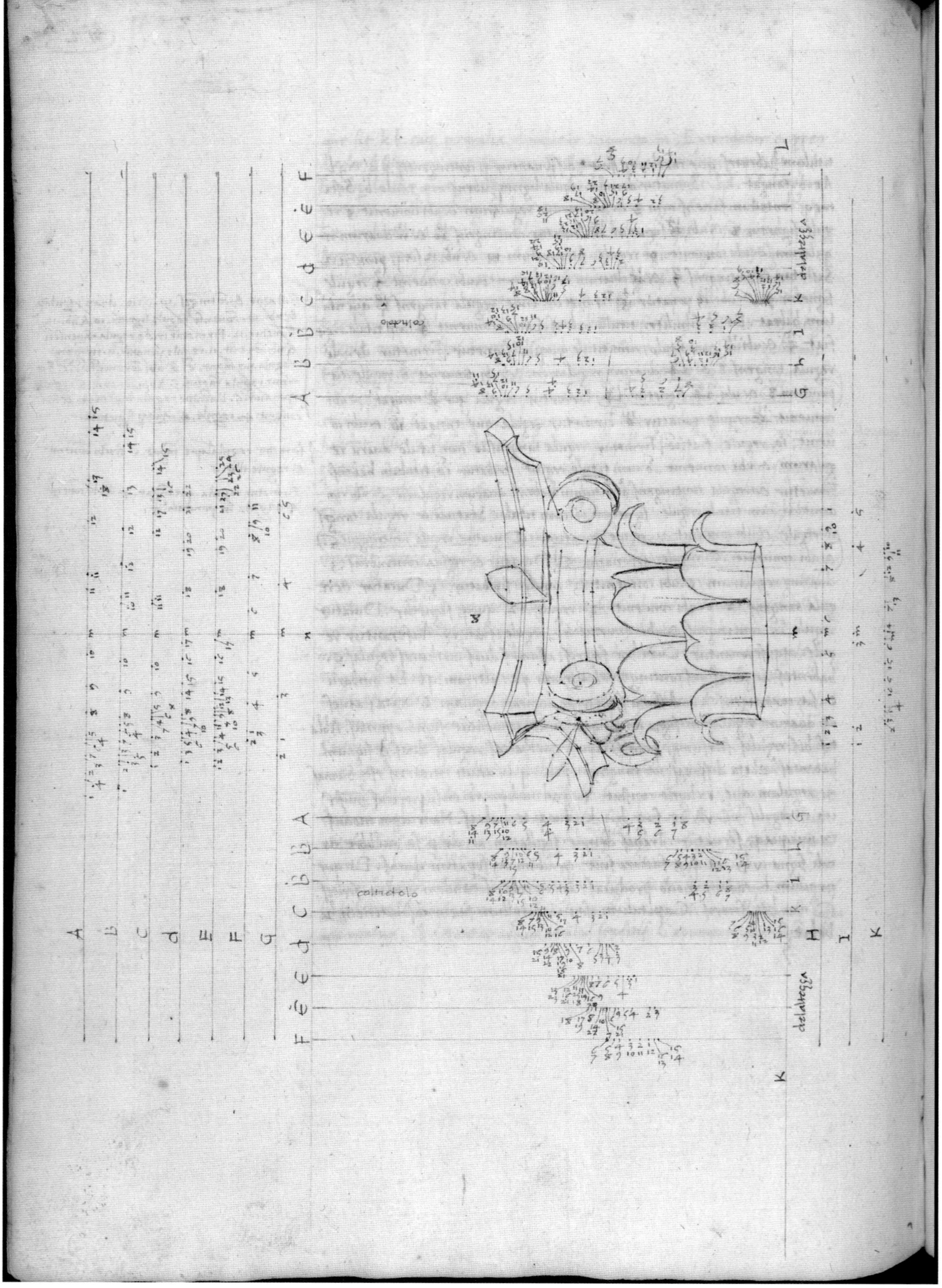

Bordeaux, Bibliothèque Municipale, Ms Latin 616, folio 72 verso.

12, 13, 14, 15. Paper ruler c is marked like this: 7, 1, 6, 2, 5, 3, 4, 8, 28, 9, 10, 11, 12, 13, 14, 15. Paper ruler d is marked like this: 2, 10, 3, 9, 4, 8, 5, 7, 6, 1, 11, 12, 13, 14, 15, 16, 17, 18, 19, 20, 21, 22. Paper ruler e is marked like this: 2, 10, 3, 9, 4, 8, 5, 7, 6, 1, 11, 13, 14, 15, 16, 17, 18, 19, 20, 21, 22, 23, 24. Paper ruler e′ is marked like this: 2, 10, 3, 9, 4, 8, 5, 7, 6, 1, 11, 12, 13, 14, 15, 16, 17, 18, 19, 20, 21, 22, 23, 24. Paper ruler f is marked like this: 2, 3, 1, 4, 5, 16, 6, 7, 8. Paper ruler g is marked like this: 2, 1, 3, 4, 5. Paper ruler h is marked like this: 2, 3, 1, 4, 5, 6, 7, 8. Paper ruler i is marked like this: 1, 2, 3, 8, 4. Paper ruler k is marked like this: 16, 1, 2, 3, 4, 5, 6, 7, 8.[298]

The Shortening of Circuit A Having shown the rulers, they only have to be arranged in the realization of the work. They will be placed where the future position of the capital will take place. In this place trace straight line KL [figure B.94], which will be divided into two equal parts at point M. Trace perpendicular line MN through M. Trace perpendicular line KP through K and perpendicular line LQ through L. Then take the two paper rulers [marked with a]: place one on KP and the other on LQ. Let point a of both lines touch line KL. Let both lines be still, fixed with wax. Now take wooden ruler A and place it on the two paper rulers in contact with points 1 and 7 of the two rulers while point M is on line MN. This way of arranging rulers must always be respected. That is, let the paper rulers be on lines KP and LQ; let the marks on the rulers touch line KL; and also point M, which is reported on the wooden ruler, touches MN. When the wooden ruler is touching points 1 and 7, wherever there is point 1 of the wooden ruler, we will mark 1. Wherever there is point 7, we will mark 7.

Moreover, keep in mind that the marks are placed [on the terminal] when the numbers of the wooden ruler contact the corresponding points of the paper rulers which the wooden ruler slightly touches. Thus place the ruler so that it goes through 6 and 2; where 6 of the wooden ruler is, mark 6. Where 2 is, mark 2. Place the ruler so that it goes through 5 and 3; where 5 of the wooden ruler is, mark 5. Where 3 is, mark 3. Place the ruler so that it goes through points 4 and 4′ of the two rulers; where 4 of the wooden ruler is, mark 4. Where 4′ is, mark 4′. Place the ruler so that it goes through 8 and 28 of the two rulers; where 8 of the wooden ruler is, mark 8. Where 28 is, mark 28. Place the ruler so that it goes through 9 of the two rulers; where 9 of the ruler is, mark 9. Place the ruler so that it goes through 10 of the two rulers; where 10 of the ruler is, mark 10. Place the ruler so that it goes through 11 of the two rulers; where 11 of the ruler is, mark 11. Place the ruler so that it goes through 12; where 12 of the ruler is, mark 12. Place the ruler so that it goes through 13; where 13 of the ruler is, mark 13. Place the ruler so that it goes through 14; where 14 of the ruler is, mark 14. Place the ruler so that it goes through 15 of the two rulers; where 15 of the ruler is, mark 15.

The Shortening of Circuit B Having taken the two rulers off and put them aside, take the two paper rulers with mark b and arrange them so that they touch KP and LQ as we said before, with letter b of both coinciding with KL [figure B.95]. Take wooden ruler B and block it so that it touches the single marks of the two rulers: make marks where they correspond with the ones of the wooden ruler. Take the rulers off and put them aside.[299]

The Shortening of Circuit B′ Take two paper rulers marked b′ and place them on KP and LQ. b′ of both rulers touches KL [figure B.96]. Take wooden ruler B′,[300] make it touch 1 and 7 of both rulers. Where 1 of the wooden ruler is, mark 1; where 7 is, mark 7. Make it touch 2 and 6 of both rulers. Where 2 is, mark 2; where 6 is, mark 6. Make it touch 3 and 5 of both rulers. Where 3 is, mark 3; where 5 is, mark 5. Make it touch 3′. Make it touch 5′.[301] Where 3′ of the ruler is, mark 3′; where 5′ is, mark 5′.[302] Make it touch 8 and 28 of both rulers; where 8 of the ruler is, mark 8. Where 28 is, mark 28. Make it touch 9. Where 9 is, mark 9. Make it touch 10. Where 10 is, mark 10. Make it touch 10′ of both rulers. Where 10′ is, mark 10′. Make it touch 12′ of both rulers: where 12′ of the ruler is, mark 12′.[303] Make it touch with 11 of the ruler. Where 11 of the ruler is, mark 11. Make it touch 12. Where 12 of the ruler is, mark 12. Make it touch 13 of the two rulers. Where 13 of the ruler is, mark 13. Make it touch 14 of the two rulers. Where 14 of the ruler is, mark 14. Make it touch 15. Where 15 of the ruler is, mark 15.

The Shortening of Circuit C Take the rulers off and put them aside. Having done this, take the two paper rulers marked c [figure B.97][304] and place one on KP and the other one on LQ; let point c of both rulers be on KL. Take ruler C and arrange it so that it touches the marks of the two paper rulers. Mark the signs where they are.

The Shortening of Circuits D, E, E′, F, G, H, I, K Take the rulers off and put them aside. Having done this, take the two paper rulers marked D [figure B.97] and place one on KP and the other on LQ as you have done for the preceding ones. Take a wooden ruler again and, as you have done for the preceding ones, you will do the same for the rulers you still have, and these are E, E′, F, G, H, I, K.[305] In each of these in fact the procedure will remain the same. Place, one by one, the marks of each circuit like the ones that form circuit A: bring them one by one. Mark them in the same way as you did for circuit B, and you will observe the same procedure for the remaining circuits. Having done this, the little head [of the column], that is, the capital, will be completed [figures B.98, B.99, B.100].[306]

Proposition VIIIa

In the preestablished terminal, assigned with the point of the eye, shorten a head in a proportional way.

Two Profiles of a Head Since we have stated, at the beginning of this book, that one must represent what will be drawn in real shape, we will then draw a head so that it is only showing one eye [figure B.101], with the profile you believe has to be shortened. Together with this profile, build another one that shows the face integrated with two eyes. This [second profile] has to have the same height of the first one and all the parts must correspond [to one another]. First, trace a straight line from the end of the head with one eye [figure B.102]. This line must touch the end of the head that has two eyes. Moreover, trace a second line, which is in the middle between the line that is above and the part above the forehead: let this be line A. Trace another line on the forehead: let this be line B. Trace another line between the forehead and nose which goes through the eye: mark it line C. Trace another line that goes through the bump on the nose: mark it line D. Trace another line that goes through the lower end of the nose: mark it line E. Trace another line through the mouth, mark it F. Trace another line between mouth and chin, mark it G. Trace line H through the extreme lower part of the chin. Let all these lines be drawn parallel to the line that touches the upper ends of both heads. Let these lines go through the same parts of the heads, through the head that shows one eye only as well as the other head that shows both eyes.[307]

Vertical Lines After, trace a line so that it touches the tip of the nose of the head with one eye and so that it is at a right angle with respect to line A: let this line be drawn with no limits [figure B.103]. Trace two more lines so that the first one touches the upper lip and the second one touches the lower lip.[308] Trace another line which goes through the bump of the nose, one that aligns with the nostrils,[309] one that aligns with the chin, one at the end of the mouth, one at the end of the forehead, another one that touches line A above the forehead, one at the beginning of the eyes, another one in the middle, another one at the end of the eyes, another one at the beginning of the throat. Let all these lines be parallel to the line that goes through the end of the nose.[310] Trace another line directly at the beginning of the ear, another one at the end of the ear, another one at the end of the jaw.[311] Besides these lines, trace a line from the contact of line A with the back of the head.

Trace another line from the contact of line B, trace another line from the contact of line C, trace another line from the contact of line D, trace another line from the contact of line E, trace another line at the end of the back of the head. Let all these lines be parallel to one another as shown in the drawing. Having done this, draw a line perpendicular to all of them that will split each one of them at point 1. This perpendicular line is MN.

How to Draw Circuit A in Real Shape Divide the head in half, which shows the entire face [figure B.104]; obtain line TV by subdividing into two parts lines A, B, C, D, E, F, G, H and by placing number 1 on all of them. Take a compass and place one of the two ends on line TV. Extend the other end until it touches line A on the outline of the head which is showing the entire face. Place with this width the fixed end of the compass on point 1 of the line which is originating from the contact—above line A—of the top of the head which is showing one eye only.[312] With the other end of the compass, mark point 13 on the line in the direction of the head, mark point 5 below.

Successively, build the outline that goes through those numbers and lines that originate from the contact of line A in the back and in the front: this will be circuit A.

How to Draw Circuit B in Real Shape Place the end of the compass on point 1 of line B which is found on line TV [figure B.105]. Extend the other end until the contact of line B with the end of the forehead. Place this width on the line that comes from the contact—above line B—on the top of the head.[313] Mark point 13 on the line in the direction of the head, mark point 5 below. Build the outline, as mentioned earlier, which goes through the marks of the lines: this is circuit B.

How to Draw Circuit C in Real Shape Place the end of the compass on point 1 of line C [of the head which shows the entire face] [figure B.106]. Extend the other end till the thickness of the nose. Then place the end of the compass on the [vertical] line that begins at the thickness of the nose: mark 22,[314] below mark 2. Then place the end of the compass on point 1 of line C and extend the other end to the beginning of the eye. Place the end of the compass on point 1 of the [vertical] line that originates at the beginning of the eye: from the upper part, toward the head, mark 21,[315] below mark 3. Place the end of the compass on point 1 of line C. Extend the other end toward the middle of the eye and with this quantity fix the point of the compass on point 1 of the [vertical] line that originates from the middle of the eye. Toward the head mark 20,[316] below mark 4. Place the end of the compass on point 1 of line C. Extend the other end toward the extremity of the eye. Go back with one of the two ends of the compass on point 1 of the [vertical] line coming from the extremity of the eye. With the other end mark point 19 on the side of the head and below [mark] point 5. Place the end of the compass on point 1 of line C. Extend the other end to the beginning of the ear.[317] Apply this extension method to the line that comes from the top of the head.[318] Mark 16 on the side of the head and mark 8 below. Place the end of the compass on point 1 of line C; extend the other end toward the [internal] extremity of the ear. Place the other end of the compass on point 1 of the line coming from the [upper] end of the ear. With the other end mark point 15 toward the head and below [mark] point 9.[319] Successively draw the outline which will be circuit C.

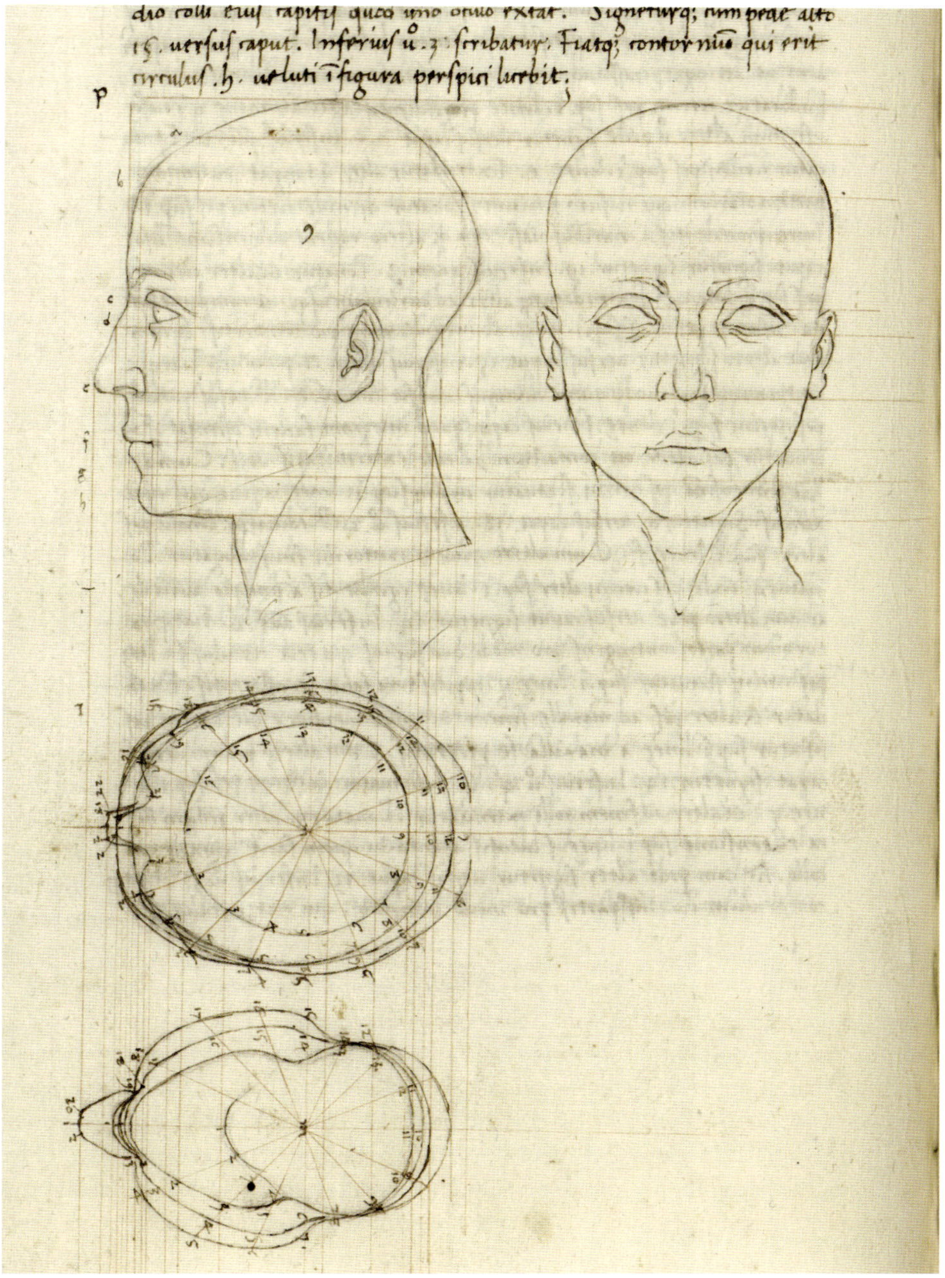

Bordeaux, Bibliothèque Municipale, Ms Latin 616, folio 74 verso.

How to Draw Circuit D in Real Shape Place the end of the compass on point 1 of line D of the head that appears with the entire face [figure B.107]. Extend the other end toward the thickness of the nose. Place the other end of the compass on point 1 of the [vertical] line coming from the beginning of the prominence of the nose. Mark point 20 toward the head and 2 below.[320] Place the end of the compass on point 1 of line D. Extend the other end till the contact with the beginning of the ear. With this quantity place the end of the compass on point 1 of the [vertical] line coming from the top of the head.[321] With the second end mark point 13 toward the head and 7 below.[322] Draw the outline which goes through the points and the lines that originate at the contact of line D: this will be circuit D.

How to Draw Circuit E in Real Shape Go back with the tip of the compass [placed] on point 1 of line E of the head which shows the entire face [figure B.108]. Extend the other end till there is contact with the thickness of the tip of the nose. With this quantity, place the end of the compass on point 1 of the line which originates from the point of the nose. With the other point mark 20 on the side of the head and 2 below.[323] Place the end of the compass on point 1 of line E. Extend the other end till there is contact with the end of the nostrils which delimit the nose. Place the end of the compass on point 1 of the line which originates from the nostrils of the nose. With the other end, with the same quantity, mark 19 on the head's side and below mark 3. Place one end of the compass on point 1 of line E. Extend the other end till the ear, and with this dimension place one end on point 1 of the line that comes from the top of the head. With the other end mark 15 on the side of the head and 5 below. Draw the outline as mentioned earlier. This will be circuit E.

How to Draw Circuit F in Real Shape Place the end of the compass again on point 1 of line F of the head which is showing the entire face [figure B.109]. Extend the other end until it touches one end of the mouth. With this dimension place the end of the compass on point 1 of the line originating at the end of the mouth. With the other end mark 18 on the side of the head and 2 below. Place one end of the compass on point 1 of line F. With the other end touch the end of the outline of the face. Place one of the two ends of the compass on point 1 of the line that comes from the top of the head.[324] With the other end mark 14 on the side of the head and 6 below. Draw the outline that joins the parts according to the way we have said. Let this be circuit F.

How to Draw Circuit G in Real Shape In the same way place the end of the compass on point 1 of line G of the head which is showing the whole face [figure B.110]. Extend the other end until it touches the end of the jaw. With this dimension place the end of the compass on point 1 of the line running from the jaw. With the other end mark 14 on the side of the head and 4 below. Place one end of the compass on point 1 of line G. With the other end touch the end of the neck. Place one of the ends of the compass on point 1 of the line which originates from the contact of line G with the neck. With the other end mark 13 on the side of the head and 5 below. Draw the outline that joins the parts according to the method we have said. Let this be circuit G.

How to Draw Circuit H in Real Shape Place the end of the compass on point 1 of line H and extend the other end until it touches the thickness of the chin [figure B.111]. Place one of the two ends of the compass on point 1 of the line originating at the throat and with the other end mark 16 on the side of the head and 2 below.[325] Place one end of the compass on point 1 of line H. With the other end touch the end of the neck. Place one of the two ends of the compass on point 1 of the line that originates from the middle of the neck[326] of the head that appears to have only one eye. With the other end mark 13 on the side of the head and 5 below. Draw the outline that corresponds to circuit H, as you can see in the picture.

The Subdivisions of the Circuits However, as we have mentioned in the beginning of the first book and this book, it is necessary that one knows the procedure of drawing in real shape, that is, to draw the extensions of the outlines so that they correspond to line A, to line B, to line C, to line D, to line E, to line F, to line G, to line H [figures B.112, B.113]. This means: as if these lines had each their own circuit,[327] as if in those positions we execute a cut [of the head]. Although one may execute multiple circuits, these may be enough. When these [circuits] have been well understood, one may introduce a multitude of circuits at one's own will.[328] If the intricate positioning of these circuits dissuades someone from executing them, he can execute them two or three at a time, at will, provided that he executes them using the same dimensions and distances with which the circuits were executed earlier. Take the center now, which corresponds to 1 and which can be found on the line that originates at the top of the head: mark point M.[329] Divide circuit A into sixteen proportionate parts.[330] The principle of subdivision begins at 1 which is on the mentioned circuit, successively marking numbers 1, 2, 3, 4, 5, 6, 7, 8, 9, 10, 11, 12, 13, 14, 15, 16. After doing this, trace the lines toward the points of divisions from M. These lines subdivide all the circuits A, B, C, D, E, F, G, H. Add six more lines to circuit C so as to have twenty-two divisions. Add four more lines to circuit D so as to have twenty divisions.[331] Add four more lines to circuit E so as to have twenty divisions. Add two more lines to circuit F so as to have eighteen divisions. In these subdivisions one will also mark the same numbers in the corresponding places: starting from these take the width and draw the elevation in the following manner.[332]

The Numbers of the Circuits on the Head Which Has One Eye Only
Circuit A and Line A Of course, place one of the two ends of a compass on the line that touches the point of the nose [in the circuit] of the head in width. Let l be this line [figure B.114]. Extend one of the two ends of the compass to point 2 of circuit A in width. When we say circuit, one must always understand the circuit of the figure in width. When one says line, it must be understood as that of the head which has one eye only, until it is otherwise specified. Then place a compass end on line A in this line's contact with line P, which is the one that touches the tip of the nose of the head with one eye only.[333] With the other end of the compass mark points 2 and 16 on line A.[334] Go back to the width. Place the end of the compass on line l. Place the other end on point 3 of circuit A. Place one of the two ends of the compass on line P and with the other end mark points 3 and 15 on line A. Place the end of the compass on line I. Place the other end on point 4 of circuit A. Place one of the two ends of the compass on line P and with the other end mark points 4 and 14 on line A. Place the end of the compass on line I. Extend the other end to point 5. With this extension place the compass on line A starting from P and mark points 5 and 13. Successively, take the quantity between lines l and 6. Place one of the two ends of the compass on P and mark points 6 and 12 on line

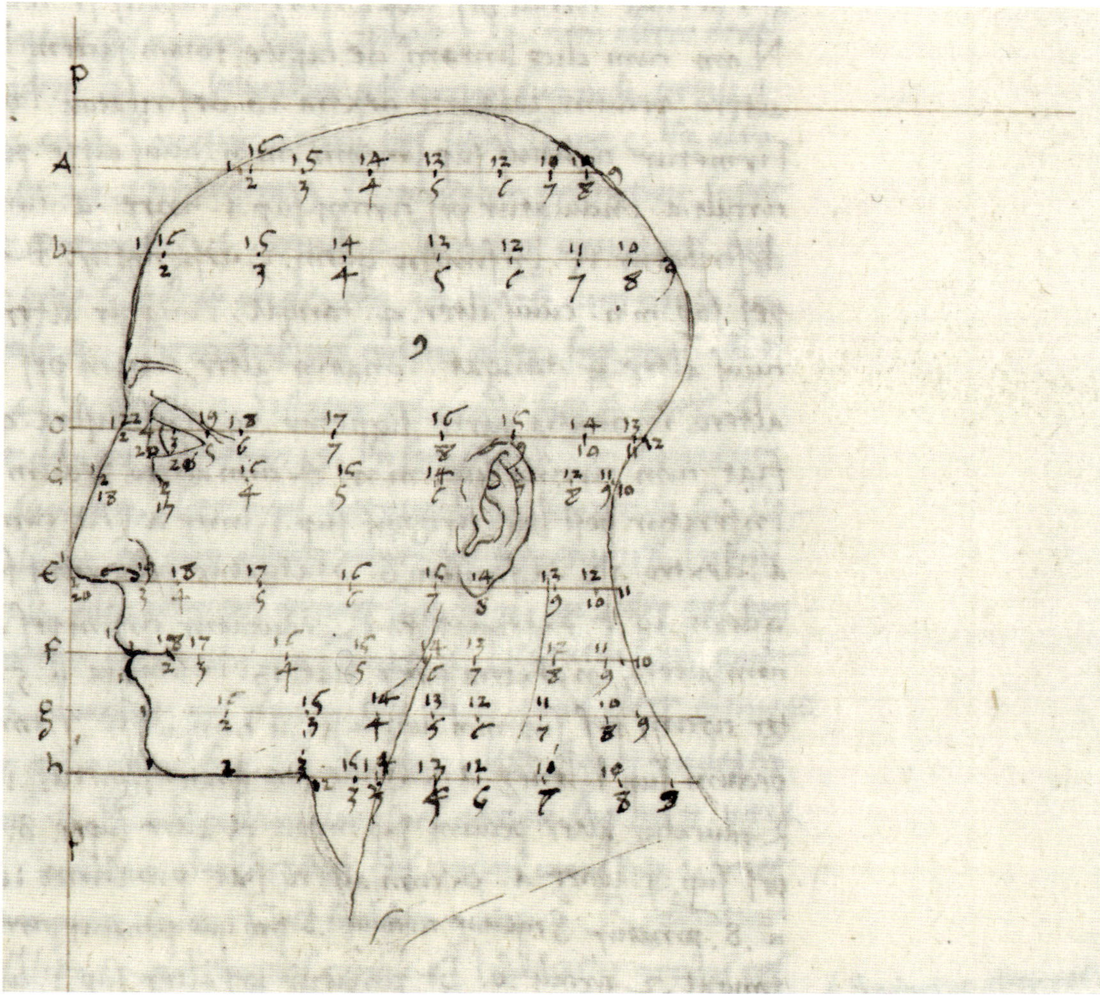

Bordeaux, Bibliothèque Municipale, Ms Latin 616, folio 76 recto.

A. Take the quantity between lines l and 7 of circuit A. Place one of the two ends of the compass on P and mark points 7 and 11 on line A. Take the quantity between lines l and 8 of circuit A. Place one of the two ends of the compass on P and mark points 8 and 10 on line A. Take the quantity between lines l and 9 of circuit A. Place one of the two ends of the compass on P and mark point 9 on line A.

Circuit B and Line B Place one of the compass's ends on line l and extend the other end to point 2 of circuit B [figure B.115]. Then place one of the ends of the compass on line P and with the other end mark points 2 and 16 on line B. Take the length between lines l and 3 of circuit B and place it on line B starting from line P: mark 3 and 15. Place one of the compass's ends on line l and extend the other end to point 4 of circuit B. Then place one of the ends of the compass on line P and with the other end mark points 4 and 14 on line B. Take the length between lines l and 5 of circuit B and place it on line B starting from line P: mark 5 and 13. Take the length between lines l and 6 of circuit B and place it on line B starting from line P: mark 6 and 12. Take the length between lines l and 7 of circuit B and place it on line B starting from line P: mark 7 and 11. Take the length between lines l and 8 of circuit B and place it on line B starting from line P: mark 8 and 10. Take the length between lines l and 9 of circuit B and place it on line B starting from line P: mark 9.

Circuit C and Line C Take the length between lines l and 2 of circuit C [figure B.116]. Place one end of the compass on line P and with the other end mark points 2 and 22 on line C. Do the same for all of circuit C till point 12. You shall have points 1, 2 and 22, 4 and 20, 3 and 21, 5 and 19, 6 and 18, 7 and 17, 8 and 16, 9 and 15, 10 and 14, 11 and 13, 12, 4′ and 20′, 3′ and 21′, 5′ and 19′ on line C.[335]

Circuit D and Line D The same procedure must be followed for circuit D [figure B.117], which will be on line D with the following numbers: 1, 2 and 20, 3 and 19, 4 and 18, 5 and 17, 6 and 16, 7 and 15, 8 and 14, 9 and 13, 10 and 12, 11.

Circuit E and Line E If you observe the same procedure for line E [figure B.117], these numbers will appear: 1′, 2 and 20, 3 and 19, 4 and 18, 5 and 17, 6 and 16, 7 and 15, 8 and 14, 9 and 13, 10 and 12, 11.[336]

Circuit F and Line F If you observe the same procedure for line F [figure B.117], these numbers will appear: 1, 1′, 1″, 2 and 18, 3 and 17, 4 and 16, 5 and 15, 6 and 14, 7 and 13, 8 and 12, 9 and 11, 10.[337]

Circuit G and Line G The line that has been marked with letter G [figure B.117] will have the following numbers: 1, 2 and 16, 3 and 15, 4 and 14, 5 and 13, 6 and 12, 7 and 11, 8 and 10, 9.

Circuit H and Line H The line that has been marked with letter H [figure B.117] will have the following numbers: 1, 2 and 16, 3 and 15, 4 and 14, 5 and 13, 6 and 12, 7 and 11, 8 and 10, 9, 1′, 2′ and 16′, 3′ and 15′.[338]

The Numbers of the Circuits on the Head Showing the Whole Face

Circuit A and Line A Now we must place these quantities on the head showing the whole face [figure B.118]. This head is split in half by line TV, which divides all the lines A, B, C, D, E, F, G, H at point 1. Line MN is perpendicular to line l and it divides all the circuits of the width at point 1. Now place one of the two compass's ends on line MN and extend the other until it touches point 2 on circuit A. Place the compass's end on point 1 of line A of the head which shows the entire face. When I say line A I mean the one on the head which is showing the entire face. With the other end mark 2 on the left and 16 on the right.[339] Place the compass's end on line MN. Extend the other end until it touches point 3 of circuit A. Place the compass's end on point 1 of line A and with the other end mark point 15 on the right and point 3 on the left. Place the compass's end on line MN till it touches point 4 of circuit A. Place the compass's end on point 1 of line A and with the other end mark point 14 on the right and point 4 on the left. Take the length from line MN to point 5 of circuit A. Place the compass's end on point 1 of line A and with the other end mark point 13 on the right and point 5 on the left. Take the length from line MN to point 6 of circuit A. Place the compass's end on point 1 of line A and with the other end mark point 12 on the right and point 6 on the left. Take the length from line MN to point 7 of circuit A. Place the compass's end on point 1 of line A and with the other end mark point 11 on the right and point 7 on the left. Take the length from line MN to point 8 of circuit A. Place the compass's end on point 1 of line A and with the other end mark point 10 on the right and point 8 on the left. Mark number 9 where you see number 1.[340]

Circuit B and Line B Next we have circuit B. Place the compass's end on line MN; the other end will touch point 2 of circuit B [figure B.119]. One of the two ends will touch point 1 of line B, the other end will mark point 16 on the right and point 2 on the left. Place the compass's end on line MN. The other end will touch point 3 of circuit B. One of the two ends will touch point 1 of line B, the other end will mark point 15 on the right and point 3 on the left. Place the compass's end on line MN. The other end will touch point 4 of circuit B. One of the two ends will touch point 1 of line B, the other end will mark point 14 on the right and point 4 on the left. Place the compass's end on line MN. The other end will touch point 5 of circuit B. One of the two ends will touch point 1 of line B, the other end will mark point 13 on the right and point 5 on the left. Place the compass's end on line MN. The other end will touch point 6 of circuit B. One of the two ends will touch point 1 of line B, the other end will mark point 12 on the right and point 6 on the left. Place the compass's end on line MN. The other end will touch point 7 of circuit B. One of the two ends will touch point 1 of line B, the other end will mark point 11 on the right and point 7 on the left. Place the compass's end on line MN. The other end will touch point 8 of circuit B. One of the two ends will touch point 1 of line B, the other end will mark point 10 on the right and point 8 on the left. Mark number 9 where you see number 1.[341]

Circuit C and Line C Next we have circuit C, and the procedure does not change [figure B.120]. Place the compass's end on line MN, the other end will touch point 2 of circuit C. One of the two ends will touch point 1 of line C, the other end will mark point 22 on the right and point 2 on the left. Place the compass's end on line MN of circuit C. The other end will touch point 3 of circuit C. One of the two ends will touch point 1 of line C, the other end will mark point 21 on the right and point 3 on the left. Place the compass's end on line MN of circuit C. The other end will touch point 4 of circuit C. One of the two ends will touch point 1 of line C, the other end will mark point 20 and point 20′ on the right and points 4 and 4′ on the left.[342] Place the compass's end on line MN of circuit C. The other end will touch point 5 of circuit C. One of the two ends of the compass will touch point 1 of line C, the other end will mark point 19 on the right and point 5 on the left. Place the compass's end on line MN of circuit C. The other end will touch point 6 of circuit C. One of the two ends of the compass will touch point 1 of line C, the other end will mark point 18 on the right and point 6 on the left. Place the compass's end on line MN of circuit C. The other end will touch point 7 of circuit C. One of the two ends of the compass will touch point 1 of line C, the other end will mark point 17 on the right and point 7 on the left. Place the compass's end on line MN of circuit C. The other end will touch point 8 of circuit C. One of the two ends of the compass will touch point 1 of line C, the other end will mark point 16 on the right and point 8 on the left. Place the compass's end on line MN of circuit C. The other end will touch point 9 of circuit C. One of the two

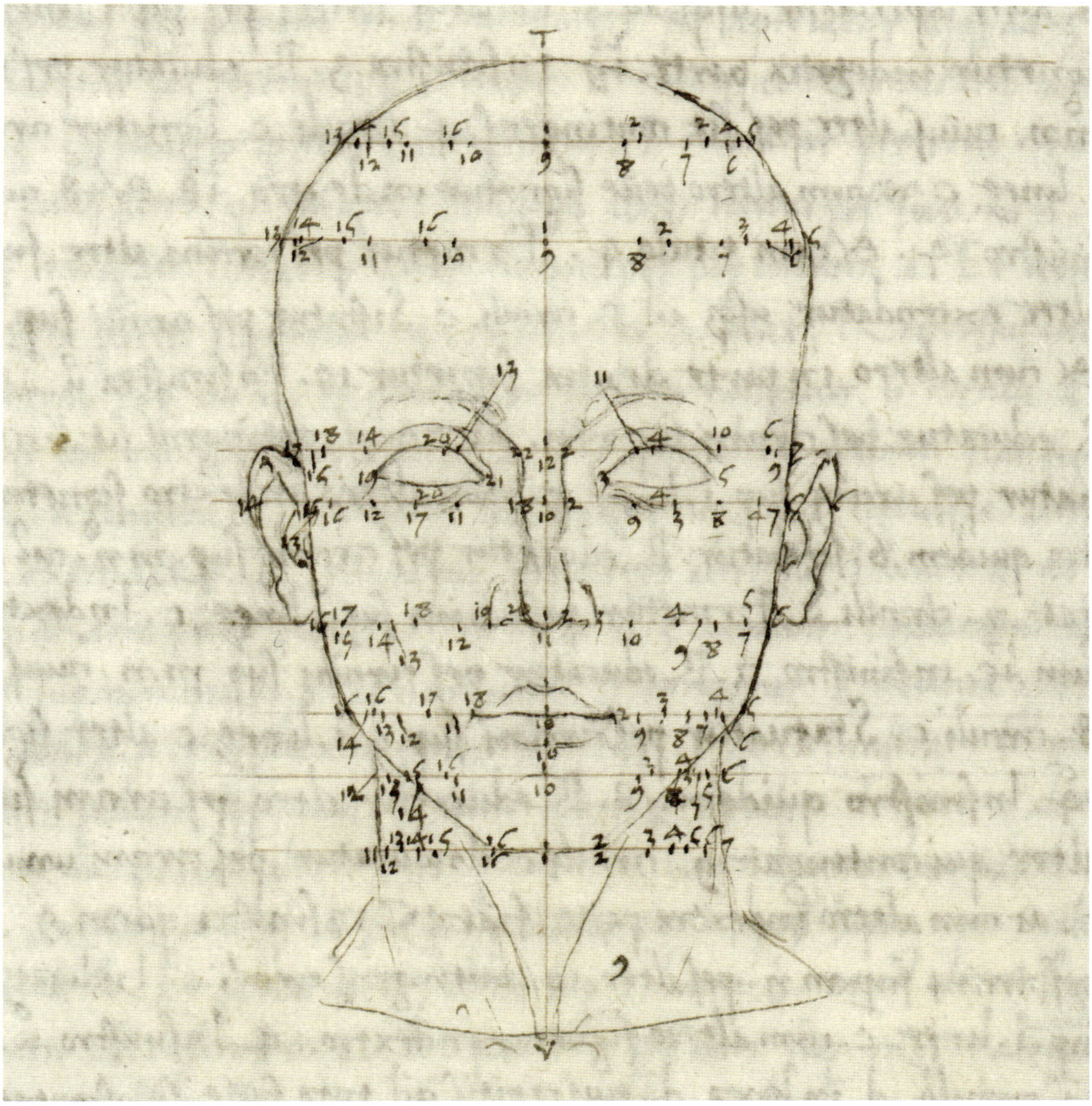

Bordeaux, Bibliothèque Municipale, Ms Latin 616, folio 77 verso.

ends of the compass will touch point 1 of line C, the other end will mark point 15 on the right and point 9 on the left. Place the compass's end on line MN of circuit C. The other end will touch point 10 of circuit C. One of the two ends of the compass will touch point 1 of line C, the other end will mark point 14 on the right and point 10 on the left. Place the compass's end on line MN of circuit C. The other end will touch point 11 of circuit C. One of the two ends of the compass will touch point 1 of line C, the other end will mark point 13 on the right and point 11 on the left. Mark number 12 where you see number 1.[343]

Circuit D and Line D Proceed in the same manner also for circuit D [figure B.121]. On line D of the head showing the whole face you shall have the following numbers: 1, 2, 3, 4, 5, 6, 7, 8, 9, 10, 11, 12, 13, 14, 15, 16, 17, 18, 19, 20.[344]

Circuit E and Line E Proceed in the same manner for circuit E and line E [figure B.121], which will have the following numbers: 1, 1′, 2, 3, 4, 5, 6, 7, 8, 9, 10, 11, 12, 13, 14, 15, 16, 17, 18, 19, 20.[345]

Circuit F and Line F Proceed in the same manner for circuit F and line F [figure B.121], which will have the following numbers: 1, 1′, 1″, 2, 3, 4, 5, 6, 7, 8, 9, 10, 11, 12, 13, 14, 15, 16, 17, 18.[346]

Circuit G and Line G Proceed in the same manner for circuit G and line G [figure B.121], which will have the following numbers: 1, 2, 3, 4, 5, 6, 7, 8, 9, 10, 11, 12, 13, 14, 15, 16.[347]

Circuit H and Line H Proceed in the same manner for circuit H and line H of the head which shows the whole face [figure B.121], which will have the following numbers: 1, 2 and 16,

3 and 15, 4 and 14, 5 and 13, 6 and 12, 7 and 11, 8 and 10, 9, 1′, 2′ and 16′, 3′ and 15′.[348]

In these conditions we will have completed the representation of the figures which have a real shape [figure B.122].[349]

The Choice of the Point of View in Plan We need to report these numbers on the rulers. Proceed first in width in the way you are used to doing [figure B.123]. Trace perpendicular line KL which is parallel to IP. Line KL will be the terminal in which we will place the wooden rulers so that they coincide with itself. Get as far as you like to see the head from a distance, and in that place mark point O where you will also place the needle with the thread.[350]

Wooden Ruler A Then take a thin wooden ruler and place it so as to make it coincide with line KL.[351] Take the thread and place it so that it touches the points on circuit A [figure B.124]. In this way, let the thread come in contact with point 14; see where it touches the ruler and mark 14. Place the thread on point 15 of circuit A, mark 15 where it touches the ruler. Place the thread on point 16 of circuit A, mark 16 where it touches the ruler. Place the thread on point 1 of circuit A, mark 1 where it touches the ruler. Place the thread on point 2 of circuit A, mark 2 where it touches the ruler. Place the thread on point 3 of circuit A, mark 3 where it touches the ruler. Place the thread on point 4 of circuit A, mark 4 where it touches the ruler. Place the thread on point 5 of circuit A, mark 5 where it touches the ruler. Place the thread on point 6 of circuit A, mark 6 where it touches the ruler. Place the thread on point 7 of circuit A, mark 7 where it touches the ruler. Place the thread on point M, mark M where it touches the ruler. After doing this, mark A on the ruler, which will be called ruler A.[352]

Wooden Ruler B Having put aside ruler A, take ruler B and place it so that it touches line KL [figure B.125]. Place the thread on point 14 of circuit B, mark 14 where it touches the ruler. Place the thread on point 15 of circuit B, mark 15 where it touches the ruler. Place the thread on point 16 of circuit B, mark 16 where it touches the ruler. Place the thread on point 1 of circuit B, mark 1 where it touches the ruler. Place the thread on point 2 of circuit B, mark 2 where it touches the ruler. Place the thread on point 3 of circuit B, mark 3 where it touches the ruler. Place the thread on point 4 of circuit B, mark 4 where it touches the ruler. Place the thread on point 5 of circuit B, mark 5 where it touches the ruler. Place the thread on point 6 of circuit B, mark 6 where it touches the ruler. Place the thread on point 7 of circuit B. You have led the thread to 7: mark 7 where the thread touches the ruler. Extend the thread to M now: mark M where it touches the ruler.[353] Put aside this ruler.[354] When the thread doesn't cut the circuit, it must not be led. Conversely the thread must be led when it is tangent to the circuit. One must stop it at the point of contact and place it on the figure of the elevation, as shown in the past [figures B.133, B.134].[355] We must mark that point on the wooden ruler. With the same procedure we must mark the points on the elevation on the paper ruler. And so we must do this for all the other circuits, whether this occurs at the beginning or at the end of a circuit.[356]

Wooden Ruler C Put aside this ruler. Take ruler C and place it so that it touches line KL [figure B.126]. You have led the thread to 16: mark 16 where the thread touches the ruler. Then place the thread on 17 of circuit C: mark 17 where the thread touches the ruler.[357] Place the thread on point 18 of circuit C, mark 18 where it touches the ruler. Place the thread on point 19 of circuit C, mark 19 where it touches the ruler. Place the thread on point 20 of circuit C, mark 20 where it touches the ruler. Place the thread on point 21 of circuit C, mark 21 where it touches the ruler. Place the thread on point 22 of circuit C, mark 22 where it touches the ruler. Place the thread on point 1, mark 1 where it touches the ruler. Proceed this way till point 9, marking the points where the thread touches the ruler. Mark point M on the ruler as well.[358]

Wooden Ruler D Once you have put aside this ruler, take ruler D and place it so that it touches line KL [figure B.127]. Place the thread on points 15, 16, 17, 18, 19, 20, 1, until point 8 of circuit D.[359] Mark all the points where the thread touches the ruler. Place the thread on M: mark M where the thread touches the ruler.

Wooden Ruler E Once you have put aside this ruler, take ruler E and place it so that it touches line KL [figure B.128]. Place the thread on points 16, 17, 18, 19, 20, 1, 1′, until point 9 and M of circuit E.[360] Mark all the points where the thread touches the ruler. Put this ruler aside.

Wooden Ruler F Take a ruler marked with F and place it so that it touches line KL [figure B.129]. Place the thread on points 15, 16, 17, 18, 1, 1′, 1″ until point 8 and above M of circuit F.[361] Mark all the points where the thread touches the ruler.

Wooden Ruler G Put aside this ruler. Take ruler G and place it so that it touches line KL [figure B.130]. Place the thread on points 14, 15, 16, 1, until point 7 and above M, marking the ruler where the thread touches the ruler. Put this ruler aside.[362]

Wooden Ruler H Take another ruler marked H and place it so that it touches line KL [figure B.131]. Place the thread on points 14, 15, 16, 1, until point 7 and above M of circuit H.[363] Place the thread on points 1′, 2′, 3′, 15′, 16′.[364] Mark all the points where the thread touches the ruler. Put this ruler aside.

All Points on the Wooden Rulers At the moment, all the widths are on the rulers [figure B.132]. On ruler A we have points 14*, 15, 16, 1, 2, 3, 4, 5, 6, 7*.[365] On ruler B we have the same points of the preceding ruler.[366] On ruler C we have

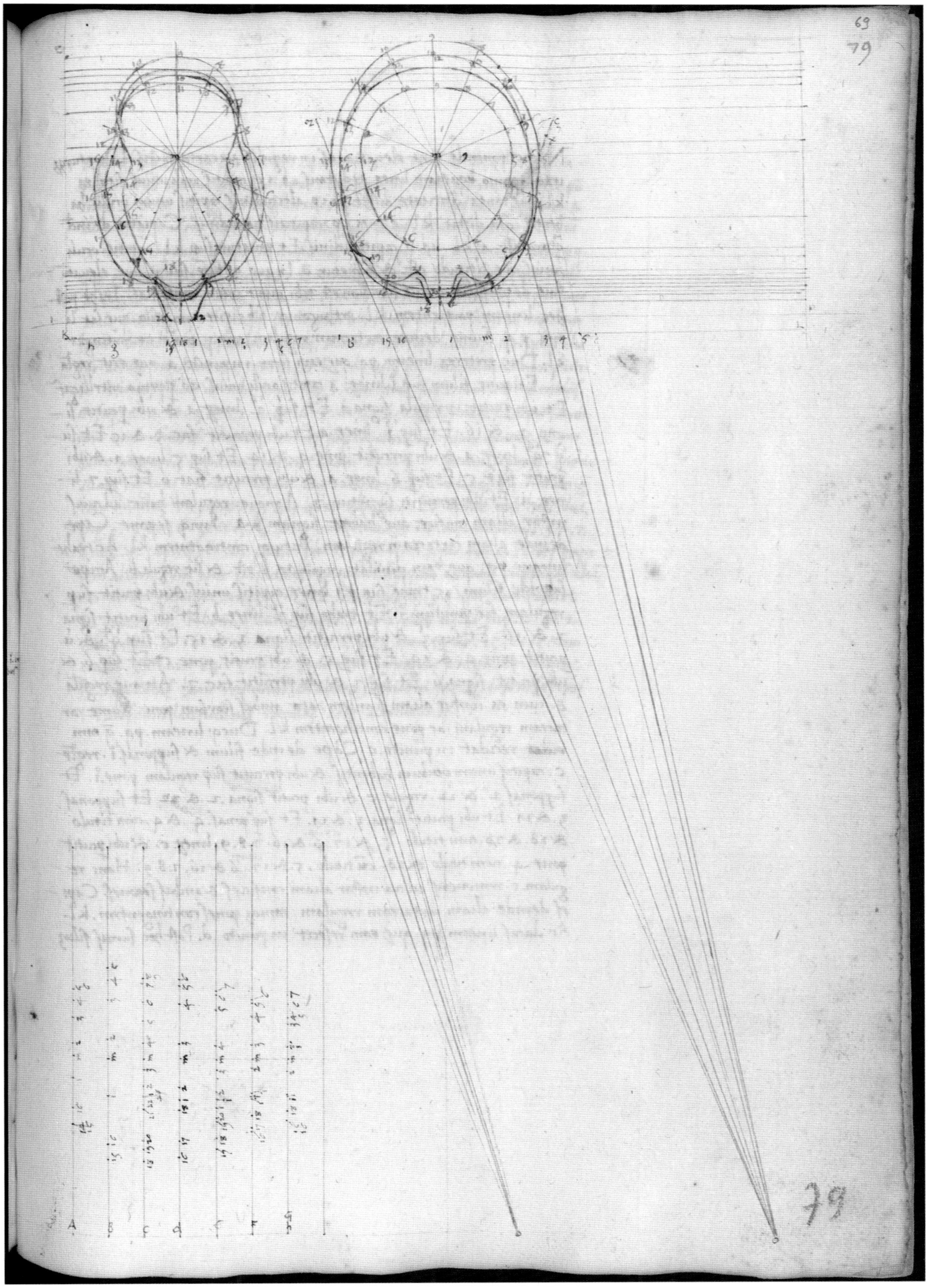

Bordeaux, Bibliothèque Municipale, Ms Latin 616, folio 79 recto.

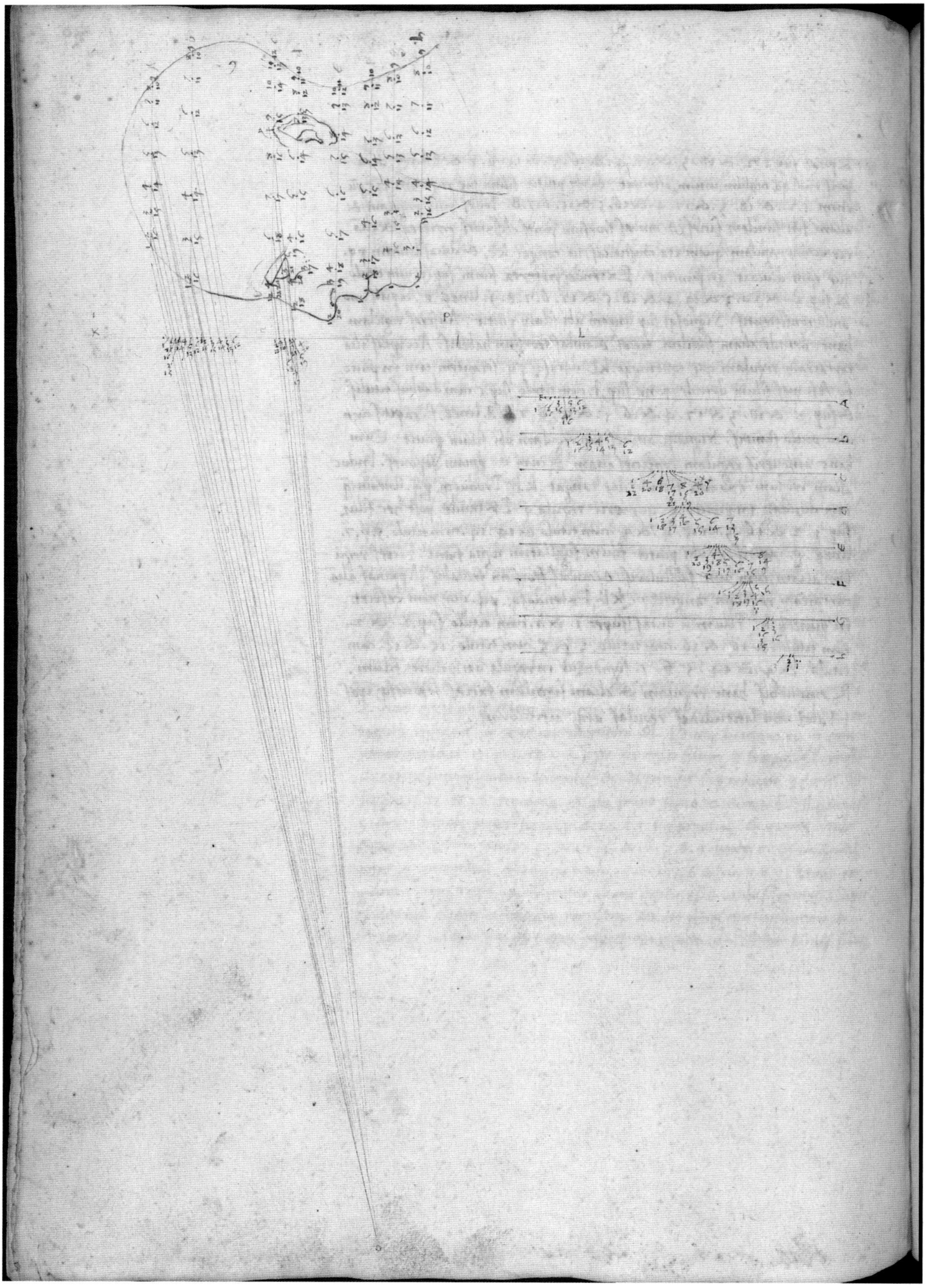

Bordeaux, Bibliothèque Municipale, Ms Latin 616, folio 80 verso.

points 17*, 18, 19, 19′, 20, 20′, 21, 21′, 22, 1, 2, 3, 3′, 4, 4′, 5, 5′, 6, 7, 8, 9*, y on the left and y on the right.[367] On ruler D we have points 17*, 17, 18, 19, 20, 1, 2, 3, 4, 5, 6, 7, 7*, z.[368] On ruler E we have points 16*, 16, 17, 18, 19, 20, 1, 1′, 2, 3, 4, 5, 6, 7, x.[369] On ruler F we have points 16*, 16, 17, 18, 1, 1′, 1″, 2, 3, 4, 5, 6.[370] On ruler G we have points 15*, 16, 1, 2, 3, 4, 4*, 6, 7.[371] On ruler H we have points 12*, 12, 13, 14, 15, 16, 1, 2, 3, 4, 5, 6, 7, 7*, 15′, 16′, 1′, 2′, 3′.[372]

The Choice of the Point of View in Elevation Now we have to place the heights on the paper rulers [figure B.135].[373] Therefore, first draw a line parallel to IP of the head with one eye only. This is line KL; it will be as far distant from line IP of the head with one eye only as line KL is distant from line IP of the head in width. Take the distance between line KL and O in the figure of the head in width and report it starting from line KL of the head with one eye only: this is the elevation.

Mark point O. Fix the thread in this location, which is low or high according to how you prefer to look at the head.[374] Then trace a line through O which divides KL orthogonally below the head in profile: this is line PQ.

Paper Rulers A Take a paper ruler and make it coincide with KL [figures B.135, B.136]. Trace line PQ which intersects line KL at point A. This will be ruler A. Place the thread on point 1 of line A of the head with one eye only, something that you must always understand. Mark point 1 where it hits the ruler. Place the thread on 2 and 16 of line A, mark 2 and 16 where it hits the ruler. Place the thread on 3 and 15 of line A, mark 3 and 15 where it hits the ruler. Place the thread on 14* of line A, mark 14* where it hits the ruler.[375] Place the thread on 4 of line A, mark 4 where it hits the ruler. Place the thread on 5 of line A, mark 5 where it hits the ruler. Place the thread on 6 of line A, mark 6 where it hits the ruler. Place the thread on 7* of line A, mark 7* where it hits the ruler.[376] Put this ruler aside and make another one with the same mark A and place it aside.[377]

Paper Rulers B Take another paper ruler and make it coincide with KL [figures B.135, B.137]. Trace line PQ which intersects line KL at point B. This will be ruler B. Place the thread on point 1 of line B of the head with one eye only. Mark point 1 where it hits the ruler. Place the thread on 2 and 16 of line B, mark 2 and 16 where it hits the ruler. Place the thread on 3 and 15 of line B, mark 3 and 15 where it hits the ruler. Place the thread on 14* of line B, mark 14* where it hits the ruler.[378] Place the thread on 4 of line B, mark 4 where it hits the ruler. Place the thread on 5 of line B, mark 5 where it hits the ruler. Place the thread on 6* of line B, mark 6* where it hits the ruler.[379] Put this ruler aside and make another one with the same mark B and place it aside.

Paper Rulers C Take another paper ruler and make it coincide with KL [figures B.135, B.137]. Trace line PQ which intersects line KL at point C. Place the thread on point 1 of line C of the head with one eye only. Mark point 1 where it hits the ruler. Place the thread on 2 and 22 of line C, mark 2 and 22 where it hits the ruler. Place the thread on 3 and 21 of line C, mark 3 and 21 where it hits the ruler. Place the thread on 4 and 20 of line C, mark 4 and 20 where it hits the ruler.[380] Place the thread on 4′ and 20′ of line C, mark 4′ and 20′ where it hits the ruler. Place the thread on 3′ and 21′ of line C, mark 3′ and 21′ where it hits the ruler. Place the thread on 5 and 19 of line C, mark 5 and 19 where it hits the ruler. Place the thread on 5′ and 19′ of line C, mark 5′ and 19′ where it hits the ruler. Place the thread on 6 and 18 of line C, mark 6 and 18 where it hits the ruler. Place the thread on 17* of line C, mark 17* where it hits the ruler. Place the thread on 8 of line C, mark 8 where it hits the ruler. Place the thread on 9* of line C, mark 9* where it hits the ruler. Place the thread on y of line C, mark y where it hits the ruler. Put this ruler aside and make another one with the same mark C and place it aside.

Paper Rulers D Take another paper ruler and make it coincide with KL [figures B.135, B.137]. Trace line PQ which intersects line KL at point D. Successively, place the thread on points 1, 2, 20, 19, 3, 4, 18, 17, 5, 17*, 6, 7, 7*, z of line D of the head with one eye only. Mark points 1, 2, 20, 19, 3, 4, 18, 17, 5, 17*, 6, 7, 7*, z where the thread hits the ruler.[381] Take away this ruler and make another one; put both rulers aside.

Paper Rulers E Take another paper ruler and make it coincide with KL [figures B.135, B.137]. Trace line PQ which intersects line KL at point E. Successively, take the thread and place it on 1, 1′, 2, 20, 19, 3, 18, 4, 17, 5, 16, 6, 16*, 7, x of line E of the head with one eye only.[382] Mark the points where the thread touches the ruler and remove it. Make a similar one and put both aside.

Paper Rulers F Take another paper ruler and make it coincide with KL [figures B.135, B.137]. Trace line PQ which intersects line KL at point F. Successively, place the thread on points 1, 1′, 1″, 2, 18, 3, 17, 4, 16, 16*, 5, 6 of line F of the head with one eye only.[383] Mark the points where the thread touches the ruler and remove it. Make a similar one and put both aside.

Paper Rulers G Take another paper ruler and make it coincide with KL [figures B.135, B.137]. Trace line PQ which intersects line KL at point G: you shall have ruler G. Successively, place the thread on 1, 2, 16, 3, 15*, 4, 4*, 5, 6, 7 of line G of the head with one eye only.[384] Mark the points where the thread touches the ruler and remove it. Make a similar one and put both aside.

Paper Rulers H Take another paper ruler and make it coincide with KL [figures B.135, B.137]. Trace line PQ which intersects line KL at point H. Successively, place the thread on points 1, 2, 16, 3, 15, 4, 14, 5, 13, 6, 12, 12*, 7, 7*, 1′, 2′, 16′, 3′, 15′.[385] Mark the points where the thread touches the ruler and remove it. Make a similar one and put both aside.[386]

The Shortening of the Head Thus far you have the rulers of width and height. Now we must place them where we want to realize the head [figure B.138]. Therefore, draw straight line KL which is greater than the head we have to represent: divide this line into equal parts at point M. Draw perpendicular MN on M. Then draw perpendicular KP on K and the second perpendicular LQ on L.

Rulers A/A/A Take the two paper rulers with mark A [figure B.138]. Place one so that it coincides with KP and the other so that it coincides with LQ. Let points A of the two rulers touch line KL. Glue them with wax so that they stay still. Take wooden ruler A and place it on the two rulers so that it passes through point 1 of both rulers. Point M touches line MN. Always remember that point M, which is present on all wooden rulers, touches line MN. Where point 1 of the wooden ruler lies, mark 1. When I say "where it lies" I mean some of the marks of the wooden ruler. Place the ruler so that it goes through 2 and 16 of the two rulers. Mark point 2 where point 2 lies, mark point 16 where point 16 lies. Place the ruler so that it goes through 3 and 15 of the two rulers. Mark point 3 where point 3 lies, mark point 15 where point 15 lies. Place the ruler so that it goes through 4 and 14 of the two rulers. Mark point 4 where point 4 lies, mark point 14 where point 14 lies. Place the ruler so that it goes through point 5 of the two rulers. Mark point 5 where point 5 lies. Place the ruler so that it goes through point 6 of the two rulers. Mark point 6 where point 6 lies. Place the ruler so that it goes through point 7 of the two rulers. Mark point 7 where point 7 lies.[387] Take the rulers off and place them aside.

Rulers B/B/B Take the two paper rulers with mark B [figures B.139, B.141].[388] Place one so that it coincides with KP and the other so that it coincides with LQ. Let points B of the two rulers touch line KL. Take wooden ruler B and place it on the two rulers so that it passes through point 1 of both rulers. Point M touches line MN. Where point 1 of the wooden ruler lies, mark 1. Place the ruler so that it goes through 2 and

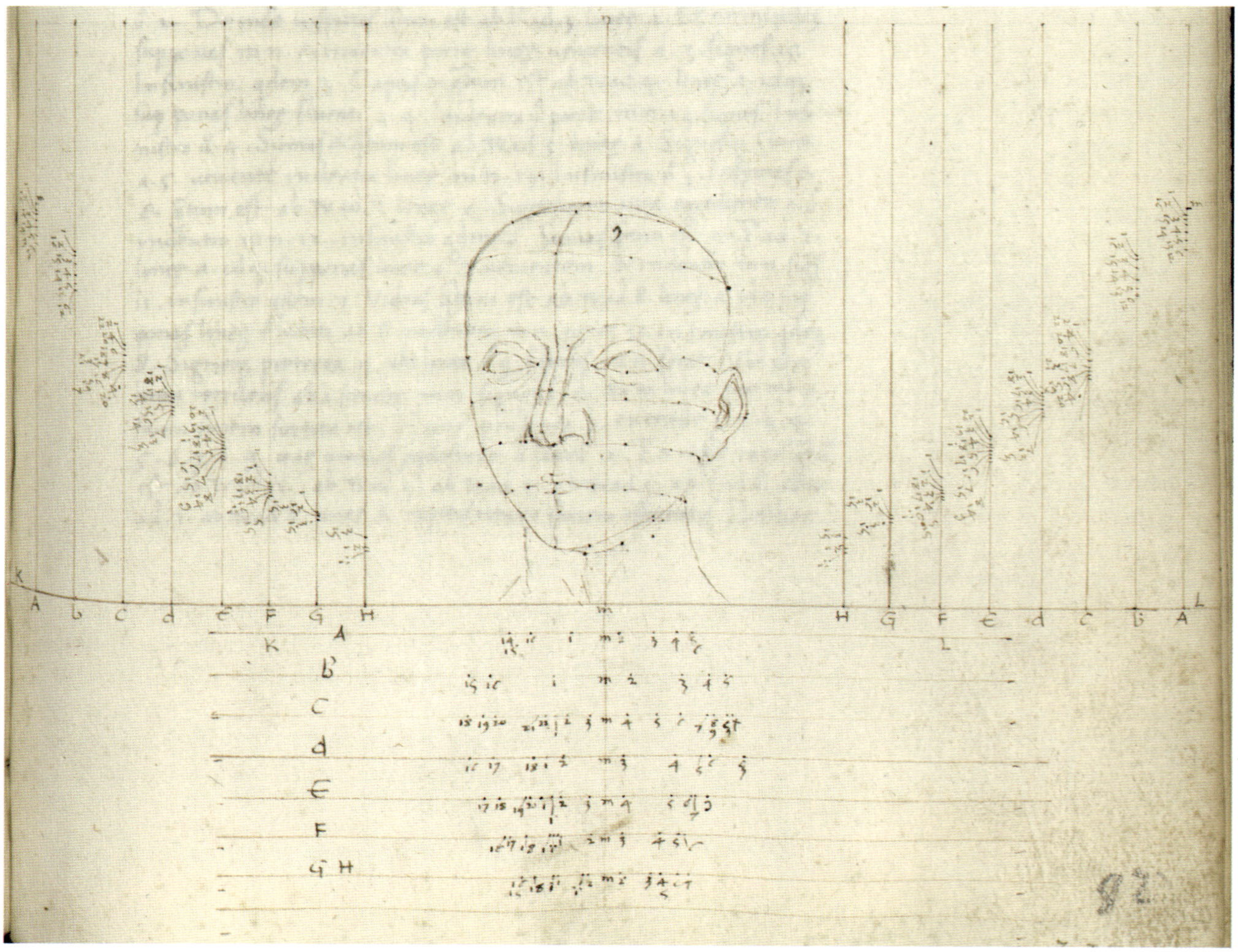

Bordeaux, Bibliothèque Municipale, Ms Latin 616, folio 82 recto.

16 of the two rulers. Mark point 2 where point 2 lies, mark point 16 where point 16 lies. Place the ruler so that it goes through 3 and 15 of the two rulers. Mark point 3 where point 3 lies, mark point 15 where point 15 lies. Place the ruler so that it goes through 4 and 14 of the two rulers. Mark point 4 where point 4 lies, mark point 14 where point 14 lies. Place the ruler so that it goes through point 5 of the two rulers. Mark point 5 where point 5 lies. Place the ruler so that it goes through point 6 of the two rulers. Mark point 6 where point 6 lies. Place the ruler so that it goes through point 7 of the two rulers. Mark point 7 where point 7 lies.[389] After completing this, take the rulers off and place them aside.

Rulers C/C/C Take the two paper rulers with mark C [figures B.139, B.141]. Place one so that it coincides with KP and the other one so that it coincides with LQ. Let points C of the two rulers touch line KL. Take wooden ruler C and place it on the two rulers so that it passes through point 1 of both rulers. Point M touches line MN. Where point 1 of the wooden ruler lies, mark 1. Place the ruler so that it goes through 2 and 22 of the two rulers. Mark point 2 where point 2 lies, mark point 22 where point 22 lies. Place the ruler so that it goes through 3 and 21 of the two rulers. Mark point 3 where point 3 lies, mark point 21 where point 21 lies. Place the ruler so that it goes through 4 and 20 of the two rulers. Mark point 4 where point 4 lies, mark point 20 where point 20 lies.[390] Place the ruler so that it goes through point 4′ and 20 of the two rulers.[391] Where point 4′ lies, mark point 4′; where point 20′ lies, mark point 20′.[392] Place the ruler so that it goes through points 5 and 19 of the two rulers.[393] Mark point 5 where point 5 lies. Mark point 19 where point 19 lies.[394] Place the ruler so that it goes through 6 and 18 of the two rulers. Mark point 6 where point 6 lies, mark point 18 where point 18 lies. Place the ruler so that it goes through point 7 of the two rulers. Mark point 7 where point 7 lies. Place the ruler so that it goes through point 17 of the two rulers. Mark point 17 where point 17 lies.[395] Place the ruler so that it goes through point 8 of the two rulers. Mark point 8 where point 8 lies. Place the ruler so that it goes through point 9 of the two rulers. Mark point 9 where point 9 lies.[396] Take the rulers off.

The Remaining Rulers Take the two paper rulers with mark D [figures B.139, B.141]. Place one so that it coincides with KP and the other one so that it coincides with LQ. Let points D of the two rulers touch line KL. Take wooden ruler D and repeat the same procedure as for the other rulers. Do the same for rulers E, F, G, H[397] which you will then place aside.

The Description of the Contours And describe the contours in good shape [figure B.139].[398] Extend the marks you reproduced with ruler A; these are 14* and 15, 15 and 16, 16 and 1, 1 and 2, 2 and 3, 3 and 4, 4 and 5, 5 and 6, 6 and 7*.[399] Extend the marks you reproduced with ruler B; these are 14* and 15, 15 and 16, 16 and 1, 1 and 2, 2 and 3, 3 and 4, 4 and 5, 5 and 6*.[400] Extend the marks you reproduced with ruler C; these are 17* and 18, 18 and 19, 19 and 20, 20 and 21, 21 and 22, 19 and 20′, 20′ and 21, 22 and 1, 1 and 2, 2 and 3, 3 and 4, 4 and 5, 3 and 4′, 4′ and 5, 5 and 6, 6 and 7, 7 and 8, 8 and 9*.[401] Extend the marks you reproduced with ruler D; these are 17* and 17, 17 and 18, 18 and 19, 19 and 20, 20 and 1, 1 and 2, 2 and 3, 3 and 4, 4 and 5, 5 and 6, 6 and 7, 7 and 7*.[402] Extend the marks you reproduced with ruler E; these are 16* and 16, 16 and 17, 17 and 18, 18 and 19, 19 and 20, 20 and 1, 1 and 1′, 1 and 2, 2 and 3, 3 and 4, 4 and 5, 5 and 6, 6 and 7.[403] Proceed for the marks reproduced with rulers F, G, and H with the same reasoning. Then draw around each mark an outline border.[404]

The Twelve Units Certainly there are twelve units [figure B.139].[405]

The Protrusions of the Head In the same manner, you will draw with outlines the protrusions, as is evident, of the eyes, the nose, the mouth, and whatever remains so that you can draw them correctly [figures B.139, B.140].[406] First the right eye presents points 19, 20, 21 in the upper part, and points 19′, 20′ and 21′ in the lower part.[407] The left eye presents points 3, 4, and 5 in the upper part and points 3′, 4′, and 5′ in the lower part.[408] The nose has the following: 19, 20, 1, 1′, 2, and 3.[409] The mouth has 18, 1, 1′, 1″, 2.[410] We must proceed this way for the remaining outlines.[411]

The Points of Tangency As mentioned earlier, it may happen that when the thread is led above the circuits, at the beginning or at the end of the object being considered, there is no correspondence to any mark; the extreme lines, in fact, that originate from the eyes surround the entire object, touching it; it happens that their mark is not found above the indicated marks but that the first line, that is, the thread, touches the circuit between points 14 and 15.

Mark point 14*.[412] We must do the same for the head that shows the whole face and the same for the head that has one eye only,[413] maintaining the already followed procedure; this implies that you have to take the length between line I(P) and 14* which you have marked on the contact of the circuit.[414] Place that length starting from line I(P) on the line corresponding to the circuit. If you are operating on circuit A on line A of the head that has one eye only, mark point 14* [figure B.133].[415] If the last line [the thread] touches the circuit between 6 and 7, mark point 7*.[416] Then look at the head which shows the whole face. Take the length between line MN and point 14* of that circuit. Place that length on the right of point 1 of the head with the whole face. On the right must be 14*, on the left must be 7*.[417] Do the same for all circuits.[418]

Proposition VIIIb

Since the head you have built till now has no inclination, it is my intention now to build that head through two movements. This means that the head is lifted up frontally, leaning on one side, and it is not equidistant from the plane of the terminal.[419] Through this procedure you will easily understand all the other head movements.

First Rotation From what was done earlier, you have the head with one eye only, which has eight transversal lines, which are A, B, C, D, E, F, G, H [figure B.142a]. Each of these lines appears with its own marks. Trace on this [head] a line that touches the point of the nose and the chin [figure B.142b]. Let this line be IP, which makes the head with one eye only lean on the side of the back of the head as much as you want it to lean.[420]

Elevation of Circuit A Also trace an unlimited line z perpendicular to line IP which touches the top of the head [figure B.143, left]. From points 1, 2, 3, 4, 5, 6, 7, 8, 9 of line A of the head with one eye only, trace lines with no end parallel to line z.[421] Successively trace perpendicularly to line z a line with a right angle which is MN. Now trace a perpendicular line which touches the point of the nose of the head with one eye only and that meets line A in a right angle: let this be line PV.[422]

Select the distance between line TV [figure B.143, right] and point 2 of line A of the first head which shows the entire face: [a distance] which we have previously stated we must delimit. Always proceed this way. Place one of the two compass ends on line MN and with the other [end] mark the line which originates from point 2 of the head with one eye only. On the right of MN mark point 16 and on the left 2.[423] Then look at the distance between TV and point 3 of line A.[424] Place the point of the compass on MN. On the line originating from 3, mark 15 on the right and 3 on the left. Take the distance between TV and point 4 of line A. Place it on the line originating from 4, mark 14 on the right and 4 on the left. Take the distance between TV and point 5 of line A. Place it on the line originating from 5, mark 13 on the right of MN and 5 on the left. Take the distance between TV and point 6 of line A. Place it on the line originating from 6, mark 12 on the right of MN and 6 on the left. Take the distance between TV and point 7 of line A. Place it on the line originating from 7, mark 11 on the right of MN and 7 on the left.

Take the distance between TV and point 8 of line A. Place it on the line originating from 8, mark 10 on the right of MN and 8 on the left. Mark 9 where the line coming from 9 intersects line MN. Mark 1 where the line coming from 1 intersects line MN.[425] This way we have obtained a [curved] line that has its own delimitation.[426]

Elevation of Circuit B Now, for what concerns line B [figure B.144], from points 1, 2, 3, 4, 5, 6, 7, 8, 9 [of the head with one eye only] trace lines that are all equidistant to line z. Take the distances between TV and 2, TV and 3, TV and 4, TV and 5, TV and 6, TV and 7, TV and 8[427] of line B of the head that shows the whole face. Successively, on the line that originates from 2, place 16 on the right of MN and 2 on the left.[428] Mark 15 on the right of MN and 3 on the left on the line originating from 3. Mark 14 on the right of MN and 4 on the left on the line originating from 4. Mark 13 on the right of MN and 13 on the left on the line originating from 5. Mark 12 on the right of MN and 6 on the left on the line originating from 6. Mark 11 on the right of MN and 7 on the left on the line originating from 7. Mark 8 on the right of MN and 10 on the left on the line originating from 8. Mark 9 where the line coming from 9 intersects line MN. Mark 1 where the line coming from 1 intersects that line [MN].

Elevations of Circuits C, D, E, F, G, H Through the same procedure also mark the following [curved] lines [figure B.145]: that is, lines C, D, E, F, G, H. When you have done this, you will have the head with the entire inclined face. The first circuit of this head is: 1, 2, 3, 4, 5, 6, 7, 8, 9, 10, 11, 12, 13, 14, 15, 16. The second circuit, which is B, is constituted by the same numbers. The circuit marked with C has numbers from 1′ to 22.[429] The circuit marked with D has numbers from 1 to 20.[430] The circuit marked with E has numbers that begin with 1′ and end with 20. The circuit marked with F has numbers that begin with 1, with 1′, with 1″ and end with 18.[431] The circuit marked with G has the following numbers: 1, 2, 3, 4, 5, 6, 7, 8, 9, 10, 11, 12, 13, 14, 15, 16.[432] The circuit marked with H has the following numbers: 1, 2, 3, 4, 5, 6, 7, 8, 9, 10, 11, 12, 13, 14, 15, 16, 1′, 2′, 3′, 15′, 16′.[433]

Second Rotation Having done this, go back to the head with one eye only [figure B.146]. From points 1, 2, 3, 4, 5, 6, 7, 8, 9 of line A of the head with one eye only, draw lines equidistant to line IP. Softly trace these and let them be thin lines. Take them below the other head so that these [lines] lie under the [other] head twice the head itself. Trace perpendicular line mN on these lines. Then take the head that has the whole face and that was built earlier. Place it above these lines so that the upper part touches line mN: so that it leans as much as you'd want it to. Because line mN is perpendicular to the lines that originate from the marks on line A of the head with one eye only, I want it [line mN] to serve to measure both heads. Thus when I say "take the quantity" to point 2 or point 3 or point 4,[434] it must be understood that the point of the compass must be placed on line mN and with the other end we must touch the [specified] mark. When I shall say "place," "mark," "let there be a point" I mean that it is on the line that comes out. In the same way, I mean that you have to place the end of the compass on line mN and that with the other end [of the compass] you will mark that quantity on the line that comes out.

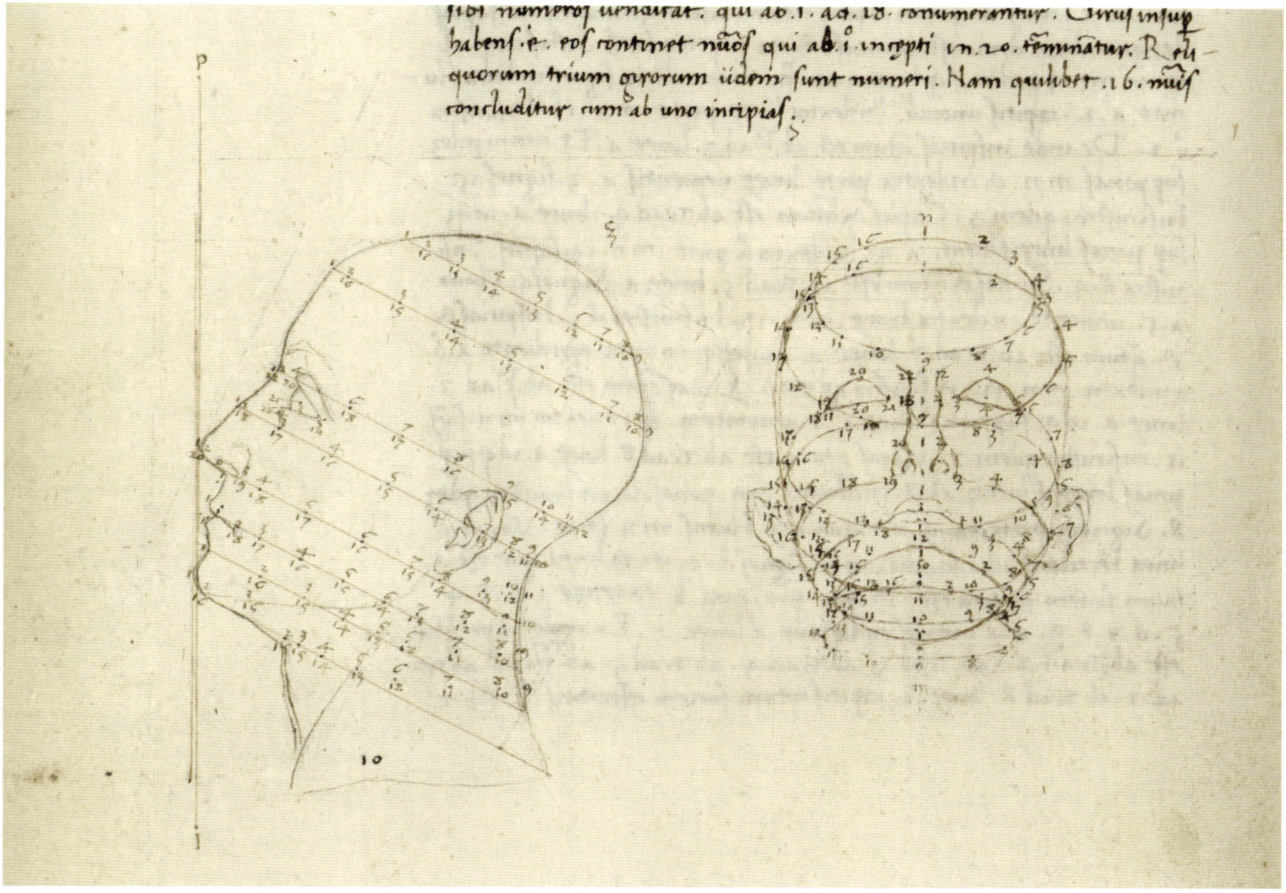

Bordeaux, Bibliothèque Municipale, Ms Latin 616, folio 83 verso.

Circuit A in Plan Take the quantity between mN and point 5 of line A [of the head that shows the whole face] [figure B.147].[435] Place it on the line originating from point 5 [of the head with one eye only] and mark 5 in plan. Take the quantity between mN and point 6 of line A. Place it on the line originating from point 6 [of the head with one eye only] and mark 6. Place the compass on line mN, place the other end at point 7 of line A. Place an end of the compass on the contact between line mN and the line originating from 7 and with the other end mark 7. Take the quantity between mN and point 8 of line A. Place it on the line originating from 8 and mark 8. Take the quantity between mN and point 9 of line A. Place it on the line originating from 9 and mark 9. Take the quantity between mN and point 10 of line A. Place it on the line originating from 8 and mark 10. Take the quantity between mN and point 11 of line A. Place it on the line originating from 7 and mark 11. Take the quantity between mN and point 12 of line A. Place it on the line originating from 12 and mark 12. Take the quantity between mN and point 13 of line A. Place it on the line originating from 13 and mark 13. Take the quantity between mN and point 14 of line A. Place it on the line originating from 14 and mark 14. Take the quantity between mN and point 15 of line A. Place it on the line originating from 3 and mark 15. Take the quantity between mN and point 16 of line A. Place it on the line originating from 2 and mark 16. Take the quantity between mN and point 1 of line A. Place it on the line originating from 1 and mark 1. Take the quantity between mN and point 2 of line A. Place it on the line originating at 2 and mark 2. Take the quantity between mN and point 3 of line A. Place it on the line originating from 3 and mark 3. Take the quantity between mN and point 4 of line A. Place it on the line originating from 4 and mark 4. Having done this, draw circuit A: join 1 with 2, 2 with 3, 3 with 4, 4 with 5, 5 with 6, 6 with 7, 7 with 8, 8 with 9, 9 with 10, 10 with 11, 11 with 12, 12 with 13, 13 with 14, 14 with 15, 15 with 16, 16 with 1. This is circuit A.

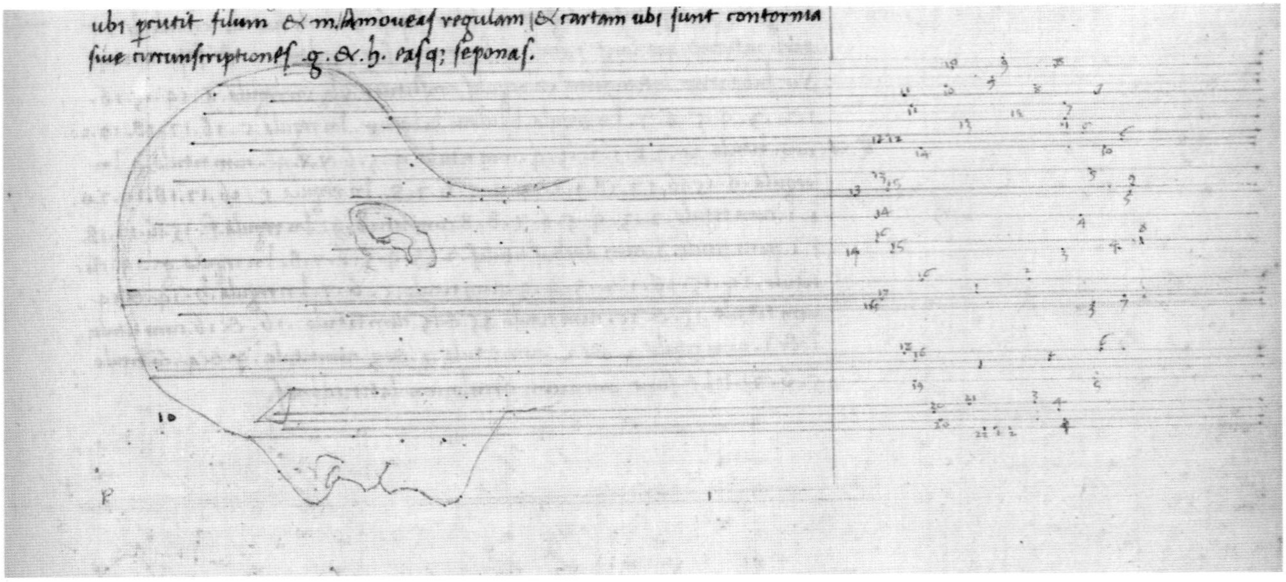

Bordeaux, Bibliothèque Municipale, Ms Latin 616, folio 86 recto.

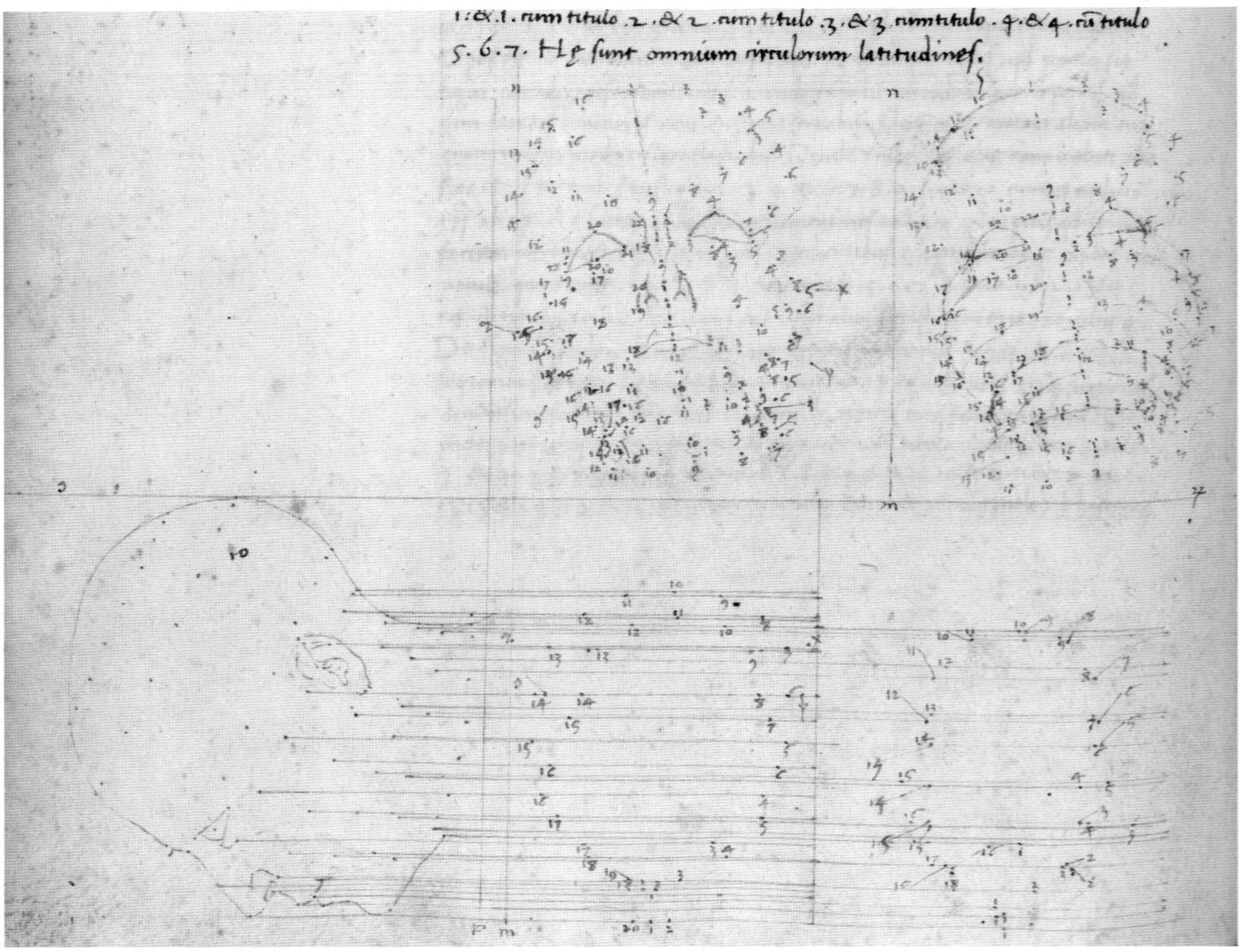

Bordeaux, Bibliothèque Municipale, Ms Latin 616, folio 86 verso.

Circuit B in Plan Now, to obtain circuit B go back to the head with one eye only [figure B.148]. Trace from points 1, 2, 3, 4, 5, 6, 7, 8, 9 of line B many lines equidistant to line IP with no end. Take the quantity between mN and point 1 of curved [line] B of the head that shows the whole face.[436] Place it on the line originating from point 1 of line B and mark 1. Take the quantity between mN and point 2 of line B. Place it on the line originating from point 2 of line B and mark 2. Take the quantity between mN and point 3 of line B. Place it on the line originating from point 3 of line B and mark 3. Take the quantity between mN and point 4 of line B. Place it on the line originating from point 4 of line B and mark 4. Take the quantity between mN and point 5 of line B. Place it on the line originating from point 5 of line B and mark 5. Take the quantity between mN and point 6 of line B. Place it on the line originating from point 6 of line B and mark 6. Take the quantity between mN and point 7 of line B. Place it on the line originating from point 7 of line B and mark 7. Take the quantity between mN and point 8 of line B. Place it on the line originating from point 8 of line B and mark 8. Take the quantity between mN and point 9 of line B. Place it on the line originating from point 9 of line B and mark 9. Take the quantity between mN and point 10 of line B. Place it on the line originating from point 8 of line B and mark 10. Take the quantity between mN and point 11 of line B. Place it on the line originating from point 11 of line B and mark 11. Take the quantity between mN and point 12 of line B. Place it on the line originating from point 6 of line B and mark 12. Take the quantity between mN and point 13 of line B. Place it on the line originating from point 5 of line B and mark 13. Take the quantity between mN and point 14 of line B. Place it on the line originating from point 4 of line B and mark 14. Take the quantity between mN and point 15 of line B. Place it on the line originating from point 3 of line B and mark 15. Take the quantity between mN and point 16 of line B. Place it on the line originating from point 2 of line B and mark 16. Now join 1 with 2, 2 with 3, 3 with 4, 4 with 5, 5 with 6, 6 with 7, 7 with 8, 8 with 9, 9 with 10, 10 with 11, 11 with 12, 12 with 13, 13 with 14, 14 with 15, 15 with 16, 16 with 1. This is circuit B.

Circuit C in Plan Now, to obtain circuit C trace from the head with one eye only from points 1, 2, 3, 4, 4′, 5, 6, 7, 8, 9 many lines with no end, all equidistant to line IP [figure B.148]. Trace them softly so that they are thin. Take the quantity between mN and point 1 of line C of the head that shows the whole face. Place it on the line originating from point 1 of line C and mark 1. Take the quantity between mN and point 2 of line C.[437] Place it on the line originating from point 2 of line C and mark 2. Take the quantity between mN and point 3 of line C. Place it on the line originating from point 3 of line C and mark 3. Take the quantity between mN and point 4 of line C. Place it on the line originating from point 4 of line C and mark 4. Take the quantity between mN and point 4′ of line C. Place it on the line originating from point 4′ of line C and mark 4′.[438] Take the quantity between mN and point 5 of line C. Place it on the line originating from point 5 of line C and mark 5. Take the quantity between mN and point 6 of line C. Place it on the line originating from point 6 of line C and mark 6. Take the quantity between mN and point 7 of line C. Place it on the line originating from point 7 of line C and mark 7. Take the quantity between mN and point 8 of line C. Place it on the line originating from point 8 of line C and mark 8. Take the quantity between mN and point x of line C. Place it on the line originating from point y of line C and mark y.[439] Take the quantity between mN and point 9 of line C. Place it on the line originating from point 9 of line C and mark 9. Take the quantity between mN and point 10 of line C. Place it on the line originating from point 10 of line C and mark 10. Take the quantity between mN and point 11 of line C. Place it on the line originating from point 11 of line C and mark 11. Take the quantity between mN and point 12 of line C. Place it on the line originating from point 12 of line C and mark 12. Take the quantity between mN and point 13 of line C. Place it on the line originating from point 11 of line C and mark 13. Take the quantity between mN and point 14 of line C. Place it on the line originating from point 10 of line C and mark 14. Take the quantity between mN and point 15 of line C. Place it on the line originating from point 9 of line C and mark 15. Take the quantity between mN and point 16 of line C. Place it on the line originating from point 8 of line C and mark 16. Take the quantity between mN and point 17 of line C. Place it on the line originating from point 7 of line C and mark 17. Take the quantity between mN and point 18 of line C. Place it on the line originating from point 6 of line C and mark 18. Take the quantity between mN and point 19 of line C. Place it on the line originating from point 5 of line C and mark 19. Take the quantity between mN and point 20 of line C. Place it on the line originating from point 4 of line C and mark 20. Take the quantity between mN and point 20′ of line C. Place it on the line originating from point 4 of line C and mark 20′. Take the quantity between mN and point 21 of line C. Place it on the line originating from point 3 of line C and mark 21. Take the quantity between mN and point C of line C. Place it on the line originating from point 2 of line C and mark 22. Having done this, you have obtained another circuit. Now join 1 with 2, 2 with 3, 3 with 4, 4 with 4′, 4 with 5, 5 with 6, 6 with 7, 7 with 8, 8 with 9, 9 with 10, 10 with 11, 11 with 12, 12 with 13, 13 with 14, 14 with 15, 15 with 16, 16 with 17, 17 with 18, 18 with 19, 19 with 20, 20 with 20′, 20 with 21, 21 with 22, 22 with 1.[440] This is circuit C. You now have three circuits. If the lines you trace in these three circuits are hard to discern because they are too many, take a clean sheet of paper and place it so that it touches line mN and it passes on line IP without touching the [two] heads at all. Place the paper well with wax. Then draw line [IP] so that it goes on the sheet of paper.[441]

Circuit D in Plan From points 1, 2, 3, 4, 5, 6, 7, 8, 9, 10[442] of line D of the head with one eye only trace many lines with no end, all equidistant to line IP [figure B.148]. Trace them softly so that they are thin. Take the quantity between mN and point 1 of line D of the head that shows the whole face. Place it on the line originating from point 1 of line D and mark 1. Take the quantity between mN and point 2 of line D. Place it on the line originating from point 2 of line D and mark 2. Execute the same procedure until point 18.[443]

Now join 1 with 2, 2 with 3, 3 with 4, 4 with 5, 5 with 6, 6 with 7, 7 with 8, 8 with 9, 9 with 10, 10 with 11, 11 with 12, 12 with 13, 13 with 14, 14 with 15, 15 with 16, 16 with 17, 17 with 18, 18 with 1.[444] Thus we have circuit D.

Circuit E in Plan From points 1, 2, 3, 4, 5, 6, 7, 8, 9, 10, 11 of line E trace many lines with no end, all equidistant to line IP [figure B.148]. Successively, take all the quantities of line E of the head which shows the entire face. Place them on the lines originating from the points, and add the numbers as you have done earlier. This will be circuit E. Now join 1 with 1′, 1 with 2,[445] 2 with 3, 3 with 4, 4 with 5, 5 with 6, 6 with 7, 7 with 8, 8 with 9, 9 with 10, 10 with 11, 11 with 12, 12 with 13, 13 with 14, 14 with 15, 15 with 16, 16 with 17, 17 with 18, 18 with 19, 19 with 20, 20 with 1.

Circuit F in Plan From points 1, 2, 3, 4, 5, 6, 7, 8, 9[446] of line F of the head with one eye only, trace many lines with no end equidistant to line IP [figure B.149]. Report the quantities of line F of the head which shows the entire face on the lines that originate from line F of the head with one eye only. Now join 1 with 1′, 1″ with 2, 2 with 3, 3 with 4, 4 with 5, 5 with 6, 6 with 7, 7 with 8, 8 with 9, 9 with 10, 10 with 11, 11 with 12, 12 with 13, 13 with 14, 14 with 15, 15 with 16, 16 with 1.[447] This is circuit F. It is necessary that you indicate with a mark the paper that you have used; this is the reason why a part of the marks are on the paper that is under, so that when it is necessary you can place it back in the same location.[448] After you have done this, you will take the paper out, you will put it aside to then use it again.

Circuit G in Plan Take another sheet of paper which you will fix in the same place [figure B.149]. Then trace line IP which will mark it like the other one.[449] From points 1, 2, 3, 4, 5, 6, 7, 8, 9 of line G trace many lines all equidistant to line IP. Report the quantities of line G of the head which shows the entire face, and place them on the lines that originate from line G of the head with one eye only. These will yield points 1, 2, 3, 4, 5, 6, 7, 8, 9, 10, 11, 12, 13, 14, 15, 16.[450] You will trace these lines as you have done in the past.[451] This is circuit G.

Circuit H in Plan Do the same thing you have done for all other marks for the marks on line H of the head with one eye only [figure B.149]. That is, trace all the parallel lines to line IP and place all the quantities on lines that originate from points of line H of the head that shows the entire face. Then join 1 with 2, 2 with 3, 3 with 4, 4 with 5, 5 with 6, 6 with 7, 7 with 8, 8 with 9, 9 with 10, 10 with 11, 11 with 12, 12 with 13, 13 with 14, 14 with 15, 15 with 16, 16 with 1, 3′ with 2′, 2′ with 1′, 1′ with 16′, 16′ with 15′.[452] Next, on the rulers, report the numbers you have found relative to these outlines or circuits.[453]

Choice of the Terminal in Plan First draw a line next to IP but not equidistant to IP [figure B.150]: let it be KL, which will be the terminal where you will place the wooden rulers.[454]

Choice of the Point of View in Plan Get as far away from this line as you would like to be distant from the head and mark point O where you should also fix the needle that carries the thread, as you have often done [figure B.151].

Wooden Rulers Place the wooden ruler so that it is tangent to KL [figure B.152]. Take the thread and in the same manner report the marks on the ruler; place the thread on M of line MN and report it on the ruler. Mark H on the ruler.[455] After removing this ruler, place ruler G. Place the thread on the marks of circuit G and report point M on ruler G and [the points] where the thread touches it. Take off the ruler and the paper where there are outlines or circuits G and H and place them aside [figure B.153]. Then take the paper on which there are the calculations of outlines D, E, F [figure B.152] and place it in the same location where you marked it so that it touches the marks you made. Take the wooden ruler marked F and place it so that it touches KL, and make sure it stays still. Place the thread on the marks of circuit F. Start where there is a contact and end when there is a contact. Report all the points on the ruler where the thread touches it and point M. Having taken this ruler off, take another wooden one marked E and proceed in a similar way. Follow the same procedure for ruler D. Thus, these lines must be put aside. Put aside the paper where you have drawn the three outlines D, E, F [figure B.153]. Place the wooden ruler marked C tangent to KL [figure B.152]. Place the thread on the marks of circuit C and on M. Mark them on the ruler. After putting this ruler aside, do the same for circuit B, and through similar procedure report the marks of circuit A on ruler A. And put these rulers aside [figure B.153].

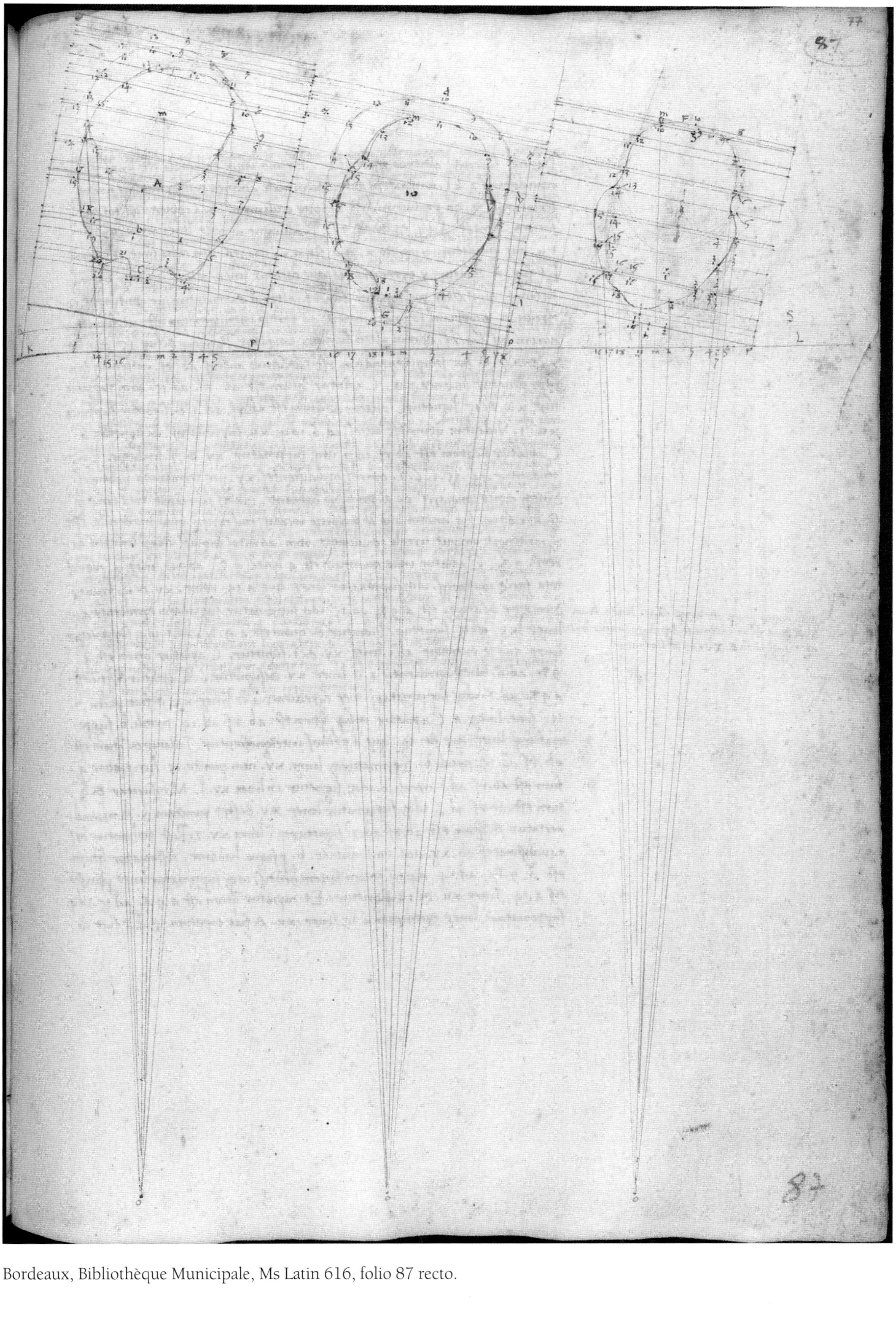

Bordeaux, Bibliothèque Municipale, Ms Latin 616, folio 87 recto.

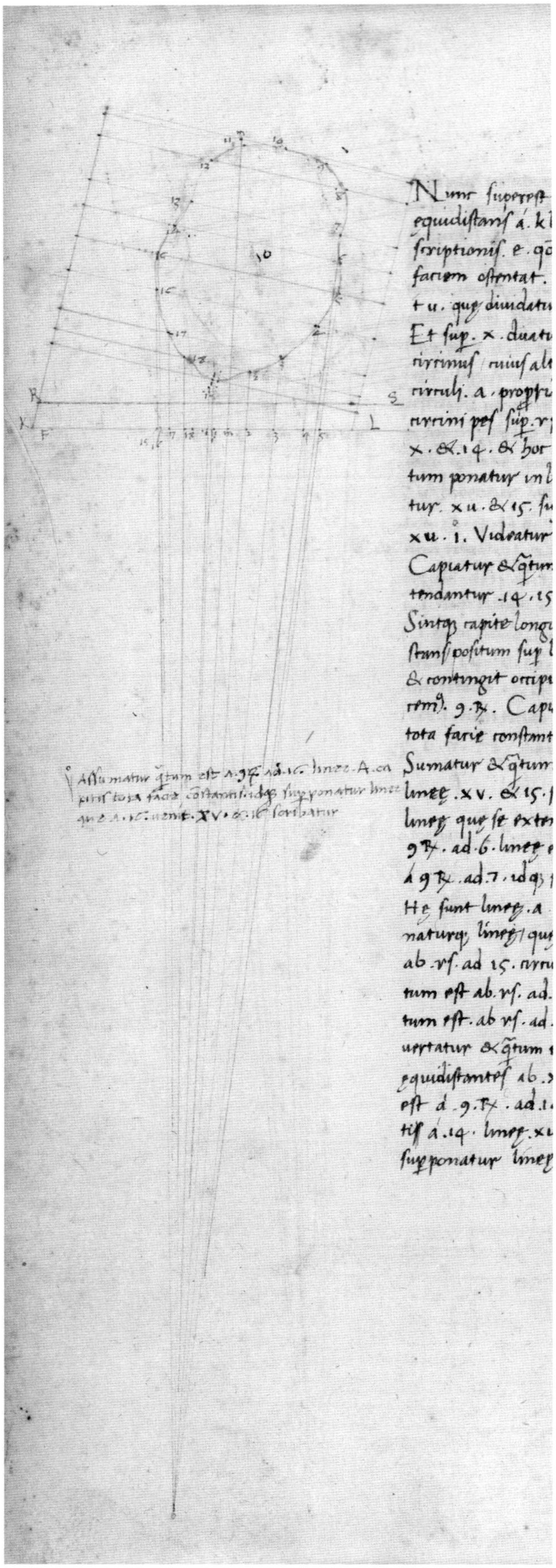

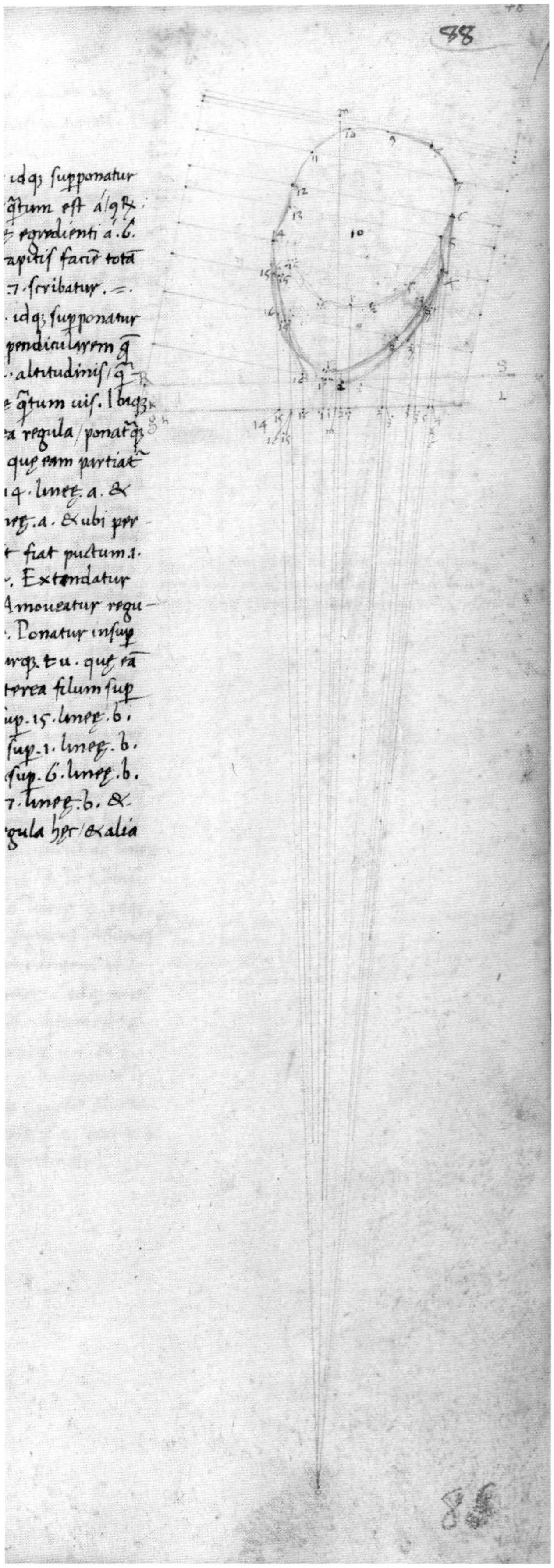

Bordeaux, Bibliothèque Municipale, Ms Latin 616, folio 87 verso.

Bordeaux, Bibliothèque Municipale, Ms Latin 616, folio 88 recto.

Numbers on Wooden Rulers In this way, you will have obtained widths on the rulers [figure B.153]. That is, on ruler A you have 14, 15, 16, 1, 2, 3, 4, 5, 6, 7; on ruler B you have the same numbers.[456] On ruler C: 16, 17, 18, 19, 20, 20′, 21, 22, 1, 2, 3, 4′, 4, 5, 6, 7, y.[457] On ruler D: 15, 16, 17, 18, 1, 2, 3, 4, 5, 6, 7, 8.[458] On ruler E: 16, 17, 18, 19, 20, 1, 1′, 2, 3, 4, 5, 6, 7, y, 8, 9.[459] On ruler F: 15, 16, 17, 18, 1, 1′, 1″, 2, 3, 4, 5, 6, 7, 8.[460] On ruler G: 14, 15, 16, 1, 2, 3, 4, 5, 6, 7.[461] On ruler H: 14, 15, 15′, 16, 16′, 1, 1′, 2, 2′, 3, 3′, 4, 5, 6, 7.[462] These are the widths of all circuits.

Heights of Points of Circuit A

First Step Now we need to find the height.[463] First draw a line parallel to KL [figure B.154] which touches the tip of the nose of the outline that is circuit E: the outline that was taken from the head with one eye only and from the head that shows the entire face. Let RS be this line. Then trace line TV, which must be divided at point X: this is VX, which is as long as the two heads or circa.[464] Trace the line perpendicular to VX up to Y; we will call this line XY. Then take the compass. Place one of the two ends on RS, and take the other end till it touches point 14 of circuit A in proper form. We must always understand this: that the end of the compass must always be placed on RS, and with the other end the mark [relative to the number in consideration] must be touched. Place the compass on XV and mark 14.[465] Take this into consideration: the quantity that you have taken from RS up to the mark on the circuit must be placed on line XV [starting from X]. Take the distance from RS to 15; place it on XV and mark 15. Take the distance from RS to 1; place it on XV and mark 1. Take the distance from RS to 6; place it on XV and mark 6. Take the distance from RS to 7; place it on XV and mark 7.

Second Step From points 14, 15, 1, 6, 7, trace lines parallel to XY with no limits. Let these lines be higher than the head and let them have A as a mark.[466] You already have the head with the entire face placed above the line that detaches from the back of the head with one eye only [figure B.154].[467] This line touches the back of the head and divides mN in right angles: we shall call this line Wp. Take the distance between line Wp and point 14 of circuit A of the head with the whole face and place this distance on the line originating from point 14 of line XV: mark 14. Take the distance between line Wp and point 15 of circuit A of the head with the whole face and place this distance on the line originating from point 15 of line XV: mark 15. Take the distance between line Wp and point 16 of circuit A of the head with the whole face and place this distance on the line originating from point 16 of line XV: mark 16. Take the distance between line Wp and point 1 and place this distance on the line originating from point 1 of line XV: mark 1. Take the distance between line Wp and point 6 and place this distance on the line originating from point 6 of line XV: mark 6. Take the distance between line Wp and point 7 and place this distance on the line originating from point 7 of line XV: mark 7. These are the points relative to A.

Heights of Points of Circuit B

First Step Take the distance from RS to point 14 of circuit B, place it on line XV and mark 14 [figure B.155]. Make a distinction between these numbers and the first ones. Take the distance from RS to point 15 of circuit B, place it on line XV and mark 15. Take the distance from RS to point 1 of circuit B, place it on line XV and mark 1. Take the distance from RS to point 6 of circuit B, place it on line XV and mark 6. Take the distance from RS to point 7 of circuit B, place it on line XV and mark 7.

Second Step Then trace all the lines parallel to line XY, which are marked on top by the letter B [figure B.155]. Take the distance between line Wp and point 14 of circuit B of the head with the entire face and place this distance on the line originating from point 14 of line XV: mark 14. Take the distance between line Wp and point 15 [of circuit B of the head with the entire face] and place this distance on the line originating from point 15 of line XV: mark 15. Take the distance between line Wp and point 1 [of circuit B of the head with the entire face] and place this distance on the line originating from point 1 of line XV: mark 1. Take the distance between line Wp and point 6 [of circuit B of the head with the entire face] and place this distance on the line originating from point 6 of line XV: mark 6. Take the distance between line Wp and point 7 [of circuit B of the head with the entire face] and place this distance on the line originating from point 7 of line XV: mark 7.[468]

Heights of Points of Circuit C

First Step Having reported the marks indicated earlier, take the compass, take the distance from RS to point 16 of circuit C, place it on line XV and mark 16 [figure B.156].[469] Take the distance from RS to point 17: place it on line XV and mark 17. Take the distance from RS to point 18 of circuit C: place it on line XV and mark 18. Take the distance from RS to point 19 of circuit C: place it on line XV and mark 19. Take the distance from RS to point 20 of circuit C: place it on line XV and mark 20. Take the distance from RS to point 20′ of circuit C: place it on line XV and mark 20′. Take the distance from RS to point 21 of circuit C: place it on line XV and mark 21. Take the distance from RS to point 22 of circuit C: place it on line XV and mark 22. Take the distance from RS to point 1 of circuit C: place it on line XV and mark 1. Take the distance from RS to point 2 of circuit C: place it on line XV and mark 2. Take the distance from RS to point 3 of circuit C: place it on line XV and mark 3. Take the distance from RS to point 4 of circuit C: place it on line XV and mark 4. Take the distance from RS to point 4′ of circuit C: place

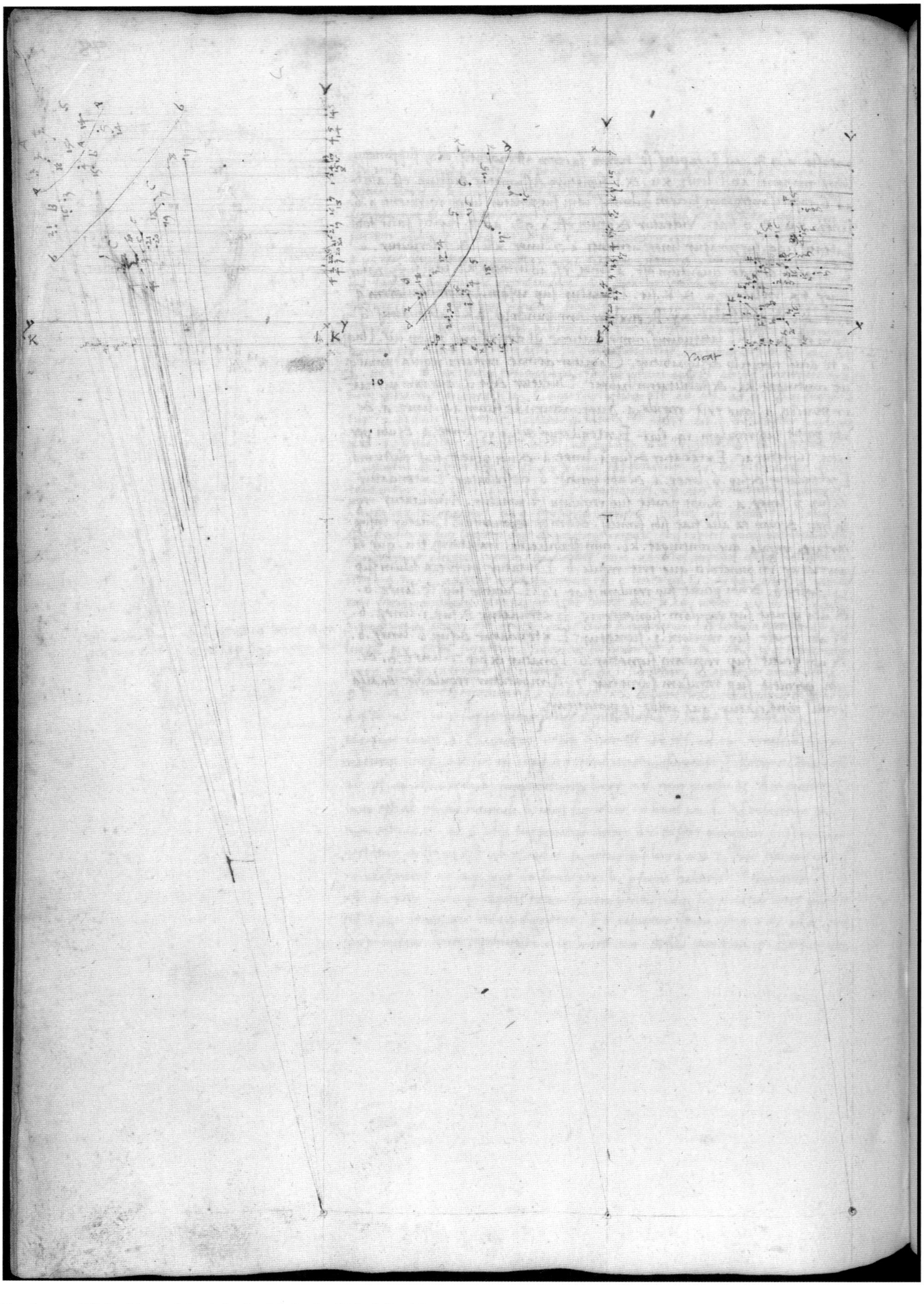

Bordeaux, Bibliothèque Municipale, Ms Latin 616, folio 88 verso.

it on line XV and mark 4′.[470] Take the distance from RS to point 5 of circuit C: place it on line XV and mark 5. Take the distance from RS to point 6 of circuit C: place it on line XV and mark 6.[471] Take the distance from RS to point 7 of circuit C: place it on line XV and mark 7.[472] Take the distance from RS to point 8 of circuit C: place it on line XV and mark 8.[473] Take the distance from RS to point y of circuit C: place it on line XV and mark y.[474]

Second Step Moreover, draw all the lines parallel to line XY with mark C. Take the distance between line Wp and point 16 of circuit C of the head with the entire face [figure B.156] and place this distance on the line originating from point 16 of line XV: mark 16.[475] Take the distance between line Wp and point 17 of circuit C of the head with the entire face and place this distance on the line originating from point 17 of line XV: mark 17. Take the distance between line Wp and point 18 of circuit C and place this distance on the line originating from point 18 of line XV: mark 18. Take the distance between line Wp and point 19 of circuit C and place this distance on the line originating from point 19 of line XV: mark 19. Take the distance between line Wp and point 20 of circuit C and place this distance on the line originating from point 20 of line XV: mark 20. Take the distance between line Wp and point 20′ of circuit C and place this distance on the line originating from point 20′ of line XV: mark 20′.[476] Take the distance between line Wp and point 21 of circuit C and place this distance on the line originating from point 21 of line XV: mark 21. Take the distance between line Wp and point 22 of circuit C and place this distance on the line originating from point 22 of line XV: mark 22.[477] Take the distance between line Wp and point 1 of circuit C and place this distance on the line originating from point 1 of line XV: mark 1. Take the distance between line Wp and point 2 of circuit C and place this distance on the line originating from point 2 of line XV: mark 2. Take the distance between line Wp and point 3 of circuit C and place this distance on the line originating from point 3 of line XV: mark 3. Take the distance between line Wp and point 4 of circuit C and place this distance on the line originating from point 4 of line XV: mark 4. Take the distance between line Wp and point 4′ of circuit C and place this distance on the line originating from point 4′ of line XV: mark 4′. Take the distance between line Wp and point 5 of circuit C and place this distance on the line originating from point 5 of line XV: mark 5. Take the distance between line Wp and point 6 of circuit C and place this distance on the line originating from point 6 of line XV: mark 6.[478] Take the distance between line Wp and point 7 of circuit C and place this distance on the line originating from point 7 of line XV: mark 7. Take the distance between line Wp and point 8 of circuit C and place this distance on the line originating from point 8 of line XV: mark 8.[479] Take the distance between line Wp and point x of circuit C and place this distance on the line originating from point x of line XV: mark x.[480]

Heights of Points of Circuit D

First Step[481] Now[482] take the distance from RS to point 15 of circuit D [figure B.157]: place it on line XV and mark 15. Then take the distance from RS to point 18 of circuit D: place it on line XV and mark 18. Take the distance from RS to point 1 of circuit D: place it on line XV and mark 1.[483] Take the distance from RS to point 2 of circuit D: place it on line XV and mark 2. Take the distance from RS to point 6 of circuit D: place it on line XV and mark 6.[484] Take the distance from RS to point 7 of circuit D: place it on line XV and mark 7. Take the distance from RS to point z of circuit D: place it on line XV and mark z.[485]

Second Step From all these marks draw lines parallel to line XY with no end [figure B.157]: let these be lines D. After doing this, take the distance between line Wp and point 15 of circuit D of the head with the entire face and place this distance on the line originating from point 15 of line XV: mark 15. Take the distance between line Wp and point 18 of circuit D and place this distance on the line originating from point 18 of line XV: mark 18. Take the distance between line Wp and point 1 of circuit D and place this distance on the line originating from point 1 of line XV: mark 1. Take the distance between line Wp and point 2 of circuit D and place this distance on the line originating from point 2 of line XV: mark 2. Take the distance between line Wp and point 6 of circuit D and place this distance on the line originating from point 6 of line XV: mark 6. Take the distance between line Wp and point 7 of circuit D and place this distance on the line originating from point 7 of line XV: mark 7. Take the distance between line Wp and point z of circuit D and place this distance on the line originating from point z of line XV: mark z.[486]

Heights of Points of Circuit E

First Step Take[487] the distance from RS to 16 of circuit E; place it on XV and mark 16 [figure B.158]. Take the distance from RS to 19 of circuit E; place it on XV and mark 19.[488] Take the distance from RS to 20 of circuit E; place it on XV and mark 20. Take the distance from RS to 1 of circuit E; place it on XV and mark 1. Take the distance from RS to 1′ of circuit E; place it on XV and mark 1′. Take the distance from RS to 2; place it on XV and mark 2. Take the distance from RS to 3 of circuit E; place it on XV and mark 3. Take the distance from RS to 7 of circuit E; place it on XV and mark 7. Take the distance from RS to x of circuit E; place it on XV and mark x.[489] Take the distance from RS to 8 of circuit E; place it on XV and mark 8. Take the distance from RS to 9 of circuit E; place it on XV and mark 9.[490]

Second Step From all these points, trace lines parallel to line XY: mark them with E [figure B.158]. Then go back to the head with the whole face. Take the distance between line Wp and point 16 of circuit E of the head that shows the entire face and place this distance on the line originating from point 16 of line XV: mark 16. Take the distance between line Wp and point 19 of circuit E and place this distance on the line originating from point 19 of line XV: mark 19. Take the distance between line Wp and point 20 of circuit E and place this distance on the line originating from point 20 of line XV: mark 20. Take the distance between line Wp and point 1 of circuit E and place this distance on the line originating from point 1 of line XV: mark 1. Take the distance between line Wp and point 1′ of circuit E and place this distance on the line originating from point 1′ of line XV: mark 1′. Take the distance between line Wp and point 2 of circuit E and place this distance on the line originating from point 2 of line XV: mark 2. Take the distance between line Wp and point 3 of circuit E and place this distance on the line originating from point 3 of line XV: mark 3. Take the distance between line Wp and point 7 of circuit E and place this distance on the line originating from point 7 of line XV: mark 7. Take the distance between line Wp and point x of circuit E and place this distance on the line originating from point x of line XV: mark x.[491] Take the distance between line Wp and point 8 of circuit E and place this distance on the line originating from point 8 of line XV: mark 8. Take the distance between line Wp and point 9 of circuit E and place this distance on the line originating from point 9 of line XV: mark 9.[492]

Heights of Points of Circuit F

First Step For circuit F, take the distance from RS to 14 of circuit F; place it on XV and mark 14 [figure B.159].[493] Take the distance from RS to 15 of circuit F; place it on XV and mark 15.[494] Take the distance from RS to 16 of circuit F; place it on XV and mark 16. Take the distance from RS to 1 of circuit F; place it on XV and mark 1. Take the distance from RS to 1′ of circuit F; place it on XV and mark 1′. Take the distance from RS to 1″ of circuit F; place it on XV and mark 1″. Take the distance from RS to 2 of circuit F; place it on XV and mark 2. Take the distance from RS to 5 of circuit F; place it on XV and mark 5. Take the distance from RS to 6 of circuit F; place it on XV and mark 6. Take the distance from RS to 7 of circuit F; place it on XV and mark 7.[495]

Second Step From all these points, trace lines parallel to line XY: mark them with F [figure B.159]. Then go back to the head with the whole face. Take the distance between line Wp and point 14 of circuit F of the head that shows the entire face and place this distance on the line originating from point 14 of line XV: mark 14. Take the distance between line Wp and point 16 of circuit F and place this distance on the line originating from point 16 of line XV: mark 16. Take the distance between line Wp and point 1 of circuit F and place this distance on the line originating from point 1 of line XV: mark 1. Take the distance between line Wp and point 1′ of circuit F and place this distance on the line originating from point 1′ of line XV: mark 1′. Take the distance between line Wp and point 1″ of circuit F and place this distance on the line originating from point 1″ of line XV: mark 1″. Take the distance between line Wp and point 2 of circuit F and place this distance on the line originating from point 2 of line XV: mark 2. Take the distance between line Wp and point 5 of circuit F and place this distance on the line originating from point 5 of line XV: mark 5. Take the distance between line Wp and point 6 of circuit F and place this distance on the line originating from point 6 of line XV: mark 6. Take the distance between line Wp and point 7 of circuit F and place this distance on the line originating from point 7 of line XV: mark 7.[496]

Heights of Points of Circuit G

First Step Then for what concerns circuit G,[497] take the distance from RS to 14 of circuit G; place it on XV and mark 14 [figure B.160]. Take the distance from RS to 15 of circuit G; place it on XV and mark 15.[498] Take the distance from RS to 16 of circuit G; place it on XV and mark 16. Take the distance from RS to 1 of circuit G; place it on XV and mark 1. Take the distance from RS to 2 of circuit G; place it on XV and mark 2. Take the distance from RS to 3 of circuit G; place it on XV and mark 3.[499] Take the distance from RS to 4 of circuit G; place it on XV and mark 4.[500] Take the distance from RS to 5 of circuit G; place it on XV and mark 5.[501] Take the distance from RS to 6 of circuit G; place it on XV and mark 6. Take the distance from RS to 7 of circuit G; place it on XV and mark 7.[502]

Second Step From all these points, trace lines parallel to line XY: mark them with G [figure B.160]. Then take the distance between line Wp and point 14 of circuit G of the head that shows the entire face and place this distance on the line originating from point 14 of line XV: mark 14.[503] Take the distance between line Wp and point 15 of circuit G and place this distance on the line originating from point 15 of line XV: mark 15.[504] Take the distance between line Wp and point 16 of circuit G and place this distance on the line originating from point 16 of line XV: mark 16. Take the distance between line Wp and point 1 of circuit G and place this distance on the line originating at point 1 of line XV: mark 1. Take the distance between line Wp and point 2 of circuit G and place this distance on the line originating from point 2 of line XV: mark 2.[505] Take the distance between line Wp and point 3 of circuit G and place this distance on the line originating from point 3 of line XV: mark 3.[506] Take the distance between line Wp and point 4 of circuit G and place this distance on the line originating from point 4 of line XV: mark 4.[507] Take the distance between line Wp and point 5 of circuit G and place this distance on the line originating from point 5 of line XV:

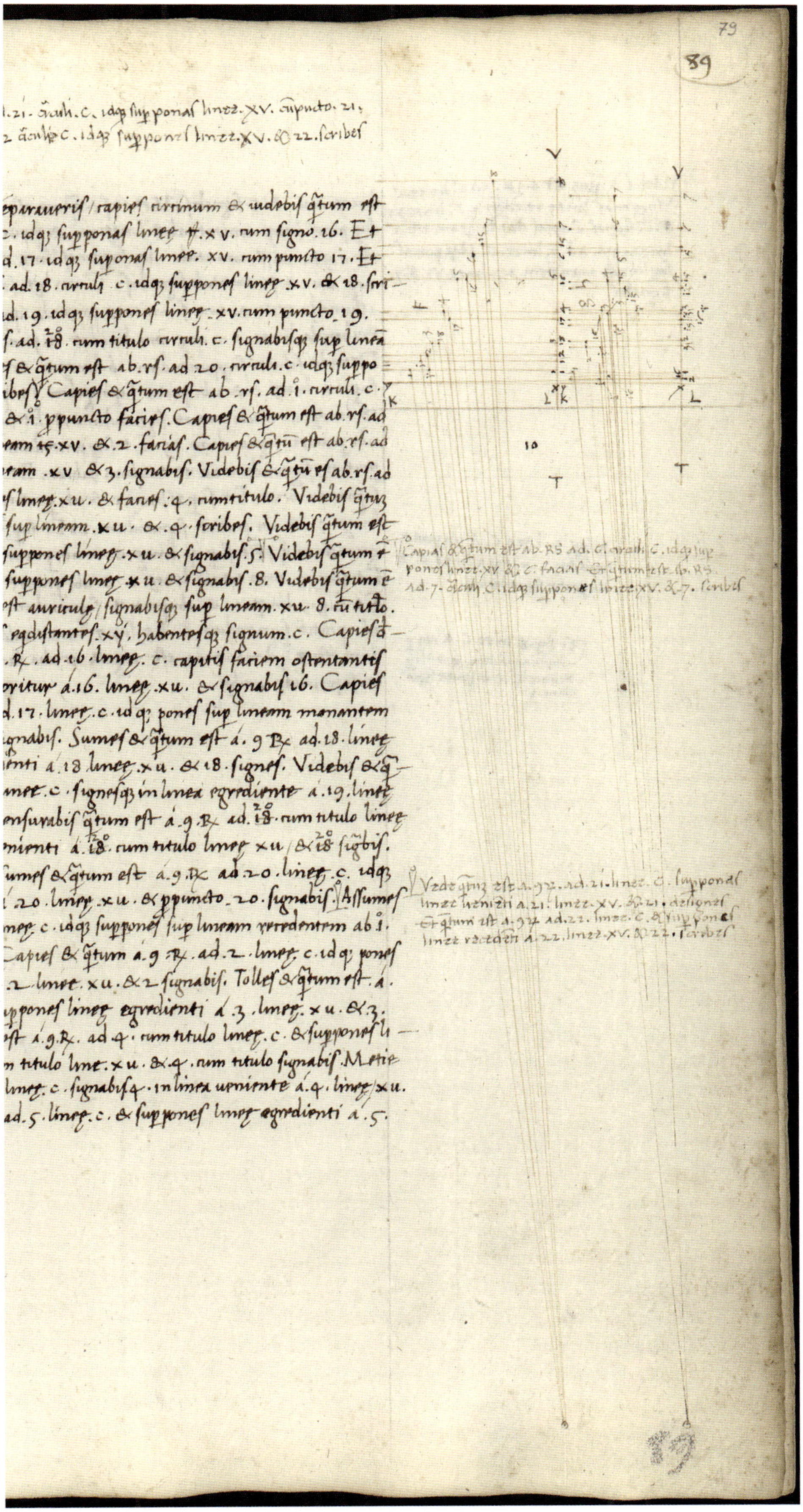

.21. circuli. c. idq; supponas lineę. XV. cu puncto. 21.
2 circuli c. idq; suppones lineę. XV. & 22. scribes

separaueris / capies circinum & uidebis quantum est
c. idq; supponas lineę .xv. cum signo. 16. Et
d. 17. idq; suponas lineę. xv. cum puncto 17. Et
ad. 18. circuli c. idq; suppones lineę .xv. & 18. scri
d. 19. idq; suppones lineę .xv. cum puncto 19.
s. ad. 18. cum titulo circuli. c. signabisq; sup linea
s & quantum est ab. rs. ad 20. circuli. c. idq; suppo
ibes. Capies & quantum est ab. rs. ad. 1. circuli. c.
& 1. p puncto facies. Capies & quantum est ab. rs. ad
eam .xv. & 2. facias. Capies & quantum est ab. rs. ad
eam .xv & 3. signabis. Videbis & quantum est ab. rs. ad
s lineę. xu. & facies. 4. cum titulo. Videbis quantum
sup lineam. xu. & 4. scribes. Videbis quantum est
suppones lineę. xu. & signabis. 5. Videbis quantum e
suppones lineę. xu. & signabis. 8. Videbis quantum e
est auriculę signabisq; sup lineam. xu. 8. cu titulo.
equidistantes. xv. habentesq; signum. c. Capies q
. R. ad. 16. lineę. c. capitis faciem ostentantis
oritur a. 16. lineę. xu. & signabis 16. Capies
d. 17. lineę. c. idq; pones sup lineam manantem
signabis. Sumes & quantum est a. 9. R. ad. 18. lineę
ienti a. 18 lineę. xu. & 18. signes. Videbis & q
lineę. c. signesq; in linea egrediente a. 19. lineę
ensurabis quantum est a. 9. R. ad. 18. cum titulo lineę
nienti a. 18. cum titulo lineę xu & 18 signabis.
Sumes & quantum est a. 9. R. ad. 20. lineę. c. idq;
20. lineę. xu. & p puncto. 20. signabis. Assumes
lineę. c. idq; suppones sup lineam recedentem ab 1.
Capies & quantum a. 9. R. ad. 2. lineę. c. idq; pones
2. lineę. xu. & 2 signabis. Tolles & quantum est. a.
uppones lineę egredienti a. 3. lineę. xu. & 3.
est a. 9. R. ad. 4. cum titulo lineę. c. & suppones li
n titulo line. xu. & 4. cum titulo signabis. Metie
lineę. c. signabisq; in linea ueniente a. 4. lineę xu.
ad. 5. lineę. c. & suppones lineę egredienti a. 5.

Bordeaux, Bibliothèque Municipale, Ms Latin 616, folio 89 recto.

mark 5.[508] Take the distance between line Wp and point 6 of circuit G and place this distance on the line originating from point 6 of line XV: mark 6. Take the distance between line Wp and point 7 of circuit G and place this distance on the line originating from point 7 of line XV: mark 7.[509]

Heights of Points of Circuit H

First Step[510] Take the distance from RS to 14 of circuit H; place it on XV and mark 14 [figure B.161].[511] Take the distance from RS to 15 of circuit H; place it on XV and mark 15.[512] Take the distance from RS to 15′ of circuit H; place it on XV and mark 15′. Take the distance from RS to 16 of circuit H; place it on XV and mark 16. Take the distance from RS to 16′ of circuit H; place it on XV and mark 16′. Take the distance from RS to 1 of circuit H; place it on XV and mark 1. Take the distance from RS to 1′ of circuit H; place it on XV and mark 1′. Take the distance from RS to 2 of circuit H; place it on XV and mark 2. Take the distance from RS to 2′ of circuit H; place it on XV and mark 2′. Take the distance from RS to 3 of circuit H; place it on XV and mark 3. Take the distance from RS to 3′ of circuit H; place it on XV and mark 3′. Take the distance from RS to 4 of circuit H; place it on XV and mark 4. Take the distance from RS to 5 of circuit H; place it on XV and mark 5. Take the distance from RS to 6 of circuit H; place it on XV and mark 6. Take the distance from RS to 7 of circuit H; place it on XV and mark 7.[513]

Second Step From all these points, trace lines parallel to line XY: mark them with H [figure B.161]. Then go back to the head with the whole face, that is, the head that has two eyes. Take the distance between line Wp and point 14 of circuit H and place this distance on the line originating from point 14 of line XV: mark 14.[514] Take the distance between line Wp and point 15 of circuit H and place this distance on the line originating from point 15 of line XV: mark 15. Take the distance between line Wp and point 15′ of circuit H and place this distance on the line originating from point 15′ of line XV: mark 15′. Take the distance between line Wp and point 16 of circuit H and place this distance on the line originating from point 16 of line XV: mark 16. Take the distance between line Wp and point 16′ of circuit H and place this distance on the line originating from point 16′ of line XV: mark 16′. Take the distance between line Wp and point 1 of circuit H and place this distance on the line originating from point 16 of line XV: mark 1. Take the distance between line Wp and point 1′ of circuit H and place this distance on the line originating from point 1′ of line XV: mark 1′. Take the distance between line Wp and point 2 of circuit H and place this distance on the line originating from point 2 of line XV: mark 2. Take the distance between line Wp and point 2′ of circuit H and place this distance on the line originating from point 2′ of line XV: mark 2′. Take the distance between line Wp and point 3 of circuit H and place this distance on the line originating from point 3 of line XV: mark 3. Take the distance between line Wp and point 3′ of circuit H and place this distance on the line originating from point 3′ of line XV: mark 3′.[515] Take the distance between line Wp and point 4 of circuit H and place this distance on the line originating from point 4 of line XV: mark 4. Take the distance between line Wp and point 5 of circuit H and place this distance on the line originating from point 5 of line XV: mark 5. Take the distance between line Wp and point 6 of circuit H and place this distance on the line originating from point 6 of line XV: mark 6. Take the distance between line Wp and point 7 of circuit H and place this distance on the line originating from point 7 of line XV: mark 7 [figure B.162].[516]

The Terminal in the New Elevation[517] Take the distance [figure B.163] from RS [of the width] to line KL [of the width] and place it on TX starting from X, on the side: mark K. Trace KL [of the height] perpendicularly to TX which [KL] will be parallel to XY.

The Point of View in the New Elevation Get as far away from line KL of the height as point O of line KL of the width [figure B.163]: with point O, above or below, however you would like it. In this point O, place the needle with thread.

Paper Ruler A Take a paper ruler and place it along KL so that it stays still [figures B.164, B.168]. Trace line TV which cuts KL at point A: you have ruler A. Then take the thread and place it on point 14 of line A: mark 14 where it intersects the ruler. Place the thread on point 15 of line A: mark 15 where it intersects the ruler. Place the thread on point 16 of line A: mark 16 where it intersects the ruler.[518] Place the thread on point 1 of line A: mark 1 where it intersects the ruler. Place the thread on point 2 of line A: mark 2 where it intersects the ruler.[519] Place the thread on point 3 of line A: mark 3 where it intersects the ruler. Place the thread on point 4 of line A: mark 4 where it intersects the ruler. Place the thread on point 5 of line A: mark 5 where it intersects the ruler. Place the thread on point 6 of line A: mark 6 where it intersects the ruler. Place the thread on point 7 of line A: mark 7 where it intersects the ruler. Take this ruler off, make a similar one, and put both aside.[520]

Paper Ruler B Take another paper ruler and place it along KL so that it stays still [figures B.164, B.168]. Trace line TV which cuts KL at point B: you have ruler B. Then take the thread and place it on point 14 of line B: mark 14 where it intersects the ruler. Place the thread on point 15 of line B: mark 15 where it intersects the ruler. Place the thread on point 16 of line B: mark 16 where it intersects the ruler.[521] Place the thread on point 1 of line B: mark 1 where it intersects the ruler. Place the thread on point 2 of line B: mark 2 where it intersects the ruler.[522] Place the thread on point 3 of line B: mark 3 where it intersects the ruler. Place the thread on point 4 of line B: mark 4 where it intersects the ruler. Place the thread on point 5 of line B: mark 5 where it intersects the ruler. Place the thread on point 6 of line B: mark 6 where it intersects the ruler. Place the thread on point 7 of line B: mark 7 where it intersects the ruler. Take this ruler off, make a similar one, and put both aside.[523]

Paper Ruler C[524] Having done this, take a paper ruler and place it along KL so that it stays still [figures B.164, B.168]. Trace line TV which cuts KL at point C: you have ruler C. Then take the thread and place it on point 17 of line C: mark 17 where it intersects the ruler. Place the thread on point 18 of line C: mark 18 where it intersects the ruler. Place the thread on point 19 of line C: mark 19 where it intersects the ruler. Place the thread on point 20 of line C: mark 20 where it intersects the ruler. Place the thread on point 20′ of line C: mark 20′ where it intersects the ruler. Place the thread on point 21 of line C: mark 21 where it intersects the ruler. Place the thread on point 22 of line C: mark 22 where it intersects the ruler. Place the thread on point 1 of line C: mark 1 where it intersects the ruler. Place the thread on point 2 of line C: mark 2 where it intersects the ruler. Place the thread on point 3 of line C: mark 3 where it intersects the ruler. Place the thread on point 4 of line C: mark 4 where it intersects the ruler. Place the thread on point 4′ of line C: mark 4′ where it intersects the ruler. Place the thread on point 5 of line C: mark 5 where it intersects the ruler. Place the thread on point 8 of line C: mark 8 where it intersects the ruler.[525] Place the thread on point y of line C: mark y where it intersects the ruler.[526] Take this ruler off, make a similar one, and put both aside.[527]

Paper Ruler D[528] Now, take into consideration the heights of line D which you will place on the ruler this way [figures B.165, B.168].[529] Take a paper ruler and place it along KL so that it stays still. Trace line TV which cuts KL at point D: you have ruler D. Then take the thread and place it on point 15 of line D: mark 15 where it intersects the ruler. Place the thread on point 18 of line D: mark 18 where it intersects the ruler. Place the thread on point 1 of line D: mark 1 where it intersects the ruler. Place the thread on point 2 of line D: mark 2 where it intersects the ruler. Place the thread on point 6 of line D: mark 6 where it intersects the ruler. Place the thread on point 8 of line D: mark 8 where it intersects the ruler.[530] Take this ruler off, make a similar one.[531]

Paper Ruler E[532] This being the situation, take a paper ruler and place it along KL so that it stays still [figures B.165, B.168]. Then trace line TV which cuts KL at point E.[533] Place the thread on point 16 of line E: mark 16 where it intersects the ruler.[534] Place the thread on point 17 of line E: mark 17 where it intersects the ruler.[535] Place the thread on point 18 of line E: mark 18 where it intersects the ruler. Place the thread on point 19 of line E: mark 19 where it intersects the ruler. Place the thread on point 20 of line E: mark 20 where it intersects the ruler. Place the thread on point 1 of line E: mark 1 where it intersects the ruler. Place the thread on point 1′ of line E: mark 1′ where it intersects the ruler.[536] Place the thread on point 2 of line E: mark 2 where it intersects the ruler.[537] Place the thread on point 3 of line E: mark 3 where it intersects the ruler. Place the thread on point 4 of line E: mark 4 where it intersects the ruler.[538] Place the thread on point 5 of line E: mark 5 where it intersects the ruler. Place the thread on point 6 of line E: mark 6 where it intersects the ruler. Place the thread on point 7 of line E: mark 7 where it intersects the ruler. Place the thread on point x of line E: mark x where it intersects the ruler.[539] Place the thread on point 8 of line E: mark 8 where it intersects the ruler. Place the thread on point 9 of line E: mark 9 where it intersects the ruler.[540] Having done this, take this ruler off, make a similar one, and put them both aside.[541]

Paper Ruler F After you've done this, take a paper ruler and place it along KL so that it stays still [figures B.166, B.168]. Trace line TV which cuts KL at point F: we shall call this ruler F. Place the thread on point 14 of line F: mark 14 where it intersects the ruler. Place the thread on point 15 of line F: mark 15 where it intersects the ruler.[542] Place the thread on point 16 of line F: mark 16 where it intersects the ruler. Place the thread on point 17 and on point 18: mark 17 and 18 where it intersects the ruler.[543] Place the thread on point 1 of line F: mark 1 where it intersects the ruler. Place the thread on point 1′ of line F: mark 1′ where it intersects the ruler. Place the thread on point 1″ of line F: mark 1″ where it intersects the ruler. Place the thread on point 2 of line F: mark 2 where it intersects the ruler. Place the thread on point 3 of line F: mark 3 where it intersects the ruler.[544] Place the thread on point 4 of line F: mark 4 where it intersects the ruler. Place the thread on point 5 of line 5: mark 5 where it intersects the ruler. Place the thread on point 6 of line F: mark 6 where it intersects the ruler. Place the thread on point 7 of line F: mark 7 where it intersects the ruler. Take this ruler off, make a similar one, and put them both aside.[545]

Paper Ruler G[546] Take a paper ruler and place it along KL so that it stays still [figures B.167, B.168]. Then trace line TV which cuts KL at point G: we will call this ruler G. Place the thread on point 10 of line G: mark 10 where it intersects the ruler.[547] Place the thread on point 11 of line G: mark 11 where it intersects the ruler. Place the thread on point 14 of line G: mark 14 where it intersects the ruler. Place the thread on point 16 of line G: mark 16 where it intersects the ruler. Place the thread on point 1 of line G: mark 1 where it intersects the ruler. Place the thread on point 2 of line G: mark 2 where it intersects the ruler. Place the thread on point 3 of line G: mark 3 where it intersects the ruler.[548] Place the thread on point 4 of line G: mark 4 where it intersects the ruler.[549] Place the thread on point 5 of line G: mark 5 where it intersects the ruler. Place the thread on point 6 of line G: mark 6 where it intersects the ruler. Place the thread on point 7 of line G: mark 7 where it intersects the ruler. Take this ruler off, make a similar one, and put them both aside.[550]

Paper Ruler H[551] Take a paper ruler and place it along KL as you have done earlier [figures B.167, B.168]. Then trace line TV which cuts KL at point H: we will call this ruler H. Place the thread on point 14 of line H: mark 14 where it intersects the ruler. Place the thread on point 15 of line H: mark 15 where it intersects the ruler. Place the thread on point 15′ of line H: mark 15′ where it intersects the ruler. Place the thread on point 16 of line H: mark 16 where it intersects the ruler. Place the thread on point 16′ of line H: mark 16′ where it intersects the ruler. Place the thread on point 1 of line H: mark 1 where it intersects the ruler. Place the thread on point 1′ of line H: mark 1′ where it intersects the ruler. Place the thread on point 2 of line H: mark 2 where it intersects the ruler. Place the thread on point 2′ of line H: mark 2′ where it intersects the ruler. Place the thread on point 3 of line H: mark 3 where it intersects the ruler. Place the thread on point 3′ of line H: mark 3′ where it intersects the ruler. Place the thread on point 4 of line H: mark 4 where it intersects the ruler. Place the thread on point 5 of line H: mark 5 where it intersects the ruler. Place the thread on point 6 of line H: mark 6 where it intersects the ruler. Place the thread on point 7 of line H: mark 7 where it intersects the ruler.[552] Take the ruler off, make a similar one, and put them both aside.[553]

Final Considerations Thus far, you have the height and the width on the rulers, although the marks on the rulers in elevation have not been reported as much as the ones on the rulers of the width. This was done for conciseness: when drawing the marks of the height,[554] only the necessary ones were reported.[555] For this reason, when you will place the wooden ruler [which is the width ruler] on the paper rulers [which are the height rulers], observe where the wooden ruler touches the paper rulers. Take into consideration this mark on the wooden ruler: whenever this contact takes place, mark a point. Do this for all the rulers. It would be best if all the marks present on the rulers of the width had a corresponding one in height. After understanding this method, through the procedure I have shown, you will be able to easily compare width and height even if you have to make more marks.

The Shortening of the Head Rulers A/A/A Trace a straight line, let it be KL: divide it at point M [figure B.169]. Trace perpendicular N on M: this is line MN. Trace perpendicular line KH on K and perpendicular LI on L. Successively, take the two paper lines marked with A; place them so that they coincide: the first with KH, the other with LI. Let point A on both lines touch KL. Take wooden ruler A; place it so that it coincides with point 1 of the two rulers. Also point M of the two rulers should always touch MN: when there is point 1 on the wooden ruler, mark 1. Take the ruler that coincides with point 15 of the two rulers. Also point M will touch line MN: where point 15 is on the ruler, mark 15. Take the ruler that coincides with point 14 of the two rulers. Also point M will touch line MN: where point 14 is on the ruler, mark 14. Take the ruler that coincides with point 6 of the two rulers. Also point M will touch line MN: where point 6 is on the ruler, mark 6. Take the ruler that coincides with point 7 of the two rulers. Also point M will touch line MN: where point 7 is on the ruler, mark 7. When I say mark 6 or 7, it will be enough to put a very small point where you place the marks for the head.

The Remaining Rulers Do the same thing for all the other rulers in moving the paper and wooden rulers as you have done till now for past figures [figure B.170]. Let the paper rulers always touch lines KH and LI, let marks B, C, D, E, F, G, H lie on KL. One must also always take into consideration that point M of the wooden rulers lies on line MN.

The Outline of All the Parts of the Head Finally, after all these things have been executed, you shall realize the outline of all the parts of the head [figure B.171], so that in all respects the eyes, the nose, the mouth, the ears will have a convenient shape, without getting further away from the established terms, as it was done when delimiting the outline of the other head [figures B.172, B.173].[556]

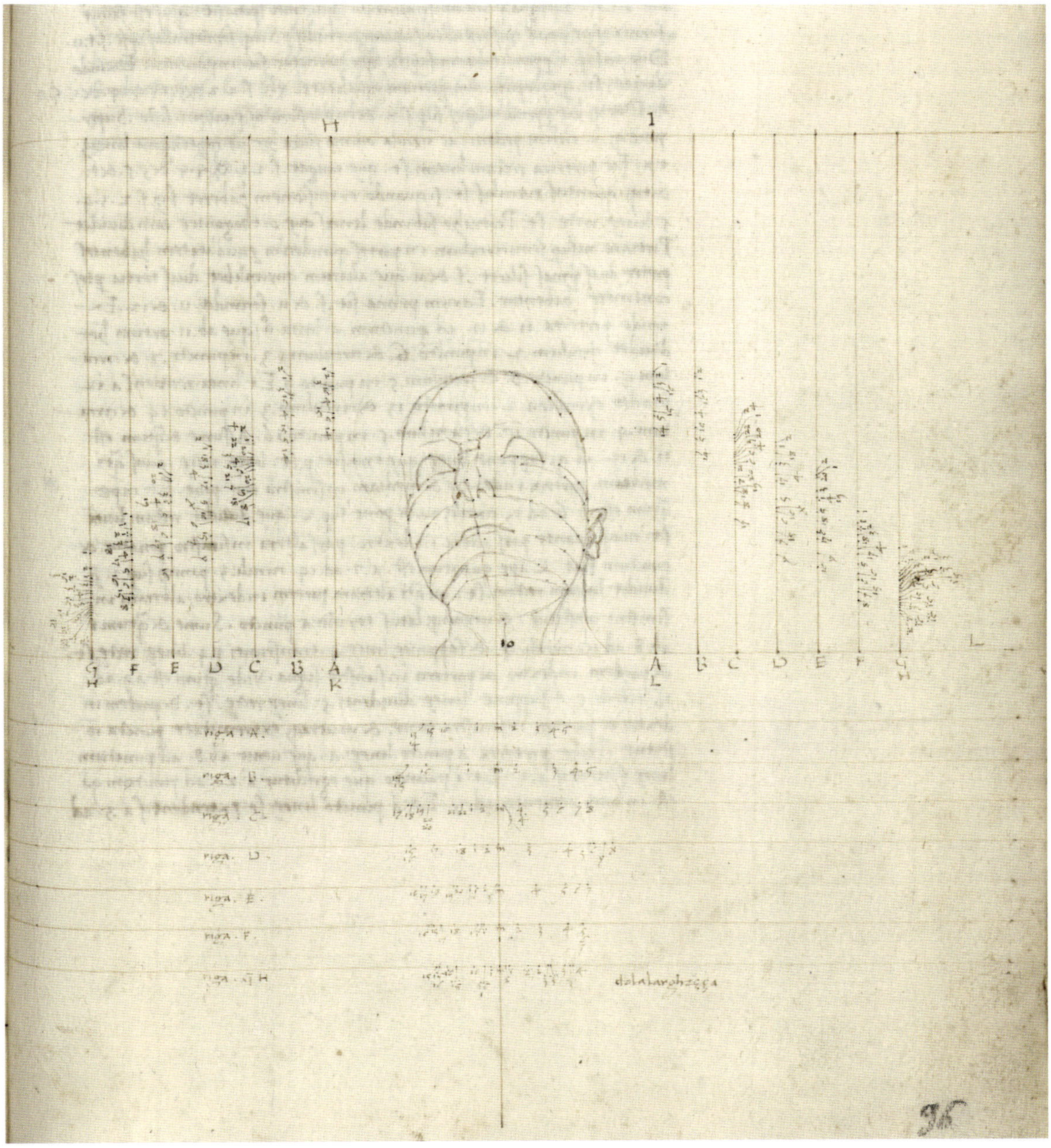

Bordeaux, Bibliothèque Municipale, Ms Latin 616, folio 96 recto.

Proposition IX

If you might need to represent appropriately a vault, what one also commonly calls a dome, and you need to divide it into squares in which you need to place the coffers, proceed in the following manner when realizing this work.

The Barrel Vault and the Arch of the Dome First draw a semicircle[557] as wide as the barrel vault or the arch of the dome you want to build [figure B.174]. This is the semicircle STV. Draw a perpendicular line to SV from T which divides SV at point X. Successively, divide ST into five equal parts, that is, S-2, 2-3, 3-4, 4-5, 5-T. Trace the perpendicular lines from these points to SX and at the meeting points mark points. Place the compass's tip on X, and from all these points draw a [portion of a] circle [from there] till you touch line TX. Moreover, draw straight line ST, which contains the subdivisions S-2, 2-3, 3-4, 4-5, 5-T [figure B.175]. Let this straight line be as long as the arch ST [figure B.174] develops points S, 2, 3, 4, 5, T in length. Trace from these points lines which divide ST orthogonally [figure B.175]. Divide a similar semicircle into fifteen equal parts [figure B.176], with the exception of the first two parts, one on the side of S and the other on the side of V, which should be two-thirds of the others. The first one [figure B.174] is S-11; the second one is 11-12. Trace lines from points 11 and 12 to point X. The line which originates at point 11 will meet circle 2 at point 6, circle 3 at point 7, circle 4 at point 8, circle 5 at point 9. The line which originates at point 12 will meet circle 2 at point 13, circle 3 at point 14, circle 4 at point 15, circle 5 at point 16.

The Slice of the Dome Take the length [of the arch] [figure B.174] between points 11 and 12 and place it [vertically] [figure B.175] on line ST: place it half above on the right and half below on the left of this line, starting from point S.[558] Take the length [figure B.174] between points 6 and 13 in circle 2 and place it [vertically] [figure B.175] on the line through point 2 that divides straight line ST, half on the right and half on the left, and place a point [at both ends]. Take the length [figure B.174] between points 7 and 14 in circle 3 and place it [vertically] [figure B.175] on the line through point 3 that divides straight line ST, half on the right and half on the left, and place a point [at both ends]. Take the length [figure B.174] between points 8 and 15 in circle 4 and place it [vertically] [figure B.175] on the line through point 4 that divides straight line ST, half on the right and half on the left, and place a point [at both ends]. Take the length [figure B.174] between points 9 and 16 in circle 5 and place it [vertically] [figure B.175] on the line through point 5 that divides straight line ST, half on the right and half on the left, and place a point [at both ends]. Then trace a line from the point on the line that originates from S to the point on the line that originates from 2. Then trace a line from the point on the line that originates from 2 to the point on the line that originates from 3. Then trace a line from the point on the line that originates from 3 to the point on the line that originates from 4. Then trace a line from the point on the line that originates from 4 to the point on the line that originates from 5. Then trace a line from the point on the line that originates from 5 to point T; you could do the same thing for the opposite side.[559]

The Seven Circles Having done this, take the perpendicular quantity between line SV [figure B.174] and point 11 which is on circle STV; place it [the perpendicular quantity] so that it divides line ST at point 1 [figure B.175]. Take the compass and trace a circle so that it touches point 1 and both lines above and below next to line ST, and let's say that these lines delimit a slice, so to speak, of the dome. Beyond this circle, toward point T trace a second circle tangent to the first one and to the two lines that delimit the slice [of the dome]. Beyond this circle, trace another circle tangent to that which precedes and to the two lines [that delimit the slice of the dome]. And proceed this way till the seventh one, as you may see in the figure.[560]

A Quarter of a Circle Having done this, with the compass take the length between S and X [figure B.174], and on a [horizontal] line draw the arch ST [figure B.177], which is a quarter of a circle and corresponds to SAT. Let A be a right angle which matches the center of the quarter-circle ST. Take the length [figure B.175] between S and 1 on line ST of the slice of the dome and place it [figure B.177] above S of the quarter of the circle and mark a point [that is, 1]. Take the length [figure B.175] from point 1 to the contact of the first circle [with the second one] and place it [figure B.177] above the point which you marked in the quarter-circle ST, and mark a point [that is, 2]. Take the length [figure B.175] of the diameter of the second circle in the slice and place it [figure B.177] above point 2 of circle ST and mark point 3. Take the length [figure B.175] of the diameter of the third circle in the slice and place it [figure B.177] above point 3 of circle ST, and mark point 4. Take the length [figure B.175] of the diameter of the fourth circle of the slice and place it [figure B.177] above point 4 of circle ST, and mark point 5. Take the length [figure B.175] of the diameter of the fifth circle in the slice of the dome and place it [figure B.177] above point 5 of circle ST, and mark point 6. Take the length [figure B.175] of the diameter of the sixth circle in the slice of the dome and place it [figure B.177] above point 6 of circle ST, and mark point 7. Take the length [figure B.175] of the diameter of the seventh circle of the slice of the dome and place it [figure B.177] above point 7 of circle ST, and mark point 8.[561]

A Second Circuit Moreover build a second circuit which goes around ST [figure B.178]. Begin beyond S, as much as you want the coffers to be leaning out, and make this circuit little by little so that it ends in T, and it is TY.[562] After that, continue the seven division points, that is, 1, 2, 3, 4, 5, 6, 7, to make eight subdivisions on circuit ST using lines parallel to AS till they intersect line AT and circuit YT.

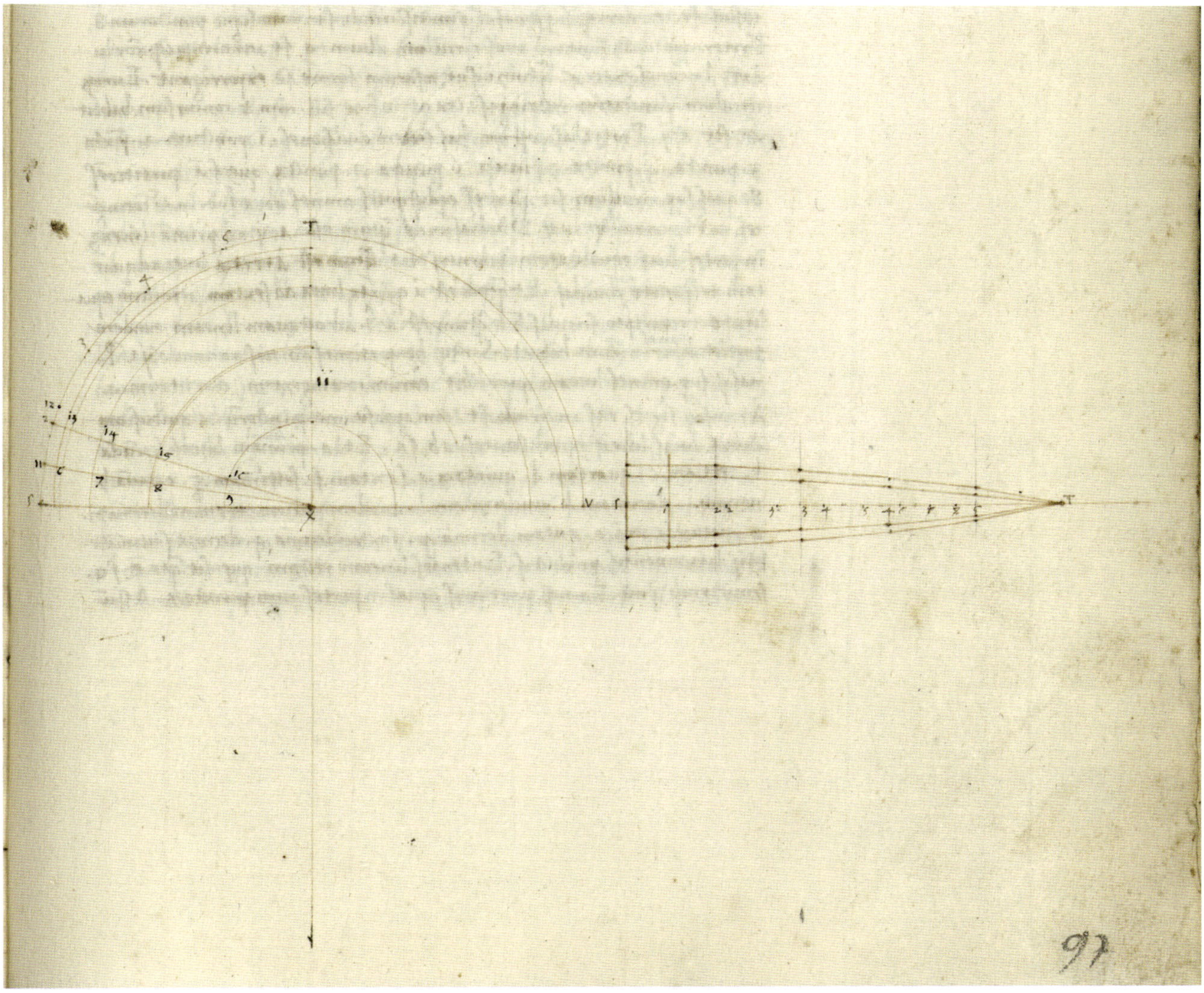

Bordeaux, Bibliothèque Municipale, Ms Latin 616, folio 97 recto.

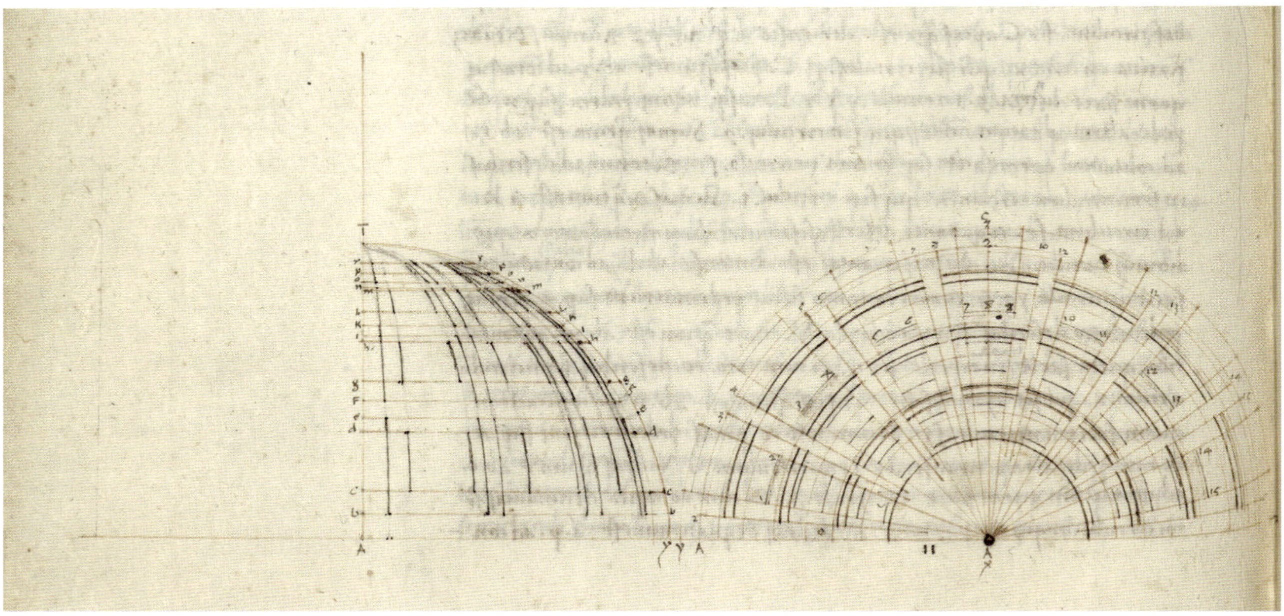

Bordeaux, Bibliothèque Municipale, Ms Latin 616, folio 98 verso.

The Subdivisions Divide the length that goes from Y to the second line [of the subdivisions] into two equal parts [figure B.179].[563] Divide the length between the third line and the fourth line into three equal parts. Divide the length between the fifth and the sixth line into three equal parts. Divide the length between the seventh and the eighth line by executing the same divisions. Let all these subdivisions be on circuit YT. Put on all points of these subdivisions a line so that it passes through point A, which is the center.[564] Moreover mark them on circle ST. Trace lines parallel to SA [figure B.180] from all these points.[565] A is the first line, B is the second one, C is the third one, D is the fourth one, E is the fifth one, F is the sixth one, G is the seventh one, H is the eighth one, I is the ninth one, K is the tenth one, L is the eleventh one, M is the twelfth one, N is the thirteenth one, O is the fourteenth one, P is the fifteenth one, Q is the sixteenth one, R is the seventeenth one.[566]

The Semicircles in Plan Having done this, draw a straight line which is as long as SV [figure B.181], the diameter of the first semicircle, and divide it [the straight line] in half at point X. Take length AS, place an end of the compass at point X of the division of the line, and with the other end trace semicircle AsT with this length. Take length AY and trace circle YzZ going around AsT.[567] Then divide semicircle YzZ into fifteen equal parts, with the exception of the first one and the last one which must be two-thirds of the others. Then trace the lines from the division points to the center X. Divide the first and the last subdivision on circle AsT into two equal parts and trace the lines that go through center X. Divide the other subdivisions on circle AsT into three equal parts and trace the lines that go through center X. Take the length of line B of the second figure [figure B.182],[568] from B to circle ST, which corresponds to the fourth part of a circle, and with that length place an end of the compass on point X and with the other end draw circle B.[569] Take the length from C to circuit YT; place an end of the compass on X and with the other end draw circle C. Take the length from D to the contact of line D on circuit YT; with this length place an end of the compass on X and with the other end draw circle D. Take the length from E to the contact of line E in circle ST; place an end of the compass on X and draw circle E. Take the length from F to the contact of line F in circle ST; place an end of the compass on X and with the other end draw circle F. Take the length from G to the contact of line G on circuit YT, place an end of the compass on X and draw circle G. Take the length from H to the contact of line H on circuit YT; place an end of the compass on X and with the other end draw circle H. Take the length from I to the contact of line I in circle ST, draw a semicircle around X;[570] this is circle I. Take the length from K to the contact of line K in circle ST, draw a semicircle around X; this is circle K. Take the length from L to the contact of line L on circuit YT; place an end of the compass on X, with the other end draw semicircle L around it. Take the length from M to the contact of line M on circuit YT, draw a semicircle around X; this is circle M. Take the length from N to the contact of line N in circle ST; place an end of the compass on X, and with the other end draw semicircle N around it. Take the length from O to the contact of line O in circle ST; place an end of the compass on X, and with the other end draw semicircle O surrounding X. Take the length from P to the contact of line P on circuit YT; draw a semicircle surrounding X. This is circle P. Take the length from Q to the contact of line Q on circuit YT; place an end of the compass on X and with the other end draw semicircle Q. Take the length from R to the contact of line R in circle ST, place an end of the compass on X and with the other end draw a semicircle. This is circle R.

The Subdivisions of the Semicircles in Plan Up until this point we have realized circles divided by lines that start at center X [figure B.182], end on semicircle AST,[571] and divide semicircle A [figure B.183], semicircle B, semicircle E, semicircle F, semicircle I, semicircle K, semicircle N, semicircle O, semicircle R with a total of 17 points of subdivisions [figure B.184].[572] The lines that start at center X and end on semicircle YzZ divide semicircle C, semicircle D, semicircle G, semicircle H, semicircle L, semicircle M, semicircle P, semicircle Q with a total of 15 points of subdivisions.[573] Having done this, you will have completed the figure of width, which shows its appropriate form [figure B.183].

Line A of the Figure of the Elevation We still need to complete the figure of the elevation: that is, AST, which is one quarter of the circle [figure B.184]. Place the compass by placing an end on line AX of the width. Extend the other end to point 2 of circle A; mark this length on line A of the elevated figure: let it be A-2. Take the length between AX and point 3 of circle A; place it on line A of the elevation and mark 3. Take the length between AX and point 4 of circle A; place it on line A of the elevation and mark 4. Take the length between AX and point 5 of circle A; place it on line A of the elevation and mark 5. Take the length between AX and point 6 of circle A; place it on line A of the elevation and mark 6. Take the length between AX and point 7 of circle A; place it on line A of the elevation and mark 7. Take the length between AX and point 8 of circle A; place it on line A of the elevation and mark 8. Take the length between AX and point 9 of circle A; place it on line A of the elevation and mark 9. Since there are seventeen numbers, we need point 1 to serve for 1 and 17,[574] 2 to serve for 2 and 16, 3 to serve for 3 and 15, 4 to serve for 4 and 14, 5 to serve for 5 and 13, 6 to serve for 6 and 12, 7 to serve for 7 and 11, 8 to serve for 8 and 10, 9 to serve only for 9: that is the end of the elevation which is represented only in half. You have line A.

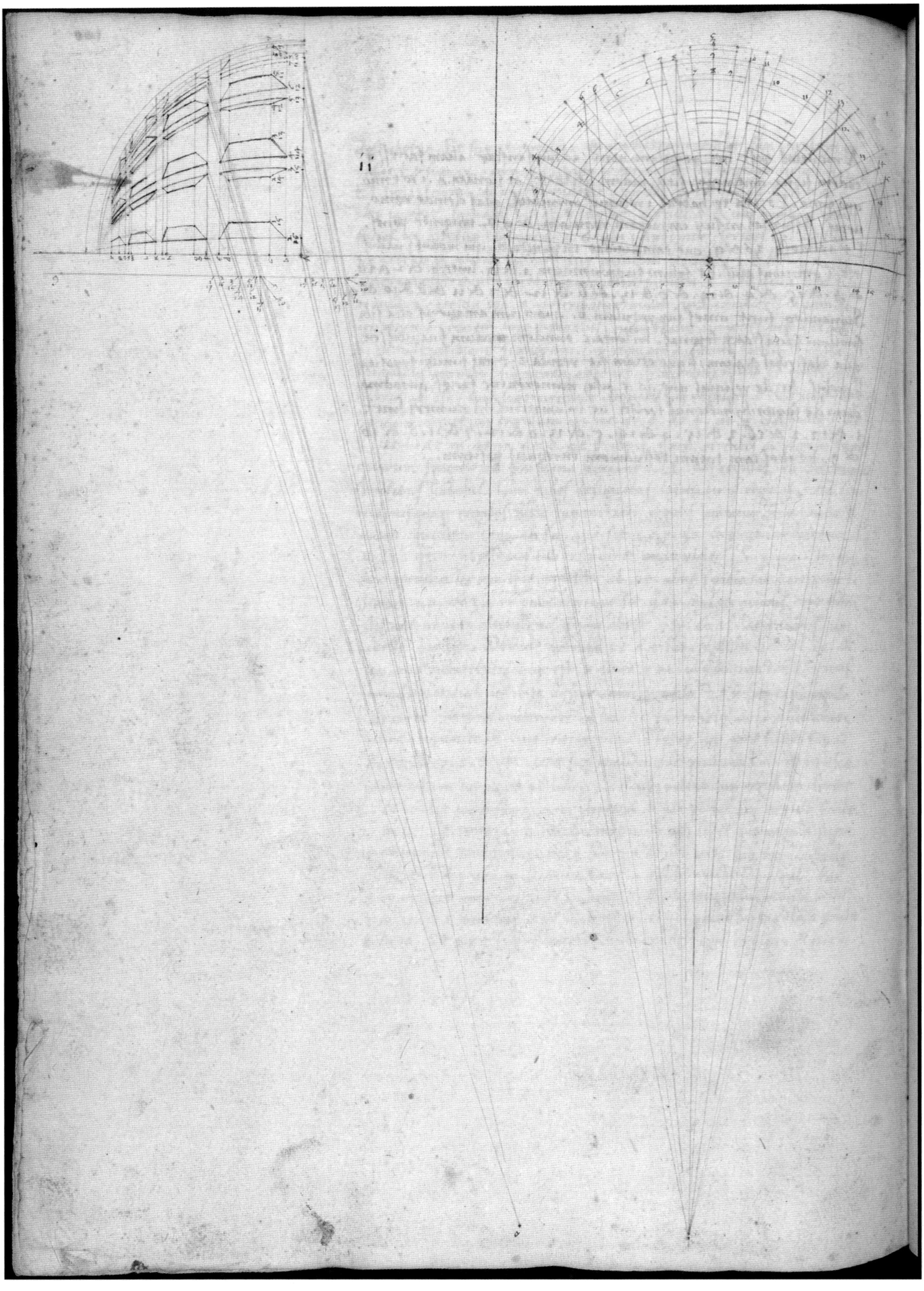

Bordeaux, Bibliothèque Municipale, Ms Latin 616, folio 100 verso.

Line B and the Remaining Ones For what concerns line B and the remaining ones, you will be proceed as in the preceding case [figure B.185]. Trace the marks of circle B on line B of the elevation. Trace the marks of circle C on line C. Proceed this way for all the remaining lines, until the end. When you have marked them, as you have till now on line A, from 1 to 17 as was said earlier, you will have the elevation [figure B.185] and the width [figure B.186] in the appropriate form [figures B.187, B.188].[575]

The Terminal and the Eye Point in Plan We must now find the shortenings on the rulers in the following manner. Draw a line parallel to line SA of the width [figure B.189]: let it be 9-R,[576] which will be the terminal, as far distant from SA as you would like it to be.[577] Stand away from 9-R at a distance you deem appropriate. To be sure about your movement, the distance at which you will place the eye O should be 10 braccia.[578] Place on it a nail and thin thread as you have often done.

Wooden Ruler A After, take a wooden ruler and place it matching 9-R as you have done earlier; place the thread on point 1 of circle A: mark point 1 where it hits the ruler [figure B.190]. Place the thread on point 2 of circle A: mark point 2 where it hits the ruler. Place the thread on point 3 of circle A: mark point 3 where it hits the ruler. Place the thread on point 4 of circle A: mark point 4 where it hits the ruler. Place the thread on point 5 of circle A: mark point 5 where it hits the ruler. Place the thread on point 6 of circle A: mark point 6 where it hits the ruler. Place the thread on point 7 of circle A: mark point 7 where it hits the ruler. Place the thread on point 8 of circle A: mark point 8 where it hits the ruler. Place the thread on point 9 of circle A: mark point 9 where it hits the ruler. Place the thread on point 10 of circle A: mark point 10 where it hits the ruler. Place the thread on point 11 of circle A: mark point 11 where it hits the ruler. Place the thread on point 12 of circle A: mark point 12 where it hits the ruler. Place the thread on point 13 of circle A: mark point 13 where it hits the ruler. Place the thread on point 14 of circle A: mark point 14 where it hits the ruler. Place the thread on point 15 of circle A: mark point 15 where it hits the ruler. Place the thread on point 16 of circle A: mark point 16 where it hits the ruler. Place the thread on point 17 of circle A: mark point 17 where it hits the ruler. Place the thread on point X which is the center: mark point M[579] where it hits the ruler. Since these points will be on circle A, mark this ruler as A.

Wooden Ruler B When you have taken it off, place on this spot a second ruler marked B [figure B.190]. Take the thread and place it on the marks of circle B [figure B.183] and report them on ruler B[580] [figure B.191] as you have done for ruler A. Mark point M on all the rulers.

Wooden Rulers C, D, E, F, G, H, I, K, L, M, N, O, P, Q, R Proceed in a similar way for the remaining circles: that is, for circle C trace the marks on ruler C, for circle D trace the marks on ruler D; do the same for circles and rulers E, F, G, H, I, K, L, M, N, O, P, Q, R.[581] Let all the rulers have mark M. Once you have marked them, put them aside. Now you have the width rulers.

The Terminal and the Eye Point in Elevation Now proceed with the paper rulers relative to the height. Trace a line parallel to SA [figure B.192] which is 20-30, as far distant from SA as you would like to stand below the vault to observe. Above this line 20-30, trace the perpendicular line parallel to AT.[582] And divide line 20-30 at point 9. This perpendicular line will indicate the terminal 9-R: let the terminal be distant from AT in height as much as 9-R of width is distant from SA in width, since the distance [between AT and 9-R] has to be equal to the distance between SA and 9-R. Stand further away from point 9 on line 20-30, at the distance that there is between 9-R in width and the point where the needle with the thread is.[583]

Paper Ruler a Place there [on point O] the needle with the thread. Take the paper ruler and place it on 9-R [figure B.193]. Extend line 20-30, which divides line 9-R at point a: this will be paper ruler a. Moreover, place the thread above 1-17 of line A: mark point 1-17 where it hits the ruler. Place the thread above 2-16 of line A: mark point 2-16 where it hits the ruler. Place the thread above 3-15 of line A: mark point 3-15 where it hits the ruler. Place the thread above 4-14 of line A: mark point 4-14 where it hits the ruler. Place the thread above 5-13 of line A: mark point 5-13 where it hits the ruler. Place the thread above 6-12 of line A: mark point 6-12 where it hits the ruler. Place the thread above 7-11 of line A: mark point 7-11 where it hits the ruler. Place the thread above 8-10 of line A: mark point 8-10 where it hits the ruler. Place the thread above 9 of line A: mark point 9 where it hits the ruler. After, you will take this ruler away and you will draw a similar one that contains the same marks you made before, which is the same length and is marked with the letter a like the previous ruler.

The Remaining Paper Rulers Do the same thing for the remaining rulers which you will take away. That is, take a paper ruler and make it coincide with 9-R. Extend line 20-30, which will cut the ruler at point b: this is a procedure that must be followed in all other cases. Place the thread on points 1-17, 2-16, 3-15, 4-14, 5-13, 6-12, 7-11, 8-10, 9 of line B [figure B.185].[584] They must all be marked on ruler b [figure B.194]. After you have taken it away, draw a similar one and take it away. Do the same thing for line C; and with this line mark a ruler which is ruler c [figure B.194], and take two rulers away. Do the same thing for all remaining lines up till line R as you have done until now with the preceding lines. Mark the following numbers on all the rulers: 1-17, 2-16, 3-15, 4-14, 5-13, 6-12, 7-11, 8-10, 9 [figure B.194]. Now you have the entire vault on the rulers.

The Shortening of the Dome We now must go back to the dome [figure B.195] and allocate it to its position,[585] where you will trace straight line SR. Trace on point S the perpendicular line ST and on point R the perpendicular line RV. Divide SR into two equal parts at point X and trace a perpendicular line through Y above X: you will have line XY.[586] Take the paper rulers marked r and make them match with ST and with RV: let point r of both lines touch line SR. Take wooden ruler R and remember that all the marks on two paper rulers have been marked here as well. Therefore when I say: trace, draw, and place a ruler I am referring to the wooden one. When I say mark the point where a certain number has a contact or where there is a mark, we are referring to the marks on the wooden ruler which we must indicate. Therefore, if the wooden ruler touches points 1-17 of the two paper rulers, we must mark point 1-17 where point 1-17 is on the wooden ruler. And always point M of the wooden ruler must touch vertical line XY.[587] Therefore place the ruler so that it touches points 1-17 on the two rulers:[588] mark point 1 where you have point 1 on the ruler; mark 17 where you have point 17. Place the ruler so that it touches points 2-16 of the two rulers: mark point 2 where you have point 2 of the ruler; mark 16 where you have point 16. Place the ruler so that it touches points 3-15 of the two rulers: mark point 3 where you have point 3 of the ruler; mark 15 where you have point 15. Place the ruler so that it touches points 4-14 of the two rulers: mark point 4 where you have point 4 of the ruler; mark 14 where you have point 14. Place the ruler so that it touches points 5-13 of the two rulers: mark point 5 where you have point 5 of the ruler; mark 13 where you have point 13. Place the ruler so that it touches points 6-12 of the two rulers: mark point 6 where you have point 6 of the ruler; mark 12 where you have point 12. Place the ruler so that it touches points 7-11 of the two rulers: mark point 7 where you have point 7 of the ruler; mark 11 where you have point 11. Place the ruler so that it touches points 8-10 of the two rulers: mark point 8 where you have point 8 of the ruler; mark 10 where you have point 10. Place the ruler so that it touches point 9 of the two rulers: mark point 9 where you have point 9 of the ruler. Remove the rulers and put them aside. Take the two paper rulers q [figure B.196] and place one on ST and the other on RV, as you have done for the others. Then take wooden ruler Q and place it on points 1-17 of the two rulers: mark point 1 where you have point 1 of the ruler; mark 17 where you have point 17. Place all the marks this way: where the numbers on the wooden ruler are, mark there[589] the same numbers. This must be done for all rulers: P, O, N, M, L, K, I, H, G, F, E, D, C, B, A [figure B.197]. Each of these rulers in fact contains the same numbers, and we can observe the same procedure in the rulers [figure B.198].[590]

The Vertical Elongations of the Points Once you have reported the marks of all the rulers, extend point 2 [figure B.199, top] of circle B toward point 2 of circle E; point 2 of circle F toward point 2 of circle I; point 2 of circle K toward point 2 of circle N; point 2 of circle O toward point 2 of circle R; point 2 of circle C toward point 2 of circle D; point 2 of circle G toward point 2 of circle H; point 2 of circle L toward point 2 of circle M; point 2 of circle P toward point 2 of circle Q.[591] After, extend point 3 of circle C toward point 3 of circle D; point 3 of circle G toward point 3 of circle H; point 3 of circle L toward point 3 of circle M; point 3 of circle P toward point 3 of circle Q. Then extend point 3 of circle B toward point 3 of circle E; point 3 of circle F toward point 3 of circle I; point 3 of circle K toward point 3 of circle N;[592] point 3 of circle O toward point 3 of circle R. Moreover, extend point 4 of circle B toward point 4 of circle E; point 4 of circle F toward point 4 of circle I; point 4 of circle K toward point 4 of circle N; point 4 of circle O toward point 4 of circle R. For the remaining ones, continue marking them, respecting the order of all seventeen circles, each with seventeen marks.

The Horizontal Elongations of the Points After this, trace lines 2-3, 4-5, 6-7, 8-9, [9-10, 11-12, 13-14, 15-16]: this is circuit B [figure B.199, top].[593] This procedure must be respected for all the other lines, and you will have completed the shortening of the vault [figure B.199, bottom], as we had previously proposed [figure B.200].[594]

Cornice Moreover, if you would like the vault to rest on the crown, commonly called the cornice, you will draw a semicircle as big as STV[595] within which you will draw the semicircles of the cornice that you intend to realize with the projections you wish [to draw]. You will then divide the semicircles into sixteen equal parts. After this, keeping the same distance and the same space interval with which you have traced the circles of the dome, extend the thread to the marks, and where the thread touches the wooden ruler, draw the points as you have done until now. In the same way, you will draw the height of the cornice in relation to a quarter of the circle, so that the width of the semicircle corresponds to the length of the projections. Allow the same thing to happen in a proportionate way in the subdivisions. Lastly, you will place the thread on the subdivisions and you will mark where thread touches the paper rulers as we have said till now.

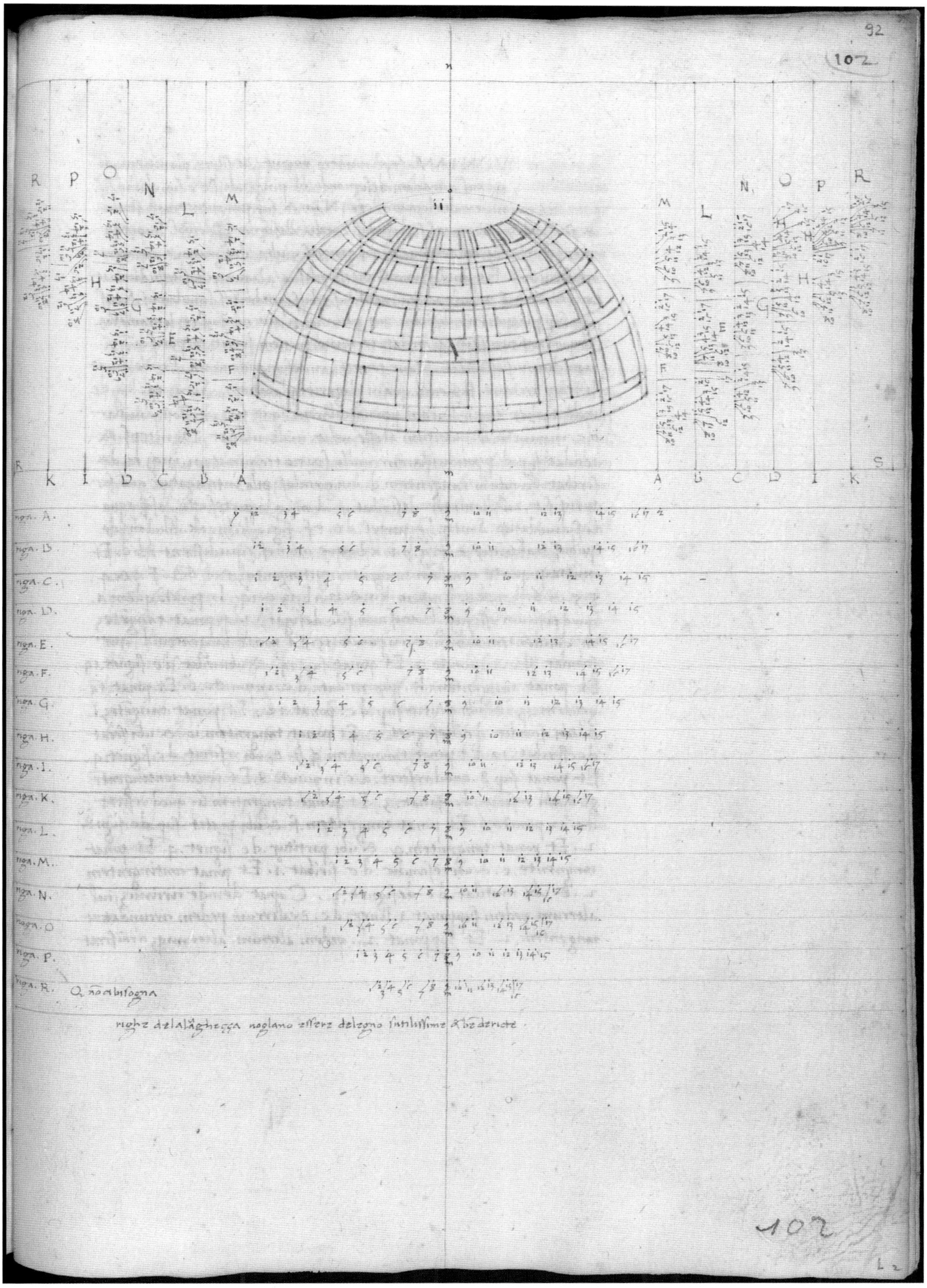

Bordeaux, Bibliothèque Municipale, Ms Latin 616, folio 102 recto.

Proposition X

Because it often happens that one will have to simulate a body above a plane, drawing it or designing it as hanging from someplace above, and because, above floors and similar shelves, we usually draw tripods and similar things on shelves, vases, chandeliers, and because we usually set them by drawing many other similar things, and we simulate rings, hooks, handles, and hanging spheres from ceilings with drawers, from shells or from volts and from other things like these that protrude: if one were to draw the sketch of a floor or anything else on a plane, a sketch that shows in a given terminal the appearance of a spherical body, one must follow this procedure to draw.

First, draw the straight line with which you will indicate the plane [figures B.201, B.202]. Let this line be BC. Draw perpendicular line AB above B, and divide BC at point D according to the distance at which you would like to observe. Then trace an unlimited perpendicular line above D on which you should draw a tangent circle at point D. Let this circle correspond to the ball which we will build. Let DM be the diameter, to be subdivided into eight equal parts. Let DE, EF, FG, GH, HI, IK, KL, LM be these parts, and draw from the points of subdivisions E, F, G, H, I, K, L lines that are parallel to BC till they touch the circle on one side and the other. Let the line through E touch it at 1 and 2; the line through F at 3 and 4; the line through G at 5 and 6; the line through H at 7 and 8; the line through I at 9 and 10; the line through K at 11 and 12; the line through L at 13 and 14. Now place the thread and the needle at point A which is above B, and let the thread go through M till it divides DC at point 15. Let the thread go through L till it divides DC at point 7. Let the thread go through 14 till it divides DC at point 14. Let the thread go through K till it divides DC at point 6′. Let the thread go through 12 till it divides DC at point 12. Let the thread go through I till it divides DC at point 5. Let the thread go through 10 till it divides DC at point 10. Let the thread go through H till it divides DC at point 4′. Let the thread go through 8 till it divides DC at point 8. Let the thread go through G till it divides DC at point 3. Let the thread go through 6 till it divides DC at point 6. Let the thread go through F till it divides DC at point 2′. Let the thread go through 4 till it divides DC at point 4. Let the thread go through E till it divides DC at point 1. Let the thread go through 2 till it divides DC at point 2.

Then take the compass, place one end on point 1 of line DC, and with the other end draw the circle that goes through 2. Place one end on point 2′ of line DC and with the other end draw the circle that goes through 4. Place one end on point 3 [of line DC] and with the other end draw the circle that goes through 6. Place one end on point 4′ [of line DC] and with the other end draw the circle that goes through 8. Place one end on point 5 [of line DC] and with the other end draw the circle that goes through 10. Place one end on point 6′ of line DC and with the other end draw the circle that goes through 12. Place one end on point 7 and with the other end draw the circle that goes through 14 of line DC: this will have to be taken into consideration for all points.

Then draw the border, which will touch all points in the external part: let the border take a good shape.[596] Such a border, executed in the assigned terminal, relates to the eye in relation to the circle [that has diameter] DM and 7-8 and that contains a spherical shape. Here is the demonstration. Point A is the eye from which you have pulled the thread which goes through M and intercepts DC at point 15. Thus 15 seems more elevated than D by length DM, as shown in the tenth proposition of *De aspectuum diversitate* of Euclid.[597] And because D-15 and DM are below the same angle and constitute two bases that present themselves to point A which is the eye, and they appear to be the same, as shown in the second proposition of the first book, it happens then that D-15 seems the same as DM which is the diameter of the maximum circle of the sphere DM-7-8. In the same manner semidiameter E-2 seems to be equal to diameter 1-2; semidiameter F-4 seems to be equal to diameter 2′-4; semidiameter G-6 seems to be equal to diameter 3-6; and in the same way with all the other semidiameters, which each appear equal to its correspondent. Thereby the complete diameters appear proportionate. The same thing will happen for the circles. Circle 7-8 in fact juxtaposes under the same angle to circle 7′-8 so that they appear to be equal. Diameter 5-6 and the corresponding diameter are under the same angle so that they appear to be equal. In the same way, diameter 3-4 and the corresponding diameter are under the same angle so that they appear to be equal to each other. In the same way, diameter 1-2 appears to be equal to 1′-2 since they are seen under the same angle, as was demonstrated in the second proposition of the first book. Length and width are presented to the eye, in the assigned terminal, as if they belonged to a spherical body.

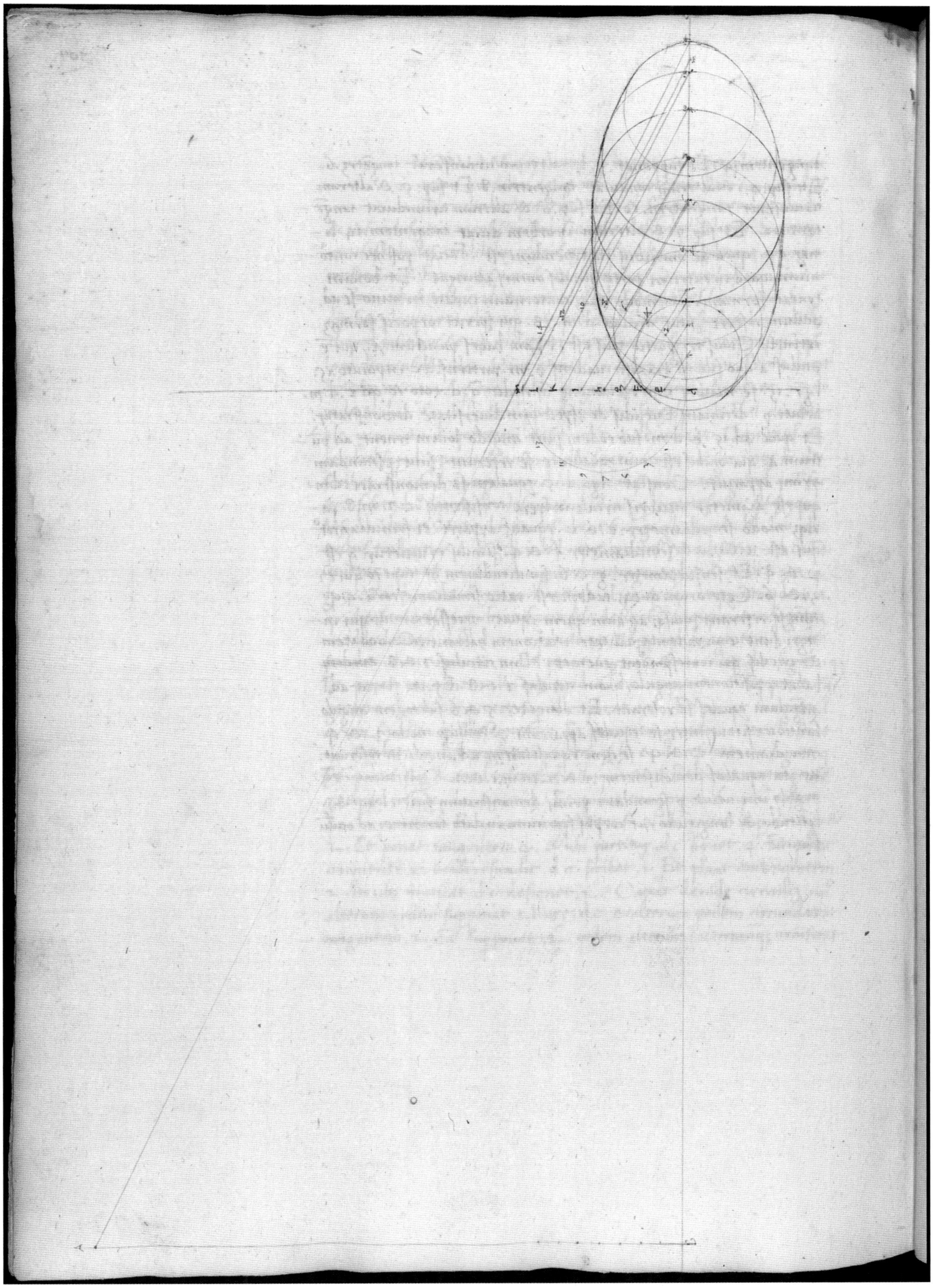

Bordeaux, Bibliothèque Municipale, Ms Latin 616, folio 103 verso.

Proposition XI

If one were to simulate a so-called fresh vase with the foot fixed on a banquet shelf or on another plane so that it appears straight as though it were real, they will draw a vase in real form.

Let the height in real form correspond to the height of the vase with the foot fixed on straight line BC [figures B.203, B.204]. Above this line trace the perpendicular line that goes through the vase and divides line BC at point D. Let this be line DO. The thickness of the margin or border of the fresh vase divides DO at point N. Let the junction line of the pedestal with the cup intersect DN at point K. Divide KN at point L and M so as to generate equal divisions. The bottom course of the foot divides DK at point E. Divide EK into five parts EF, FG, GH, HI, IK, as you can see in the figure. Then trace a line parallel to BC from E which ends on the side of C at point 1. Trace a line parallel to BC from F which touches the border at point 2. Trace a line parallel to BC from G which touches the border at point 3. Trace a line parallel to BC from H which touches the border at point 4. Trace a line parallel to BC from I which touches the border at point 5. Trace a line parallel to BC from K which touches the border at point 6. Trace a line parallel to BC from L which touches the border at point 7. Trace a line parallel to BC from M which touches the border at point 8. Obtain from N the parallel which touches the border at point 9. Obtain from O the parallel which ends at point 10, so that all these parallel lines have contact [with the outline of the vase] on the side of C. Now trace, as high as you want to observe above B, perpendicular line AB on B. Place thread and needle in this point [A] and extend the thread to line DC so that it goes through E of line DO, and where it touches DC mark point P. Extend the thread so that it goes through F, and where it touches DC mark point Q. Extend the thread so that it goes through G, and where it touches DC mark point R. Extend the thread so that it goes through H, and where it touches DC mark point S. Extend the thread so that it goes through I, and where it touches DC mark point T. Extend the thread so that it goes through K, and where it touches DC mark point V. Extend the thread so that it goes through L, and where it touches DC mark point X. Extend the thread so that it goes through M, and where it touches DC mark point Y. Extend the thread so that it goes through N, and where it touches DC mark point z. Extend the thread so that it goes through O, and where it touches DC mark point Z. Extend the thread so that it goes through 1, and where it touches DC mark point 11. Extend the thread so that it goes through 2, and where it touches DC mark point 12. Extend the thread so that it goes through 3, and where it touches DC mark point 13. Extend the thread so that it goes through 4, and where it touches DC mark point 14. Extend the thread so that it goes through 5, and where it touches DC mark point 15. Extend the thread so that it goes through 6, and where it touches DC mark point 16. Extend the thread so that it goes through 7, and where it touches DC mark point 17. Extend the thread so that it goes through 8, and where it touches DC mark point 18. Extend the thread so that it goes through 9, and where it touches DC mark point 19. Extend the thread so that it goes through 10, and where it touches DC mark point 20. To report the width or the thickness of the border of the vase, extend the thread so that it goes through 9*, and where it touches DC mark point 21.[598] Then place an end of the compass on point D, and with the other one trace a circle that touches the extreme points of the pedestal from one side and the other. Place one end of the compass on point P, and with the other end trace the circle which goes through point 11. Place one end of the compass on point Q, and with the other end trace the circle which goes through point 12. Place one end of the compass on point R, and with the other end trace the circle which goes through point 13. Place one end of the compass on point S, and with the other end trace the circle which goes through point 14. Place one end of the compass on point T, and with the other end trace the circle which goes through point 15. Place one end of the compass on point V, and with the other end trace the circle which goes through point 16. Place one end of the compass on point X, and with the other end trace the circle which goes through point 17. Place one end of the compass on point Y, and with the other end trace the circle which goes through point 18. Place one end of the compass on point z, and with the other end trace the circle which goes through point 19. Place one end of the compass on point Z, and with the other end trace the circle which goes through point 20. Place one end of the compass on point Z*, and with the other end trace the circle which goes through point 21.[599] After this, draw the border according to the circles that are in the fresh vase which will seem to stand on the shelf, and emerging from the assigned terminal, as we had we said we would do in the past. If we want the vase to be subdivided into as many faces as you wish, you will realize it through the addition of circles to the internal side, according to the length you desire it to be.

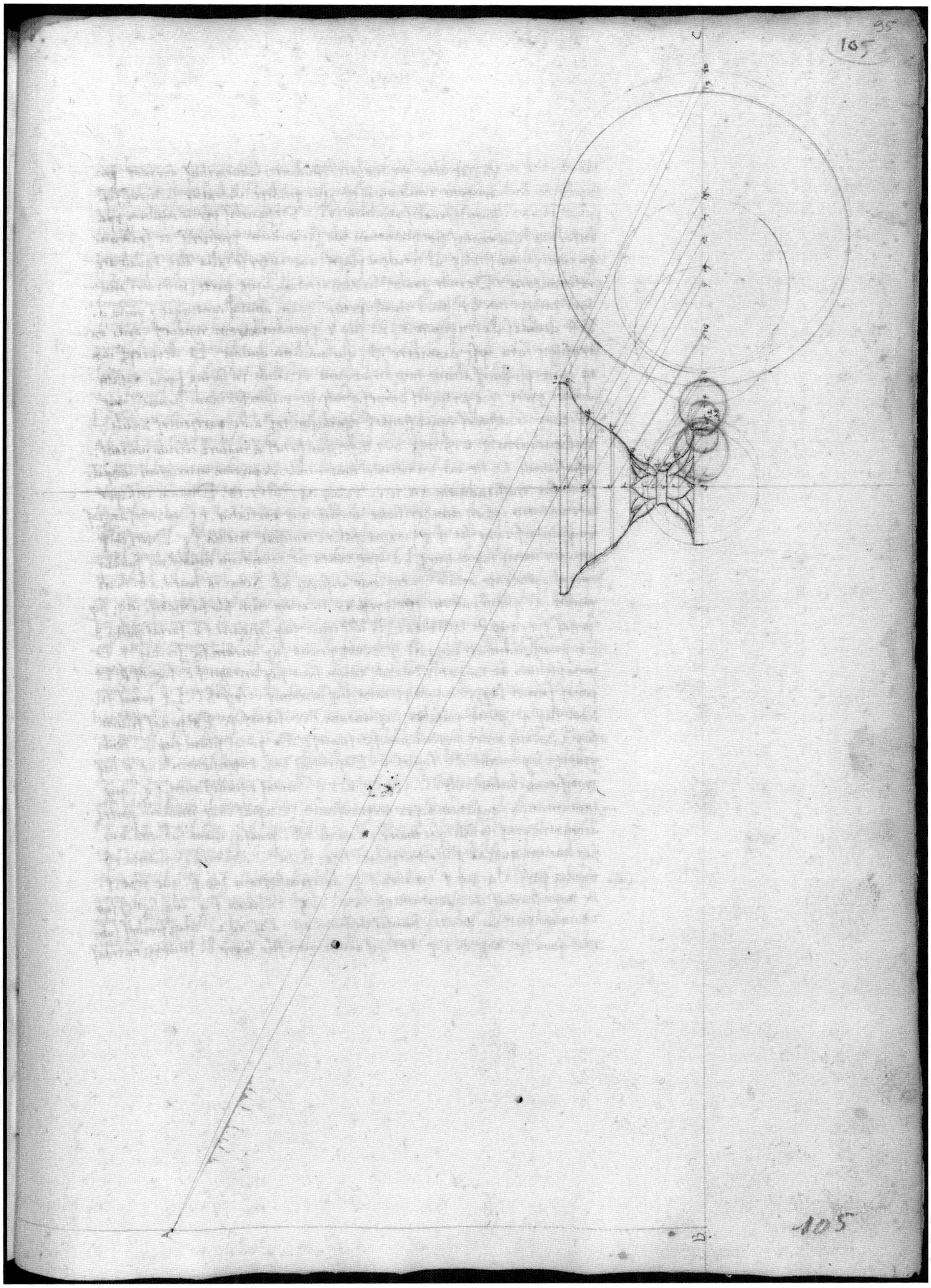

Bordeaux, Bibliothèque Municipale, Ms Latin 616, folio 105 recto.

Proposition XII

When there is a ring hanging from an elevated position like a coffered ceiling, you will proceed this way.

First you will draw a ring in real form, as big as you would like it to be [figures B.205, B.206]. Then you will divide it into sixteen parts, or if you want in as many parts more; make sure they correspond to one another. Then trace a straight line that touches the inferior part of the ring: this line will be called BC, and it will be tangent to the ring at point D. You will then divide DC at point E and you will trace the perpendicular line on E; it will be as high as the diameter which divides the ring. Next to this perpendicular line you will then trace another line as distant as the thickness of the ring in real form.[600] Trace lines parallel to BC, from the sixteen divisions of the ring in real form, which divide the height of the ring at points E, 1, 2, 3, 4, 5, 6, 7, 8. These divisions derive from the larger circle of the ring. Those that are on the smaller circle and occupy the inner side [of the ring] will intersect the height at points 11, 12, 13, 14, 15, 16, 17, 18. Moreover, divide the width by cutting it in the middle; execute this cut with right angles at FG so that point E is in the middle of FG.[601] You will also trace a line perpendicular to BC from B; this is as high as you desire to be observing the ring below the coffered ceiling. Let this line be AB, and place thread and needle on point A which is the eye. Make the thread go through 8 on FG: where it intersects fC mark point 8. Make the thread go through 7 on FG: where it intersects fC mark point 7. Make the thread go through 6 on FG: where it intersects fC mark point 6. Make the thread go through 5 on FG: where it intersects fC mark point 5. Make the thread go through 4 on FG: where it intersects fC mark point 4. Make the thread go through 3 on FG: where it intersects fC mark point 3. Make the thread go through 2 on FG: where it intersects fC mark point 2. Trace lines parallel to FG going through points 2, 3, 4, 5, 6, 7, 8, and let them cross [i.e., extend] above and below line EC. Take half of the diameter of the greater circle of the ring and place it below line BC on the line that starts from f, and obtain f-26; take half the length between 4 and 5 on line FG, place it on the line that starts from f, and obtain f-20. Take the quantity between 4 and 3 on line FG, place it below 20, and obtain 20-22.

Take the quantity between 3 and 2, place it below 22, and obtain 22-24. Then take needle with thread and place the needle on B. Extend the thread above the line that comes out of 8 until it touches the line coming out of 20 of f-26: where it touches the line coming out of 8, mark 34. Extend the thread so that it goes through 22: where it touches the line coming out of 2, mark 22. Where it touches the line coming out of 7, mark 32. Extend the thread so that it goes through 24: where it touches the line coming out of 3, mark 24. Where it touches the line coming out of 6, mark 30. Extend the thread so that it goes through 26: where it touches the line coming out of 4, mark 26. Where it touches the line coming out of 5, mark 28. See what the distance between f and 20 is, place it on F's side so that you have f-21. Place the distance between 2 and 22 on 2 and mark 23; place the distance between 3 and 24 on 3 and mark 25; place the distance between 4 and 26 on 4 and mark 27; place the distance between 5 and 28 on 5 and mark 29. And place the distance between 6 and 30 on 6 and mark 31. And place the distance between 7 and 32 on 7 and mark 33. Then trace lines 20-22, 22-24, 24-26, 26-28, 28-30, 30-32, 32-34, [and] 21-23, 23-25, 25-27, 27-29, 29-31, 31-33, 33-35. This will be the first circuit. Now for the second, smaller circuit, trace the subdivisions parallel to BC: let these divide FG at points 11, 12, 13, 14, 15, 16, 17, 18. Place needle with thread on point A and lead the thread through 18; where it hits fC, mark point 18. Lead the thread through 17; where it hits line fC, mark point 17. Lead the thread through 16; where it hits line fC, mark point 16. Lead the thread through 15; where it hits line fC, mark point 15. Lead the thread through 14; where it hits line fC, mark point 14. Lead the thread through 13; where it hits line fC, mark point 13. Lead the thread through 12; where it hits line fC, mark point 12. Lead the thread through 11; where it hits line fC, mark point 11. Then from these marks on line fC trace the parallel lines to FG till near the already executed circuit. Then take half of the diameter of the circle internal to the ring and place it below f: obtain f-46. Take the width between f and 14 on line FG and place it on 26 to obtain 26-40. Take the width between f and 13 and place it on 26 to obtain 26-42. Take the width between f and 12 and place it on 26 to obtain 26-44. Then take needle with thread and place it on point B. Lay the thread till it touches point 40. When it touches the line coming from 18, mark point 54. Extend the thread till it touches point 42. Where it touches the line coming from 12, mark point 42. Where it touches the line coming from 17, mark point 52. Extend the thread till it touches point 44. Where it touches the line coming from 13, mark point 44. Where it touches the line coming from 16, mark point 50. Extend the thread till it touches point 46. Where it touches the line coming from 14, mark point 46. Where it touches the line coming from 15, mark point 48. After this you will have to take all these measurements and report them above. Take the width between 11 and 40 and place it on 11: mark 41 and you will obtain 40-41. Take the width between 12 and 42 and place it on 12: mark 43 and you will obtain 42-43. Take the width between 13 and 44 and place it on 13: mark 45 and you will obtain 44-45. Take the width between 14 and 46 and place it on 14: mark 47 and you will obtain 46-47. Take the width between 15 and 48 and place it on 15: mark 49 and you will obtain 48-49. Take the width between 16 and 50 and place it on 16: mark 51 and you will obtain 50-51. Take the width between 17 and 52 and place it on 17: mark 53 and you will obtain 52-53. Take the width between 18 and 54 and place it on 18: mark 55 and you will obtain 54-55. Then trace lines 40-42, 42-44, 44-46, 46-48, 48-50, 50-52, 52-54, 54-55, 55-53, 53-51, 51-49, 49-47, 47-45, 45-43, 43-41, 41-40. This is the internal circuit of the ring. Now we need to eliminate the angles of the circuits to obtain the ring we had proposed. The end.[602]

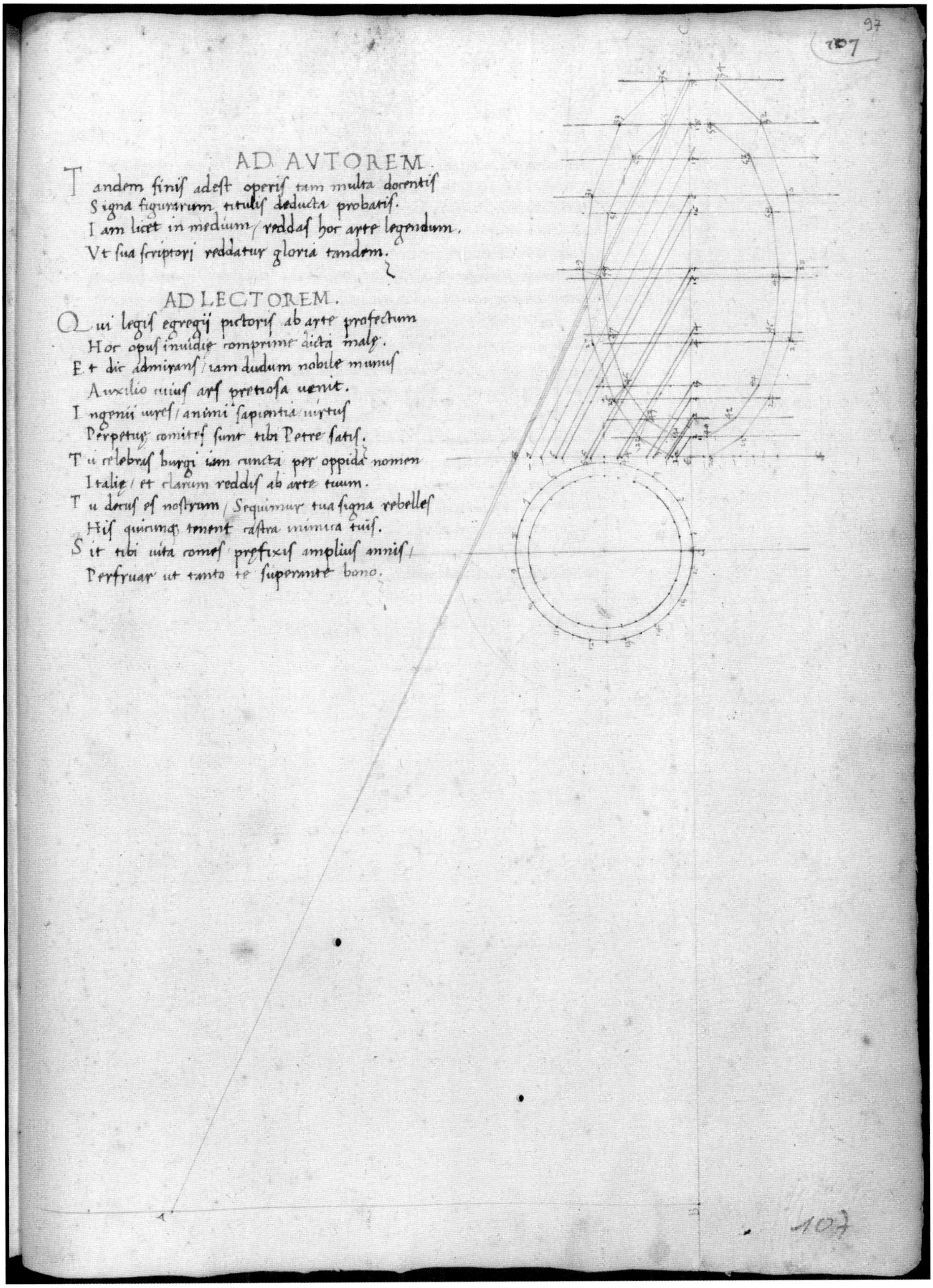

AD AVTOREM.

Tandem finis adest operis tam multa docentis
Signa figurarum titulis deducta probatis.
Iam licet in medium, reddas hoc arte legendum.
Vt sua scriptori reddatur gloria tandem.

AD LECTOREM.

Qui legis egregij pictoris ab arte profectum
Hoc opus invidie comprime dicta male.
Et dic admirans, iam dudum nobile munus
Auxilio cuius ars pretiosa uenit.
Ingenii vires, animi sapientia, virtus
Perpetue comites sunt tibi Petre satis.
Tu celebras burgi iam cuncta per oppida nomen
Italie, et clarum reddis ab arte tuum.
Tu decus es nostrum, Sequimur tua signa rebelles
His quicumque tenent castra inimica tuis.
Sit tibi uita comes prefixis amplius annis,
Perfruar ut tanto te superante bono.

107

Bordeaux, Bibliothèque Municipale, Ms Latin 616, folio 107 recto.

To the Author[603]

Thus we are at the end of a work that teaches numerous images of figures which are deduced from proven demonstrations. Now you can put the work in common in order to read it according to the rules of art, so as to offer the writer appropriate glory.

To the Reader

You reading this work which was generated by the art of an illustrious painter, hold back the words of evil envy and admire, at once, the noble effort. With the help of this work, a precious art has reached ingenuous men. Let knowledge of the soul and virtue always be sufficiently your companions, oh! Piero. You are already diffusing the name of Borgo[604] throughout all the cities of Italy, and you are making your name illustrious through art. You are our ornament: we follow your footprints, unruly to those who are hostile to your teachings. Let life be for you a fellow-traveler more than the established years! I will rejoice like you that you have reached such happiness.

Appendix A

Drawings for the Introduction

All images in this section were created by Rocco Sinisgalli to faithfully illustrate the geometric and scientific processes practiced by Piero della Francesca.

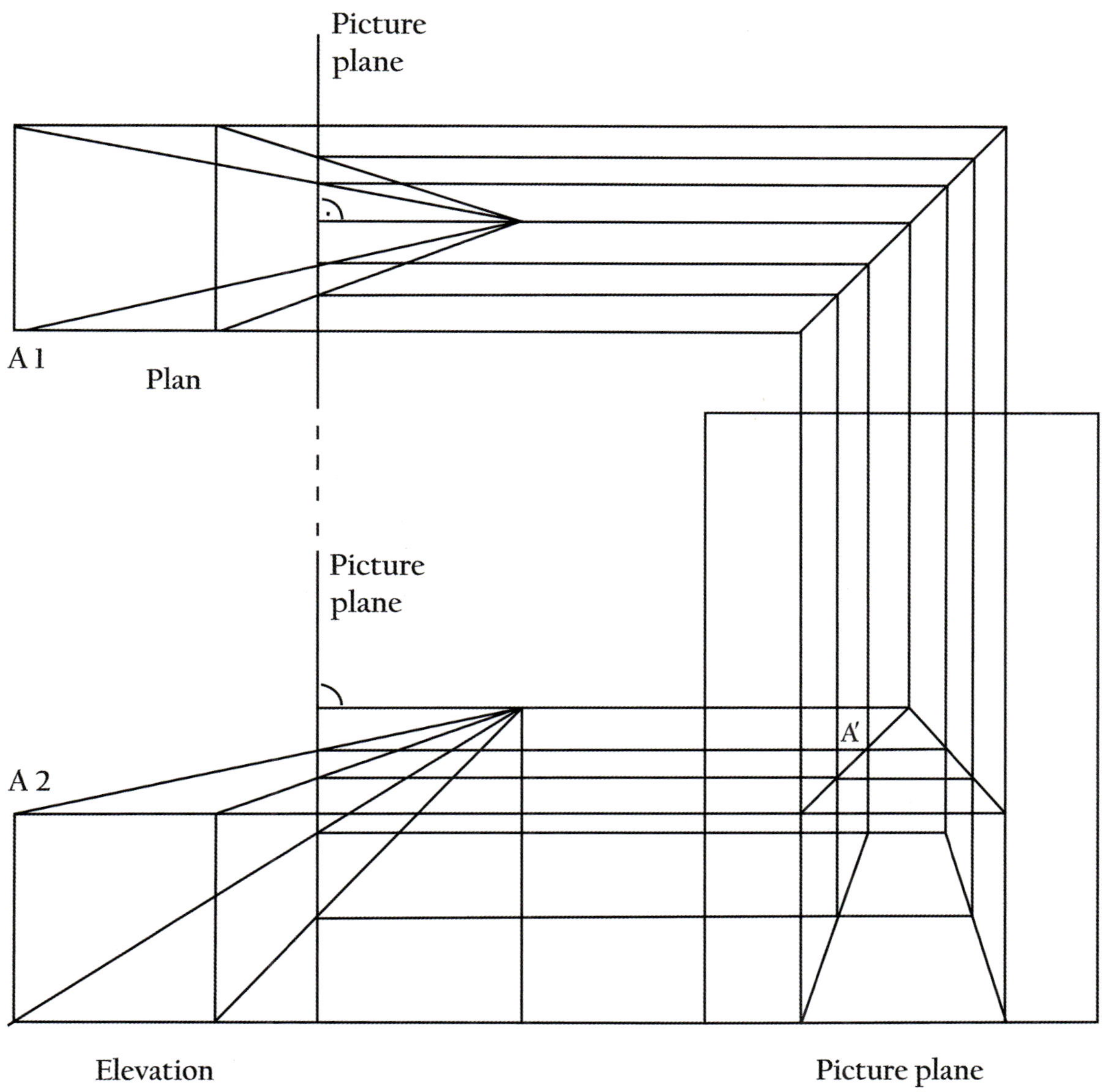

Figure A.1
Brunelleschi's "legitimate construction."

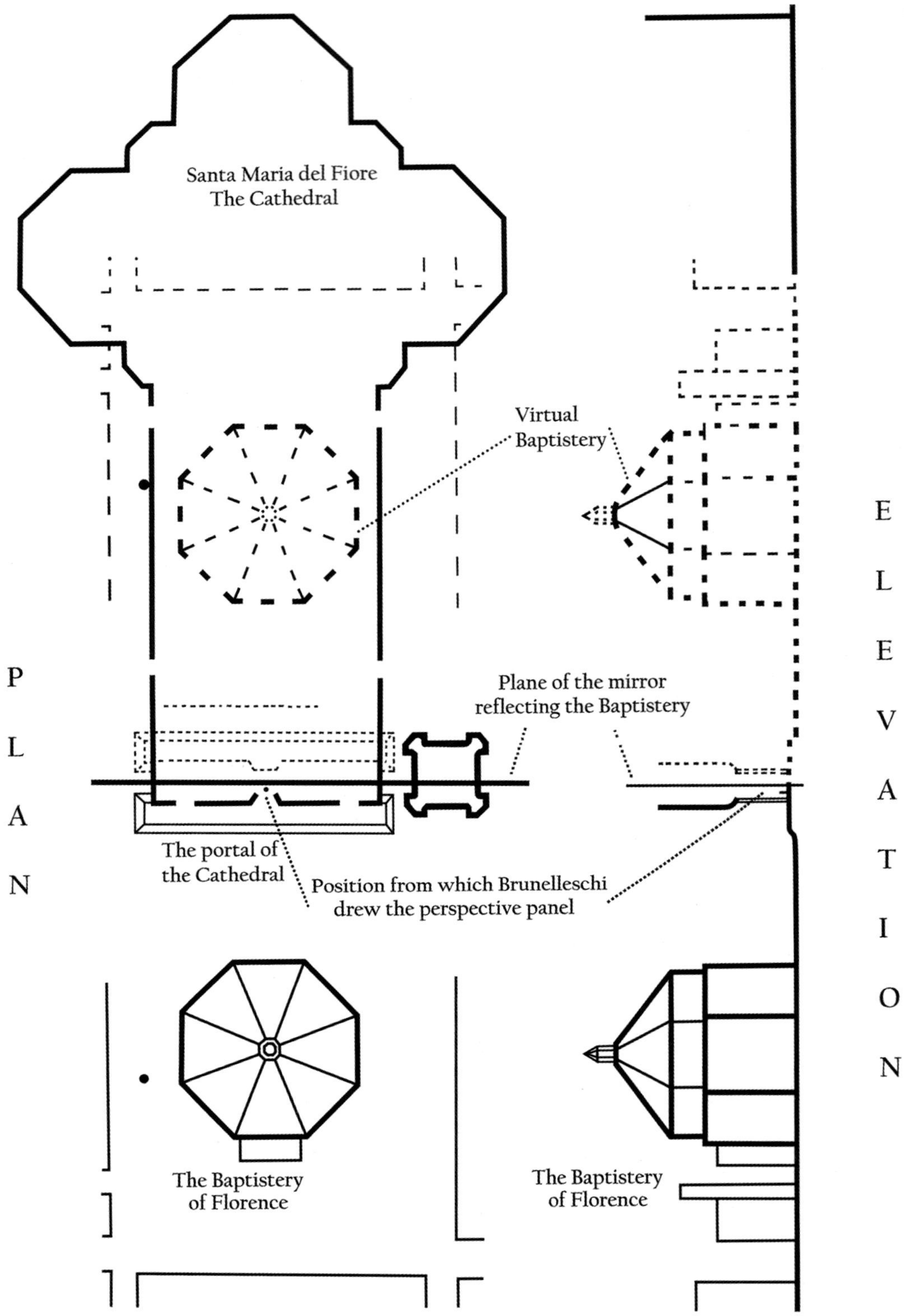

Figure A.2
The Baptistery reflected in a mirror, phase 1:
Brunelleschi observes the Baptistery reflected in a mirror.
The virtual Baptistery can be seen inside Santa Maria del Fiore.

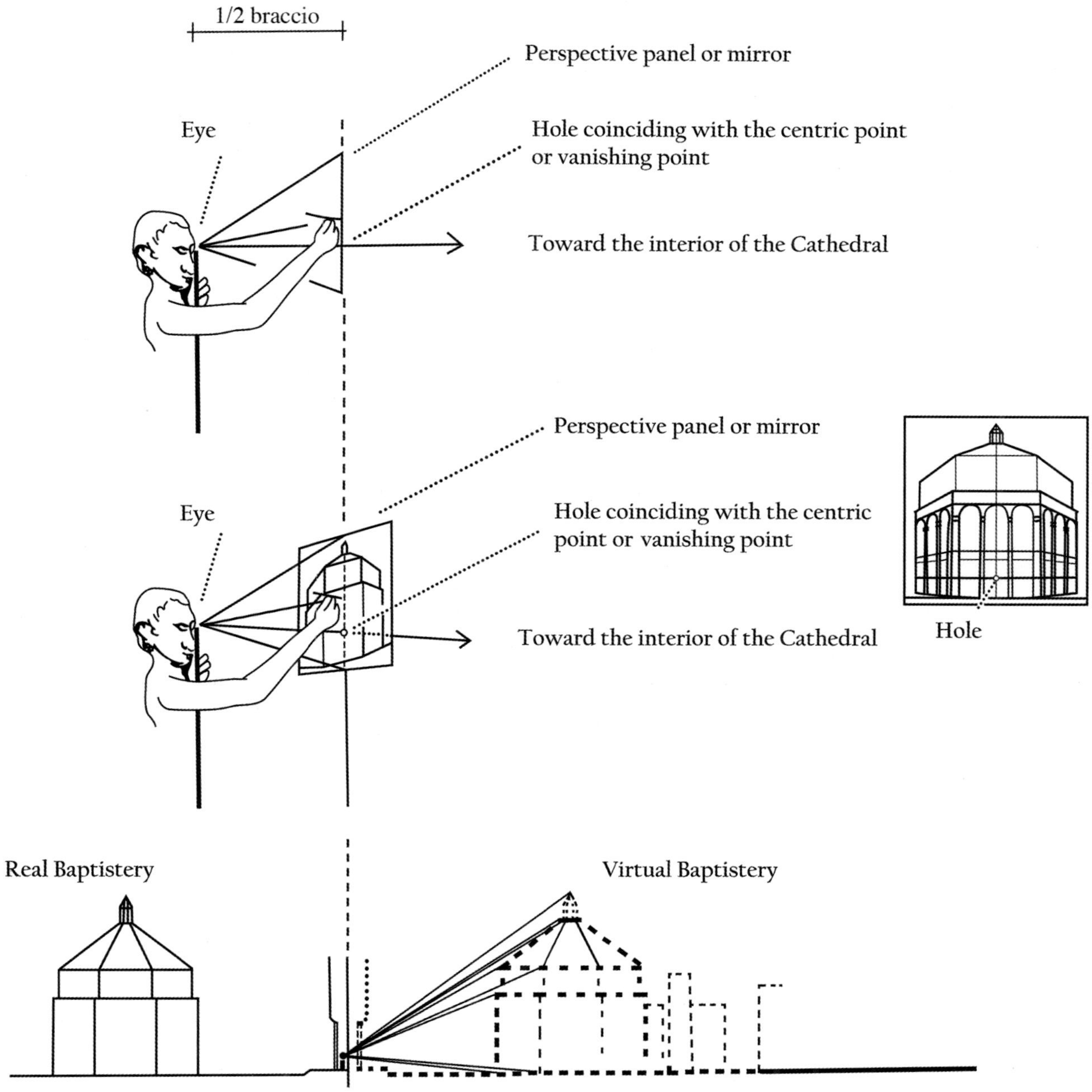

Figure A.3
The Baptistery reflected in a mirror, phase 2: Brunelleschi traces the Baptistery directly onto the mirror or copies it onto a panel beside the mirror, holding his eye fixedly on the target and the mirror at the appropriate distance of half a braccio.

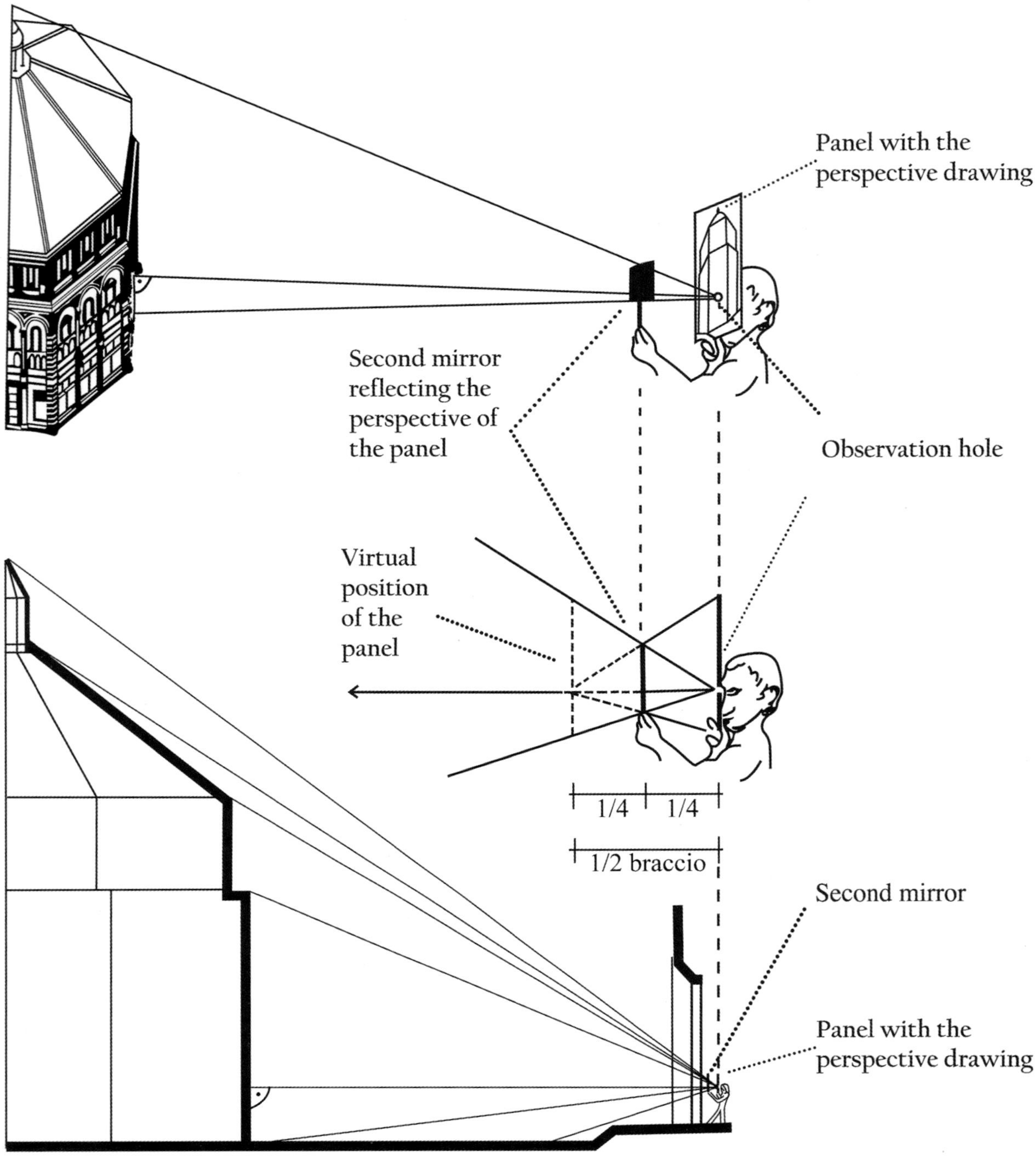

Figure A.4
The Baptistery reflected in a mirror, phase 3: through a hole made in the panel, Brunelleschi verified with a second mirror, which he removed from time to time, the perfect coincidence of the real view with that reflected on the panel.

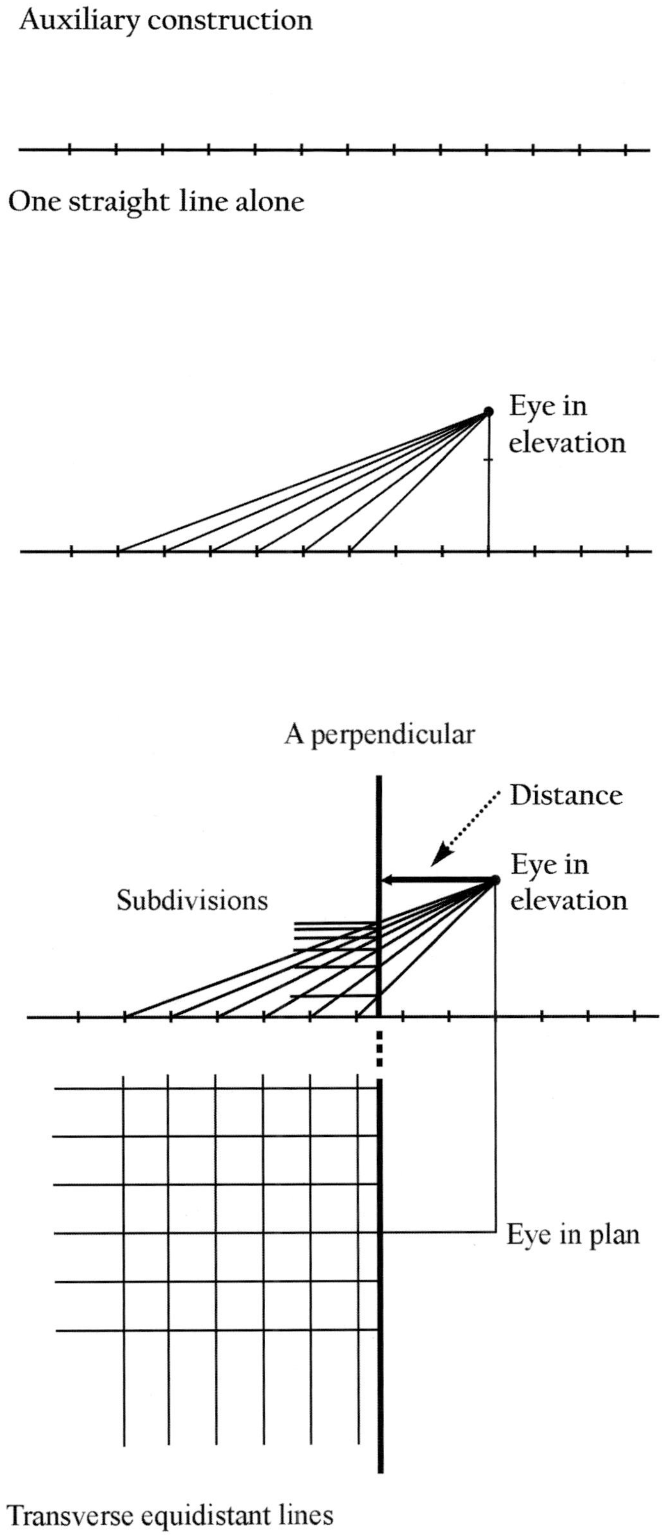

I have a small surface on which I trace one straight line alone, which I divide into those parts according to which the baseline of the quadrangle has been divided.

Hence I place a single point above, with respect to this line, as far up as the centric point in the [rectangular] quadrangle is distant from the subdivided baseline, and from this point I trace the single lines to the single subdivisions of this exact line.

I then establish how much I want the distance between the observer's eye and the painting to be; and, having fixed there the position of the cut, I obtain, by means of a perpendicular line, as mathematicians say, the cut of all lines that this [perpendicular] will have met.

This perpendicular line, then, will give me, through its subdivisions, the limits of every distance that must occur among the transverse, equidistant lines of the pavement; on this condition I have it that all parallels of the pavement are represented.

Figure A.5
Alberti's perspective, 1.

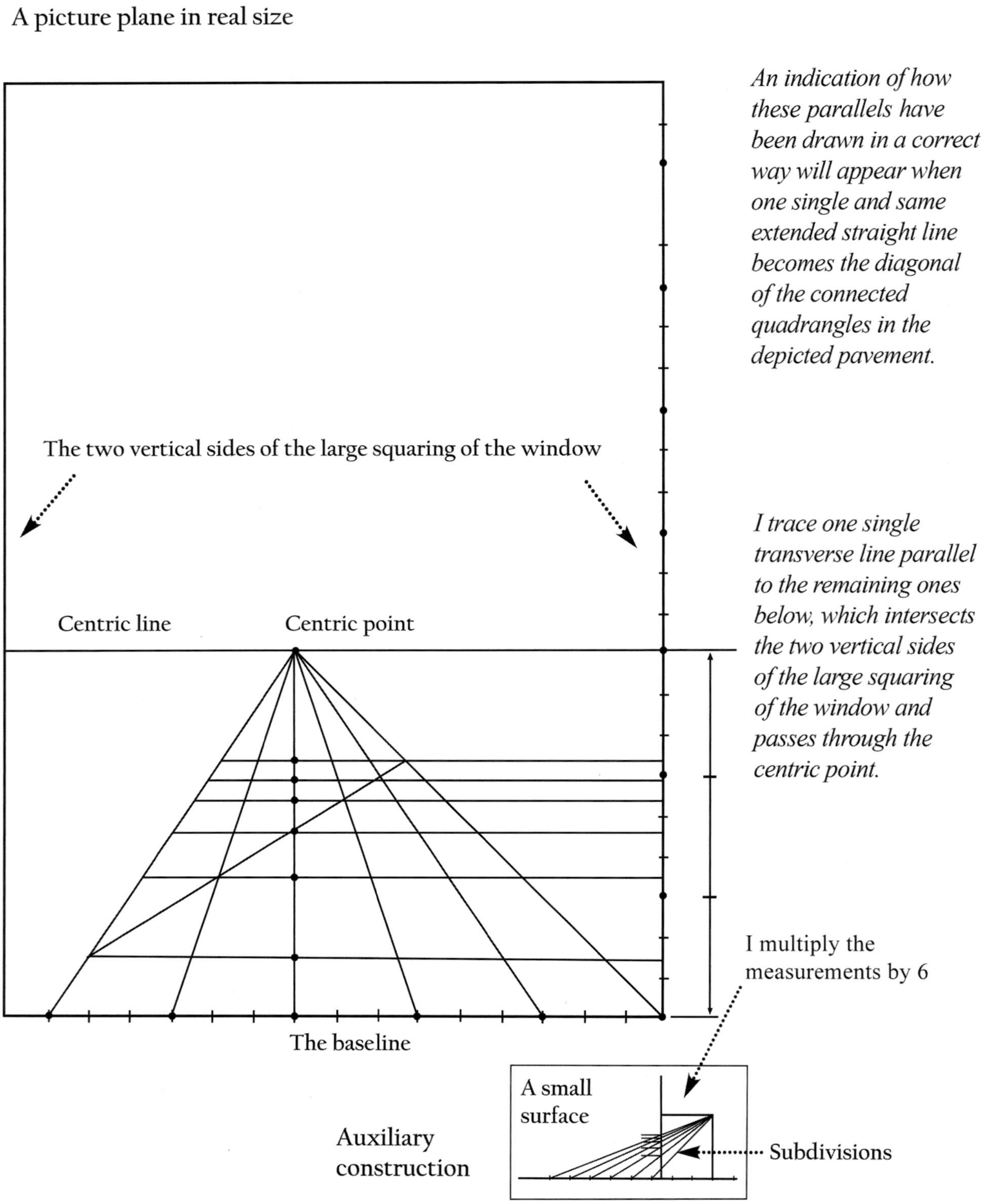

Figure A.6
Alberti's perspective, 2.

The surfaces of buildings
and of colossal objects

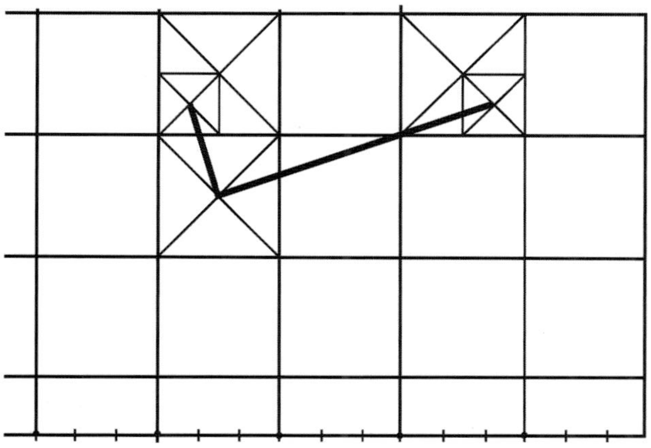

On the pavement, therefore, represented by parallels, one needs to construct the wings of the walls and whatever surfaces of this kind that we have called surfaces that lean upon a side.

Pavement in plan

Picture plane or painting

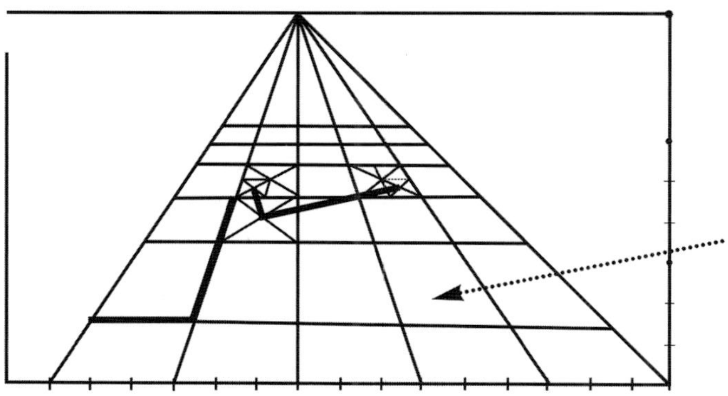

Pavement represented in painting

Squared bodies with right angles

Surfaces A and B form right angles;
the angle condition of the surfaces is implicit

Of any body squared with right angles, one cannot see, with a single glance, in relation to the ground, more than two joined surfaces that lean upon a side.

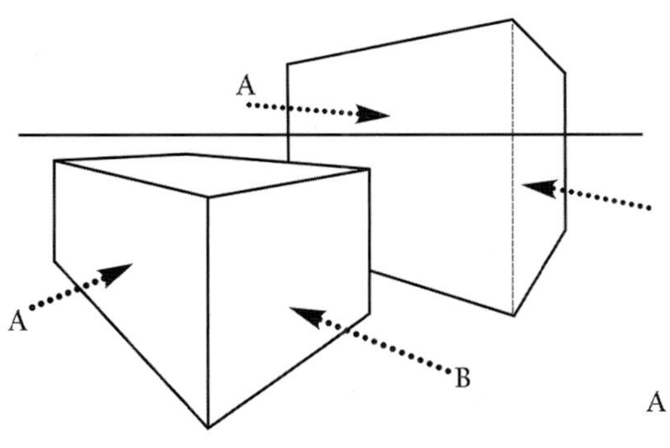

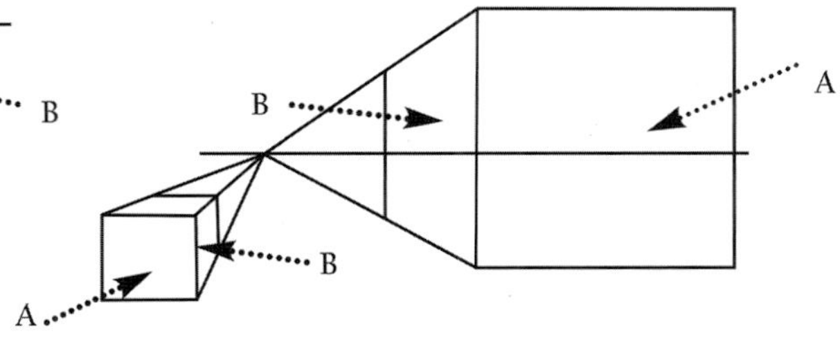

Figure A.7
Alberti: the drawing of profiles of very great surfaces, 1.

The surfaces of buildings and of colossal objects

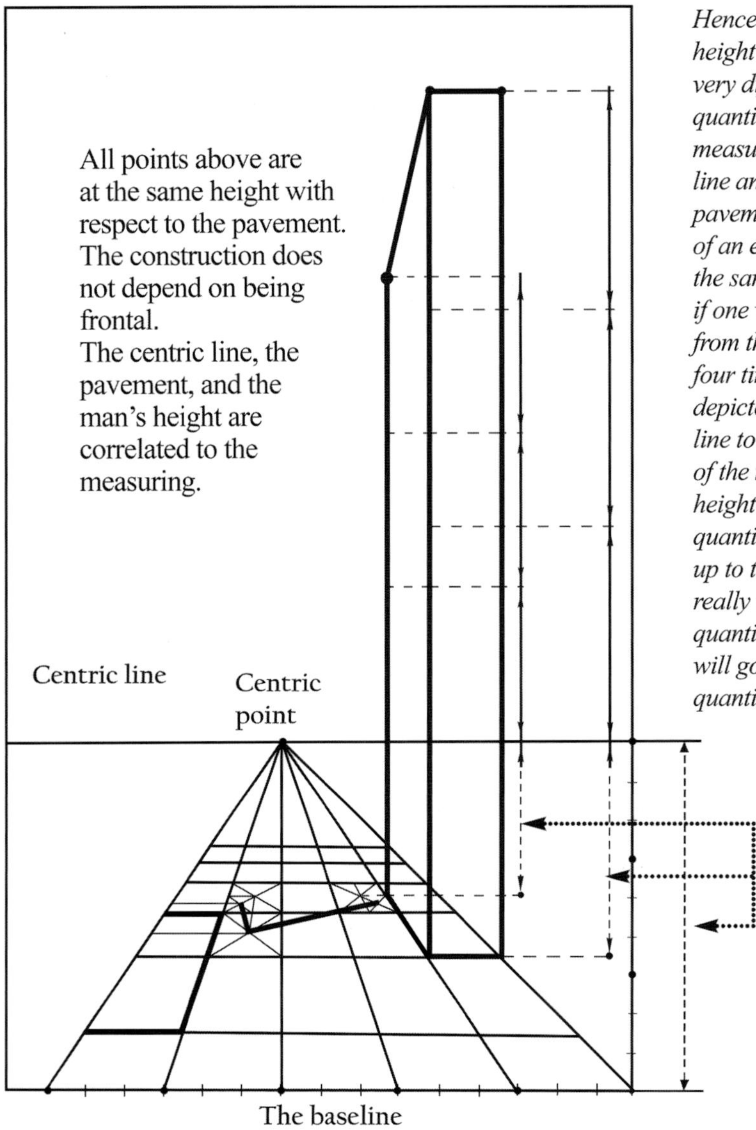

Hence, then, I also attain the height of the surfaces in a not very difficult way. That whole quantity, in fact, which is the measurement between the centric line and that place on the pavement from which the mass of an edifice rises, will maintain the same dimension. That is why, if one wants such a quantity, from the ground to the top, to be four times the height of the depicted man, and the centric line to have been placed at three of the man, certainly then the height will be the braccia of the quantity from the lower extremity up to the centric line. You who really wish to increase that quantity up to twelve braccia will go upward three times the quantity from the lower extremity.

Figure A.8
Alberti: the drawing of profiles of very great surfaces, 2.

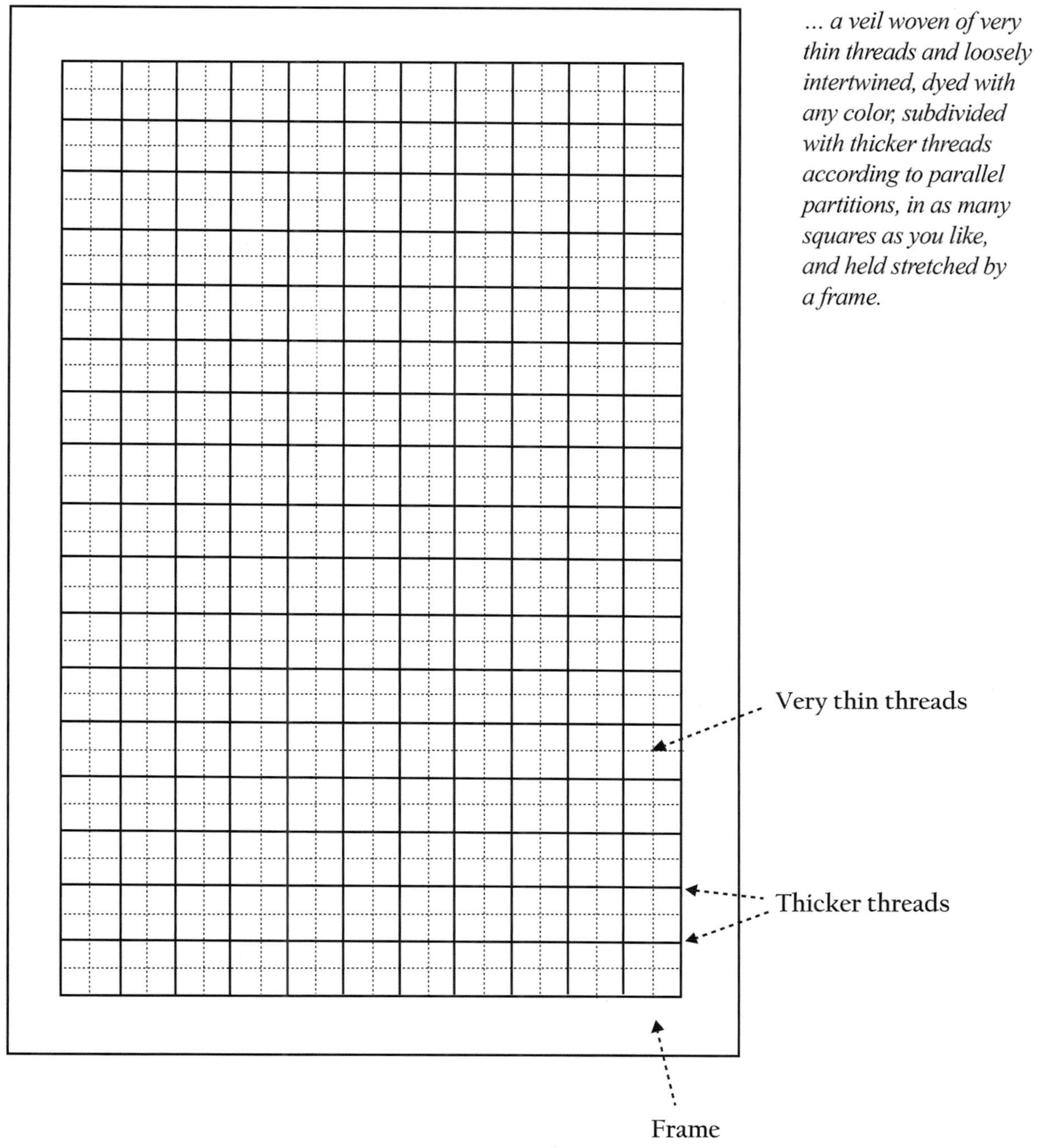

Figure A.9
Alberti's veil.

And so the veil will guarantee this not negligible advantage which I have spoken of: that the object always stays the same with respect to view.

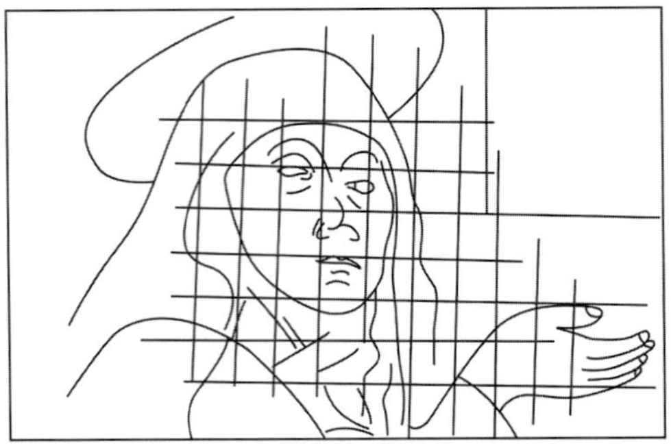

Facsimile drawing of *The Trinity* (reticulated nets), by Masaccio, Santa Maria Novella, Florence, circa 1425

A further advantage consists of the fact that in a panel to be painted, the positions of the edges and the limits of the surfaces can easily be established in very precise locations. In fact, when you observed over here in a parallel [plane] the forehead, in the very next the nose, in the near one the cheeks, in the lower the chin, and all things of that kind, situated in their own places, in the same way there, you will have placed in the best manner all the things on a panel or on a wall also subdivided by corresponding parallels.

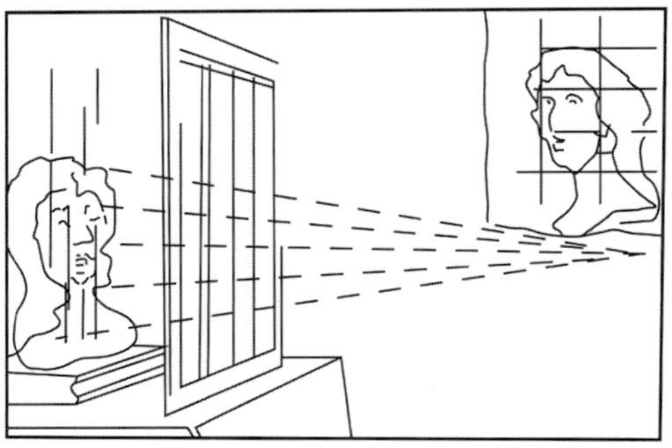

Abraham Bosse, *Les pratiques par figures*, 1665

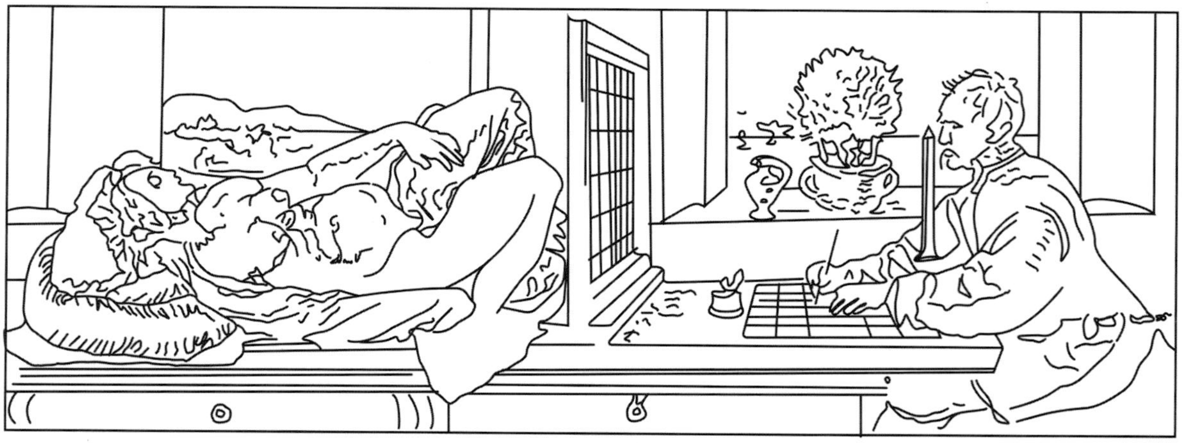

Dürer, *Institutionum geometricarum libri quatuor* (1532)

Figure A.10
The veil and the drawing of profiles of small surfaces.

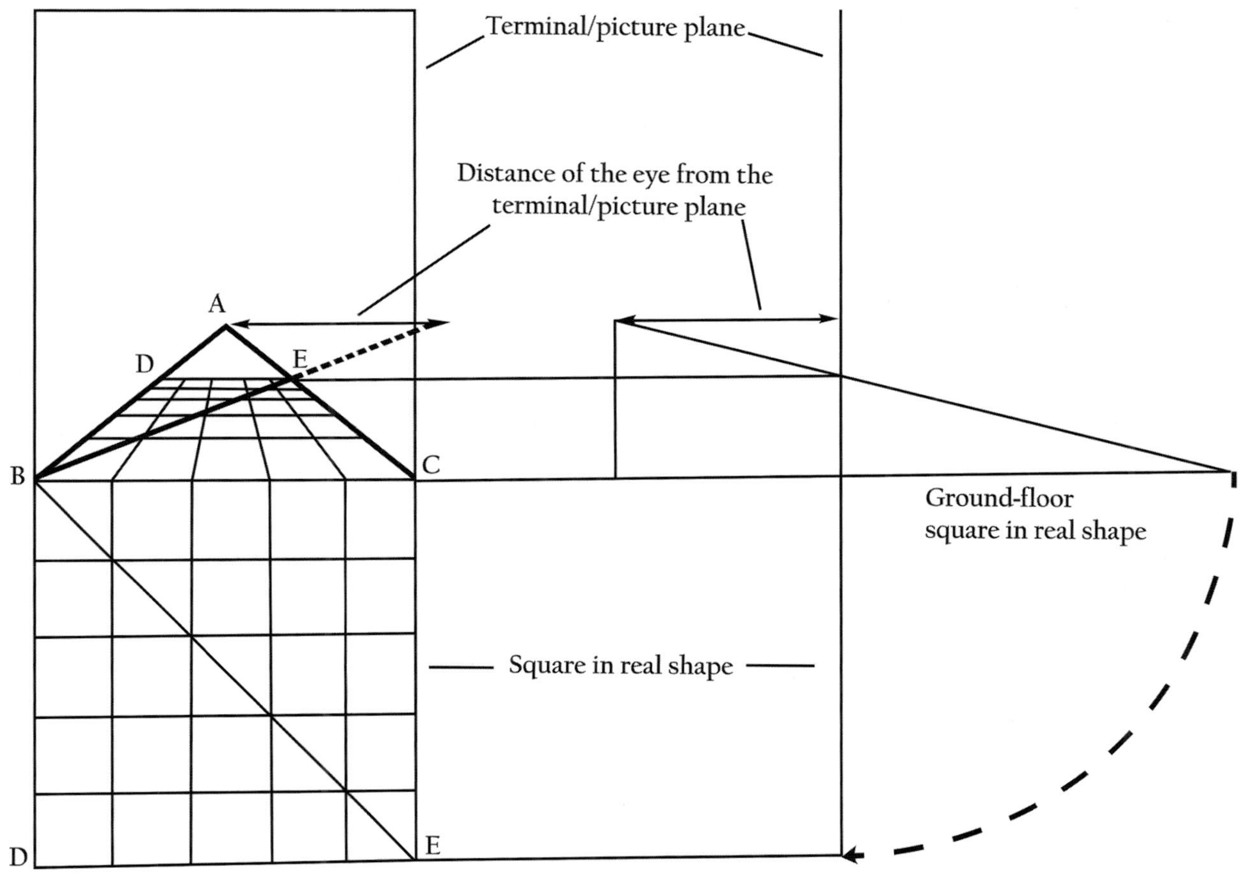

Figure A.11
De prospectiva pingendi: first method, 1.

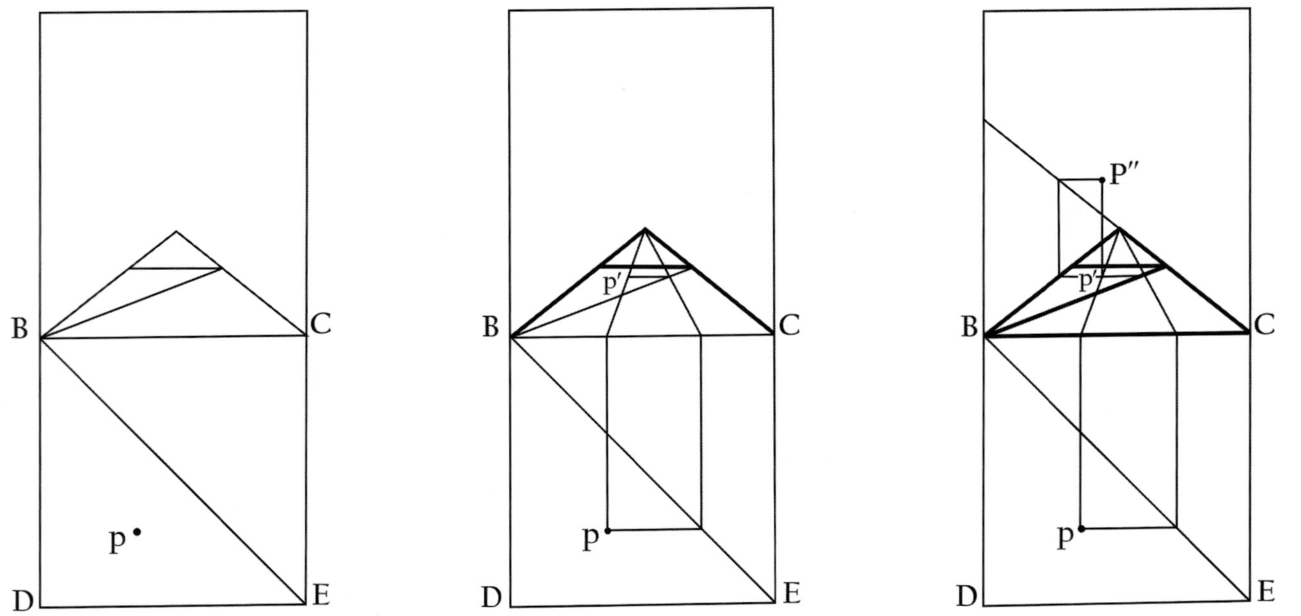

Figure A.12
De prospectiva pingendi: first method, 2.

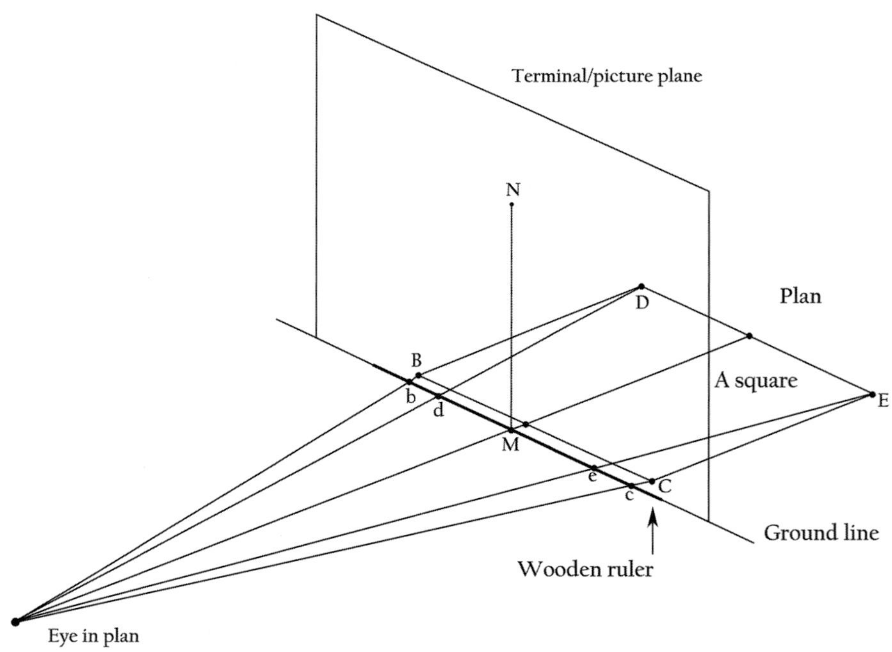

Figure A.13
De prospectiva pingendi: second method, wooden-ruler abscissa.

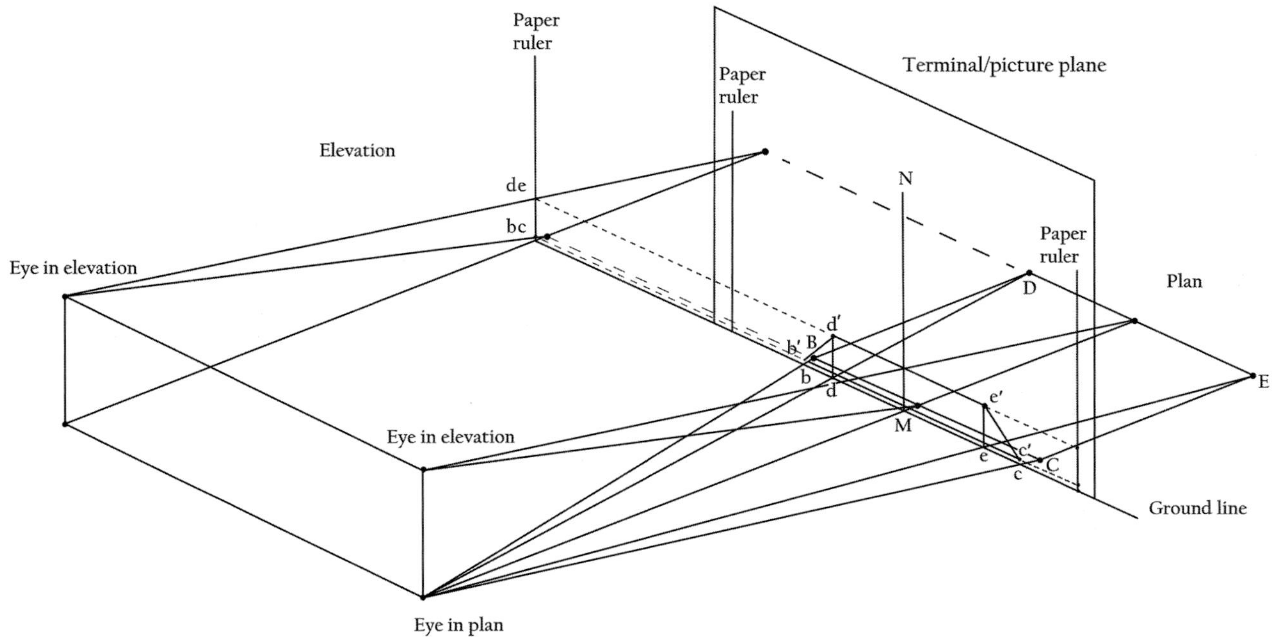

Figure A.14
De prospectiva pingendi: second method, paper-ruler ordinate.

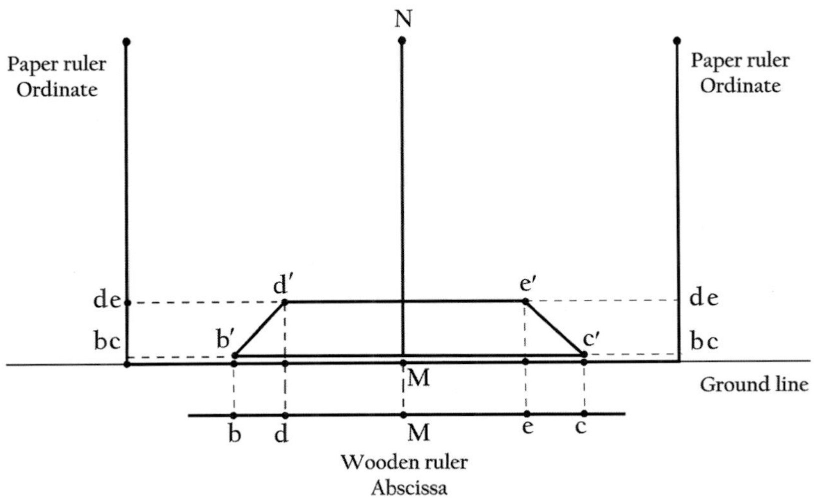

Figure A.15
De prospectiva pingendi: second method.

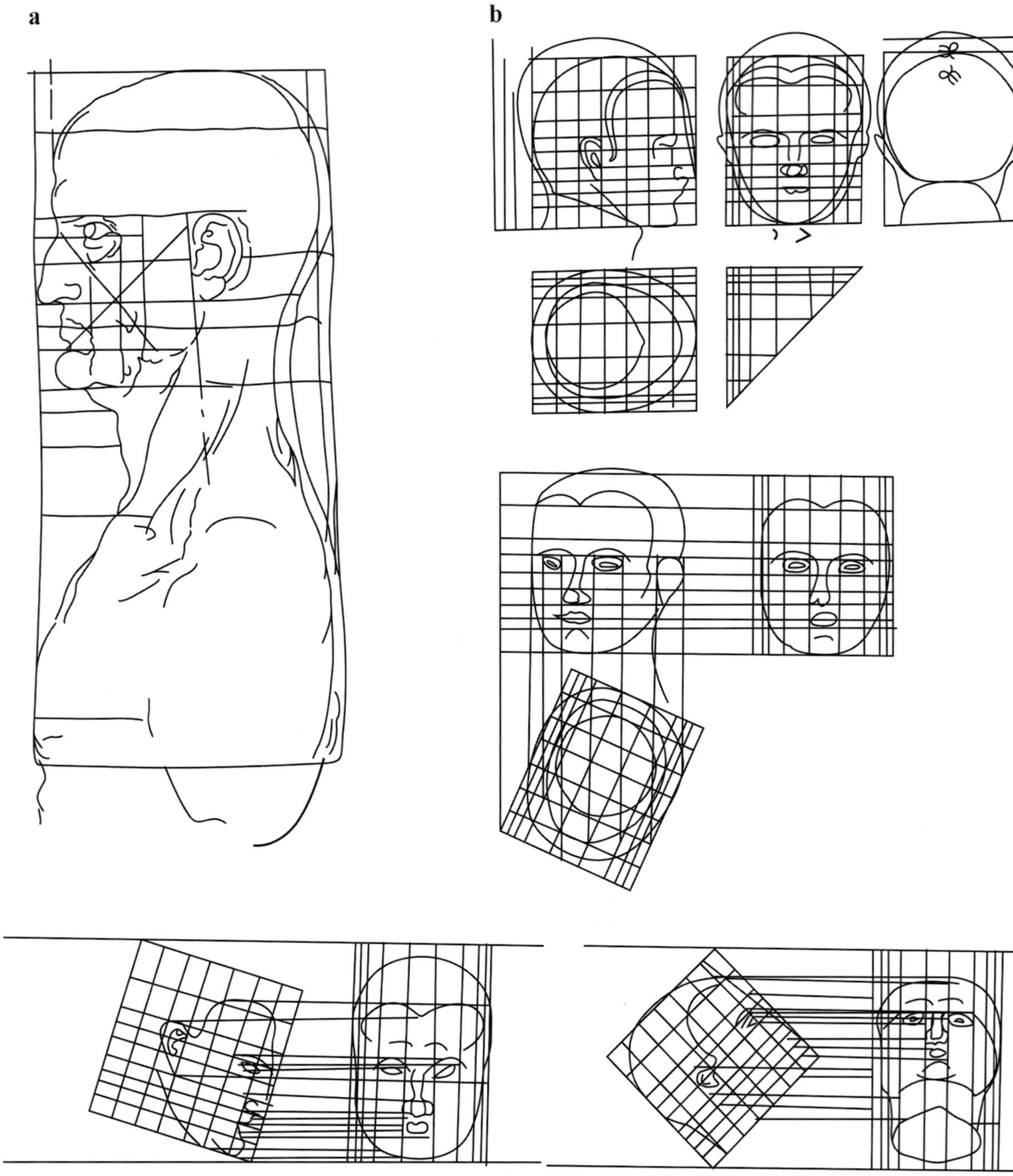

Figure A.16
Leonardo's head and Dürer's head. (*a*) Leonardo, bust of a man seen in profile, Galleria dell'Accademia, Venice (1490); (*b*) Dürer, *De symmetria partium humanorum corporum* (1531).

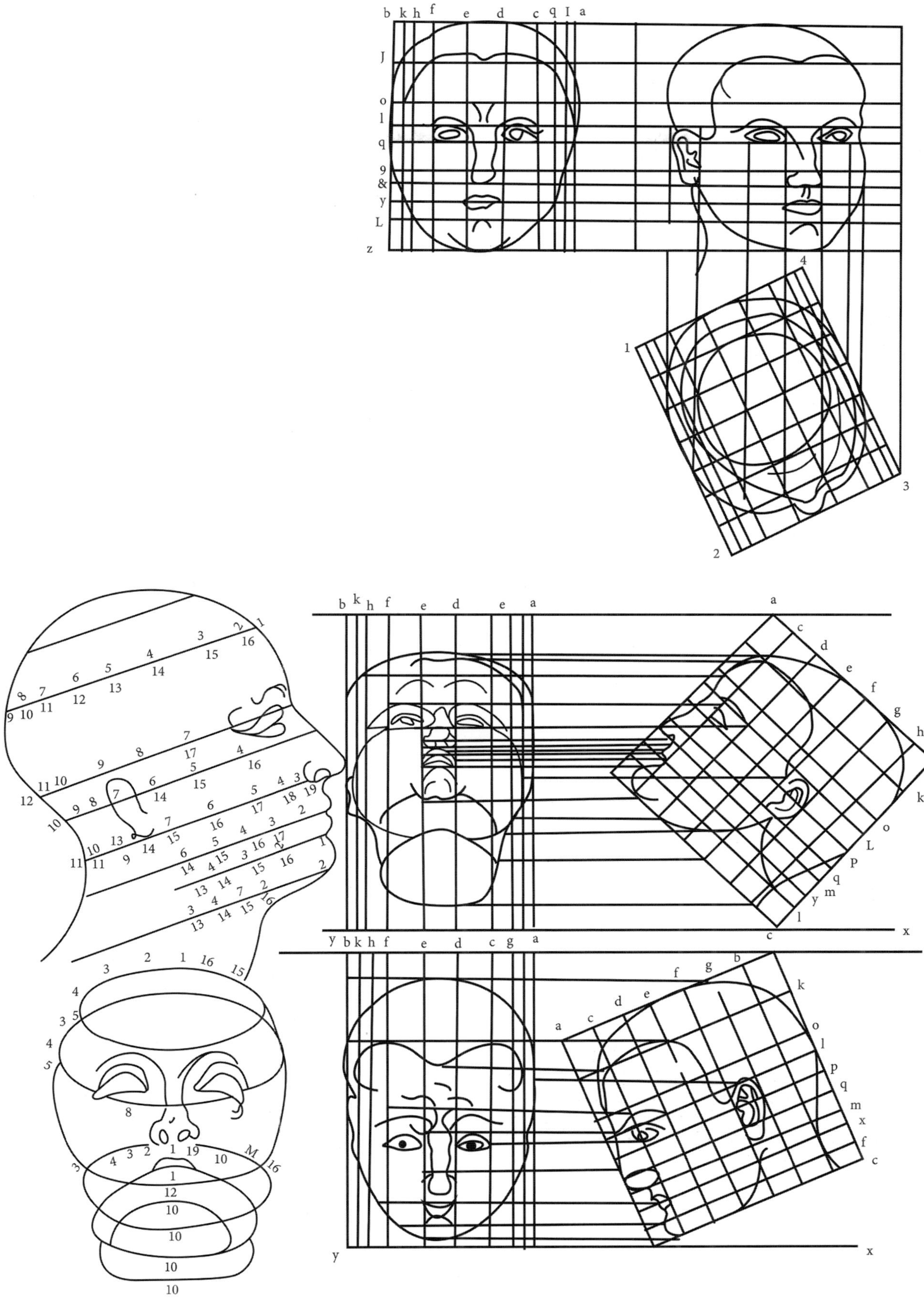

Figure A.17
Barbaro's head, from *La pratica della prospettiva* (1569).

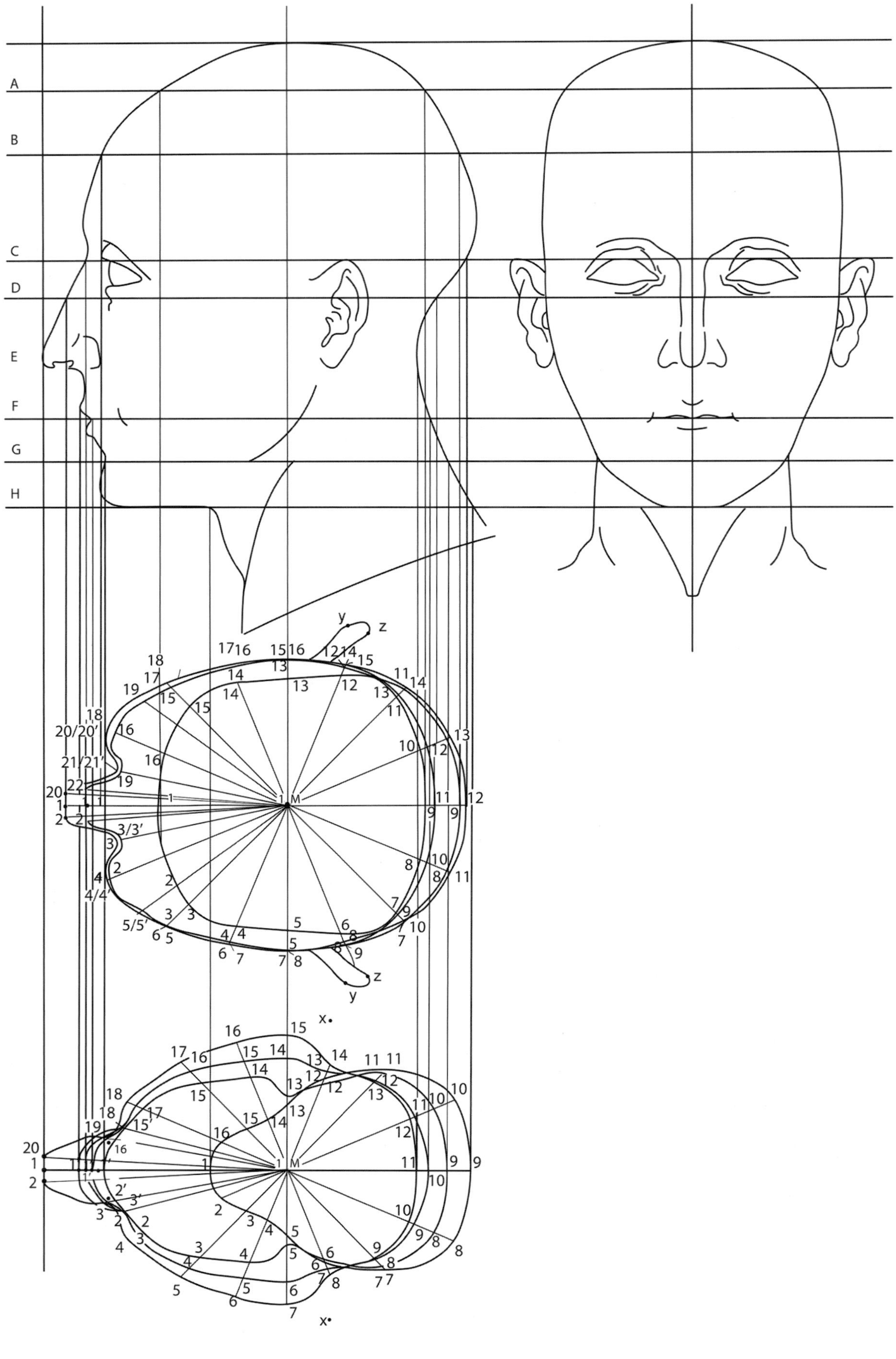

Figure A.18
The numbers in plan and the reconstruction of the circuits.

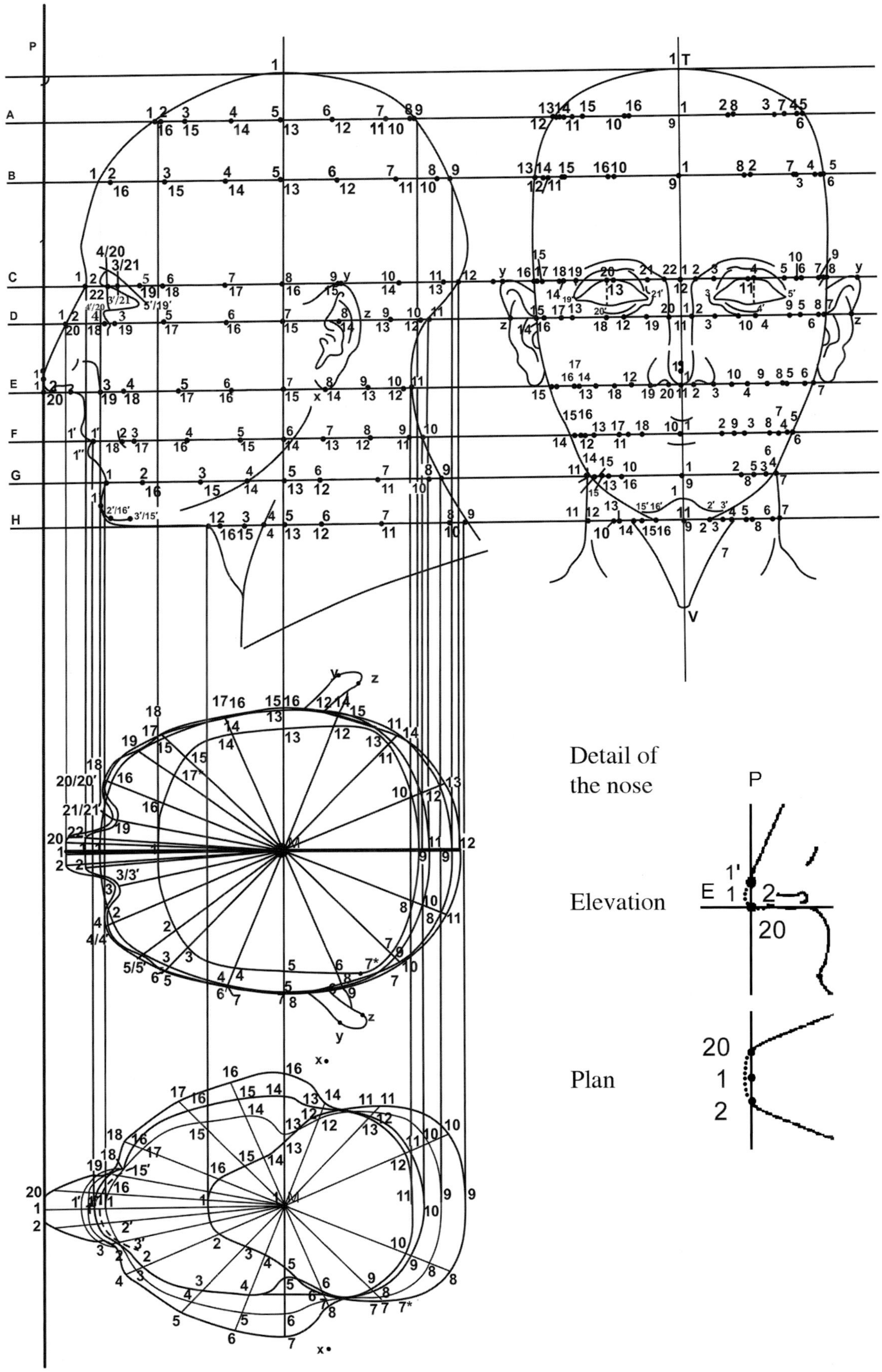

Figure A.19
The numbers on the elevations.

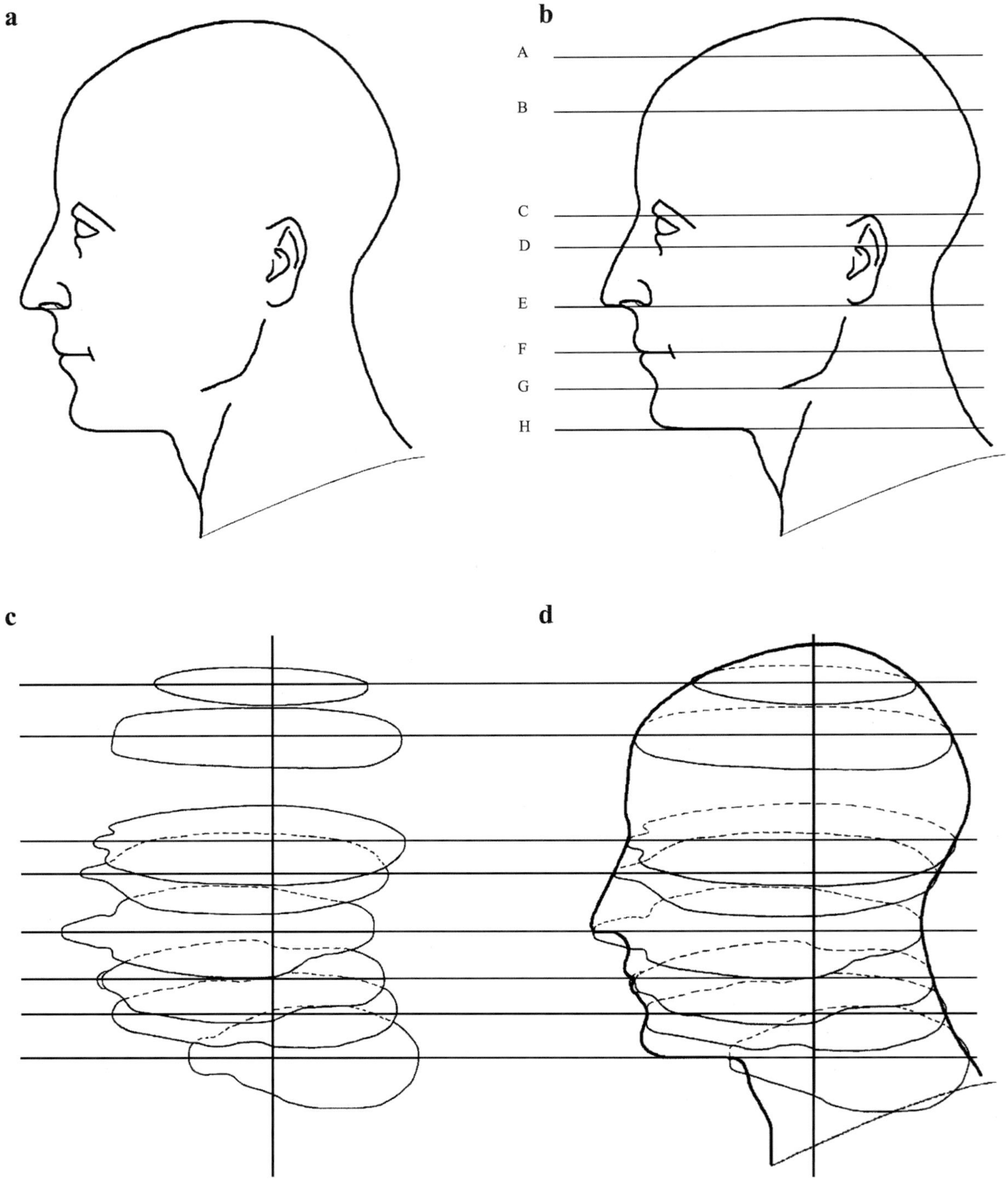

Figure A.20
A multitude of circuits at one's own will.

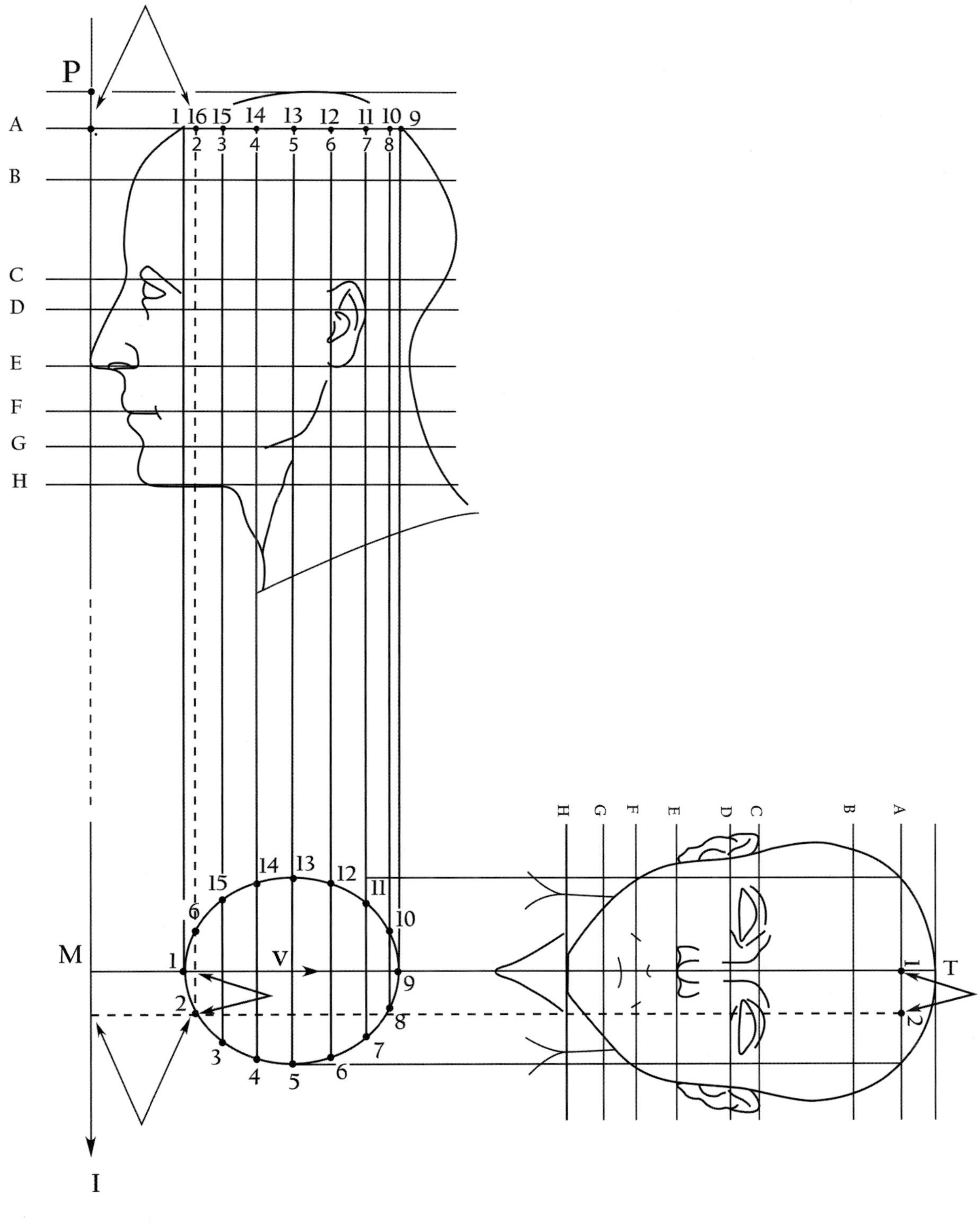

Figure A.21
The orthogonal axes.

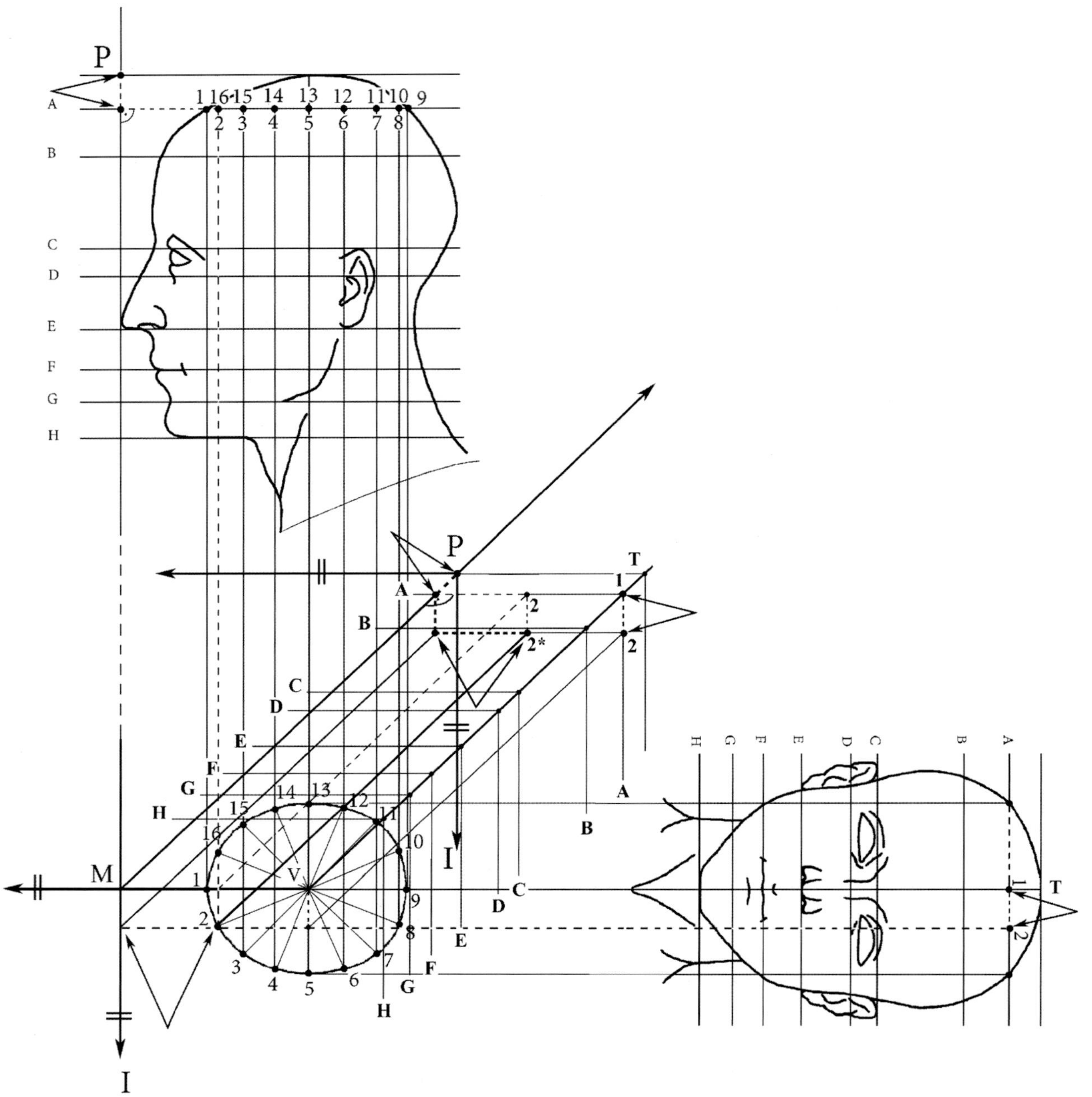

Figure A.22
A point in space.

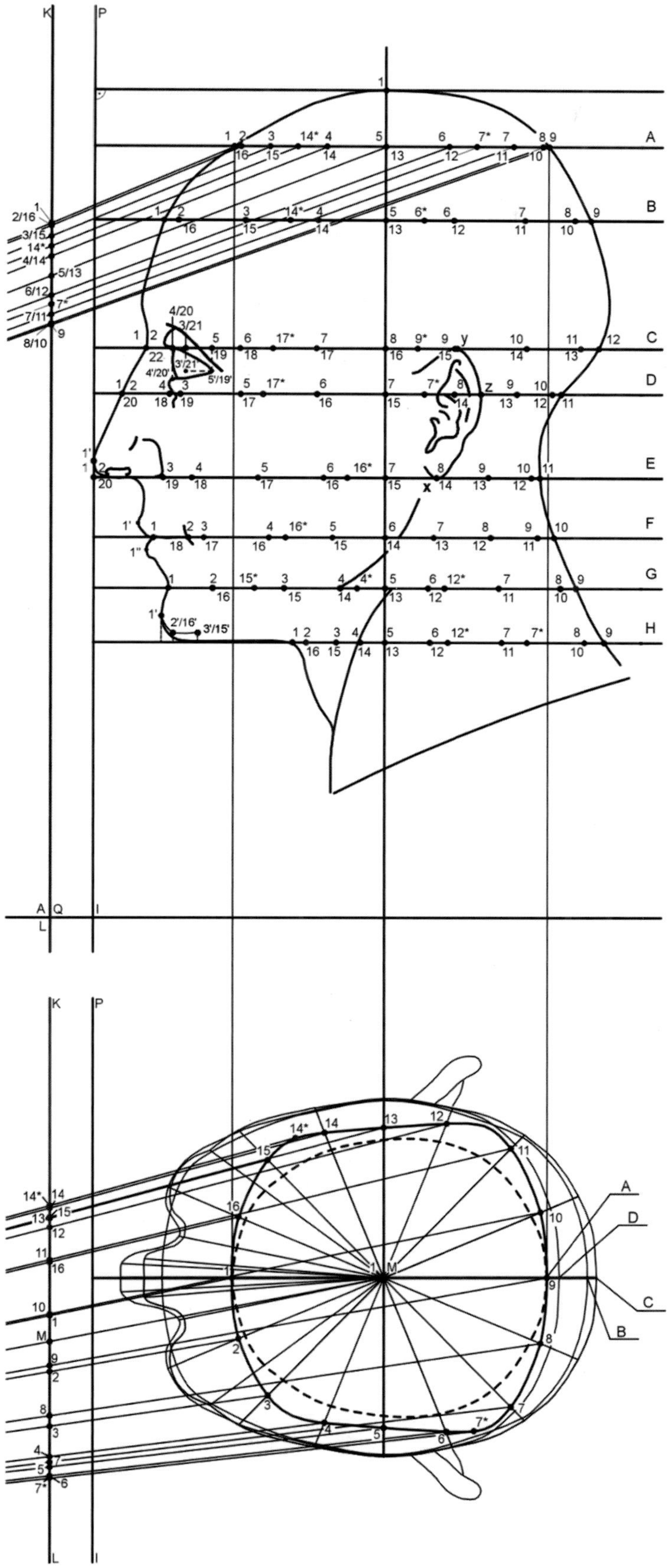

Figure A.23
The remade circuit A.

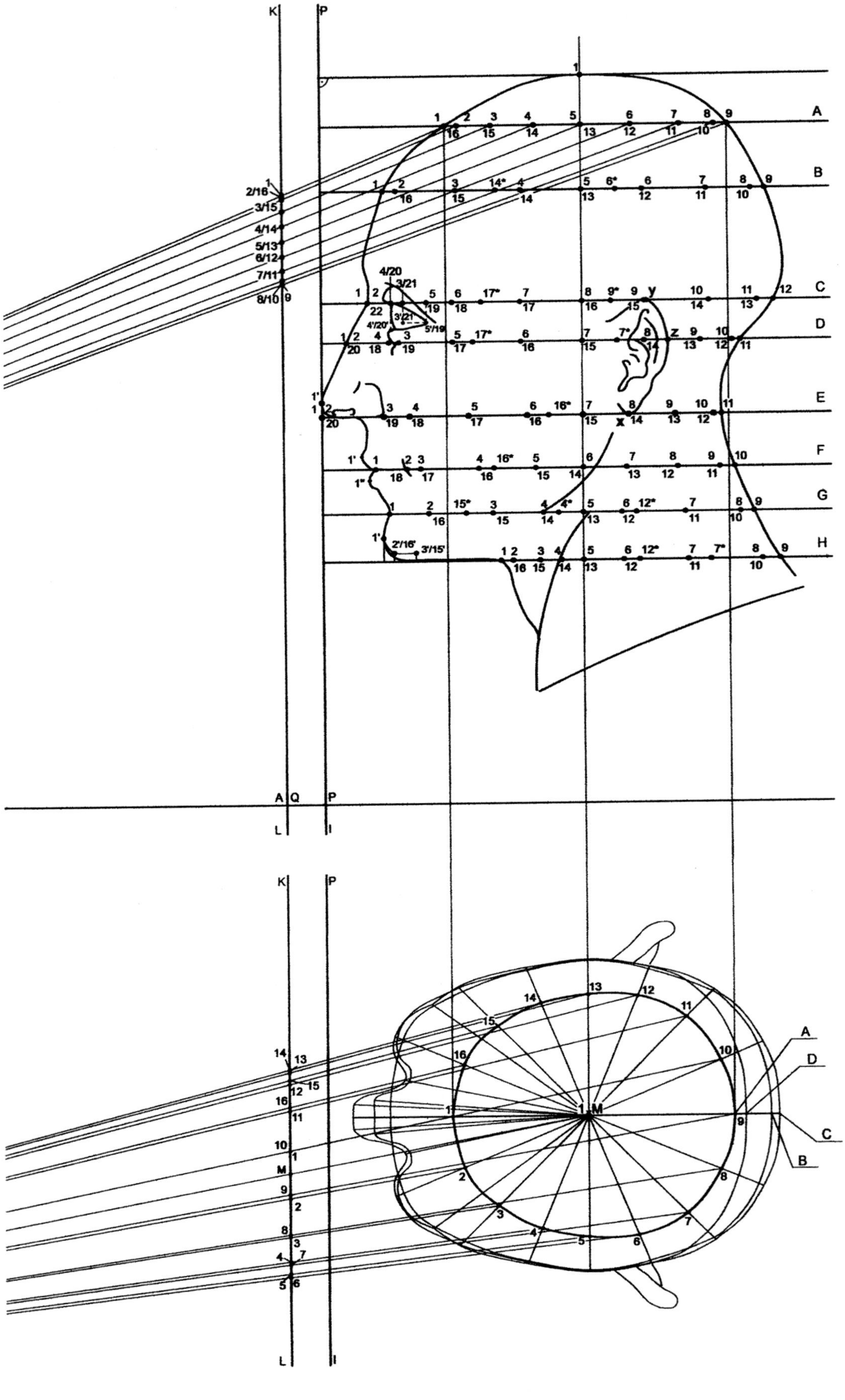

Figure A.24
Circuit A compliant with the original.

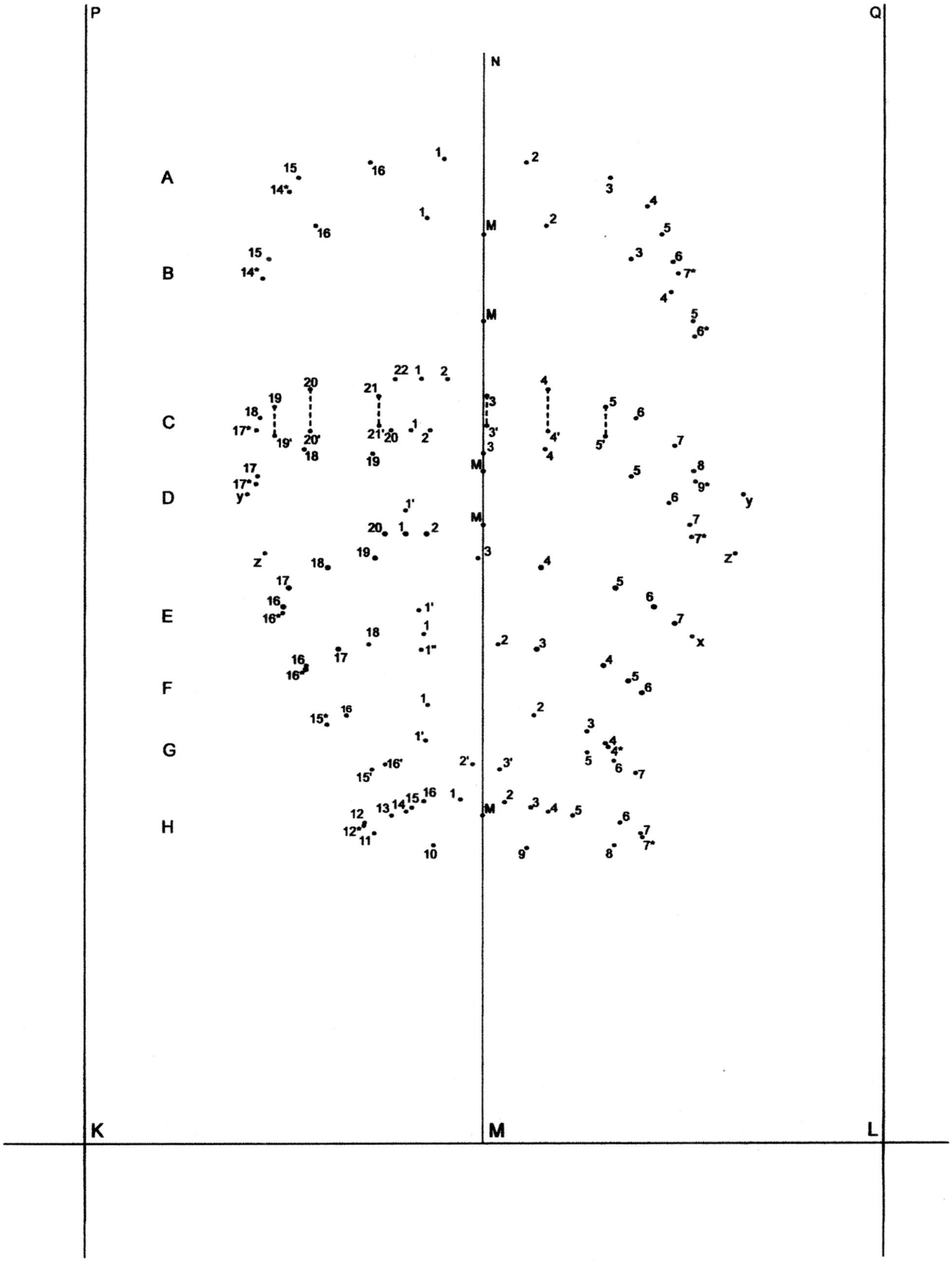

Figure A.25
Images of all the visible points of the circuits.

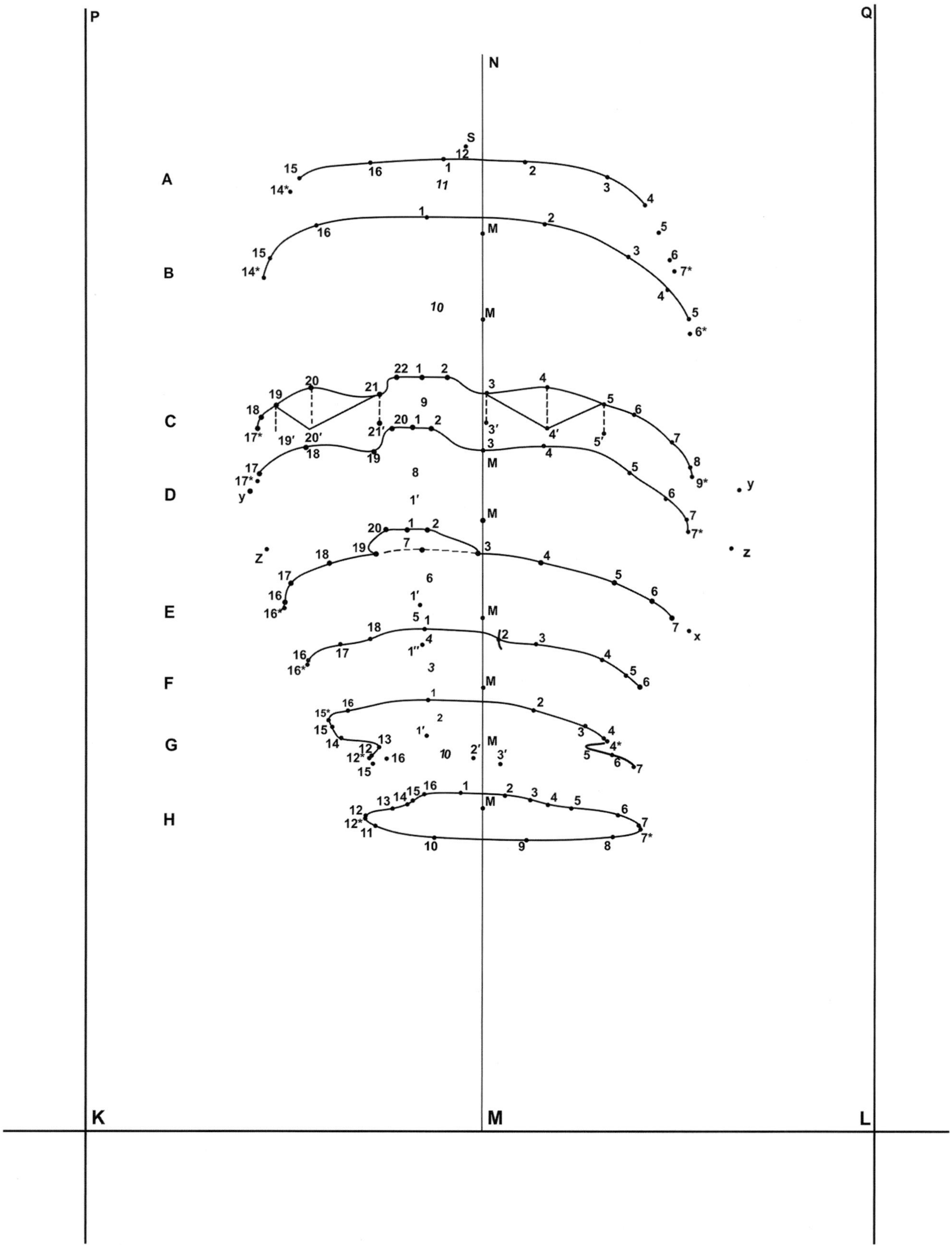

Figure A.26
The connected dots of all the circuits.

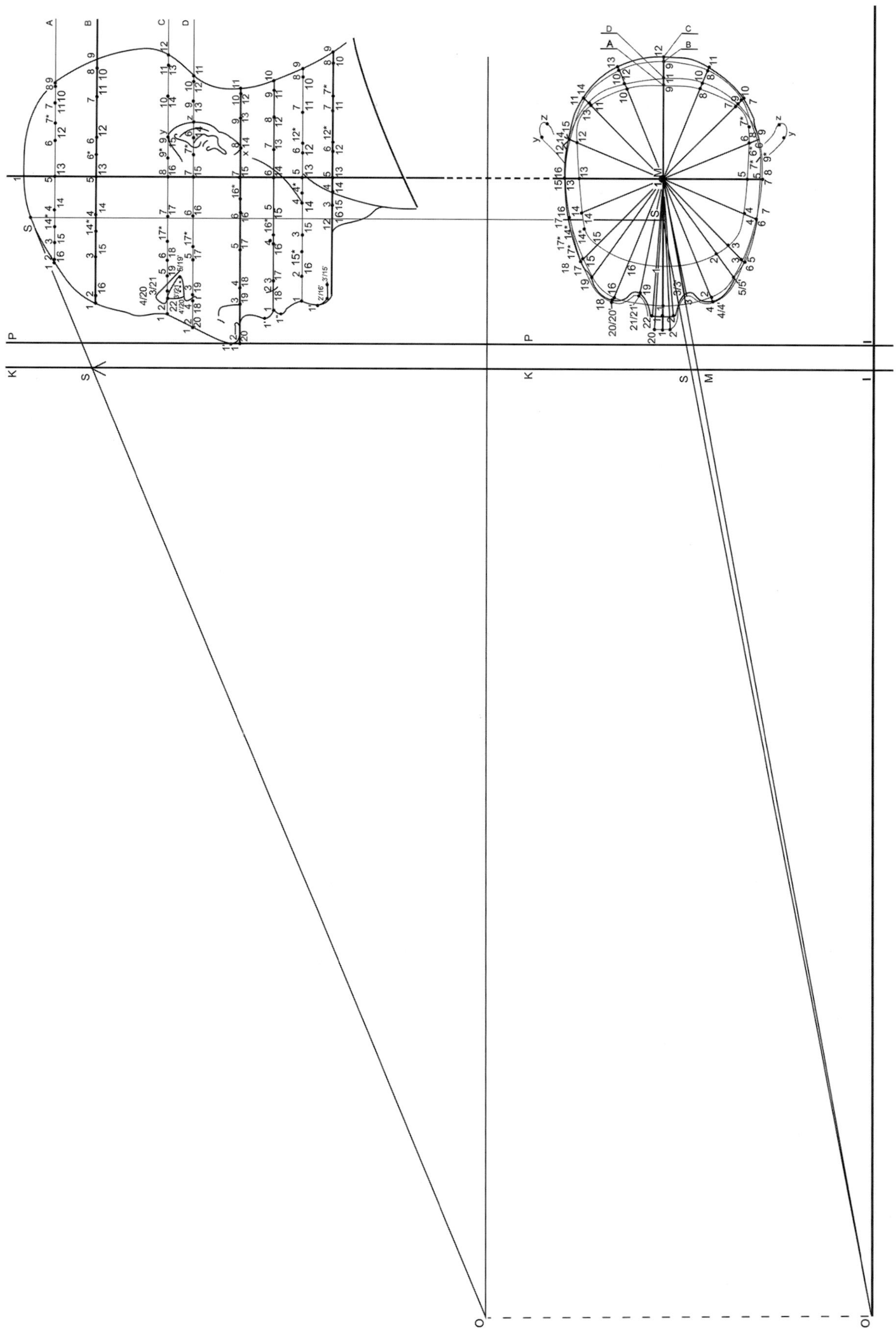

Figure A.27
The tangent from the eye to the vertical section of the head.

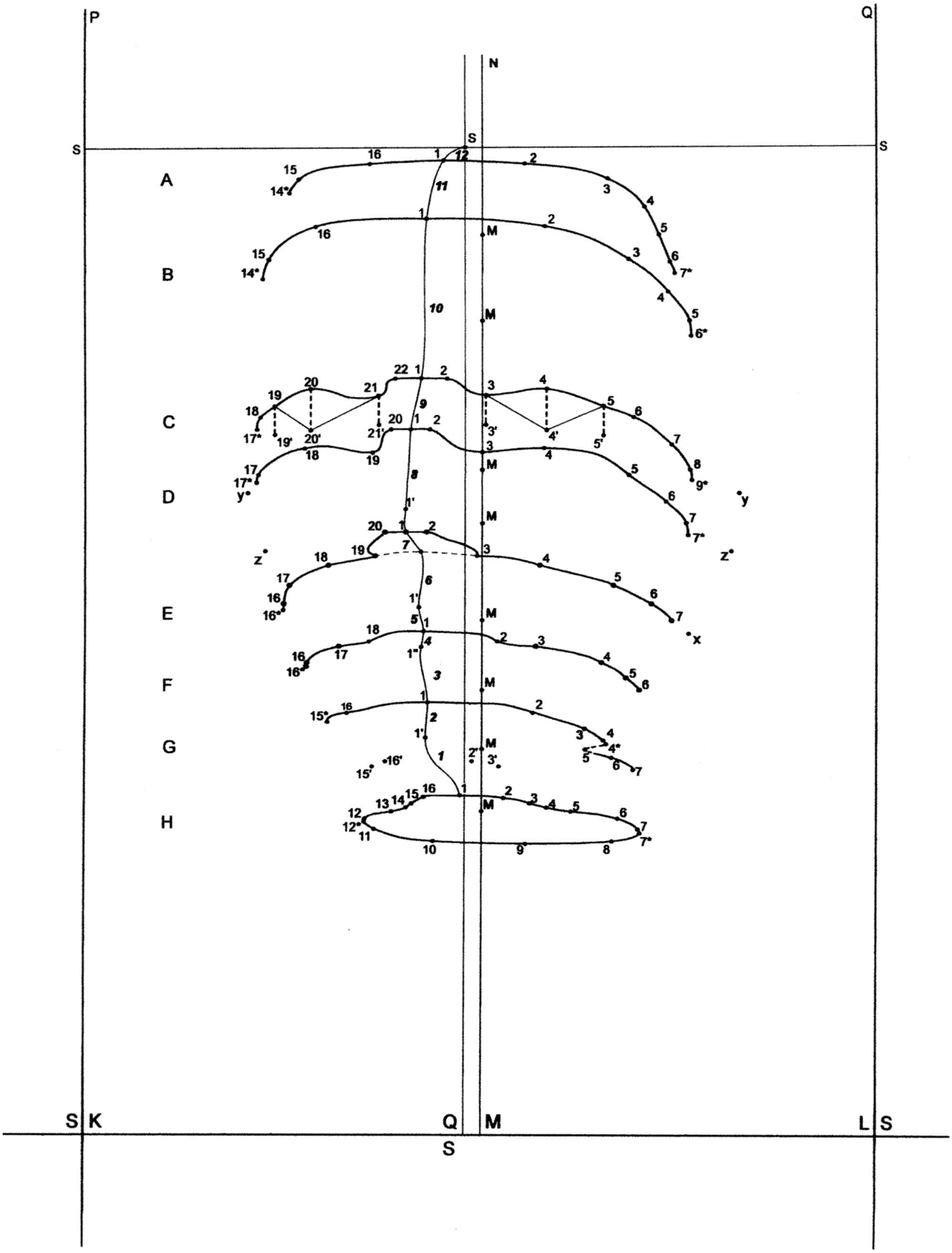

Figure A.28
The twelve parts of the vertical section of the head.

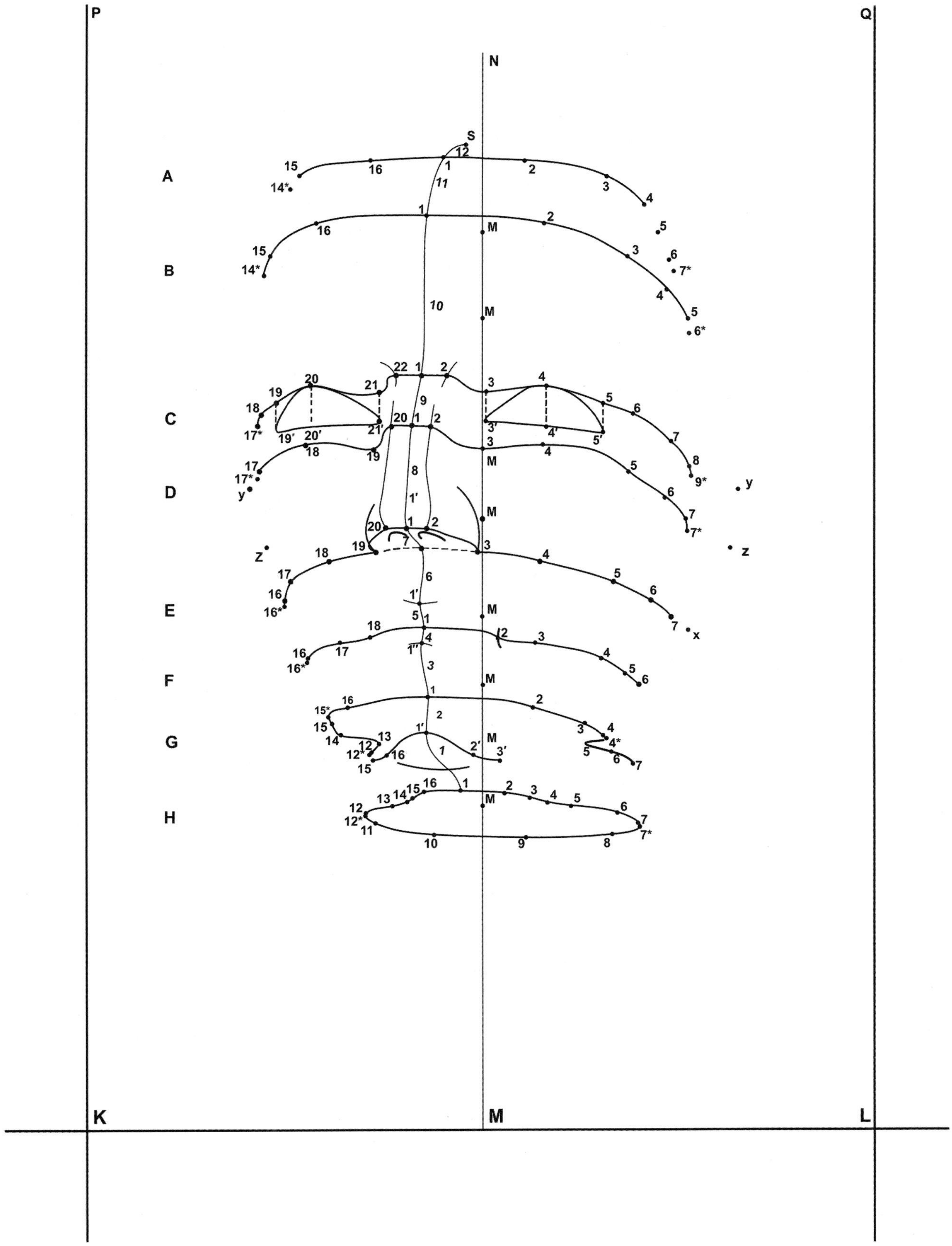

Figure A.29
The protrusions of the head.

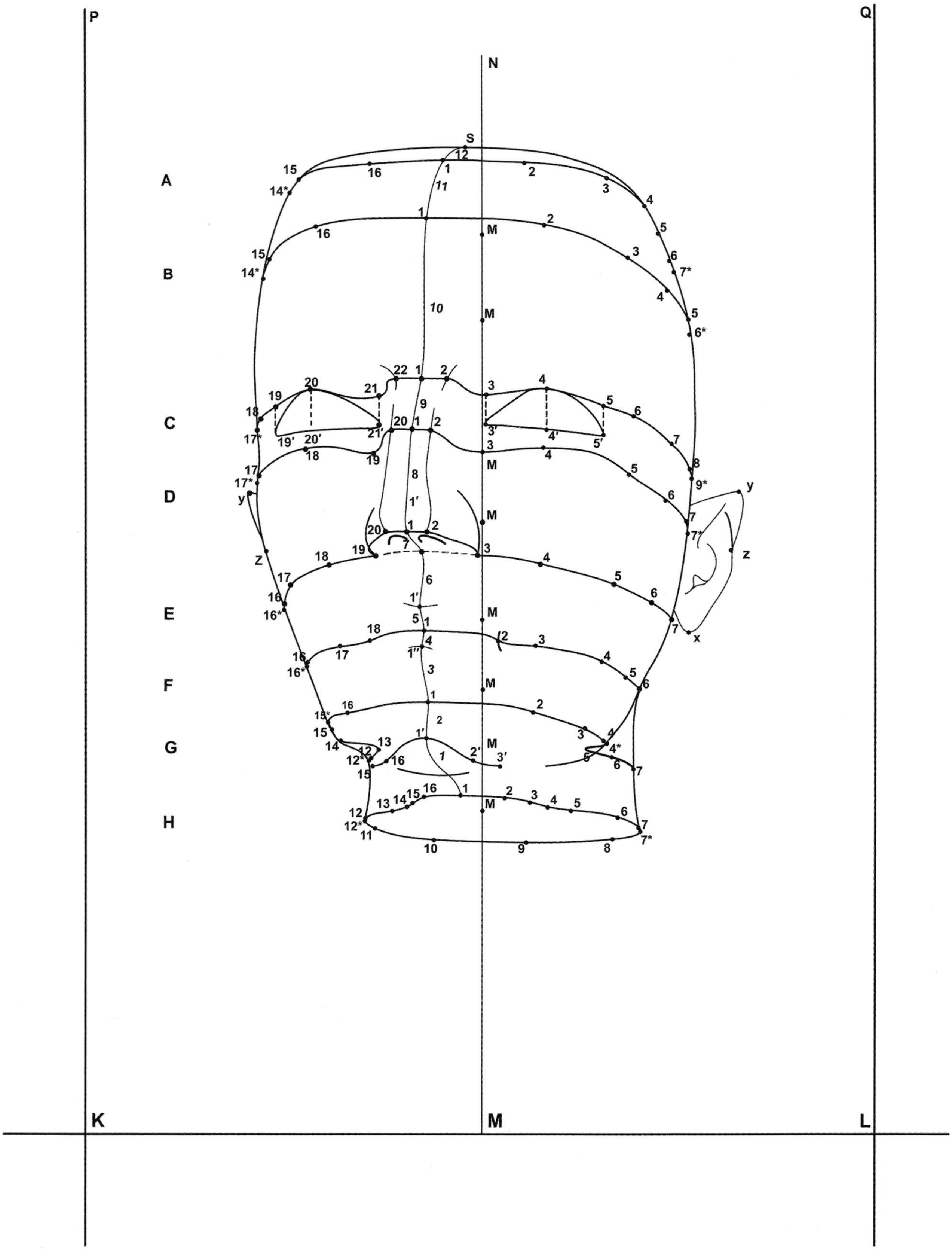

Figure A.30
The remaining outlines.

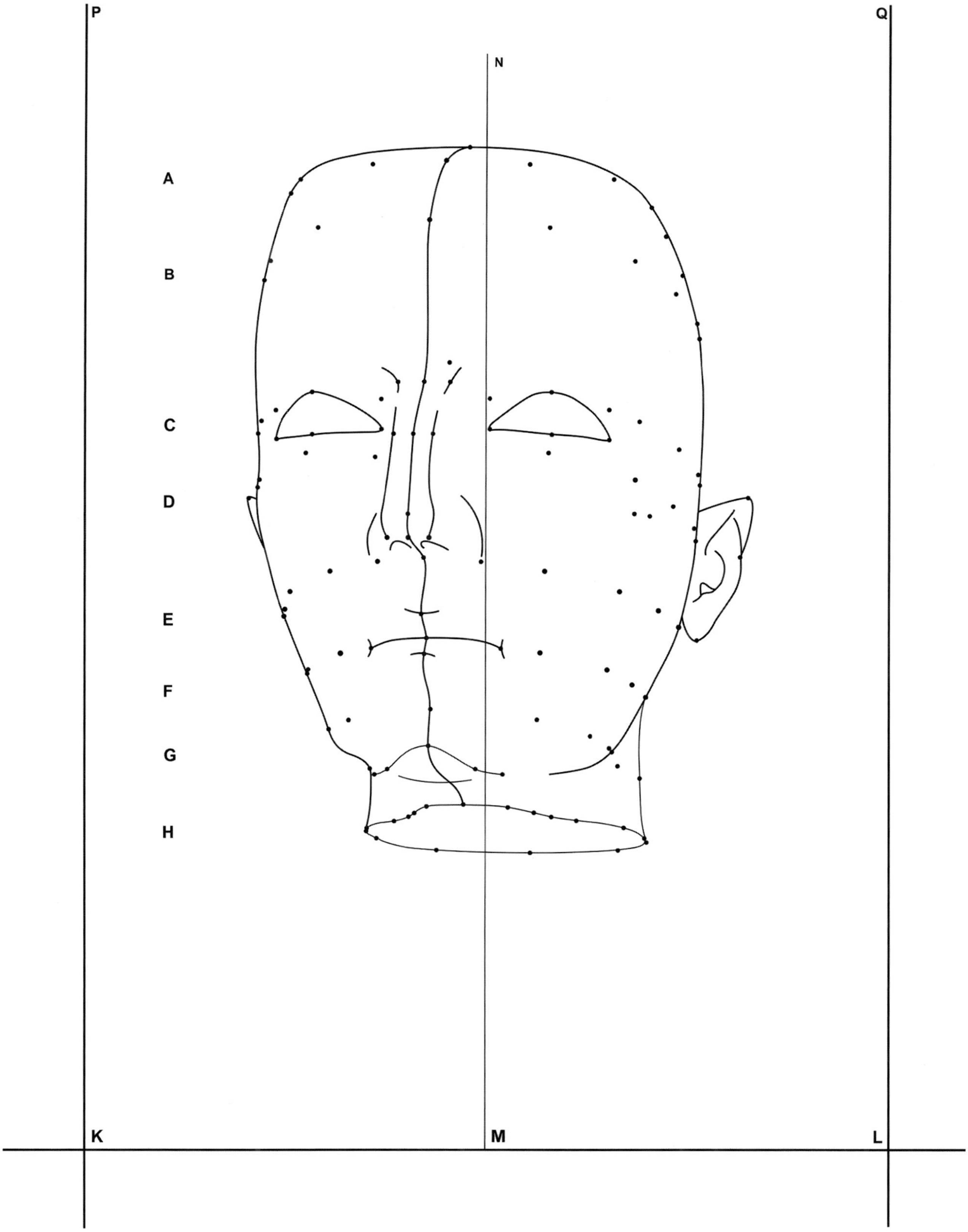

Figure A.31
First head: the perspective with the remade circuit A.

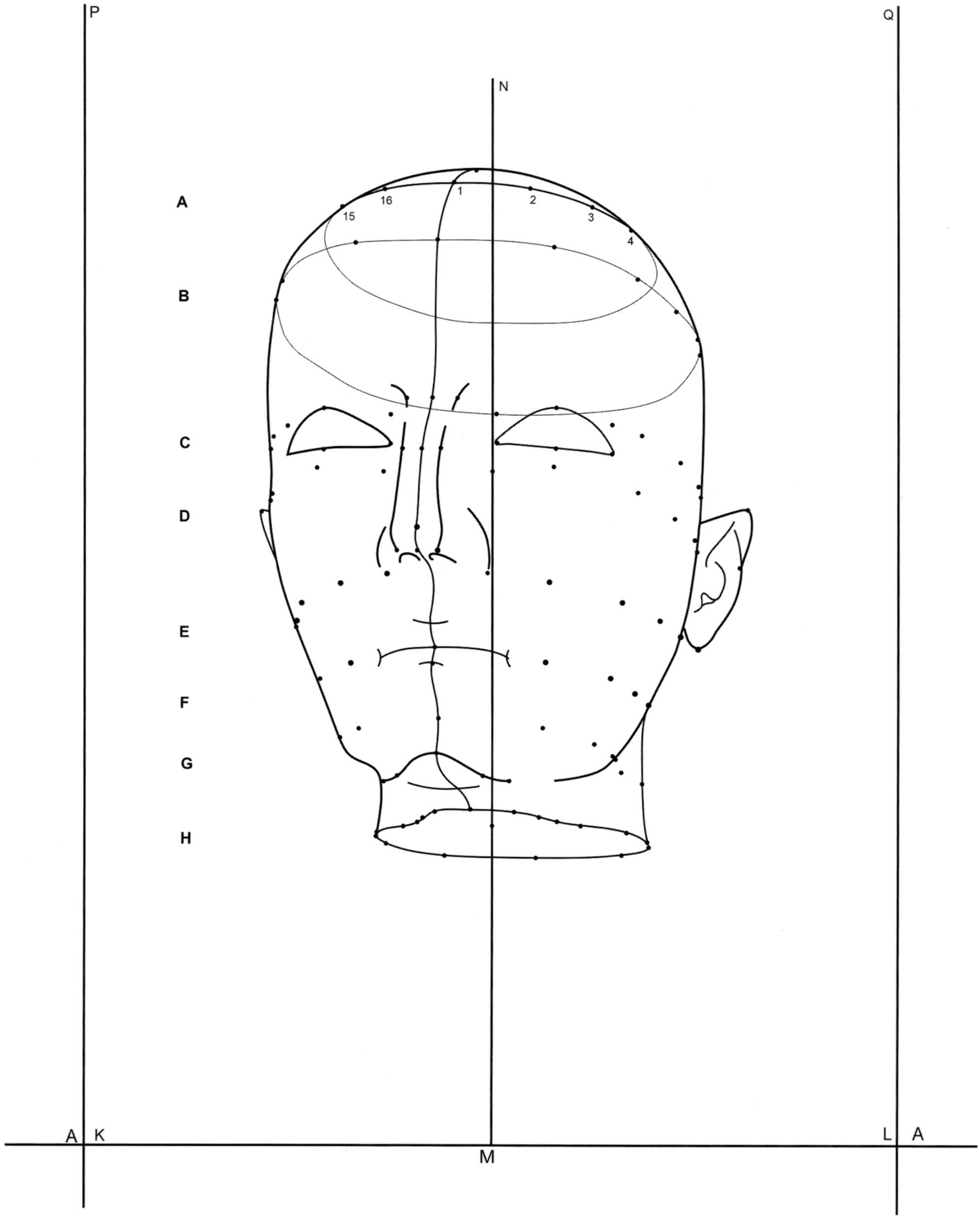

Figure A.32
First head: the outline of circuit A compliant with the original.

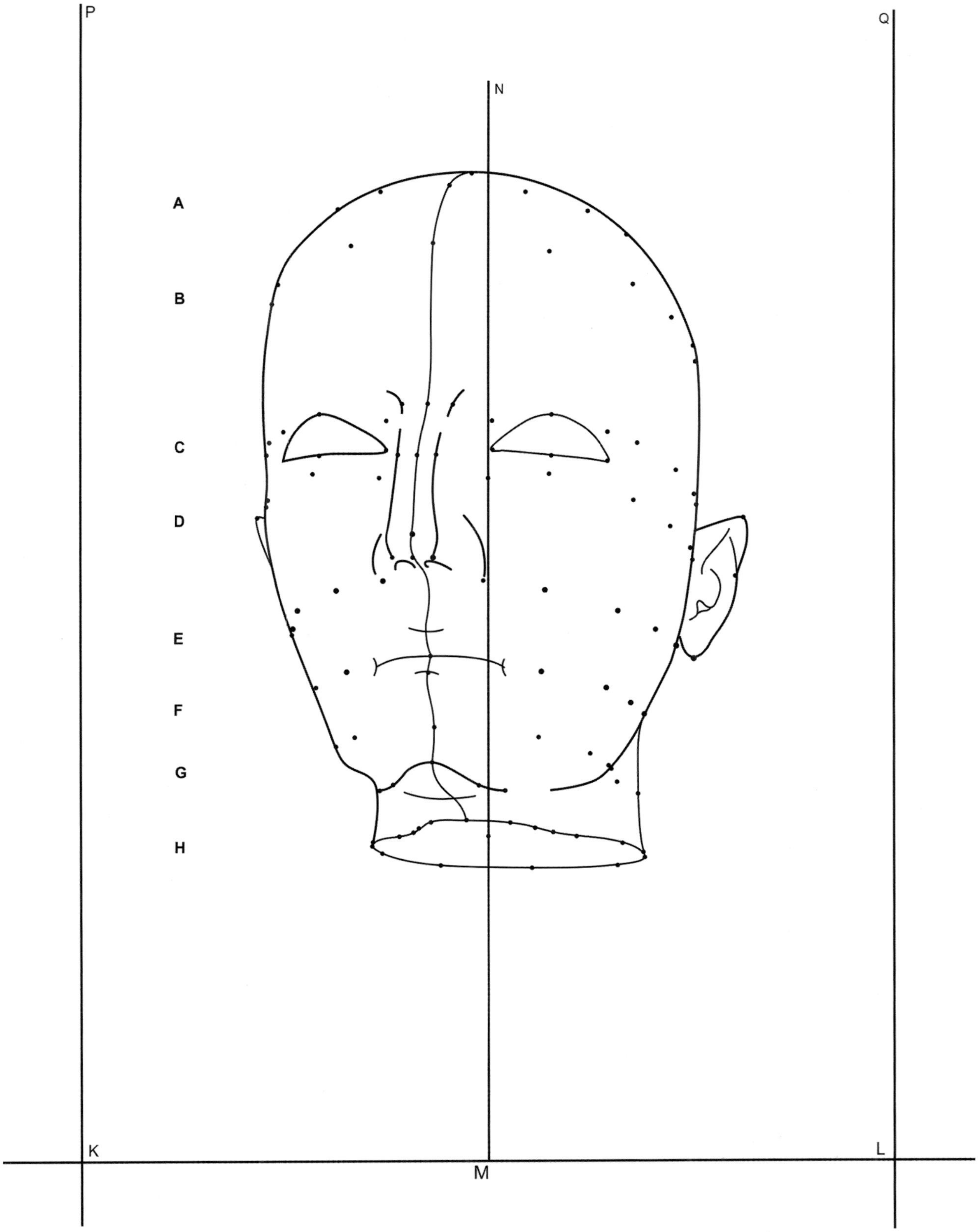

Figure A.33
First head: the perspective with circuit A compliant with the original.

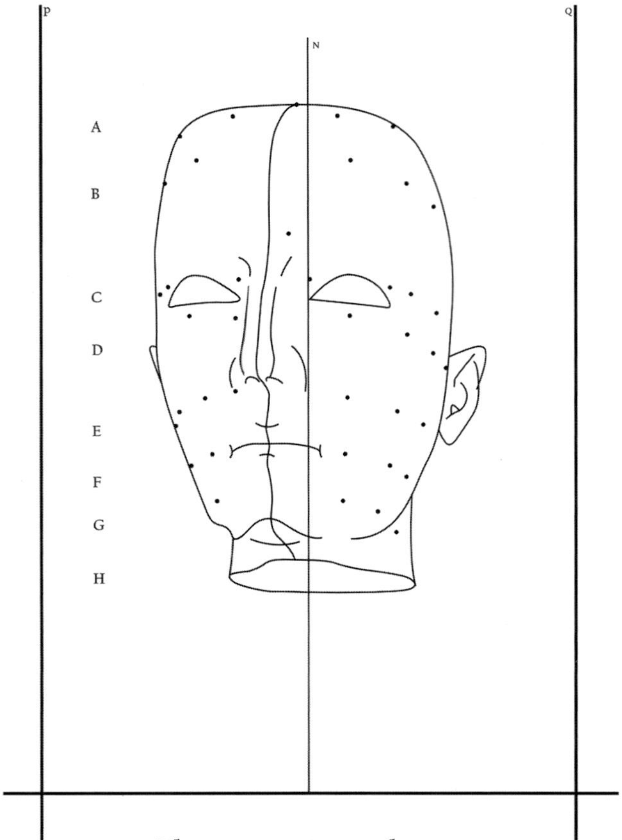

The perspective with
the remade circuit A

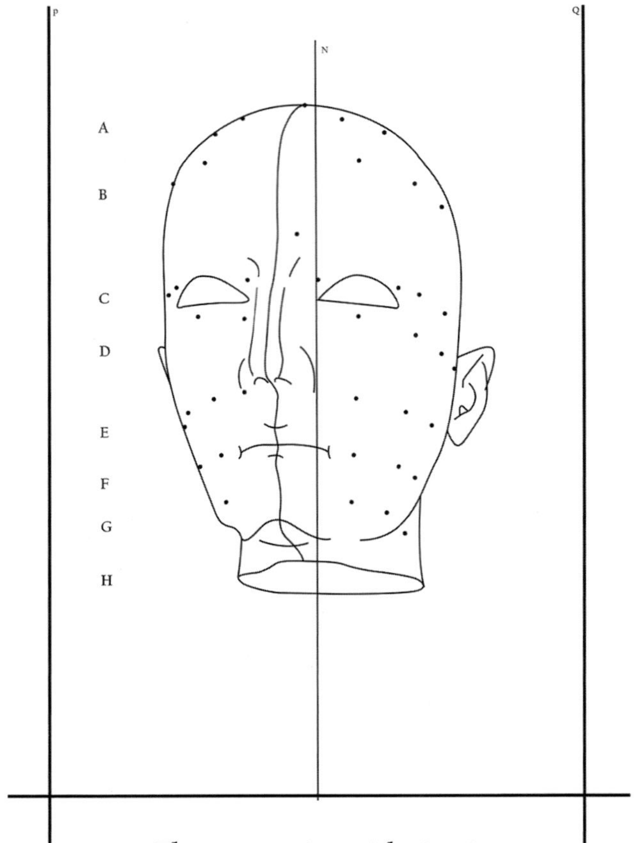

The perspective with circuit
A compliant with the original

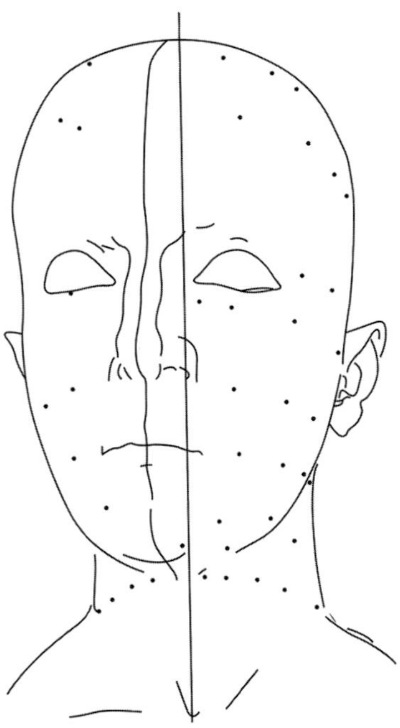

Parma manuscript, folio 66 verso

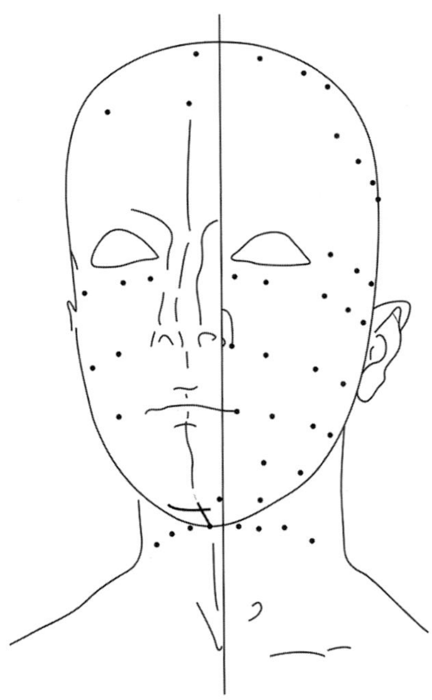

Reggio Emilia manuscript, folio 76 recto

Figure A.34
Comparisons of perspectives.

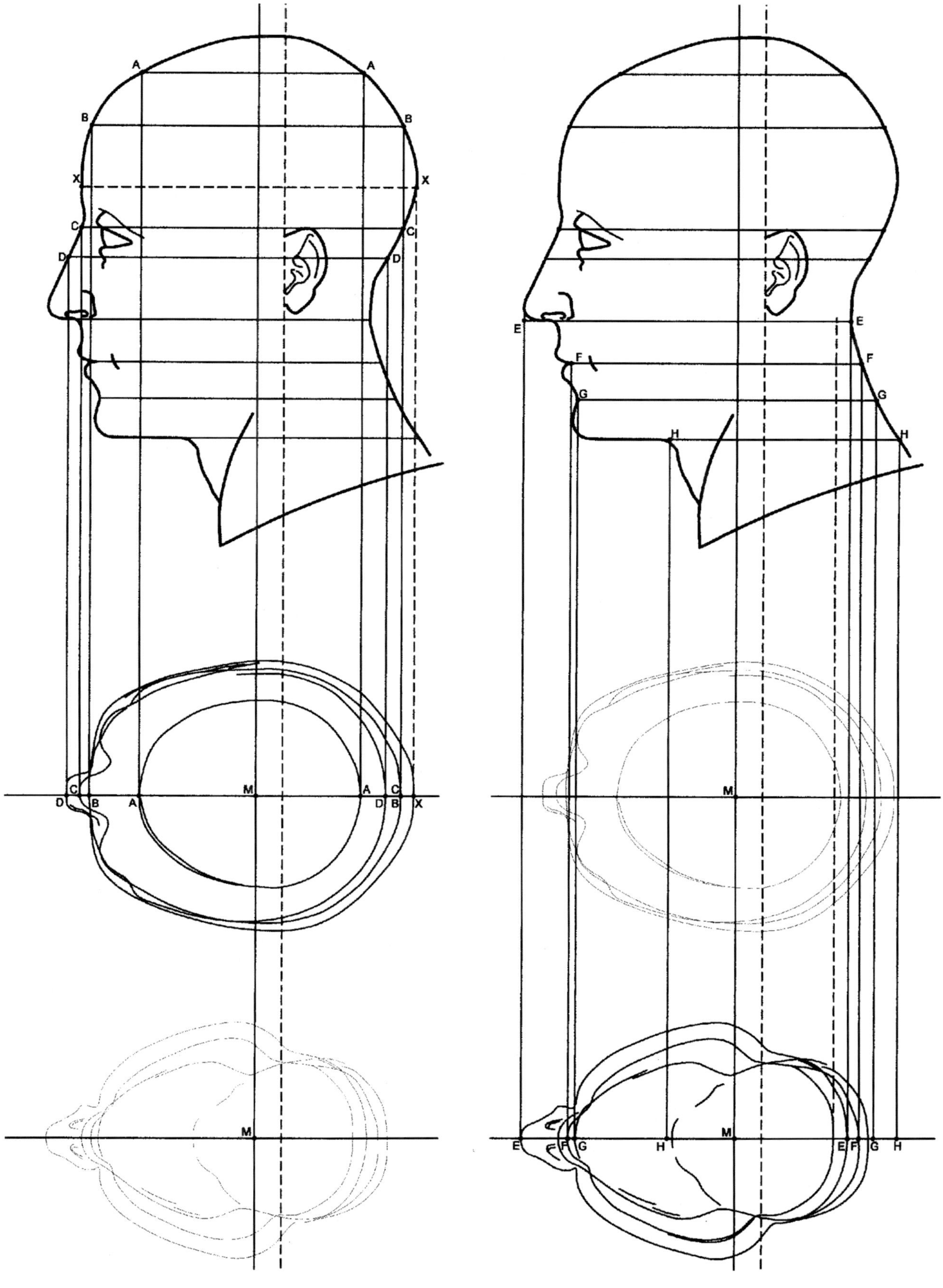

Figure A.35
Parma manuscript, folio 59 verso: incongruences.

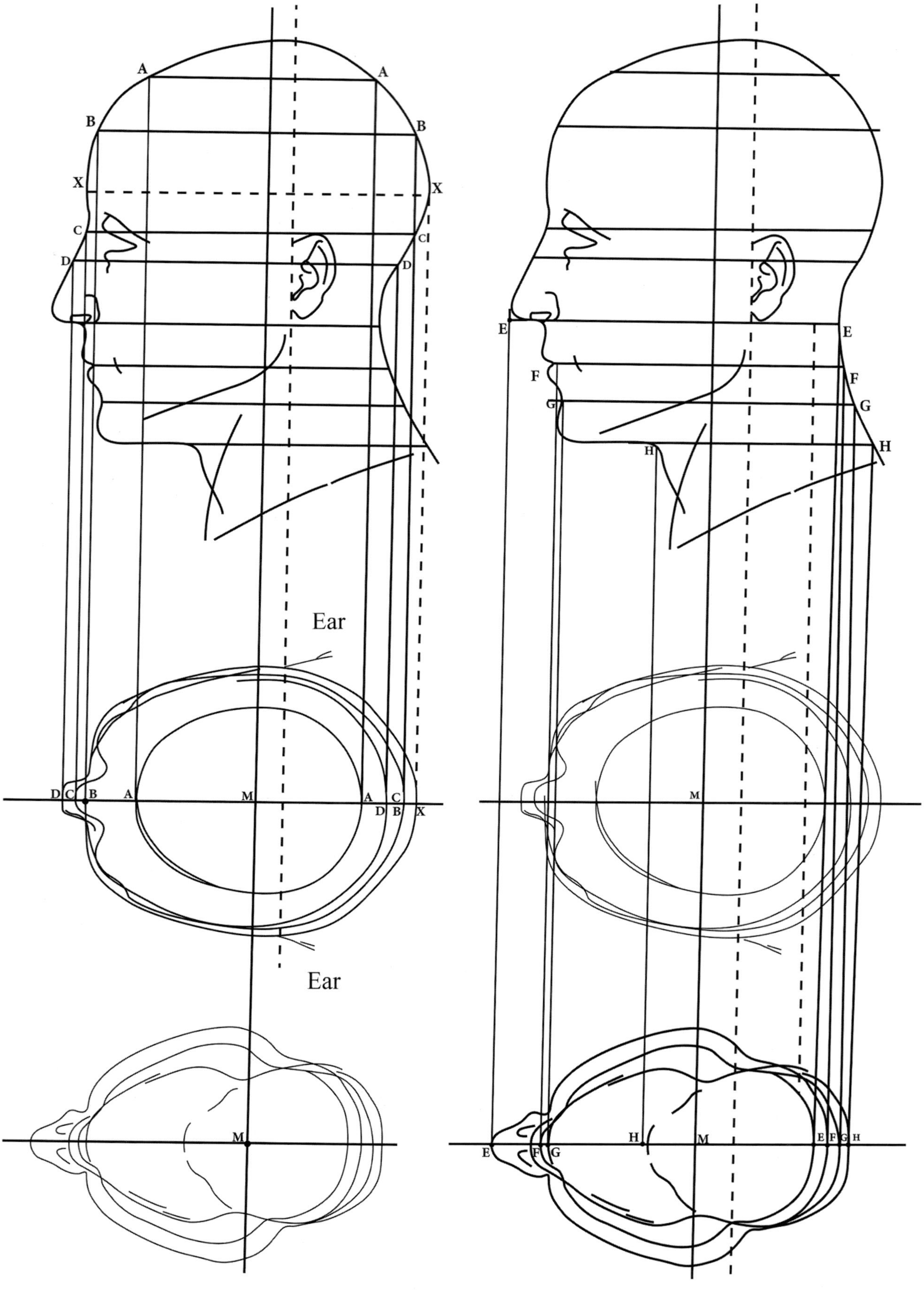

Figure A.36
Parma manuscript, folio 61 recto: incongruences.

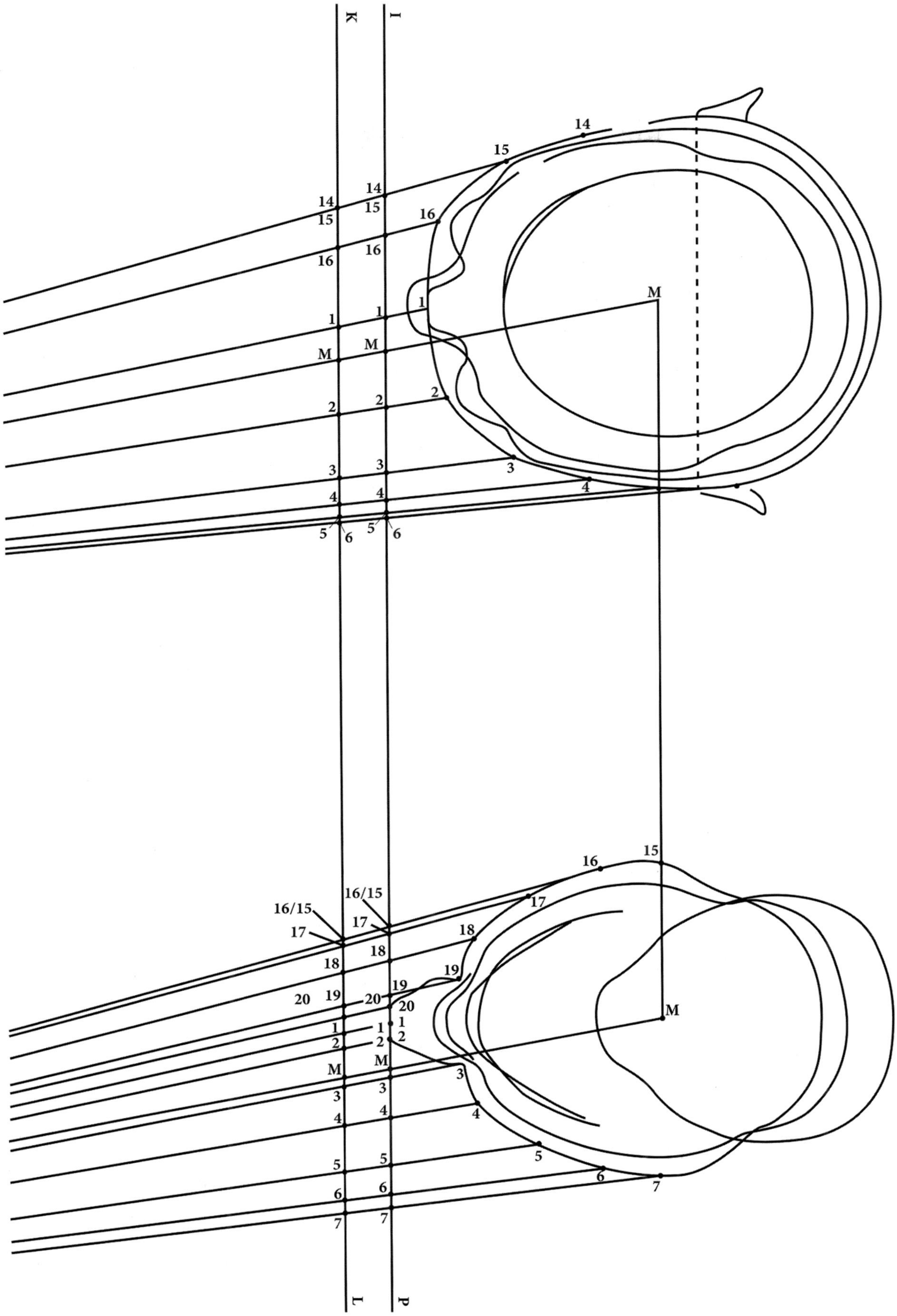

Figure A.37
Parma manuscript, folio 63 verso: incongruences.

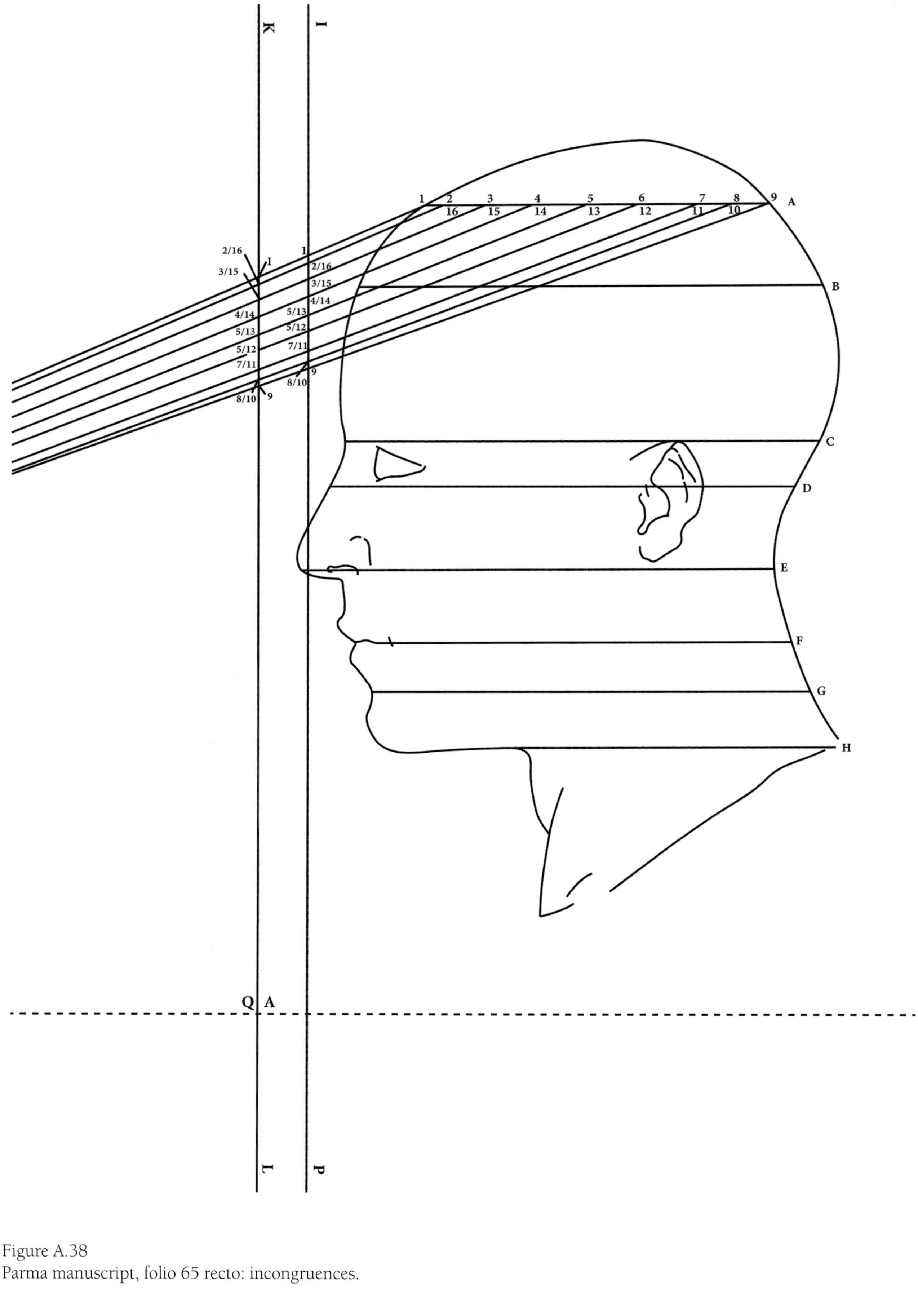

Figure A.38
Parma manuscript, folio 65 recto: incongruences.

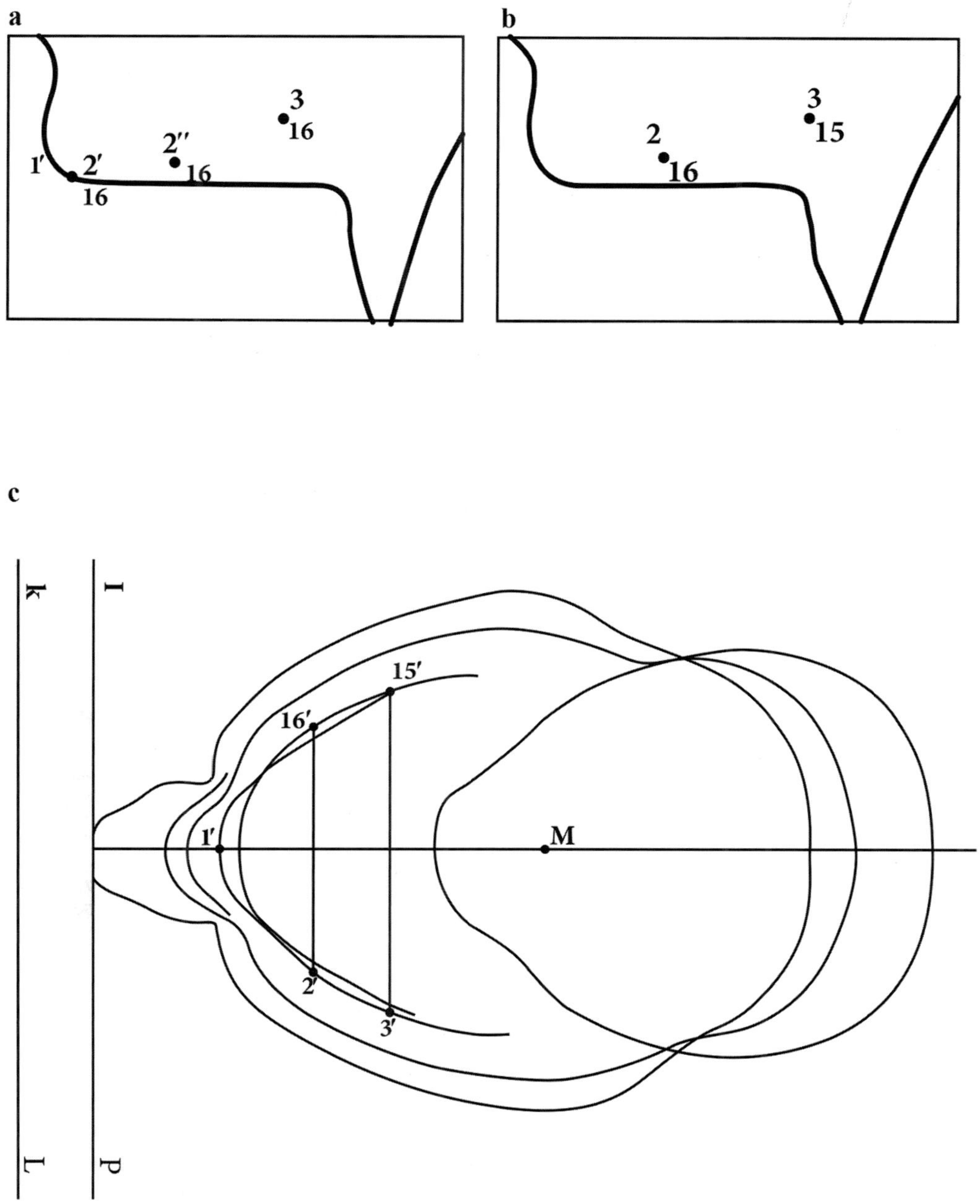

Figure A.39
Parma manuscript, folio 61 recto (*a*); Reggio Emilia manuscript,
folio 70 recto (*b*); Parma manuscript, folio 63 verso (*c*): incongruences.

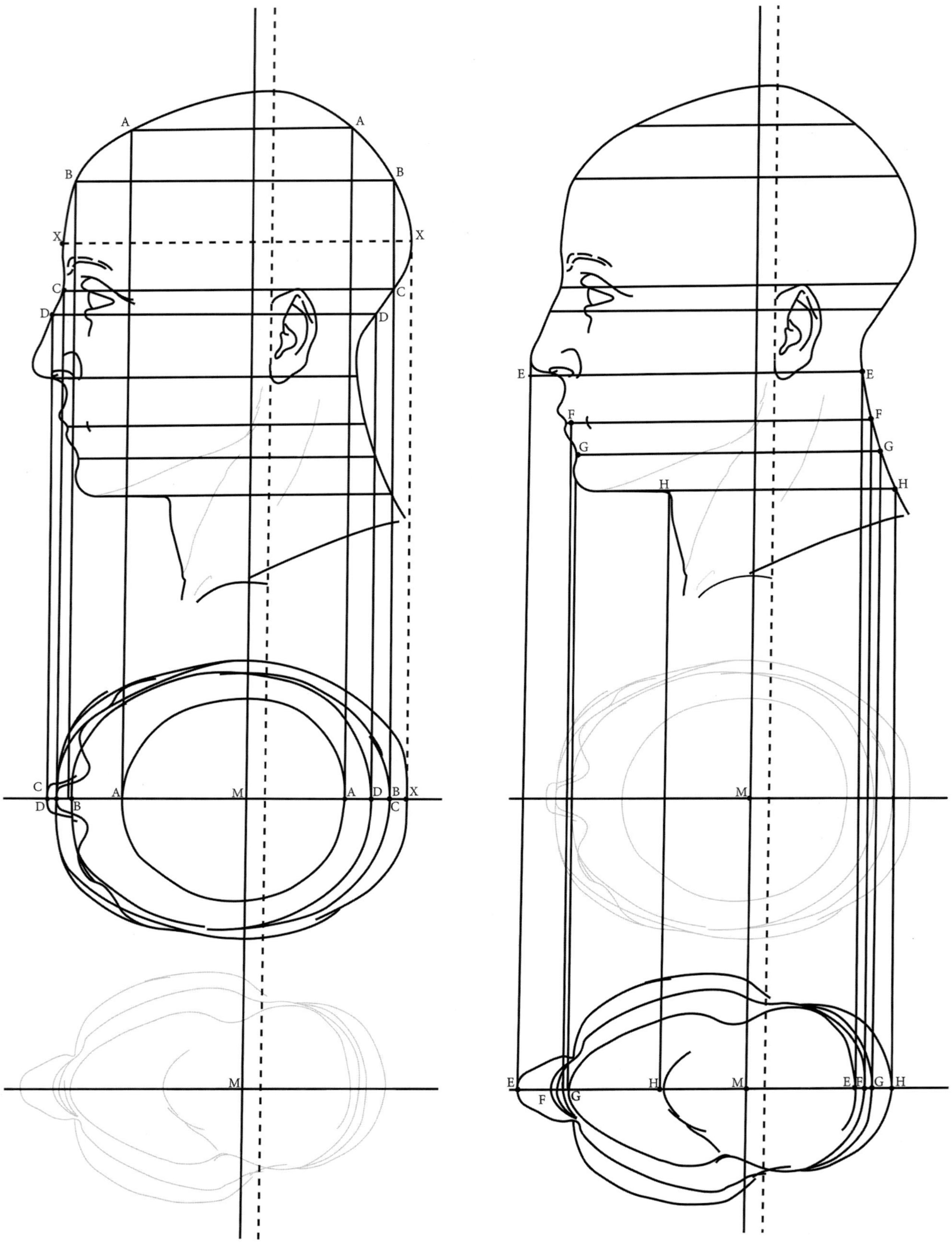

Figure A.40
Bordeaux manuscript, folio 74 verso: incongruences.

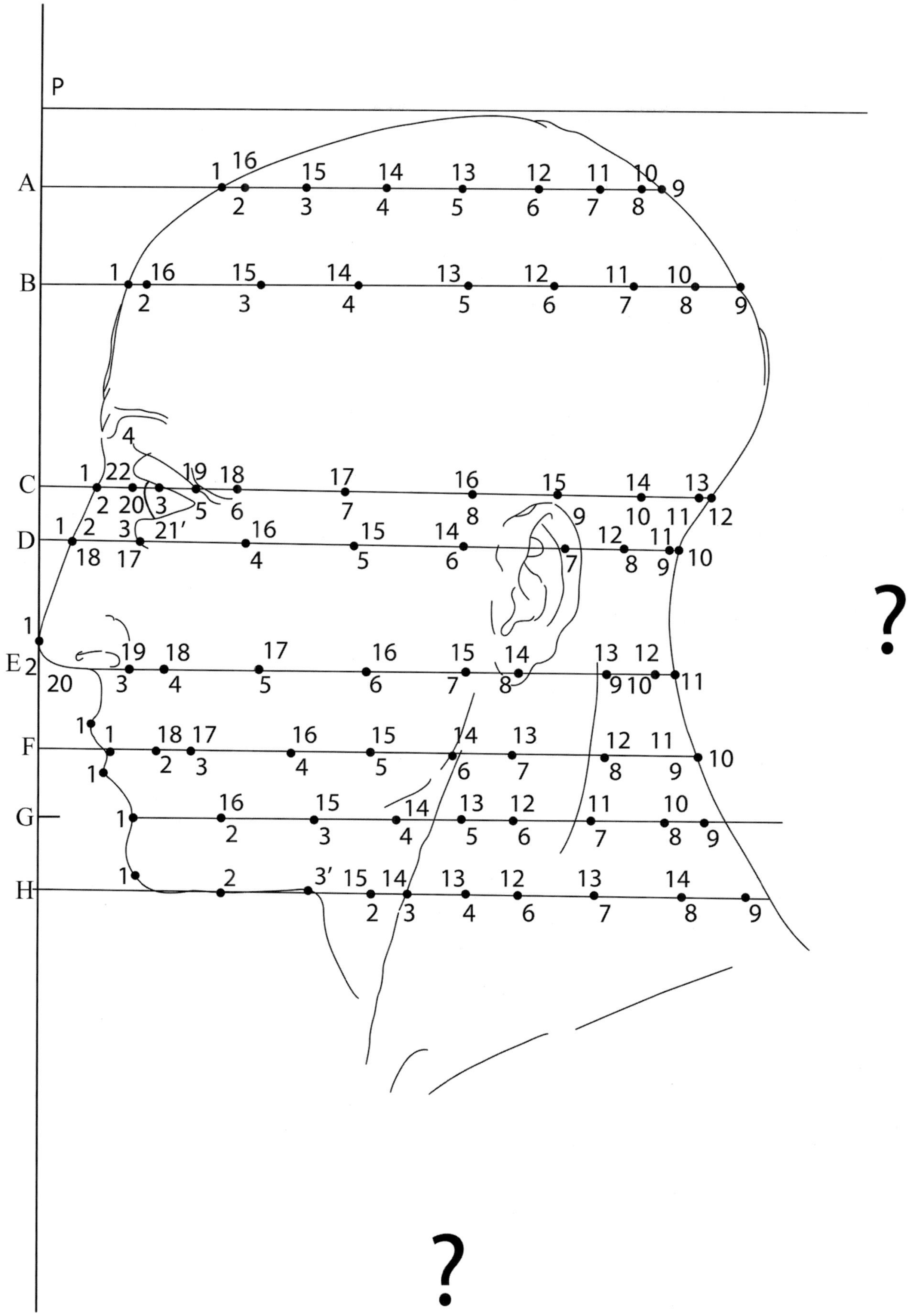

Figure A.41
Bordeaux manuscript, folio 76 recto: incongruences.

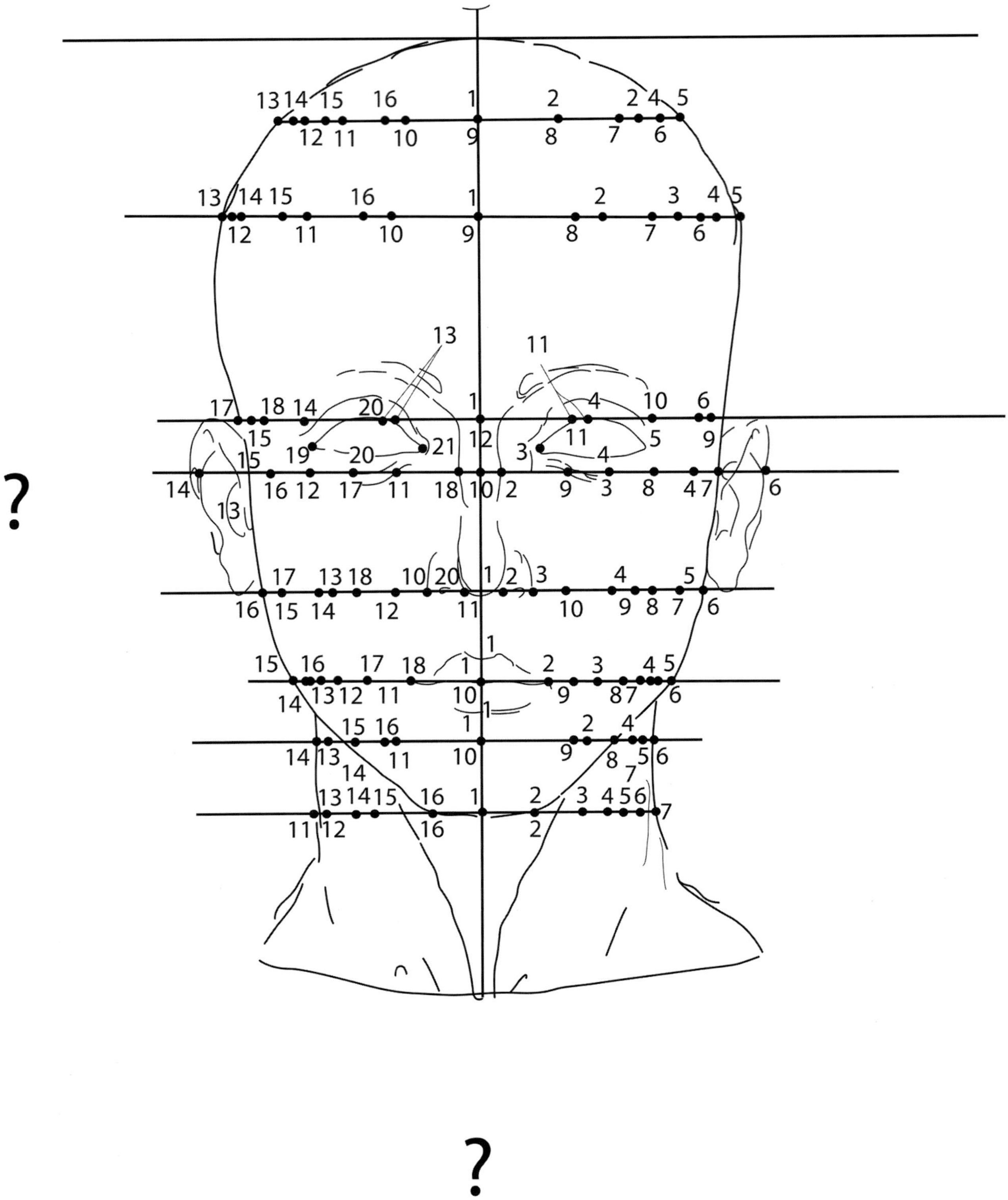

Figure A.42
Bordeaux manuscript, folio 77 recto: incongruences.

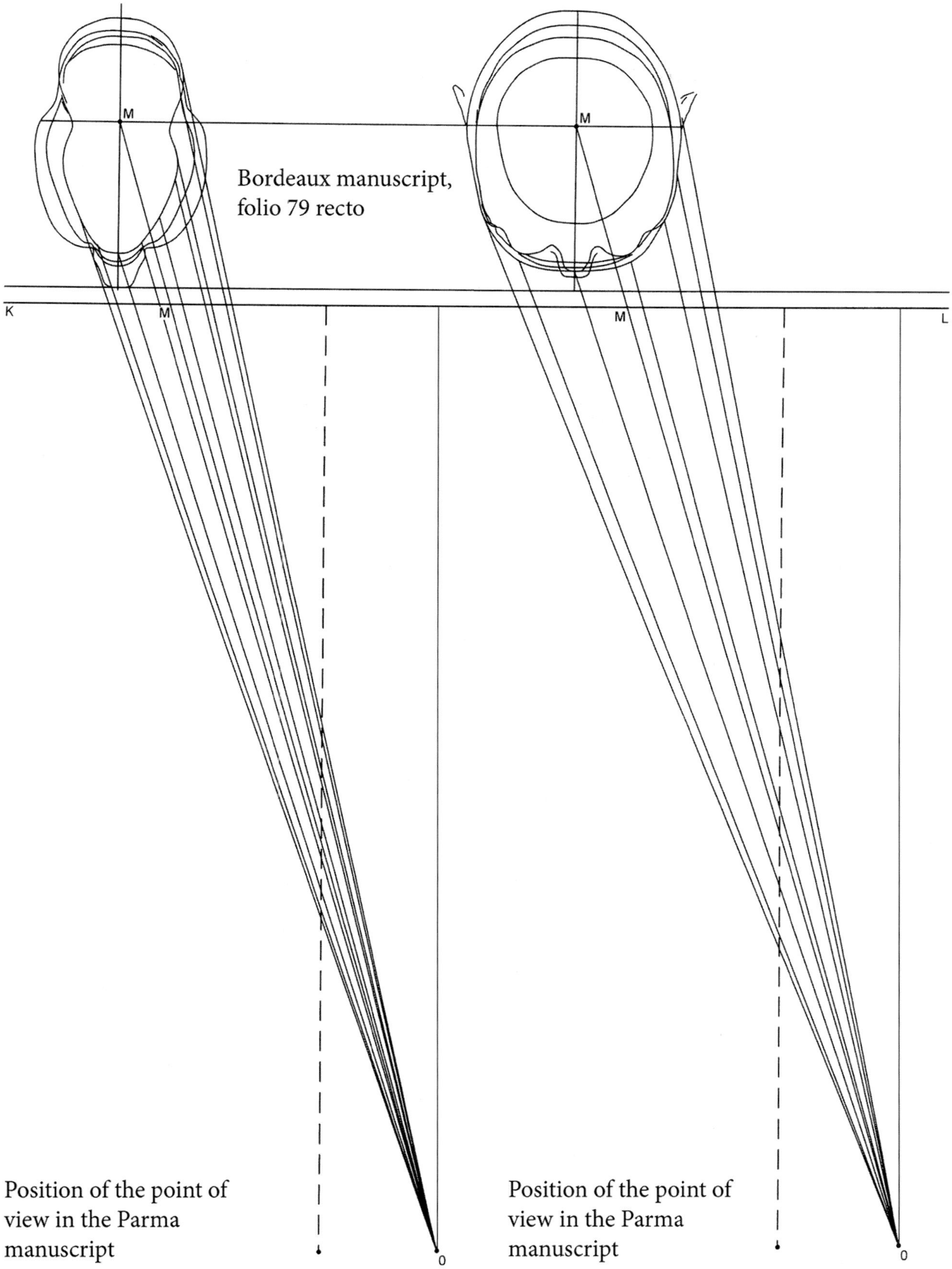

Figure A.43
Vernacular manuscripts and Latin manuscripts: comparisons of points of view in plan.

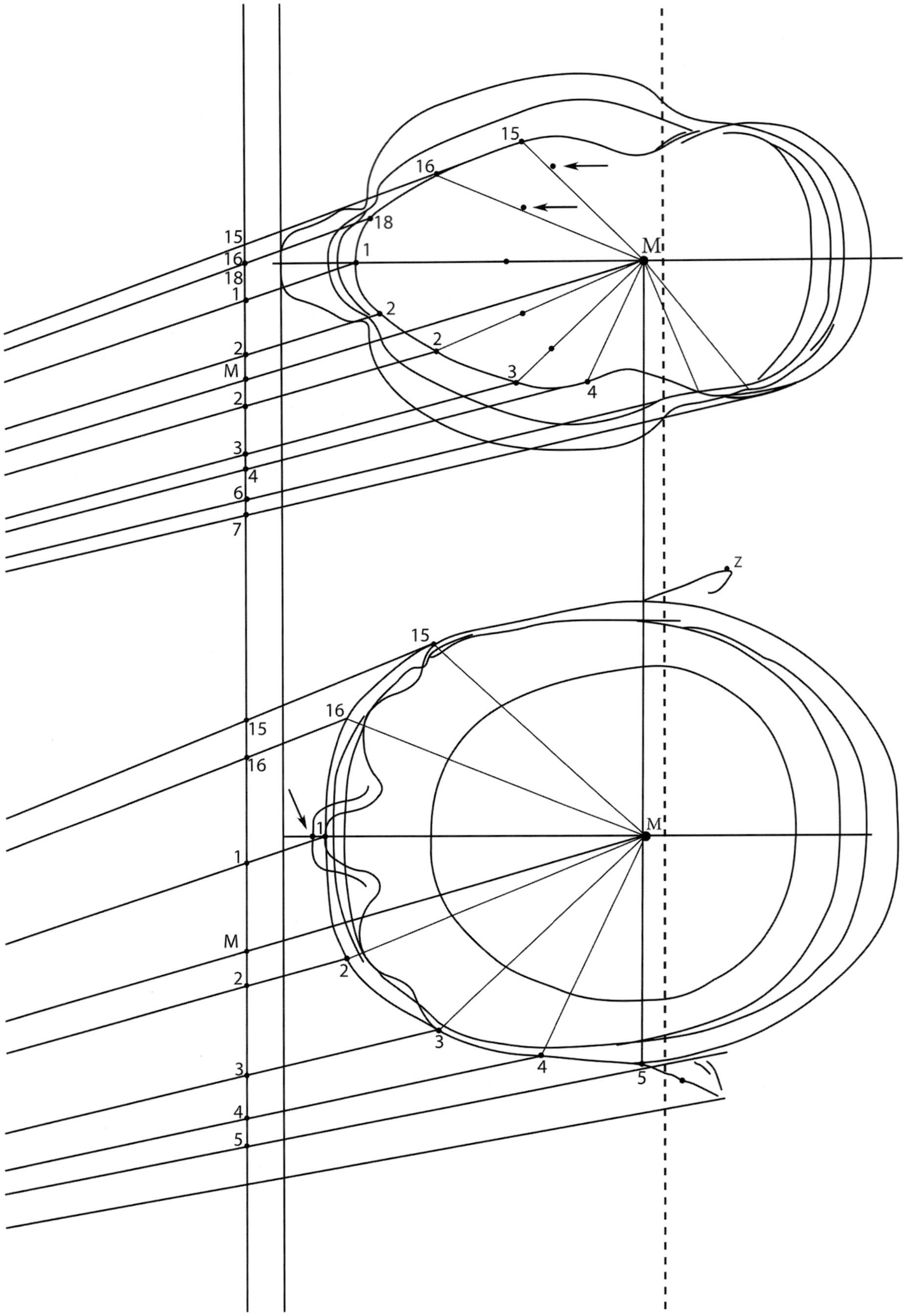

Figure A.44
Bordeaux manuscript, folio 79 recto: incongruences.

Parma manuscript,
folio 65 recto

Bordeaux manuscript,
folio 80 verso

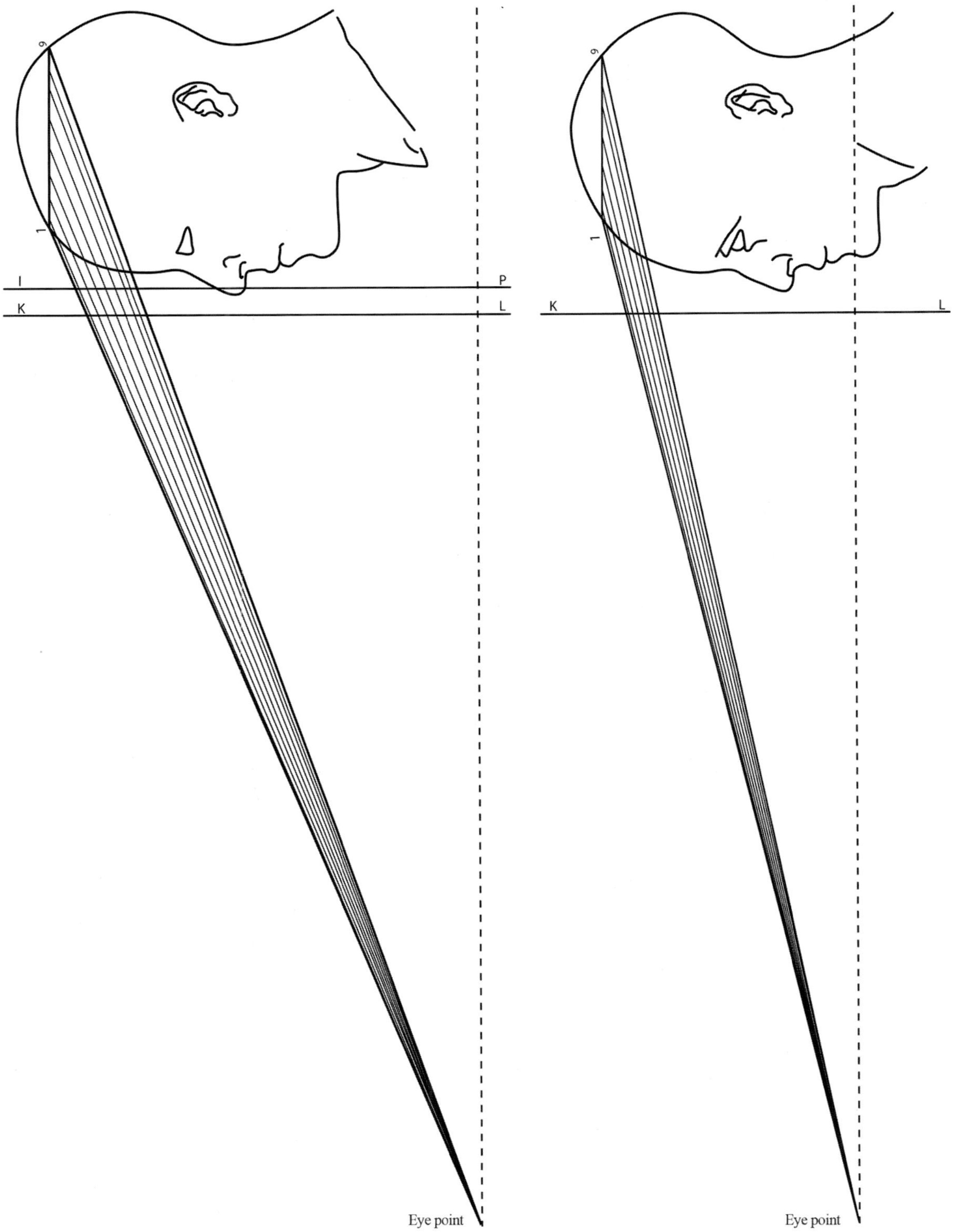

Figure A.45
Vernacular manuscripts and Latin manuscripts: comparisons of points of view in elevation.

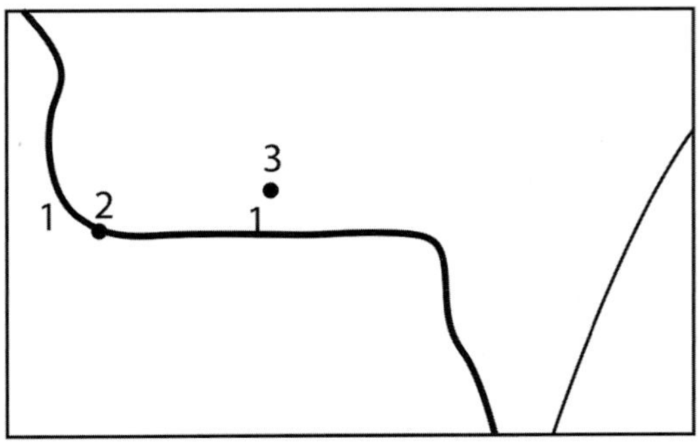

Bordeaux manuscript,
folio 80 verso

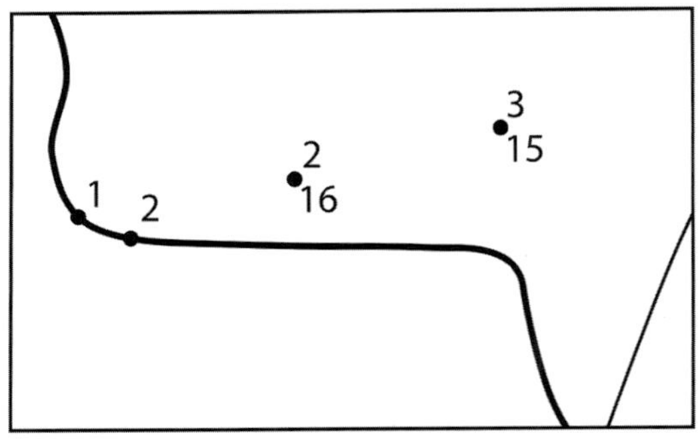

Milan manuscript,
folio 87 recto

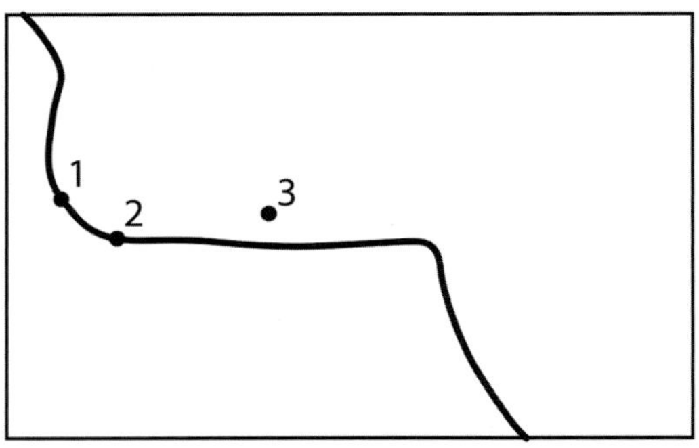

London manuscript,
folio 80 verso

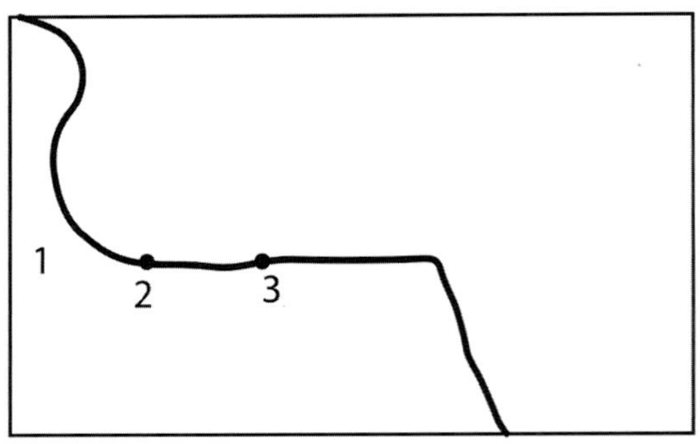

Paris manuscript,
folio 62 recto

Figure A.46
The bow of the chin as it appears in the Latin manuscripts: incongruences.

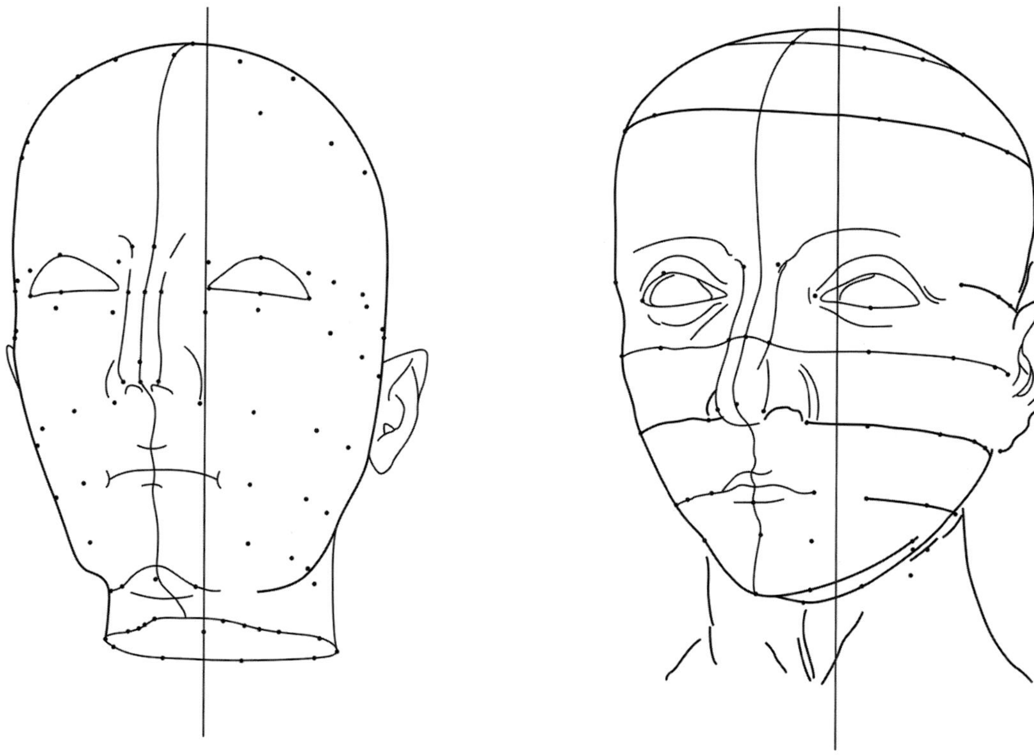

Vernacular manuscripts: the perspective with circuit A compliant with the original

Bordeaux manuscript, folio 82 recto: the perspective

Figure A.47
Vernacular manuscripts and Latin manuscripts: comparison of perspectives.

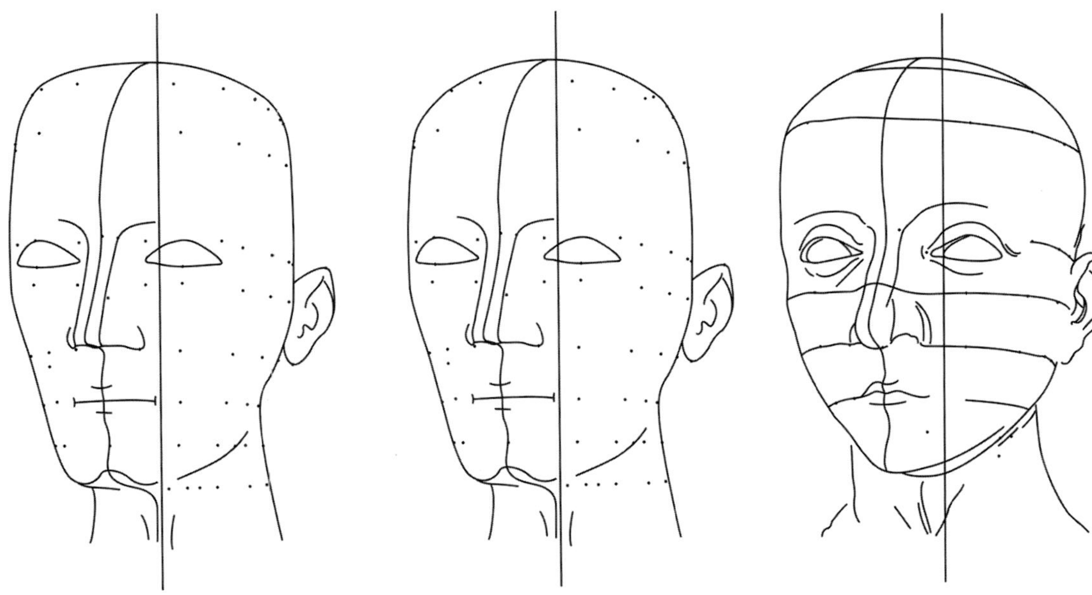

The remade circuit A

Circuit A compliant with the original

Bordeaux manuscript, folio 82 recto: the perspective

Figure A.48
The first two perspectives of the Latin head compared to the perspective of the Bordeaux manuscript.

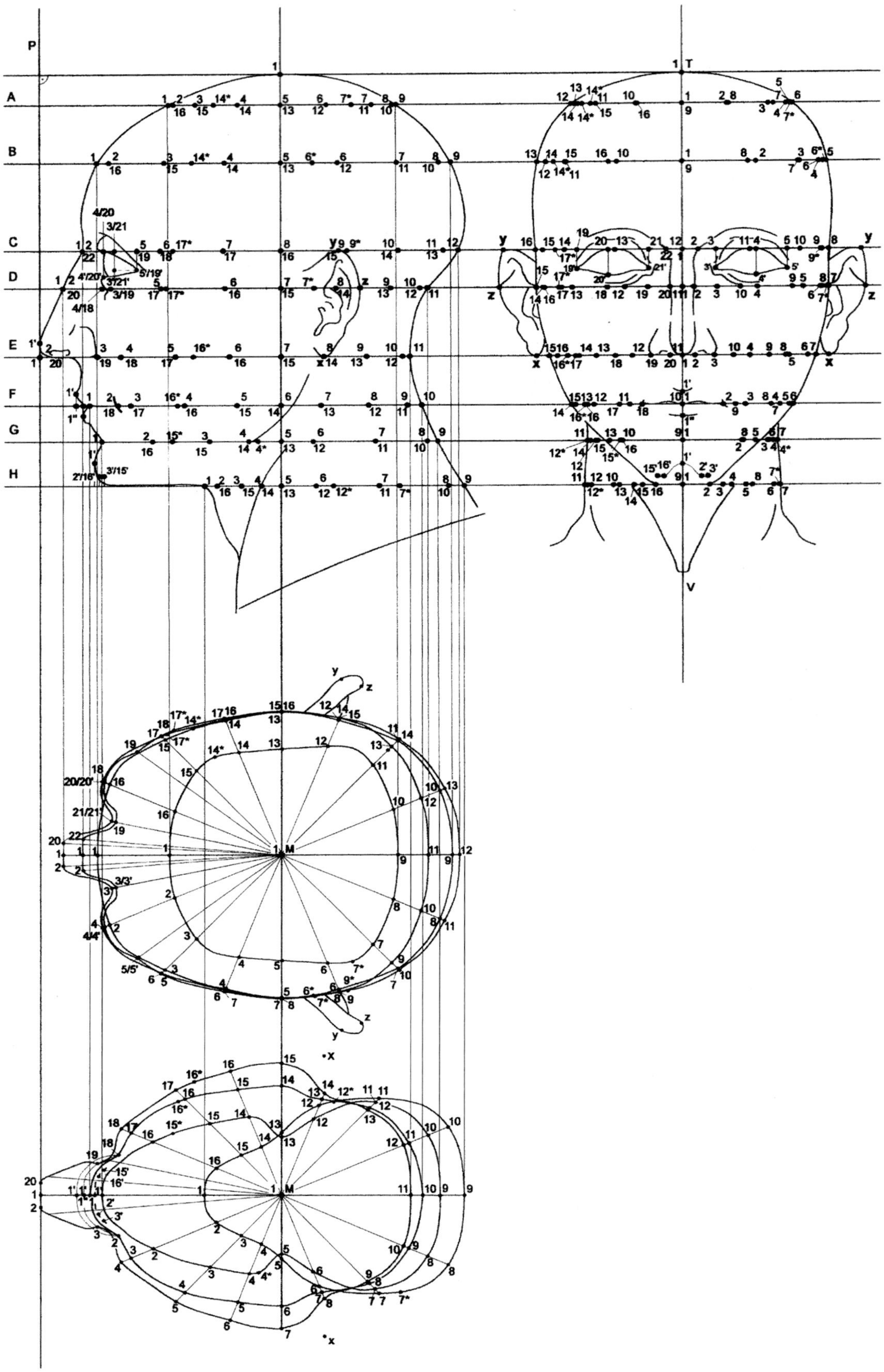

Figure A.49
The reconstruction of the new head designed in accordance with the Latin manuscripts.

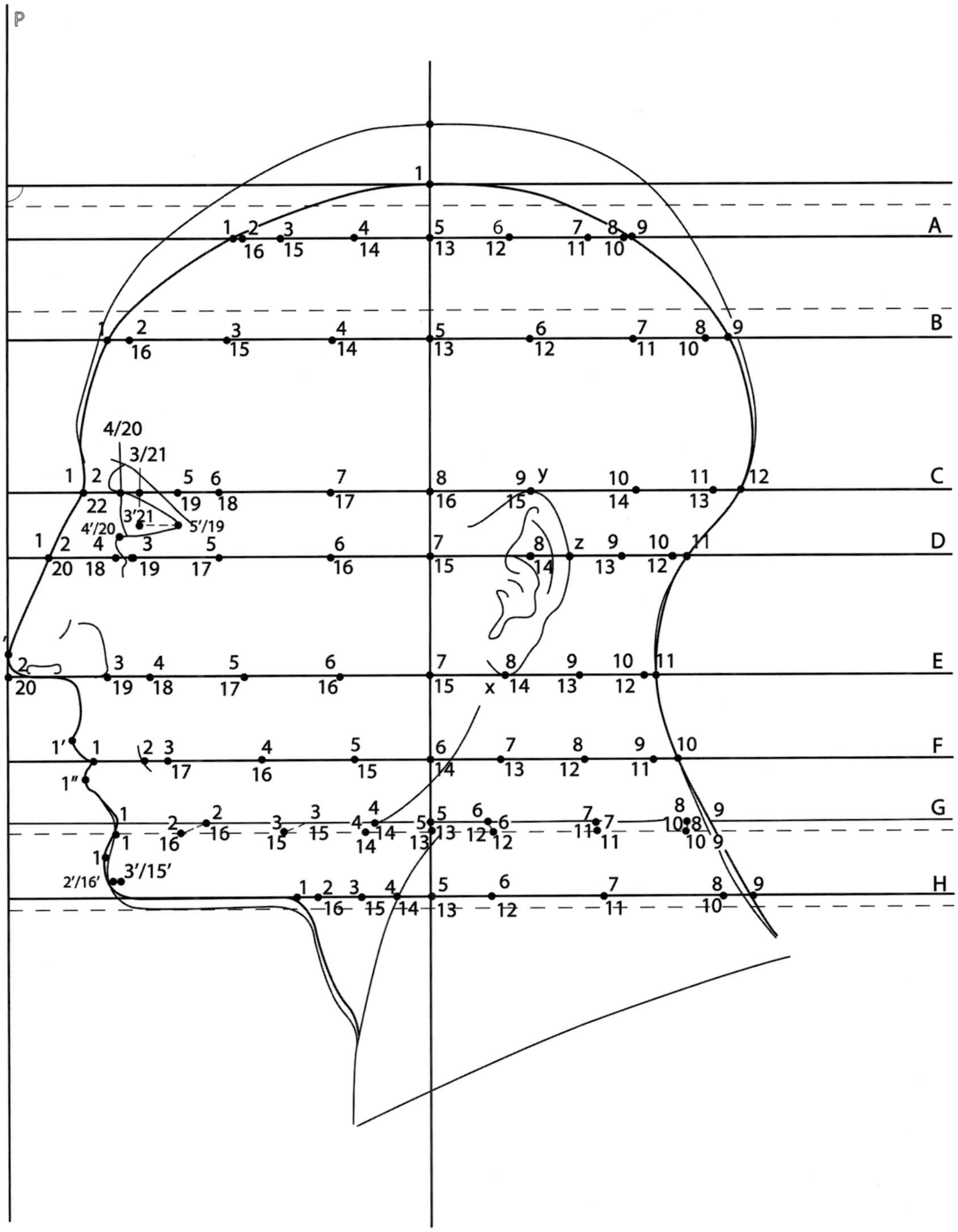

Figure A.50
Detail of the new head in elevation.

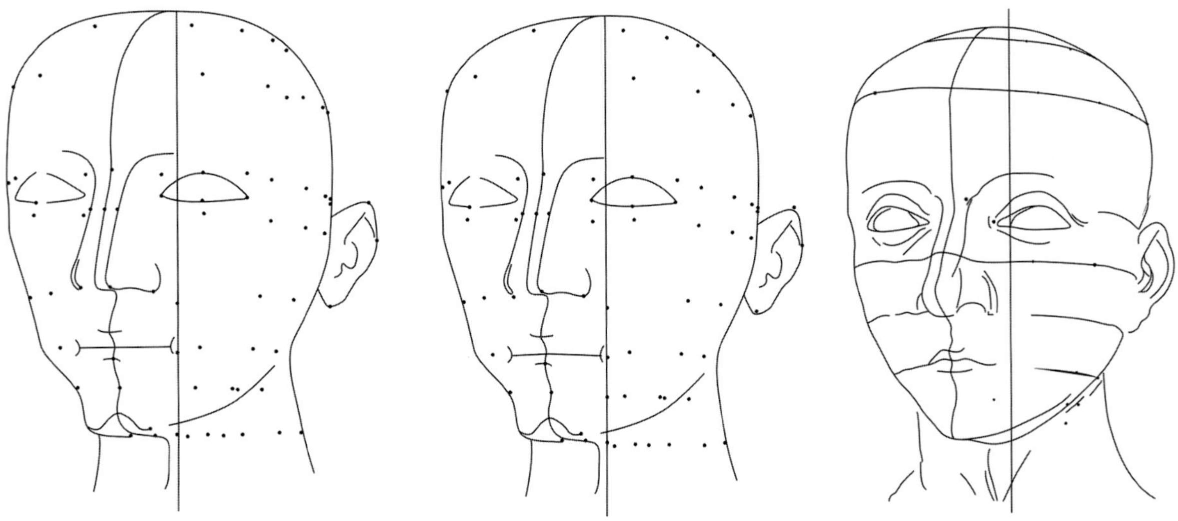

The remade circuit A

The circuit A compliant with the original

Bordeaux manuscript, folio 82 recto: the perspective

Figure A.51
The second two perspectives of the new head compared to the perspective of the Bordeaux manuscript.

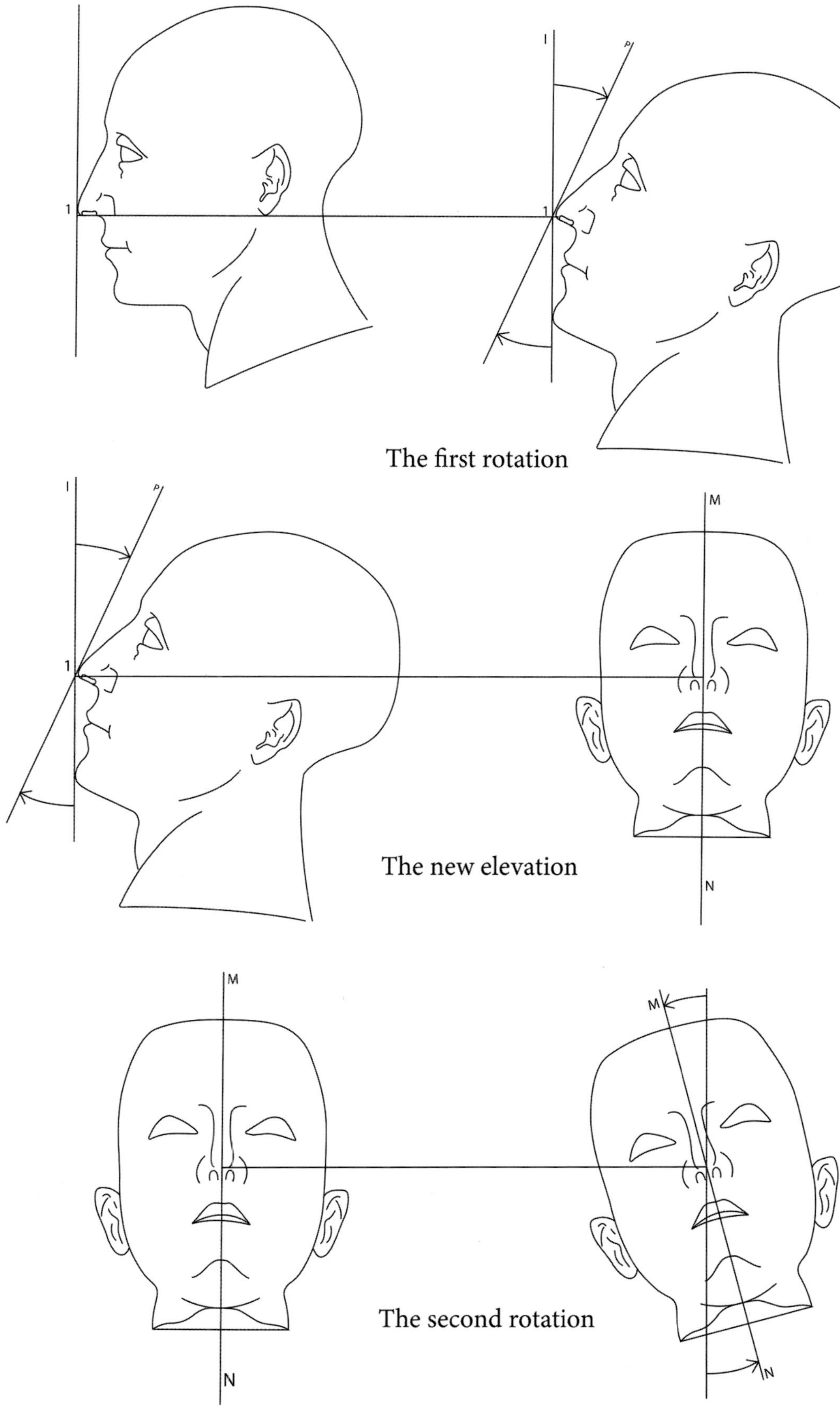

Figure A.52
Second head: significant scientific operations.

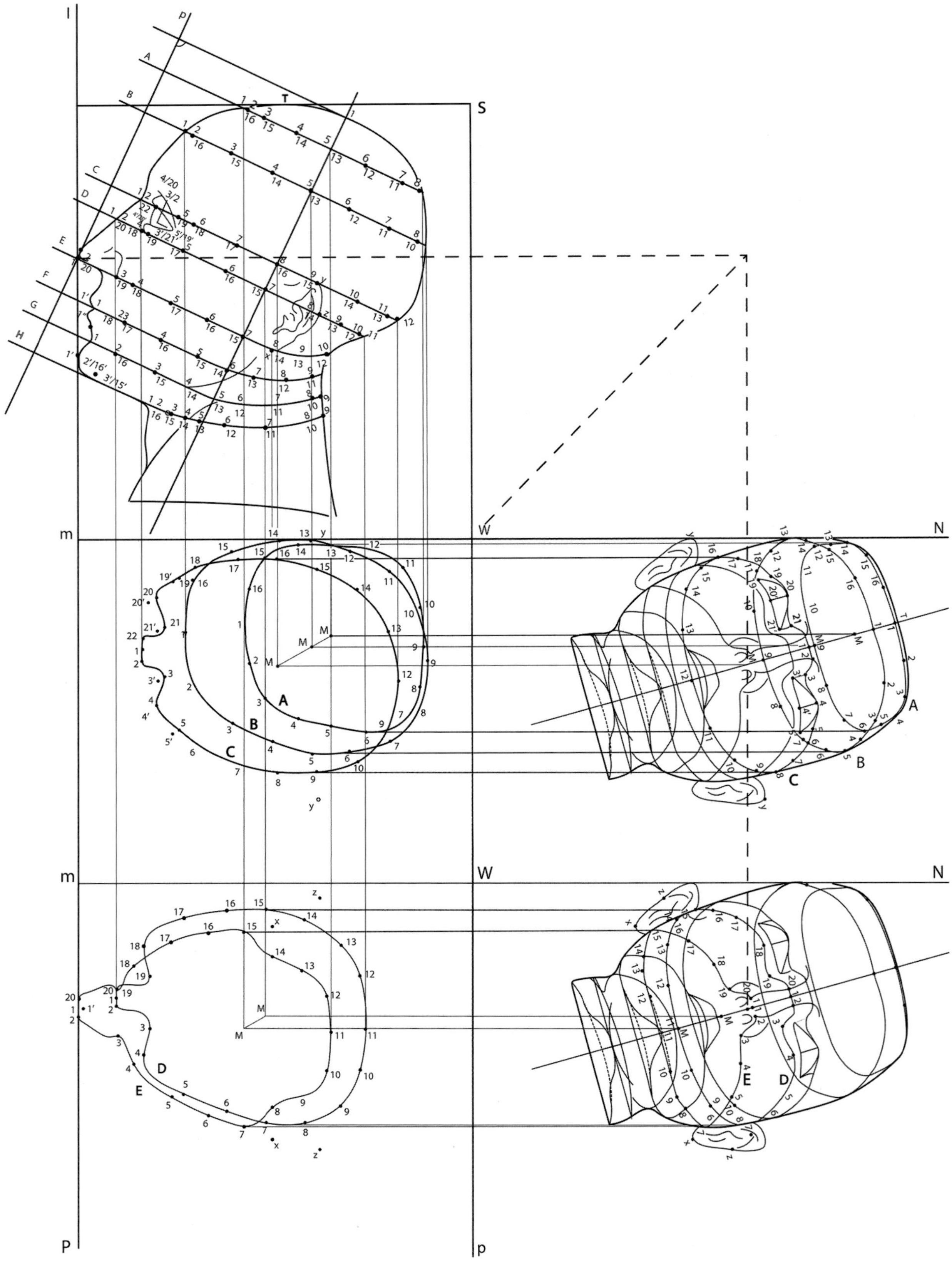

Figure A.53
The new plans with projections of circuits A, B, C, D, E.

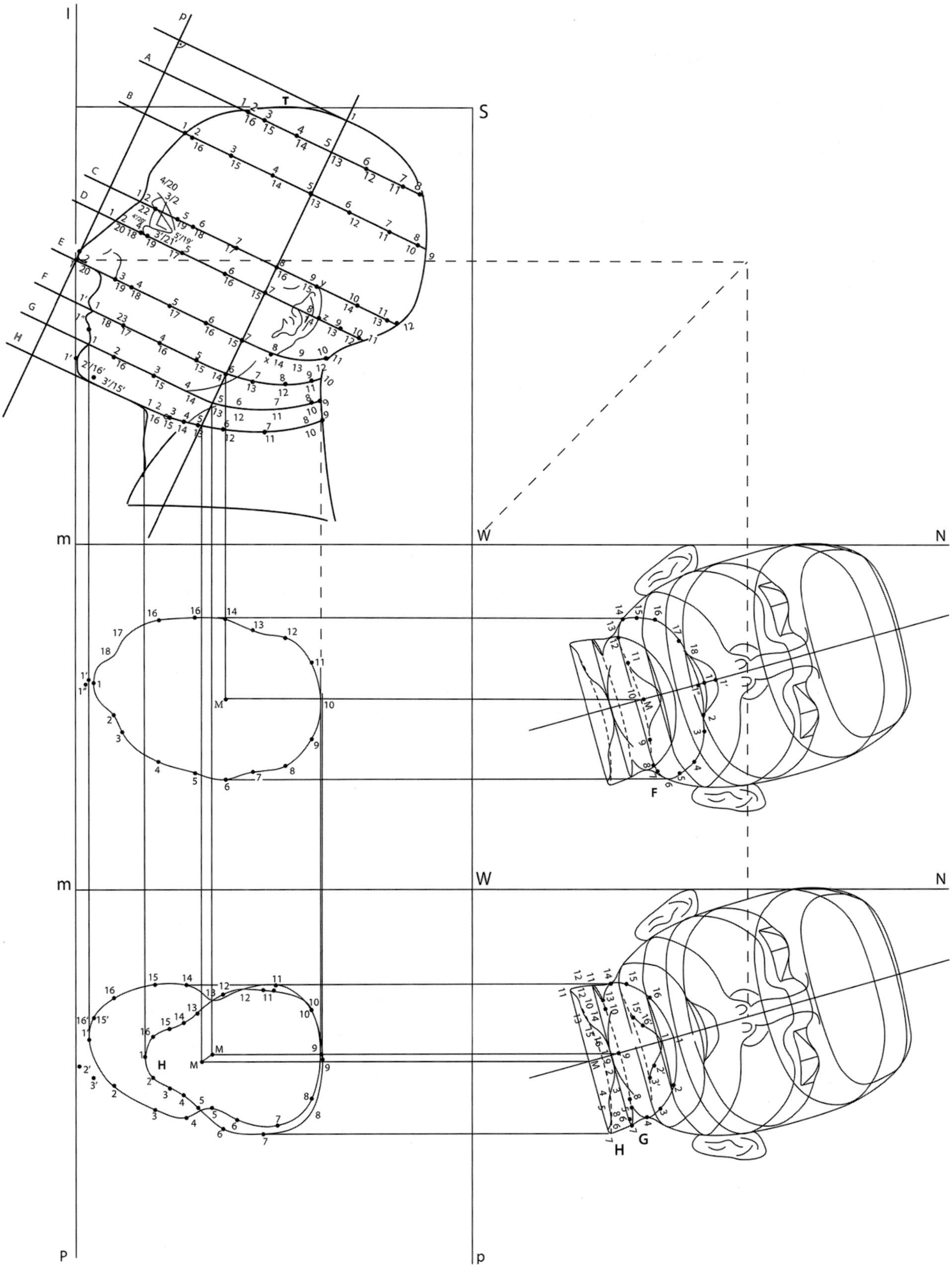

Figure A.54
The new plans with projections of circuits F, G, H.

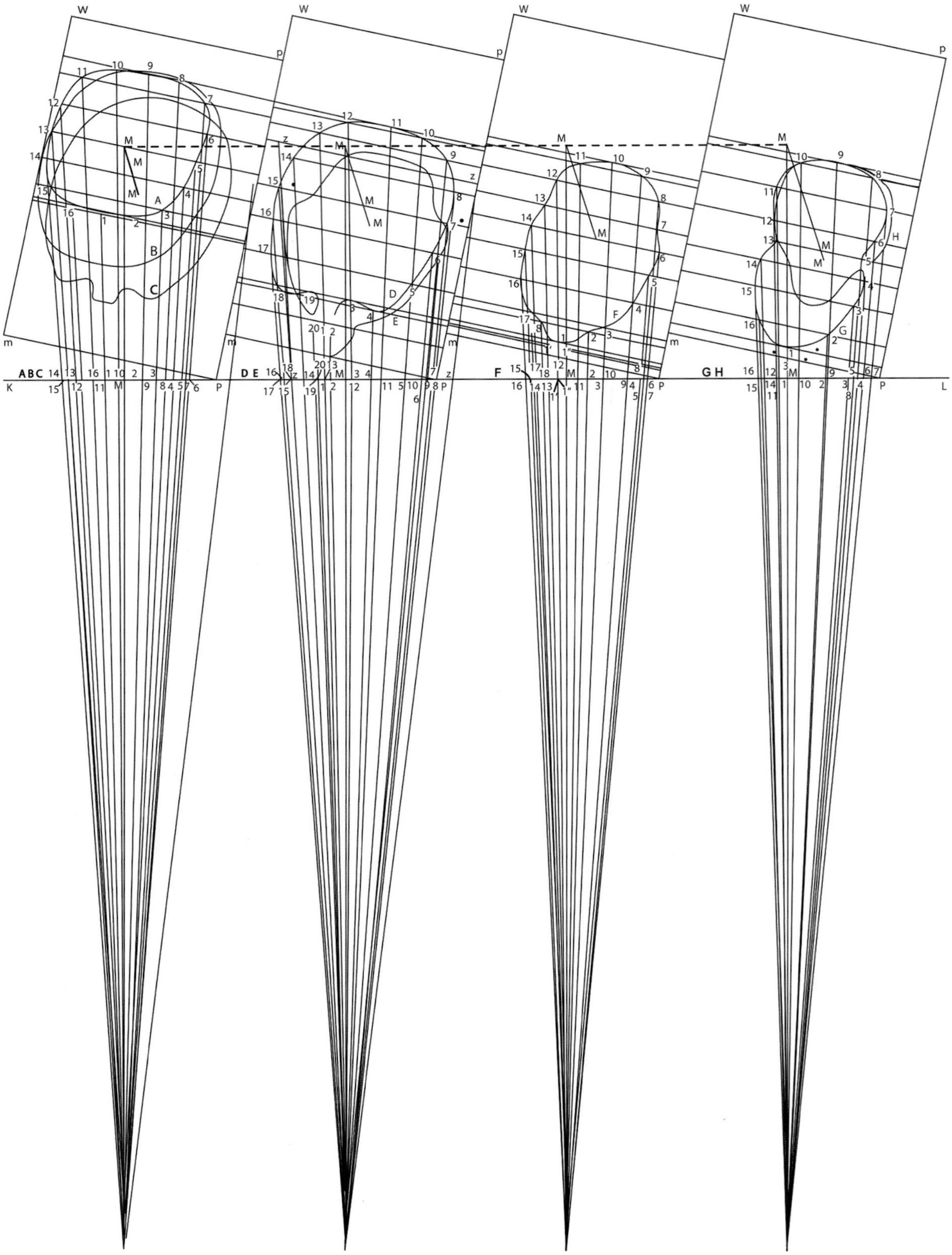

Figure A.55
The new plans of the circuits, with the terminal, the point of view, and the wooden rulers.

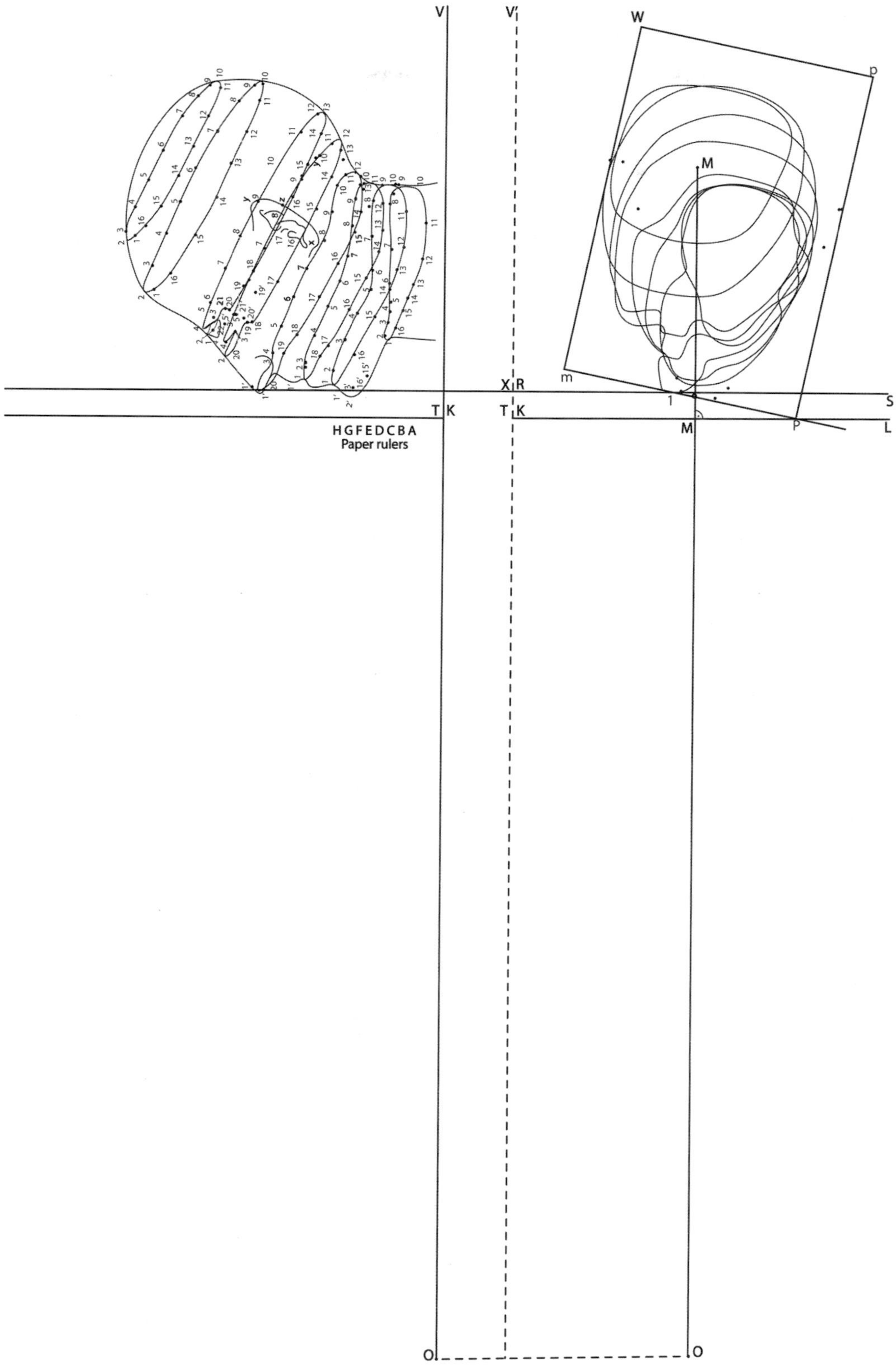

Figure A.56
The new elevations of the circuits with the heights of all points, with the terminal and the point of view in plan and elevation.

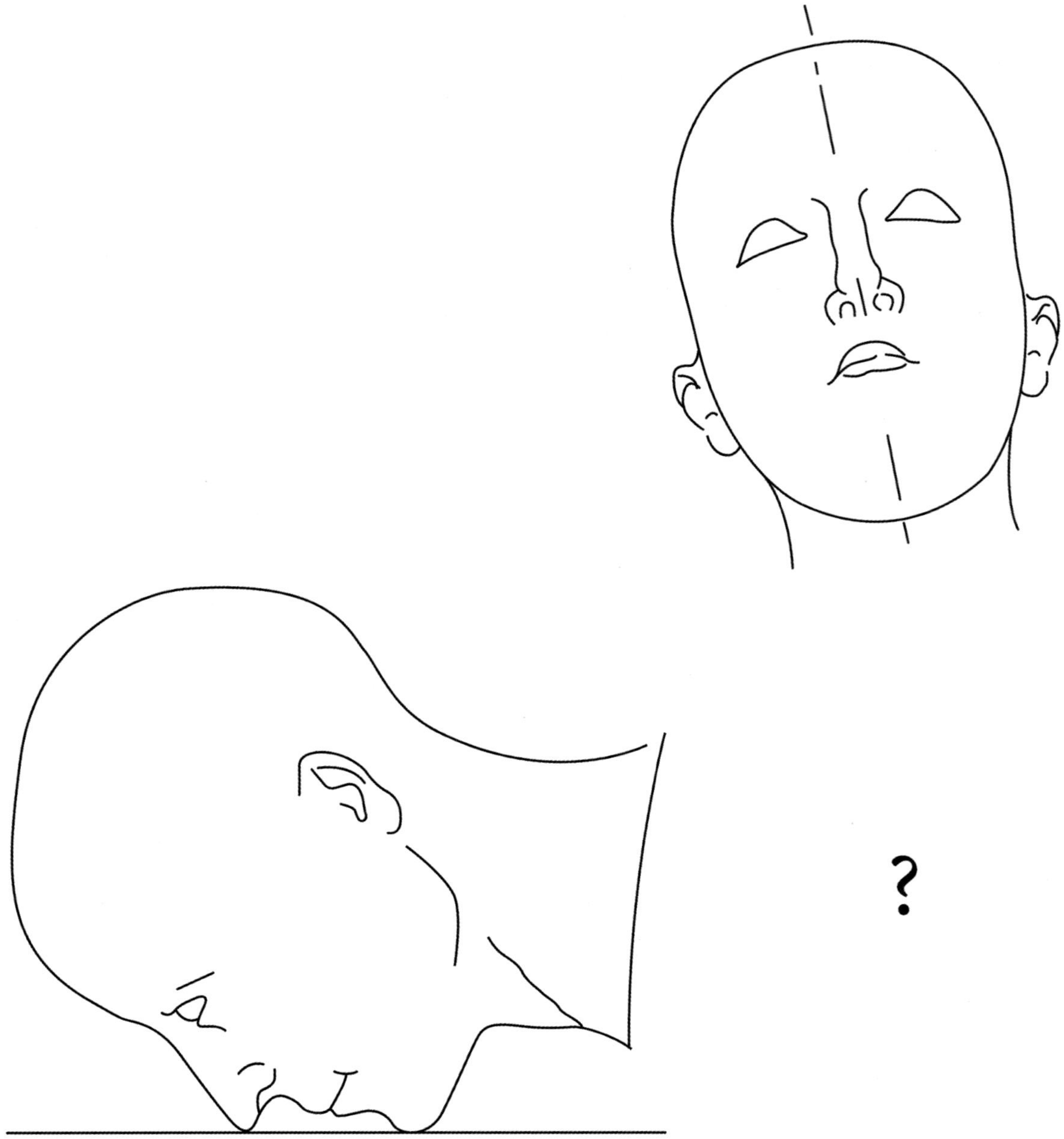

Figure A.57
Parma manuscript, folio 71 recto; Reggio Emilia manuscript, folio 81 recto: incongruences.

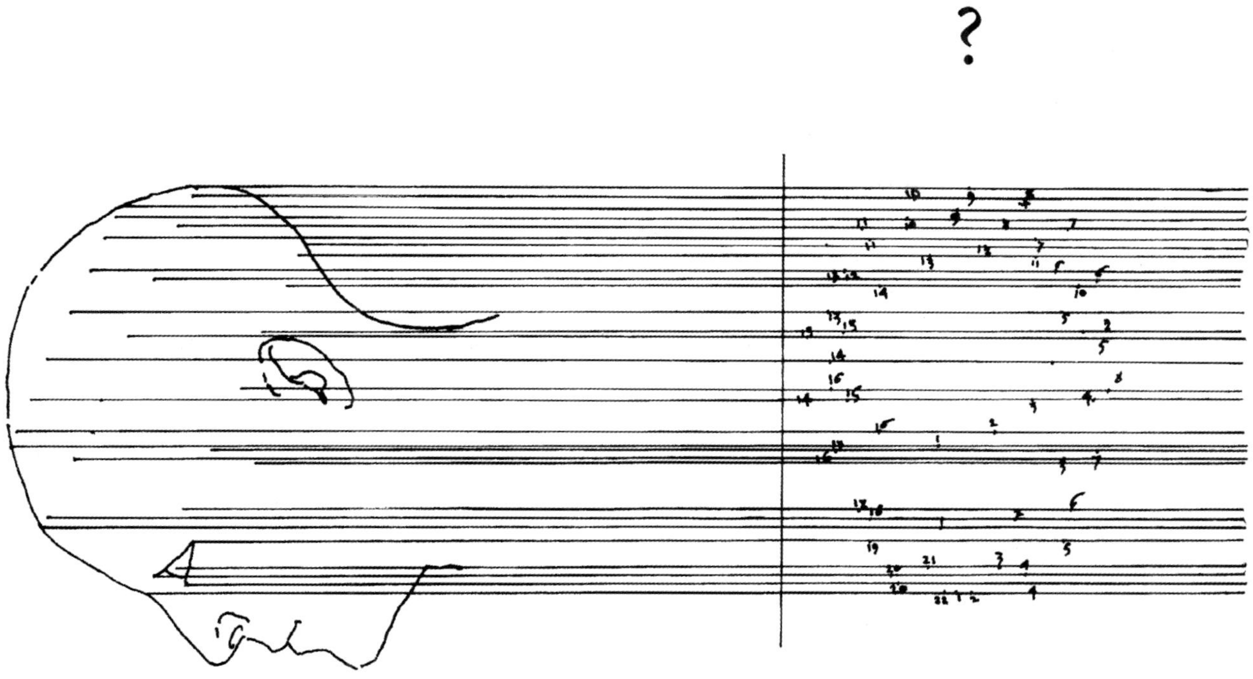

Figure A.58
Bordeaux manuscript, folio 86 recto: incongruences.

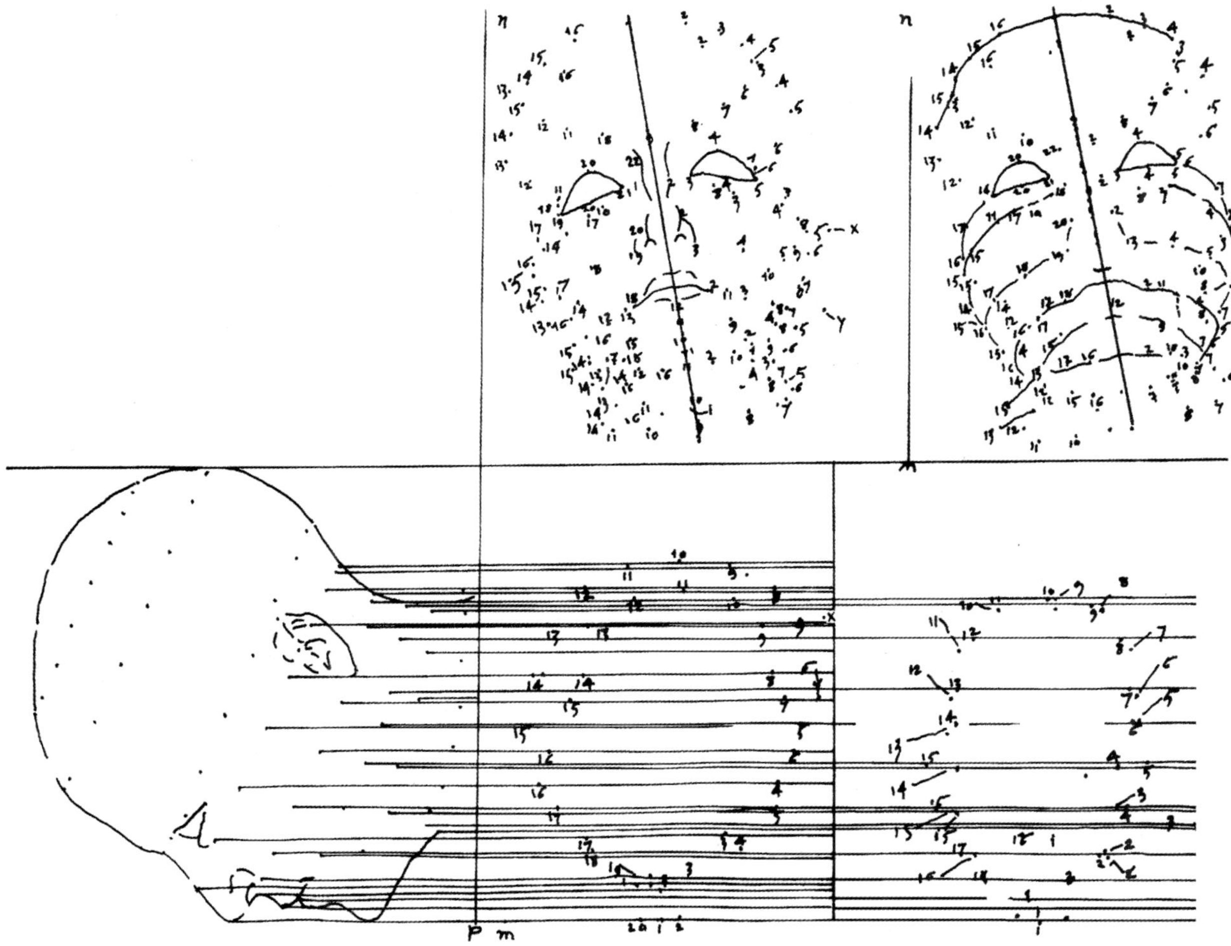

Figure A.59
Bordeaux manuscript, folio 86 verso: incongruences.

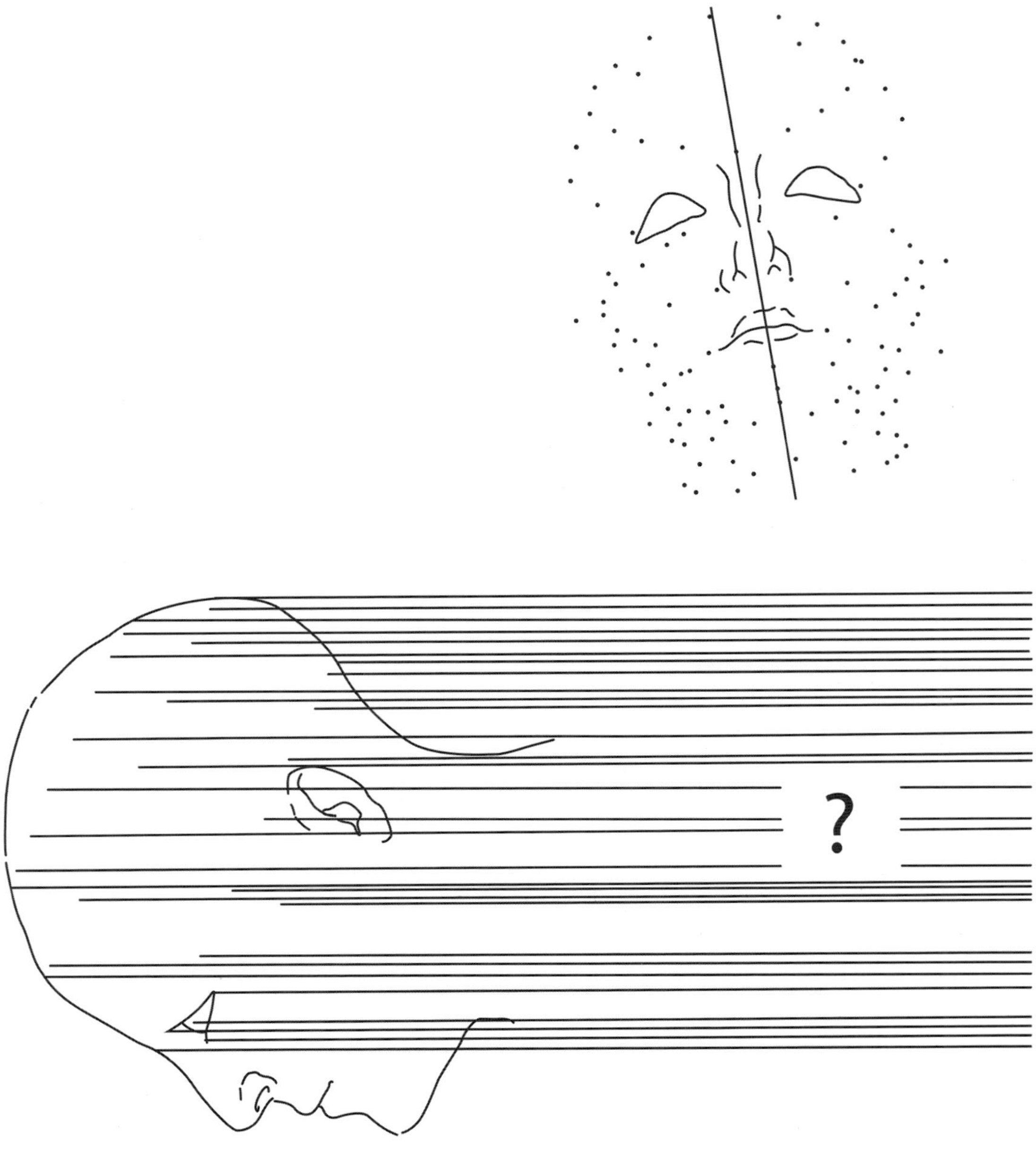

Figure A.60
Milan manuscript, folio 93 verso: incongruences.

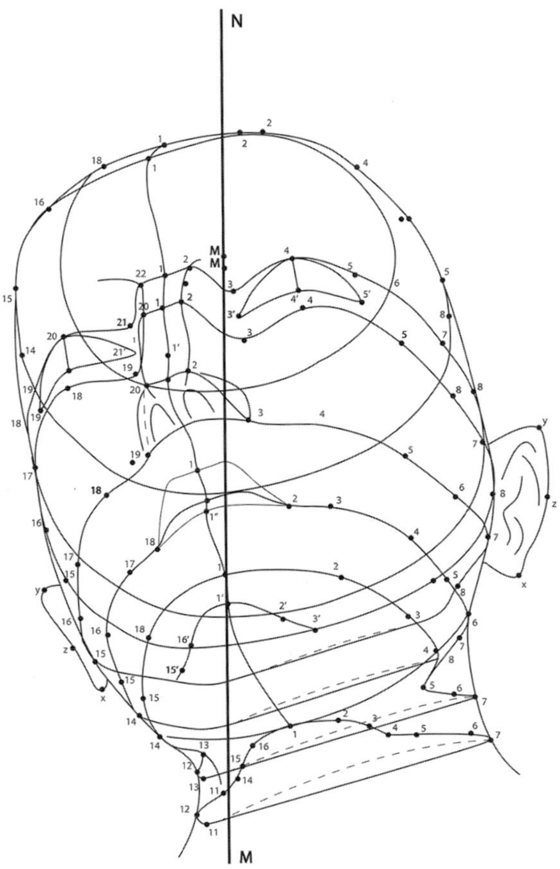

The perspective with the
outlines of all the circuits

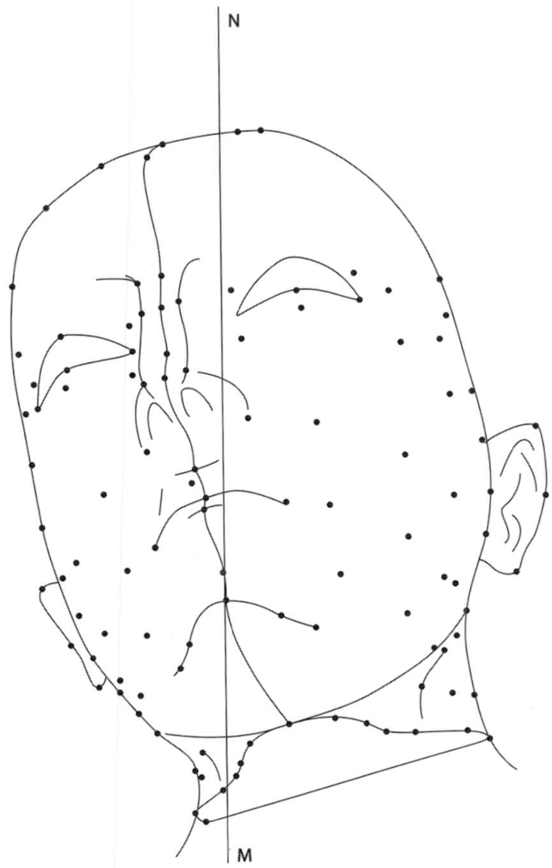

The perspective without
the outlines of the circuits

Figure A.61
Second head: perspective with circuit A compliant
with the original and the protruding ears.

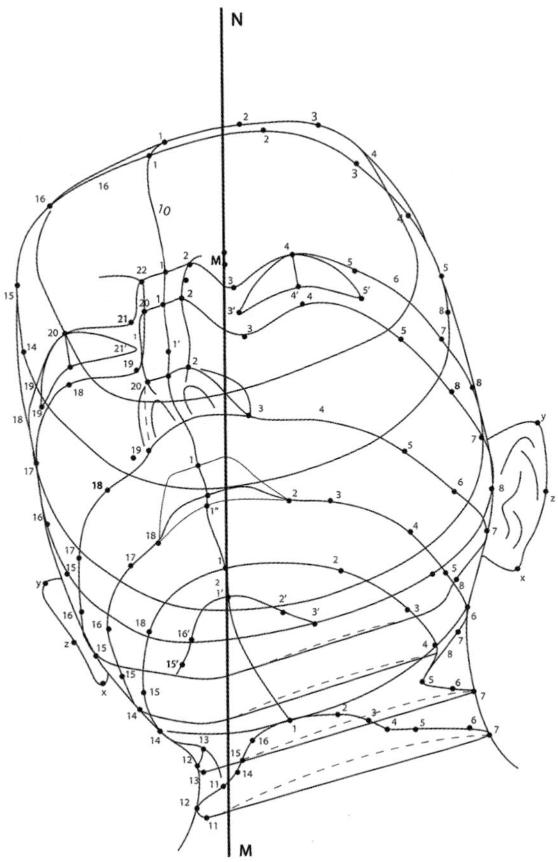

The perspective with the outlines of all the circuits

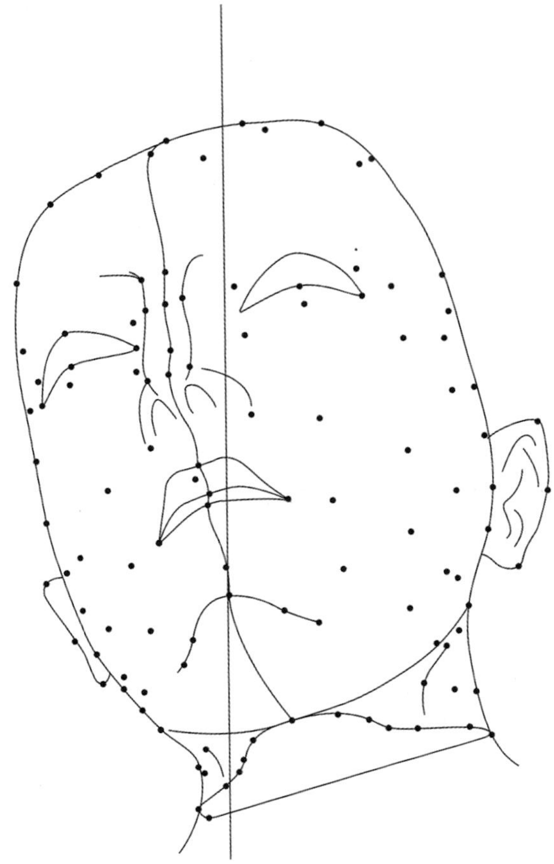

The perspective without the outlines of the circuits

Figure A.62
Second head: perspective with the remade circuit A and the protruding ears.

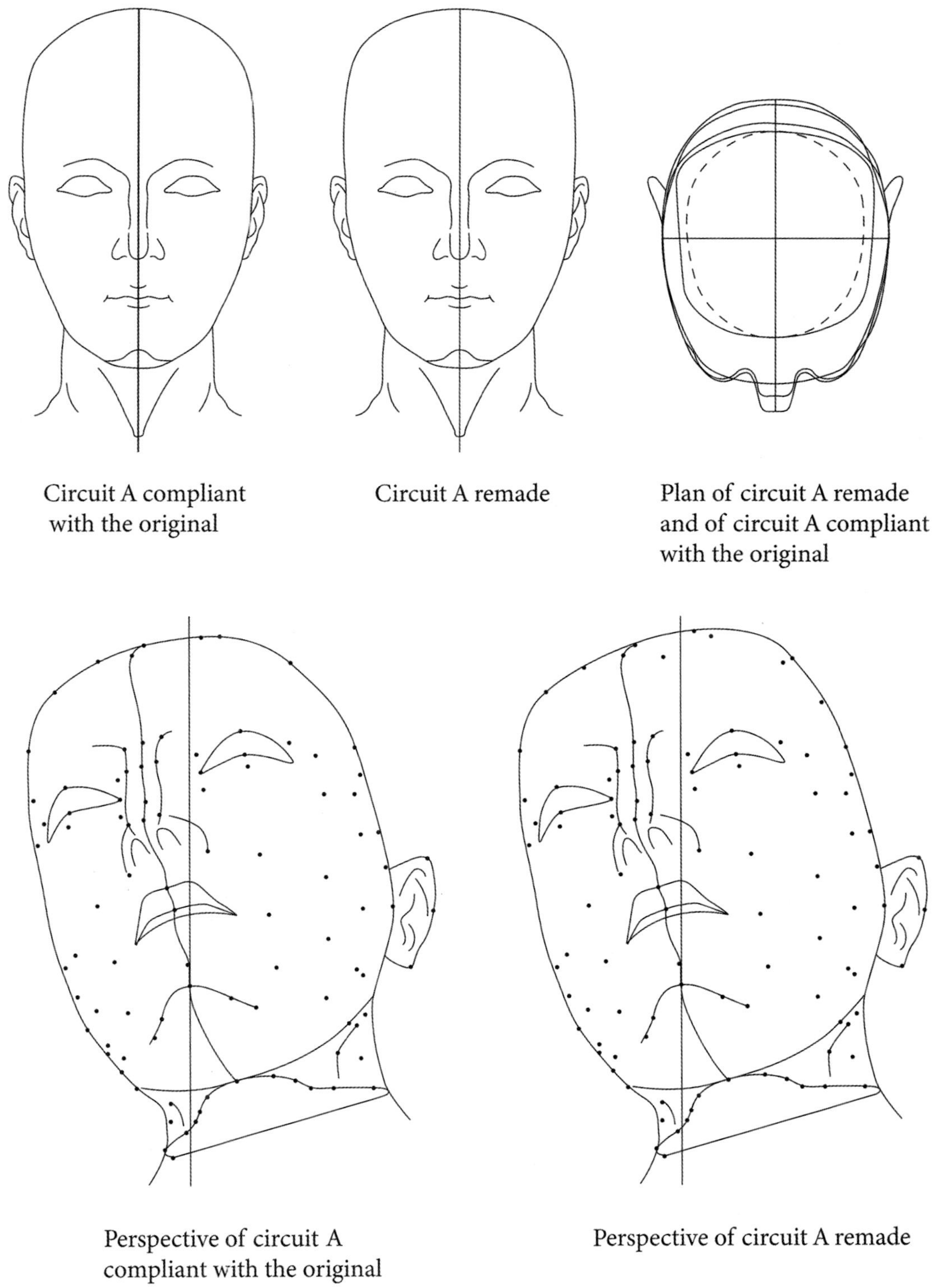

Figure A.63
Ears less protruding.

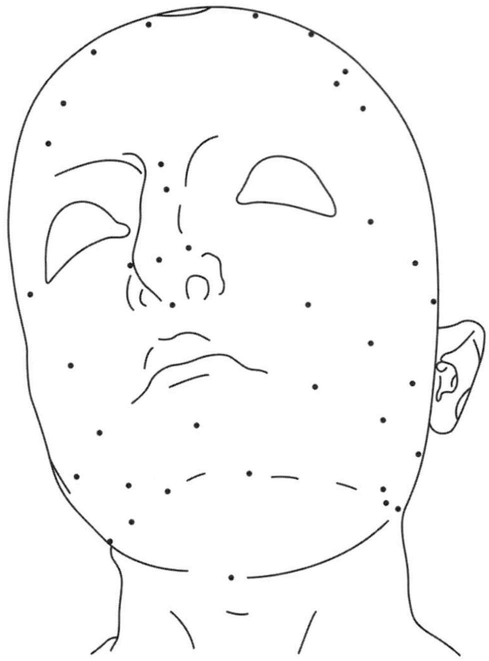

Parma manuscript,
folio 76 verso

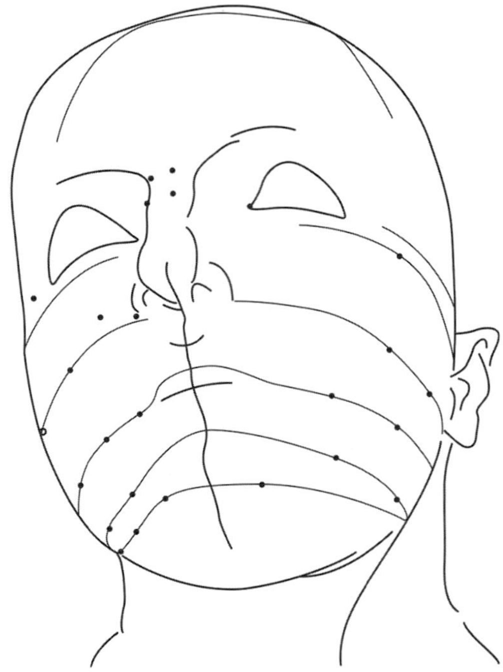

Bordeaux manuscript,
folio 96 recto

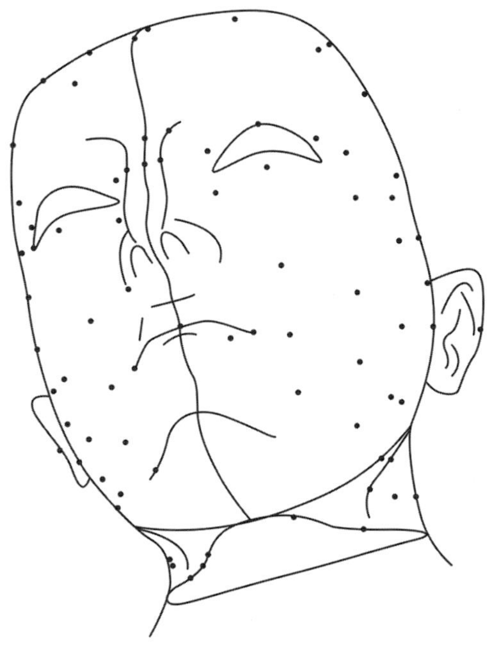

Circuit A compliant with the
original: protruding ears

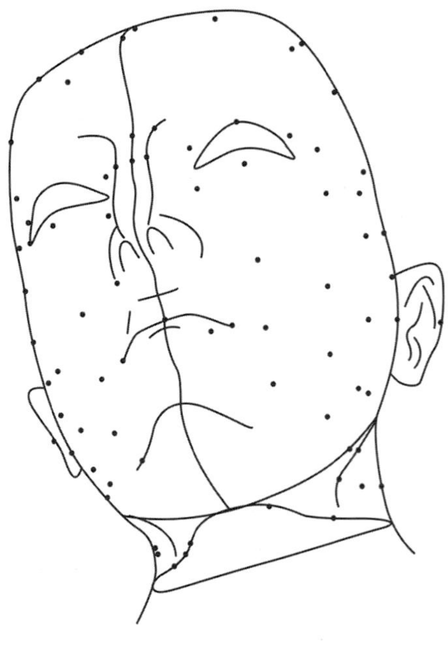

Circuit A compliant with the
original: ears less protruding

Figure A.64
Second head: comparisons of perspectives.

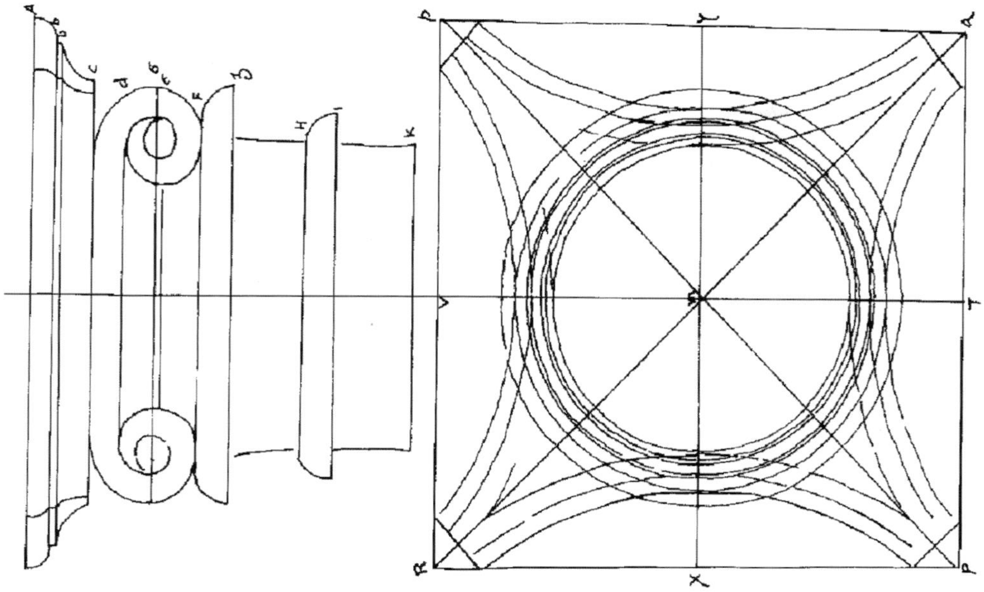

Figure A.65
Parma manuscript, folio 52 recto: incongruences.

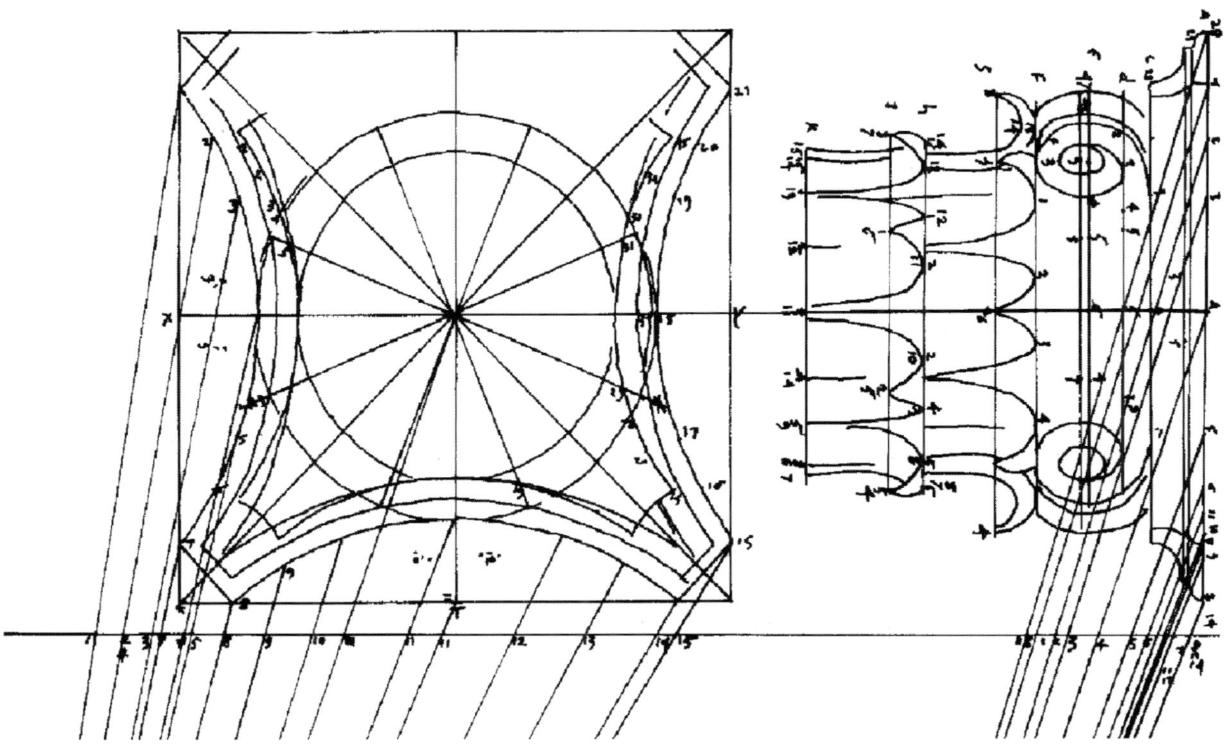

Figure A.66
Parma manuscript, folio 54 verso: incongruences.

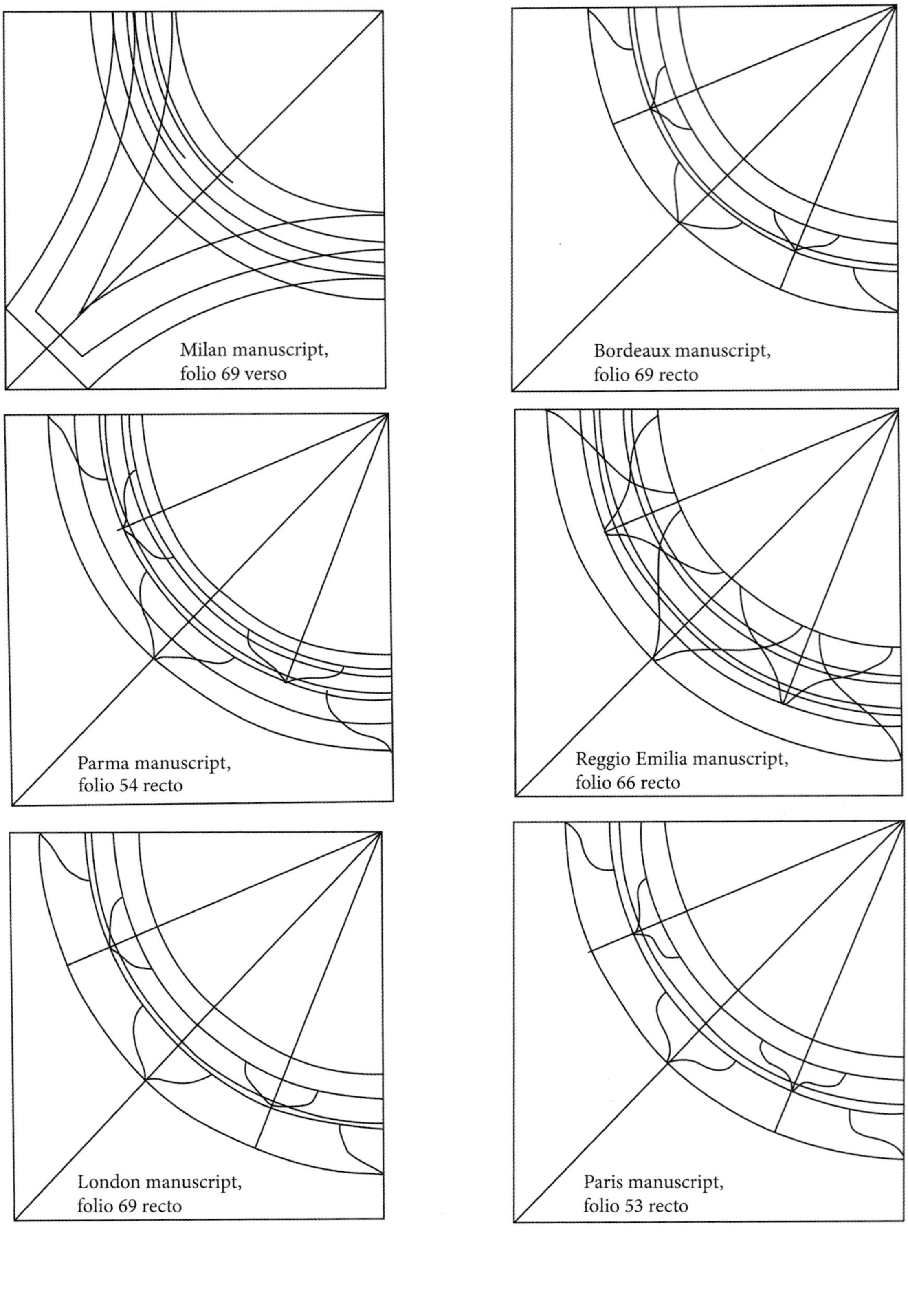

Figure A.67
Representation of the leaves: incongruences.

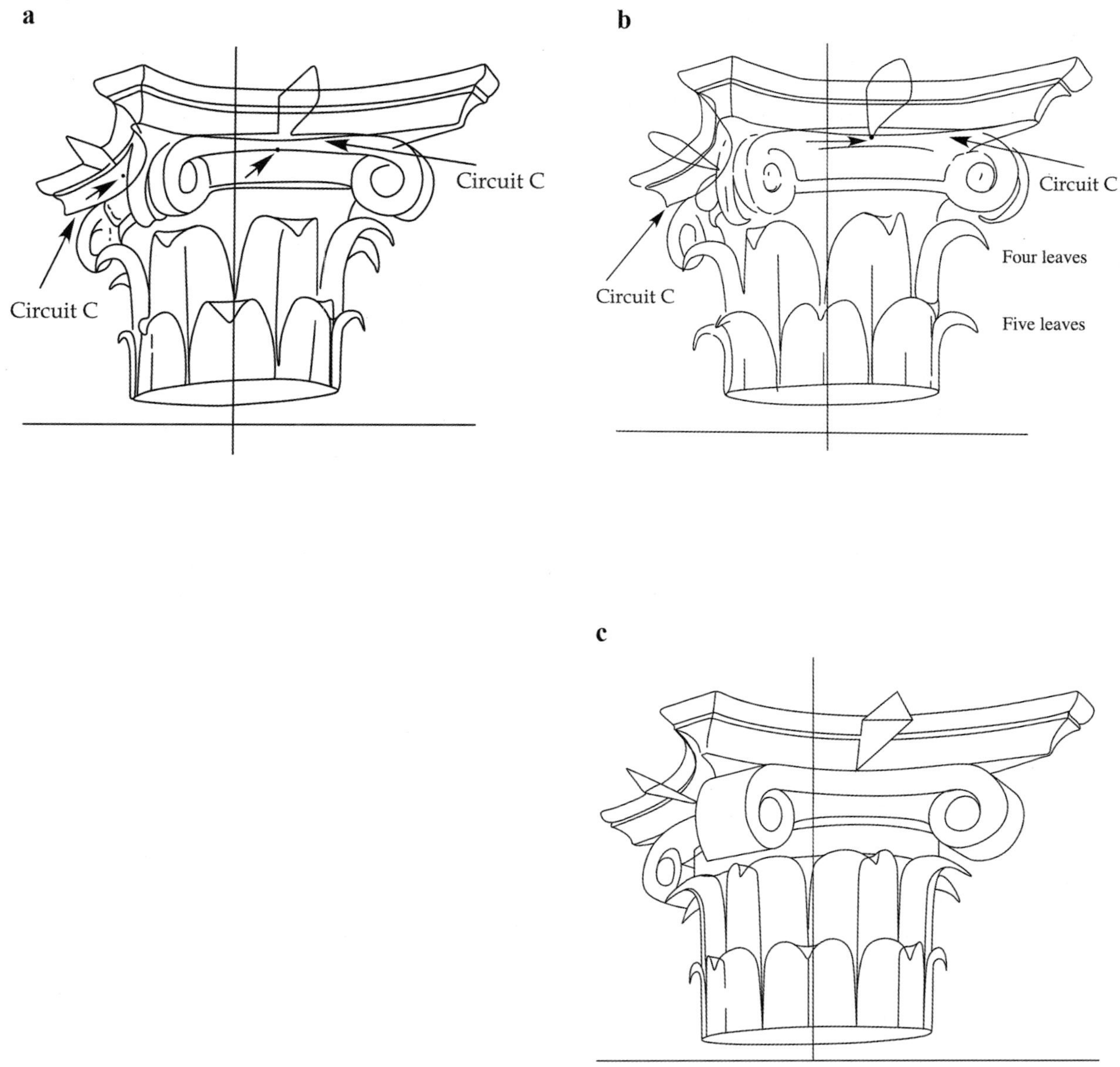

Figure A.68
Perspective of the capital: incongruences. Parma manuscript, folio 57 recto (*a*); Bordeaux manuscript, folio 72 verso (*b*); the reconstruction (*c*).

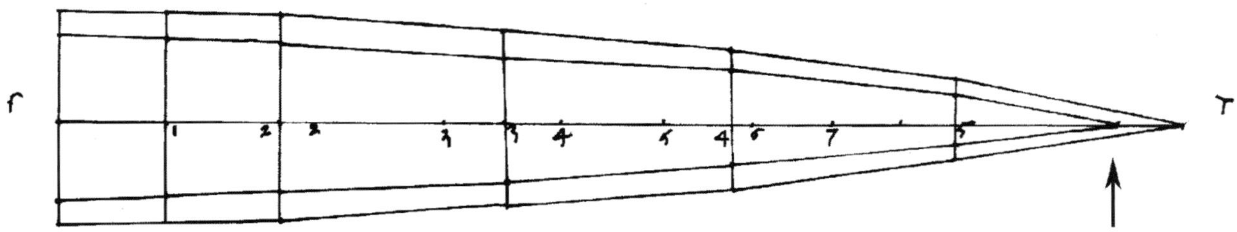

Figure A.69
Parma manuscript, folio 77 verso: incongruences.

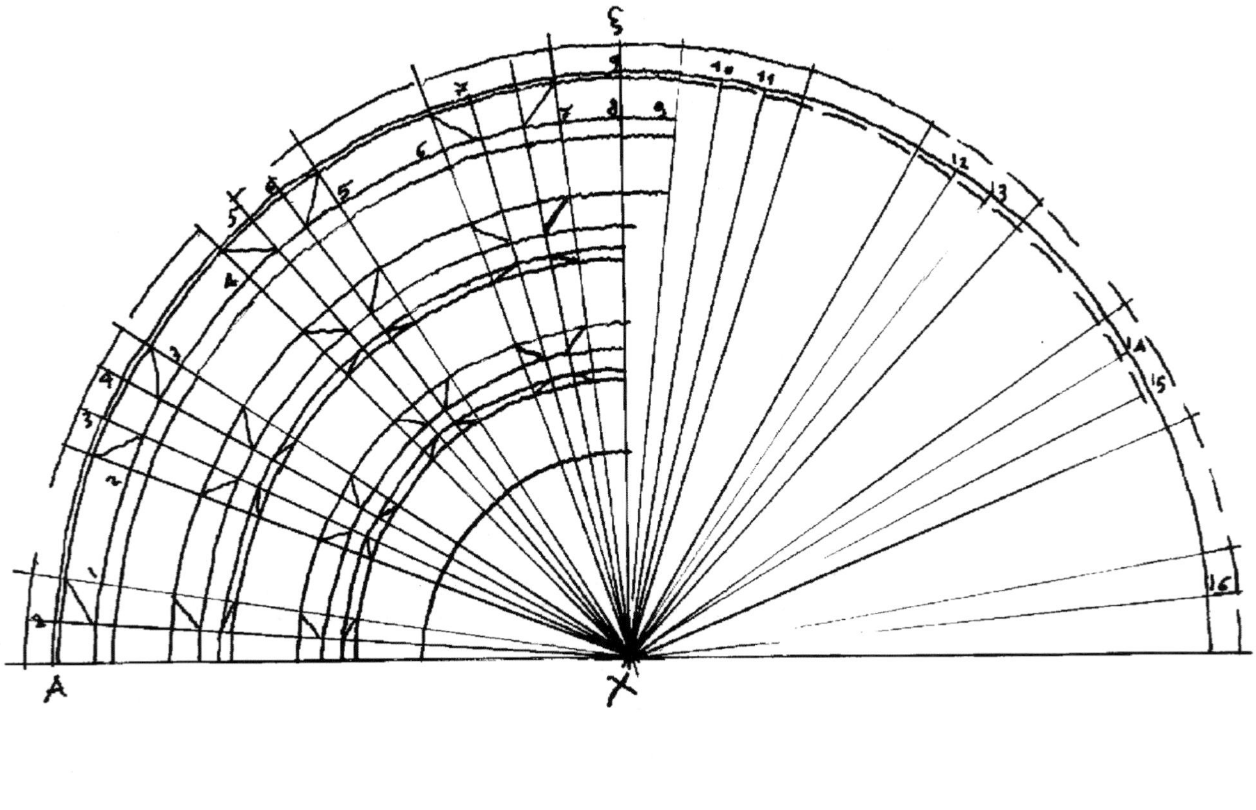

Figure A.70
Parma manuscript, folio 78 verso: incongruences.

Parma manuscript,
folio 78 verso

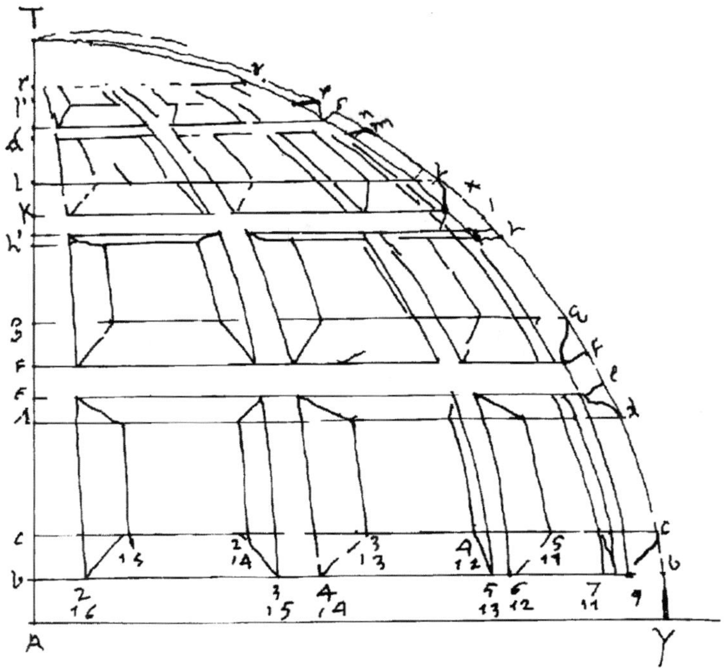

Parma manuscript,
folio 80 verso

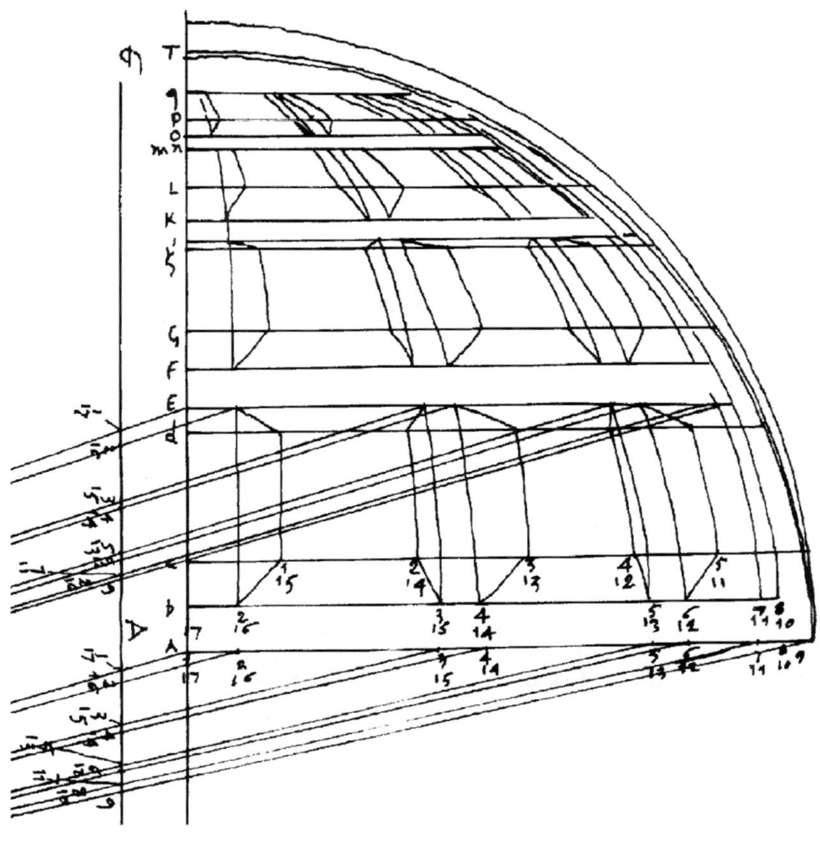

Figure A.71
The two elevations of the dome in the Parma manuscript: incongruences.

Parma manuscript, folio 78 verso

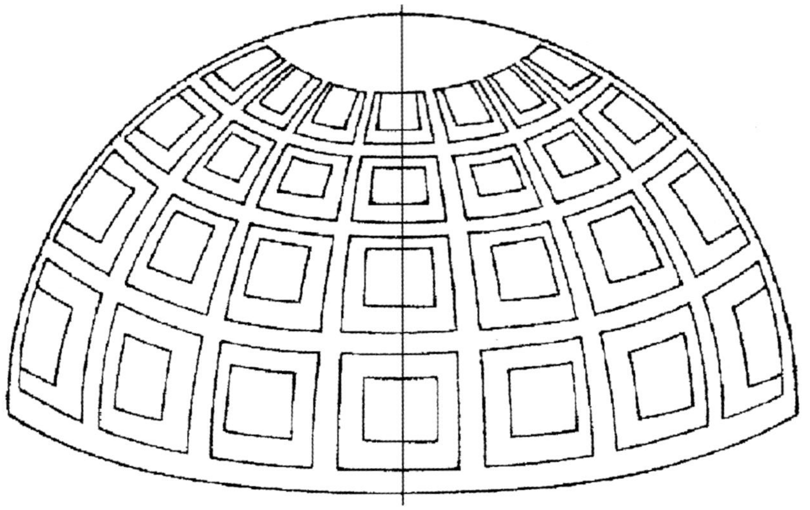

The reconstruction of the dome

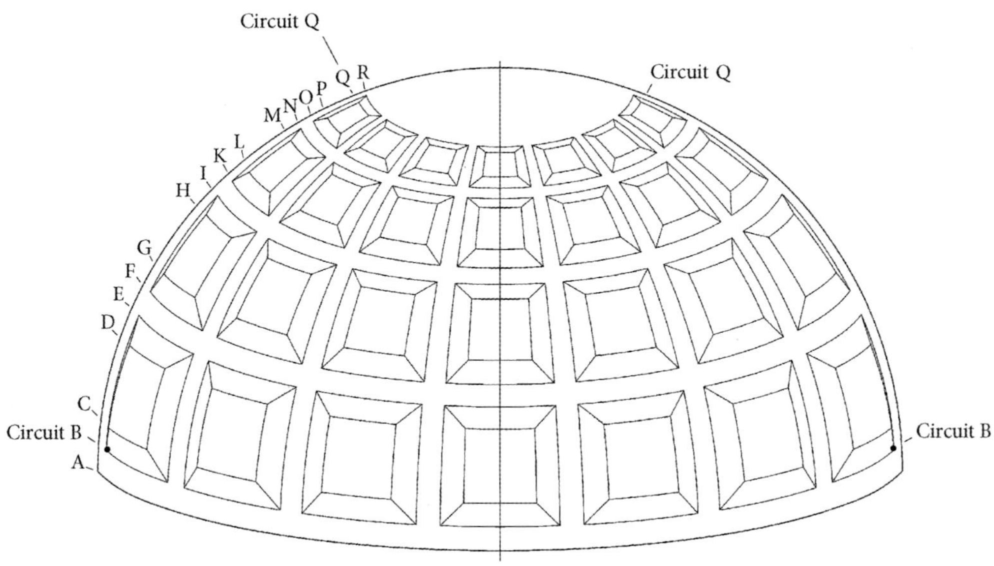

Figure A.72
Perspective of the dome: incongruences.

Appendix B

Drawings for Piero's Text

All images in this section were created by Rocco Sinisgalli to faithfully illustrate the geometric and scientific processes practiced by Piero della Francesca.

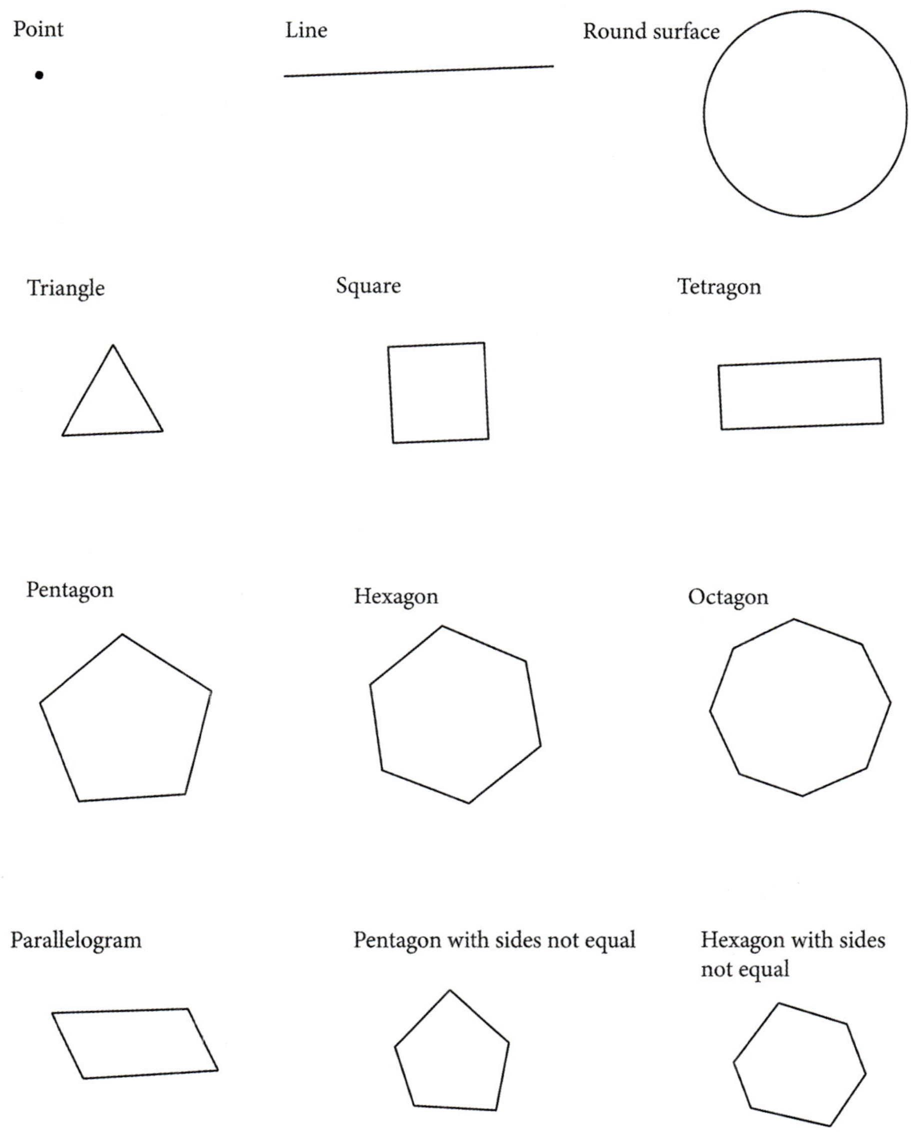

Figure B.1
Book 1, introduction.

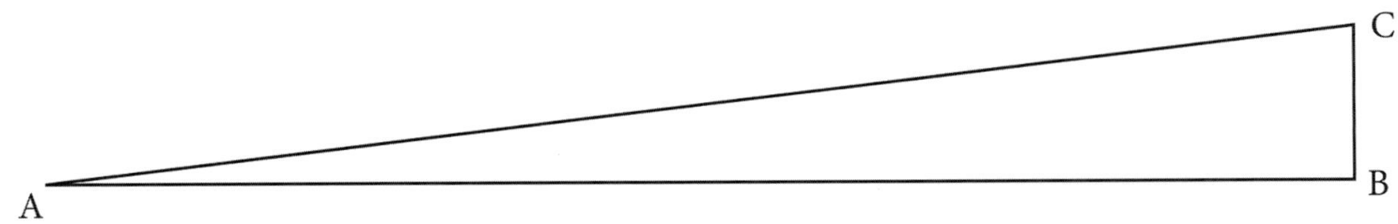

Figure B.2
Book 1, Proposition I.

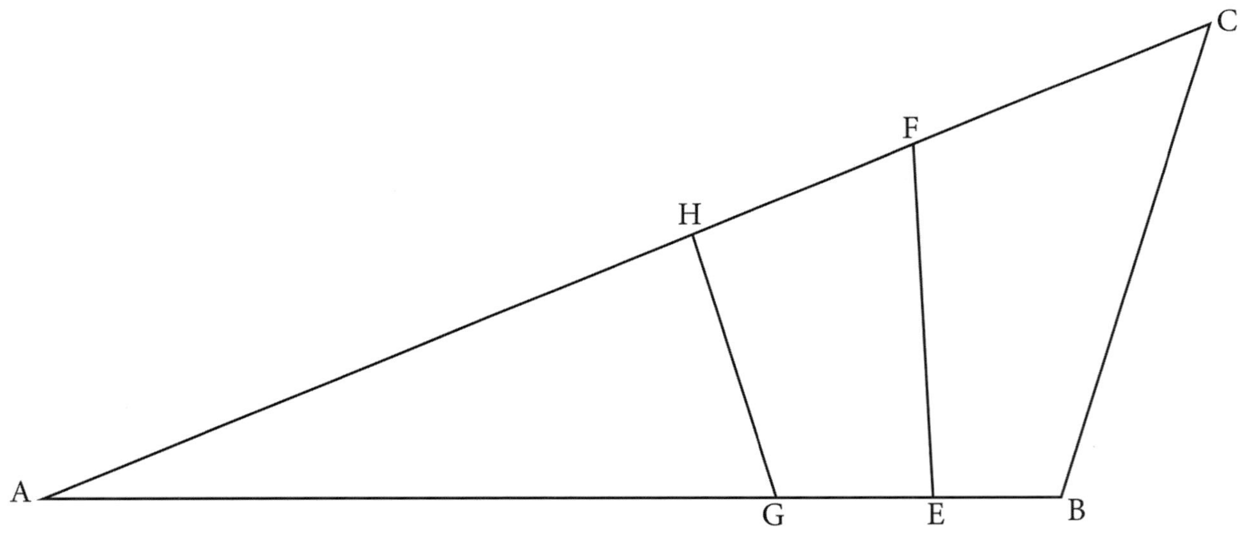

Figure B.3
Book 1, Proposition II.

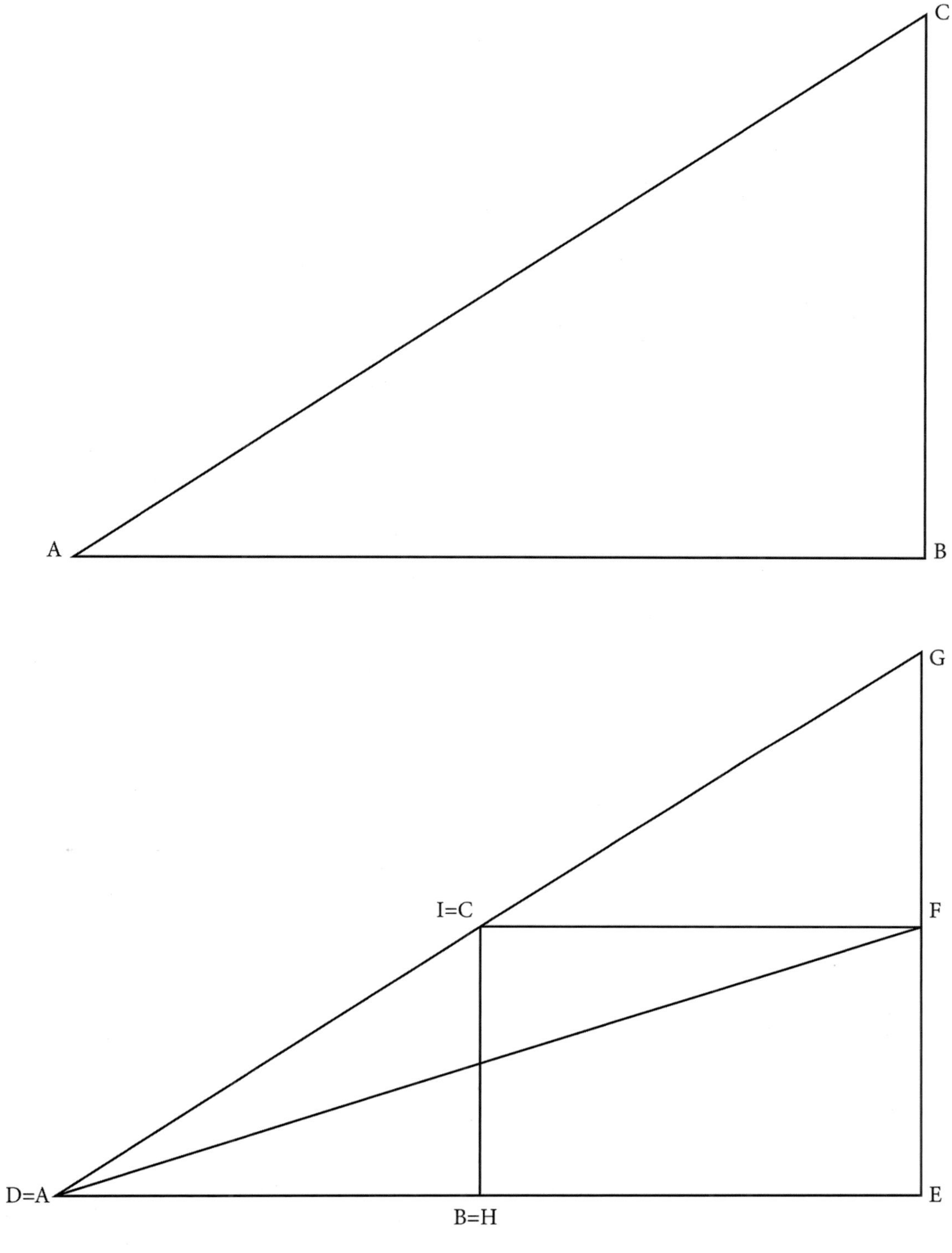

Figure B.4
Book 1, Proposition III.

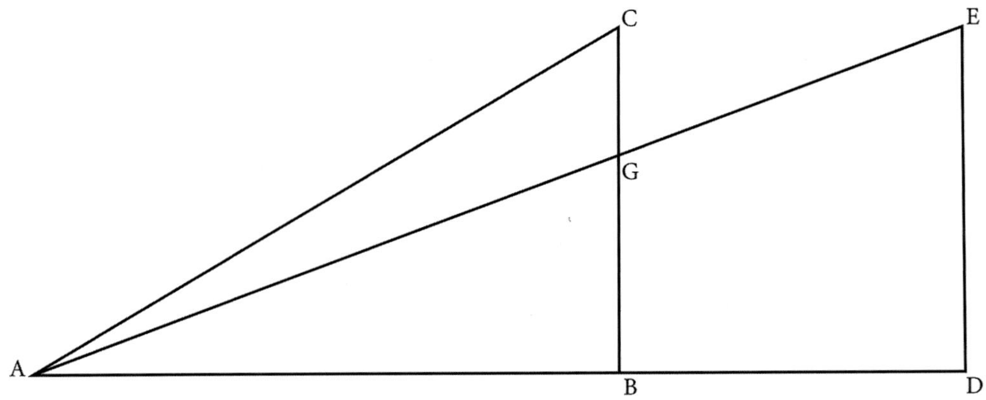

Figure B.5
Book 1, Proposition IV.

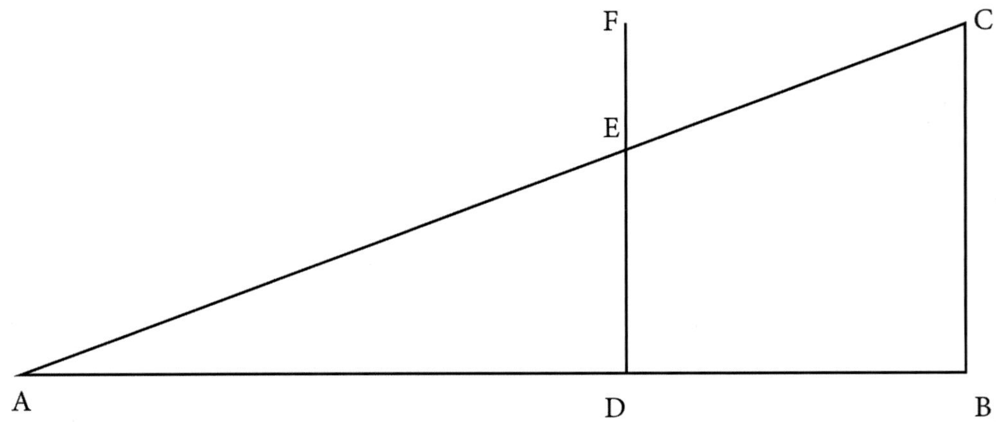

Figure B.6
Book 1, Proposition V.

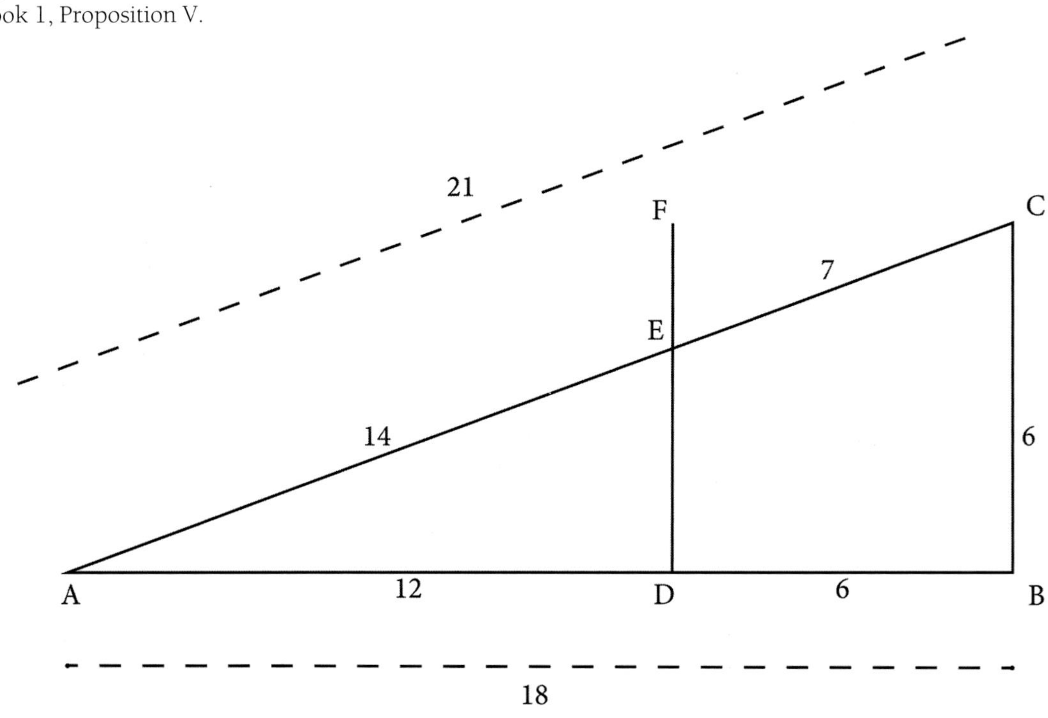

Figure B.7
Book 1, Proposition V.

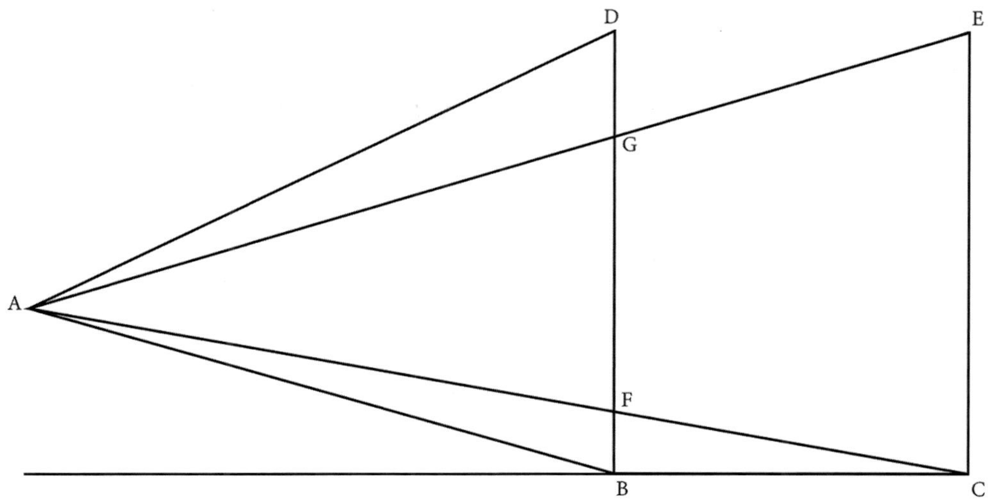

Figure B.8
Book 1, Proposition VI.

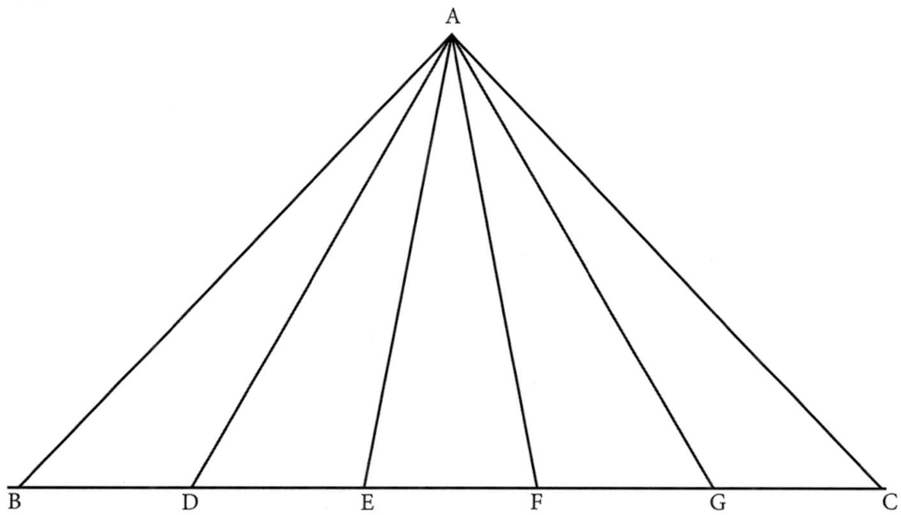

Figure B.9
Book 1, Proposition VII.

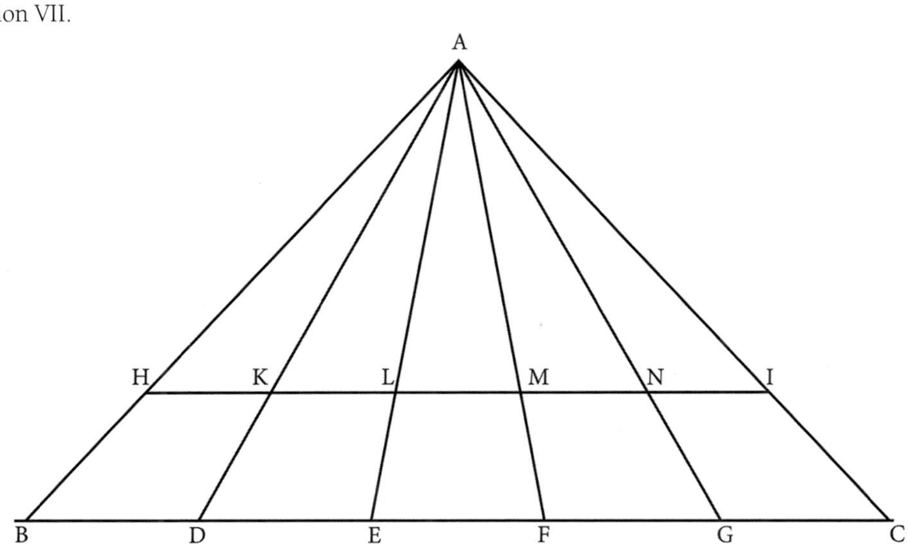

Figure B.10
Book 1, Proposition VIII.

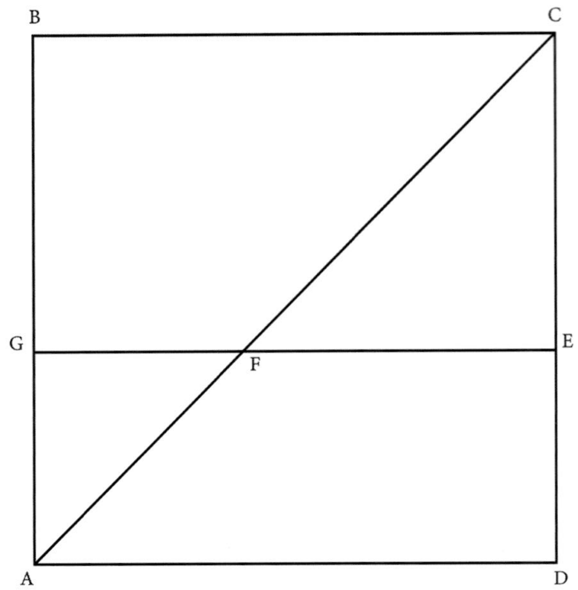

Figure B.11
Book 1, Proposition IX.

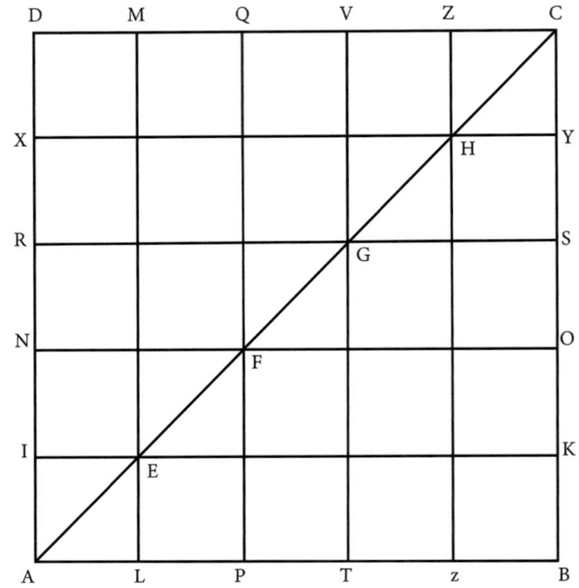

Figure B.12
Book 1, Proposition X.

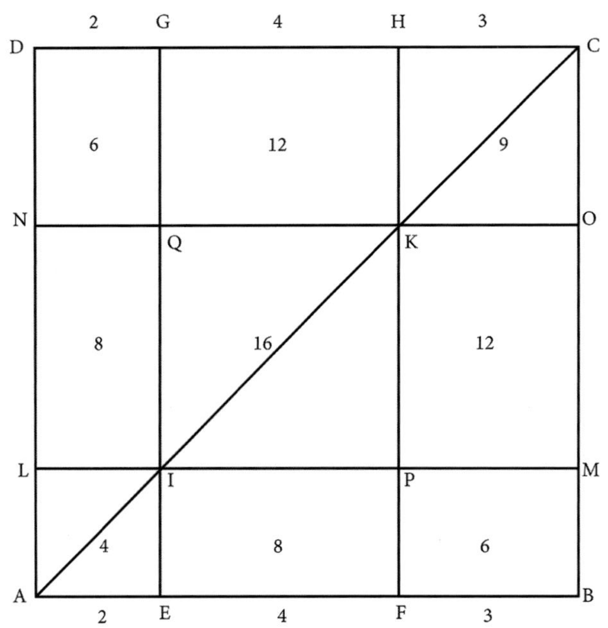

Figure B.13
Book 1, Proposition XI.

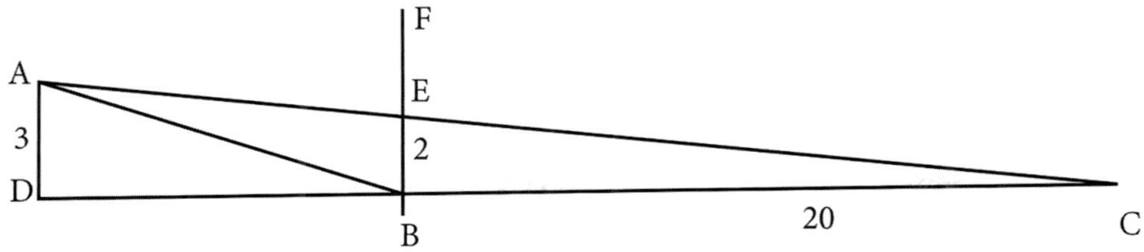

Figure B.14
Book 1, Proposition XII.

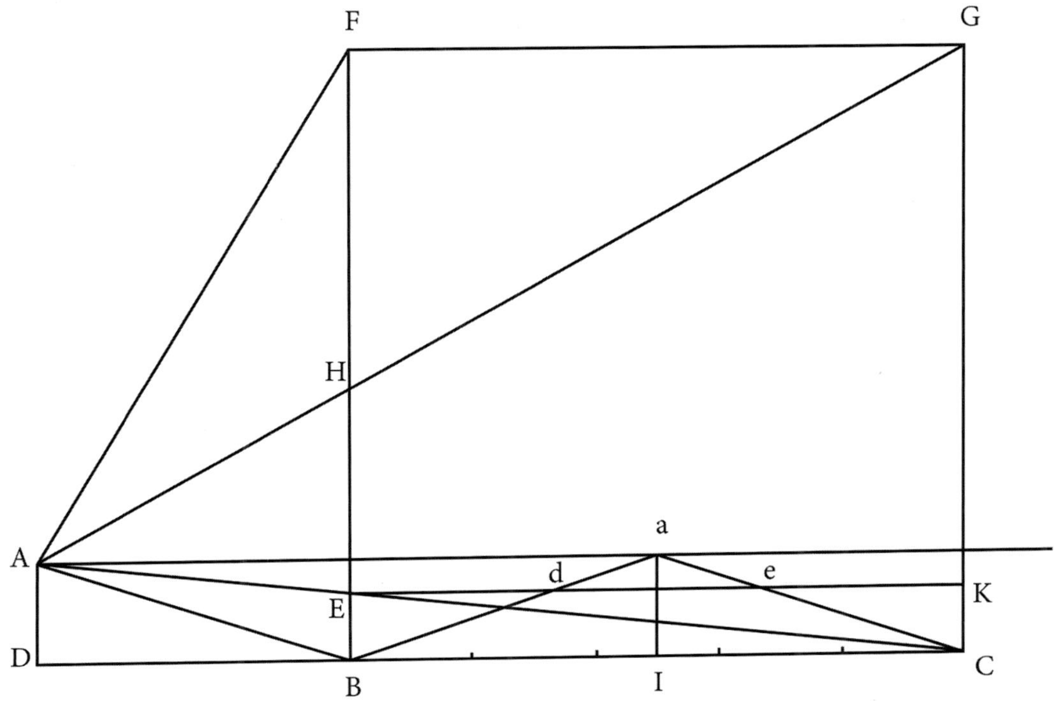

Figure B.15
Book 1, Proposition XIII.

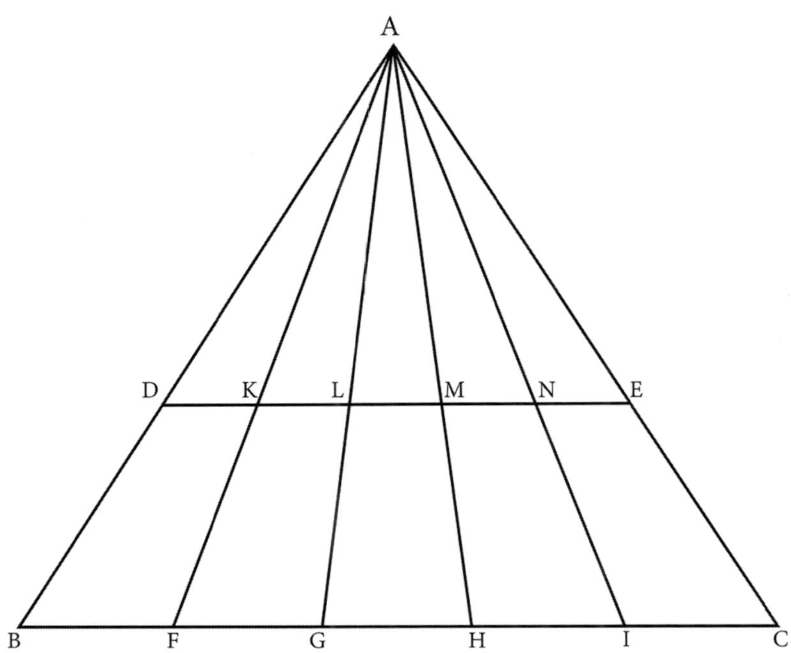

Figure B.16
Book 1, Proposition XIV.

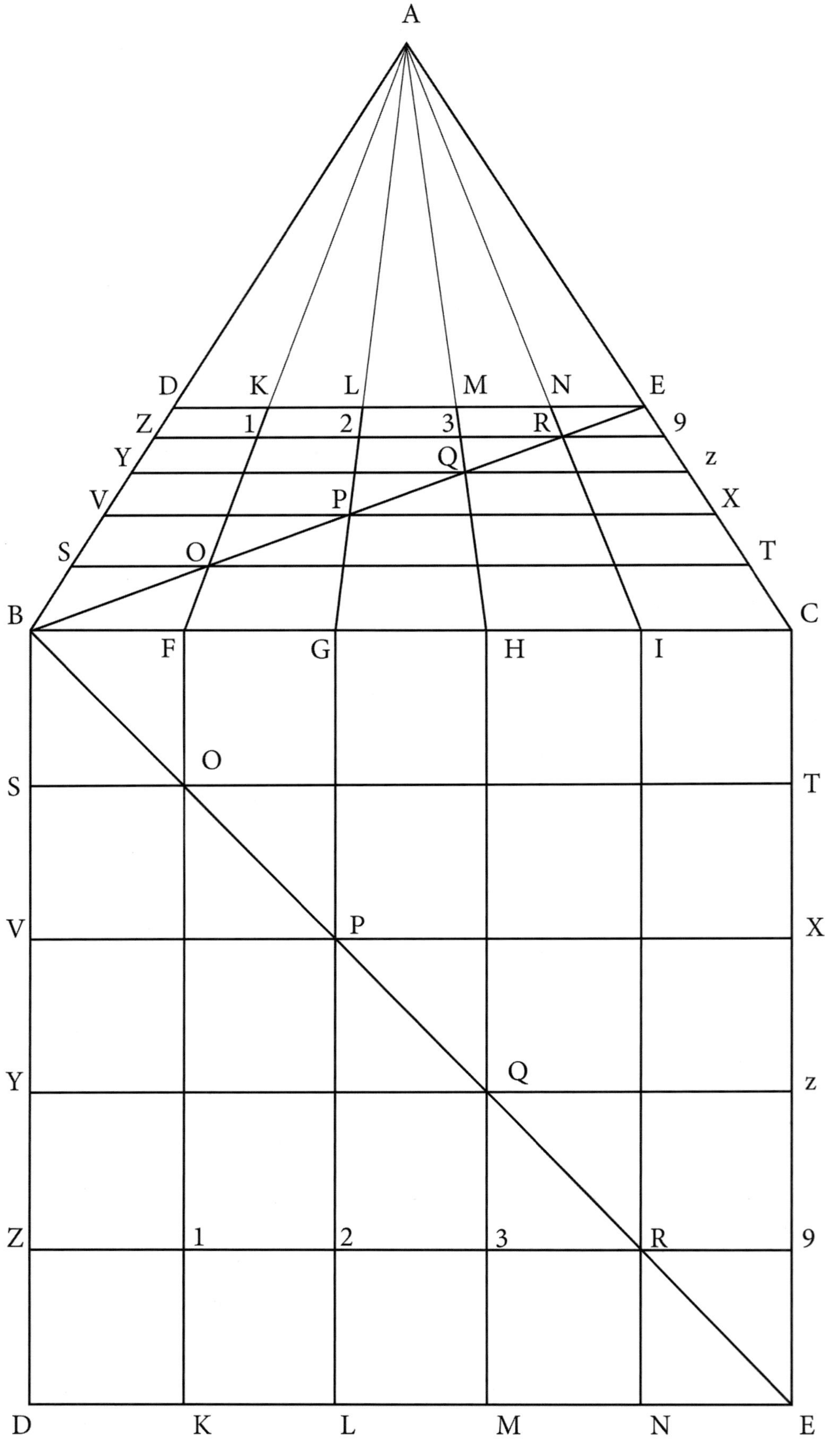

Figure B.17
Book 1, Proposition XV.

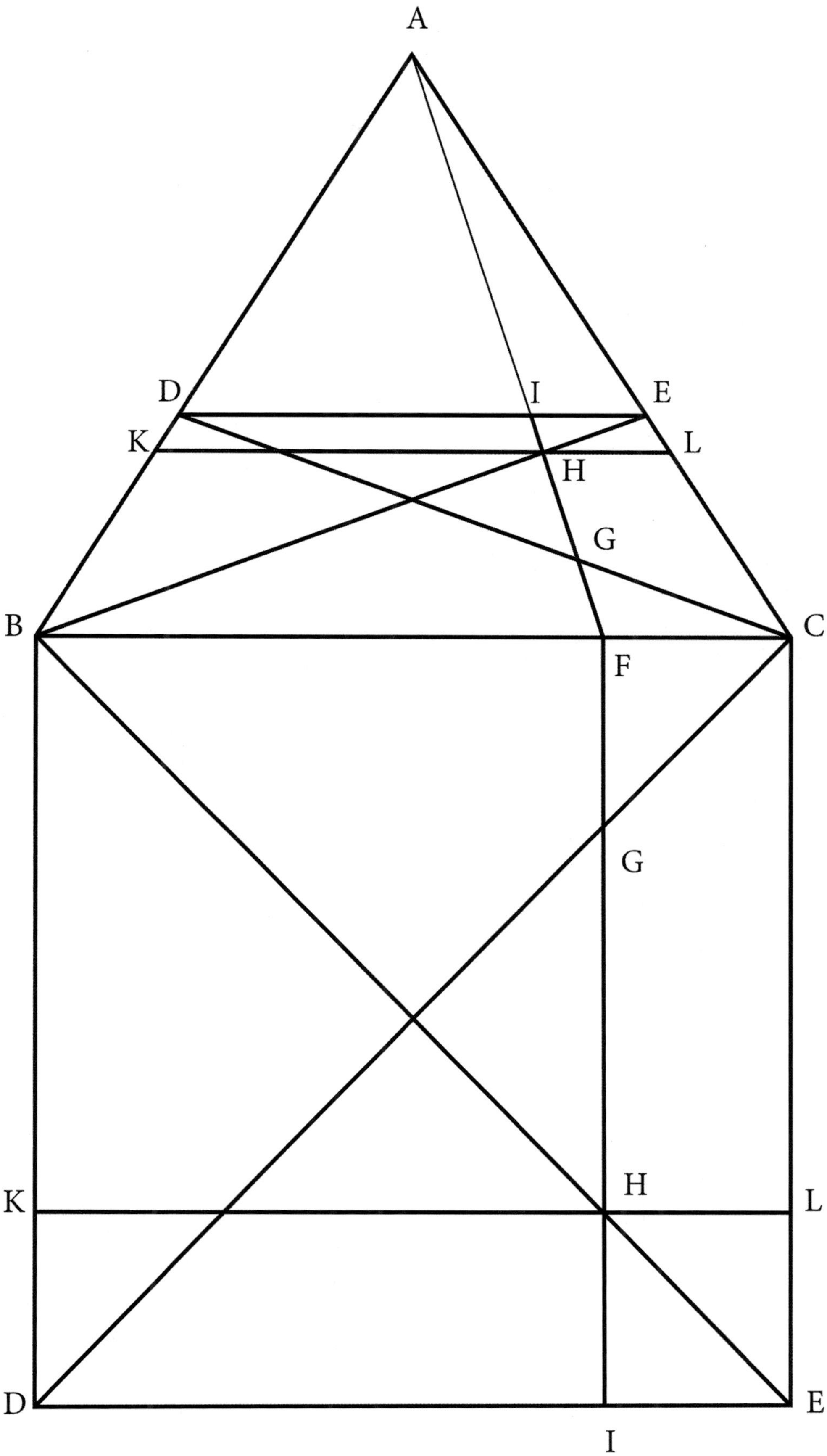

Figure B.18
Book 1, Proposition XV.

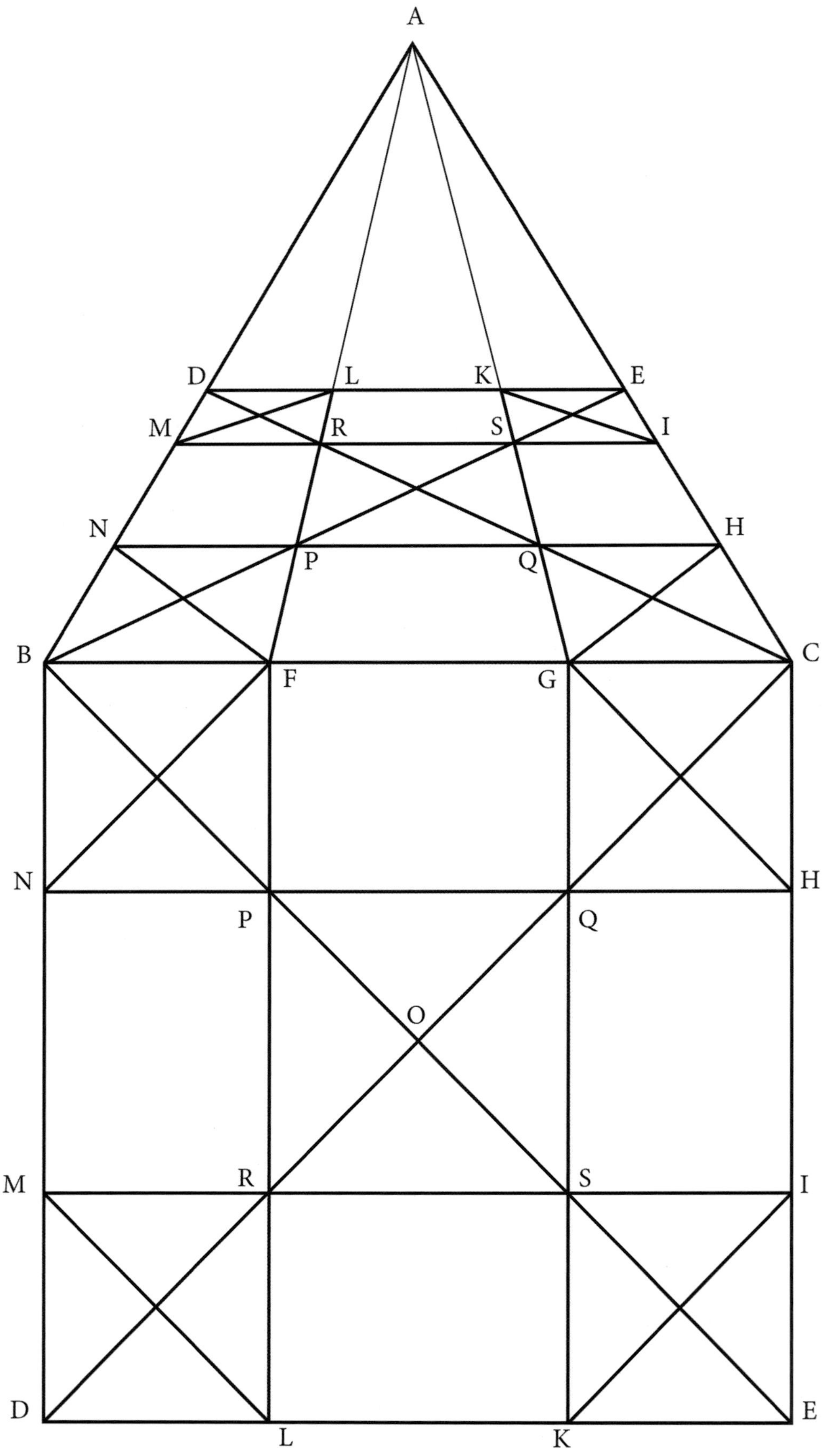

Figure B.19
Book 1, Proposition XVI.

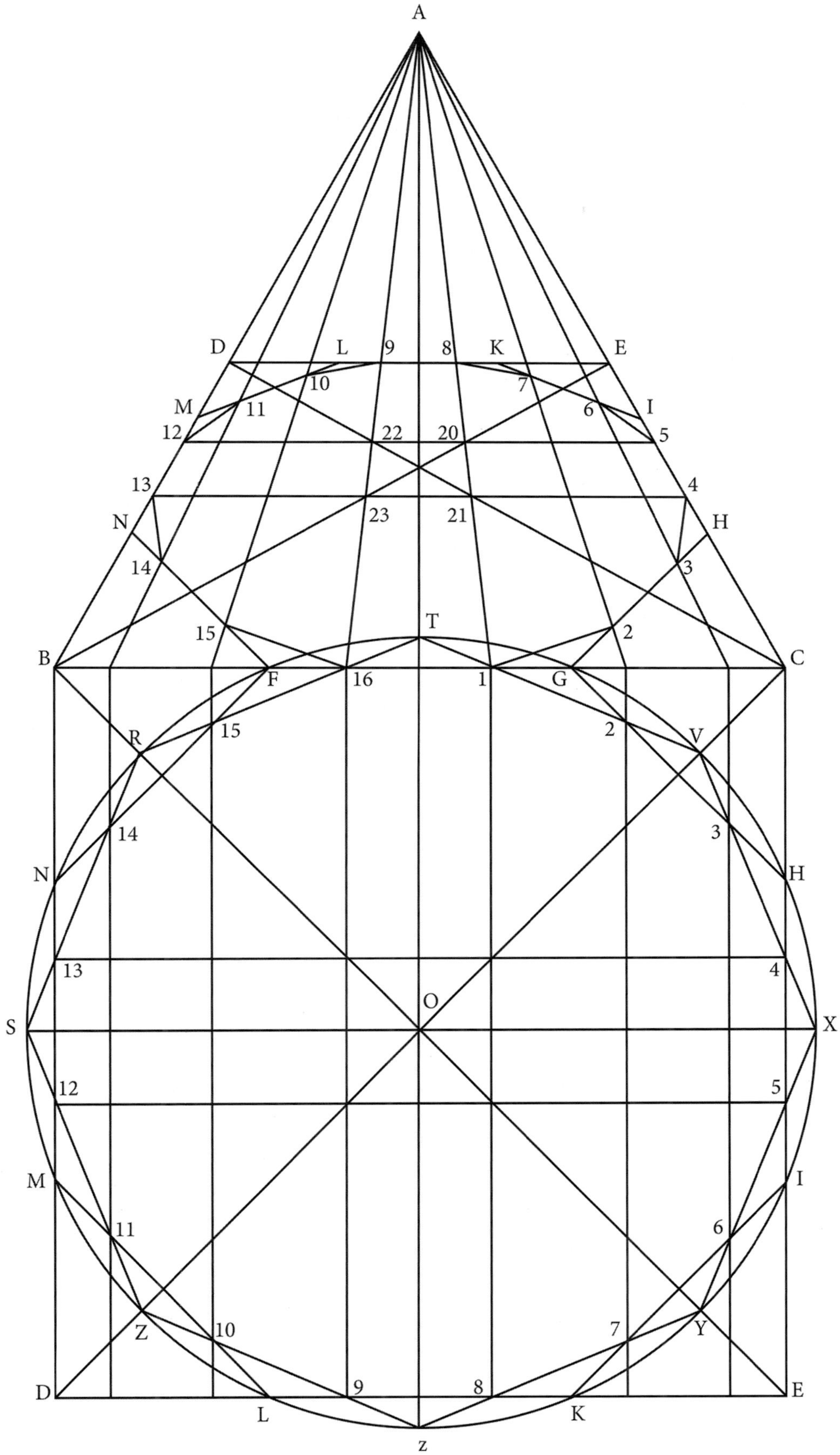

Figure B.20
Book 1, Proposition XVII.

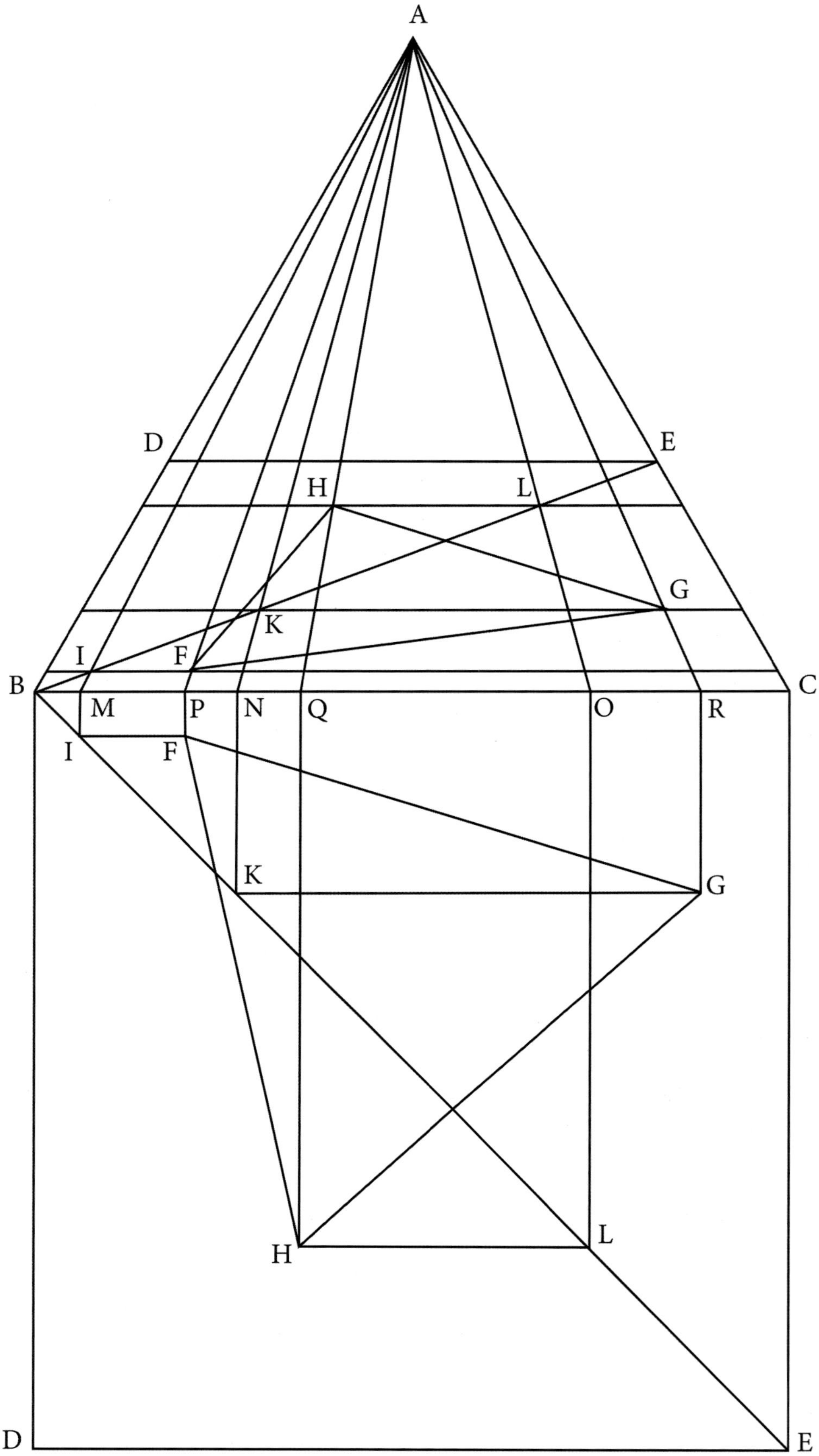

Figure B.21
Book 1, Proposition XVIII.

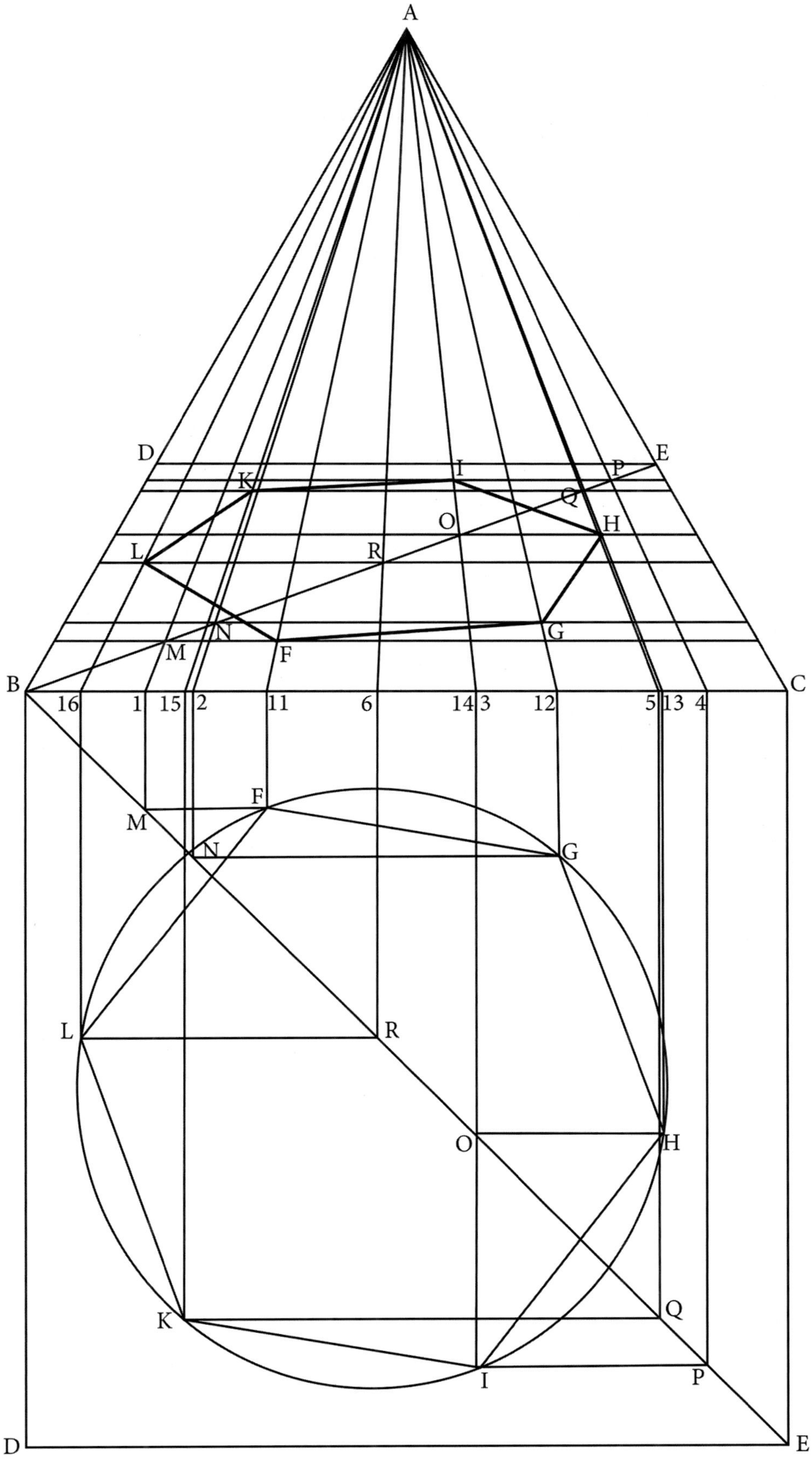

Figure B.22
Book 1, Proposition XIX.

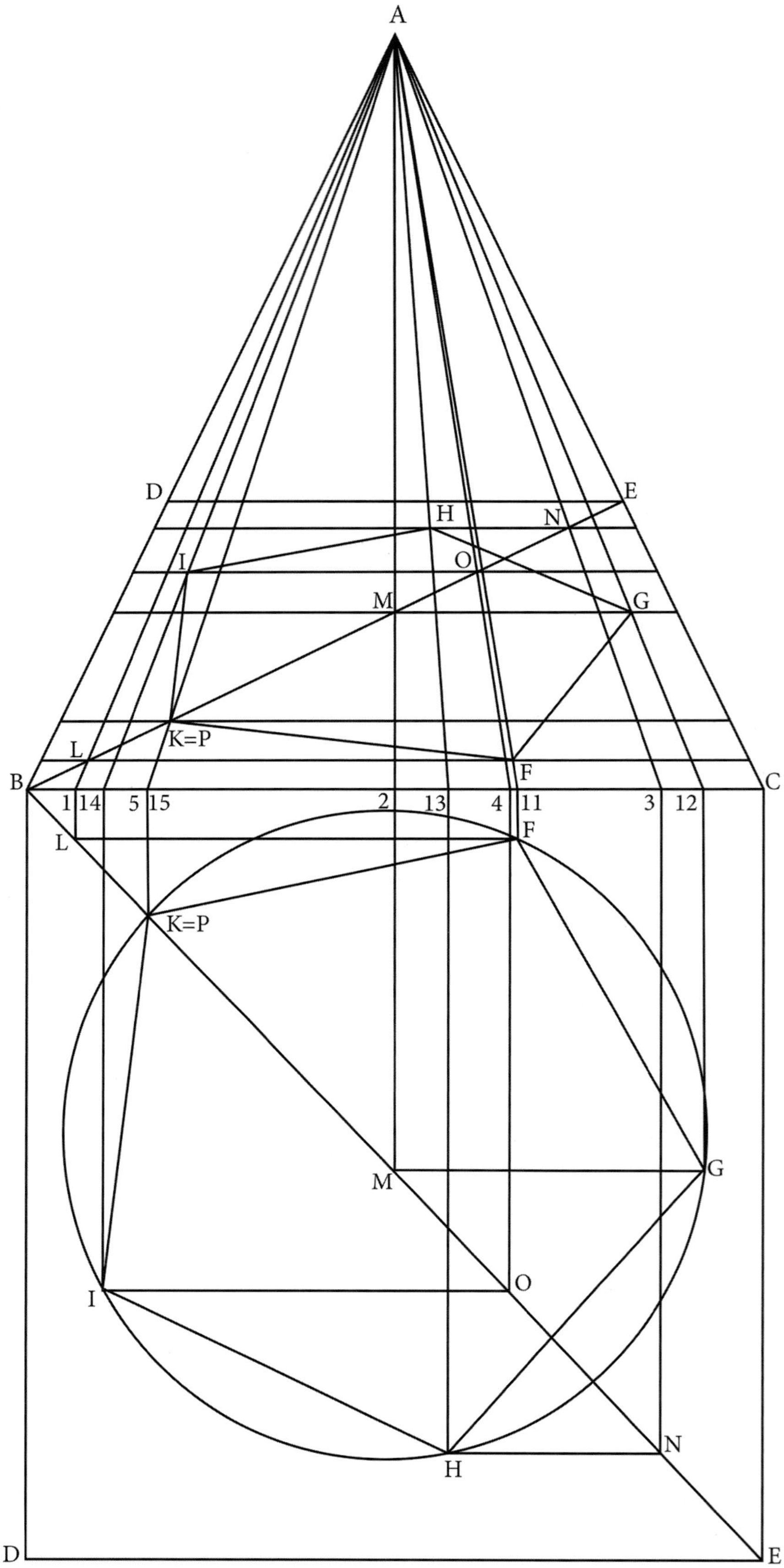

Figure B.23
Book 1, Proposition XX.

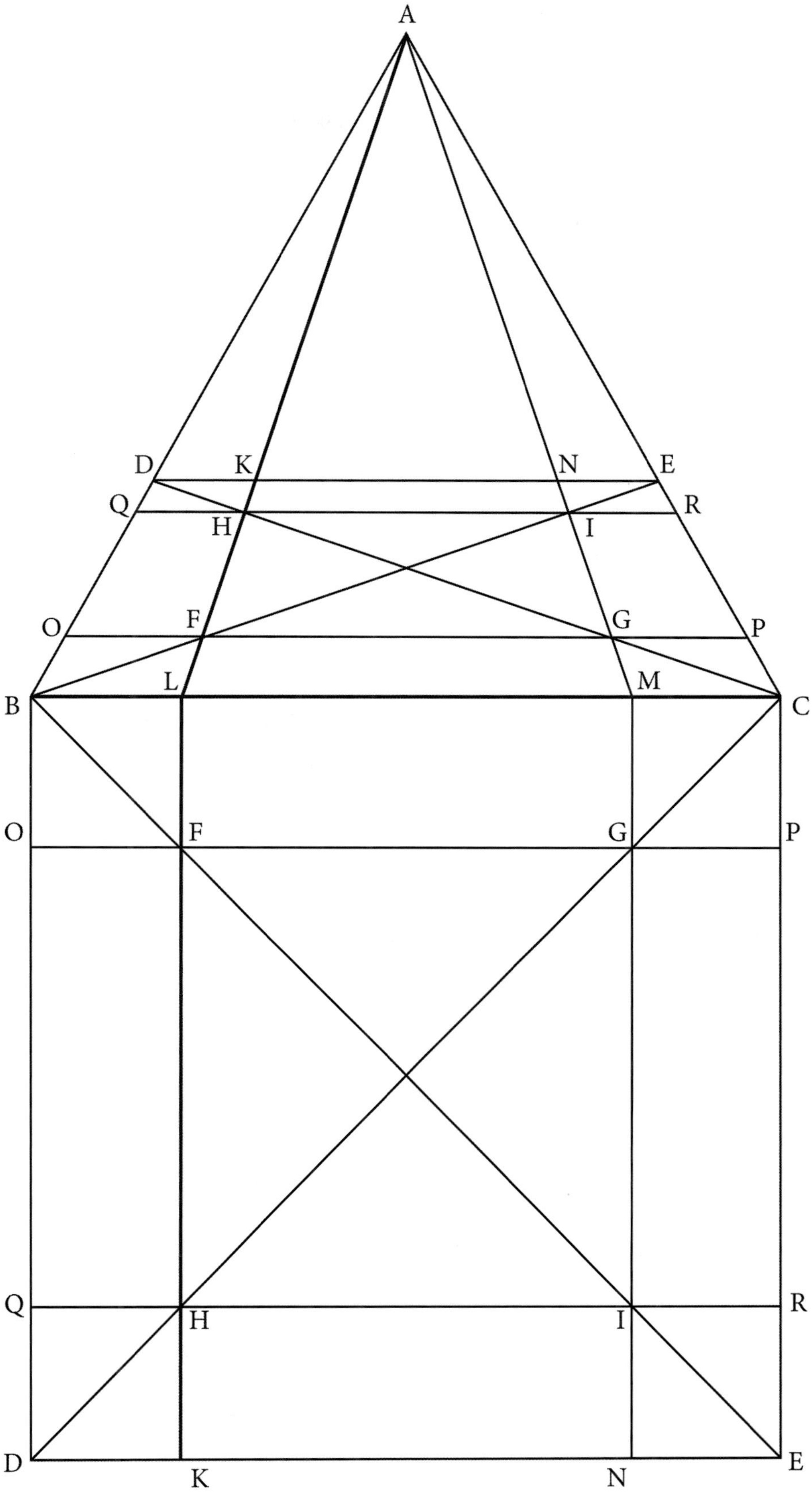

Figure B.24
Book 1, Proposition XXI.

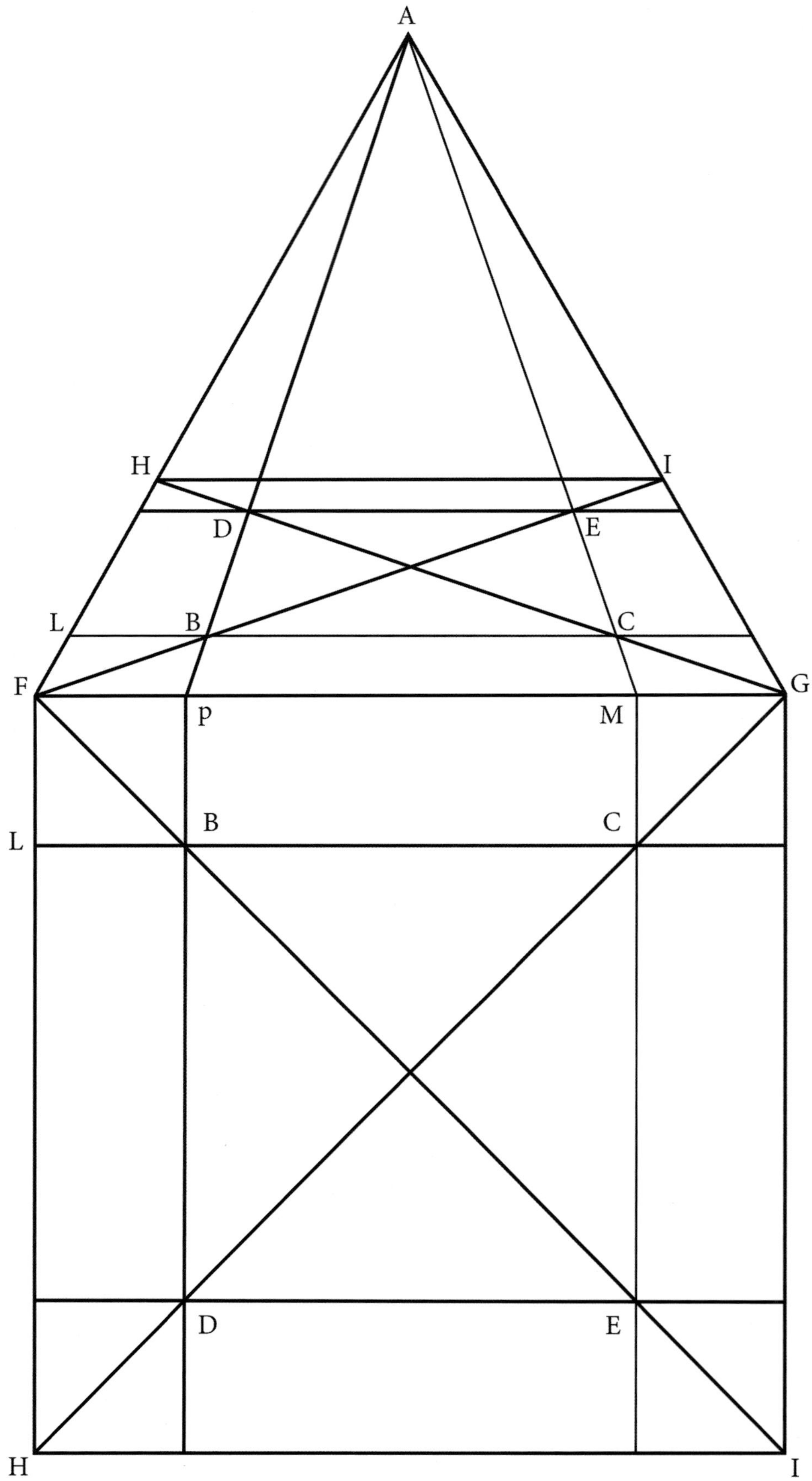

Figure B.25
Book 1, Proposition XXII.

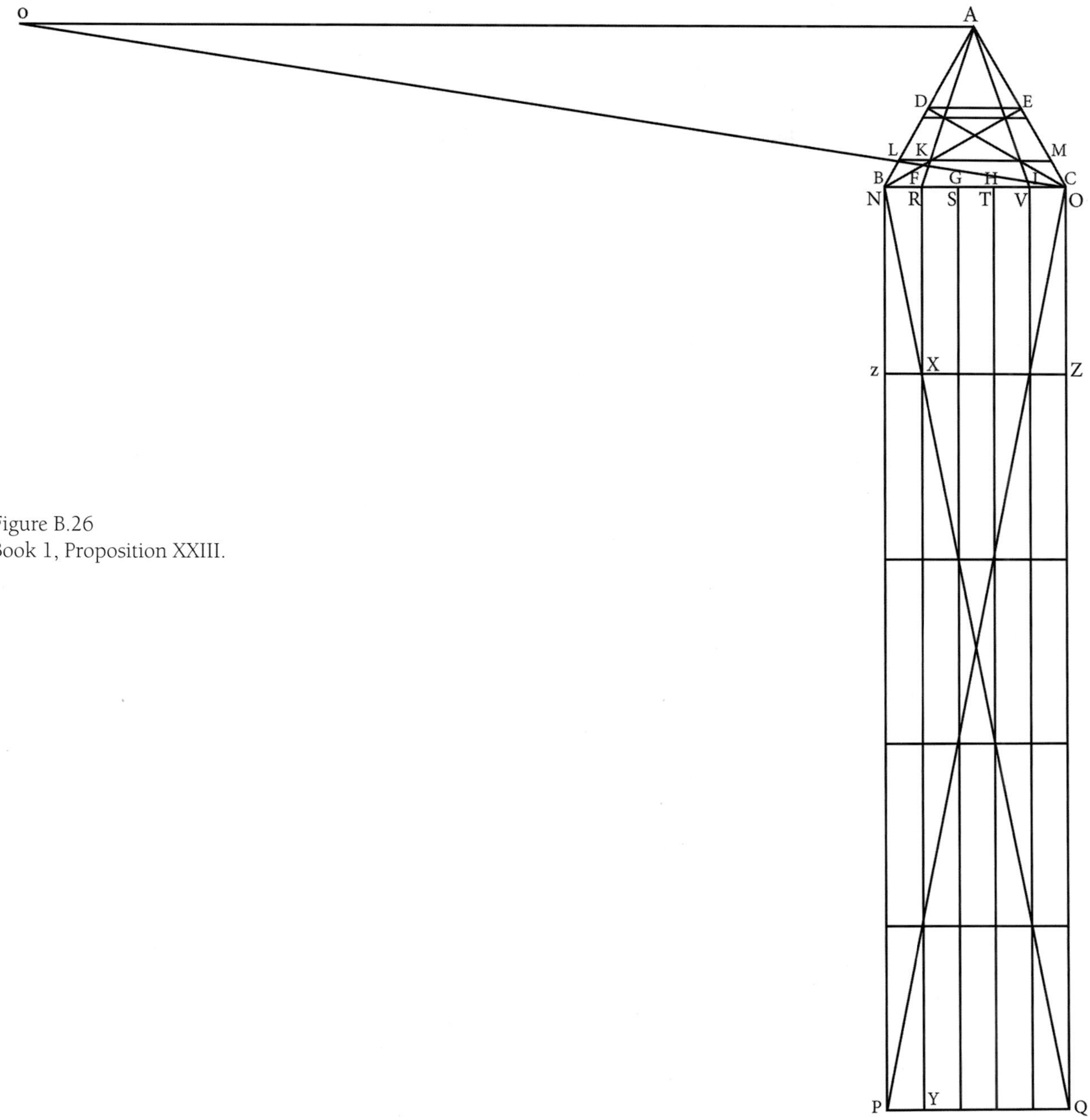

Figure B.26
Book 1, Proposition XXIII.

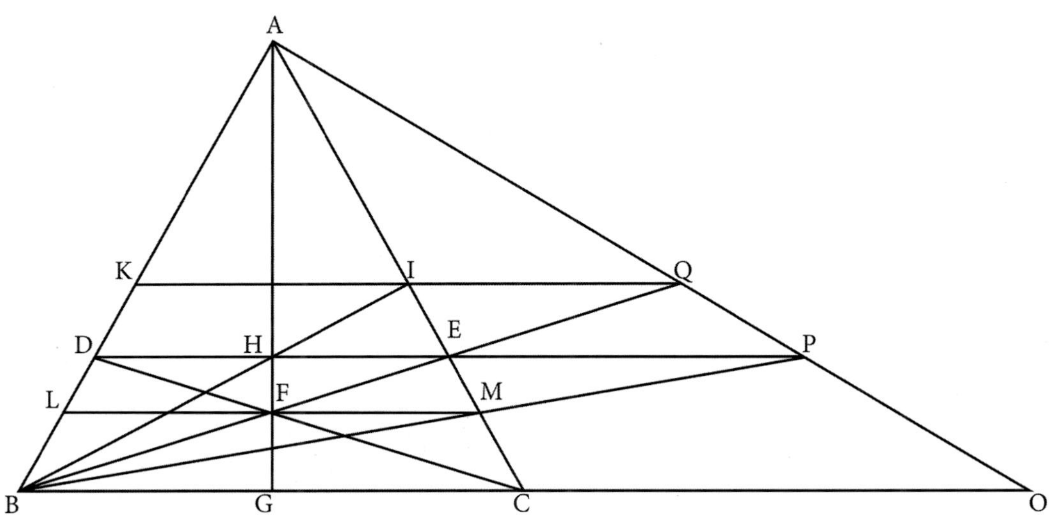

Figure B.27
Book 1, Proposition XXIV.

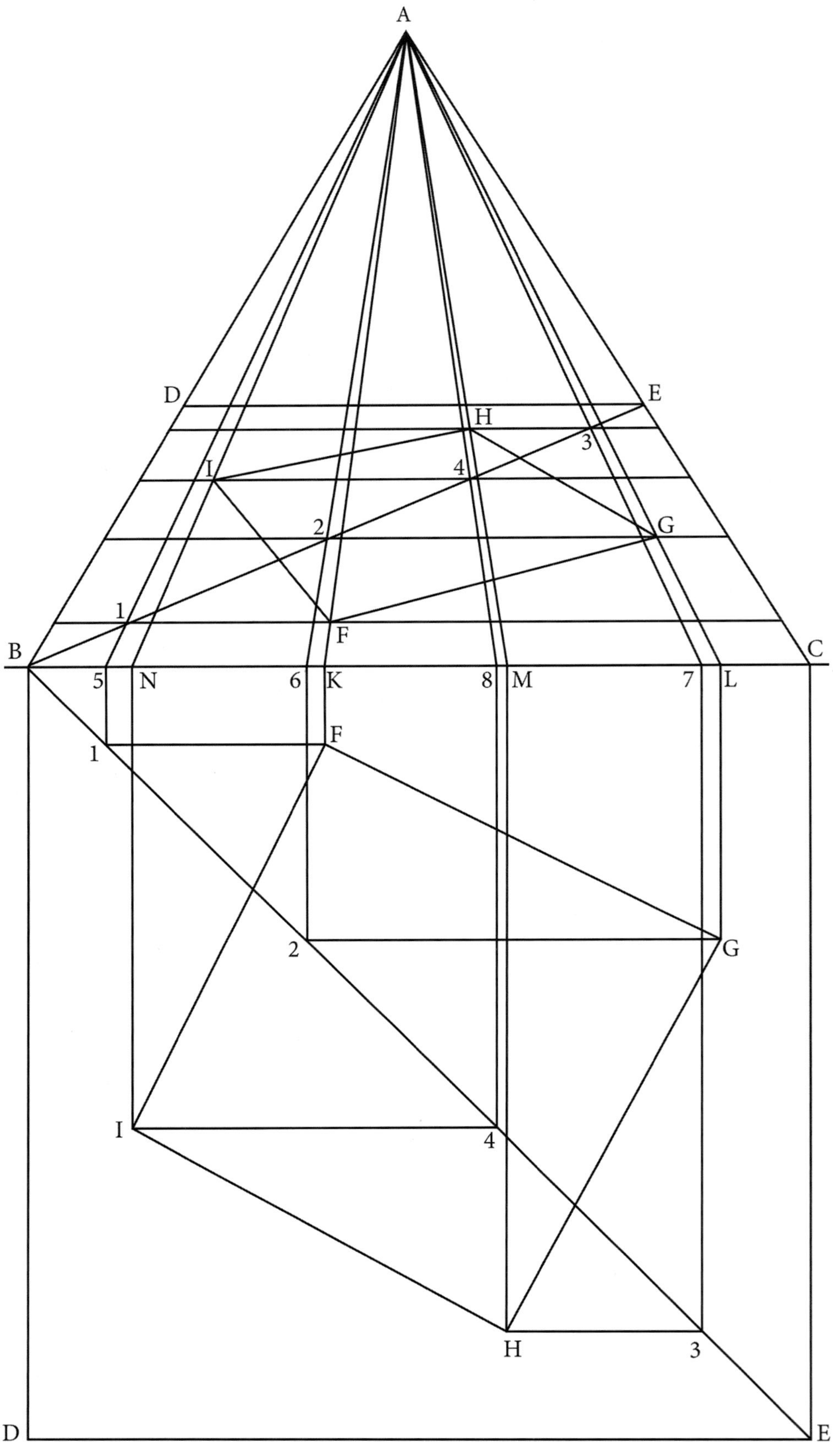

Figure B.28
Book 1, Proposition XXV.

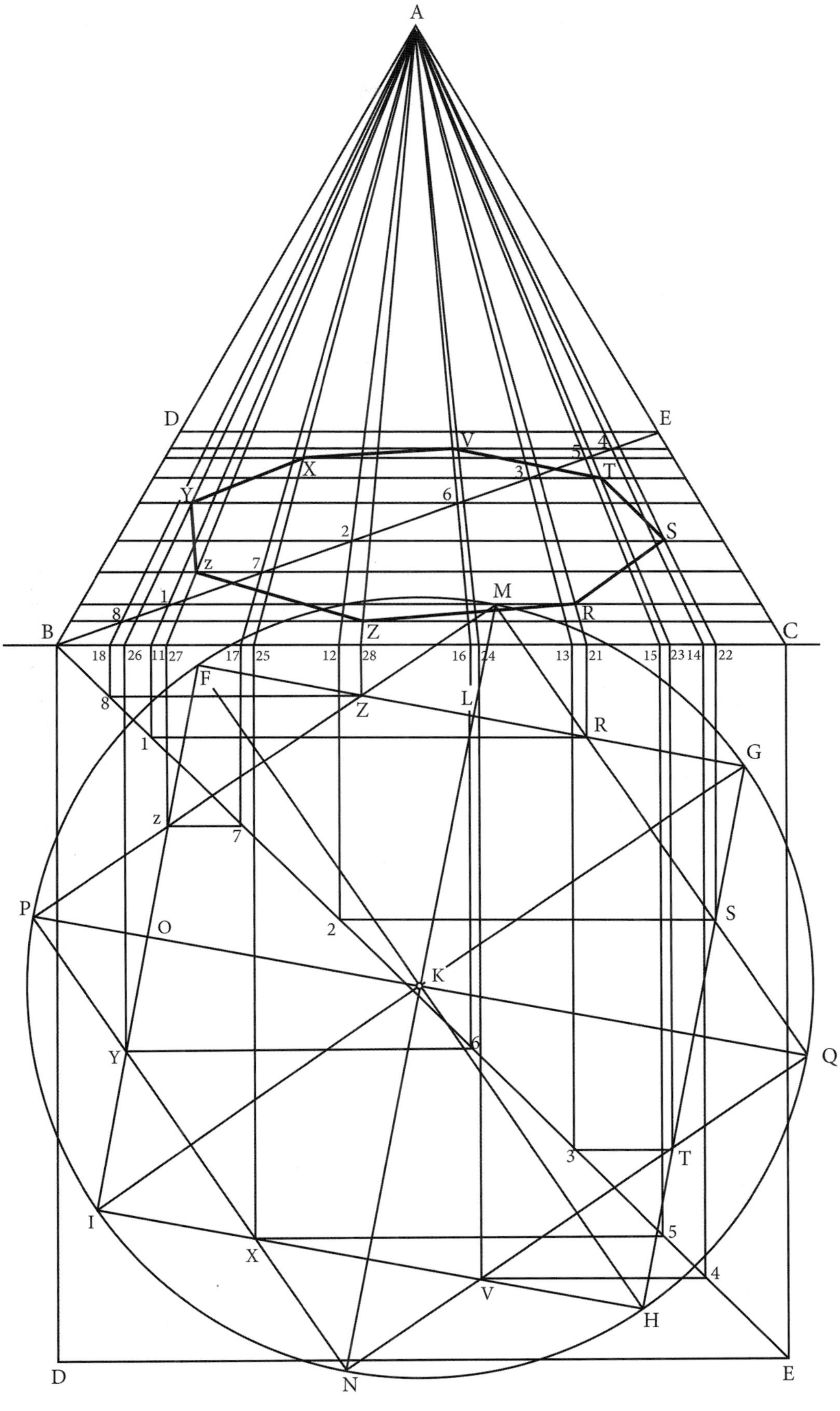

Figure B.29
Book 1, Proposition XXVI.

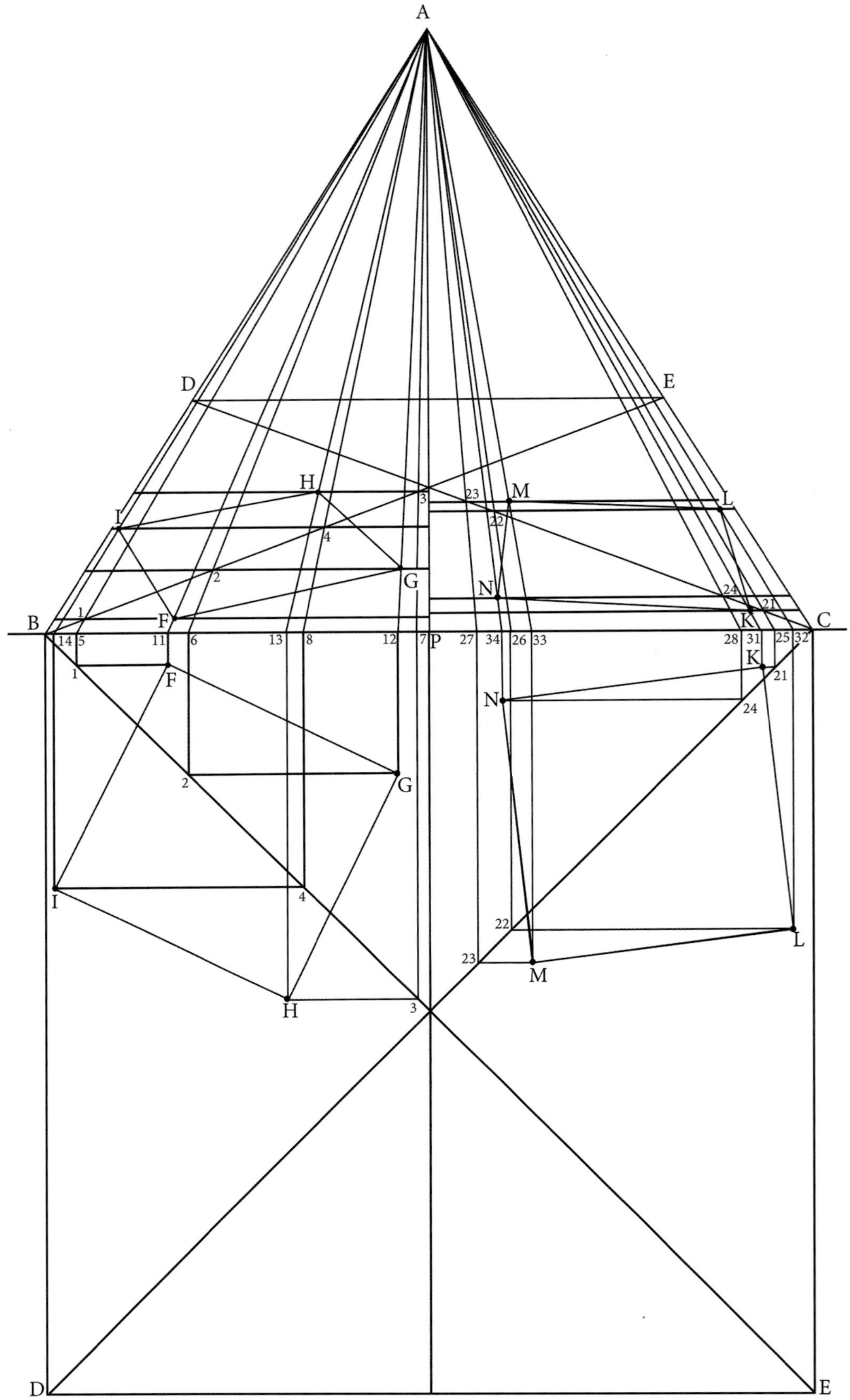

Figure B.30
Book 1, Proposition XXVII.

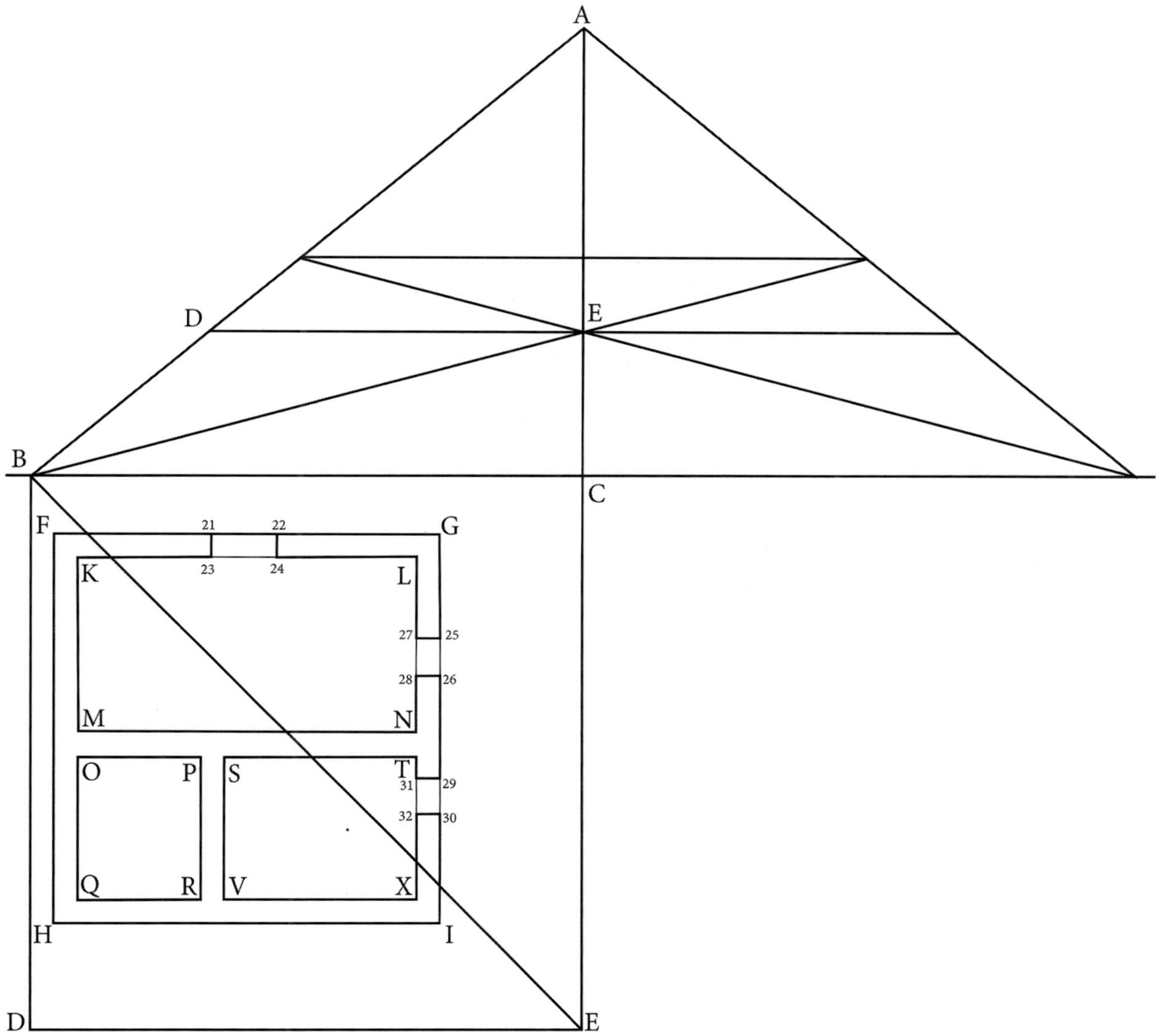

Figure B.31
Book 1, Proposition XXVIII.

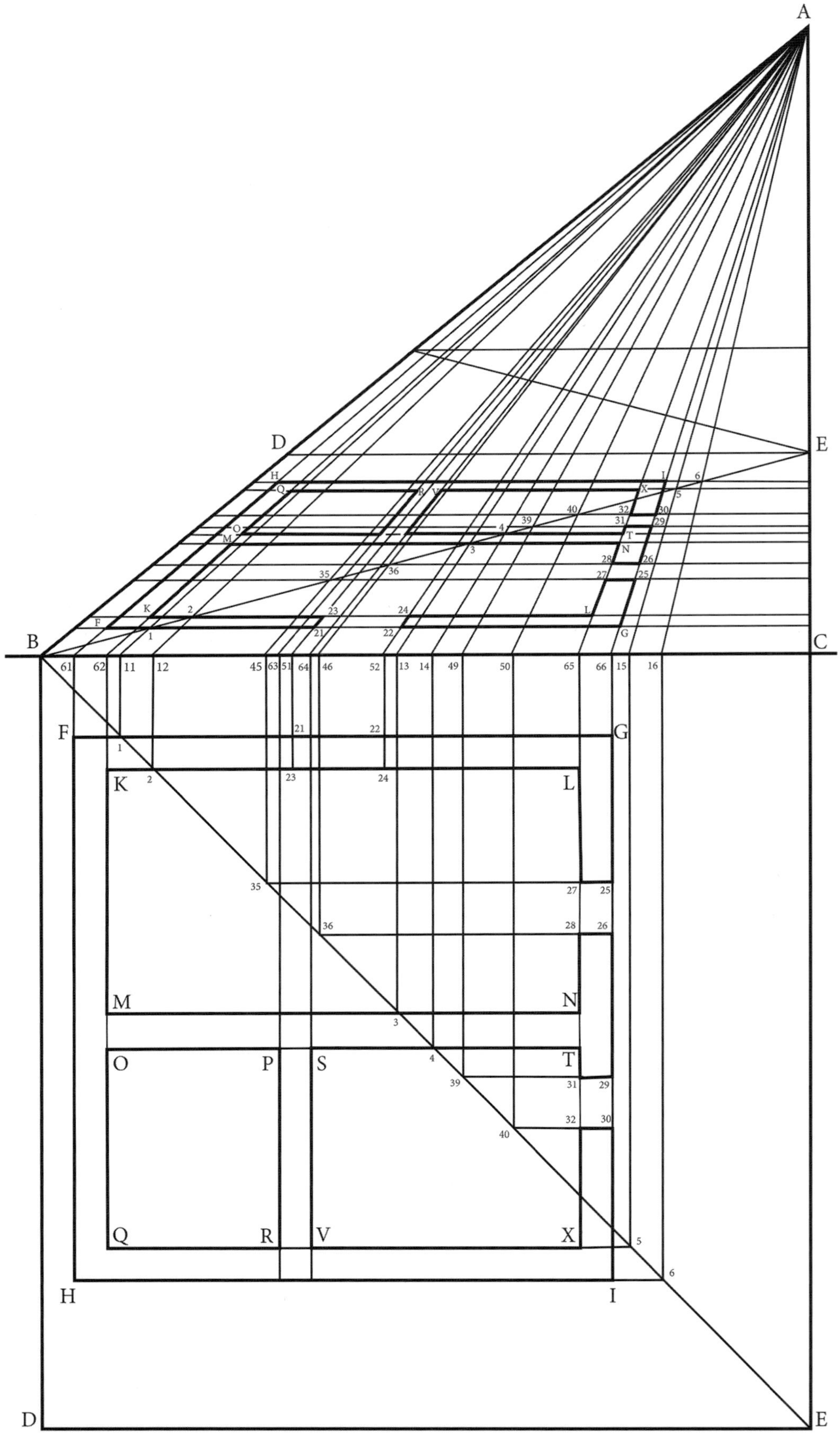

Figure B.32
Book 1, Proposition XXVIII.

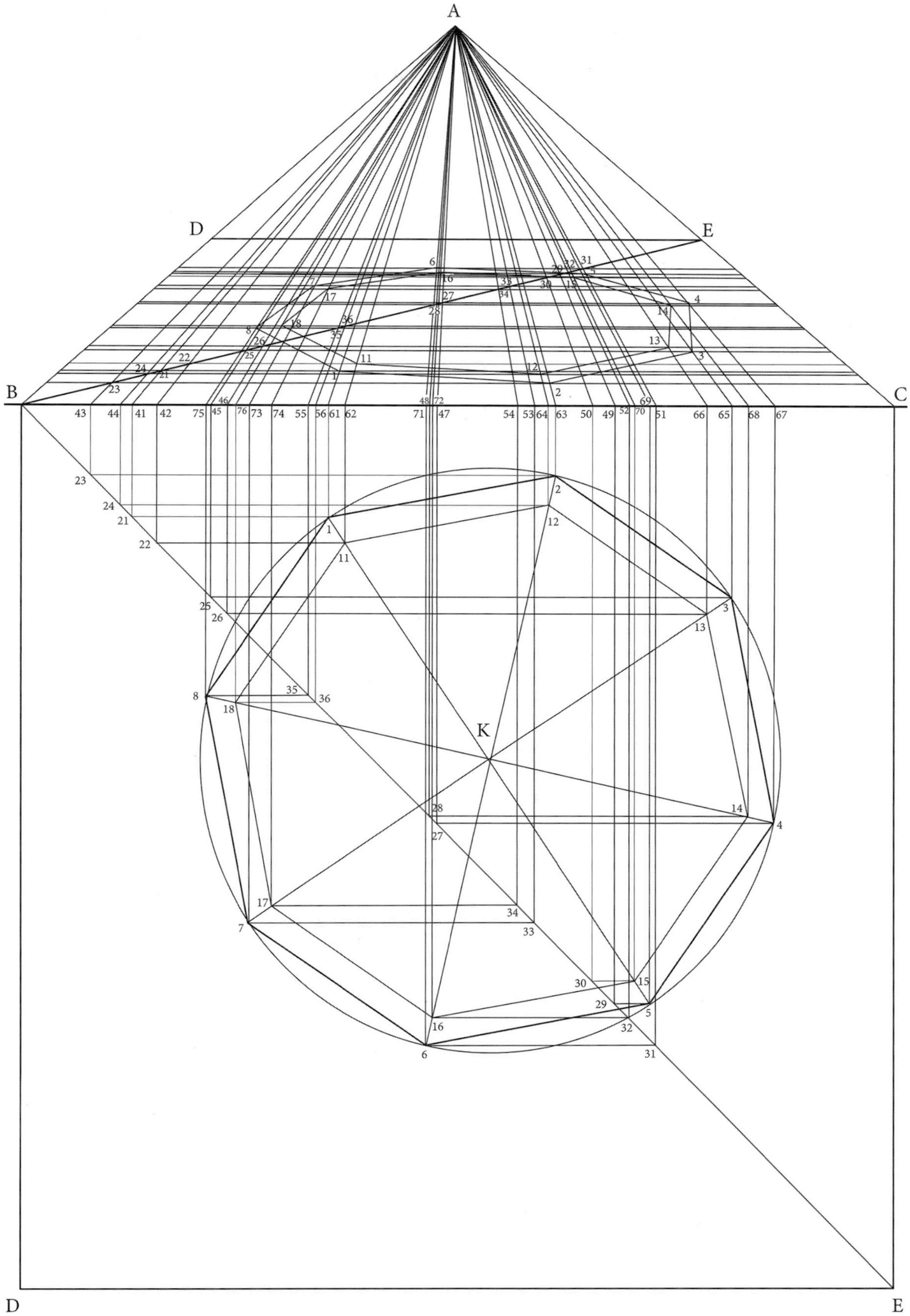

Figure B.33
Book 1, Proposition XXIX.

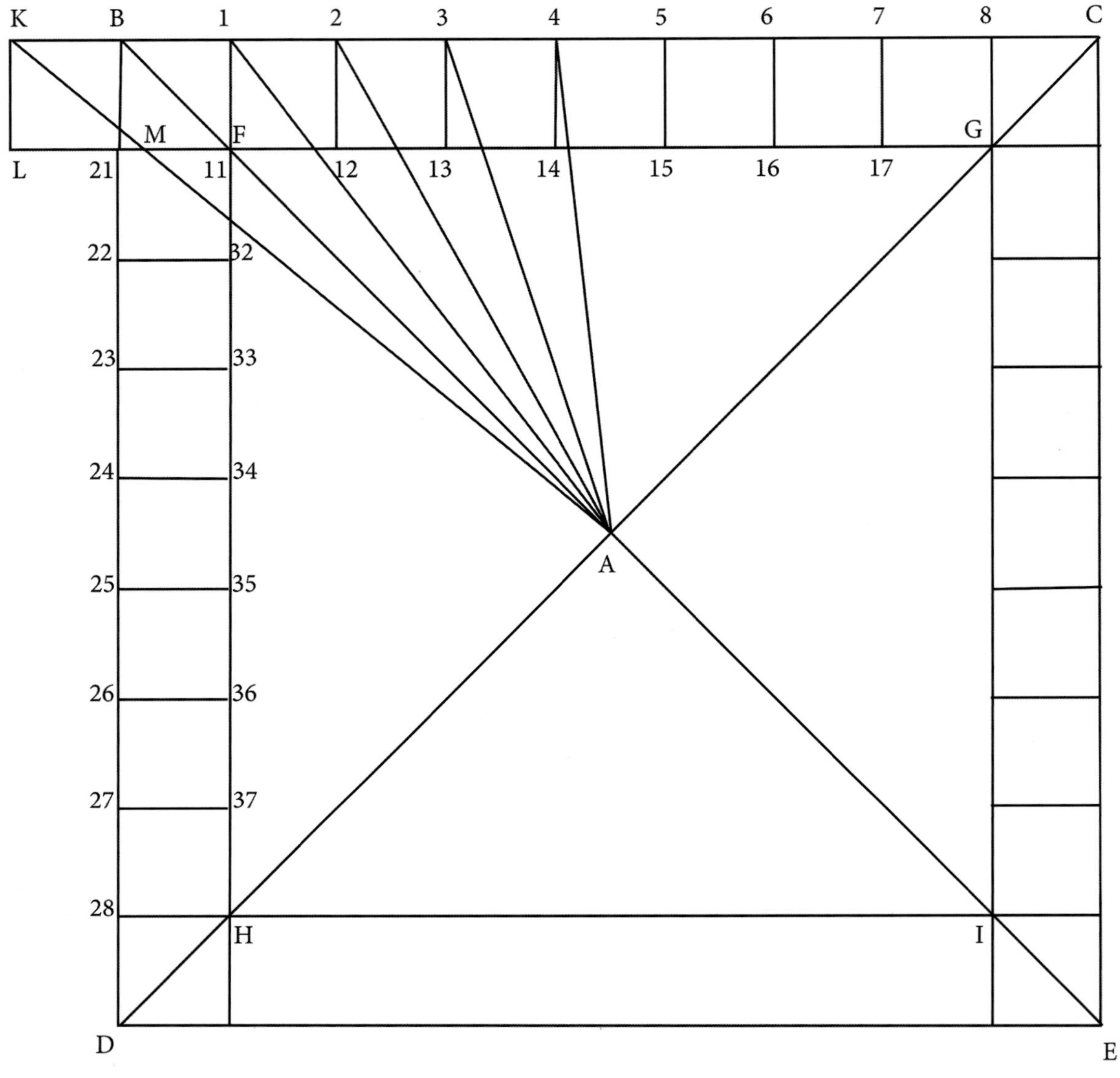

Figure B.34
Book 1, Proposition XXX.

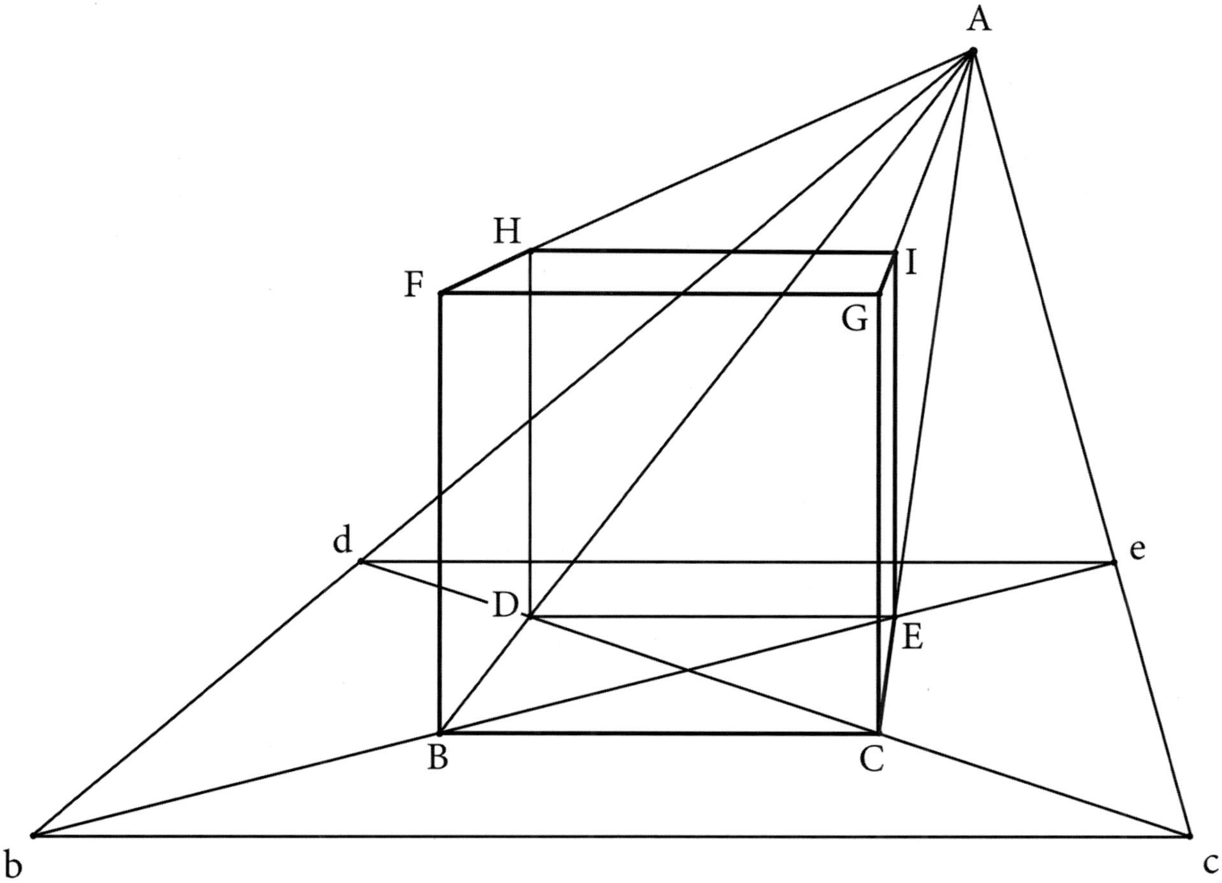

Figure B.35
Book 2, Proposition I.

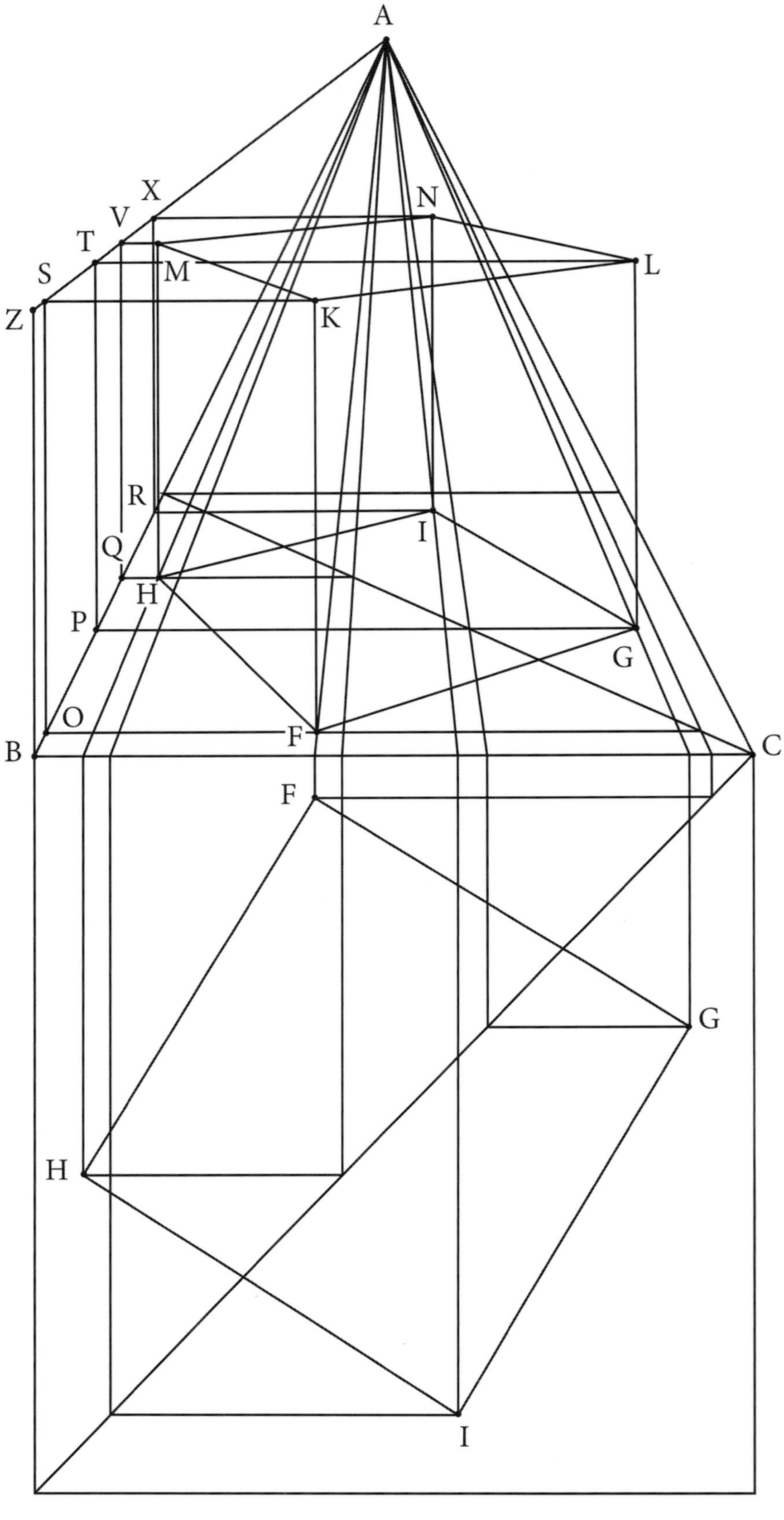

Figure B.36
Book 2, Proposition I.

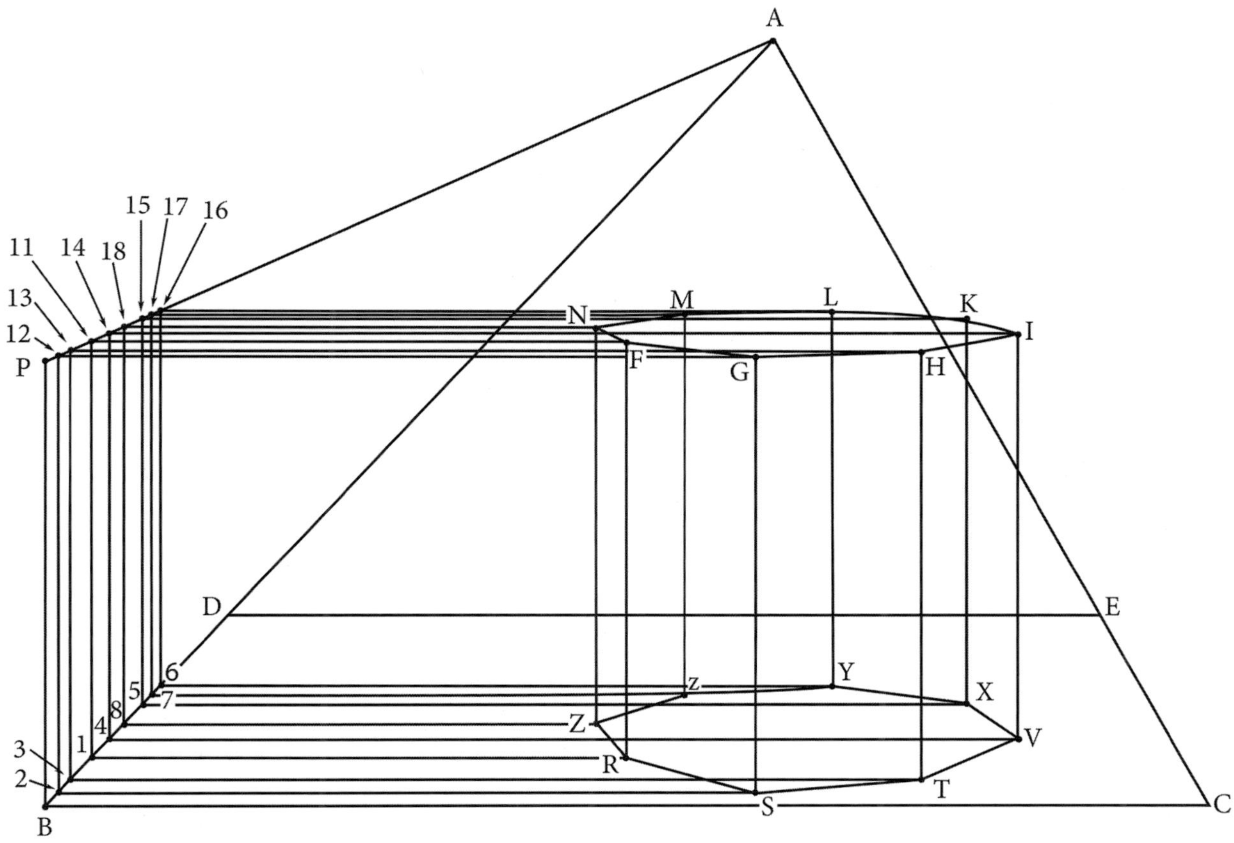

Figure B.37
Book 2, Proposition II.

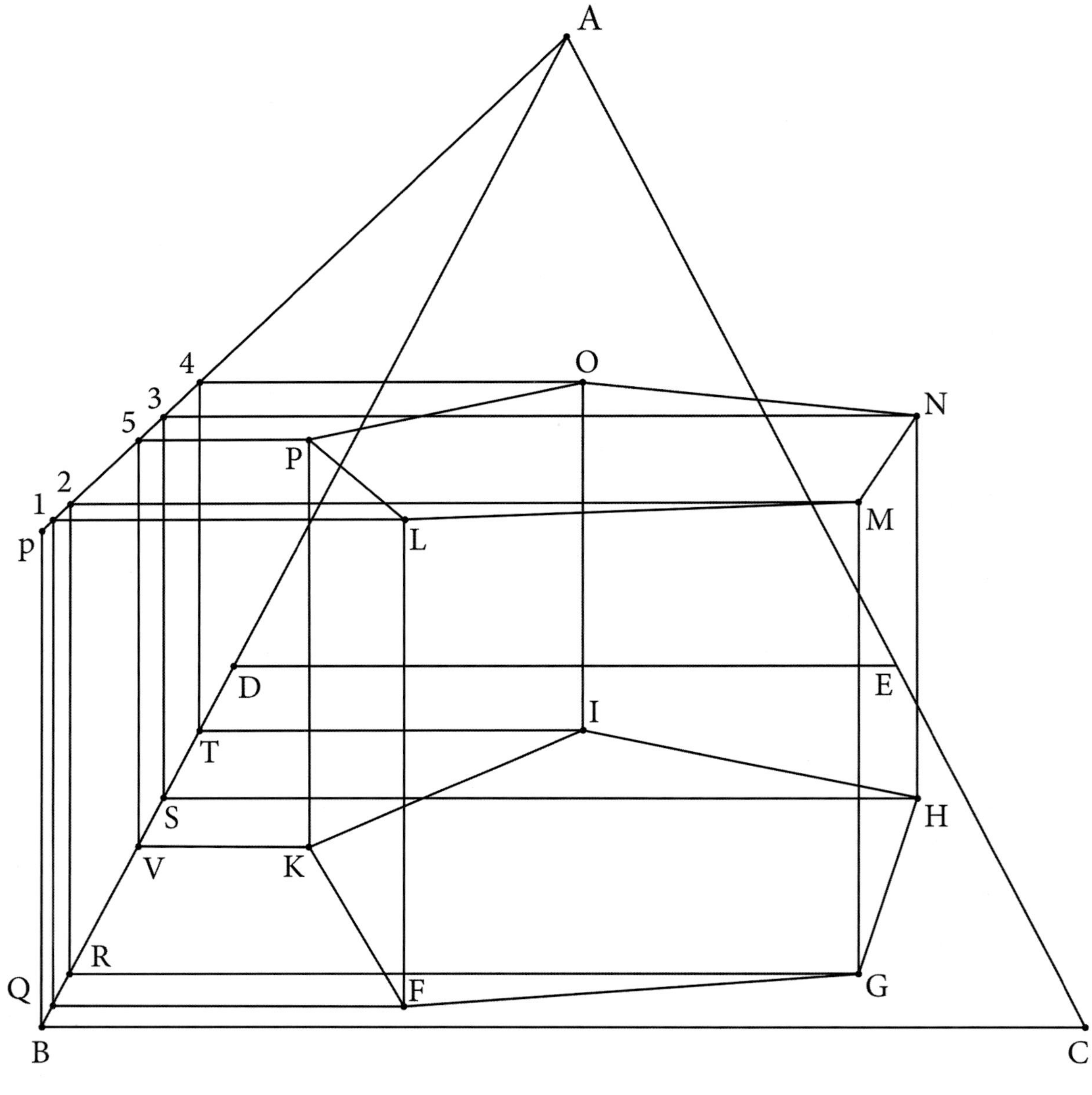

Figure B.38
Book 2, Proposition III.

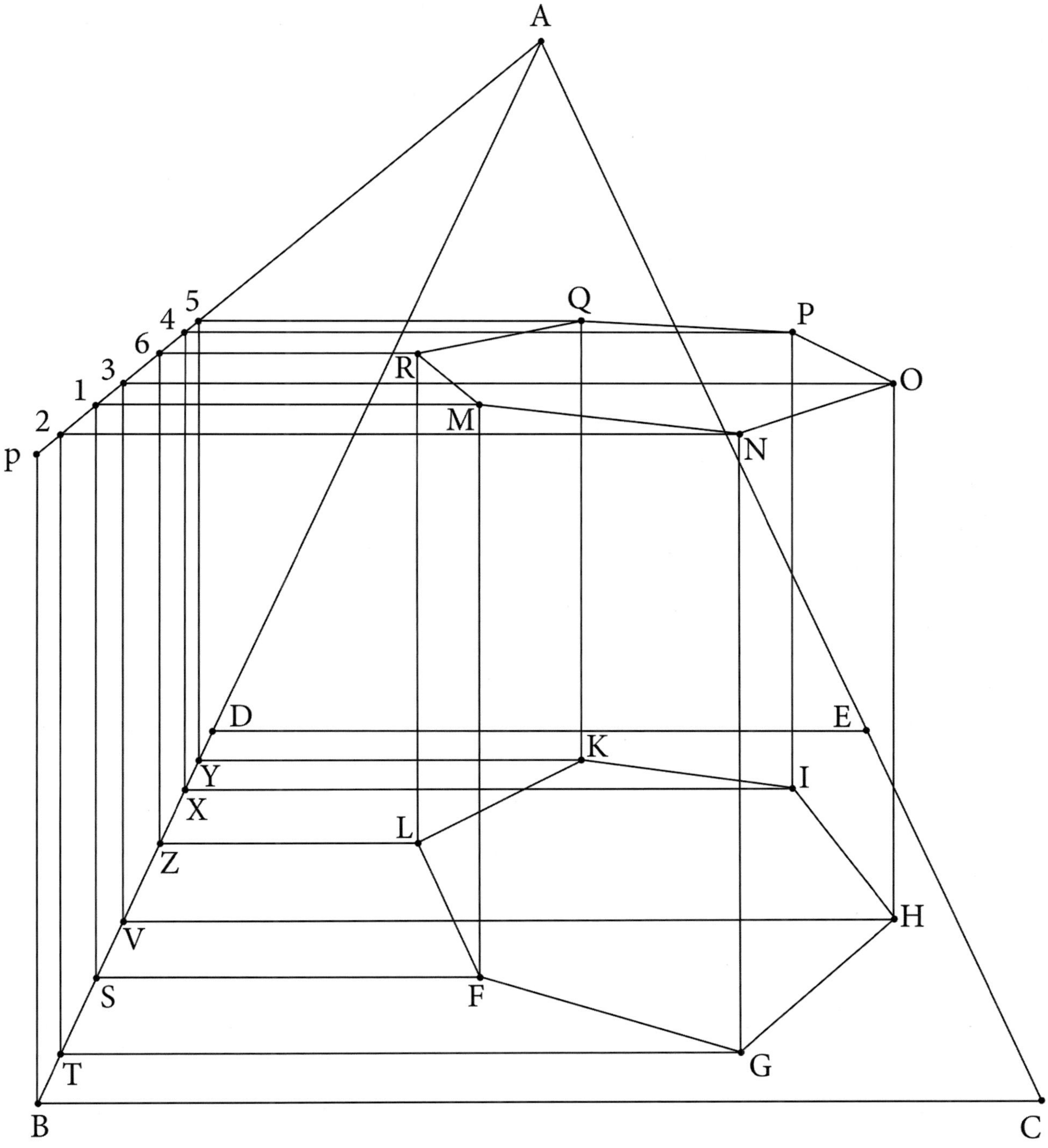

Figure B.39
Book 2, Proposition IV.

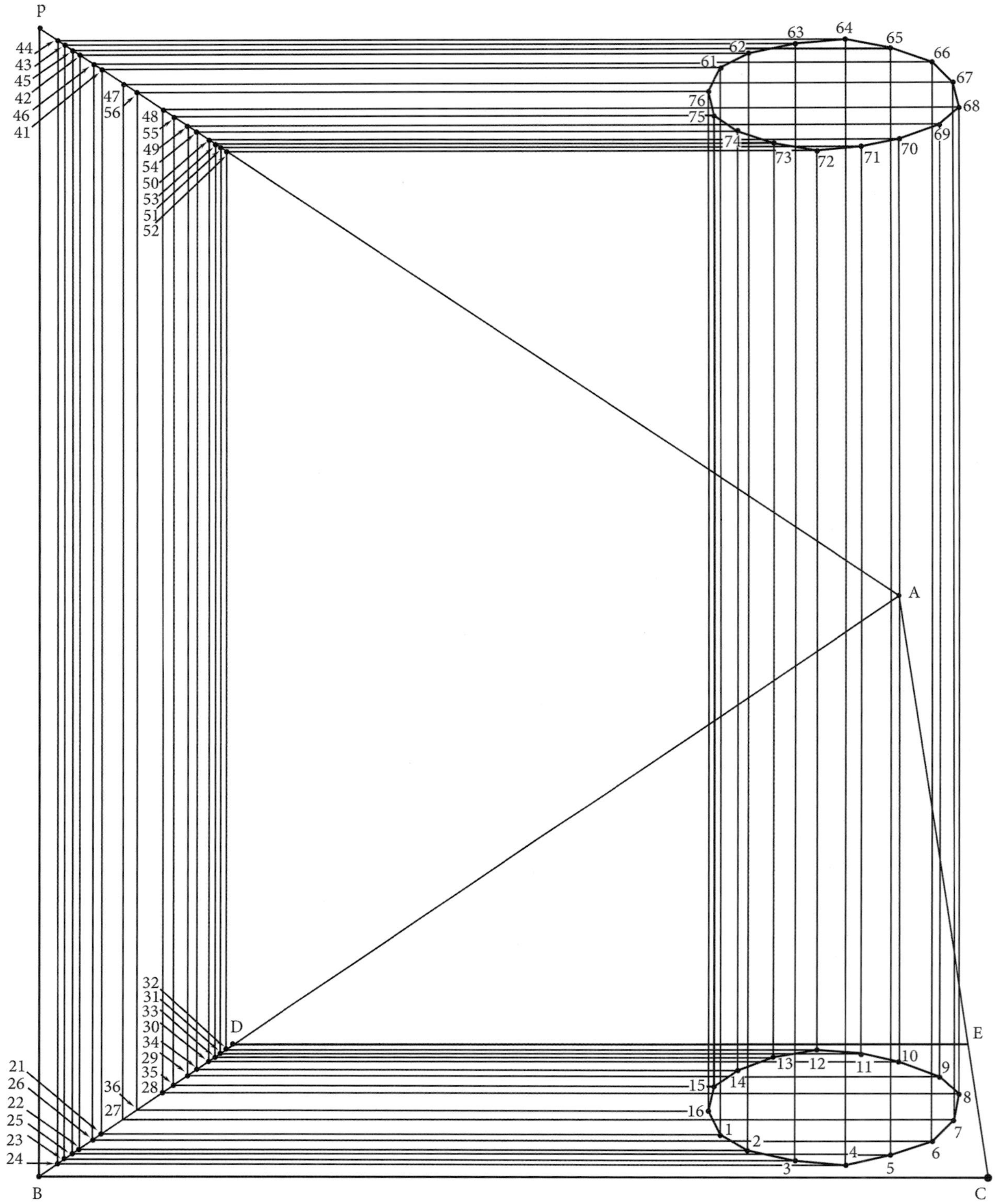

Figure B.40
Book 2, Proposition V.

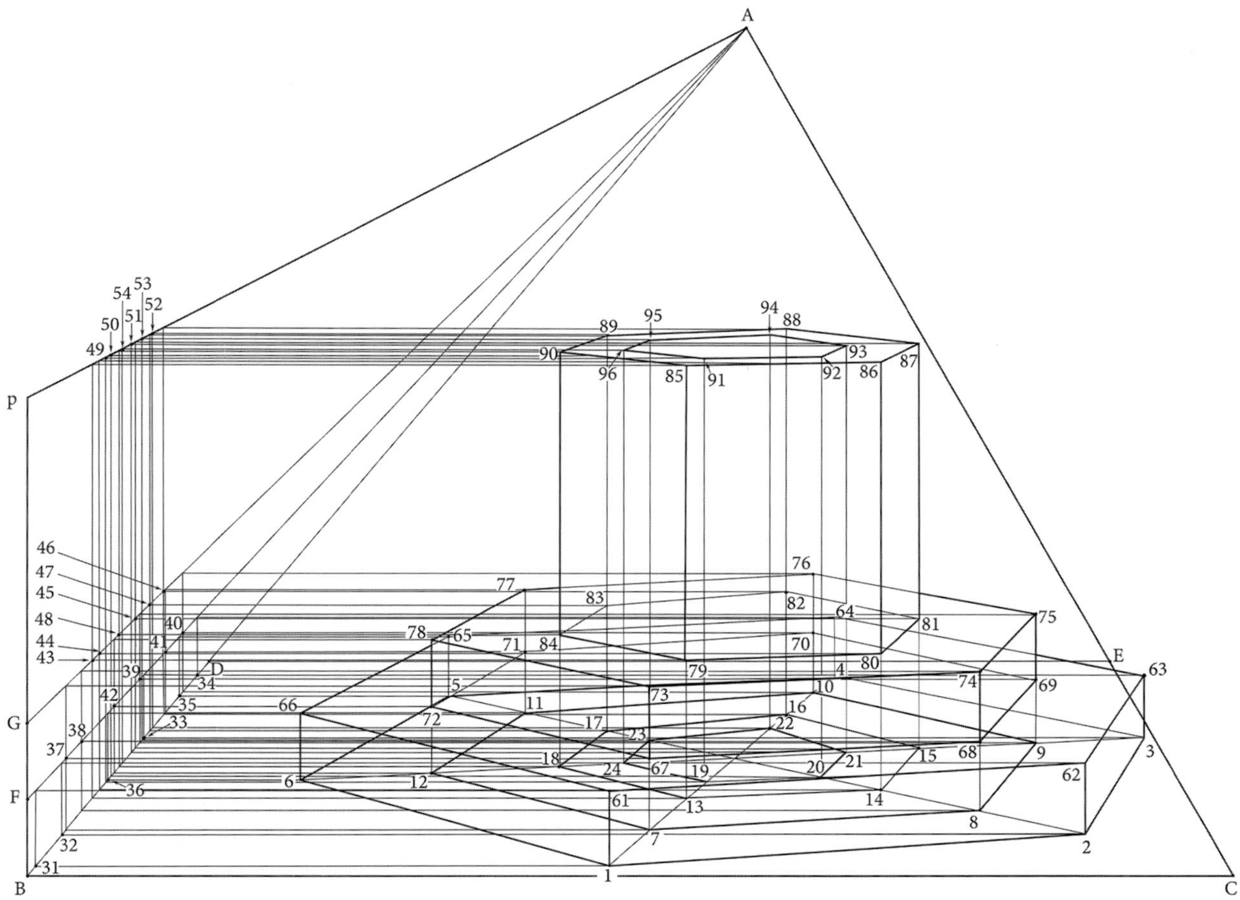

Figure B.41
Book 2, Proposition VI.

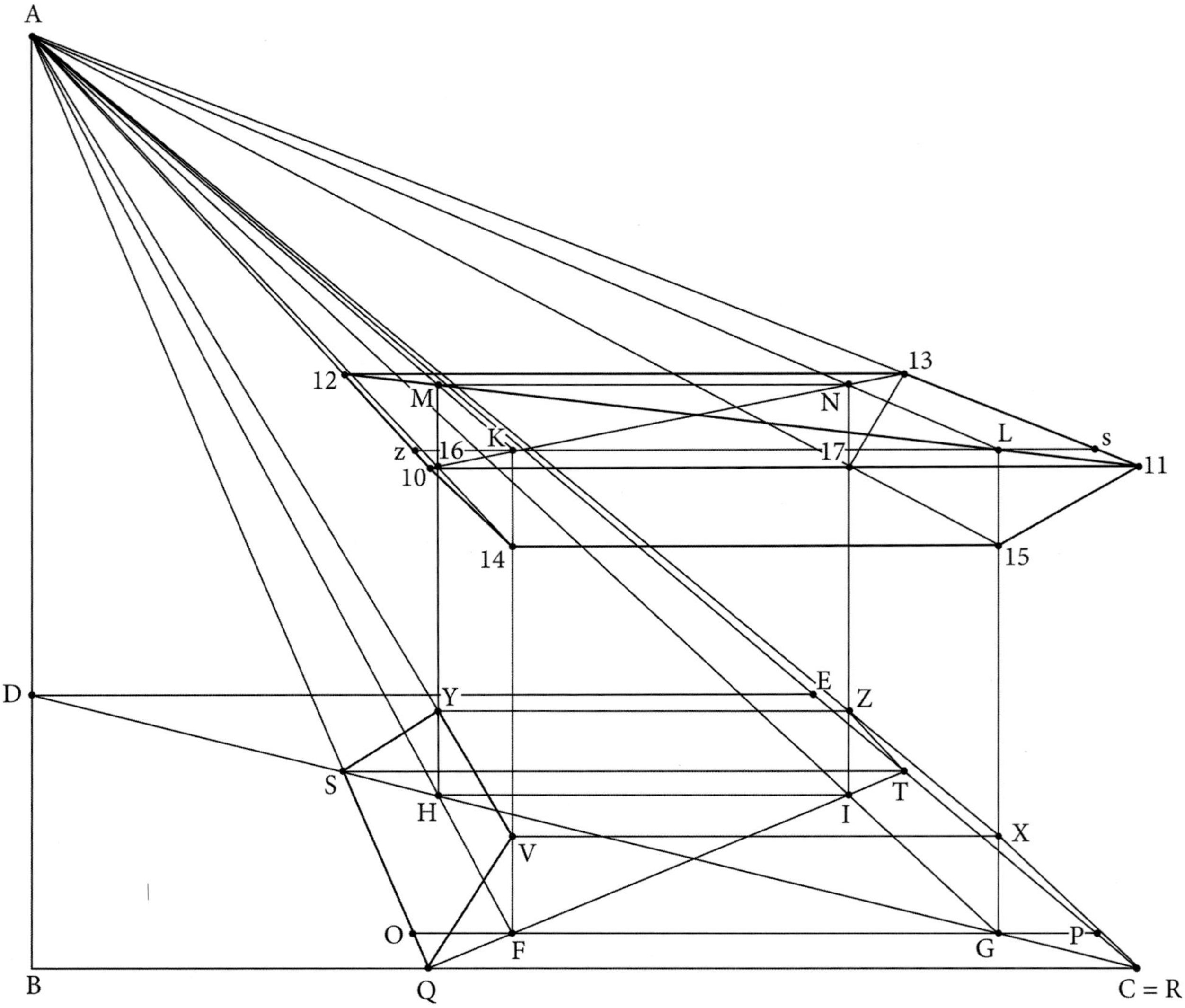

Figure B.42
Book 2, Proposition VII.

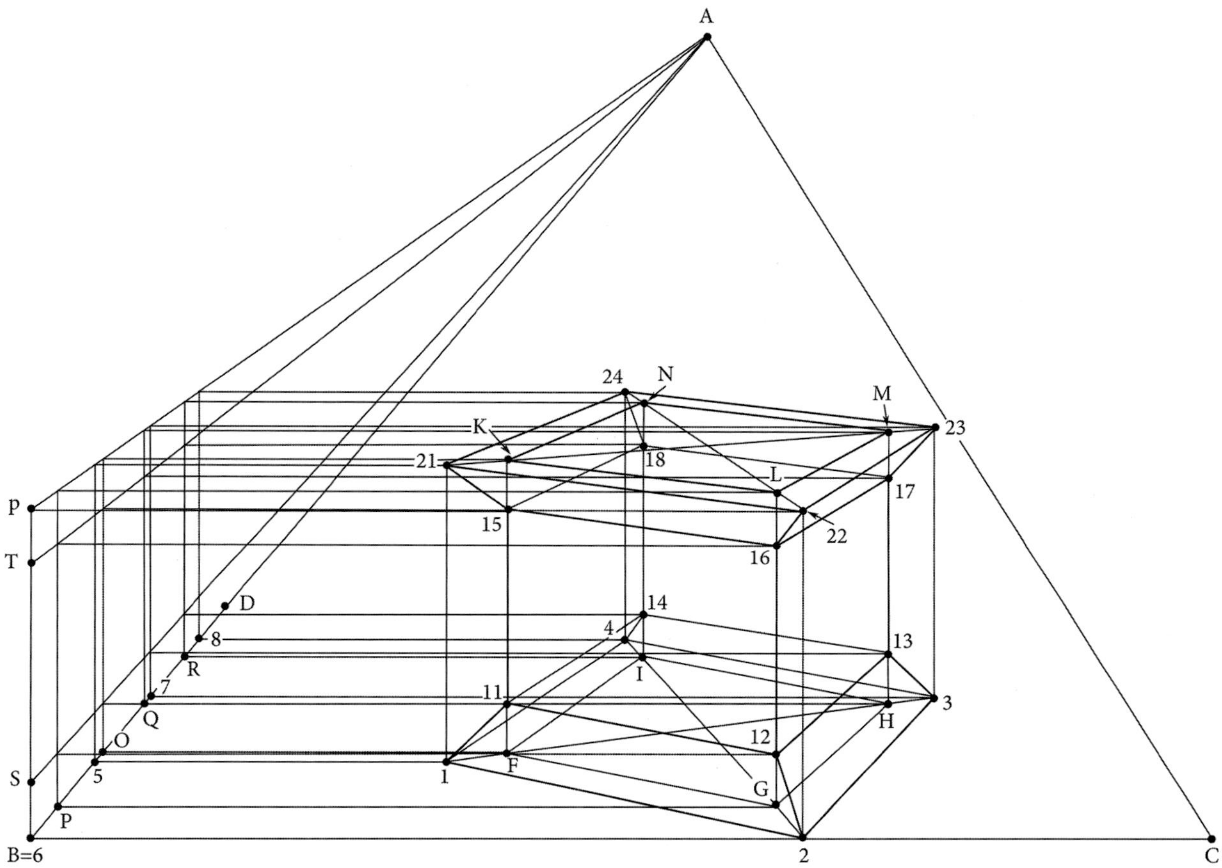

Figure B.43
Book 2, Proposition VII.

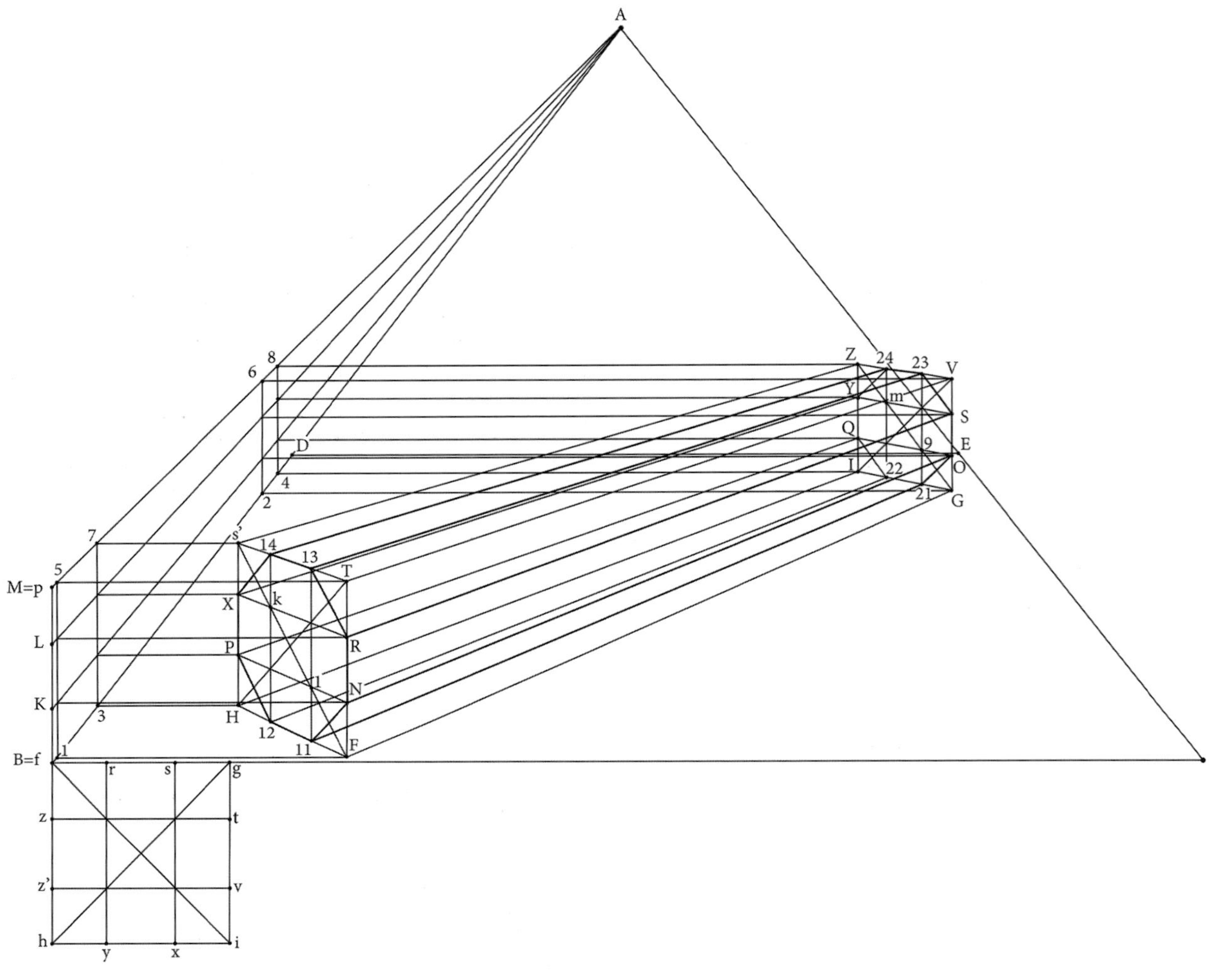

Figure B.44
Book 2, Proposition VIII.

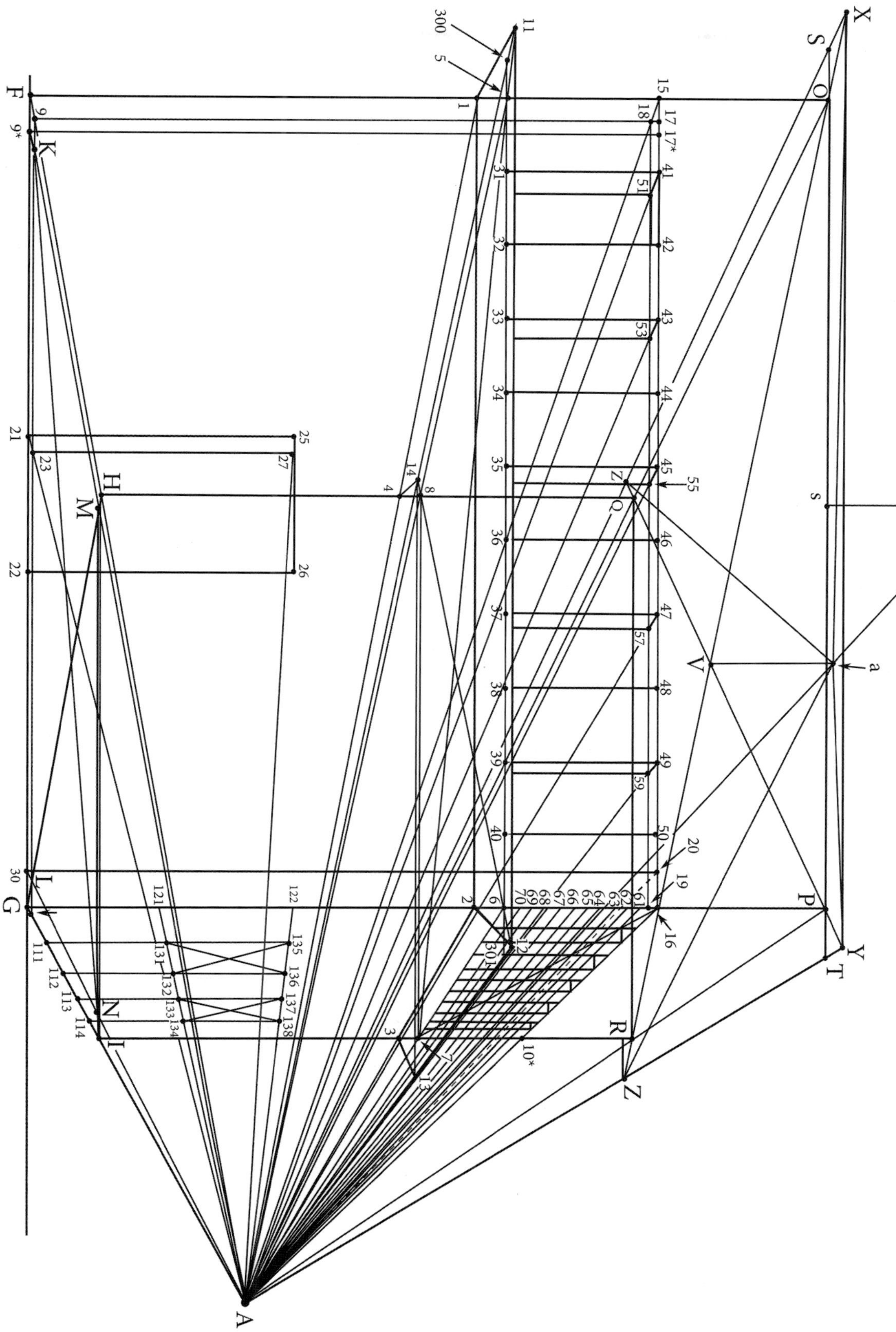

Figure B.45
Book 2, Proposition IX.

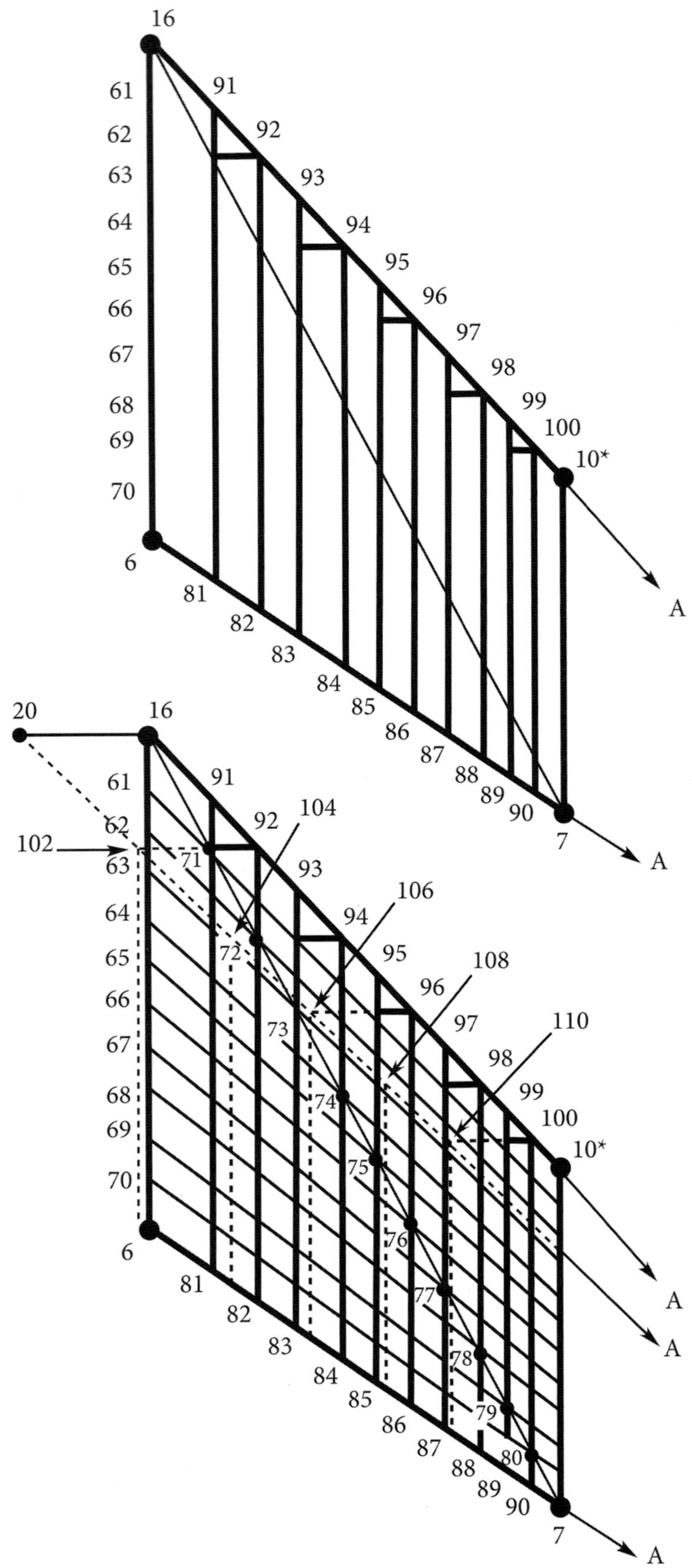

Figure B.46
Book 2, Proposition IX.

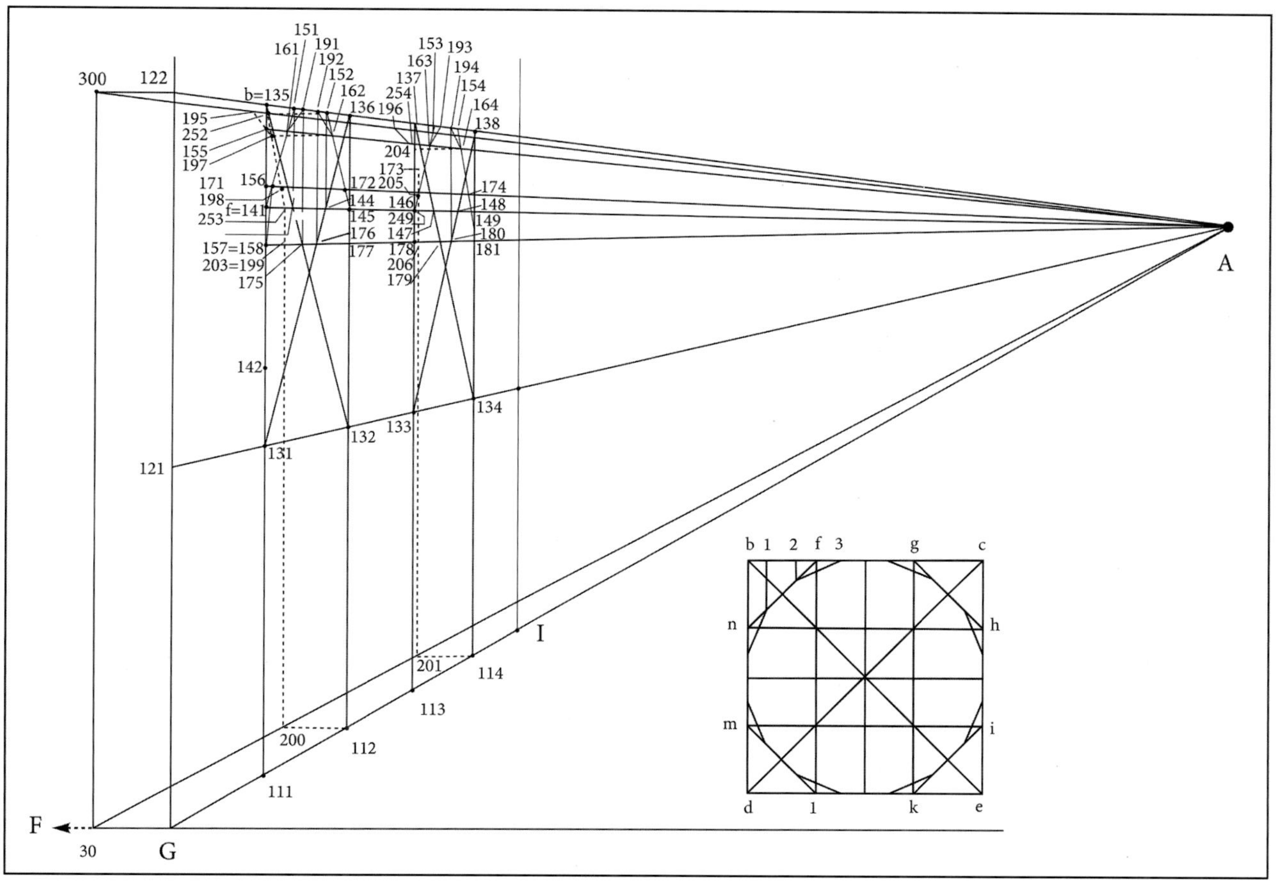

Figure B.47
Book 2, Proposition IX.

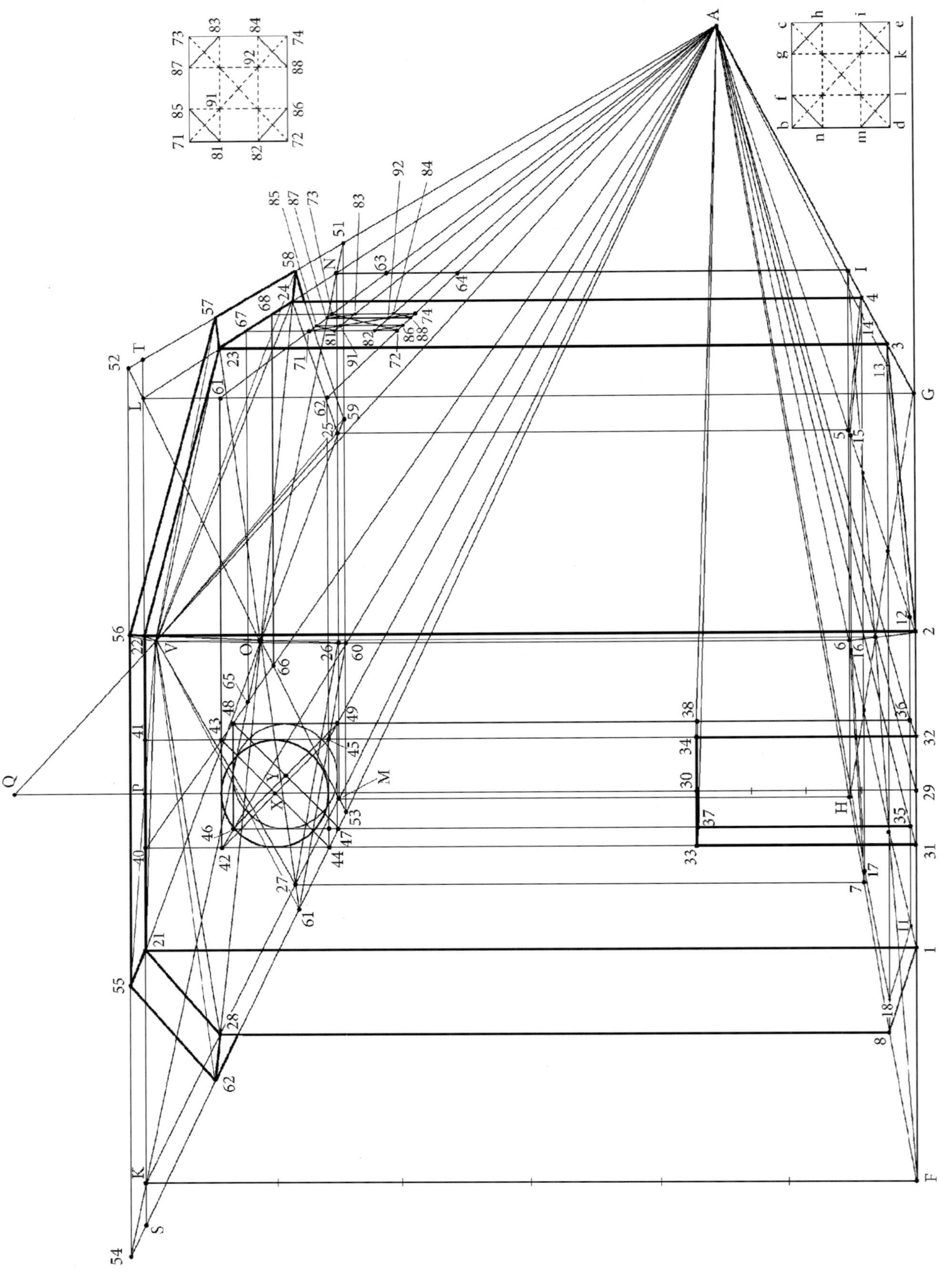

Figure B.48
Book 2, Proposition X.

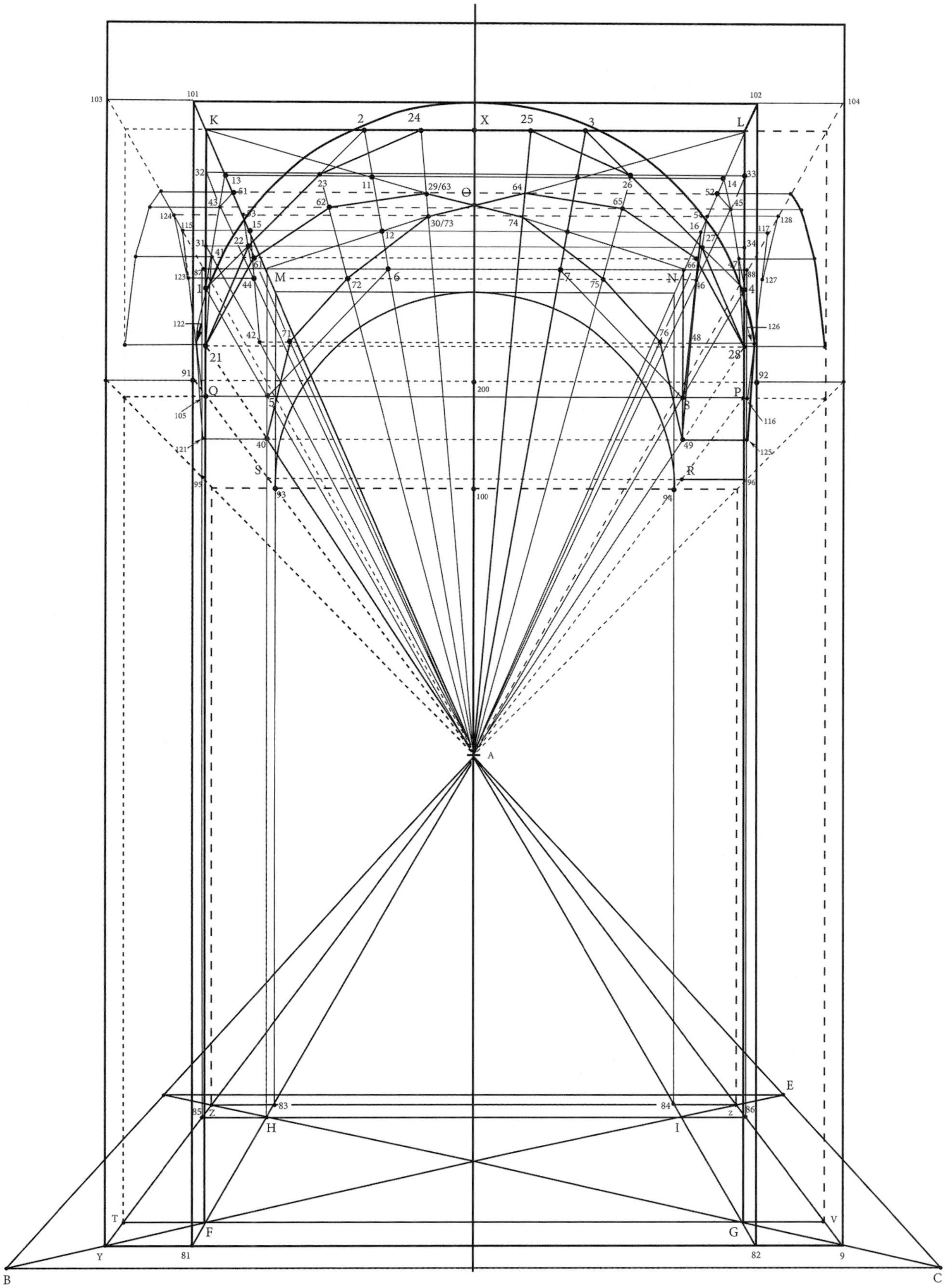

Figure B.49
Book 2, Proposition XI.

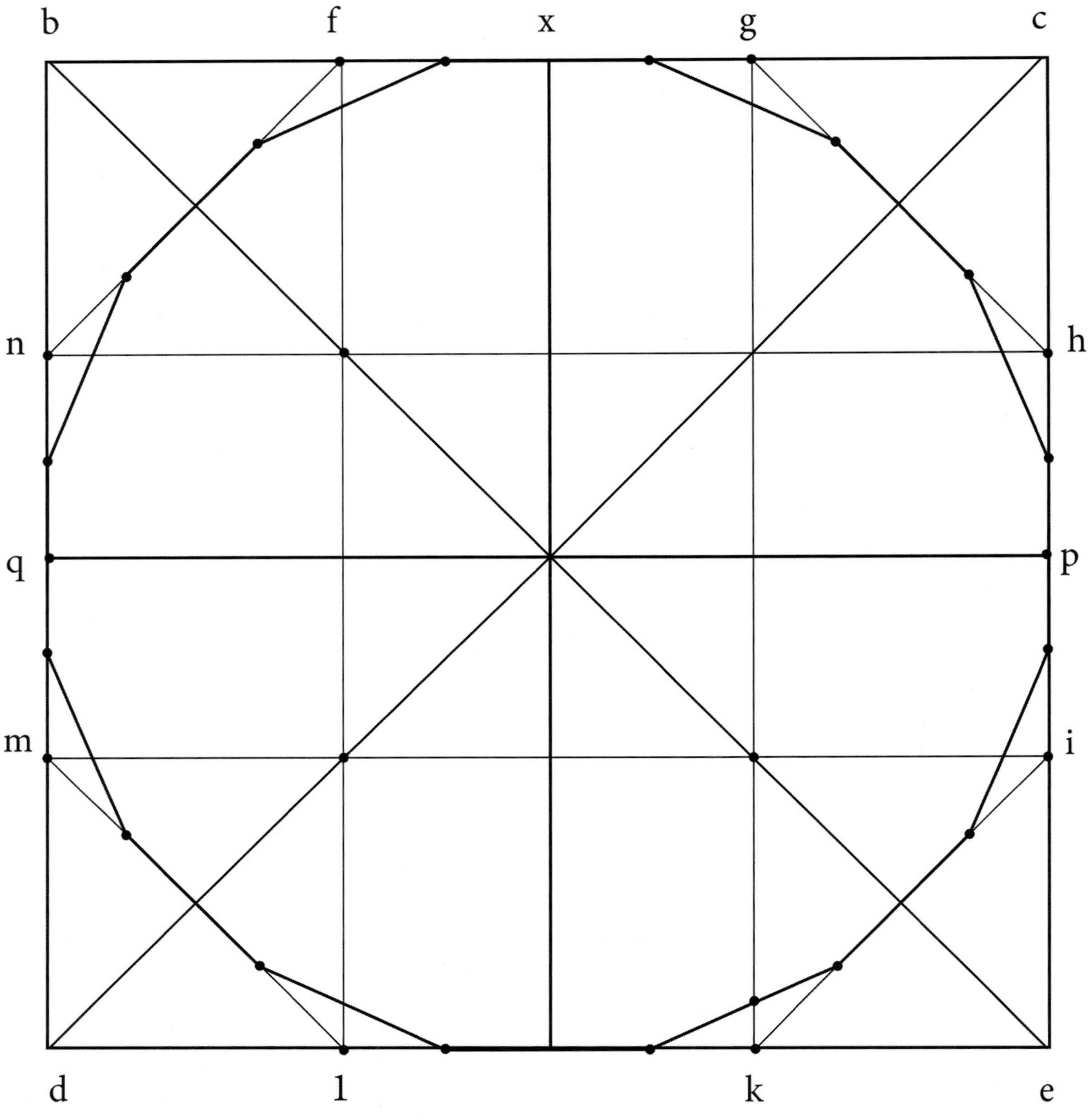

Figure B.50
Book 2, Proposition XI.

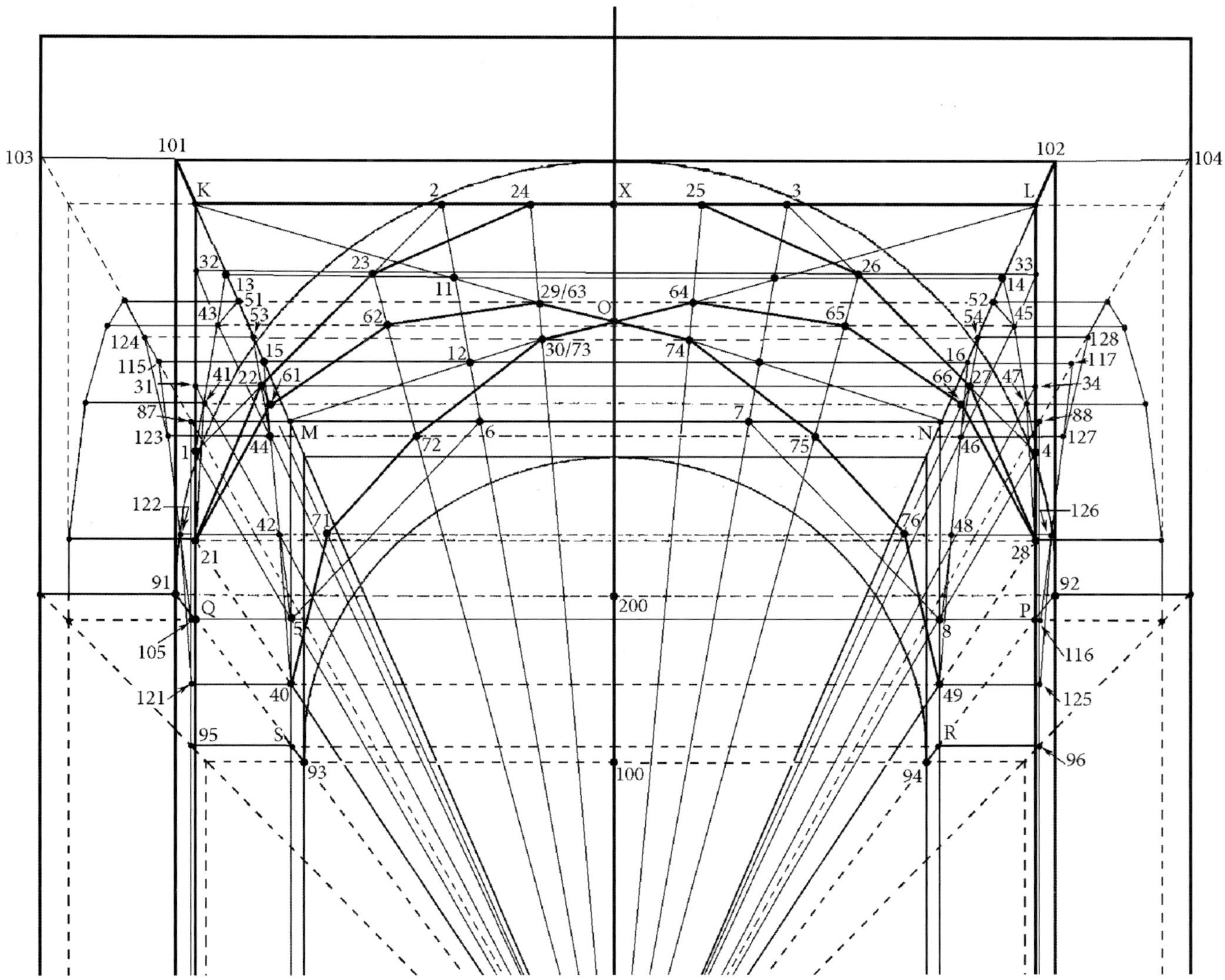

Figure B.51
Book 2, Proposition XI.

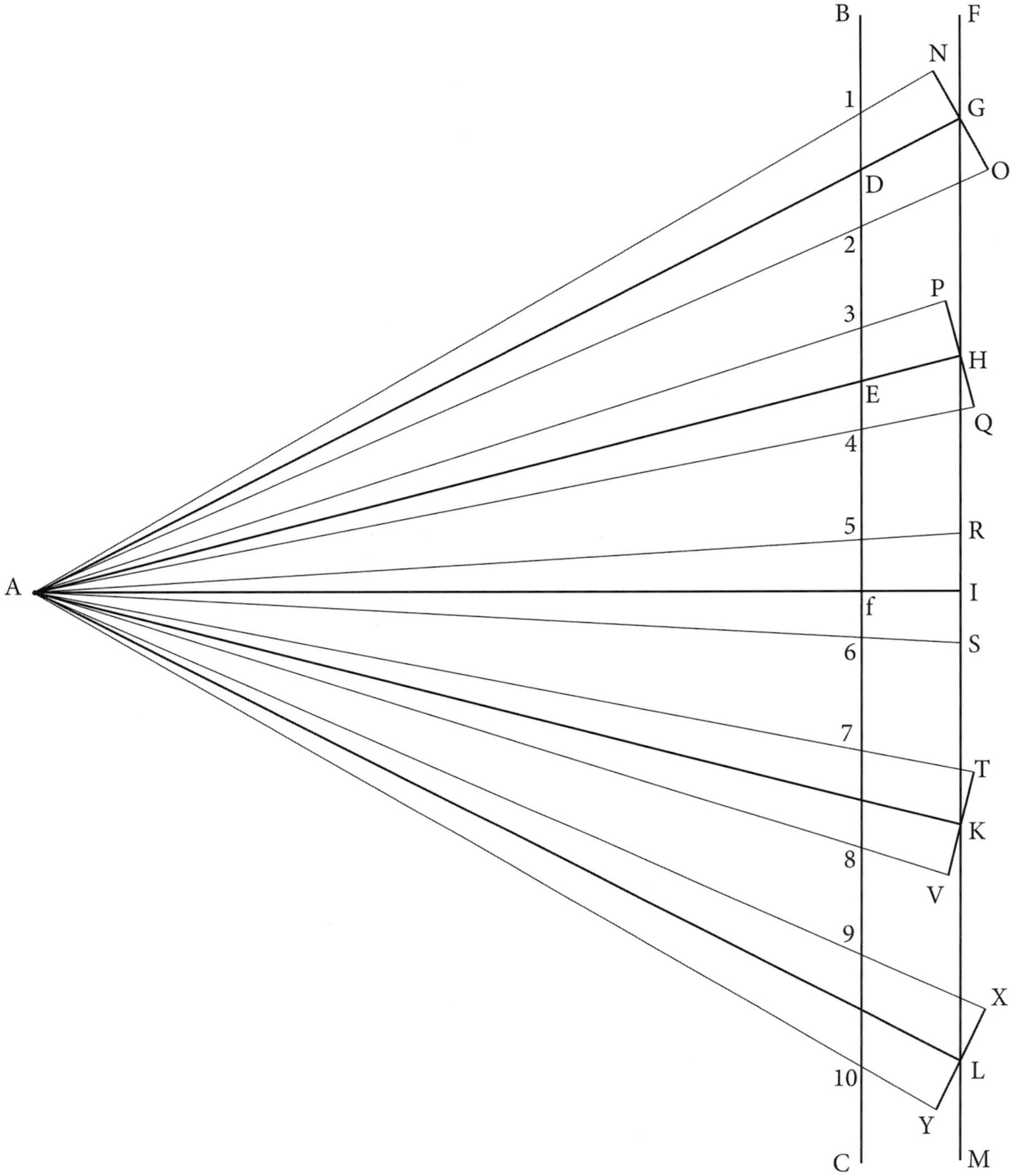

Figure B.52
Book 2, Proposition XII.

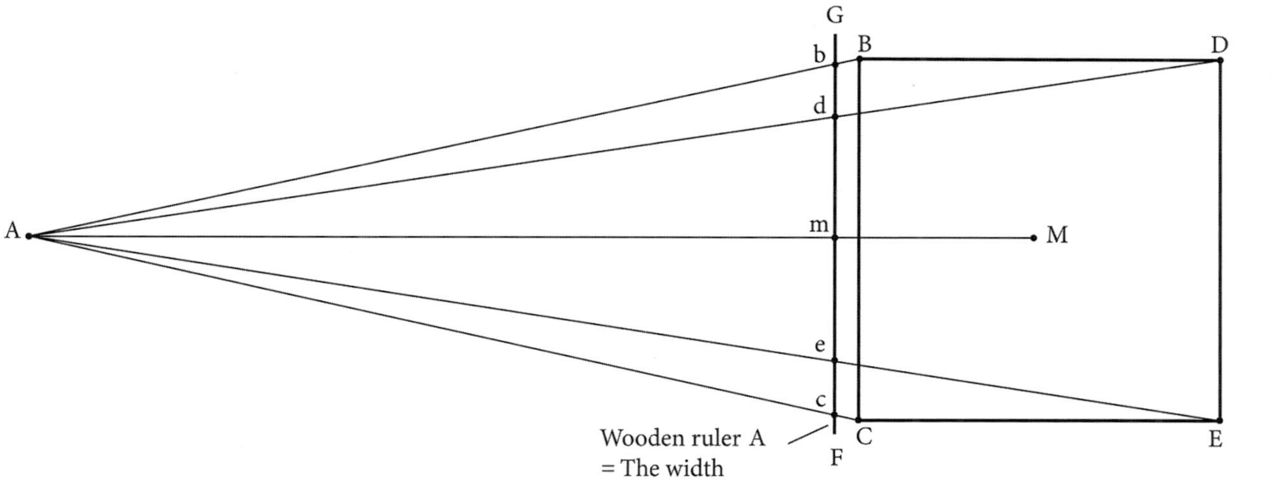

Figure B.53
Book 3, Proposition I.

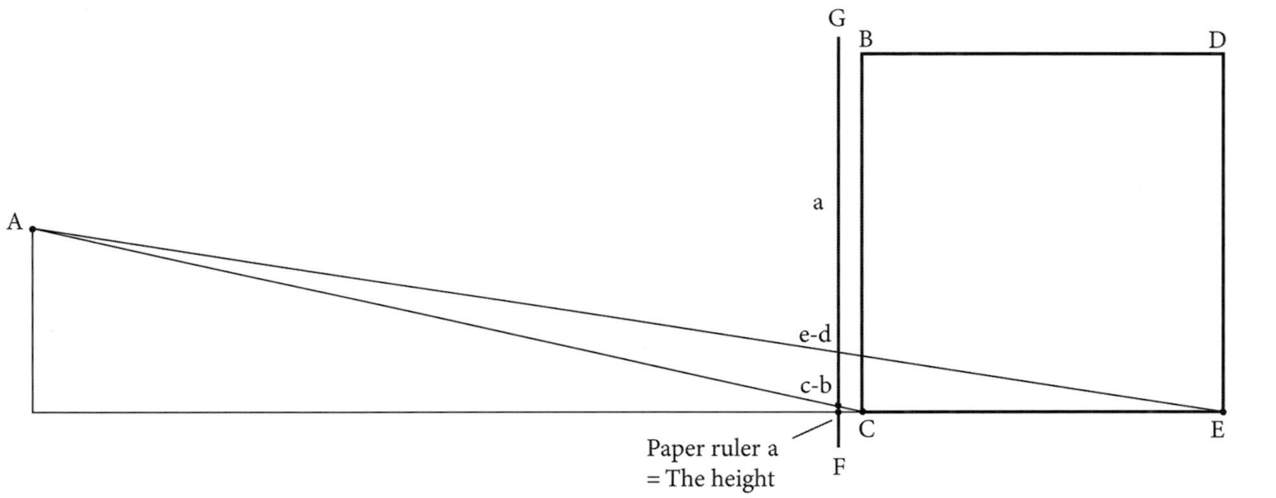

Figure B.54
Book 3, Proposition I.

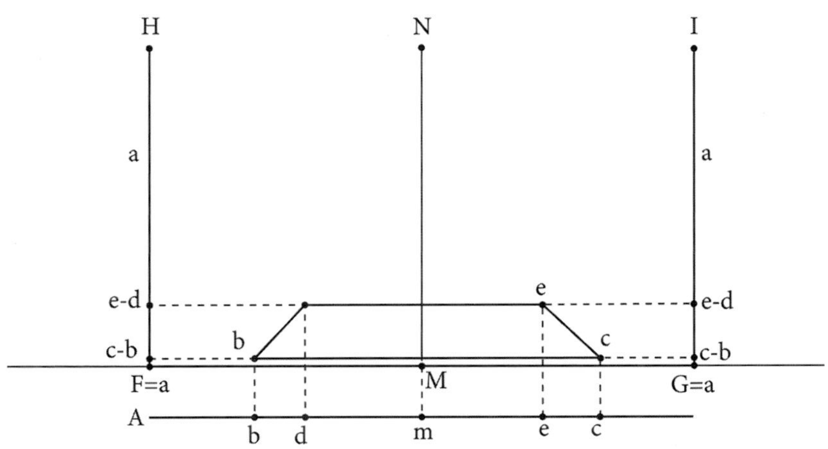

Figure B.55
Book 3, Proposition I.

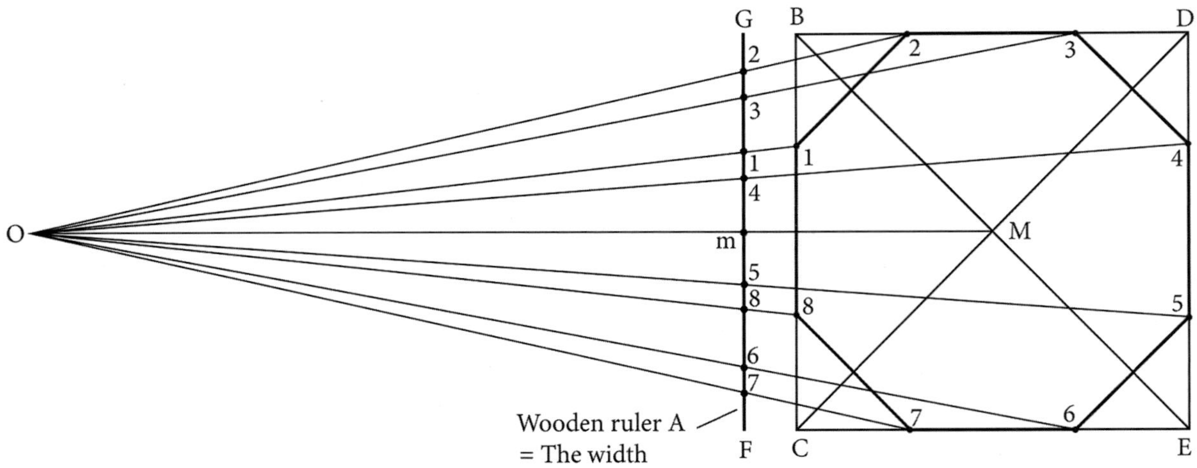

Figure B.56
Book 3, Proposition II.

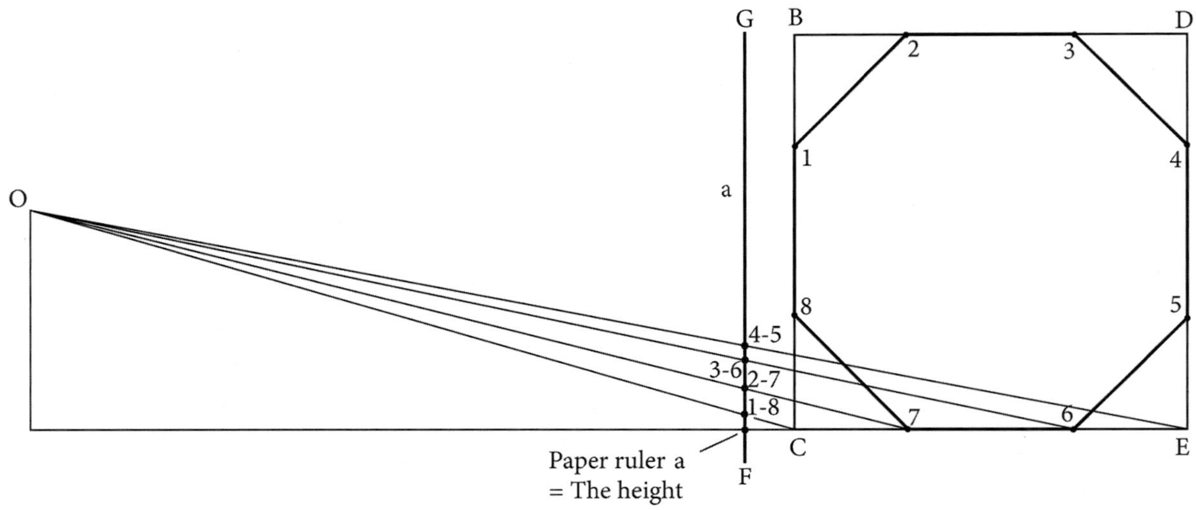

Figure B.57
Book 3, Proposition II.

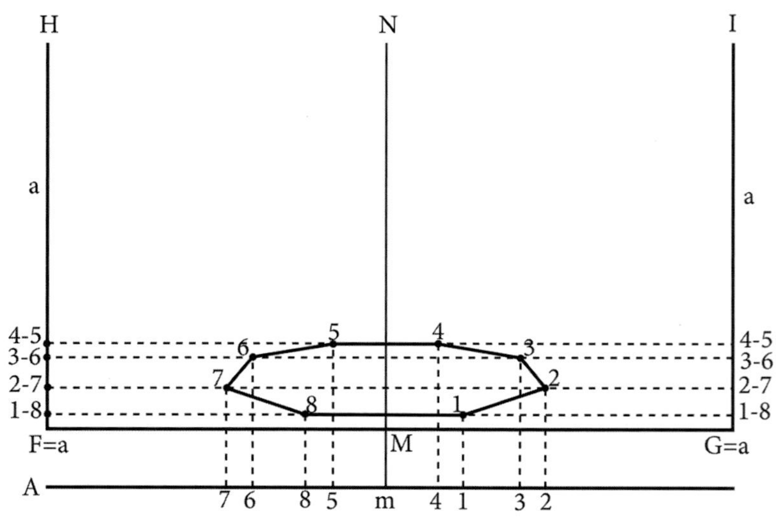

Figure B.58
Book 3, Proposition II.

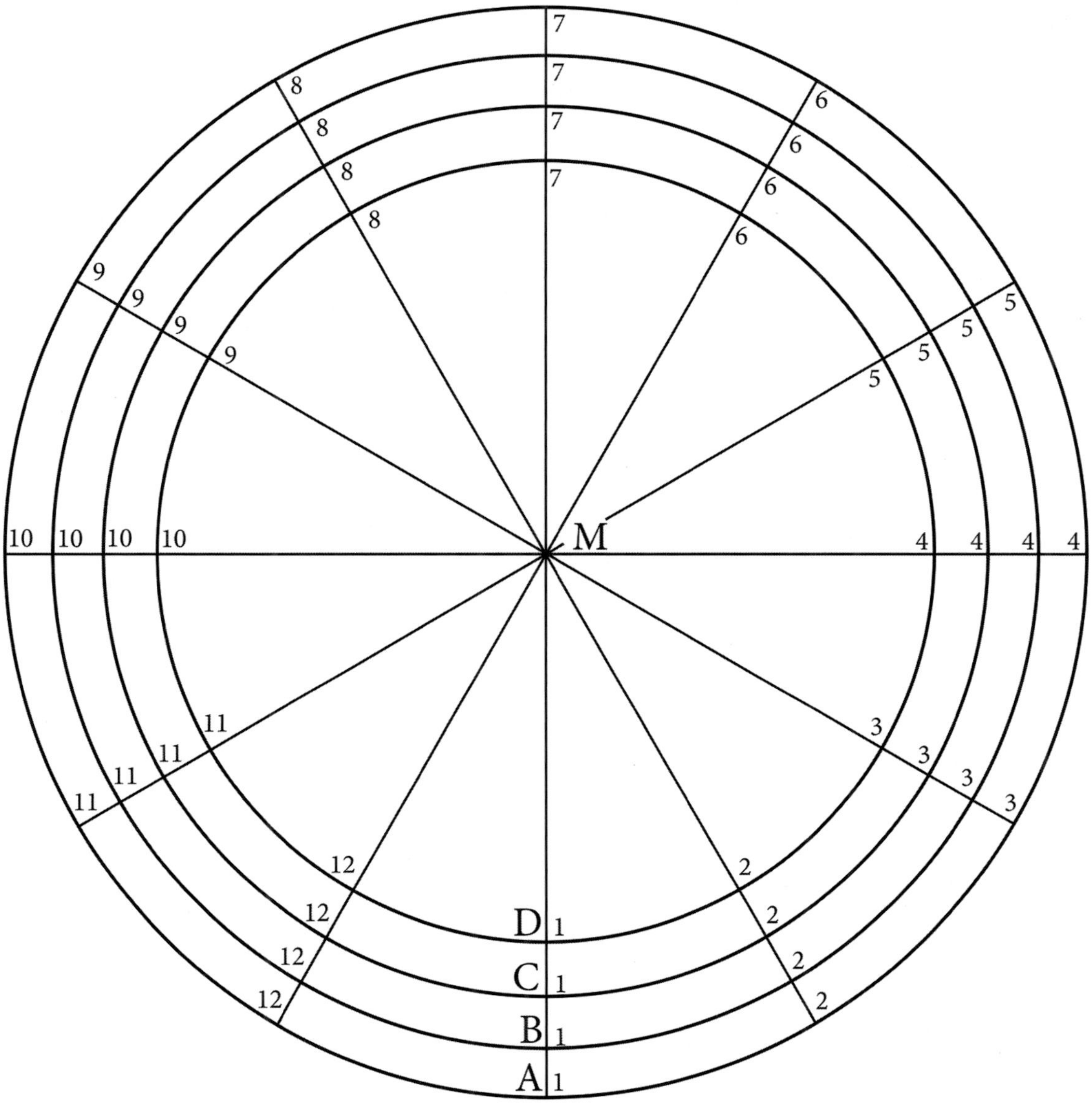

Figure B.59
Book 3, Proposition III.

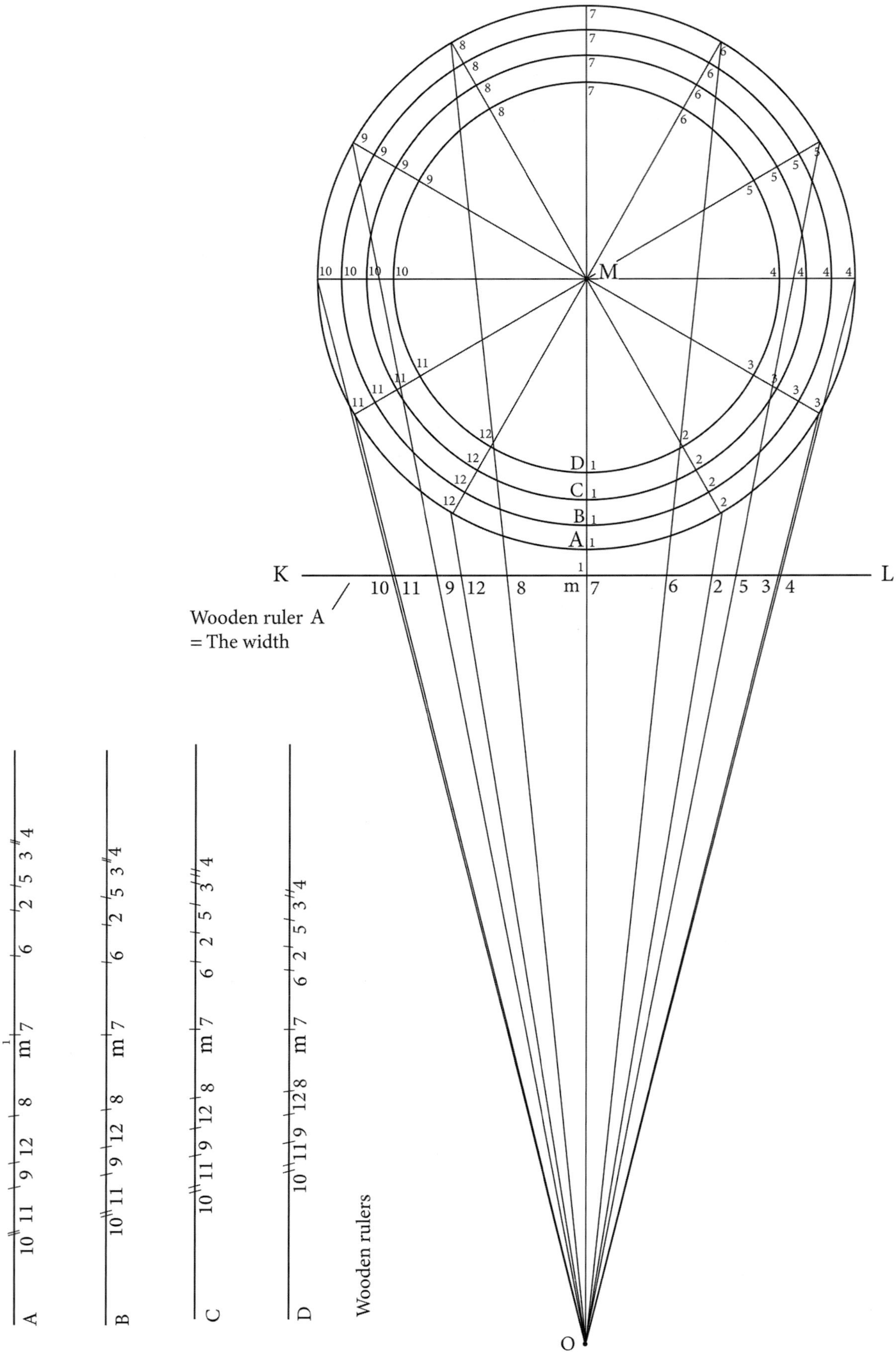

Figure B.60
Book 3, Proposition III.

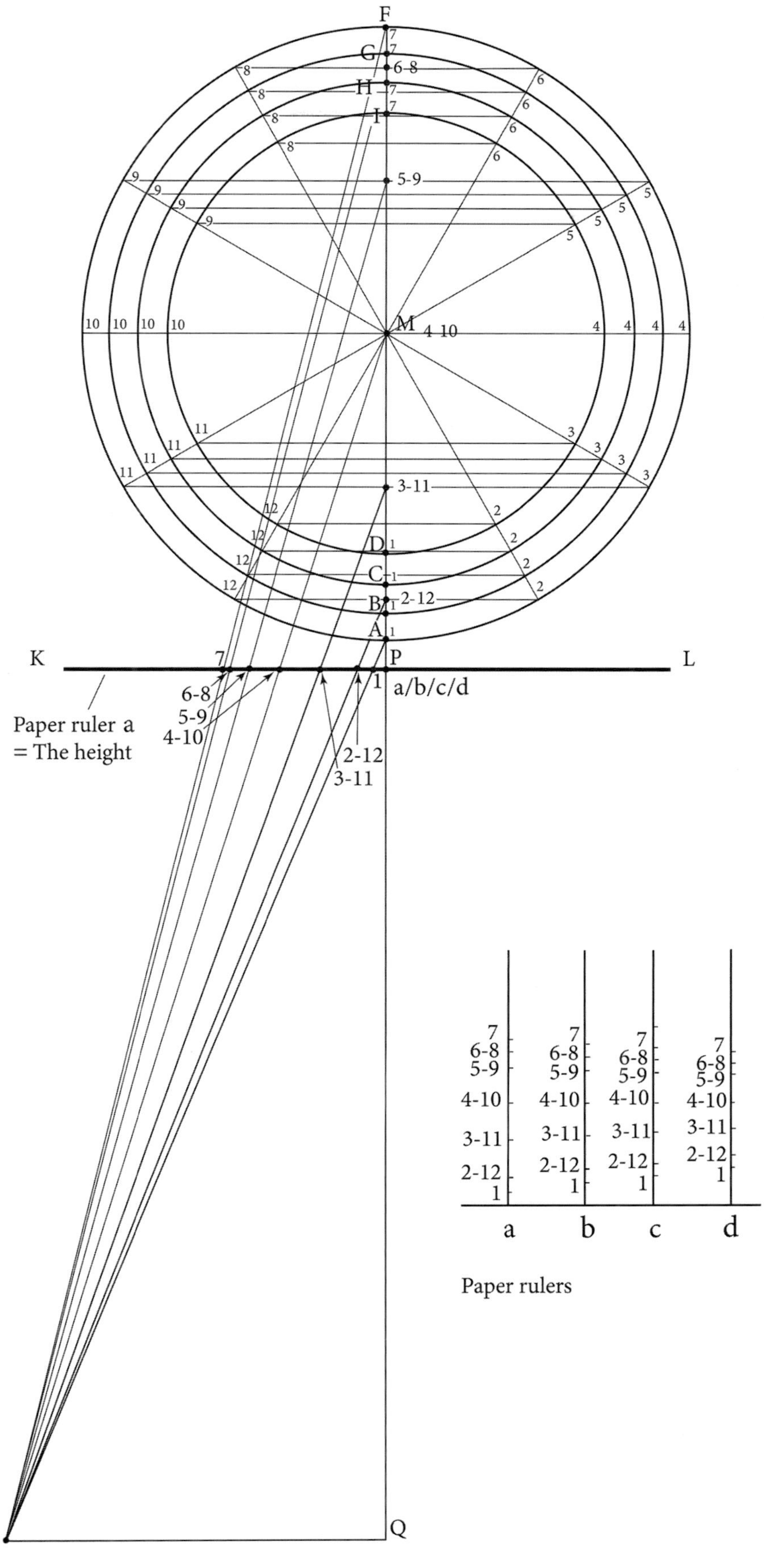

Figure B.61
Book 3, Proposition III.

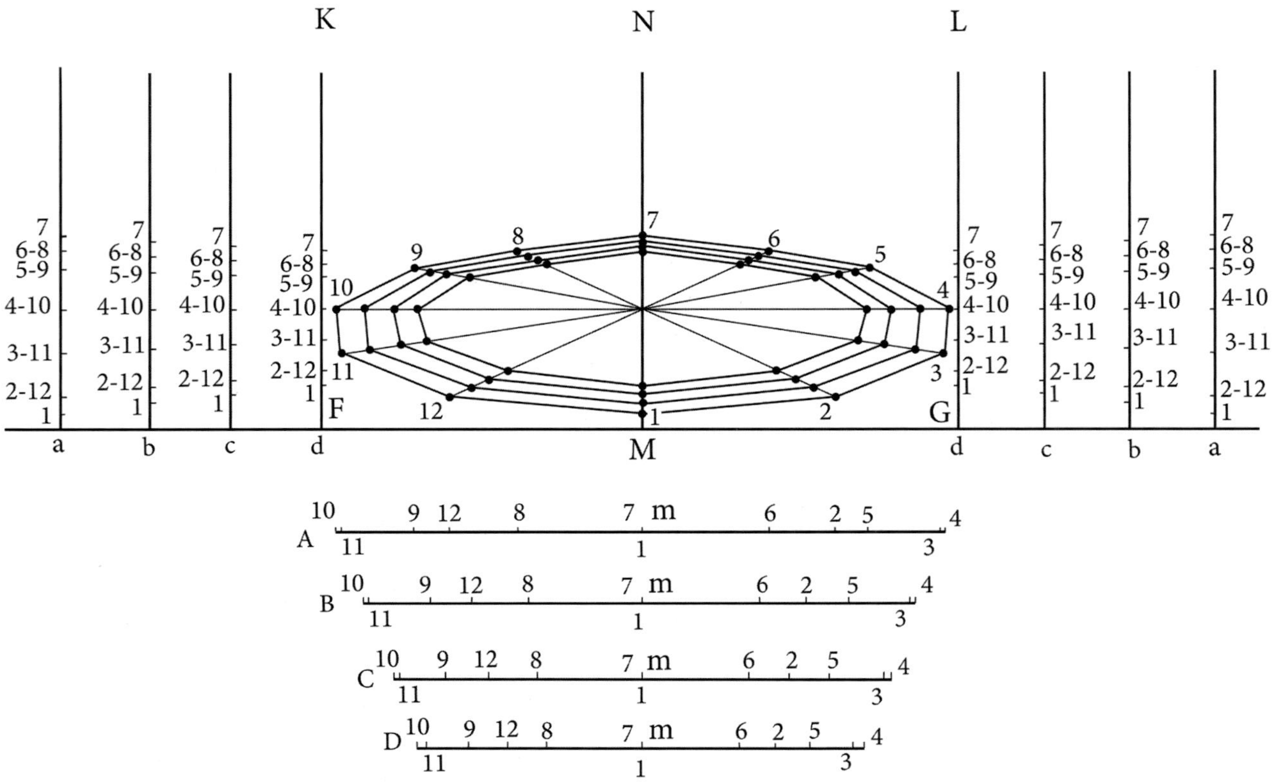

Figure B.62
Book 3, Proposition III.

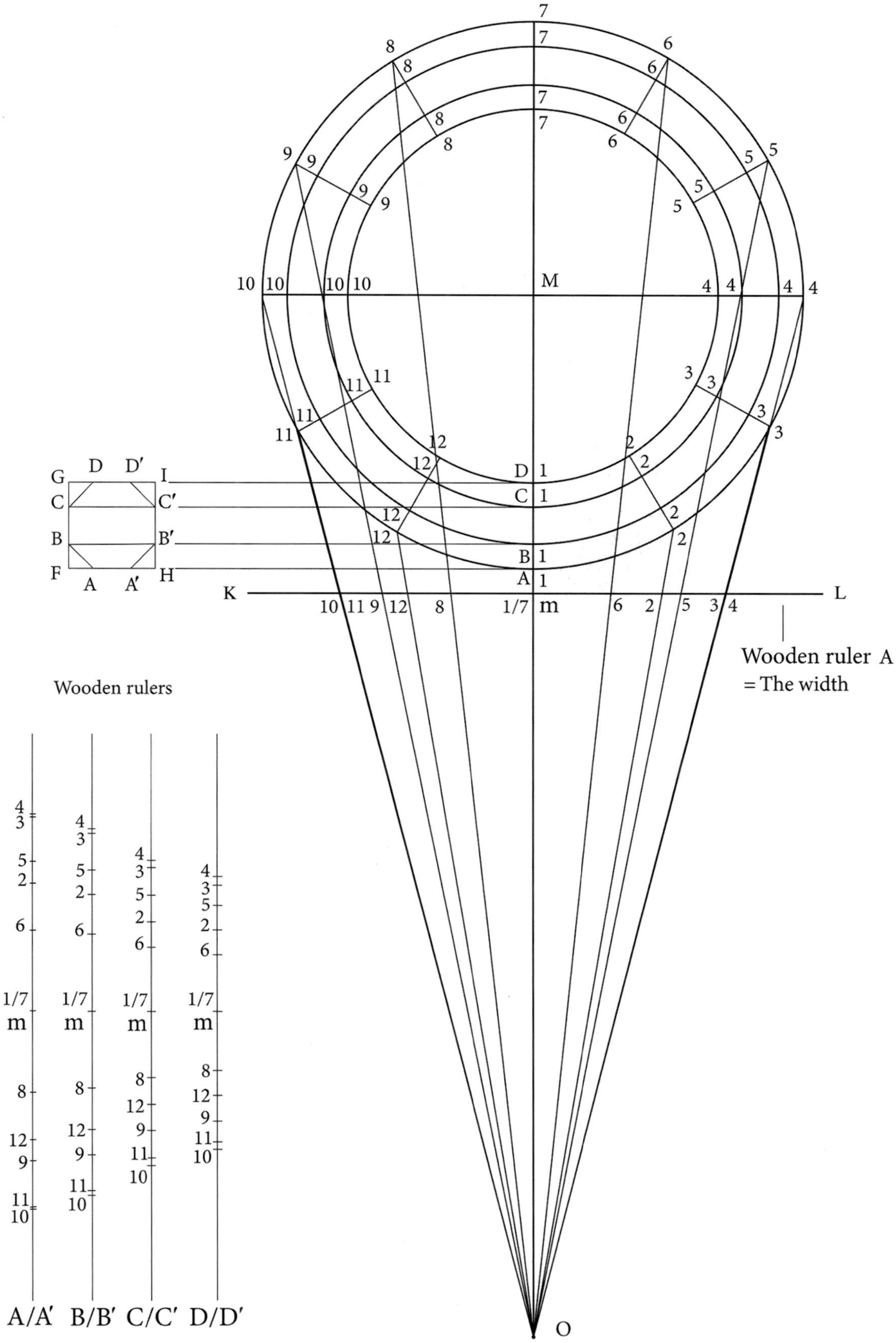

Figure B.63
Book 3, Proposition IV.

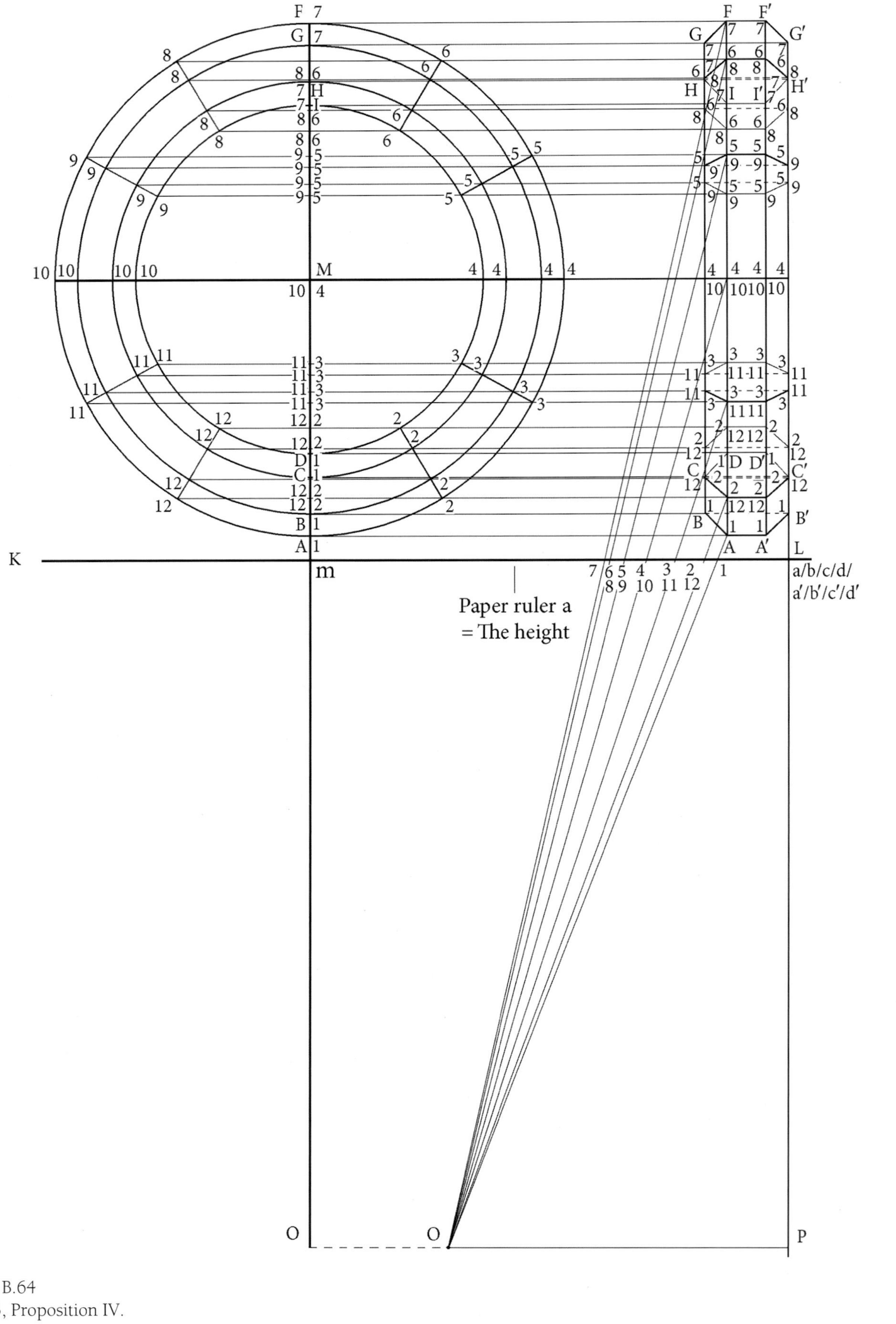

Figure B.64
Book 3, Proposition IV.

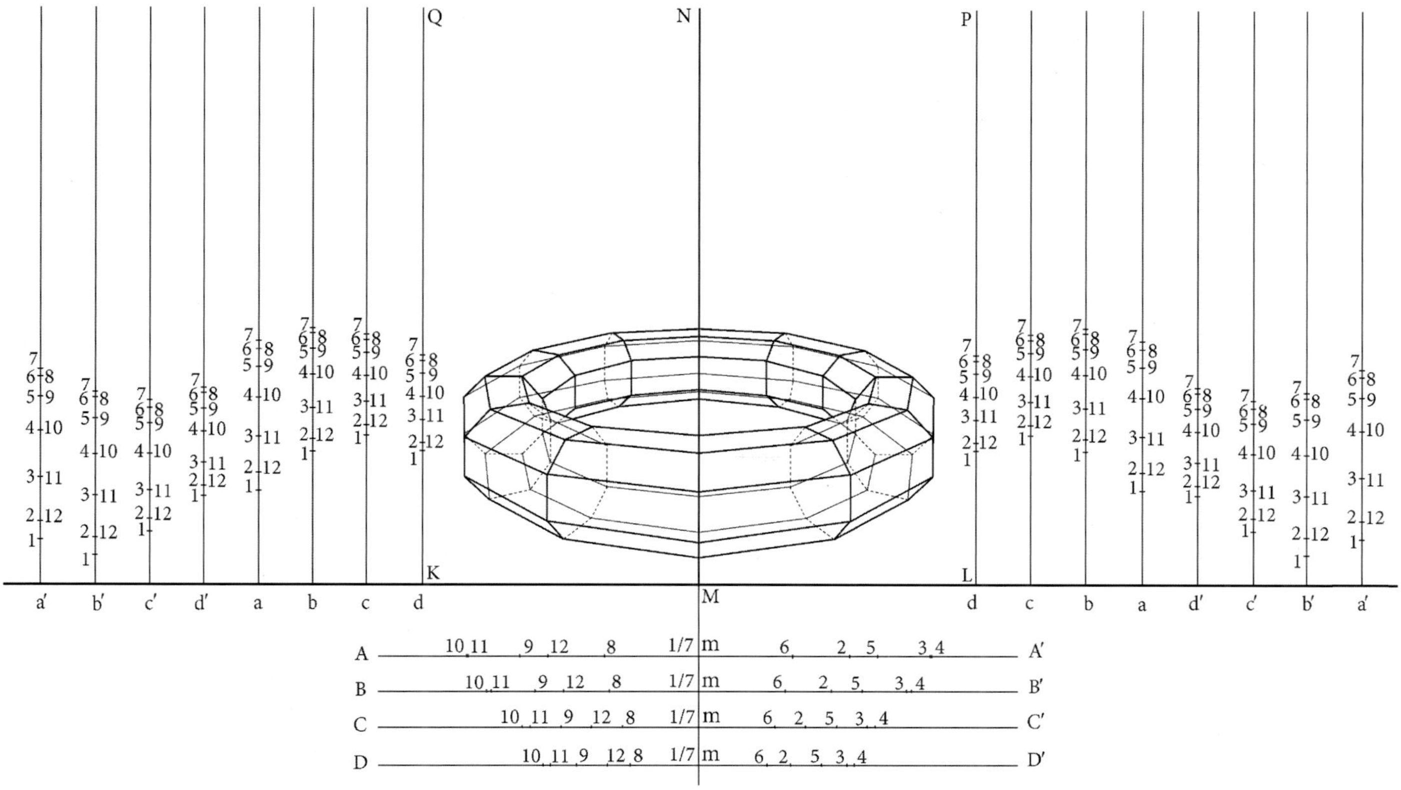

Figure B.65
Book 3, Proposition IV.

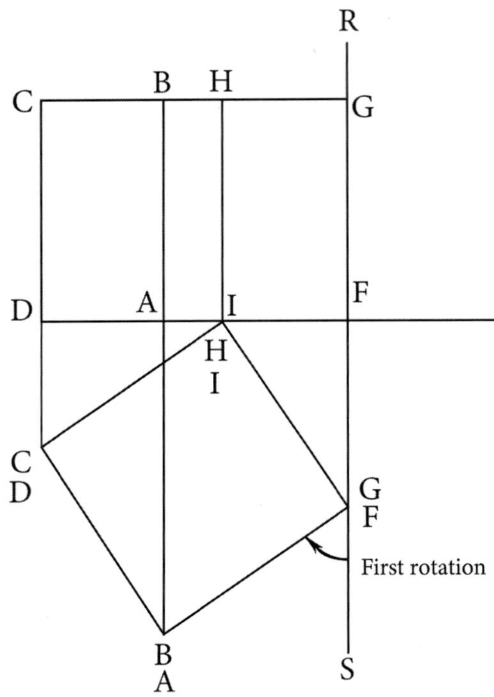

Figure B.66
Book 3, Proposition V.

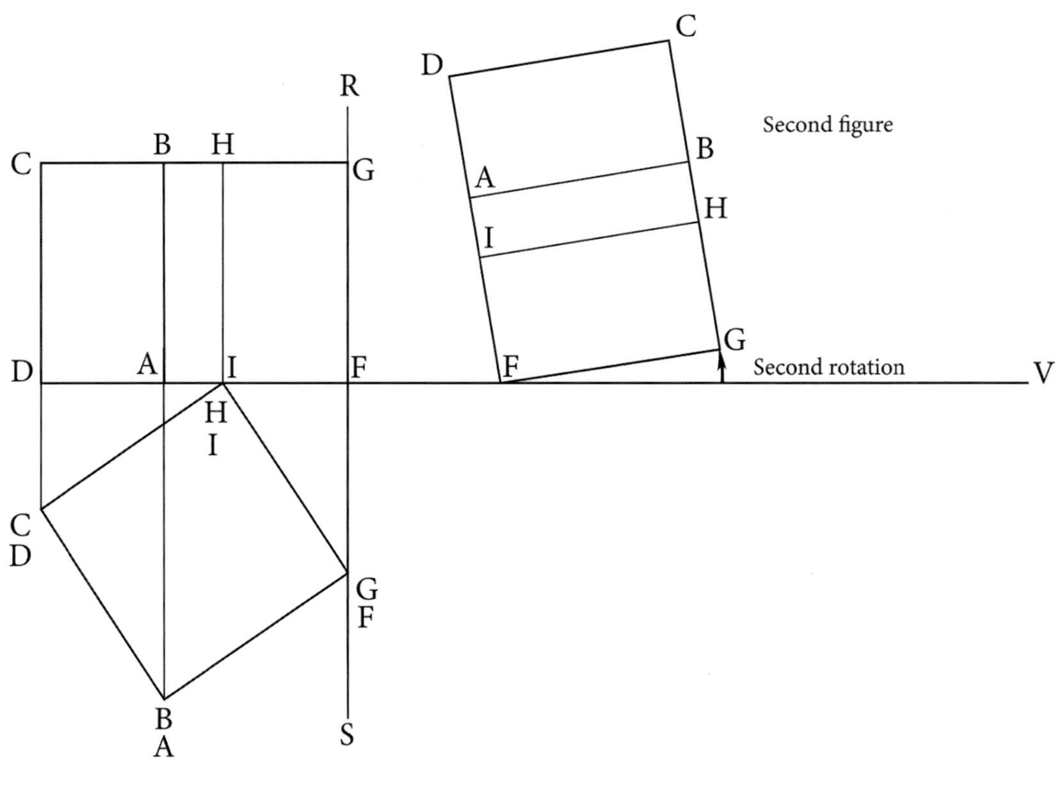

Figure B.67
Book 3, Proposition V.

Second figure

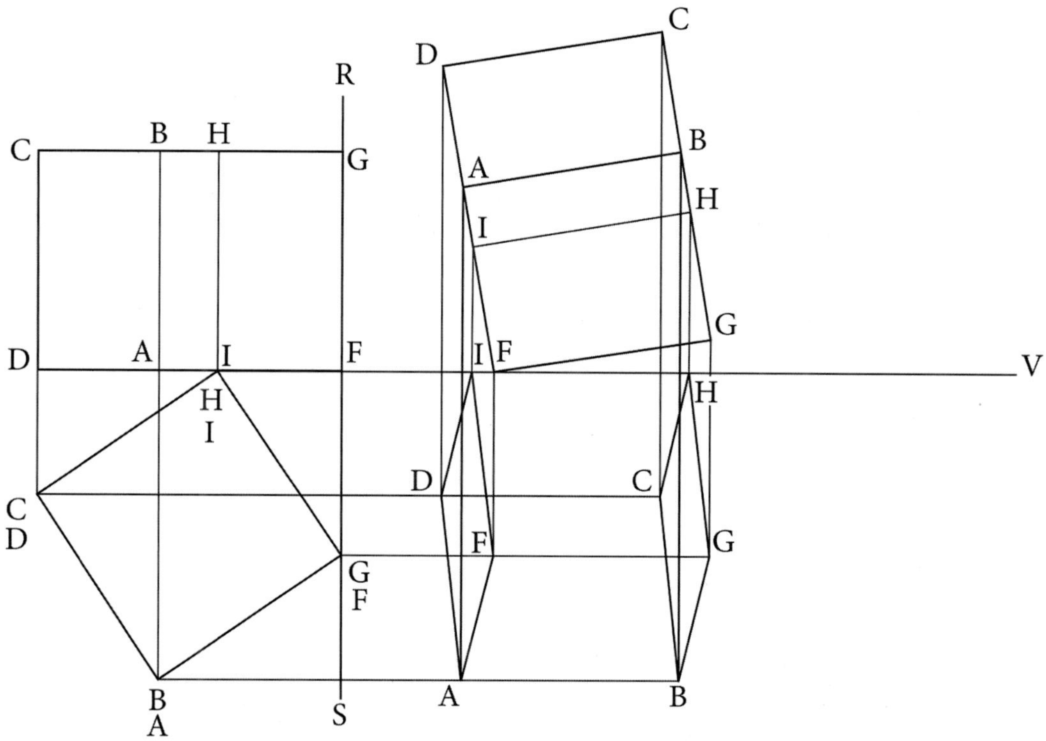

Third figure

Figure B.68
Book 3, Proposition V.

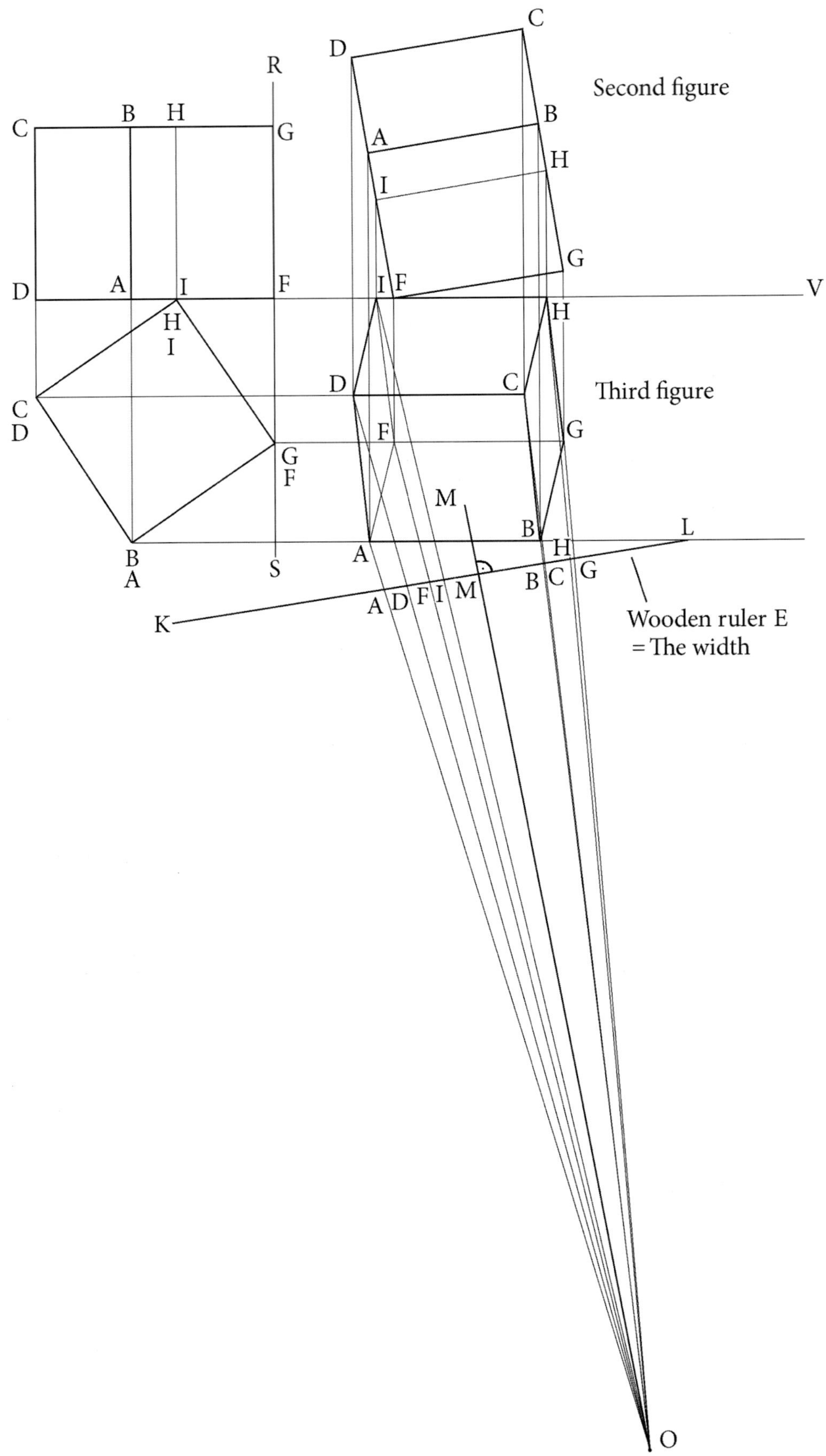

Figure B.69
Book 3, Proposition V.

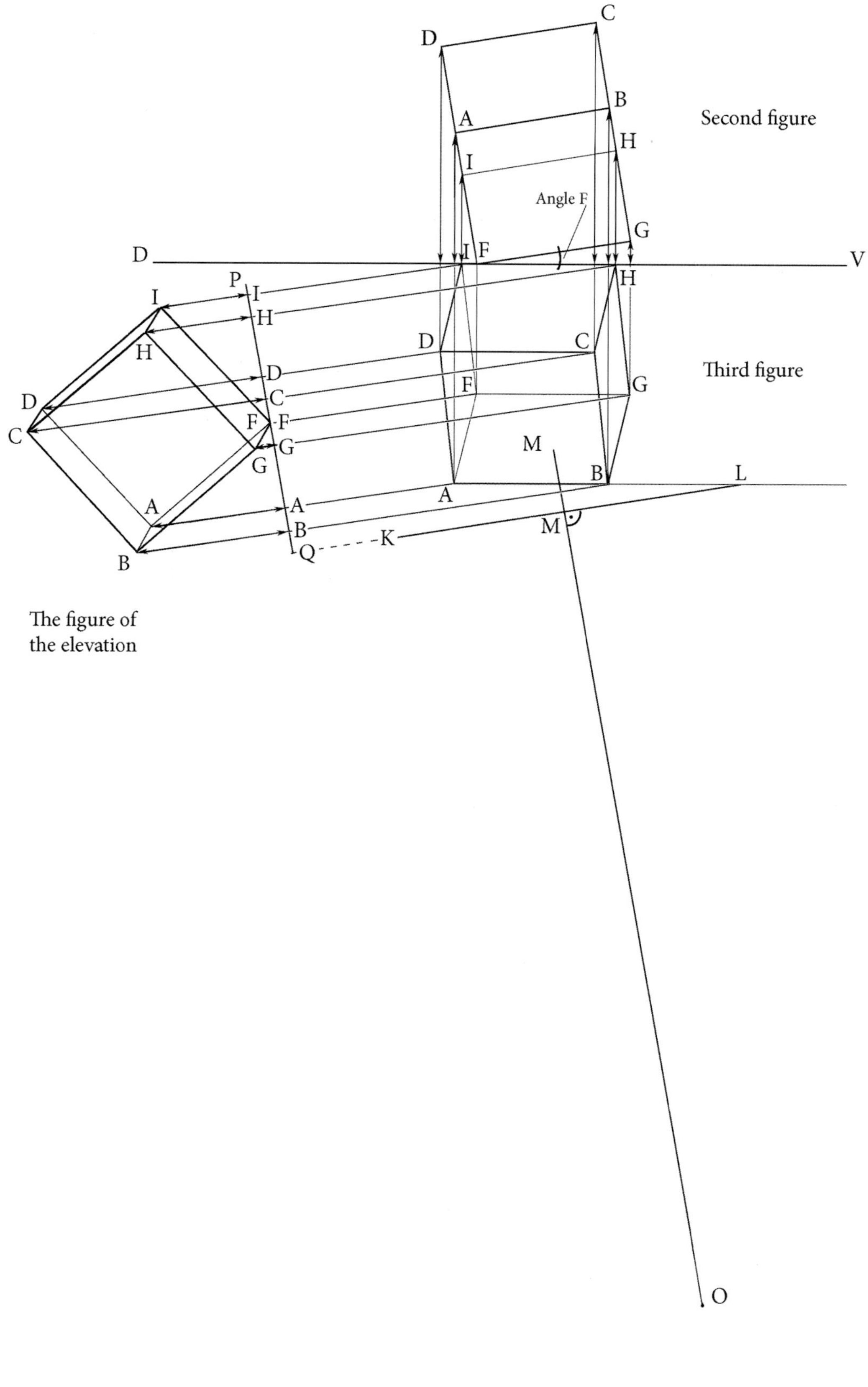

Figure B.70
Book 3, Proposition V.

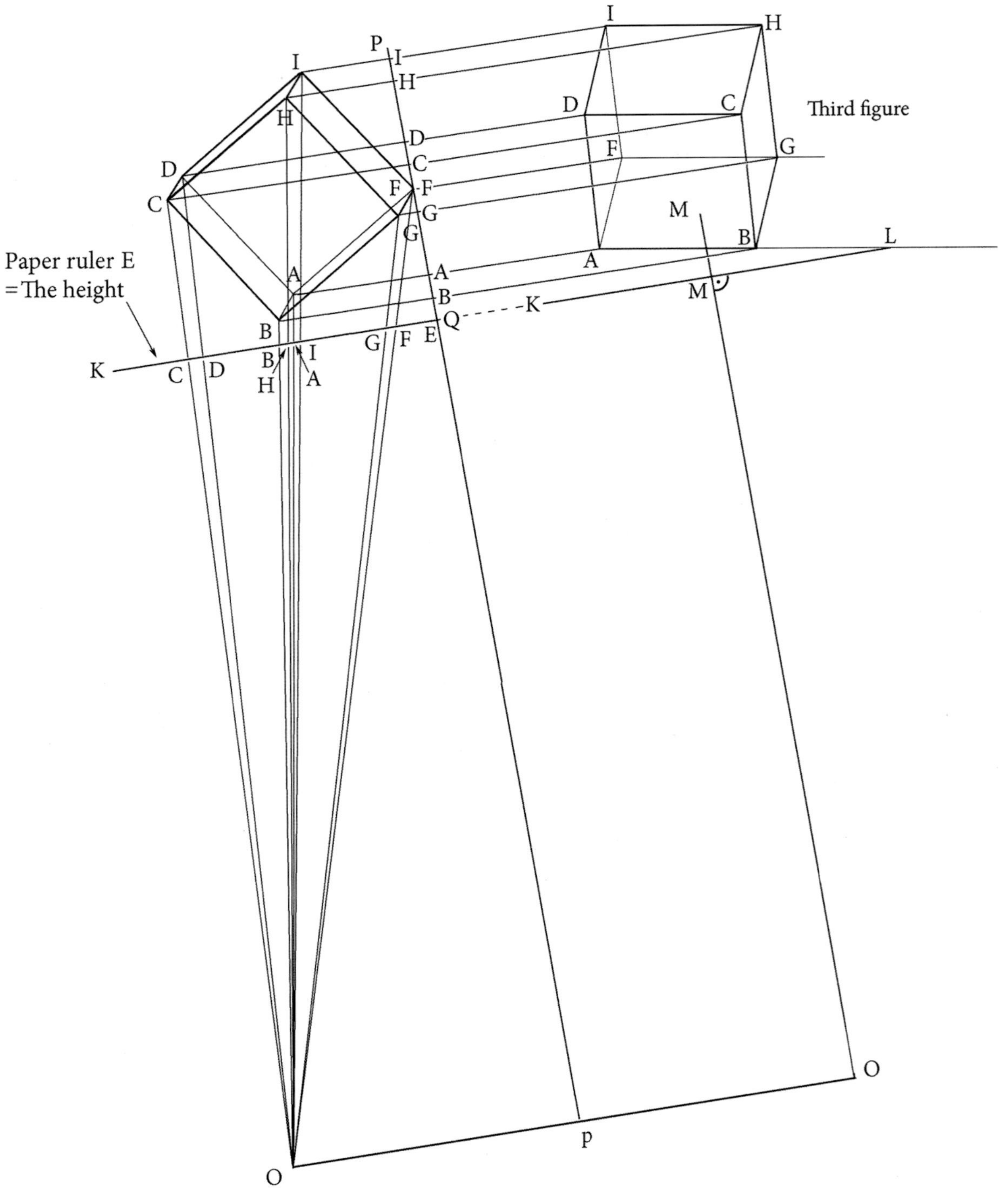

Figure B.71
Book 3, Proposition V.

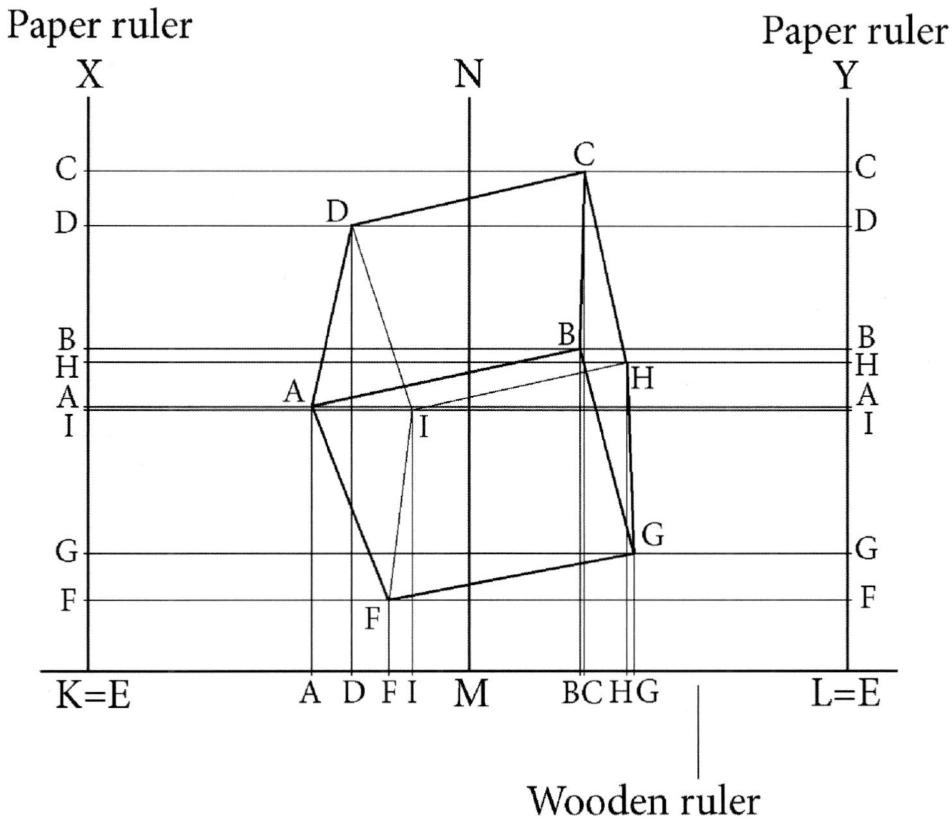

Figure B.72
Book 3, Proposition V.

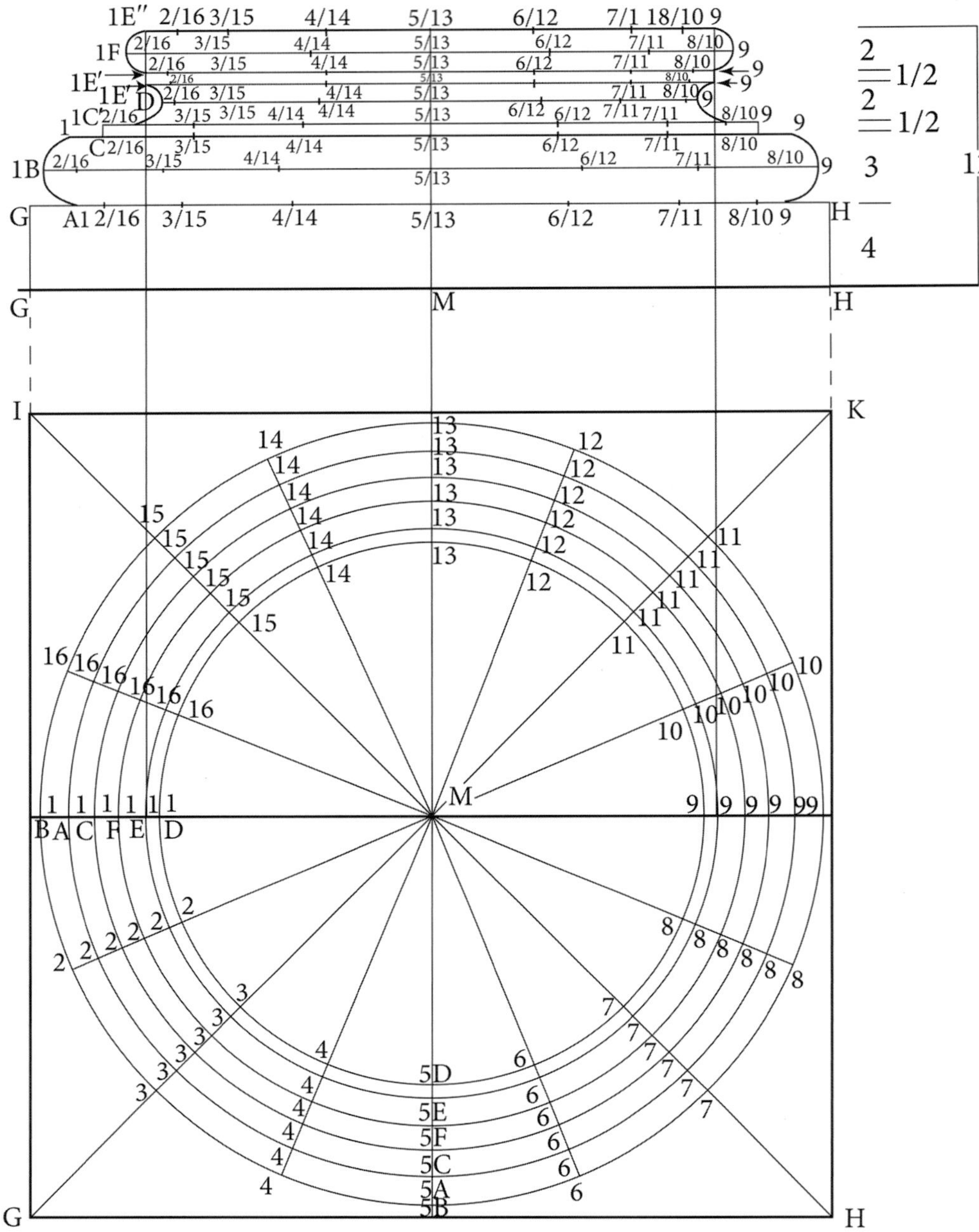

Figure B.73
Book 3, Proposition VI.

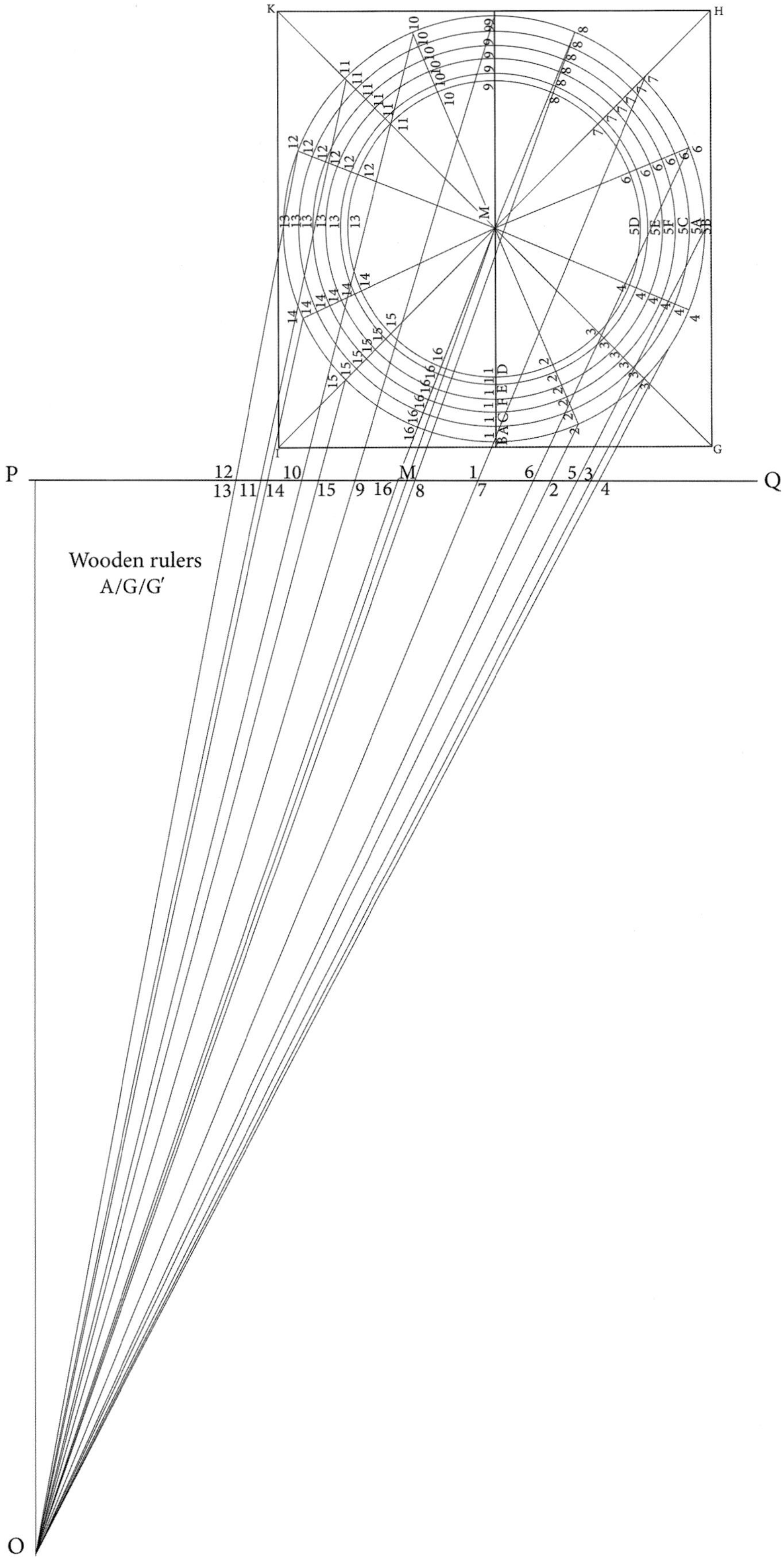

Figure B.74
Book 3, Proposition VI.

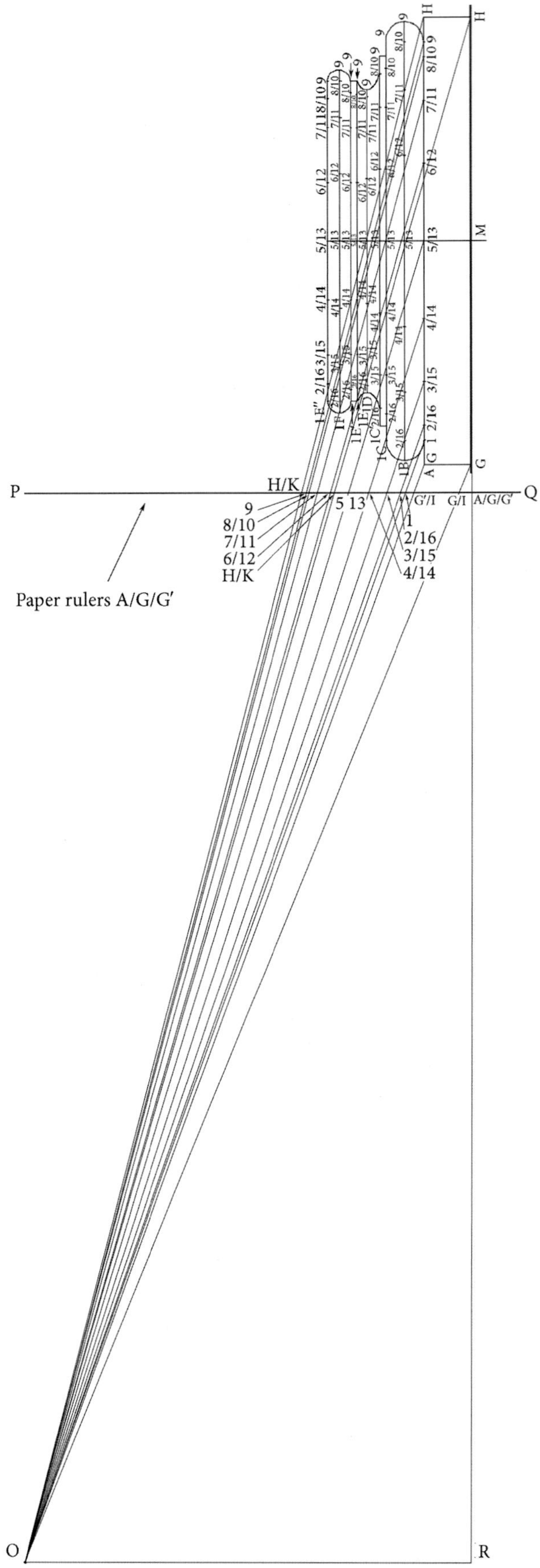

Figure B.75
Book 3, Proposition VI.

Paper rulers

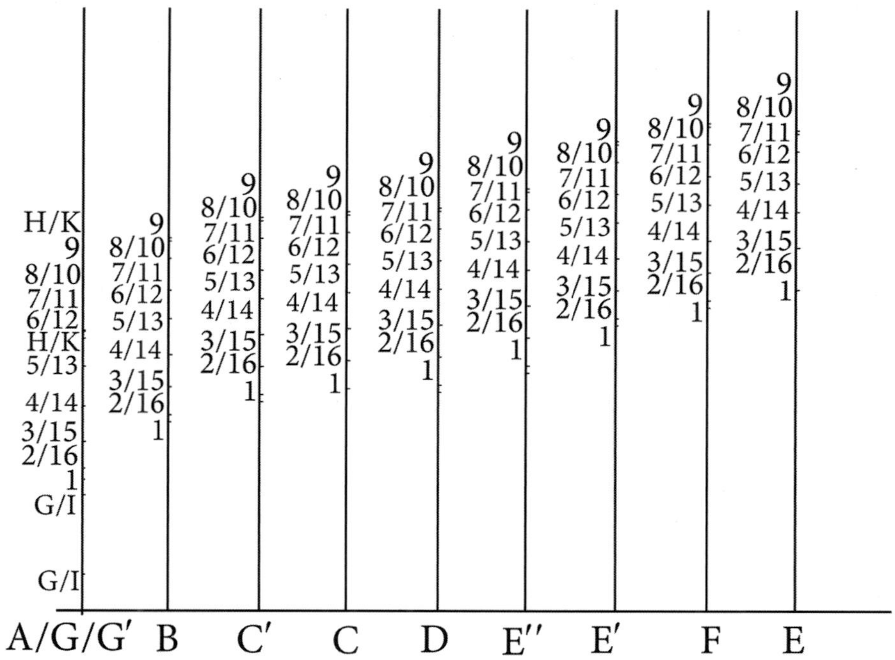

Wooden rulers

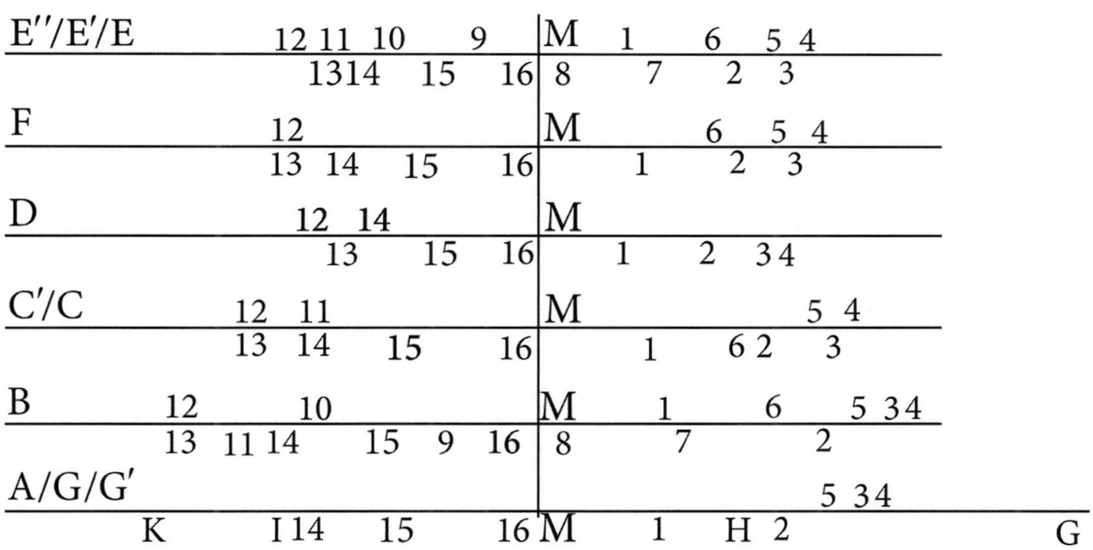

Figure B.76
Book 3, Proposition VI.

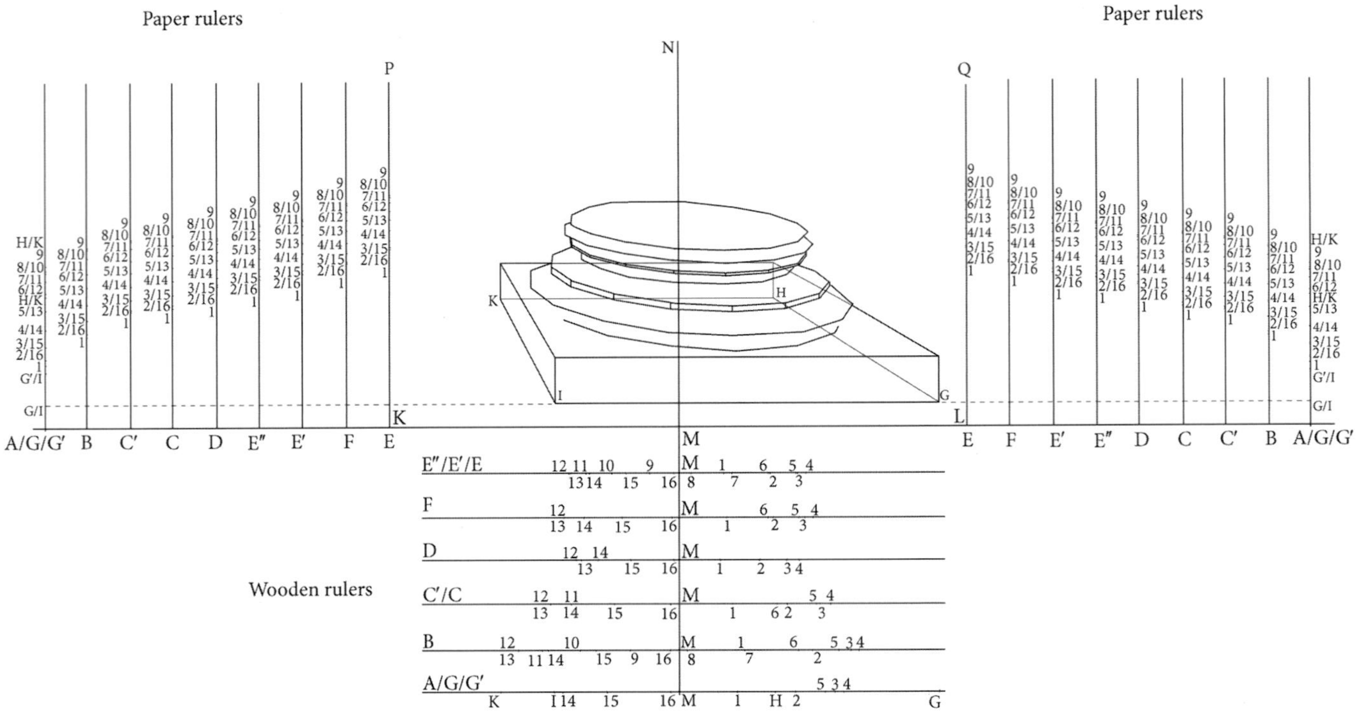

Figure B.77
Book 3, Proposition VI.

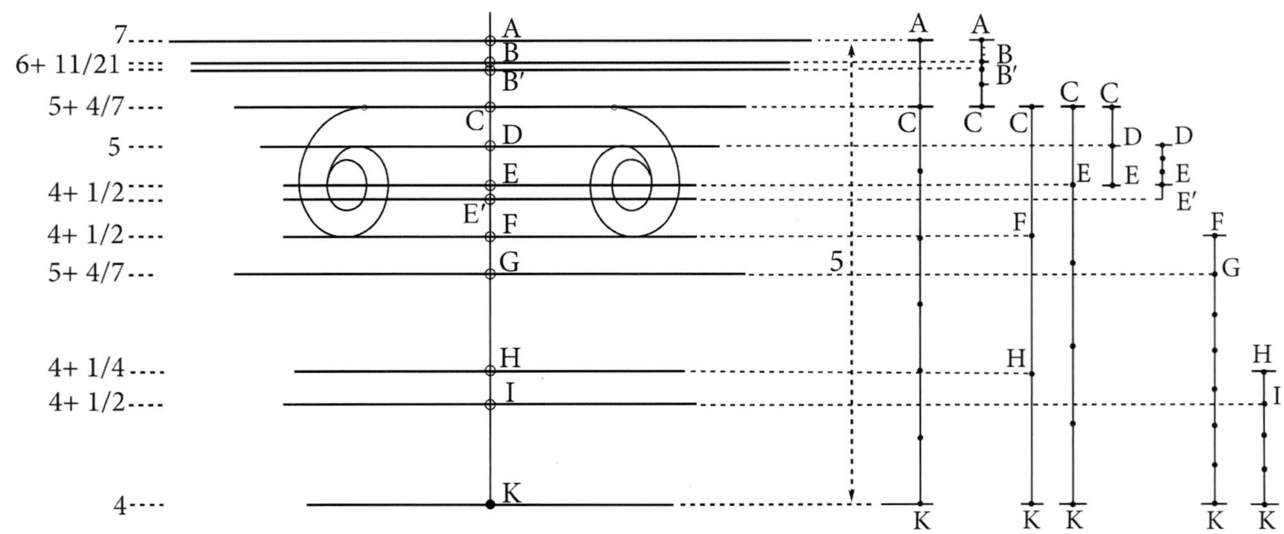

Figure B.78
Book 3, Proposition VII.

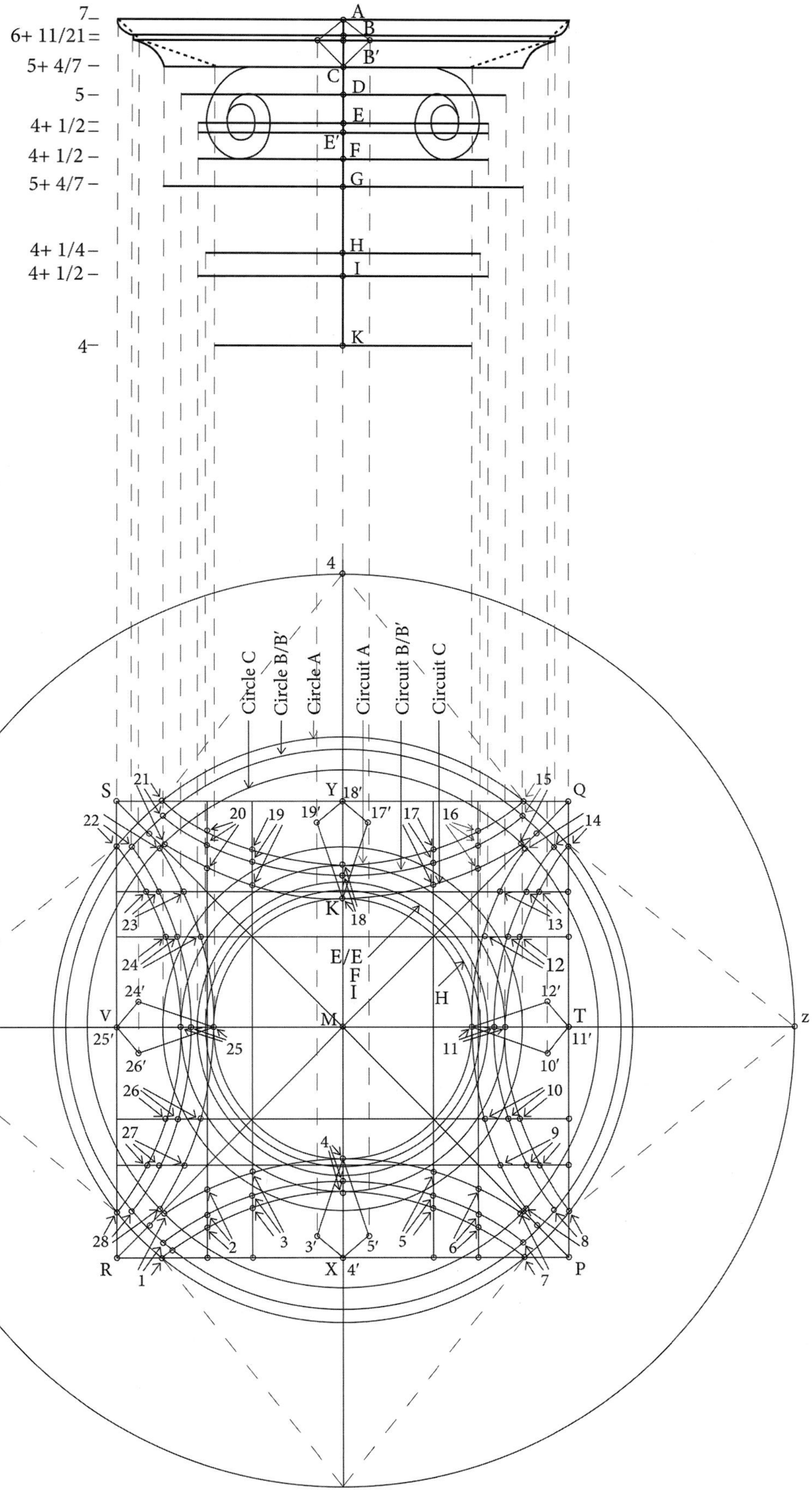

Figure B.79
Book 3, Proposition VII.

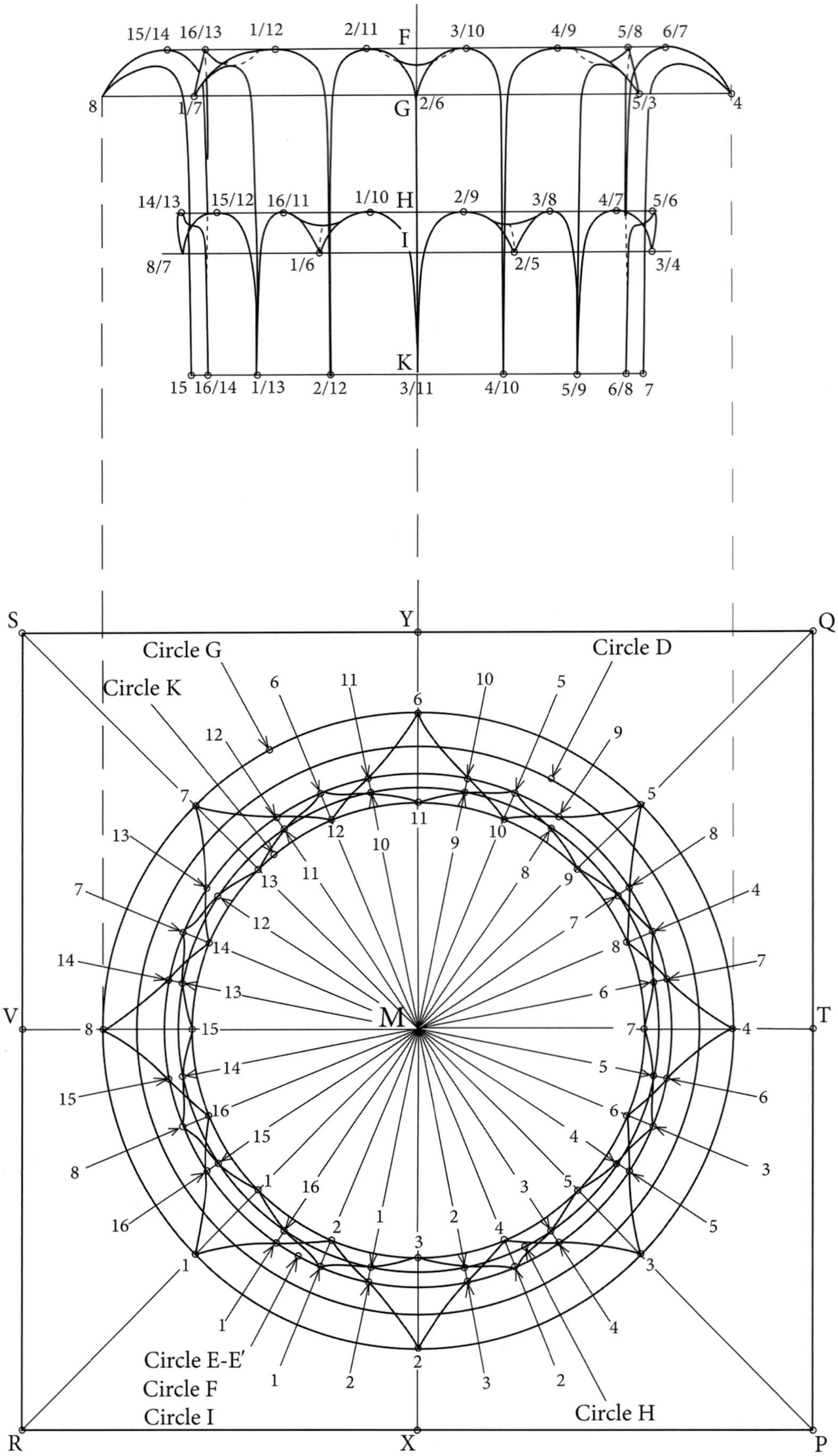

Figure B.80
Book 3, Proposition VII.

Circuit E′

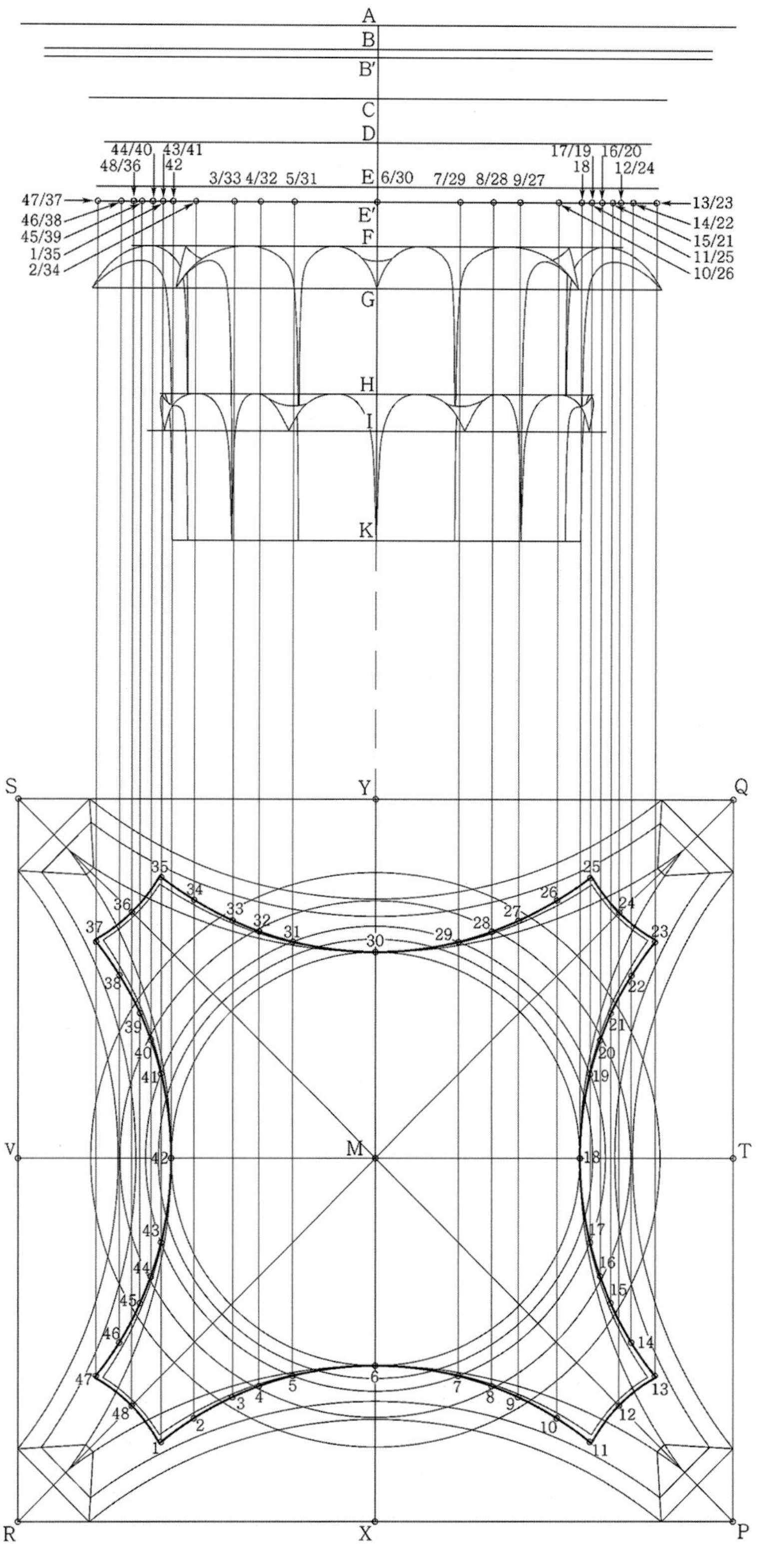

Figure B.81
Book 3, Proposition VII.

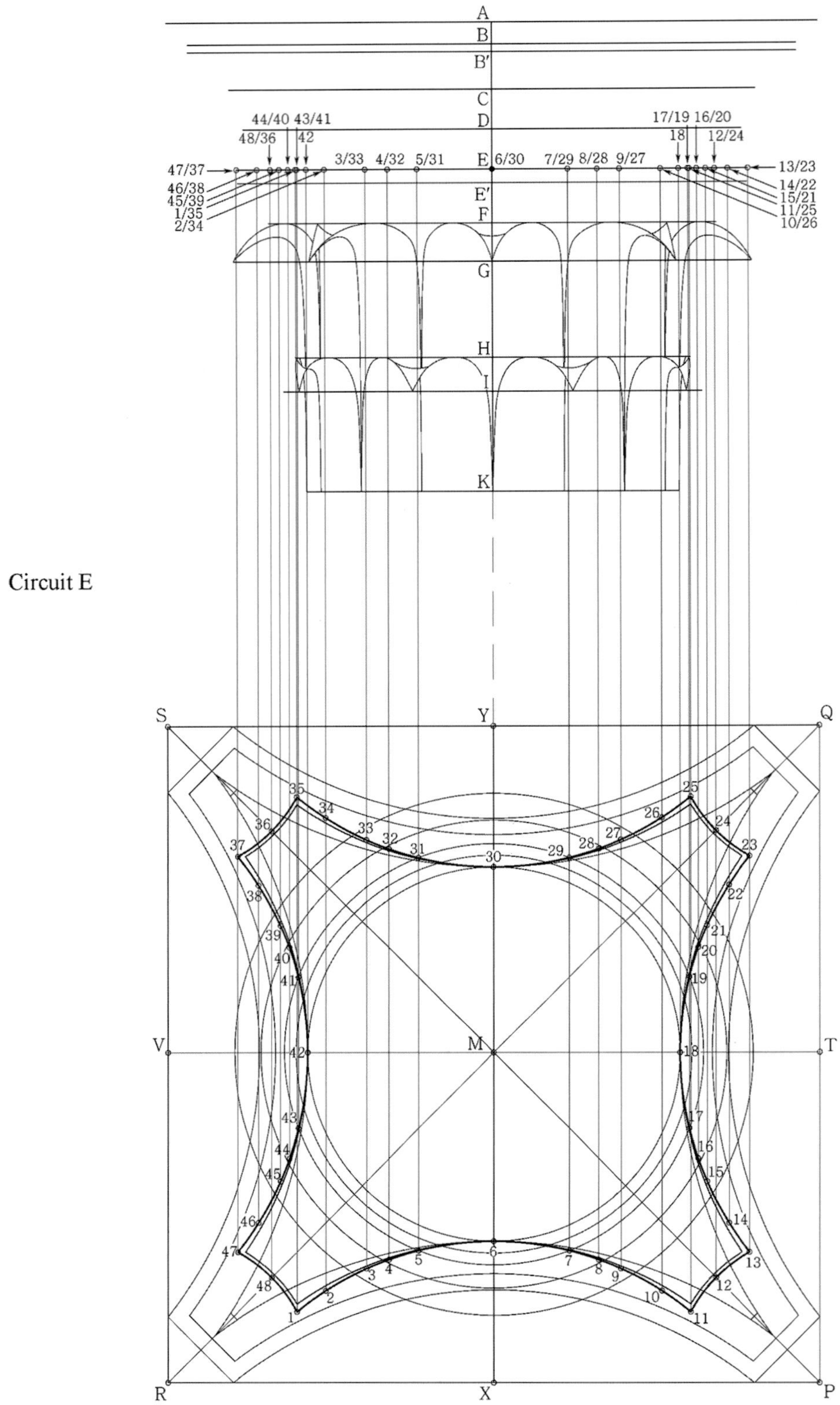

Figure B.82
Book 3, Proposition VII.

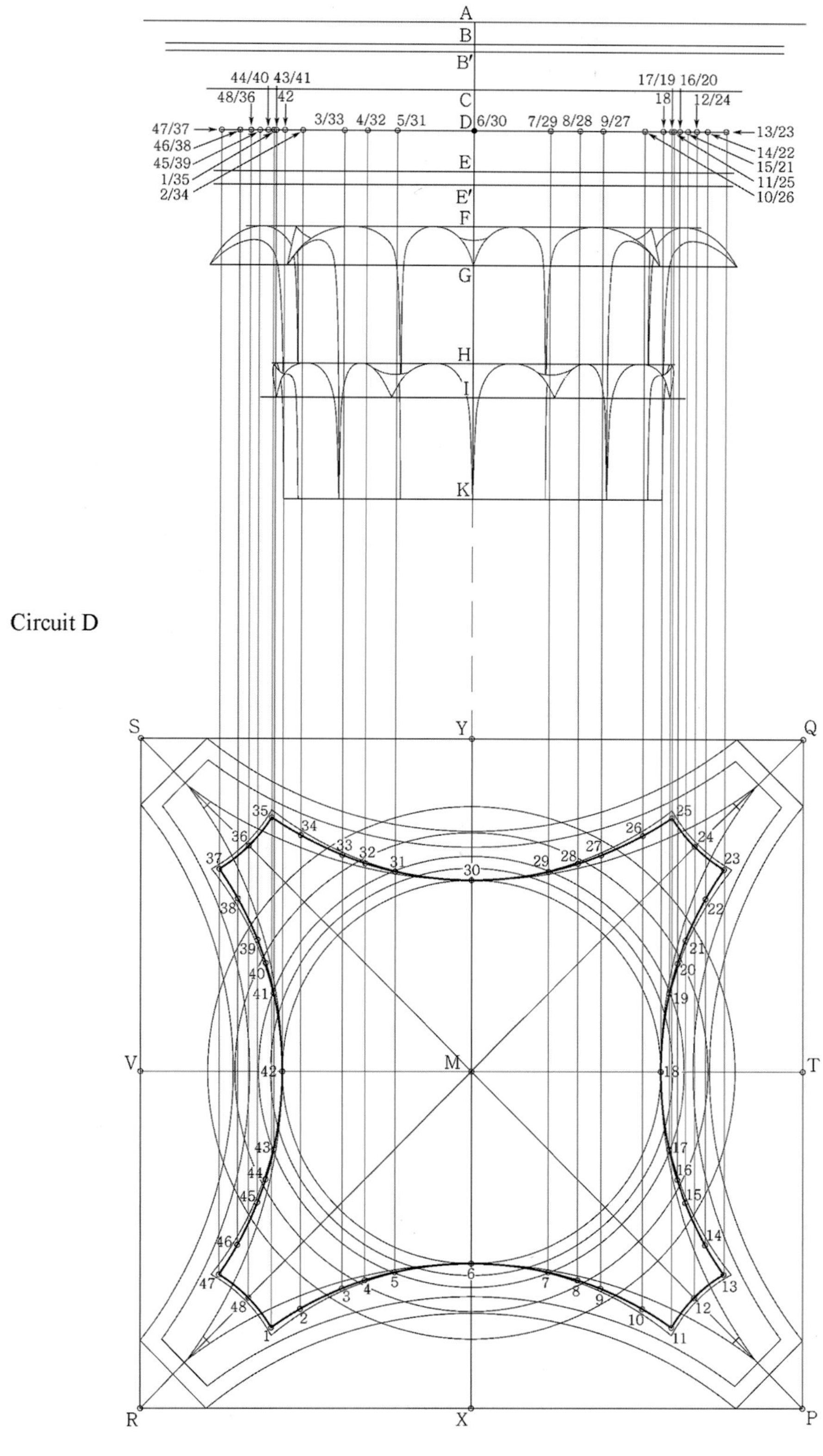

Figure B.83
Book 3, Proposition VII.

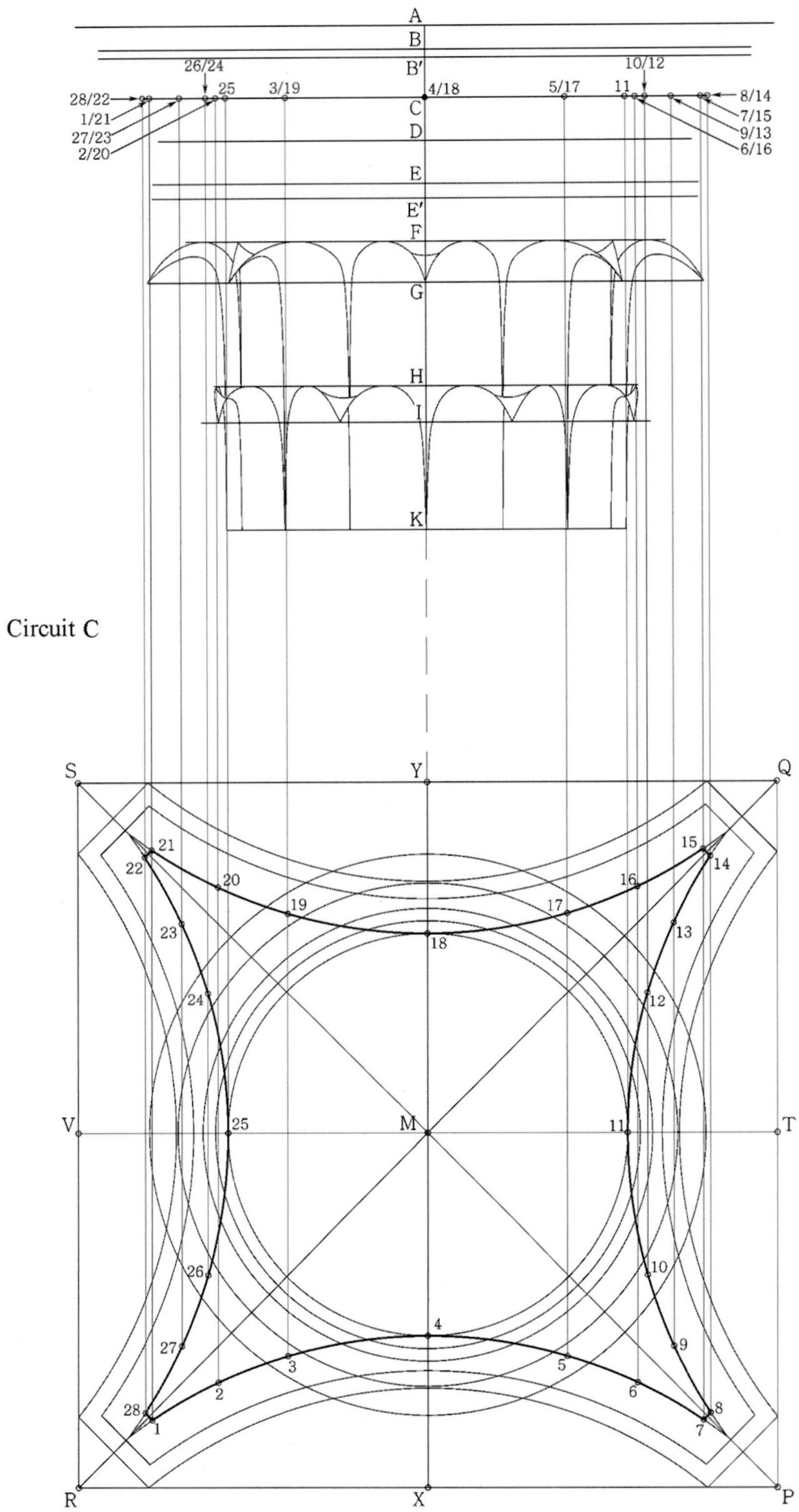

Figure B.84
Book 3, Proposition VII.

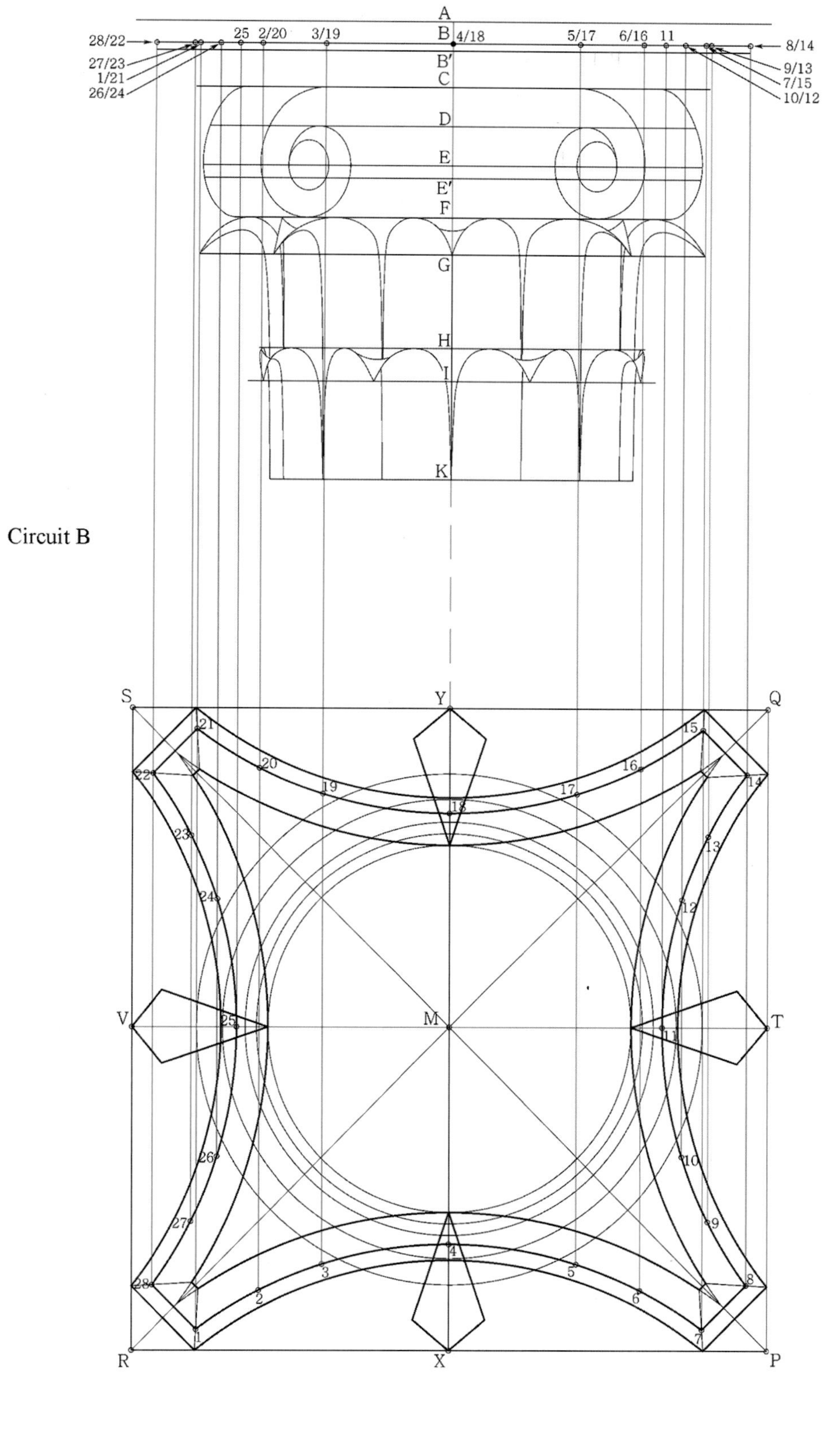

Figure B.85
Book 3, Proposition VII.

Circuit B′

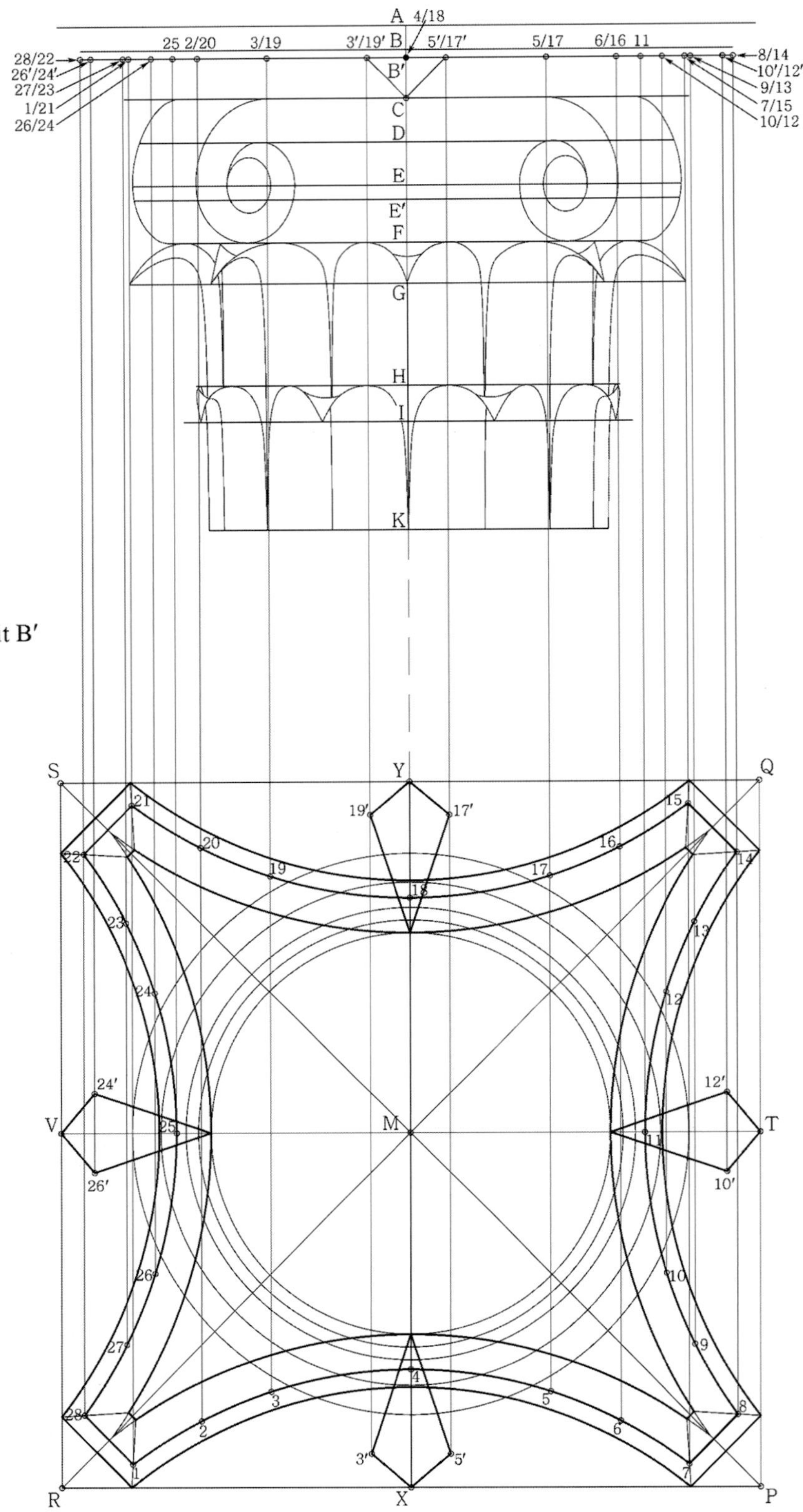

Figure B.86
Book 3, Proposition VII.

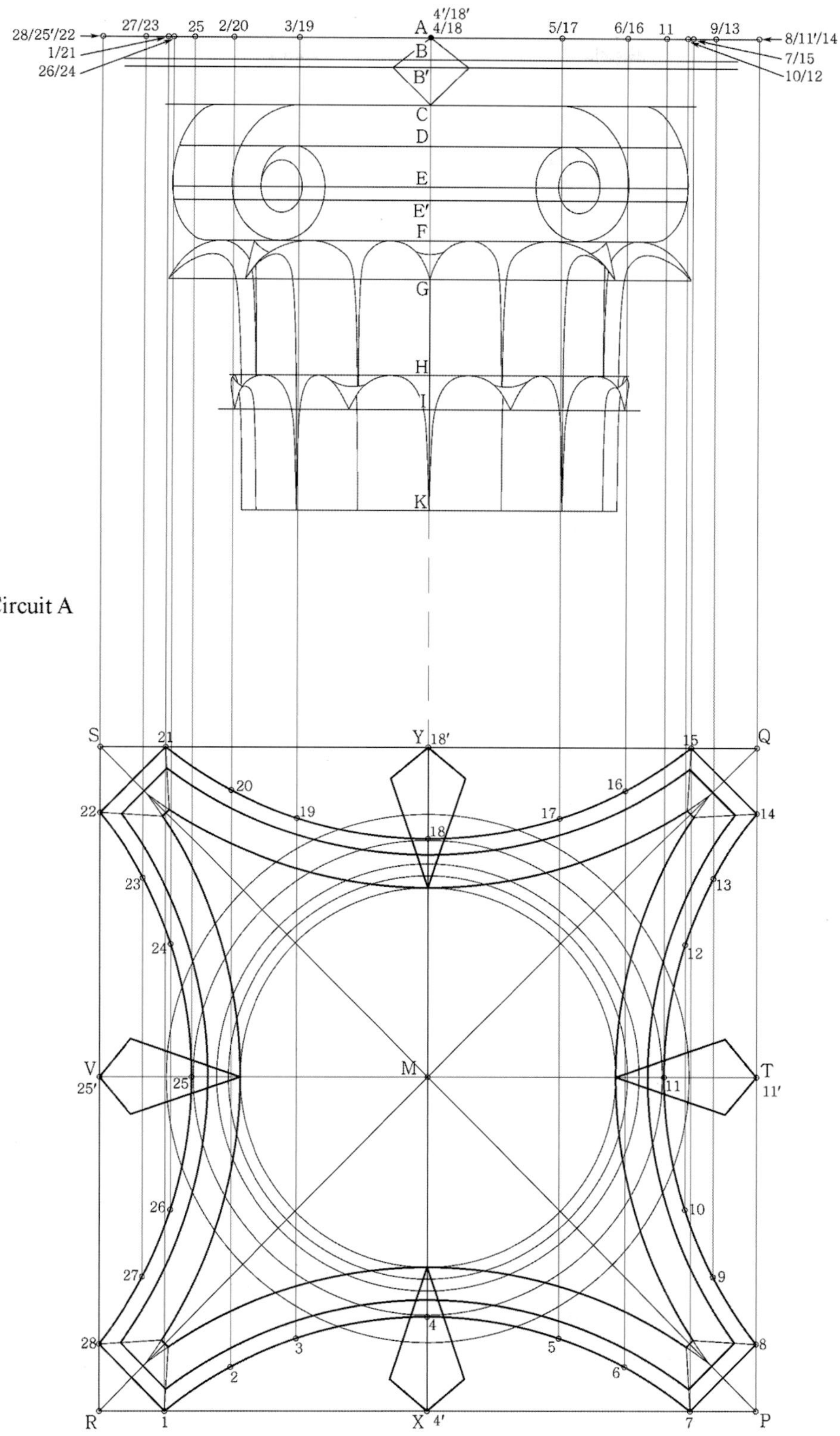

Figure B.87
Book 3, Proposition VII.

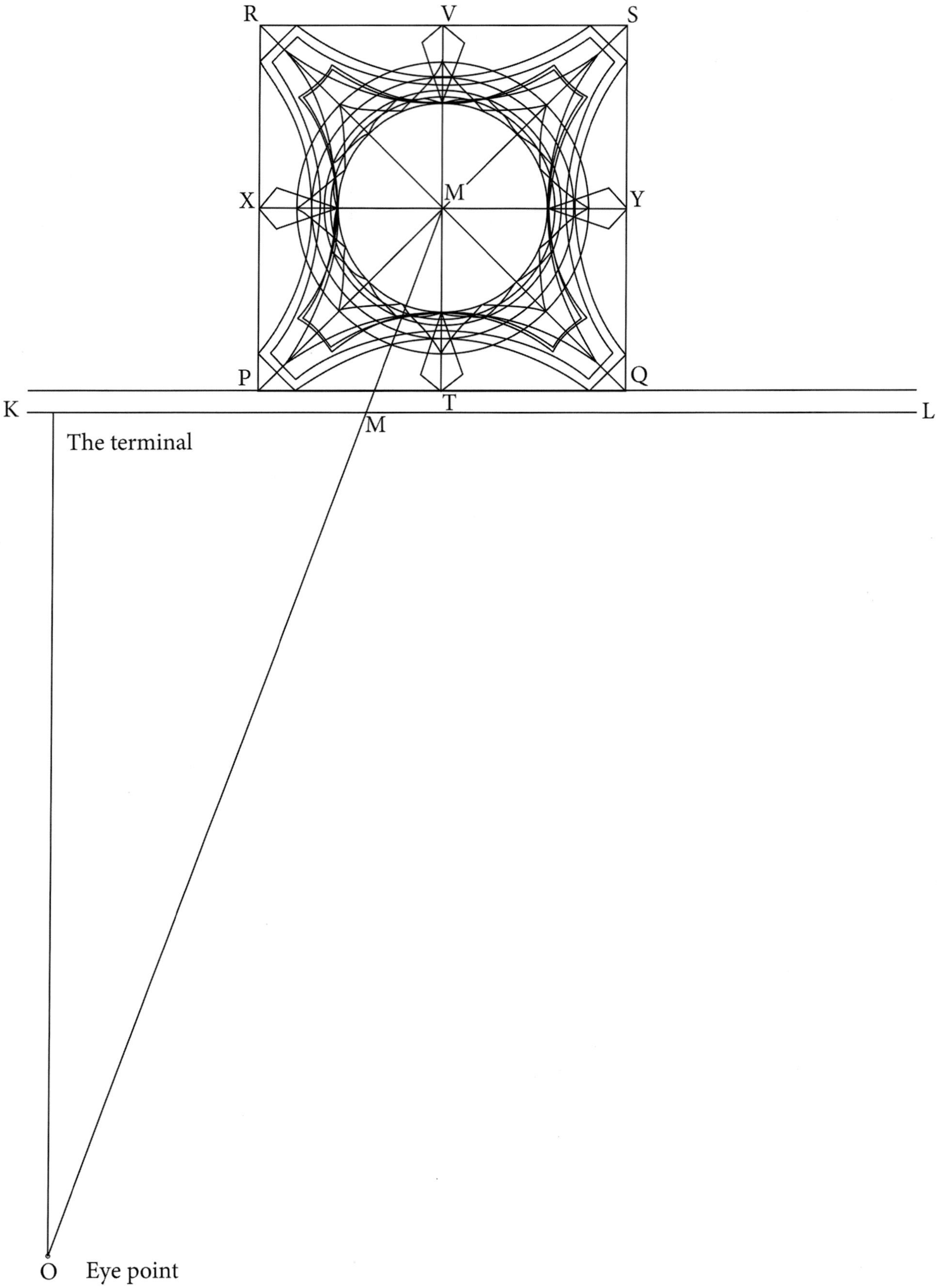

Figure B.88
Book 3, Proposition VII.

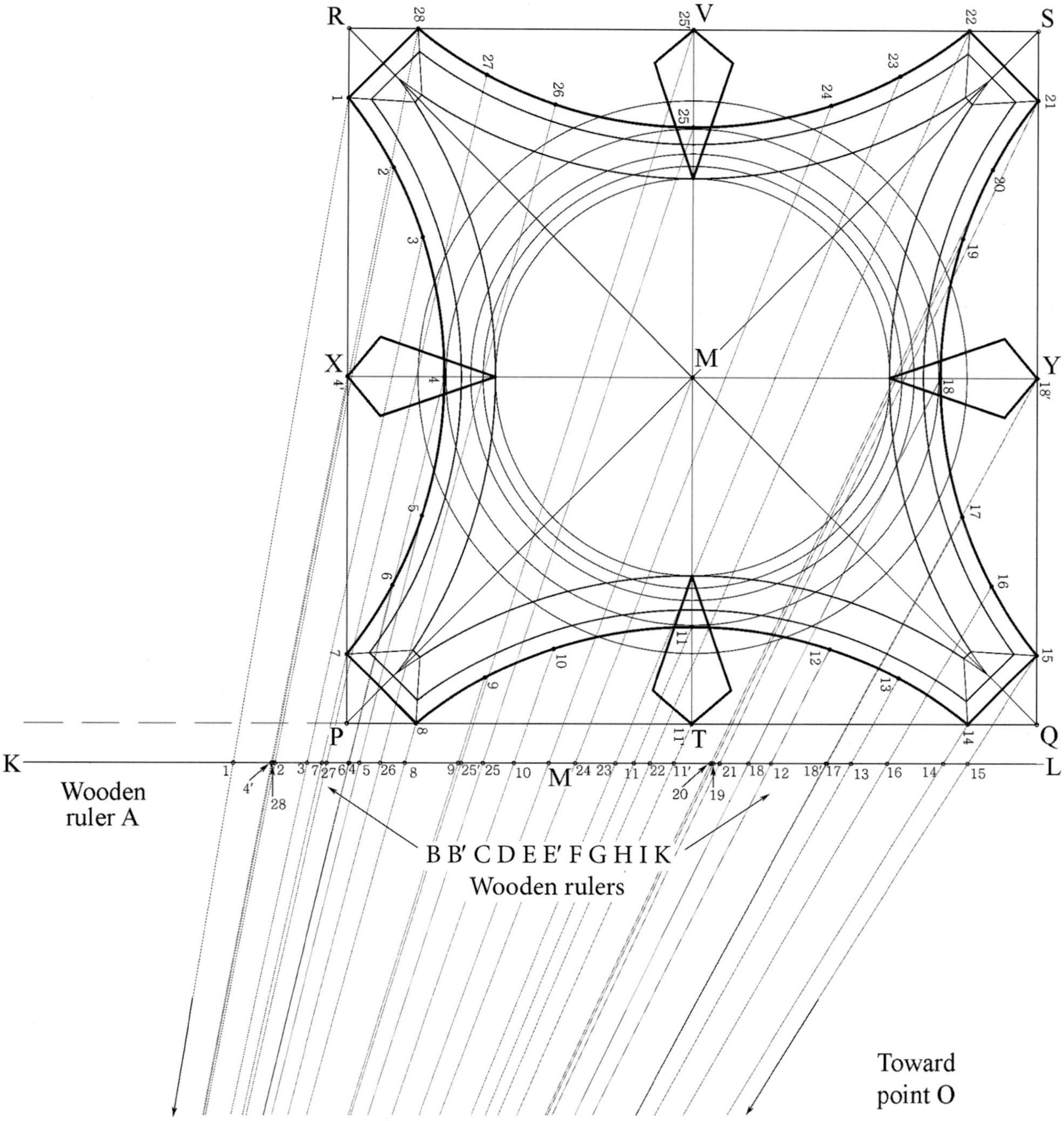

Figure B.89
Book 3, Proposition VII.

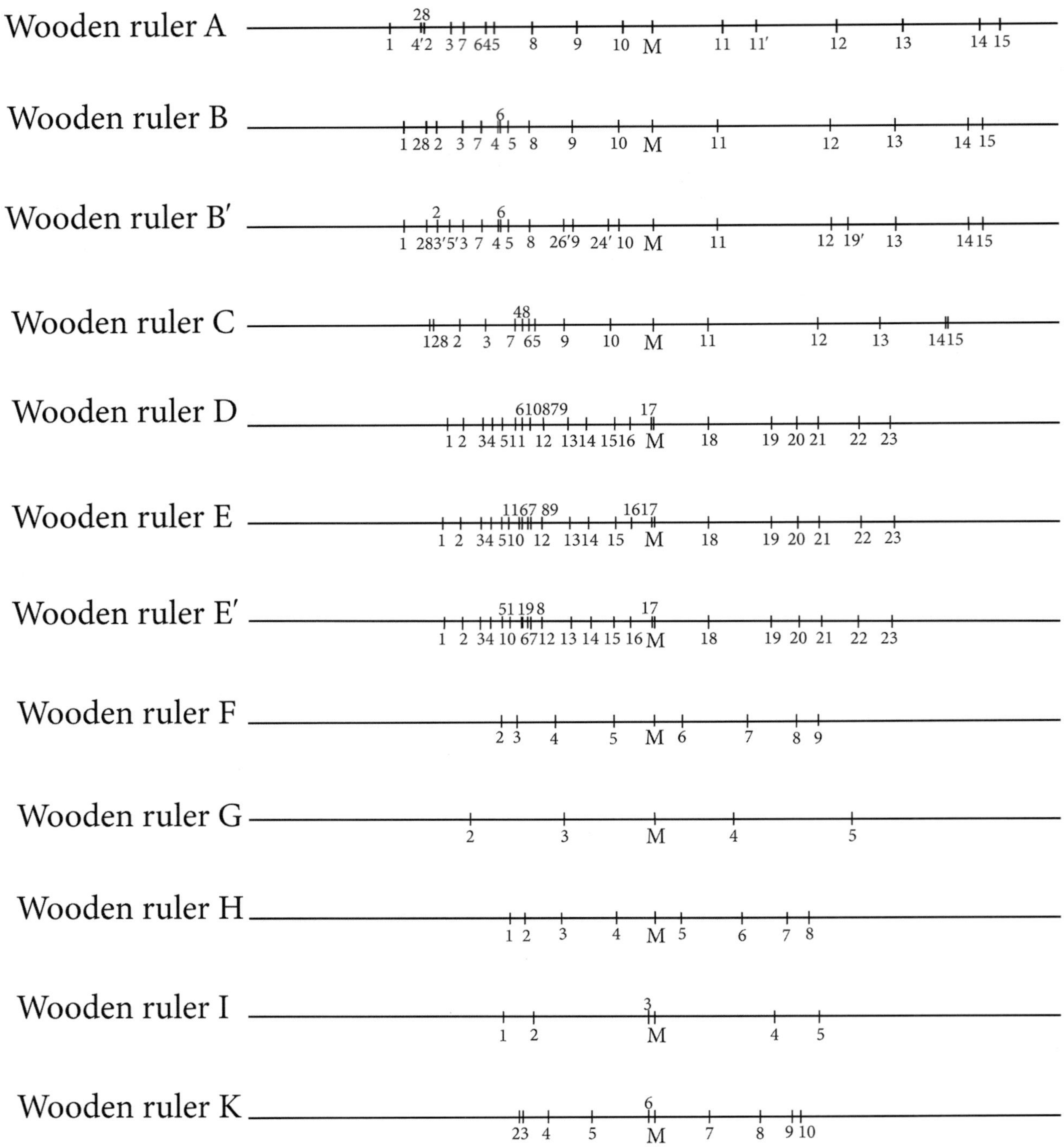

Figure B.90
Book 3, Proposition VII.

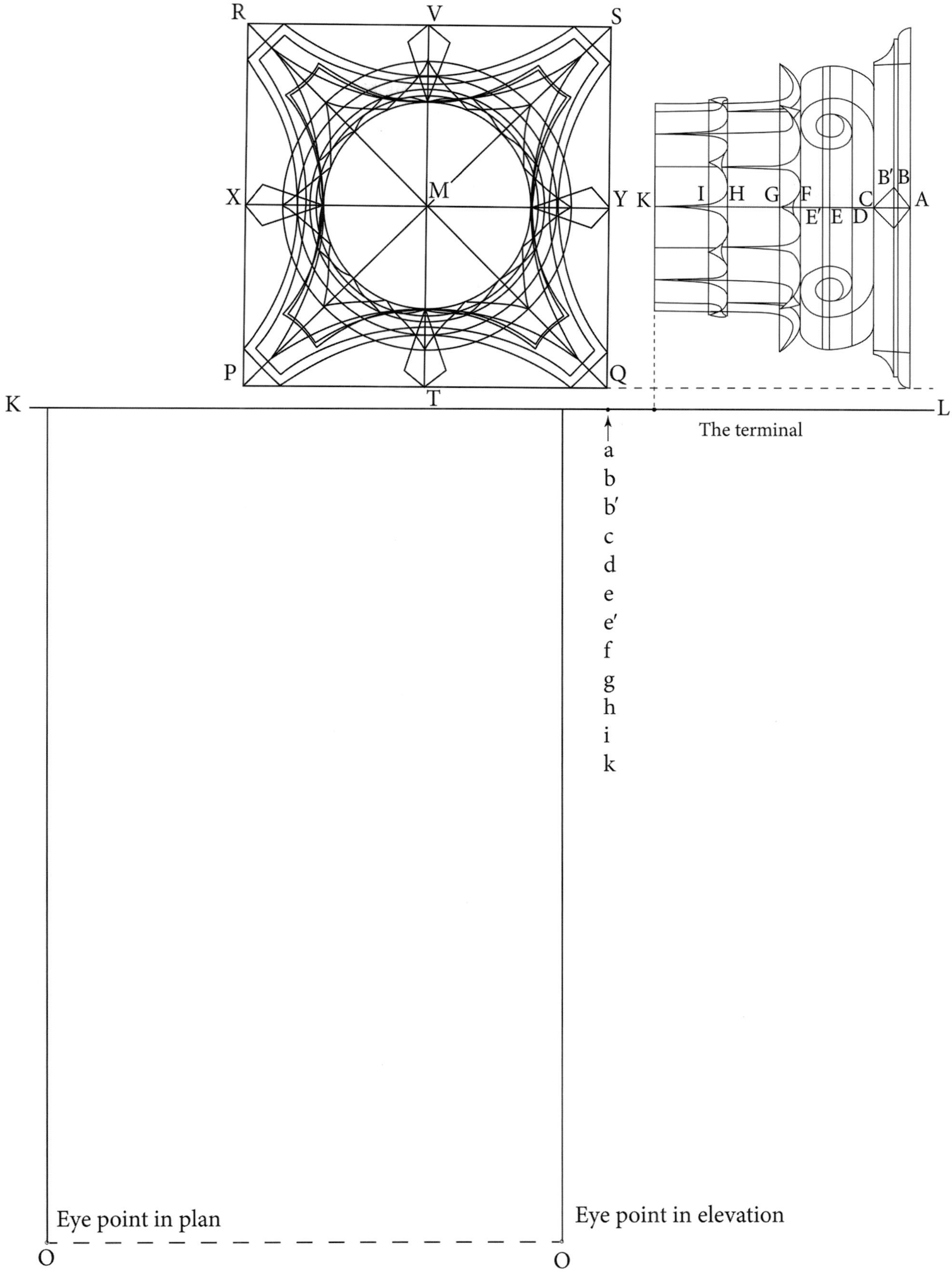

Figure B.91
Book 3, Proposition VII.

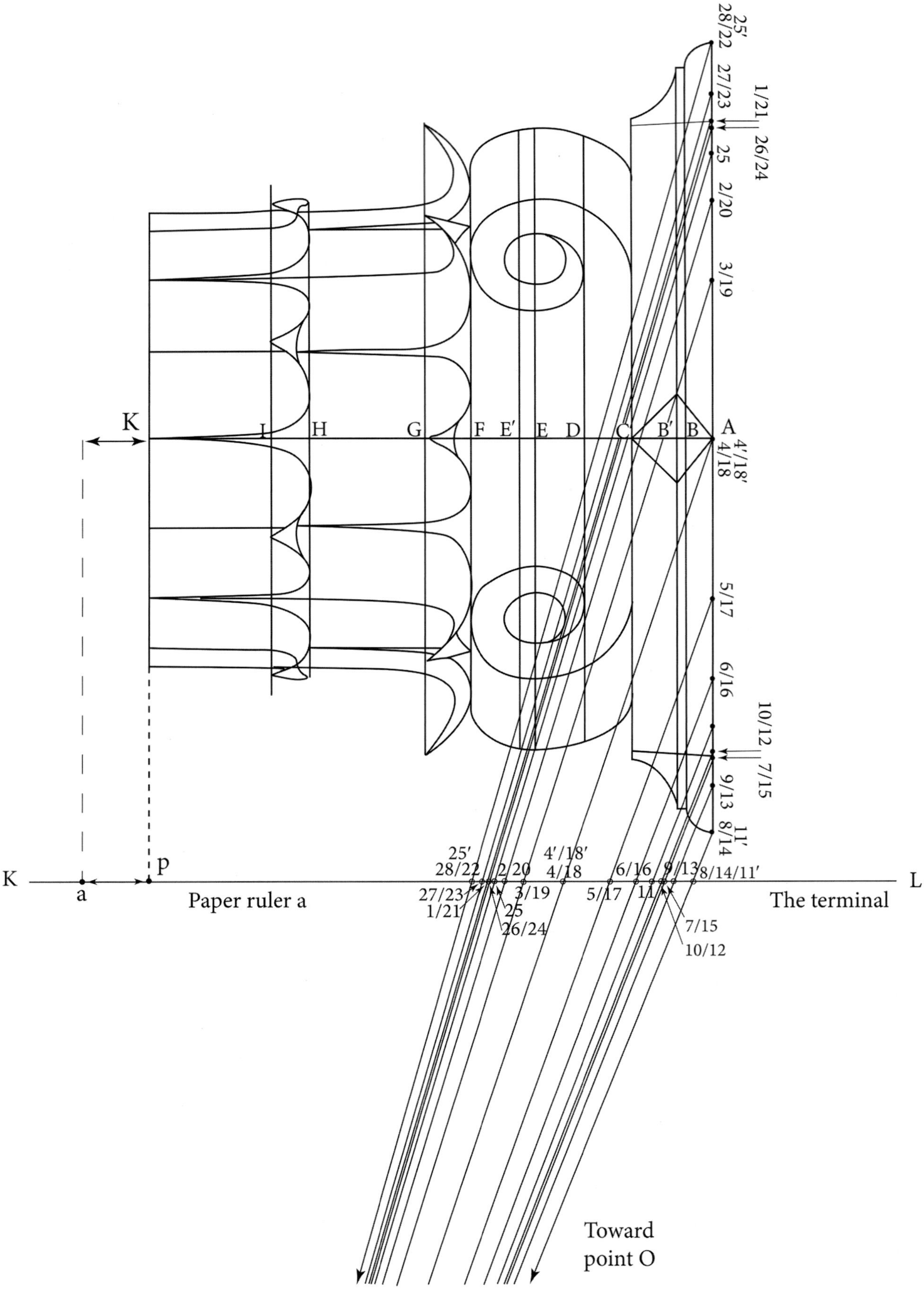

Figure B.92
Book 3, Proposition VII.

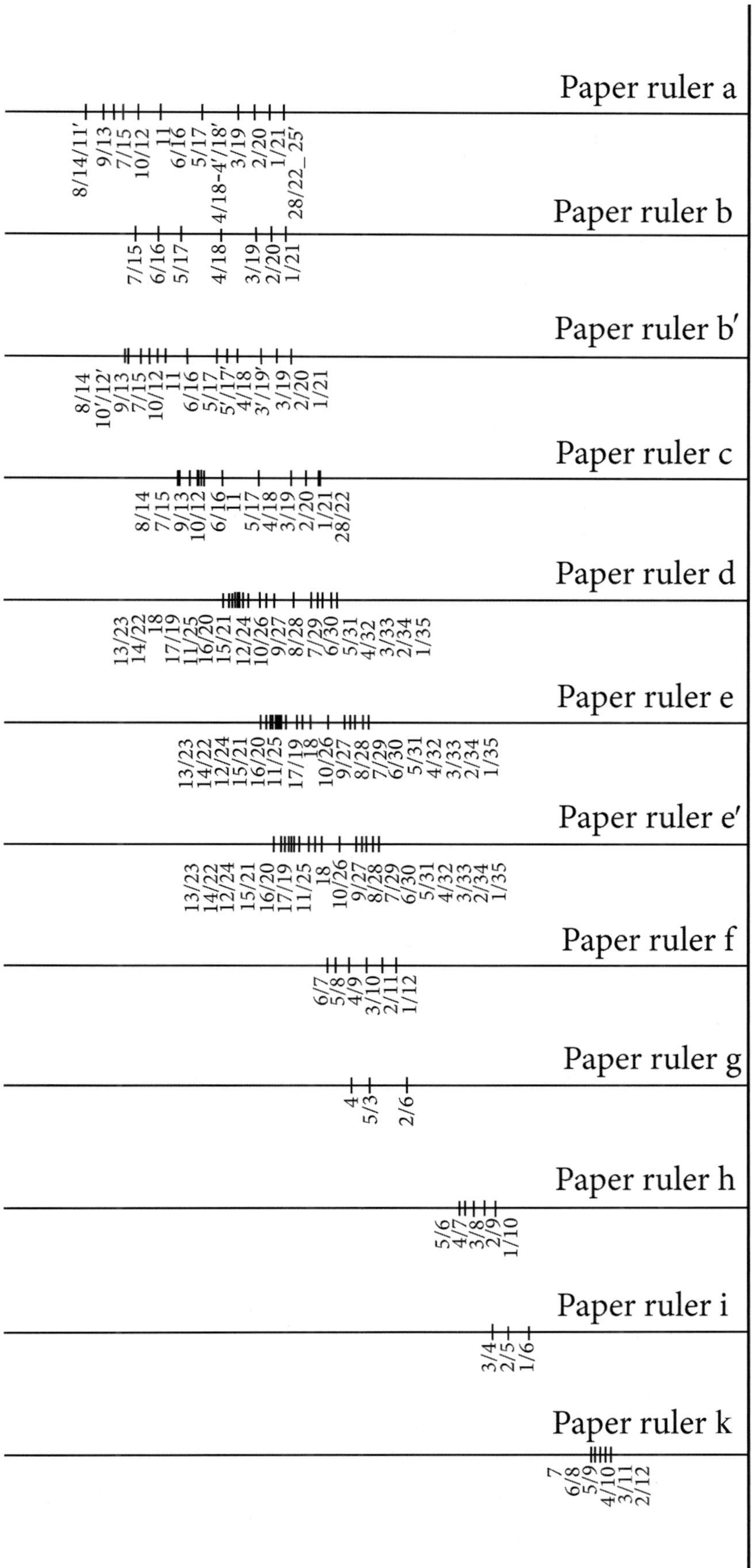

Figure B.93
Book 3, Proposition VII.

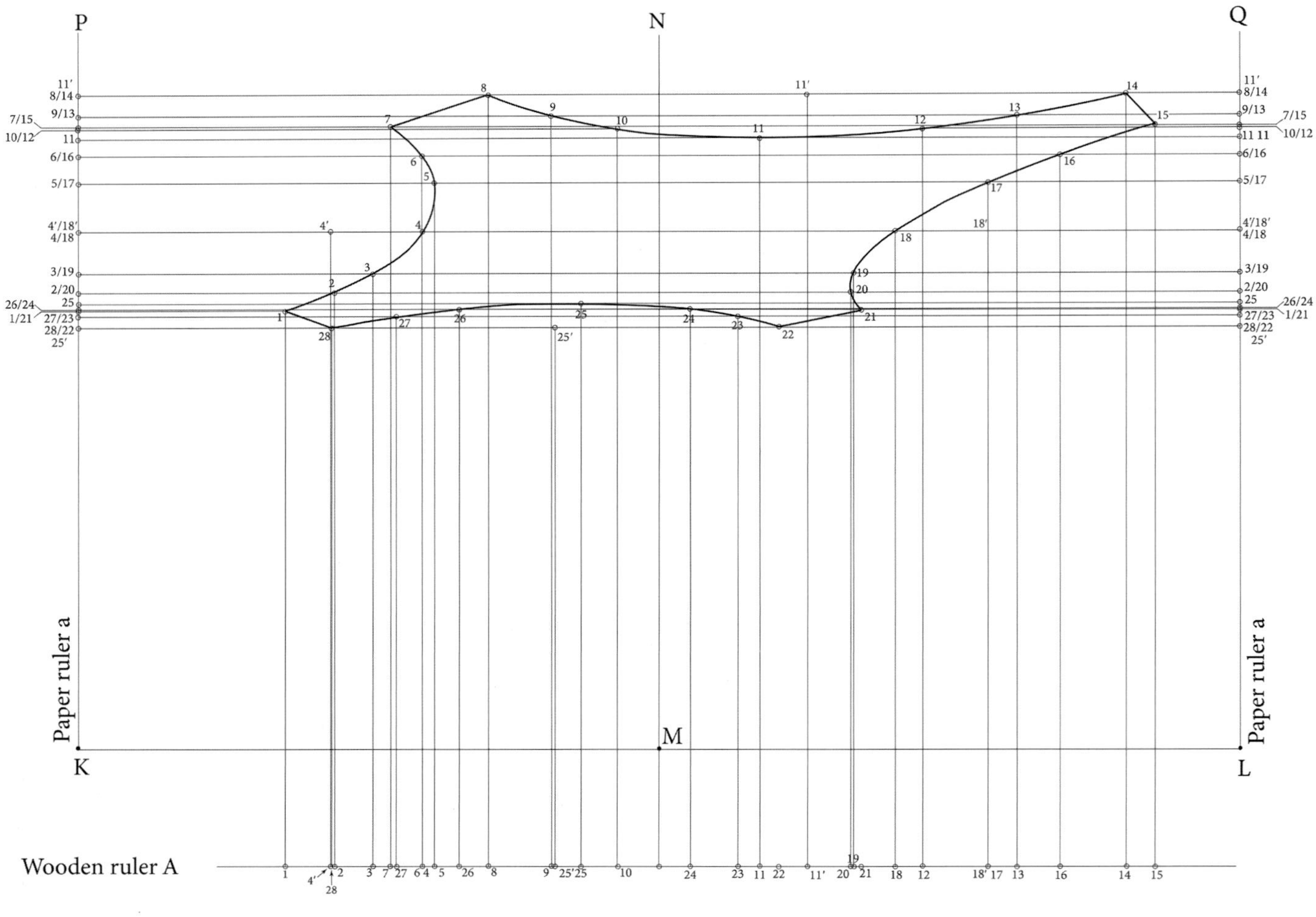

Figure B.94
Book 3, Proposition VII.

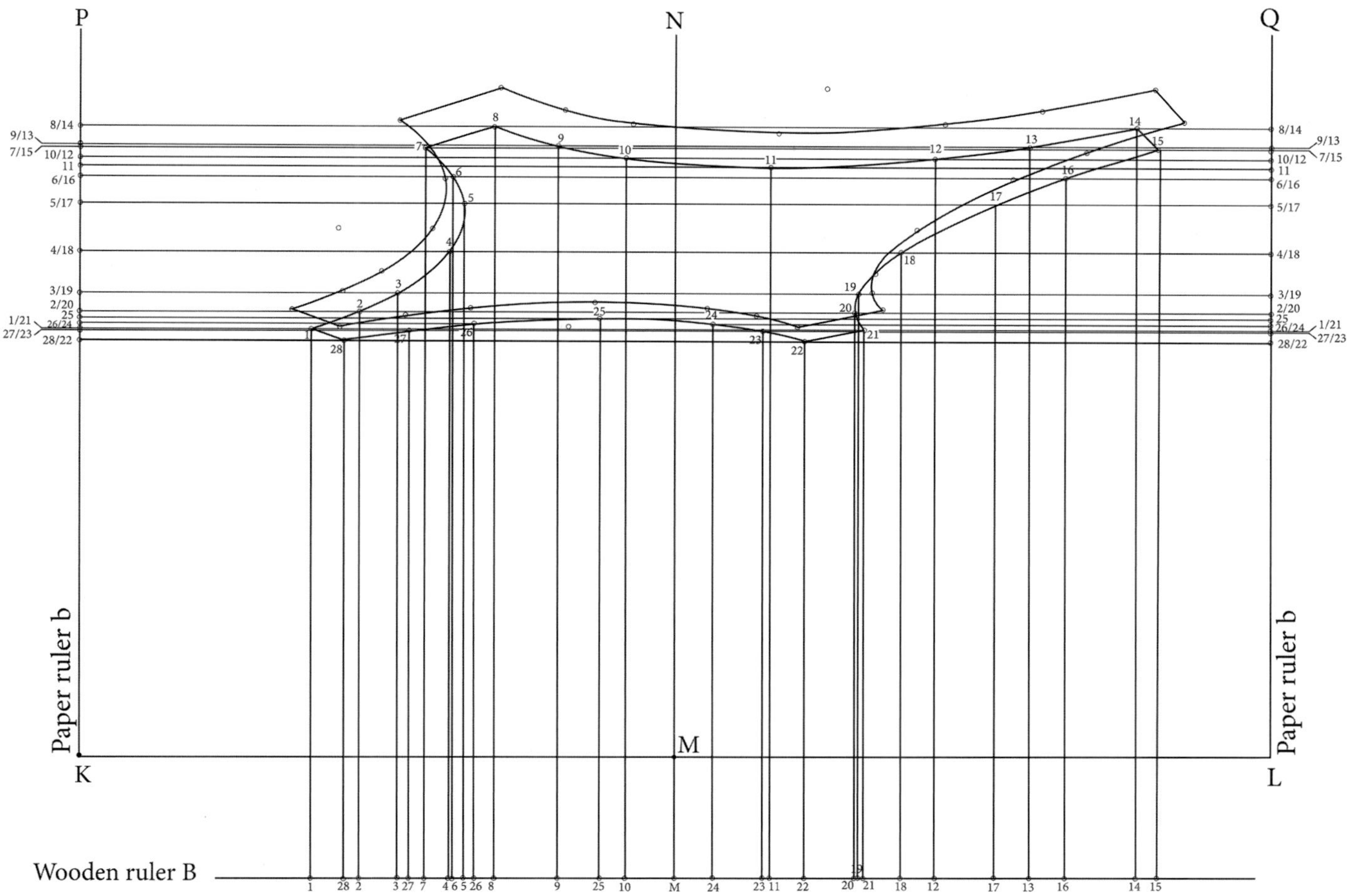

Figure B.95
Book 3, Proposition VII.

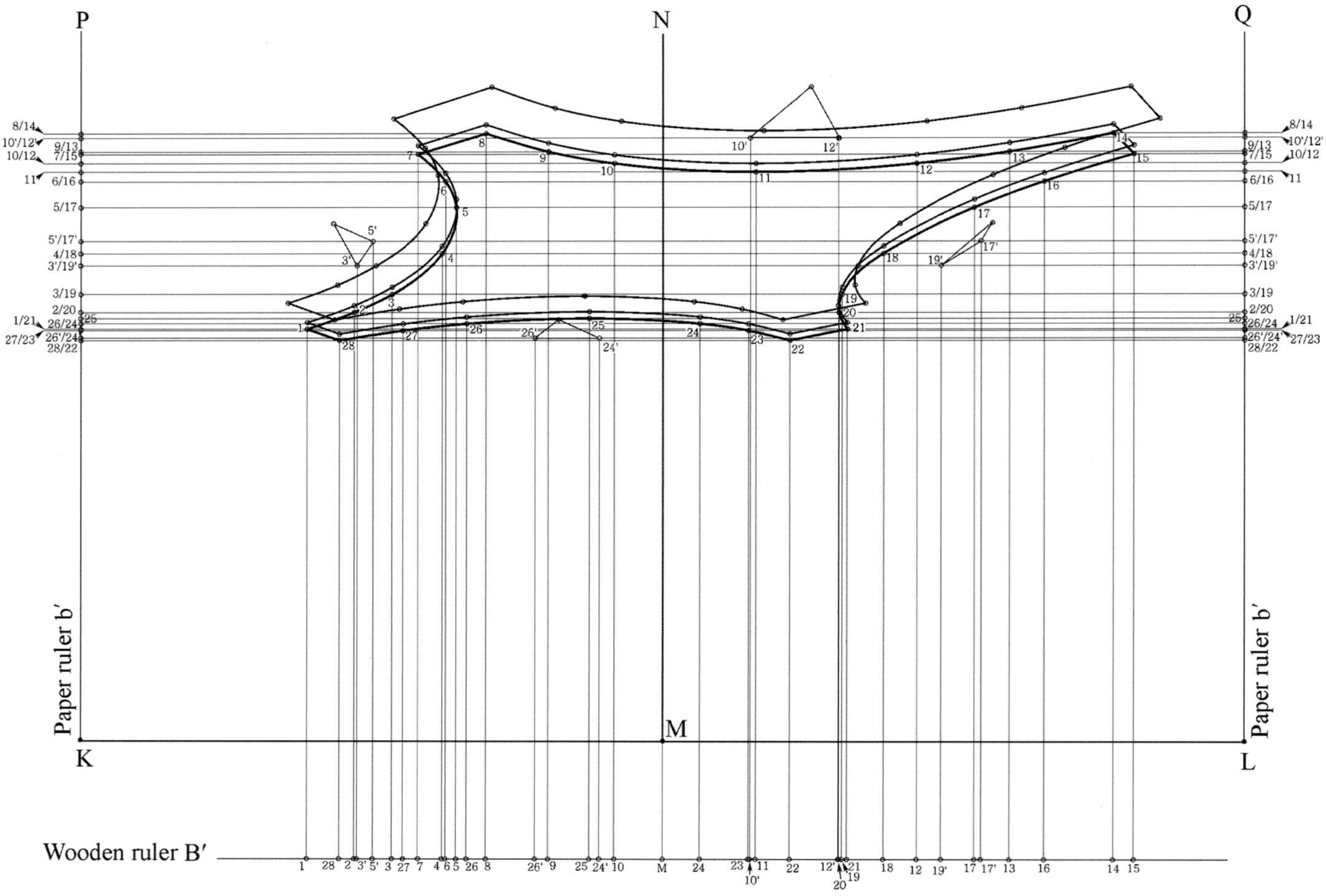

Figure B.96
Book 3, Proposition VII.

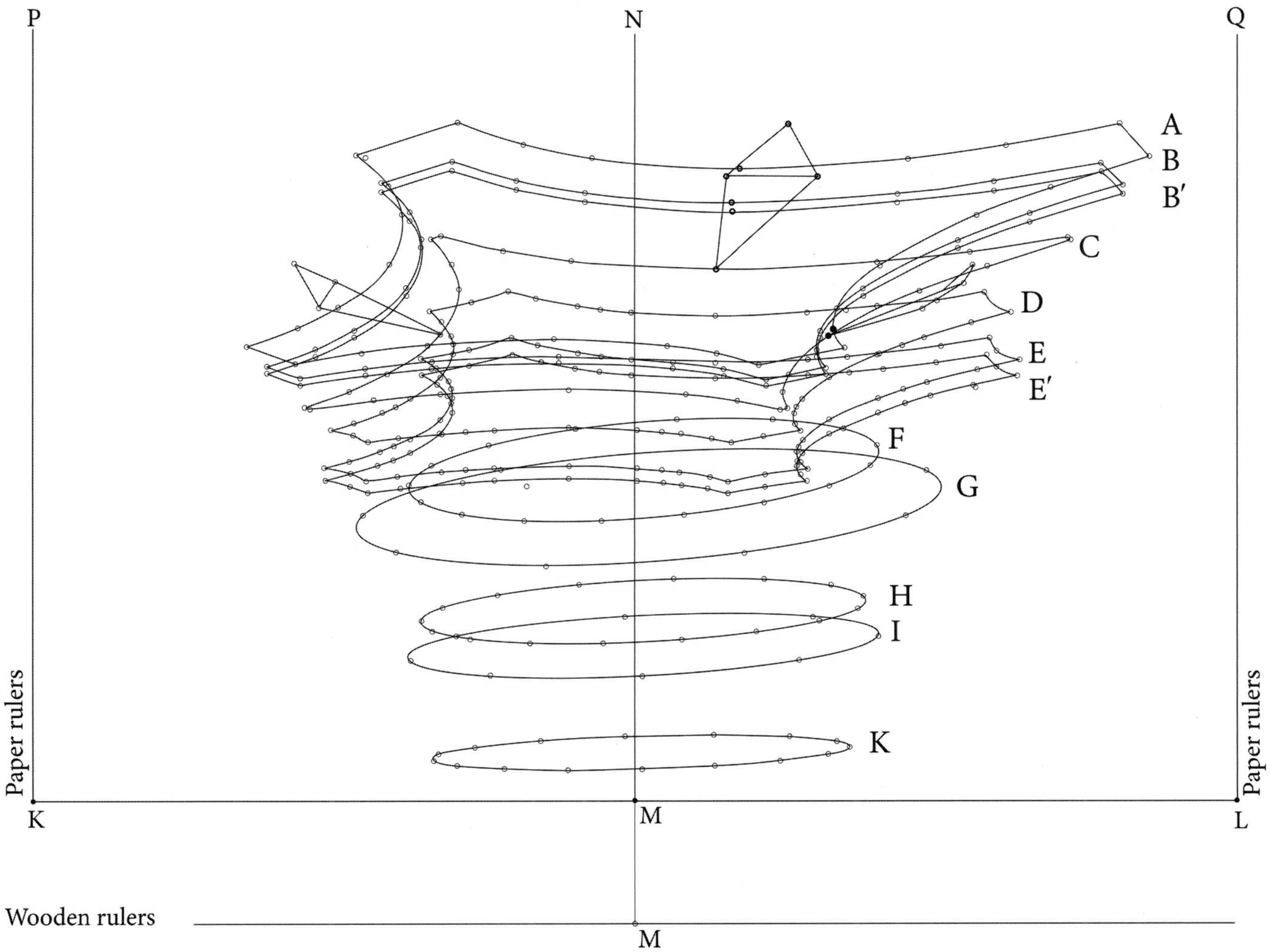

Figure B.97
Book 3, Proposition VII.

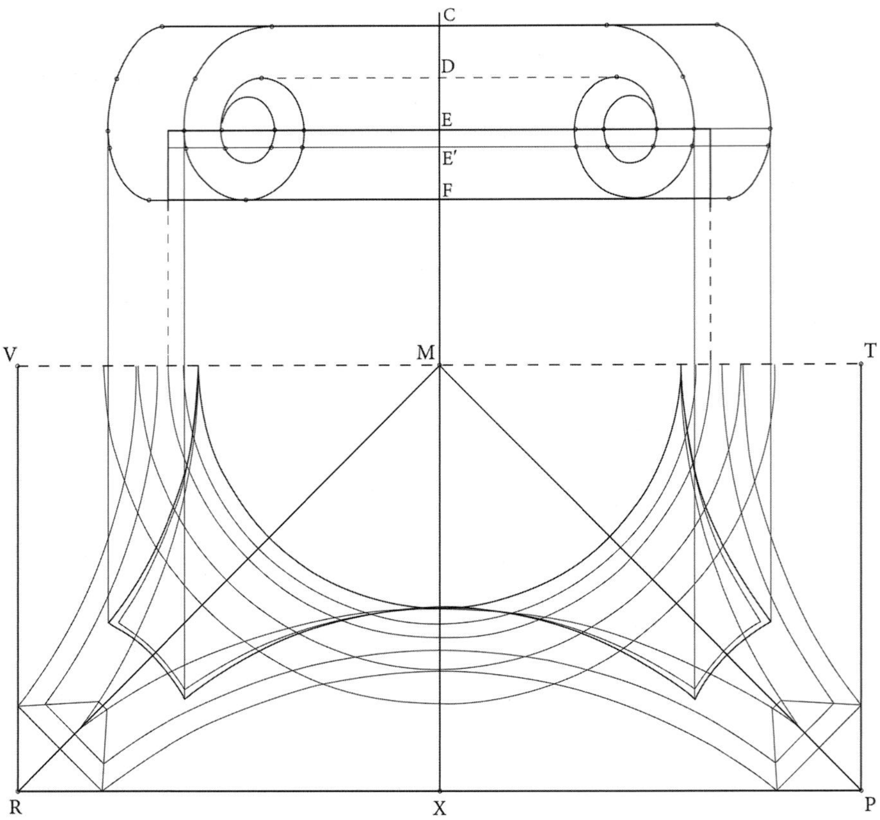

Figure B.98
Book 3, Proposition VII.

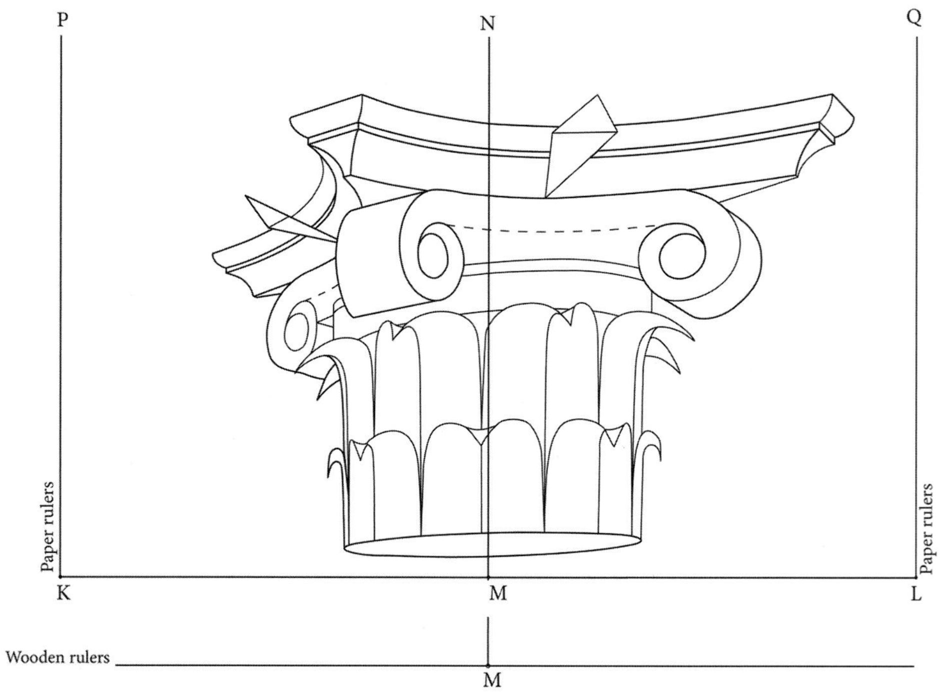

Figure B.99
Book 3, Proposition VII.

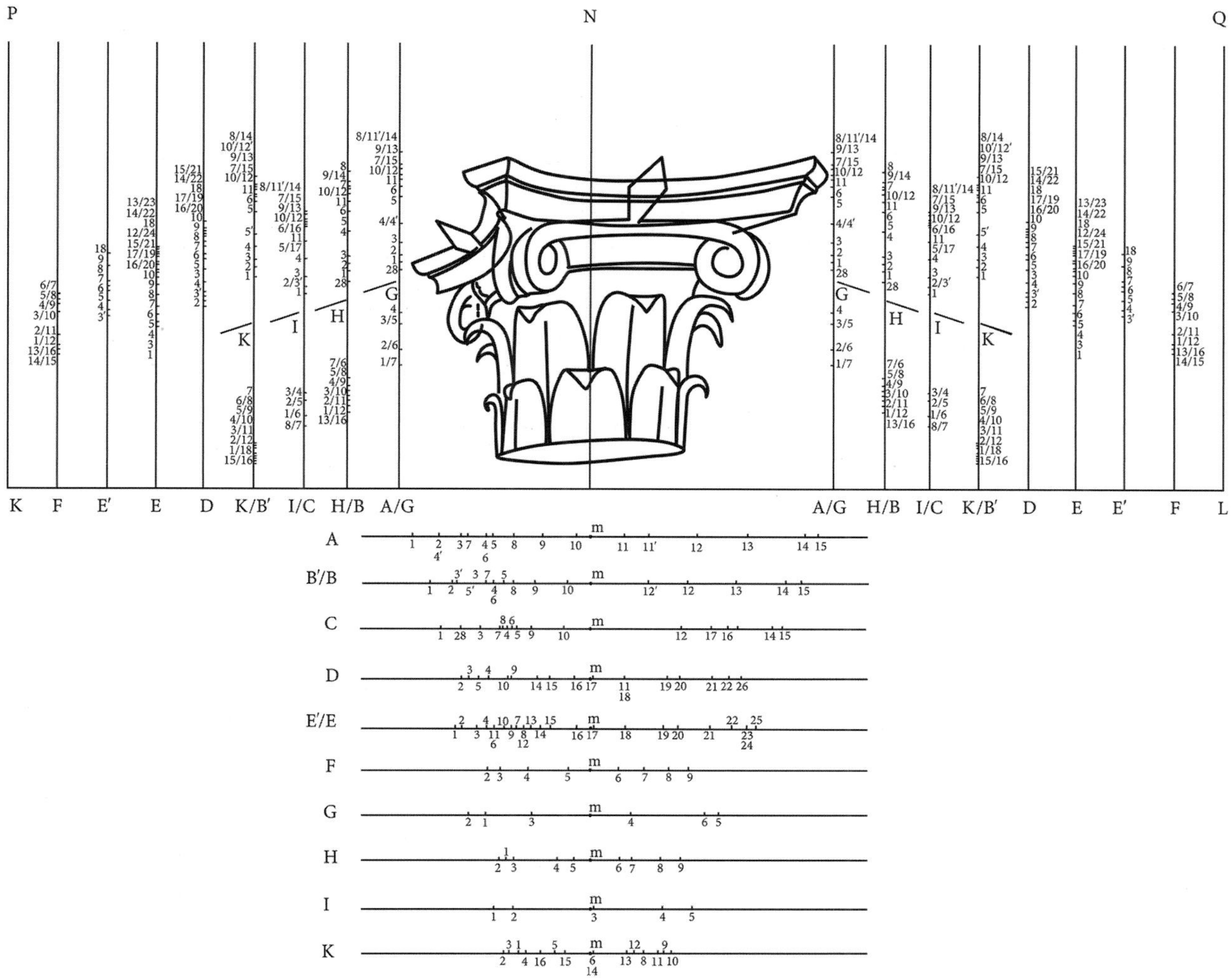

Figure B.100
Book 3, Proposition VII.

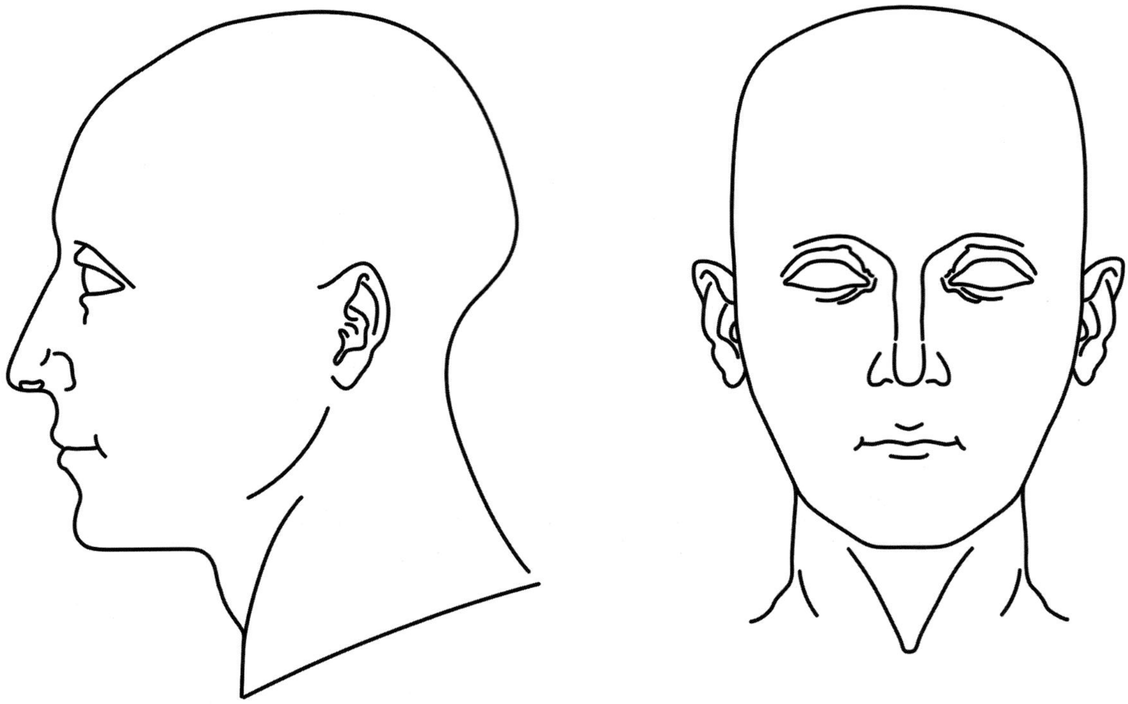

Figure B.101
Book 3, Proposition VIII.

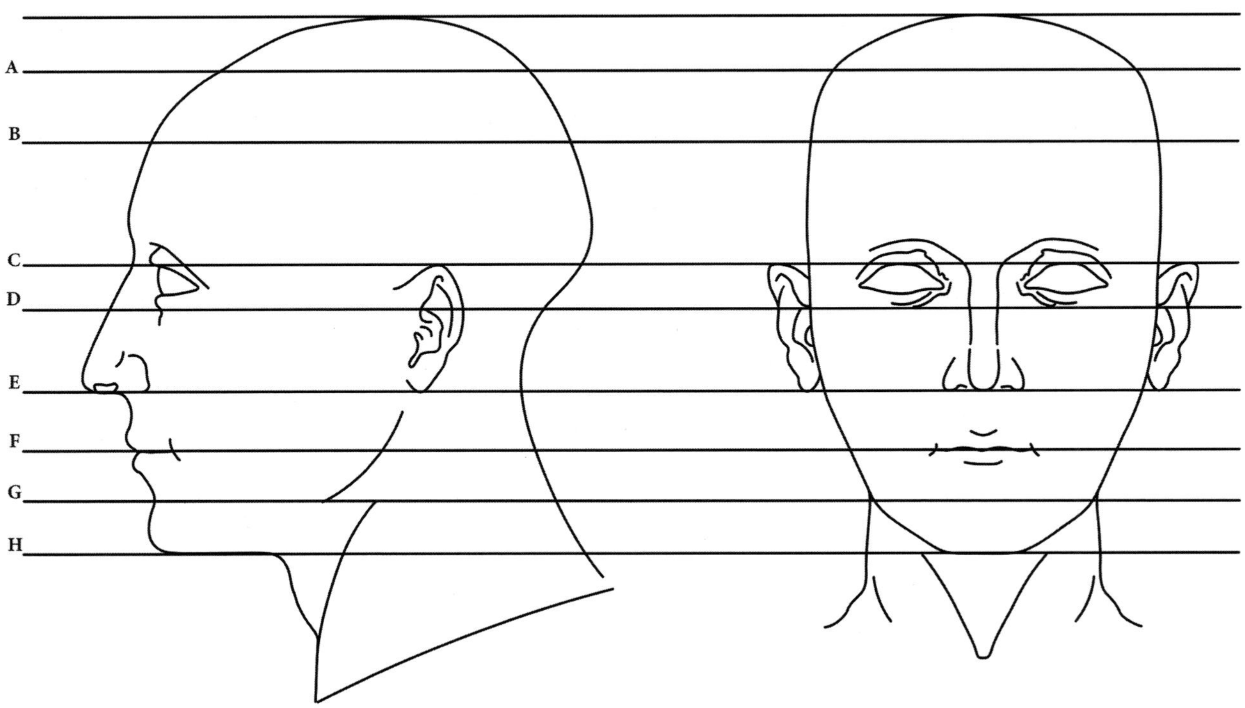

Figure B.102
Book 3, Proposition VIII.

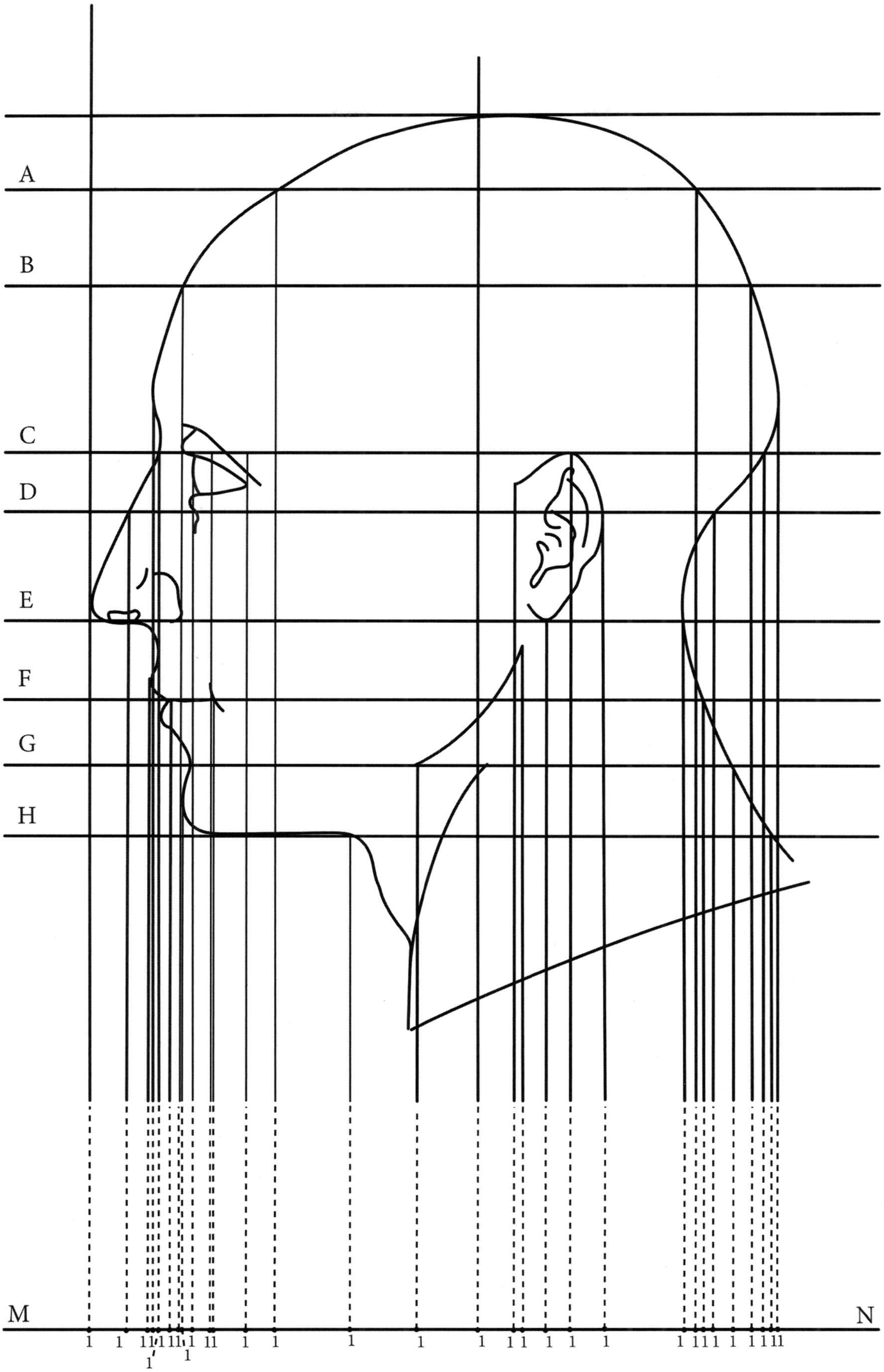

Figure B.103
Book 3, Proposition VIII.

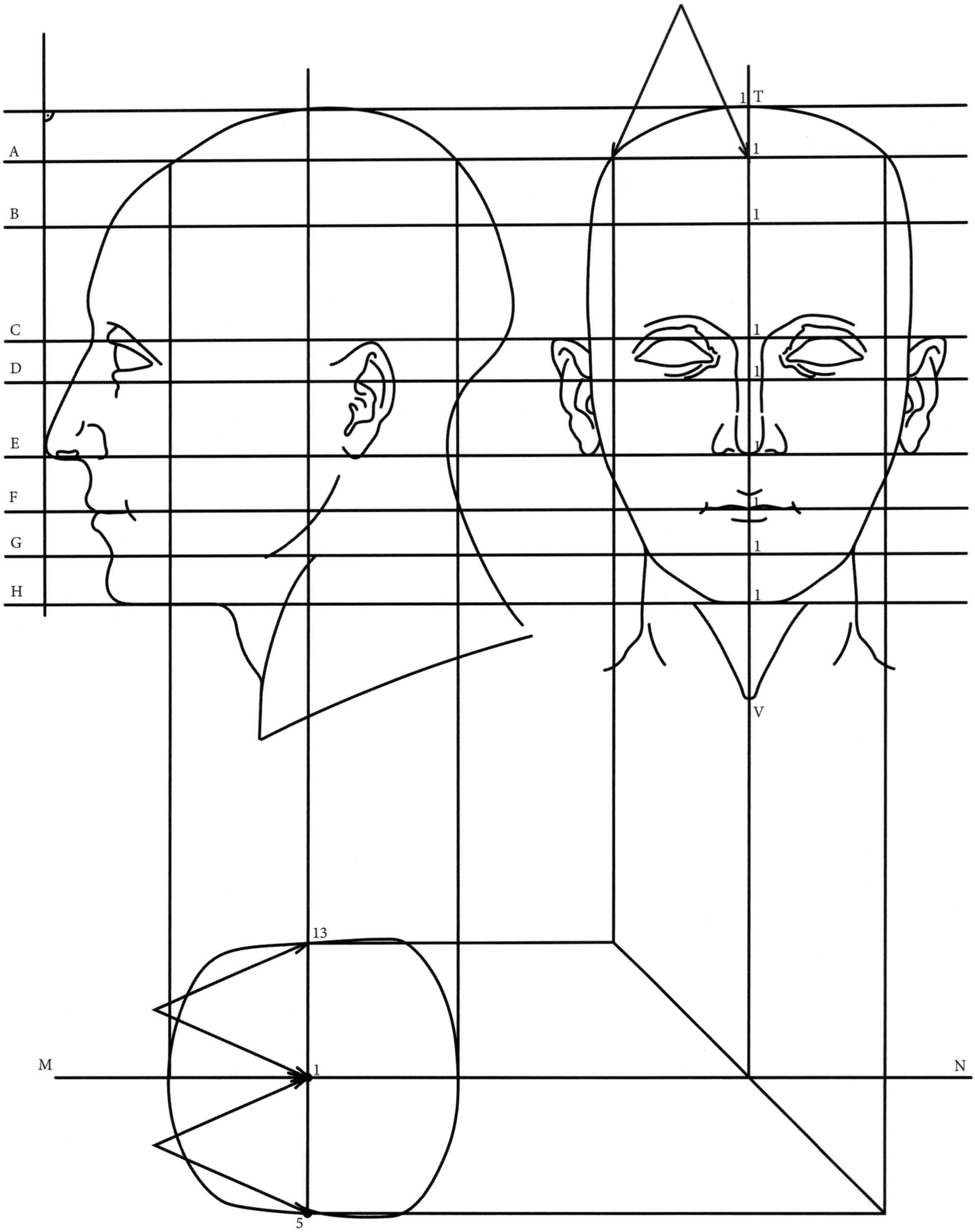

Figure B.104
Book 3, Proposition VIII.

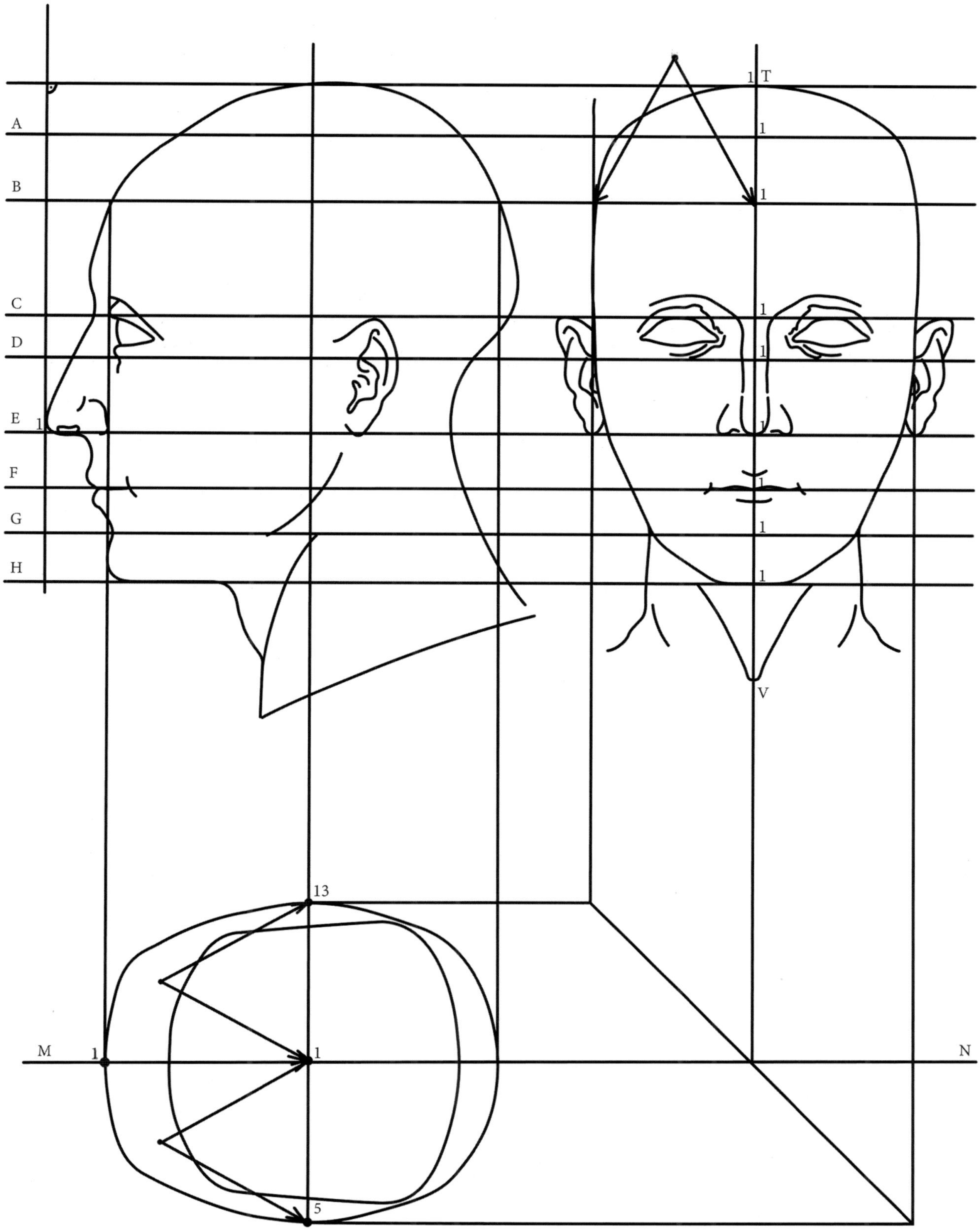

Figure B.105
Book 3, Proposition VIII.

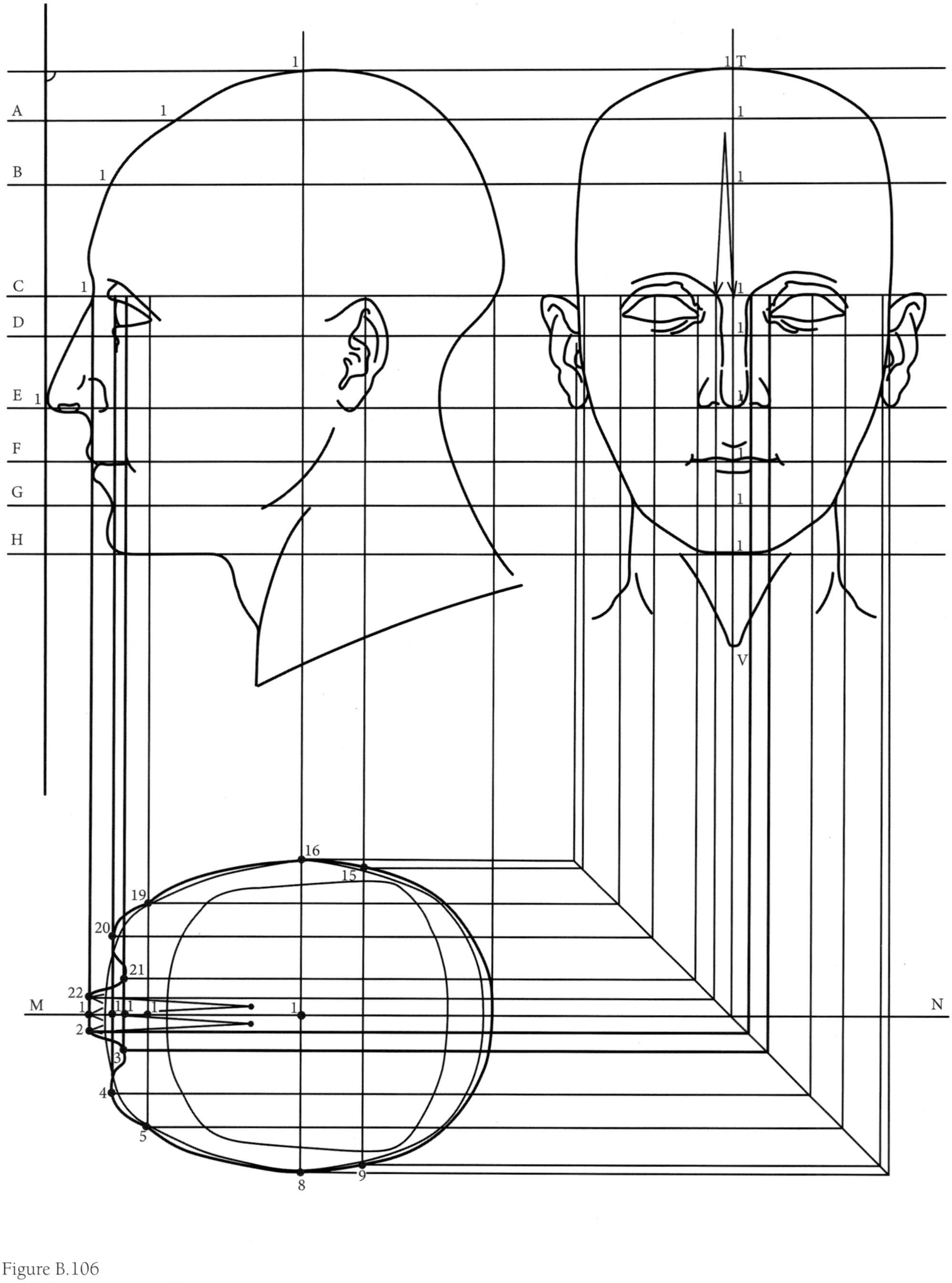

Figure B.106
Book 3, Proposition VIII.

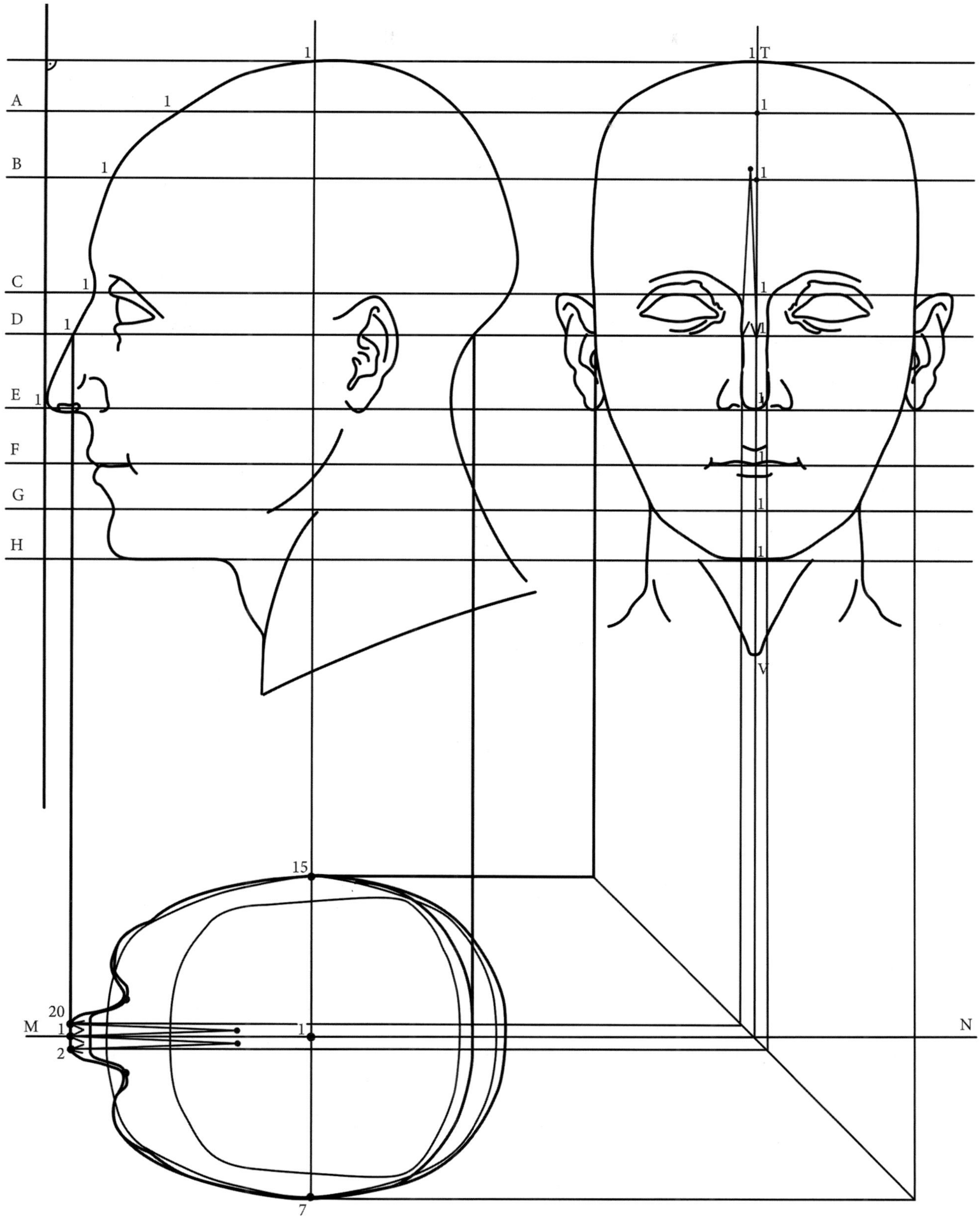

Figure B.107
Book 3, Proposition VIII.

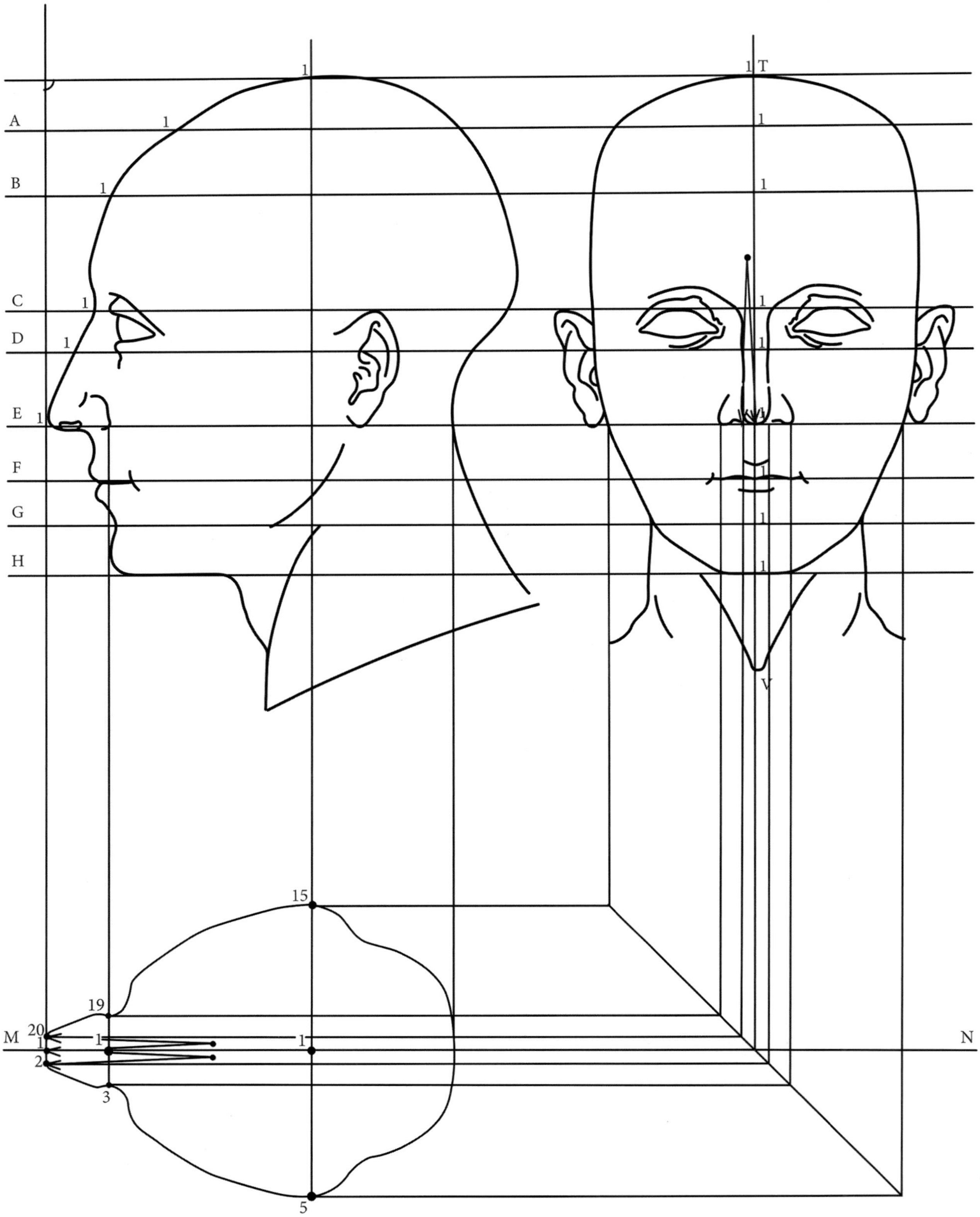

Figure B.108
Book 3, Proposition VIII.

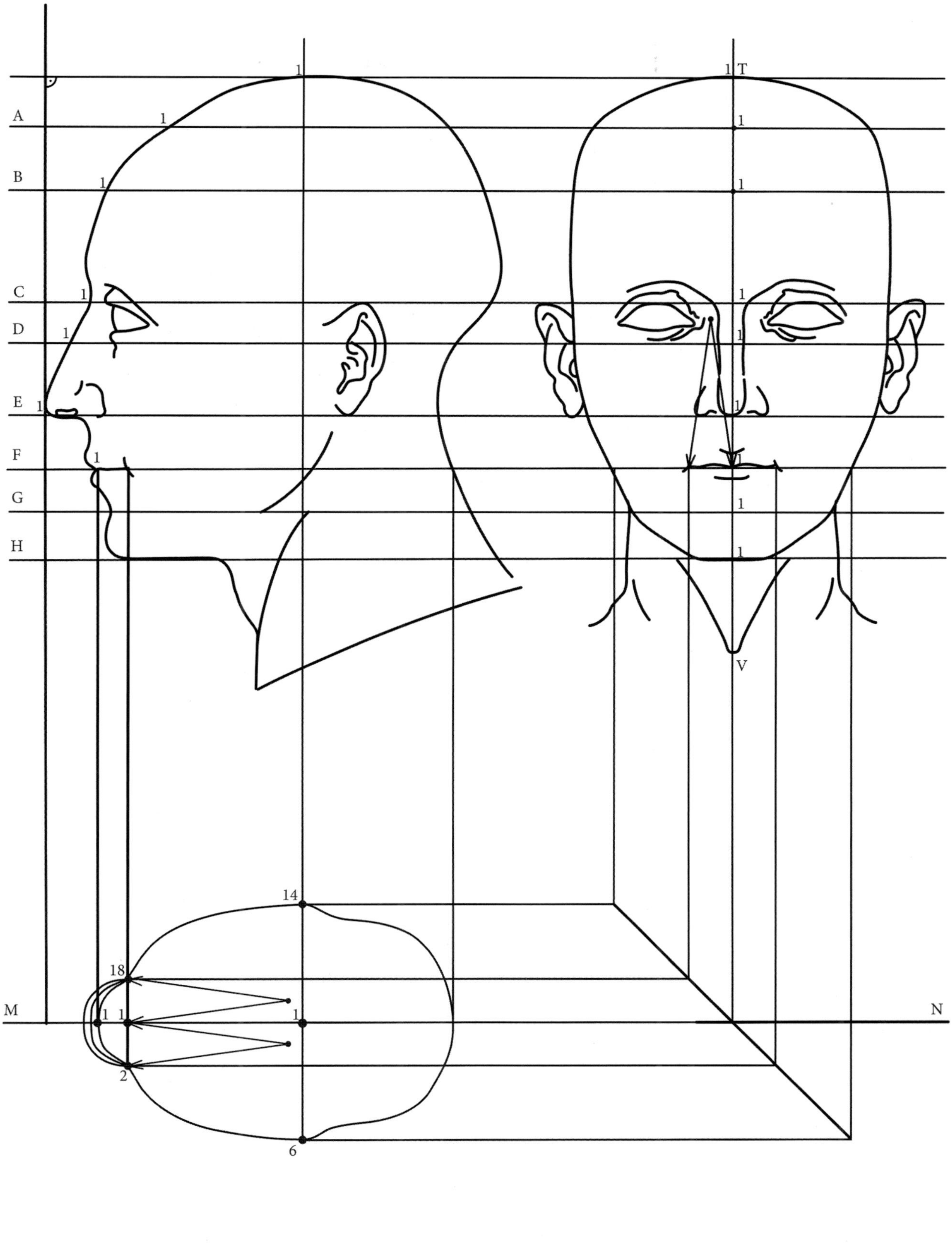

Figure B.109
Book 3, Proposition VIII.

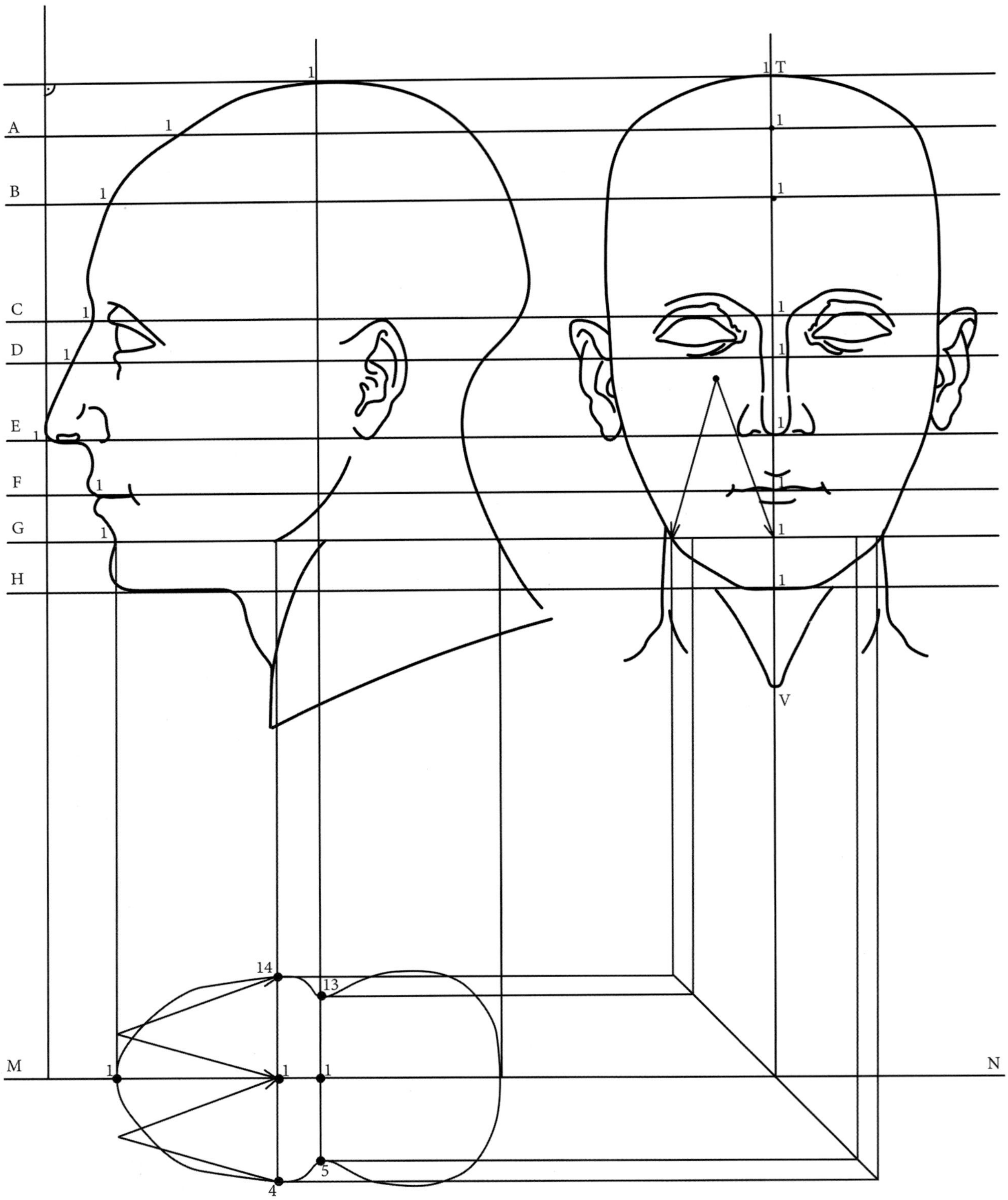

Figure B.110
Book 3, Proposition VIII.

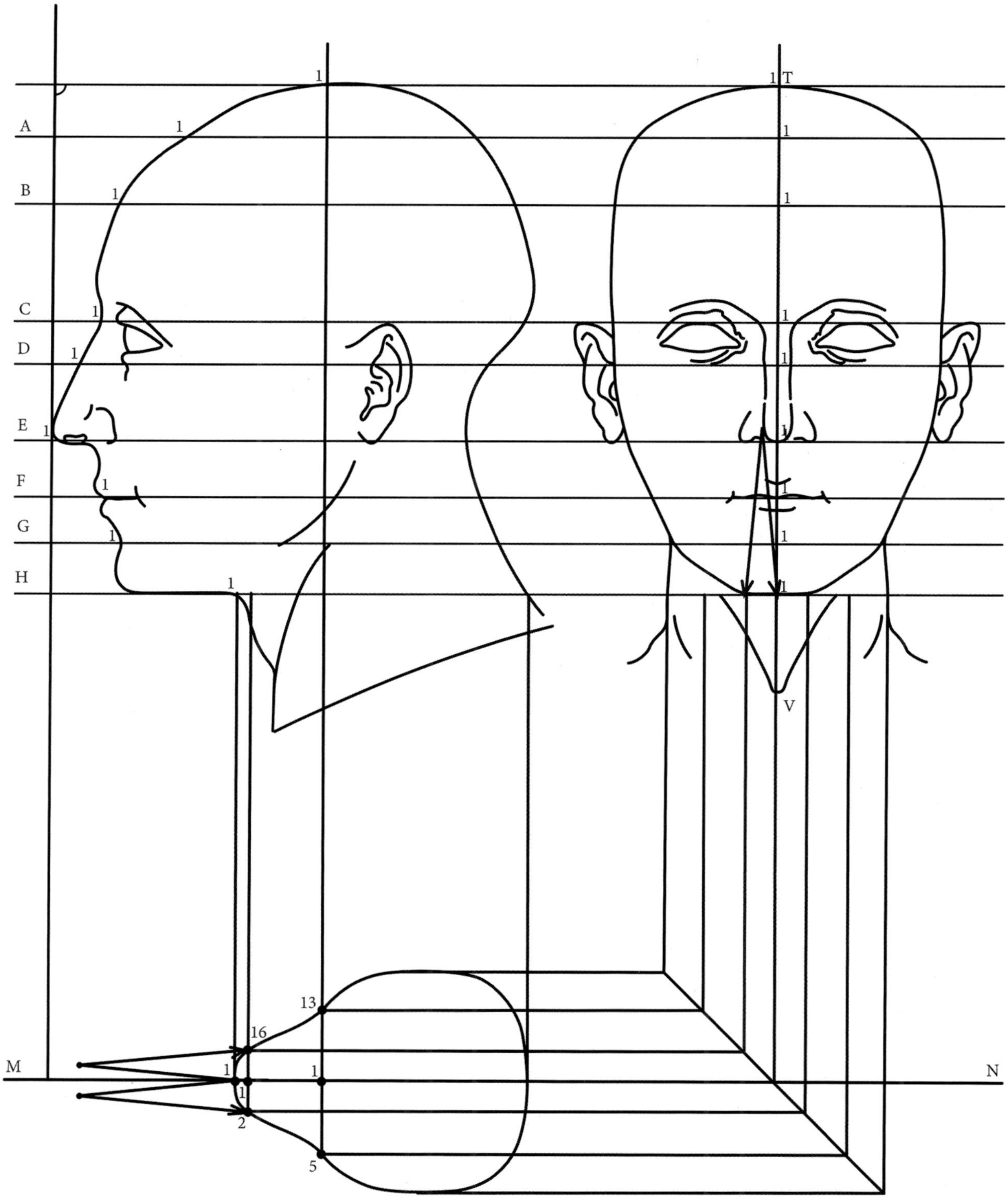

Figure B.111
Book 3, Proposition VIII.

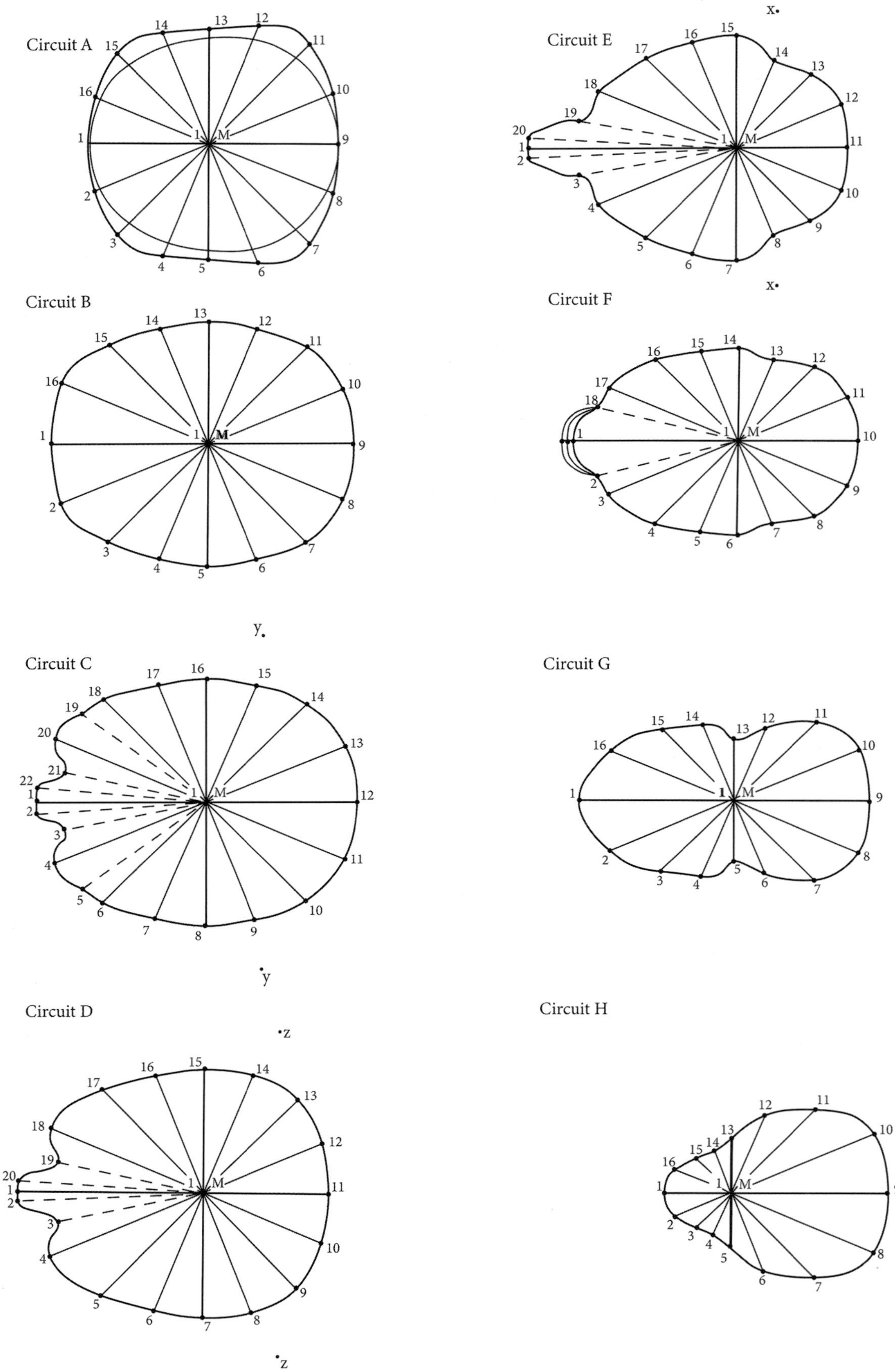

Figure B.112
Book 3, Proposition VIII.

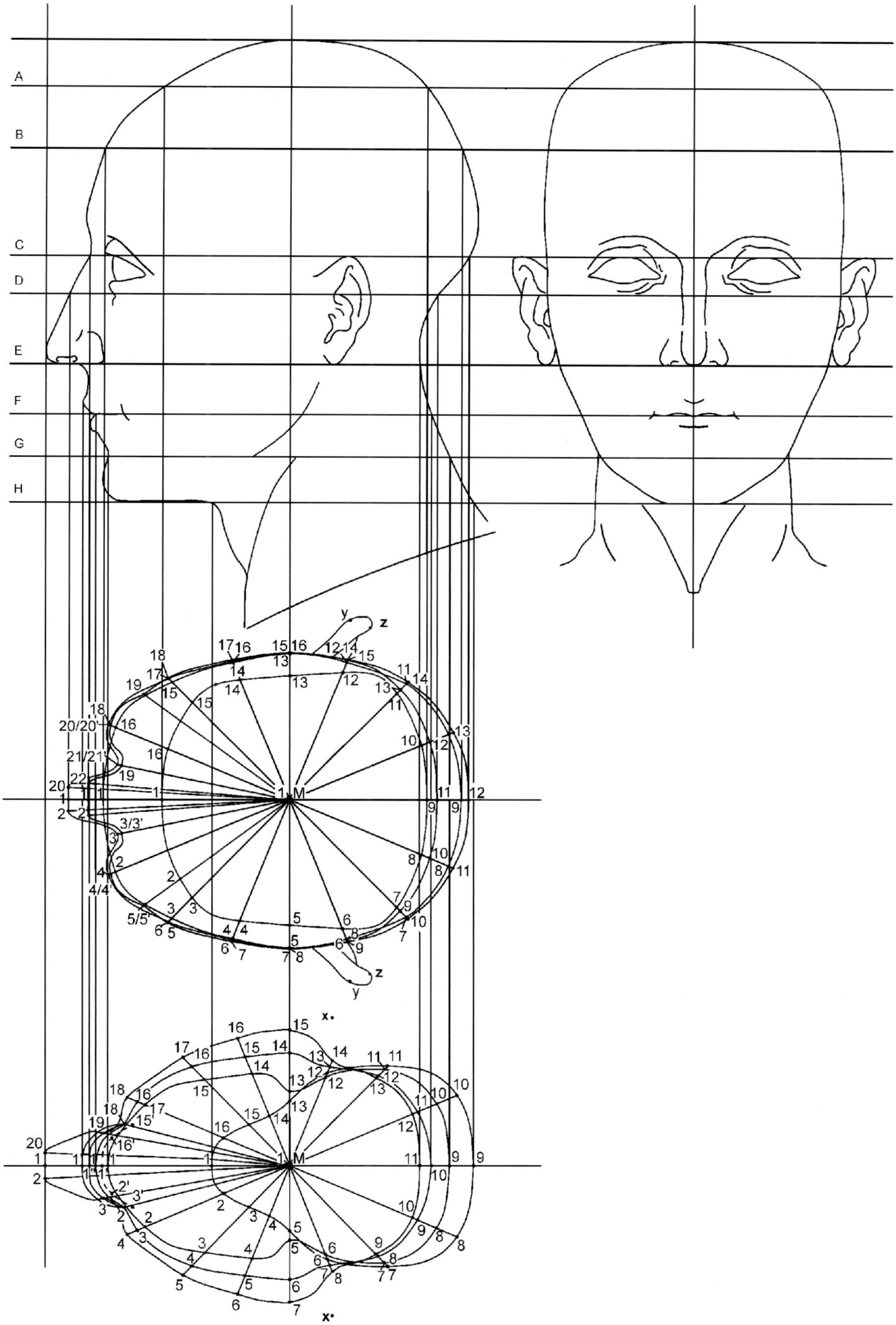

Figure B.113
Book 3, Proposition VIII.

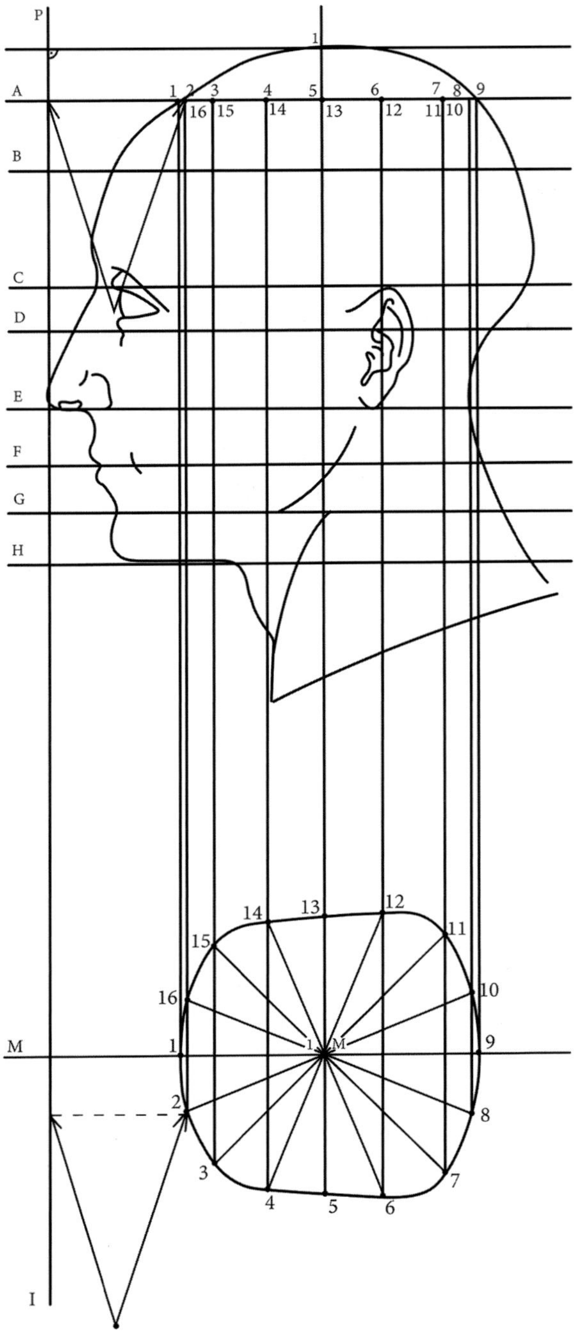

Figure B.114
Book 3, Proposition VIII.

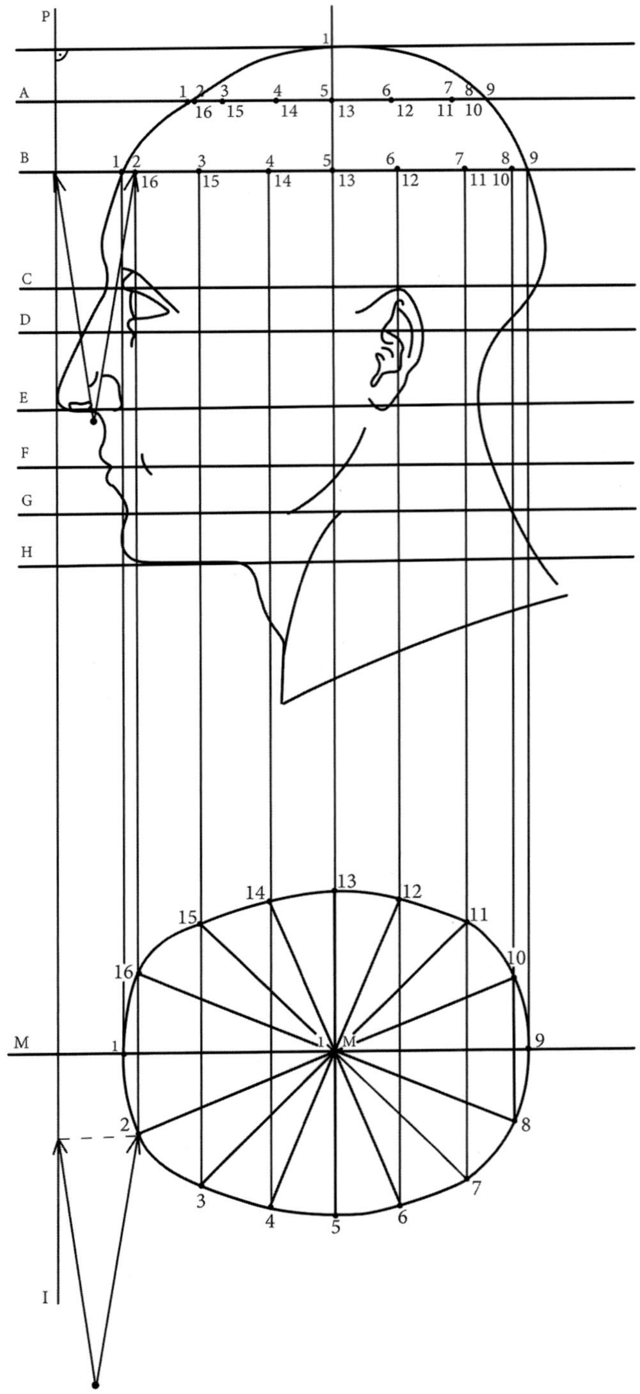

Figure B.115
Book 3, Proposition VIII.

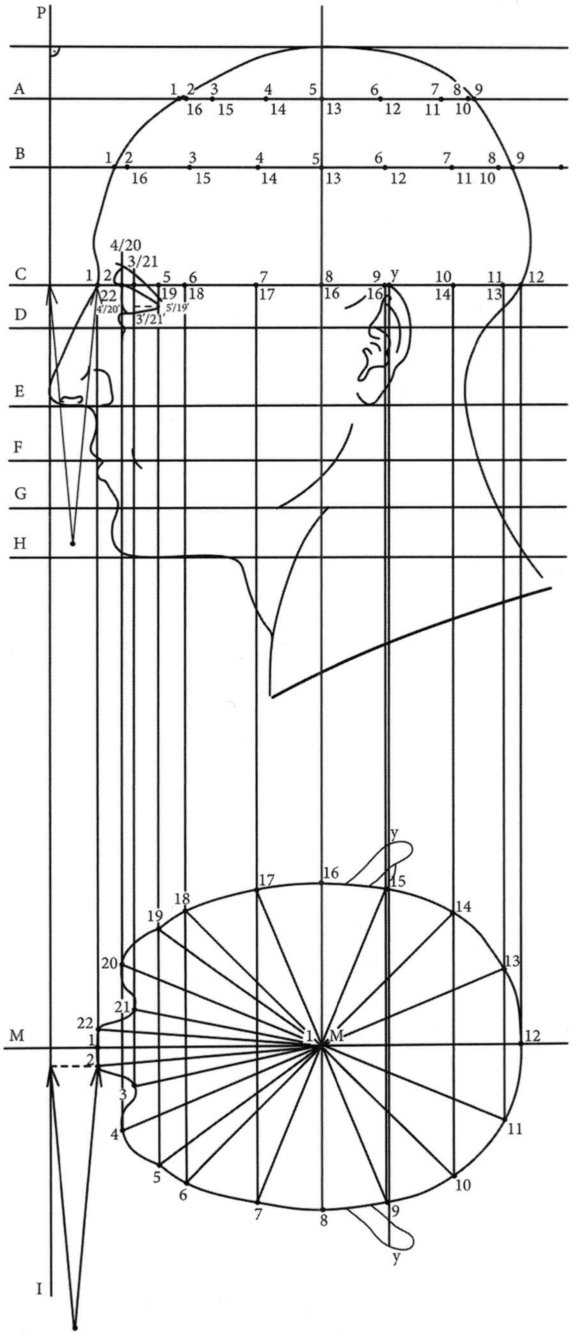

Figure B.116
Book 3, Proposition VIII.

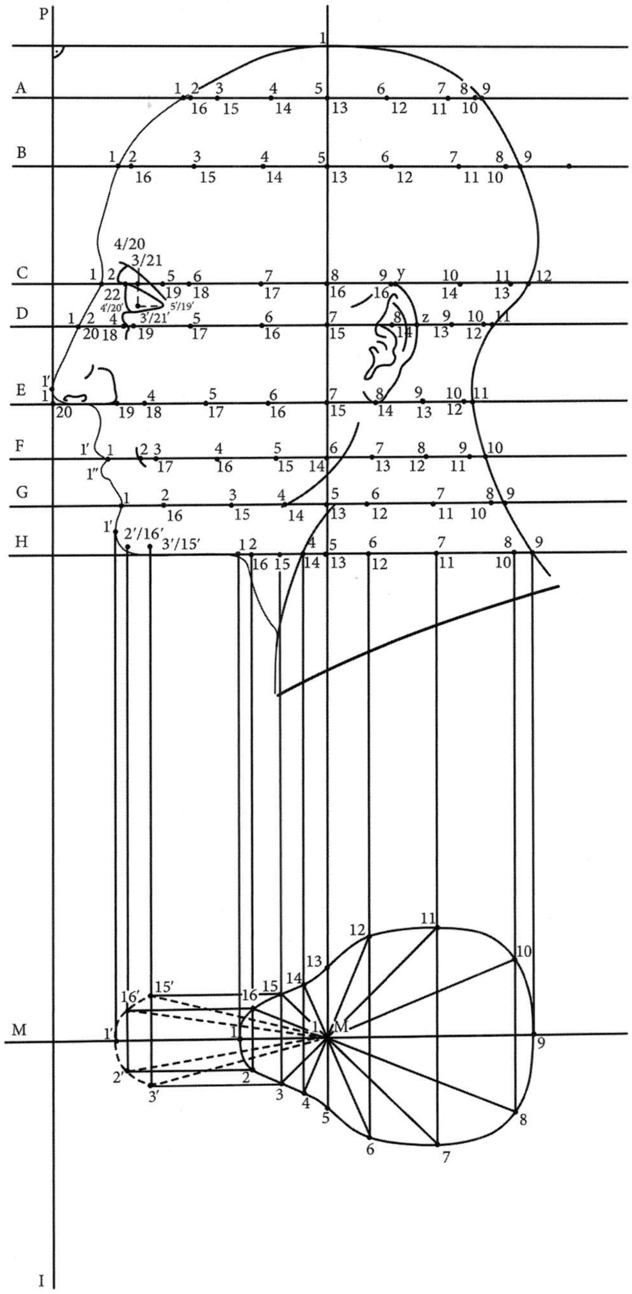

Figure B.117
Book 3, Proposition VIII.

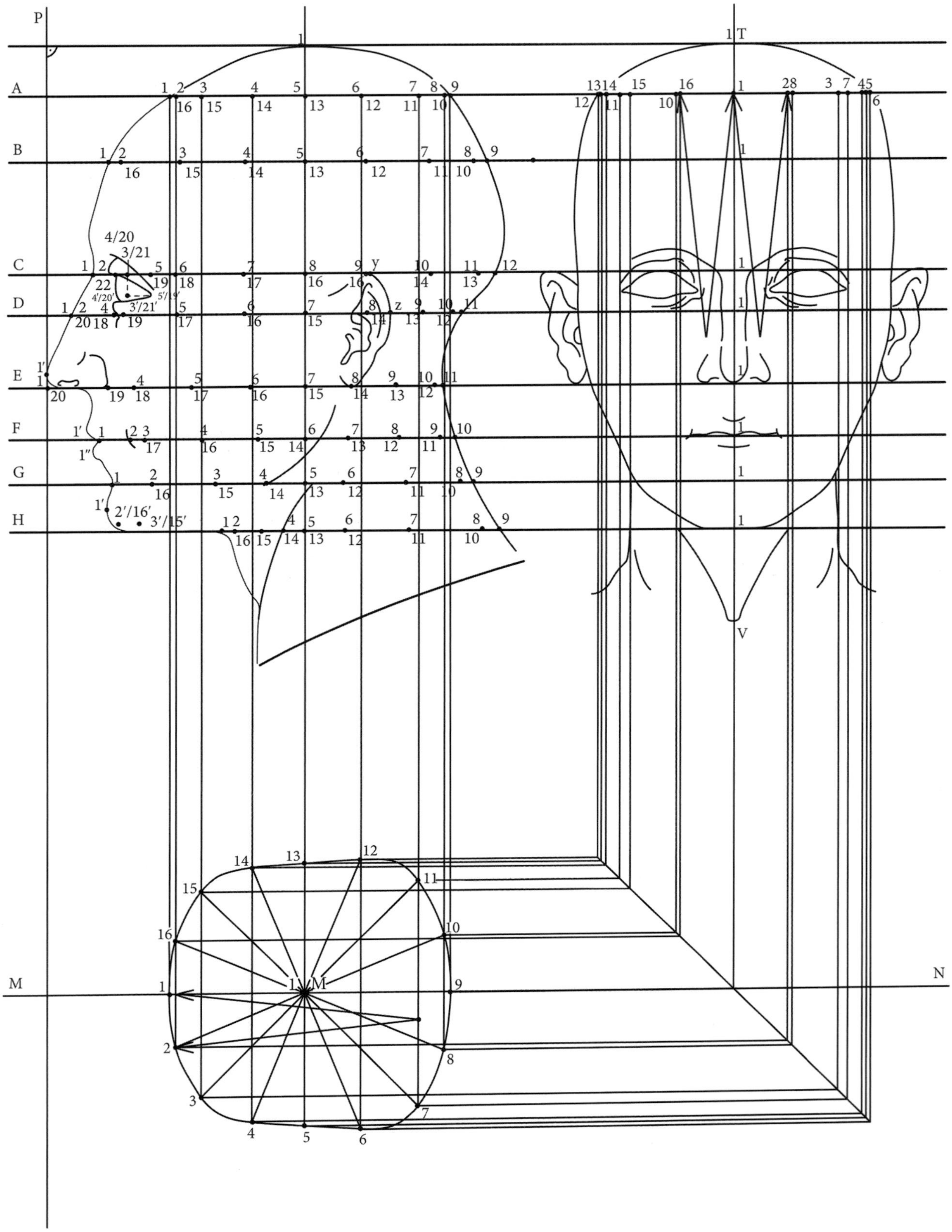

Figure B.118
Book 3, Proposition VIII.

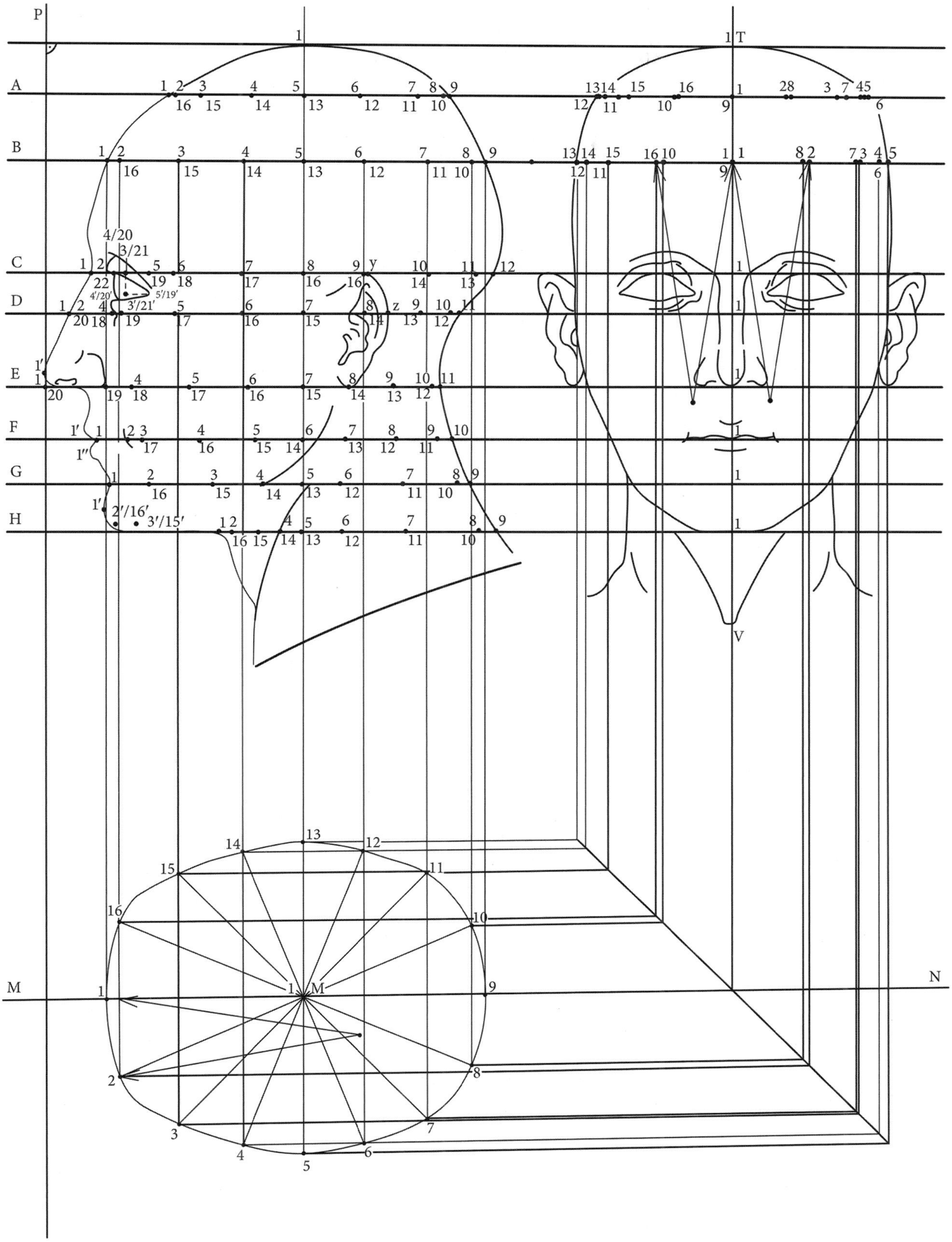

Figure B.119
Book 3, Proposition VIII.

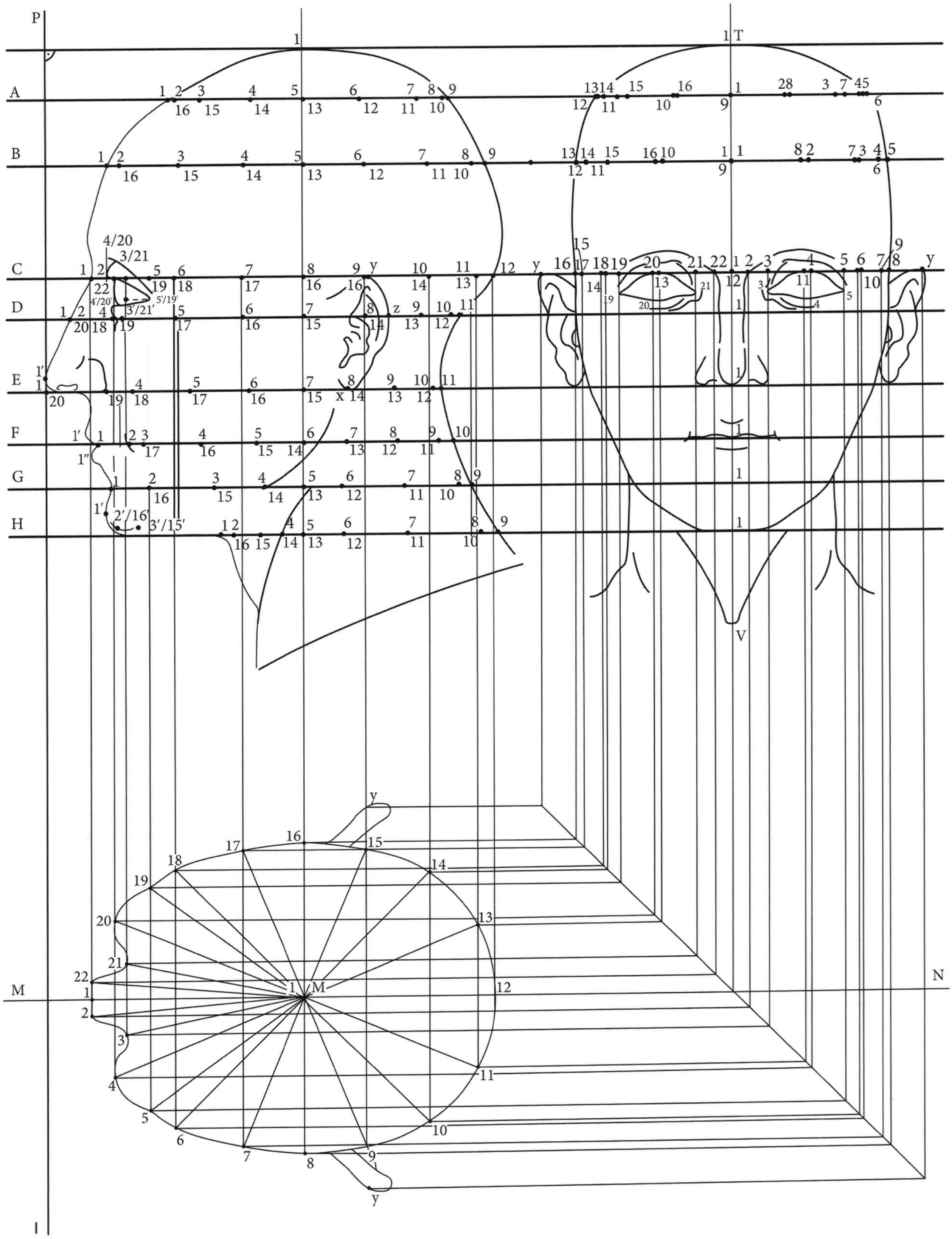

Figure B.120
Book 3, Proposition VIII.

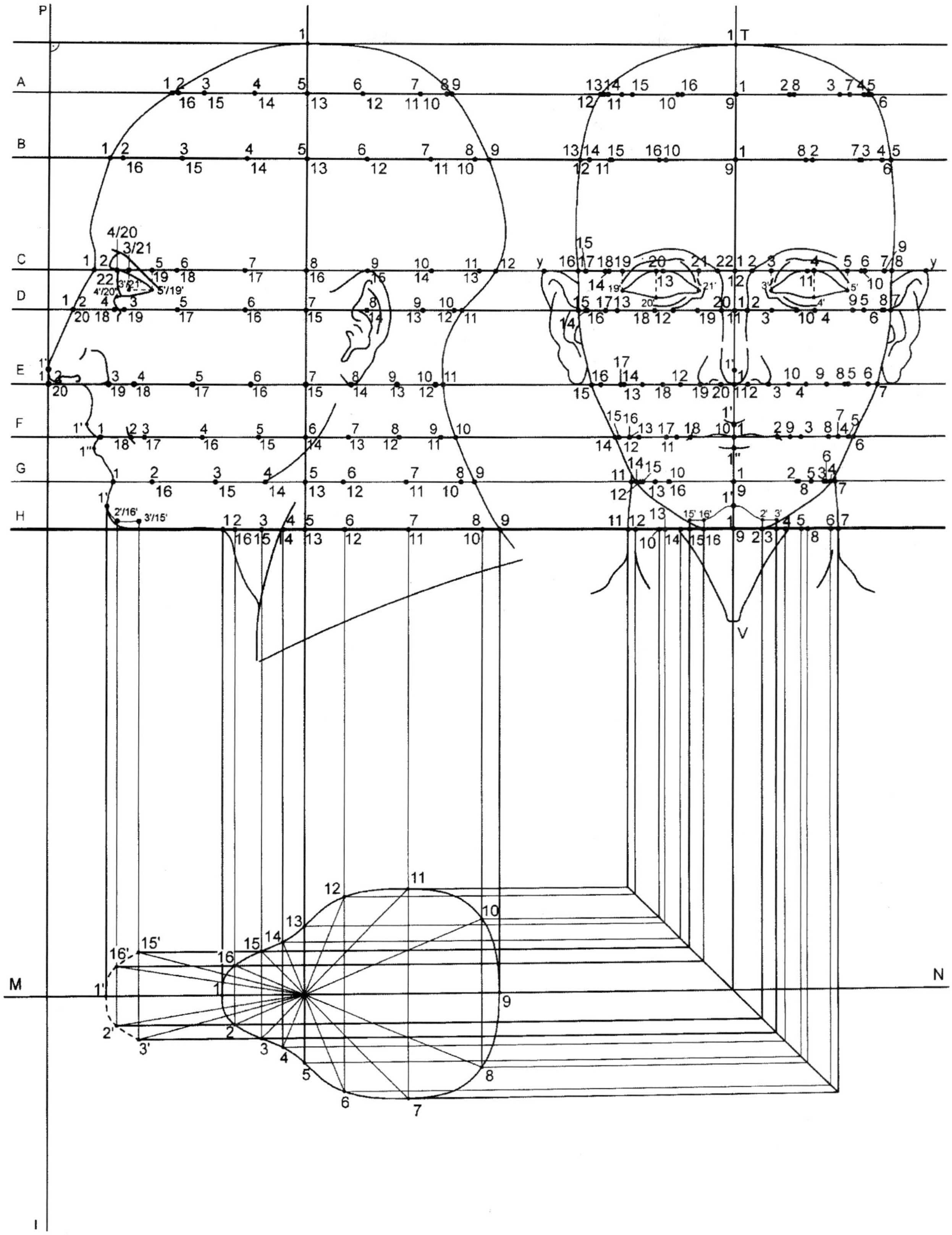

Figure B.121
Book 3, Proposition VIII.

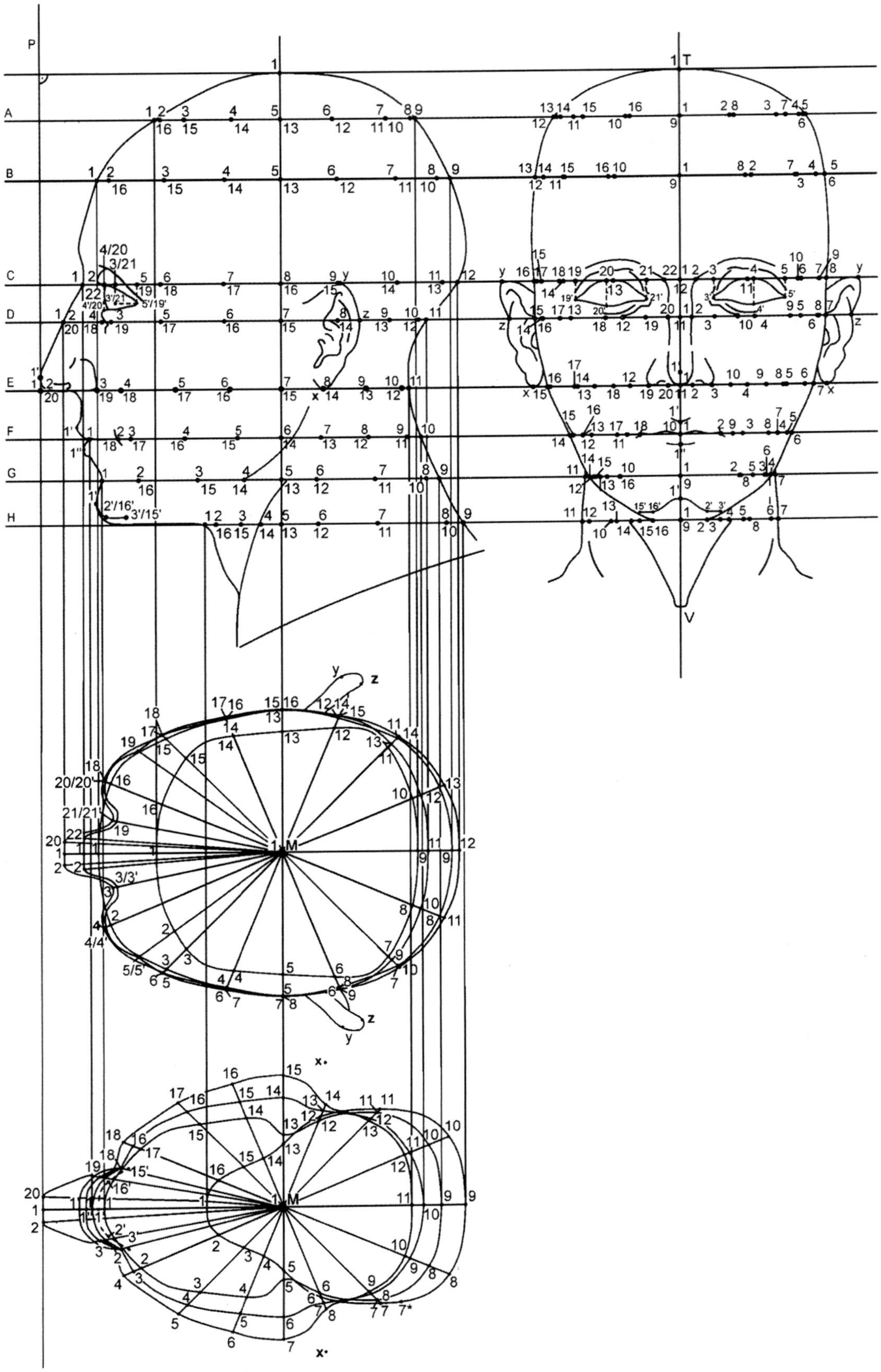

Figure B.122
Book 3, Proposition VIII.

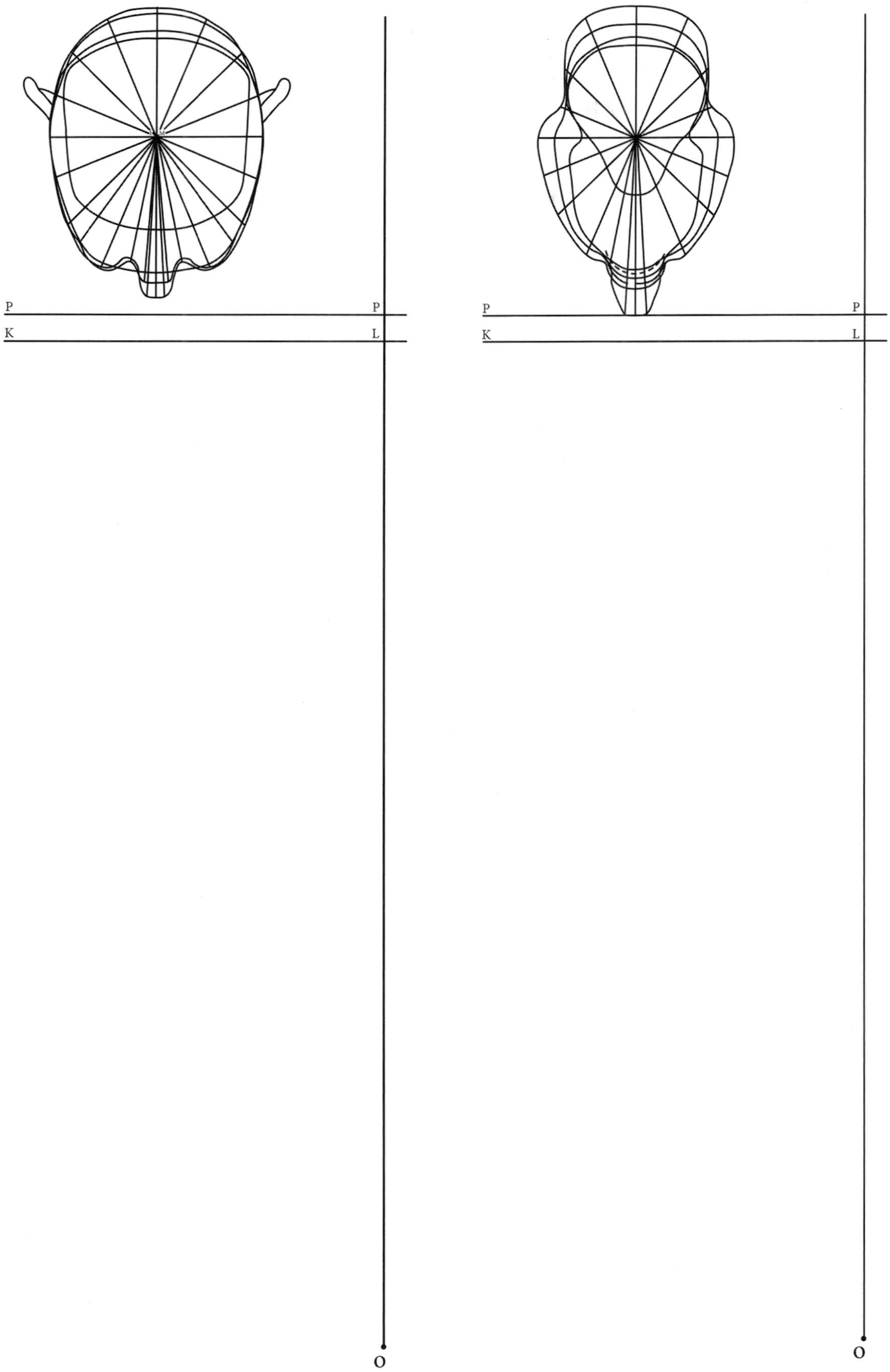

Figure B.123
Book 3, Proposition VIII.

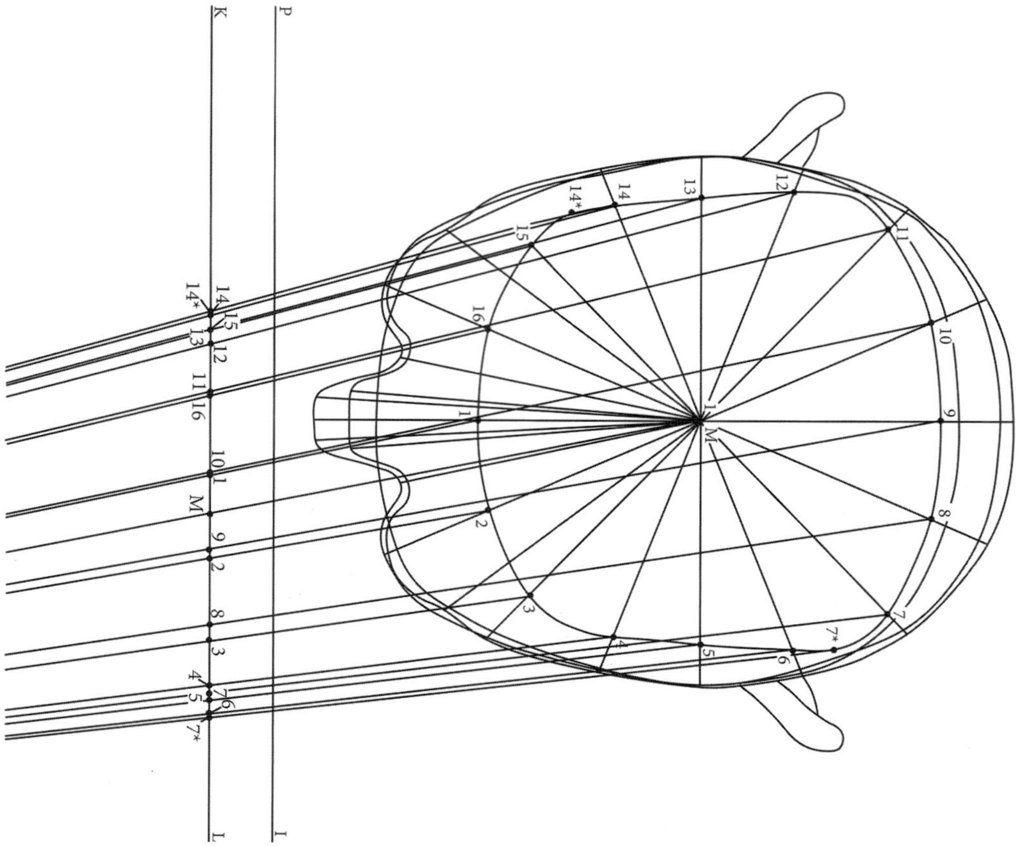

Figure B.124
Book 3, Proposition VIII.

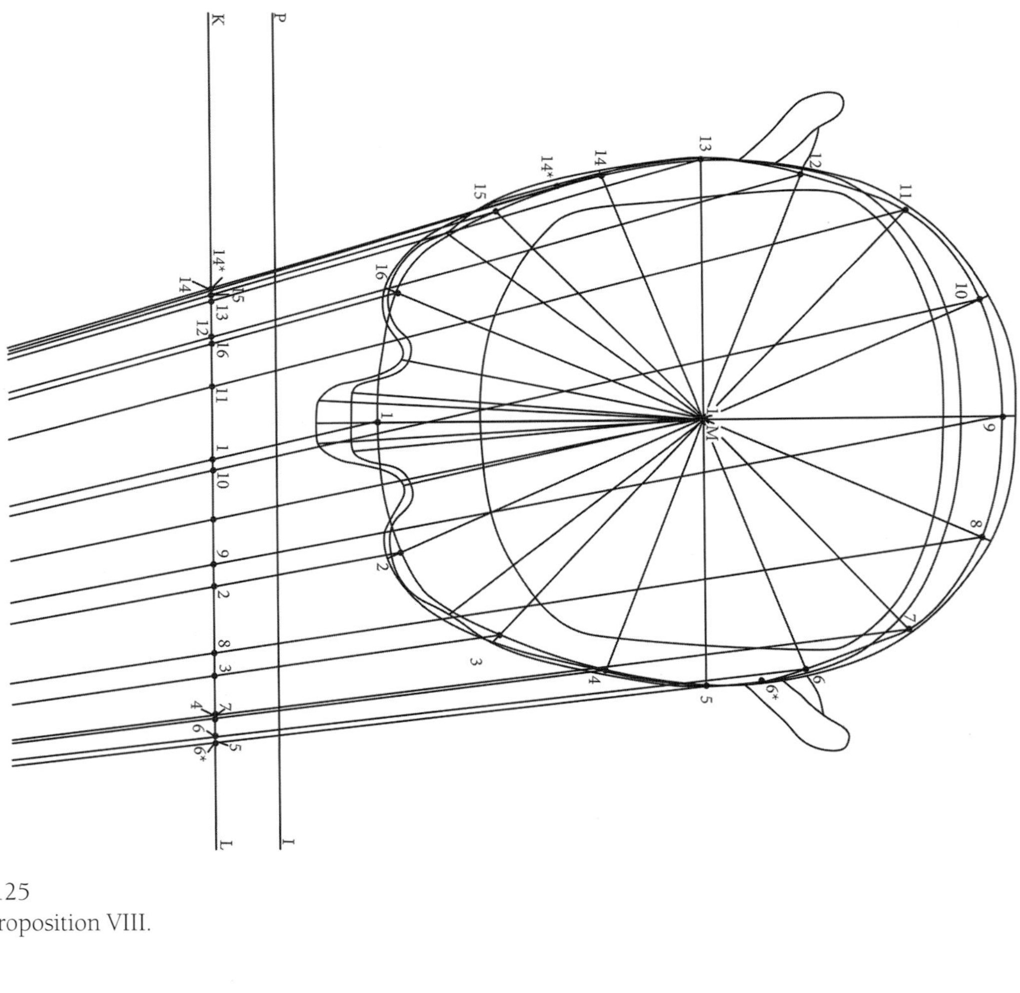

Figure B.125
Book 3, Proposition VIII.

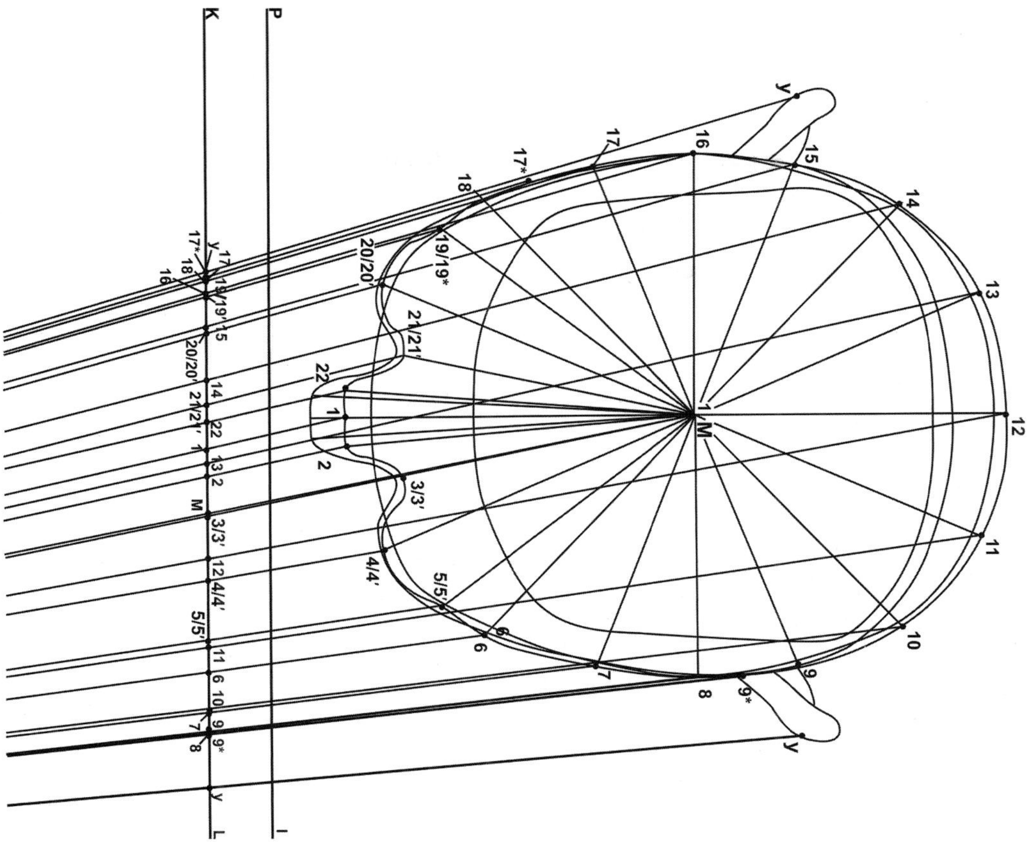

Figure B.126
Book 3, Proposition VIII.

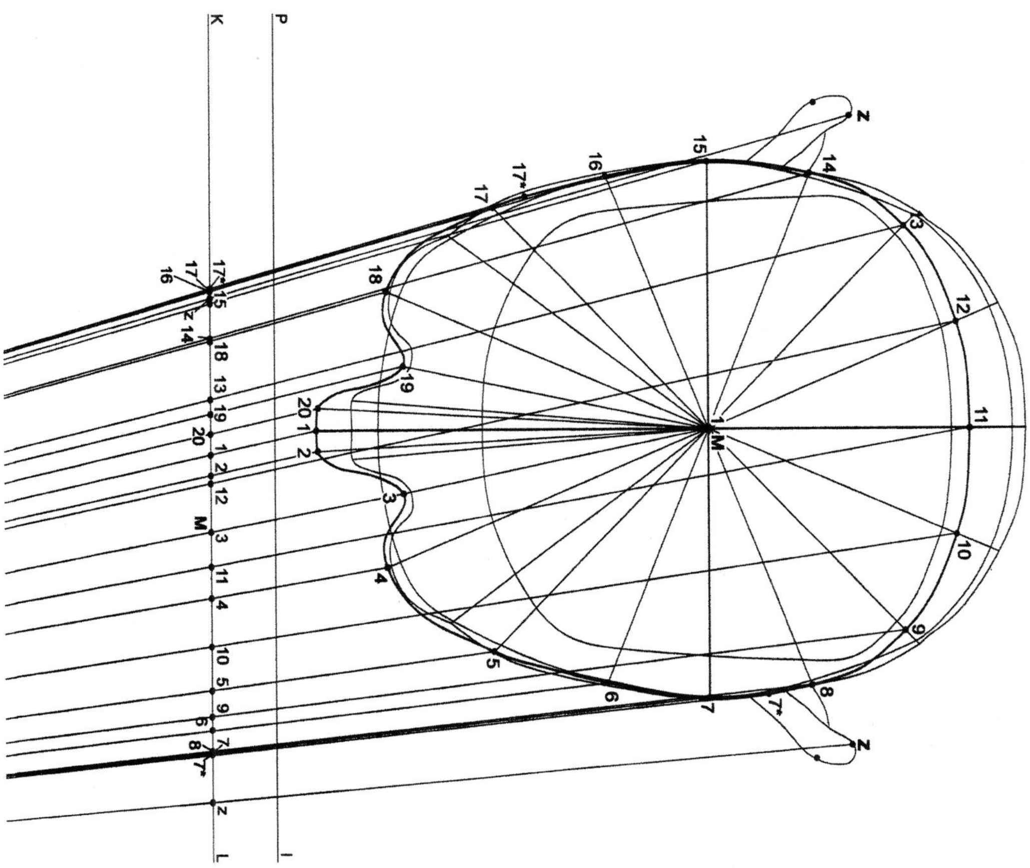

Figure B.127
Book 3, Proposition VIII.

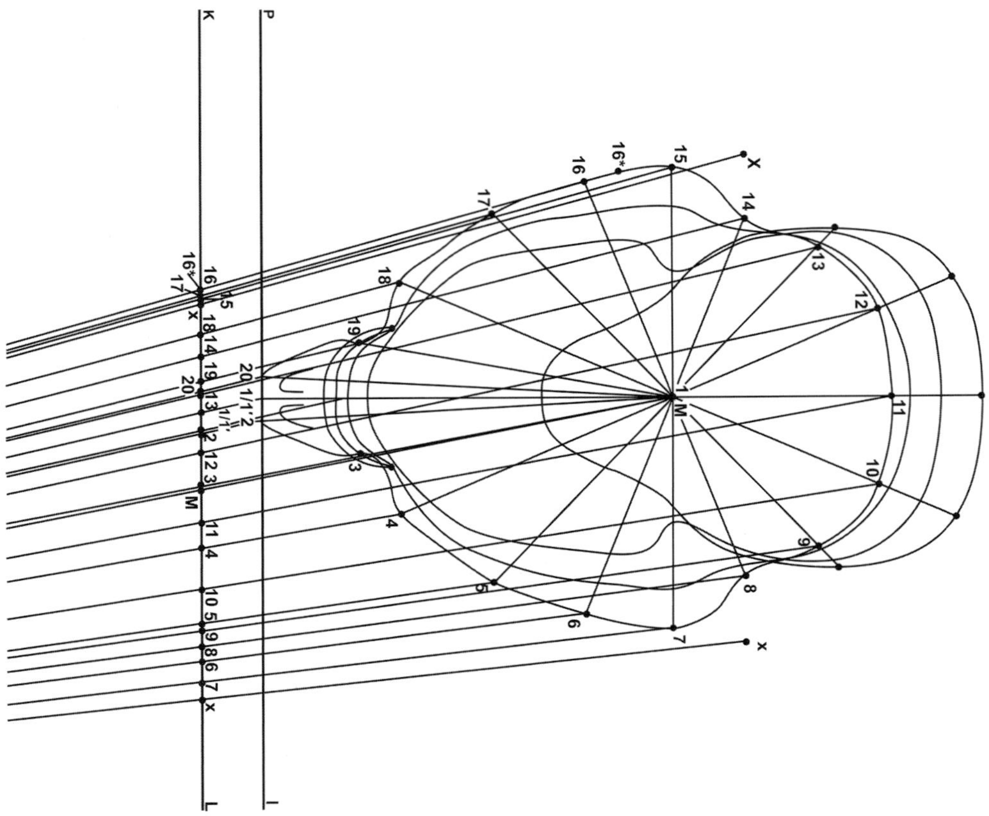

Figure B.128
Book 3, Proposition VIII.

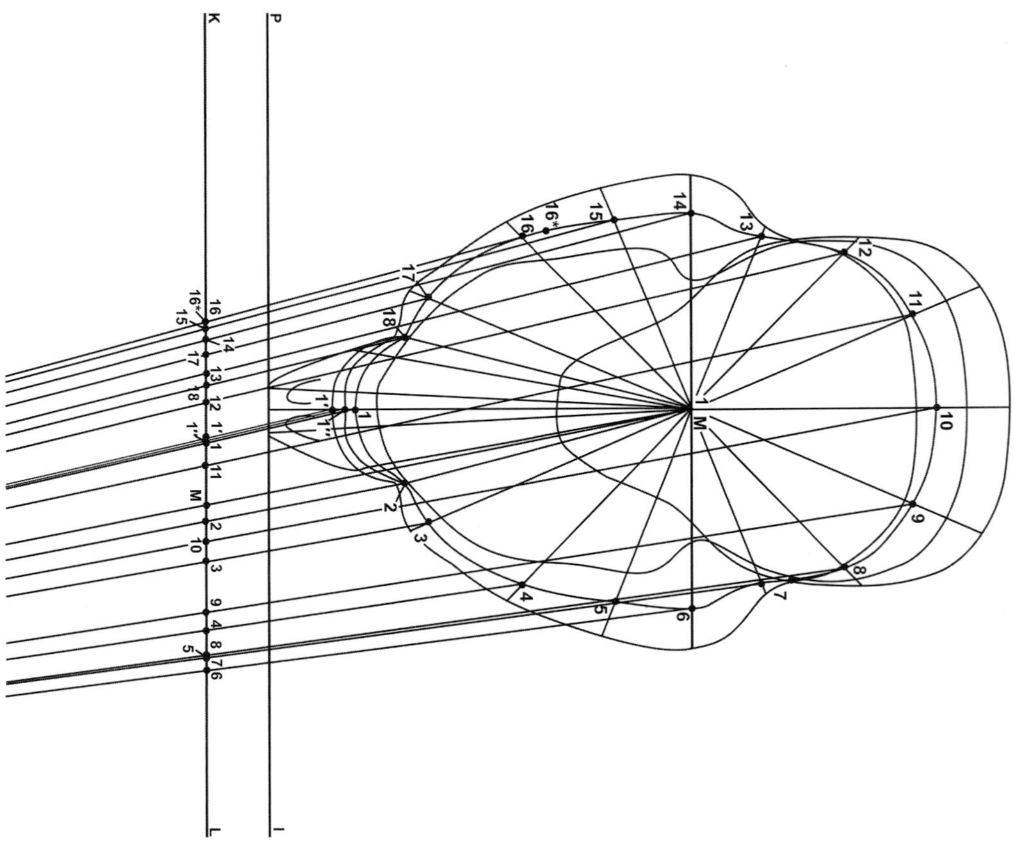

Figure B.129
Book 3, Proposition VIII.

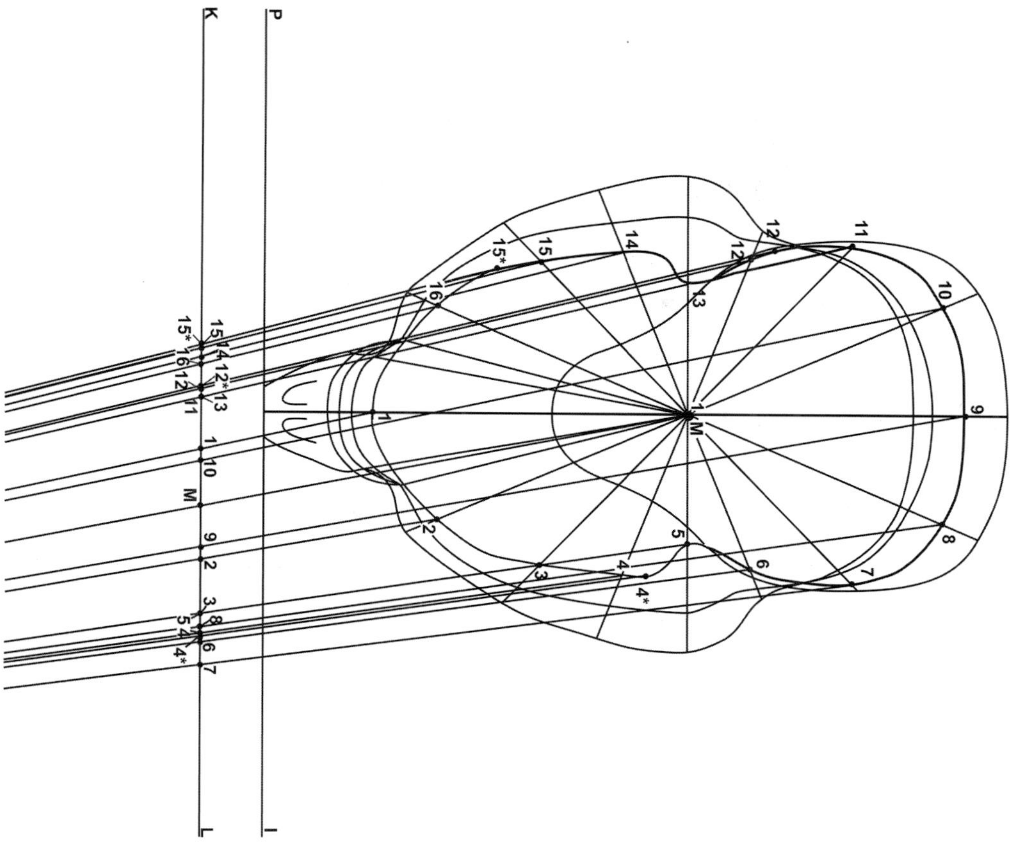

Figure B.130
Book 3, Proposition VIII.

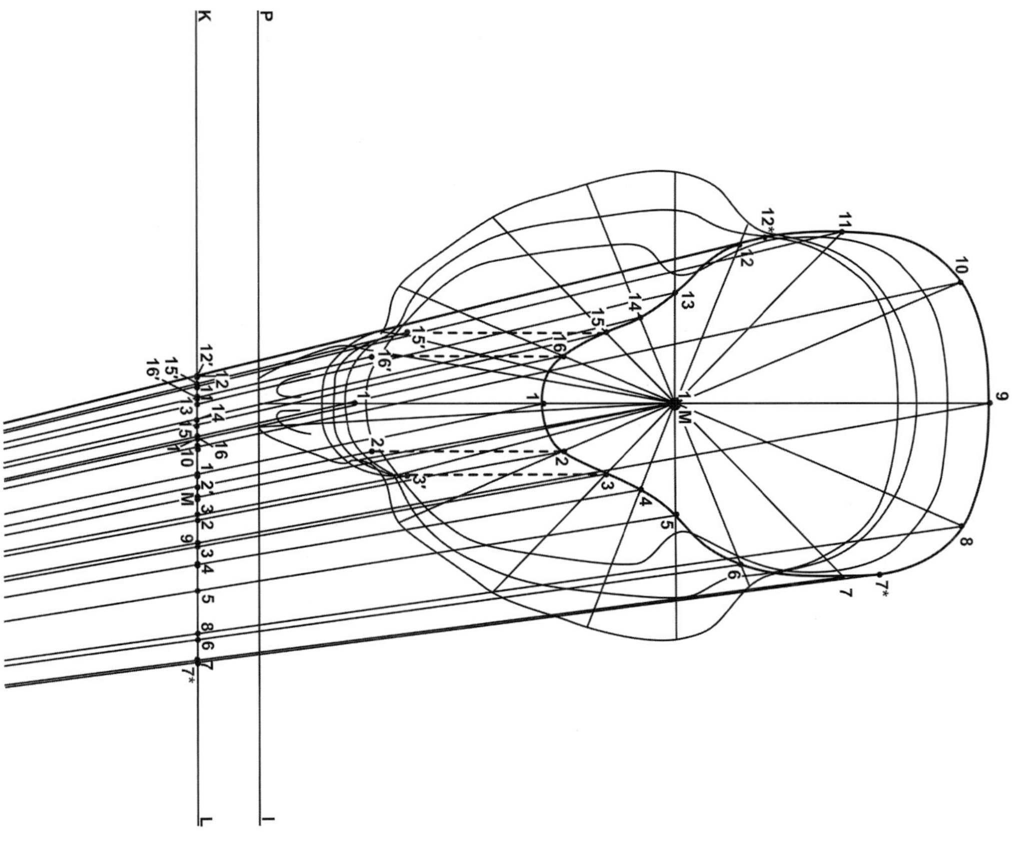

Figure B.131
Book 3, Proposition VIII.

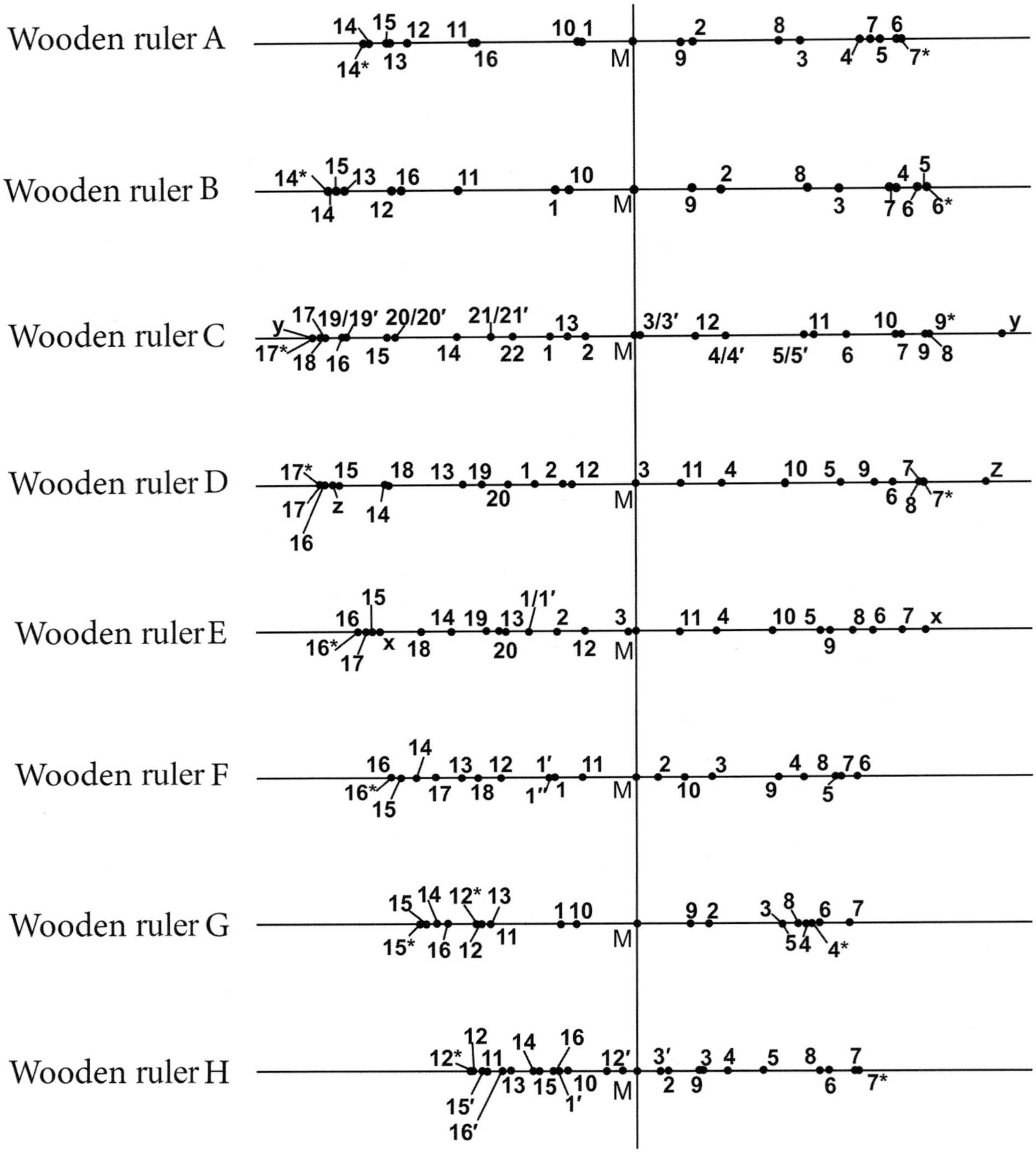

Figure B.132
Book 3, Proposition VIII.

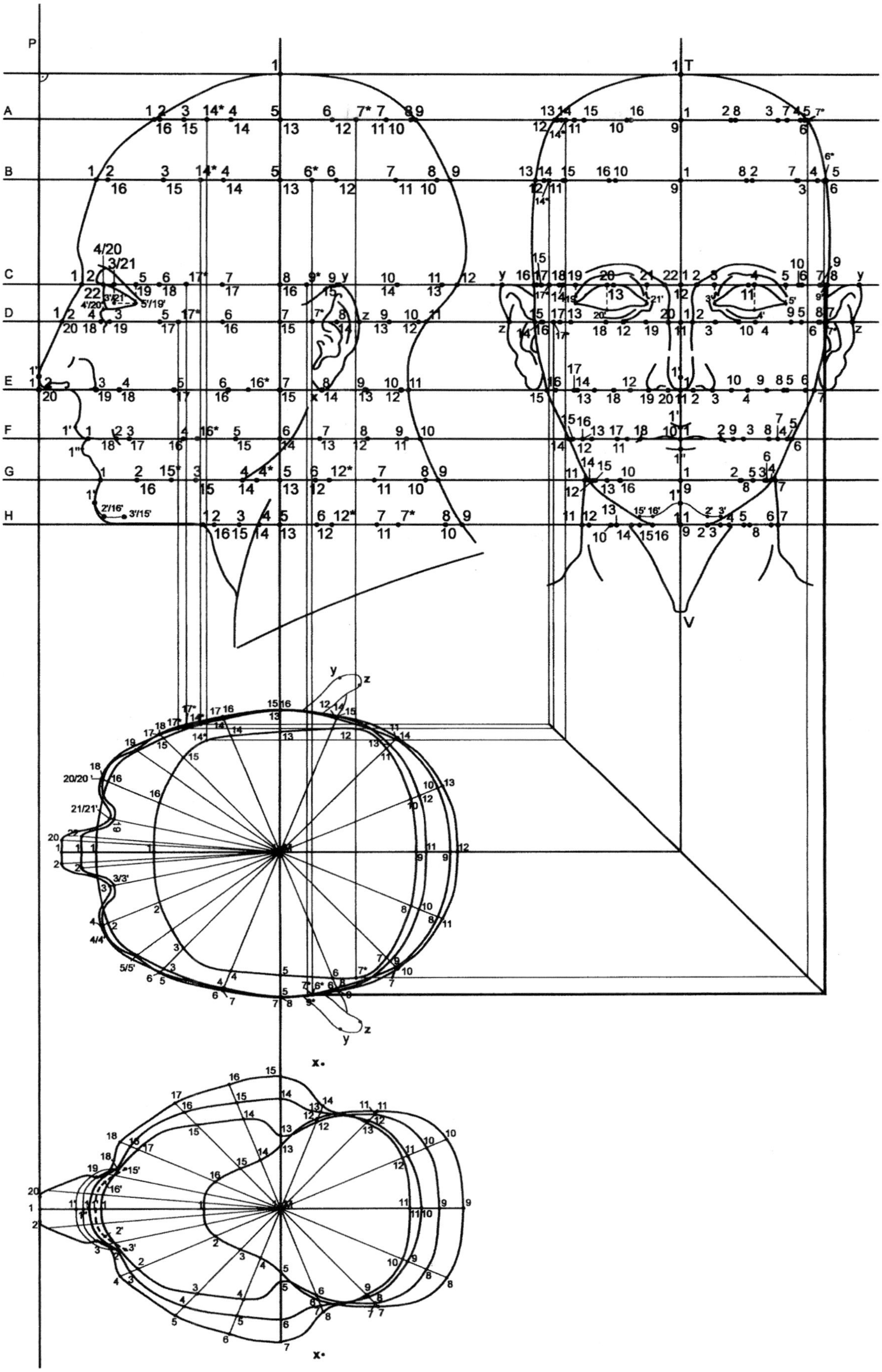

Figure B.133
Book 3, Proposition VIII.

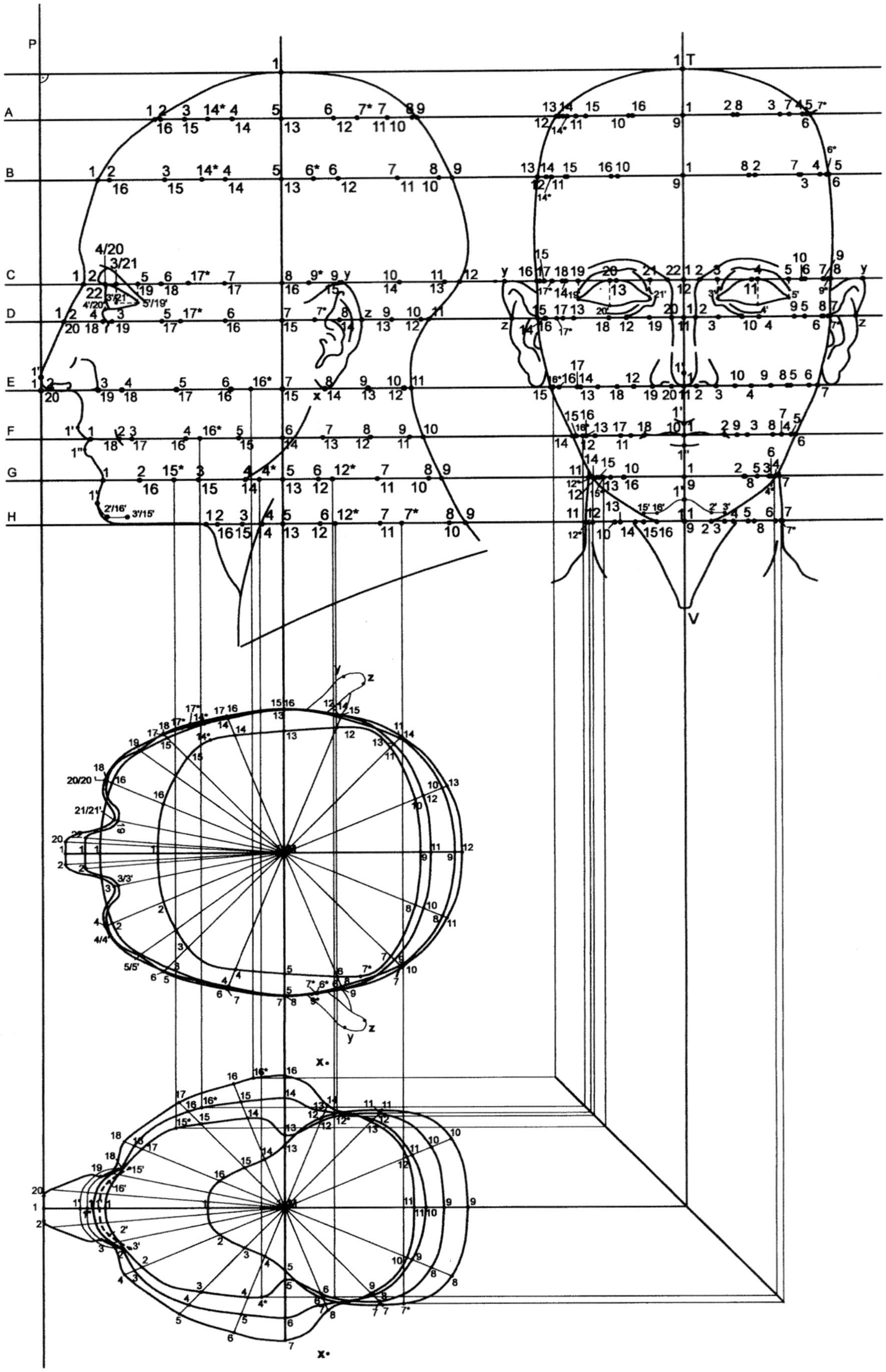

Figure B.134
Book 3, Proposition VIII.

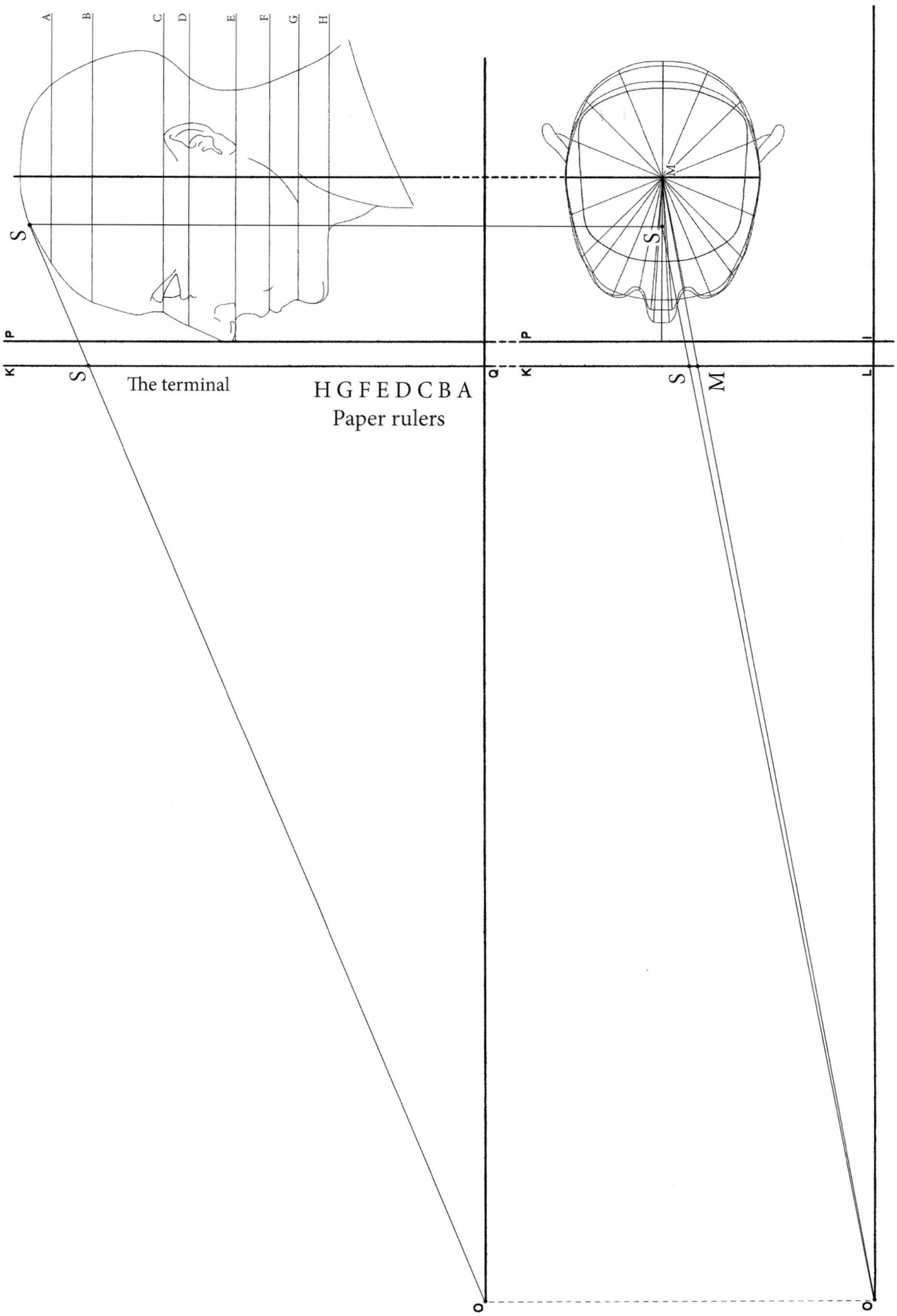

Figure B.135
Book 3, Proposition VIII.

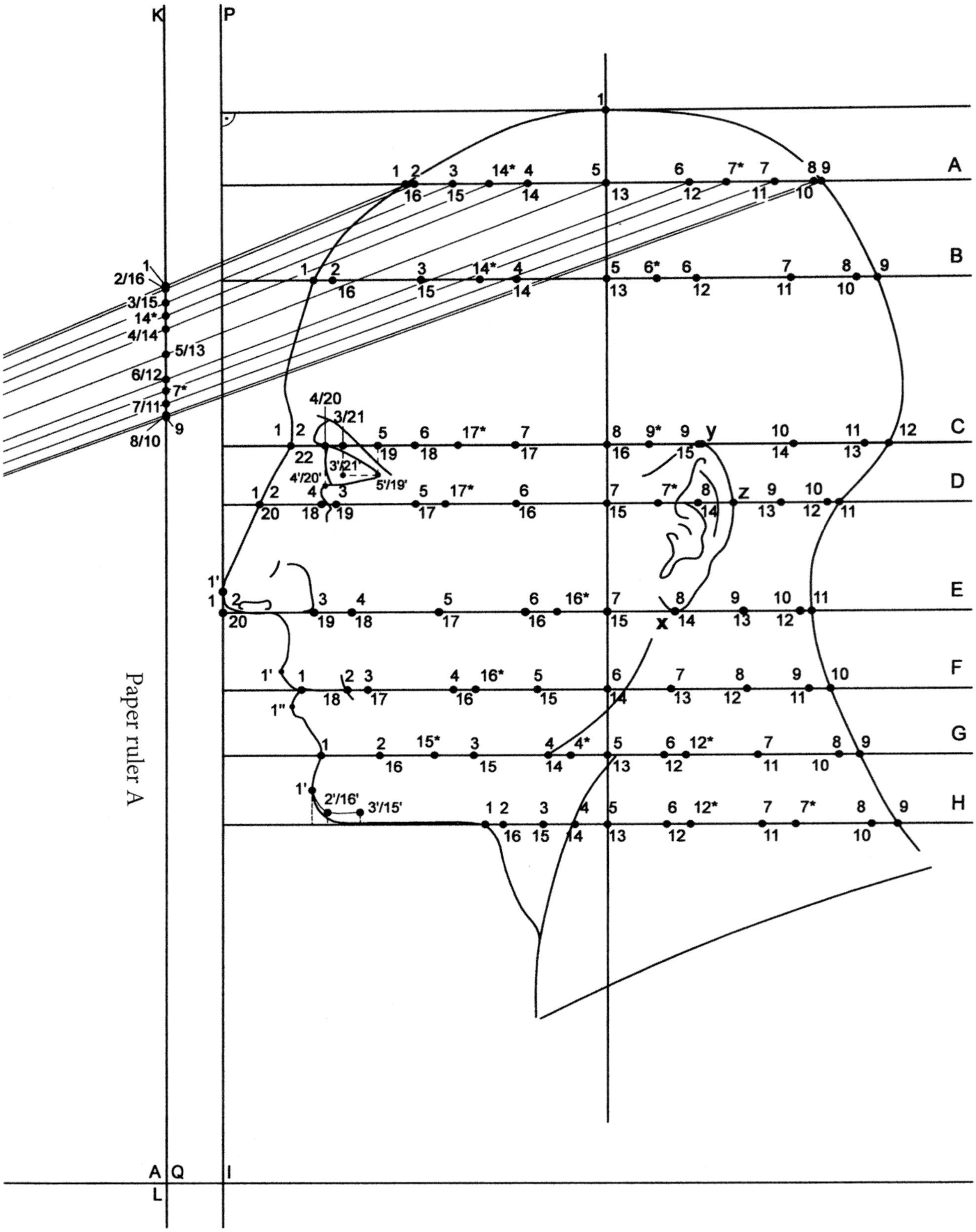

Figure B.136
Book 3, Proposition VIII.

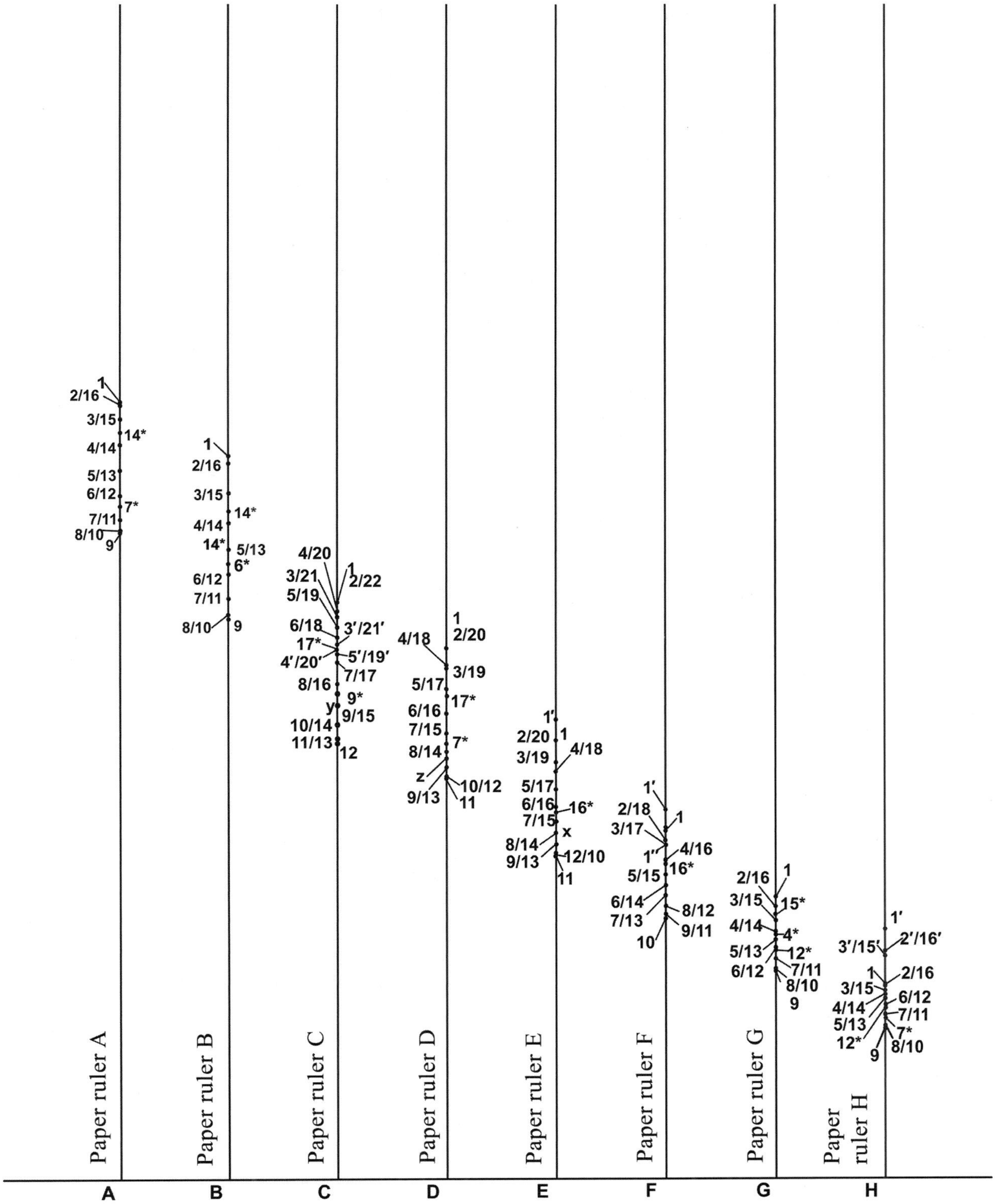

Figure B.137
Book 3, Proposition VIII.

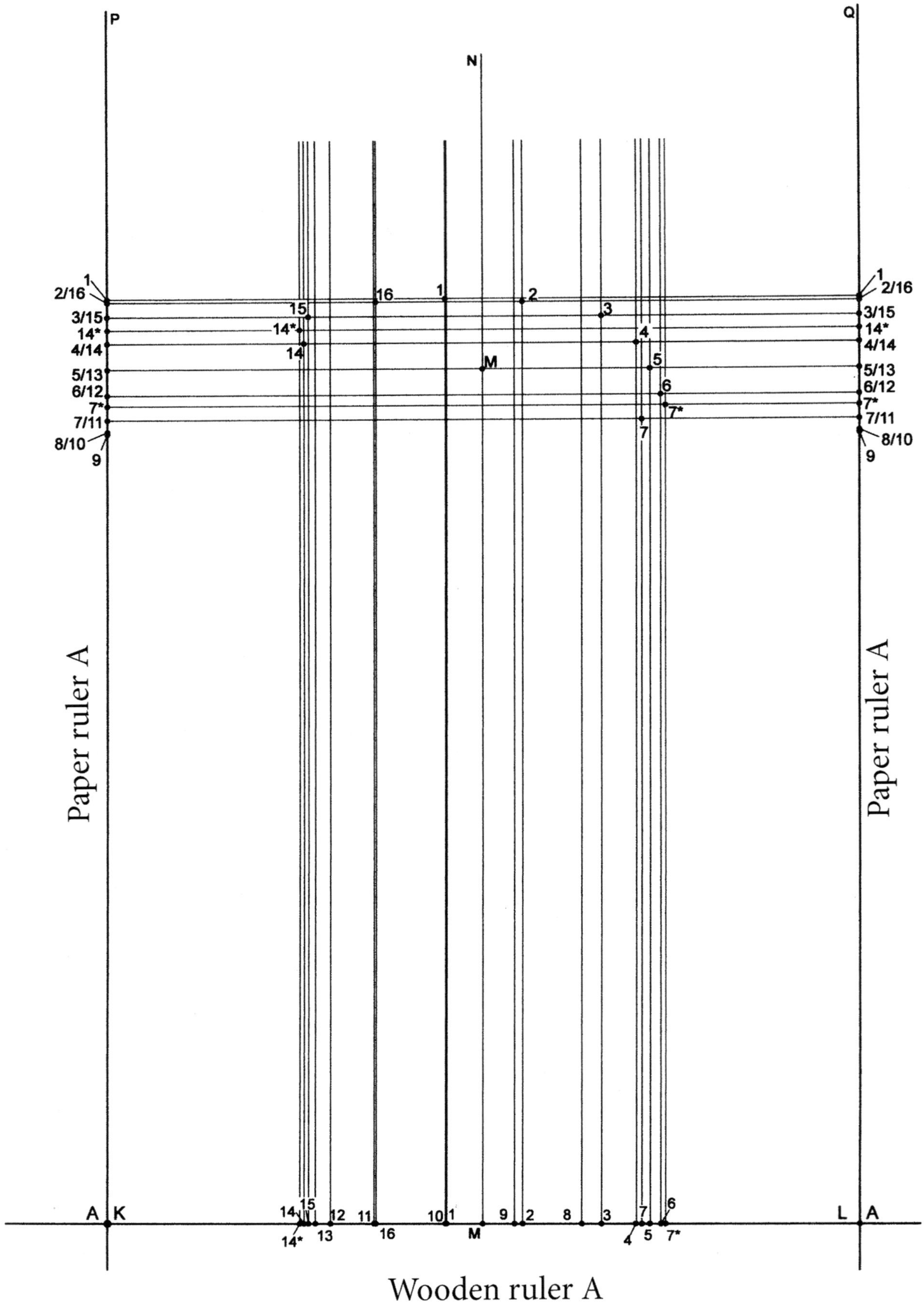

Figure B.138
Book 3, Proposition VIII.

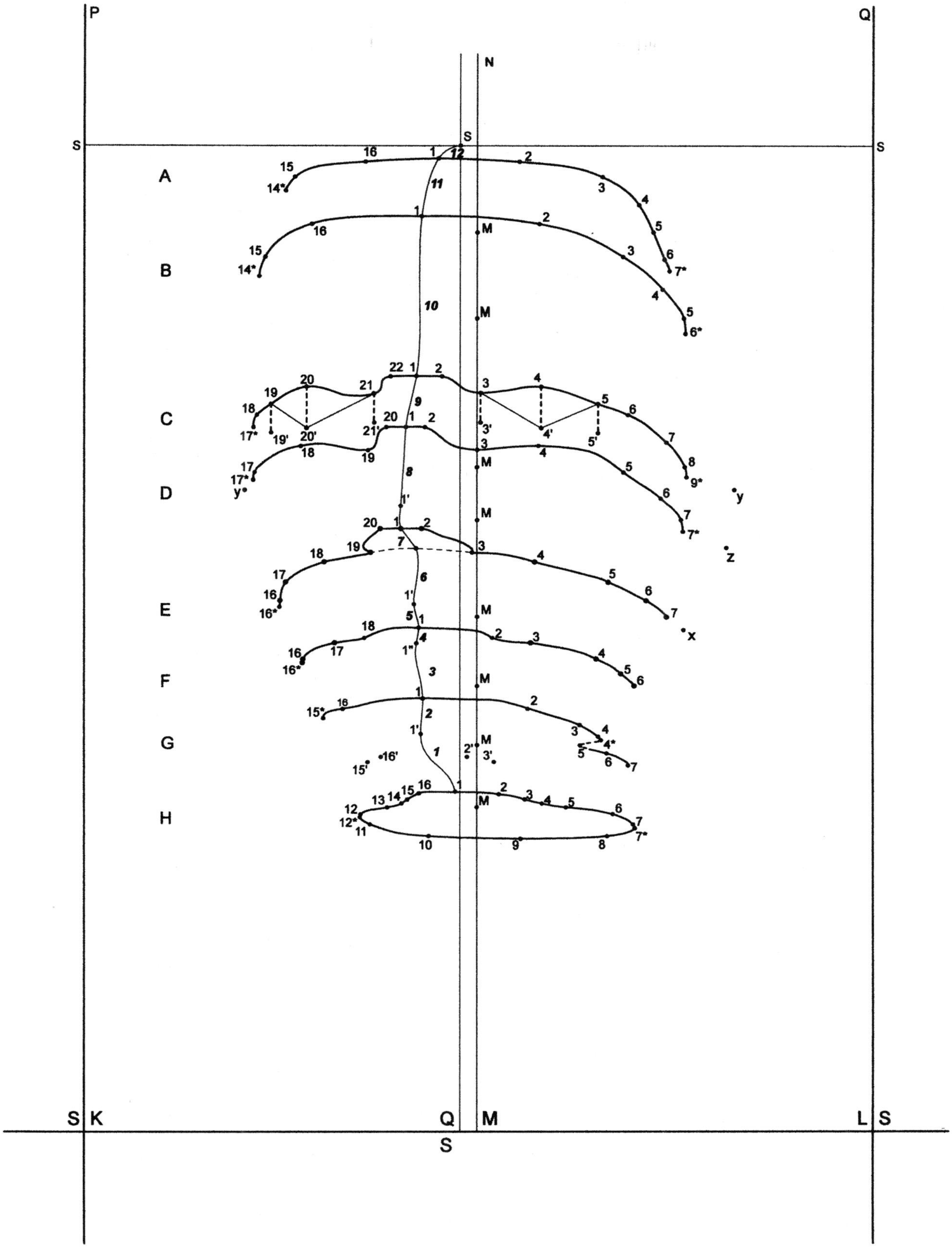

Figure B.139
Book 3, Proposition VIII.

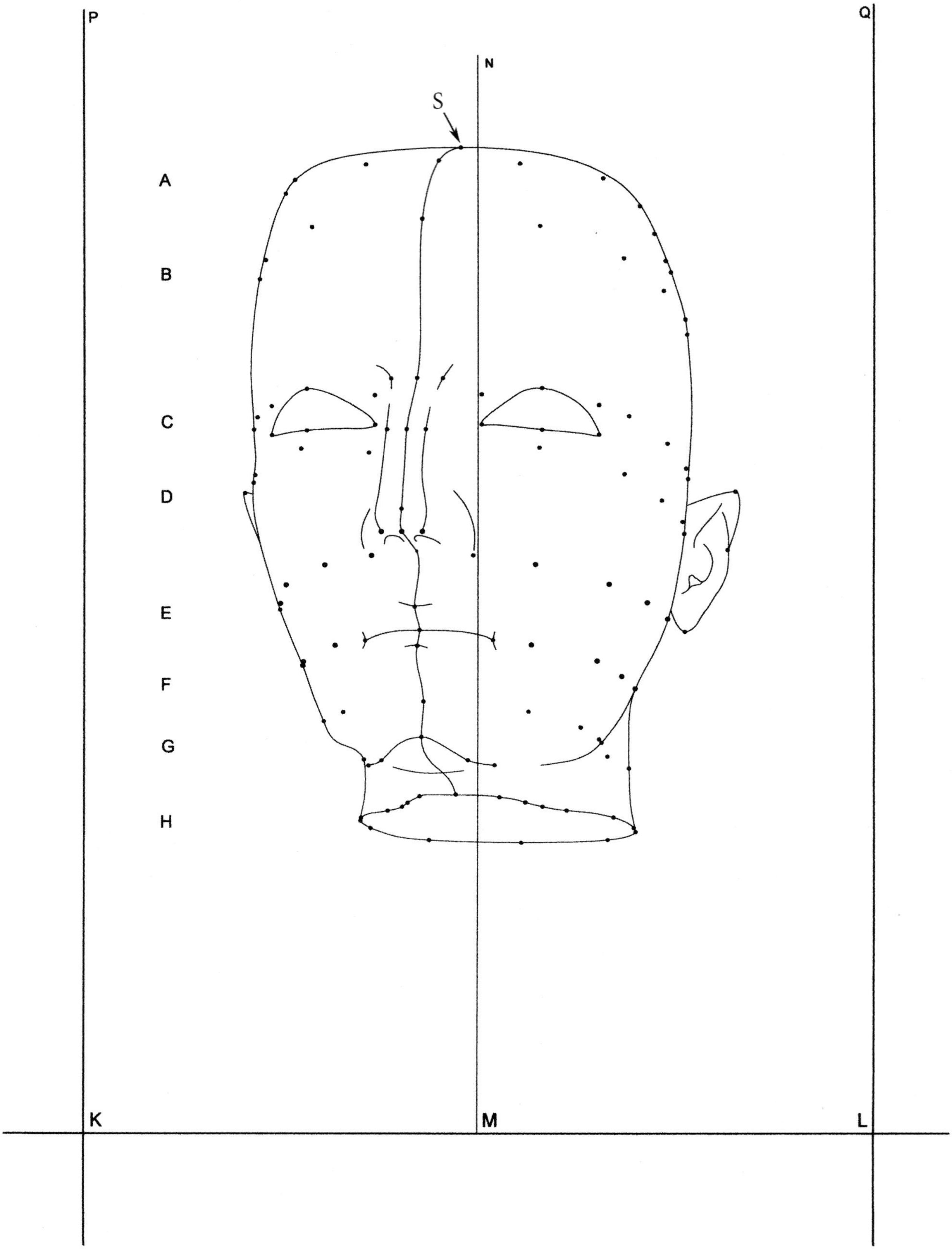

Figure B.140
Book 3, Proposition VIII.

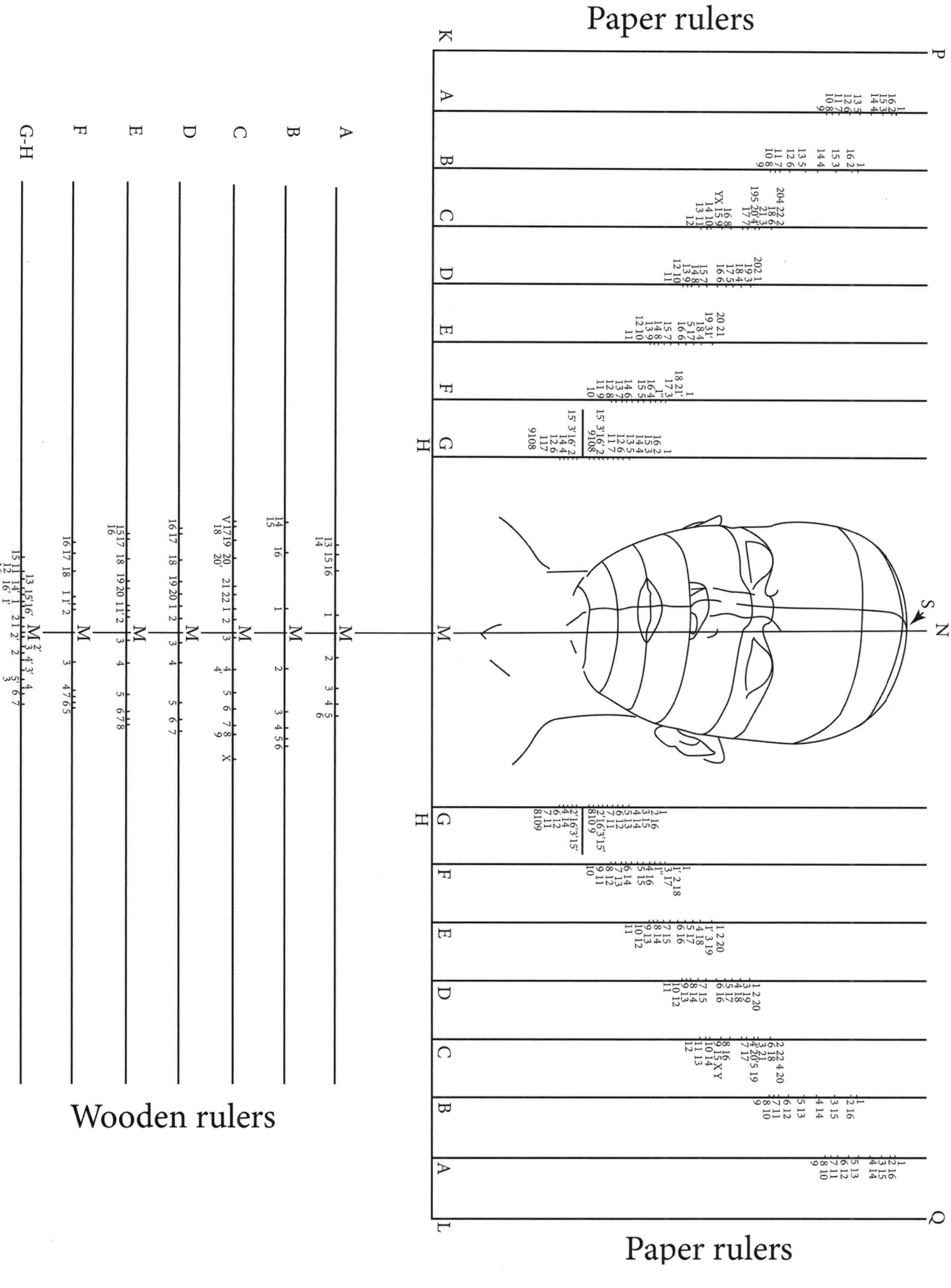

Figure B.141
Book 3, Proposition VIII.

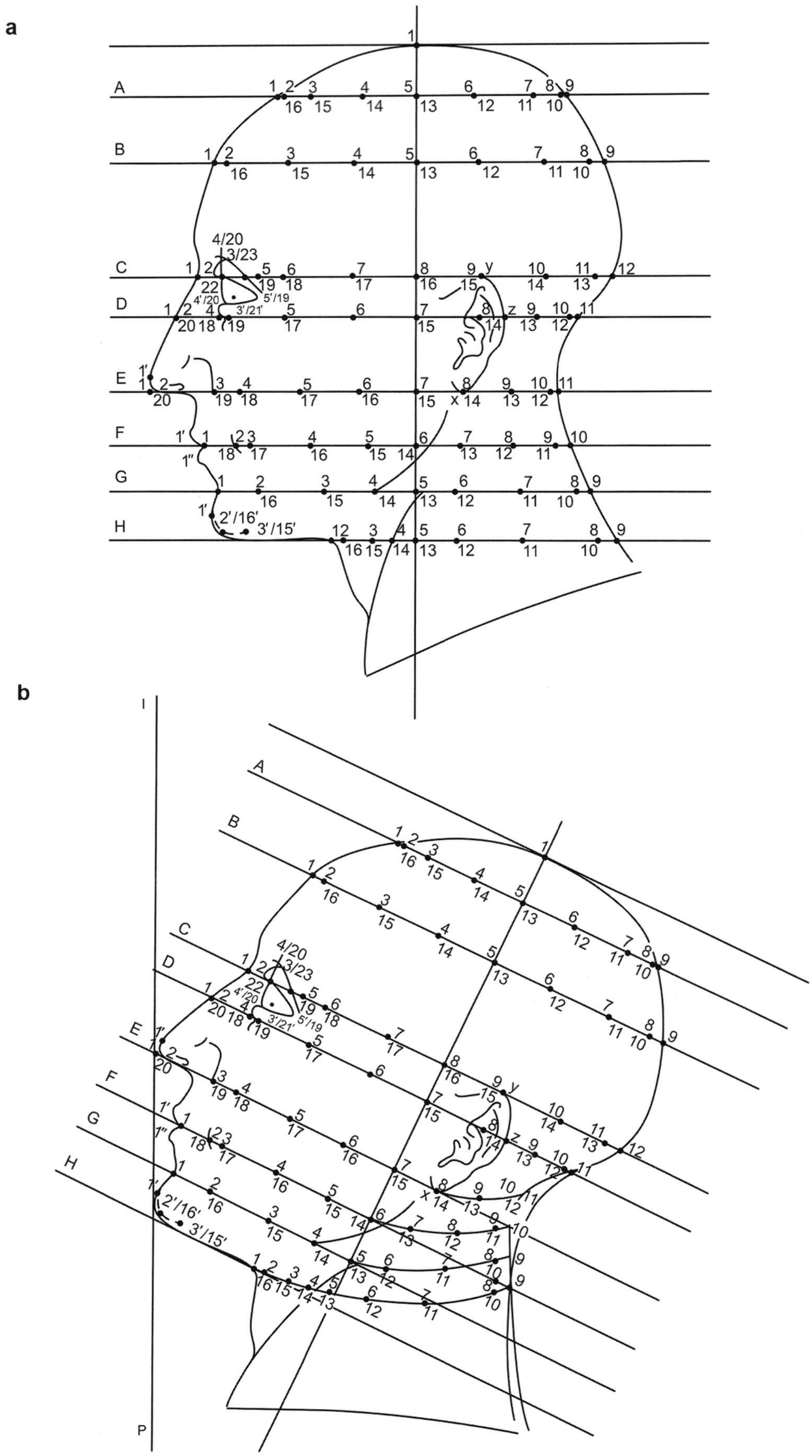

Figure B.142
Book 3, Proposition VIII.

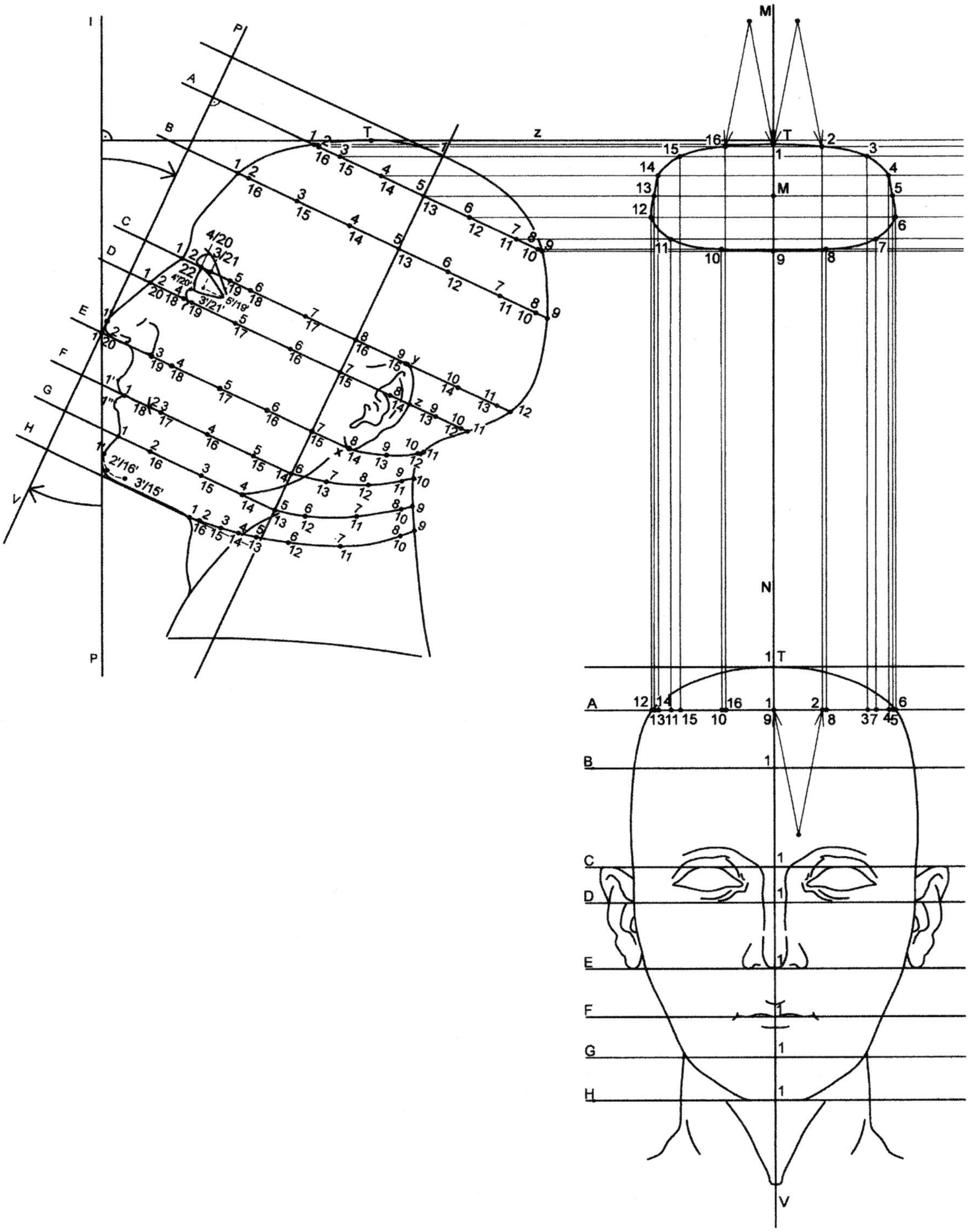

Figure B.143
Book 3, Proposition VIII.

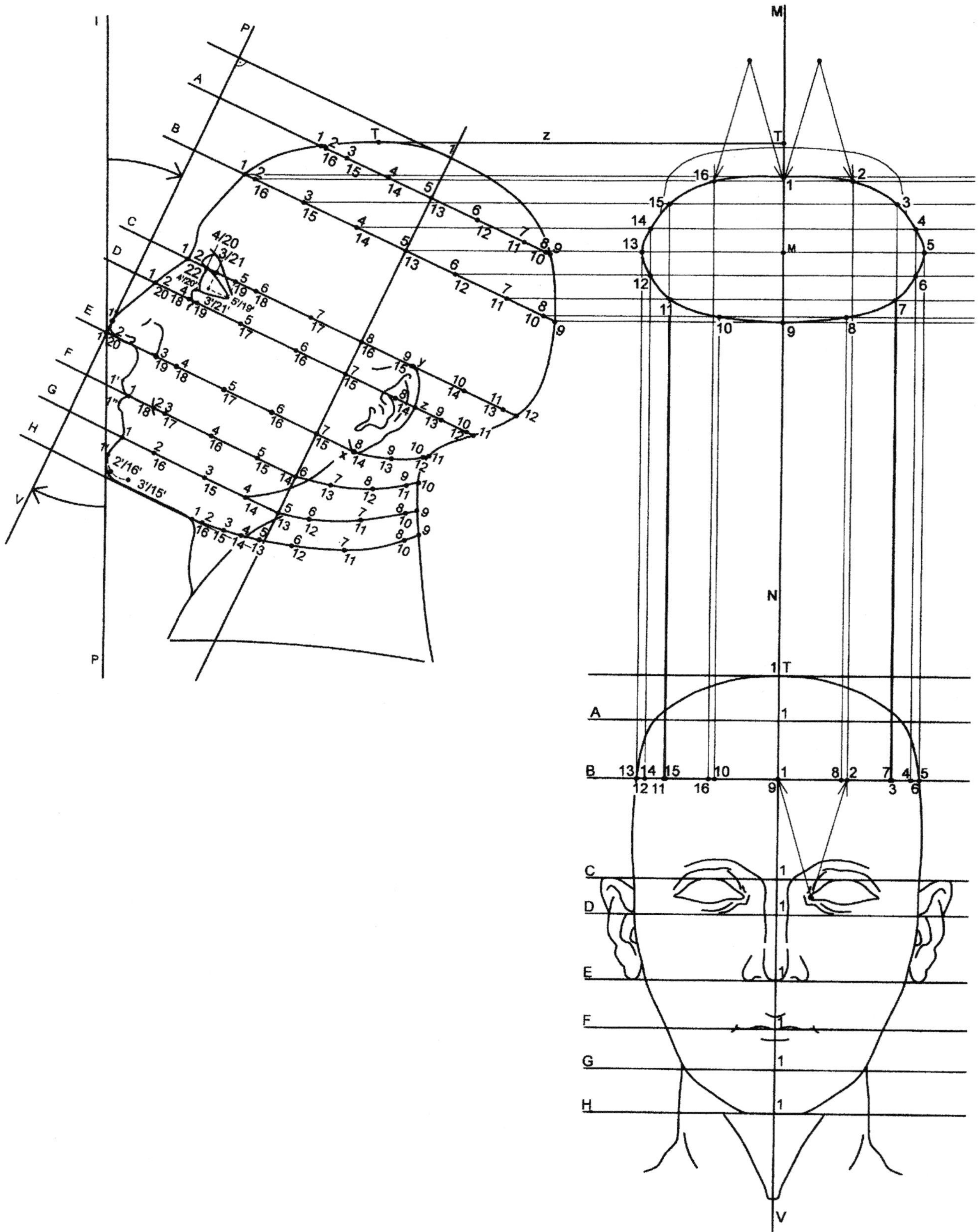

Figure B.144
Book 3, Proposition VIII.

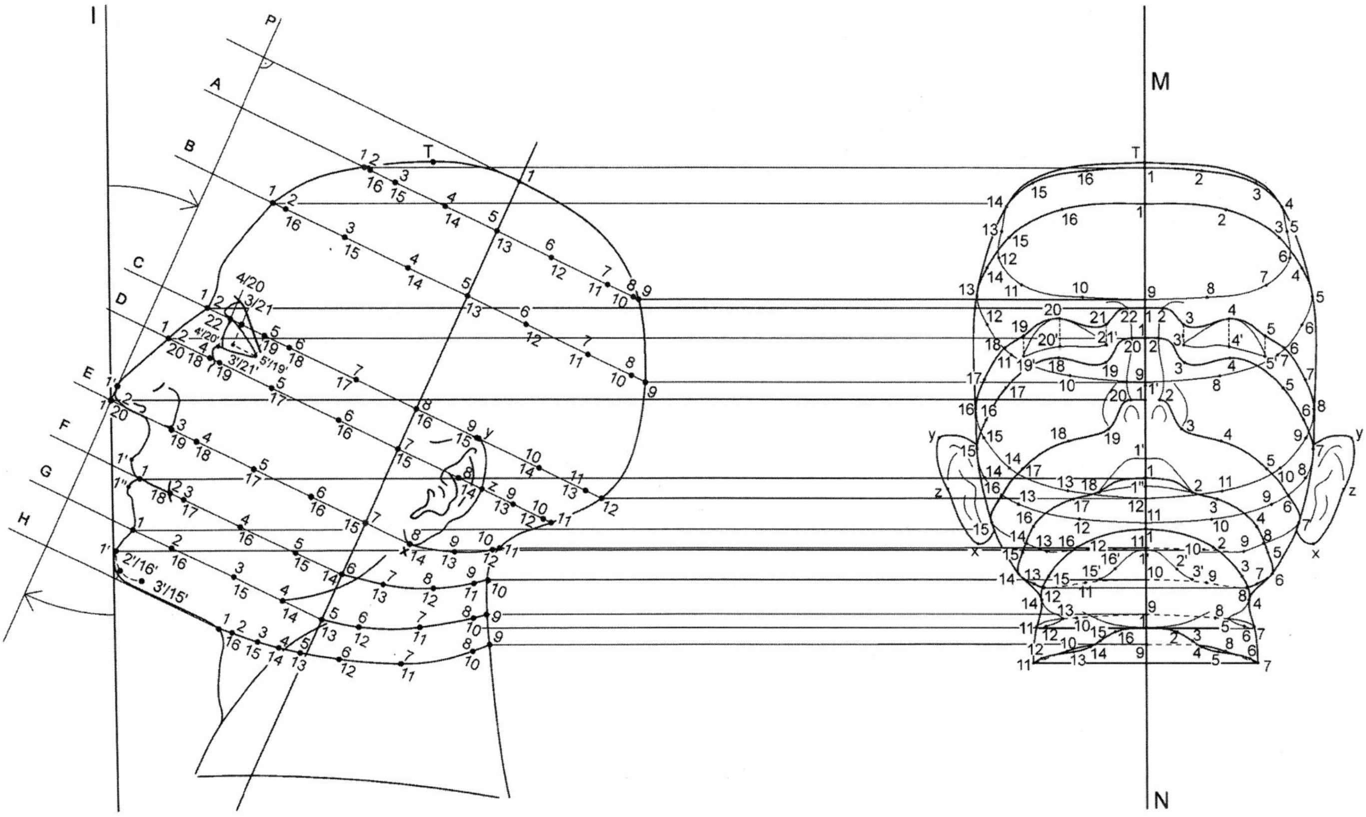

Figure B.145
Book 3, Proposition VIII.

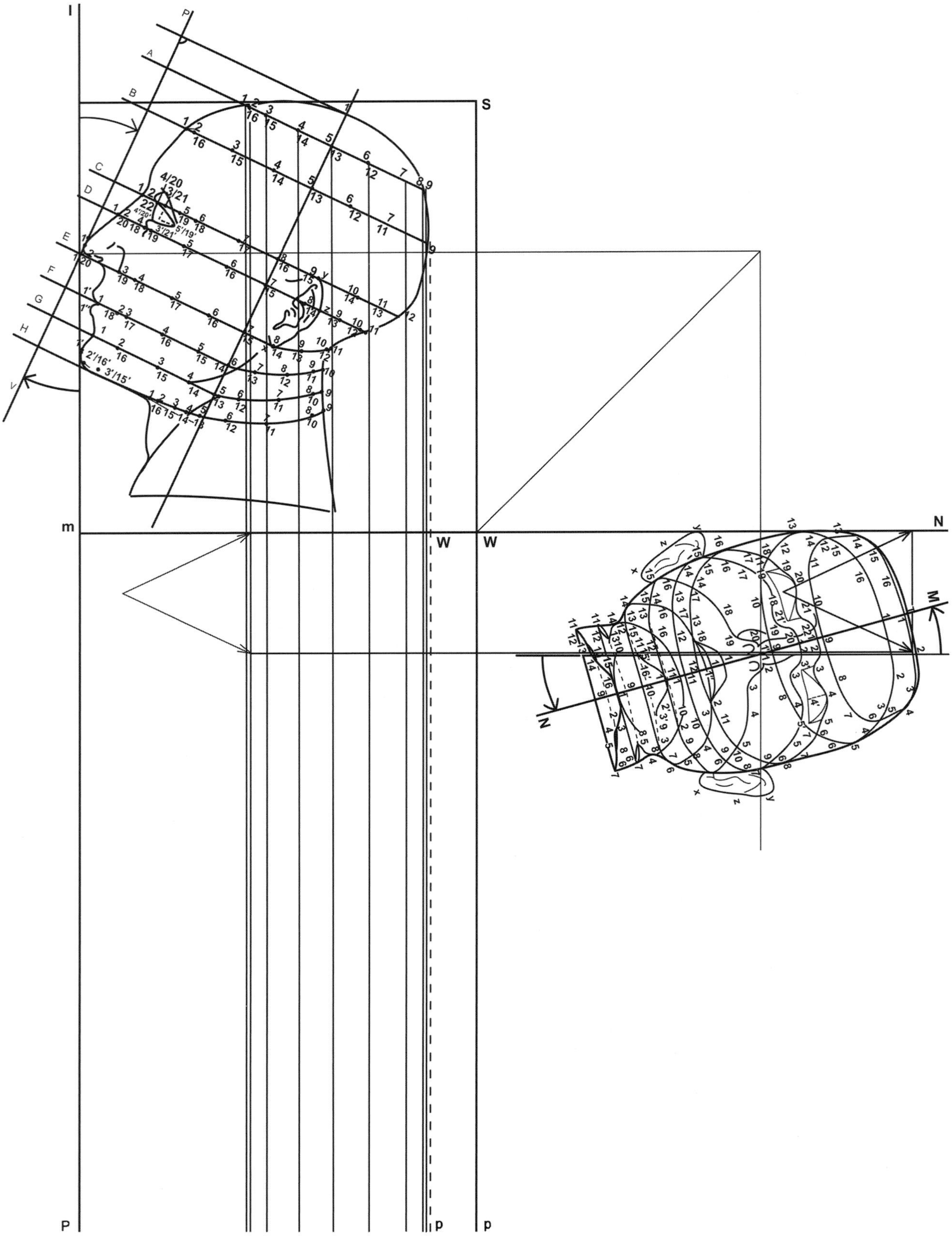

Figure B.146
Book 3, Proposition VIII.

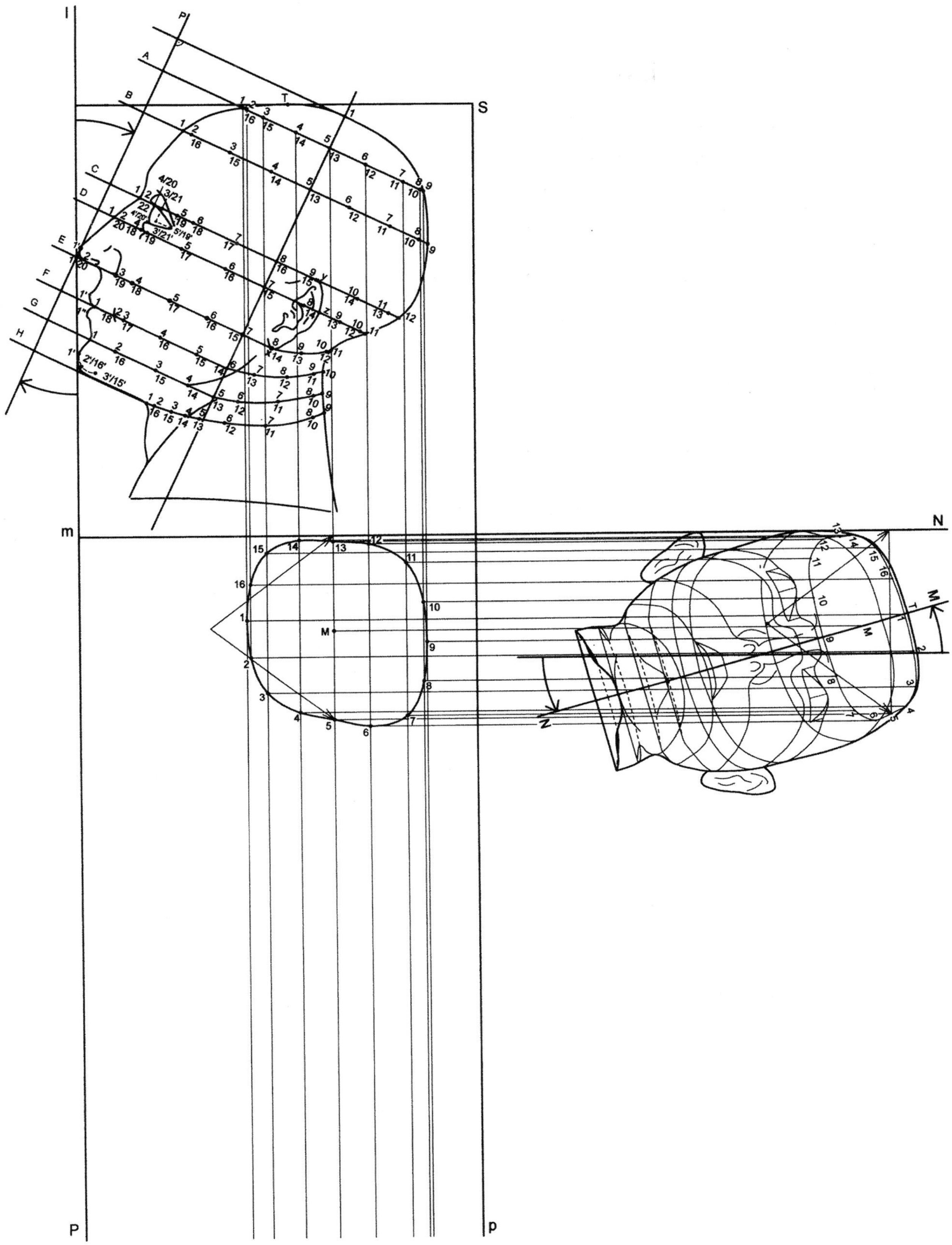

Figure B.147
Book 3, Proposition VIII.

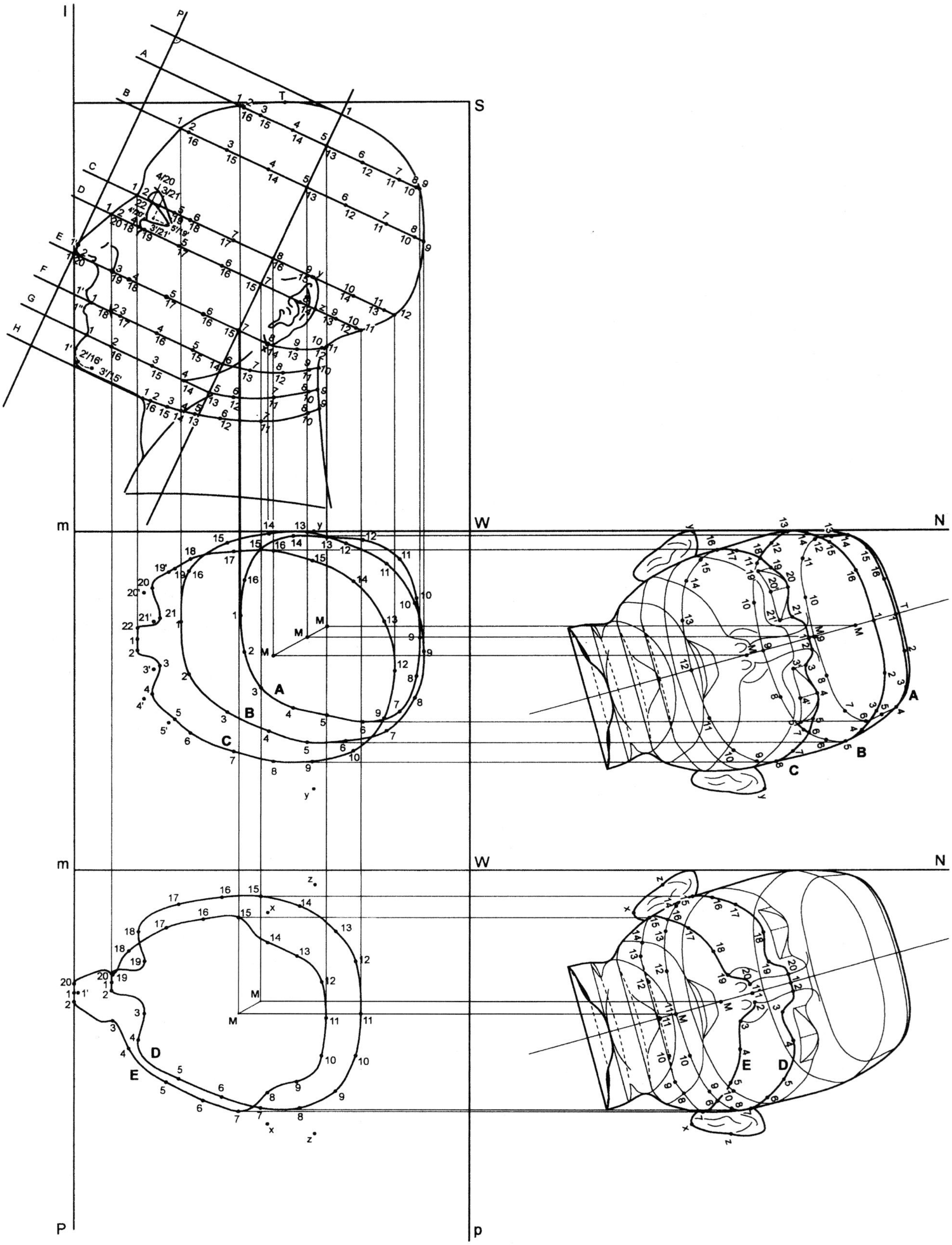

Figure B.148
Book 3, Proposition VIII.

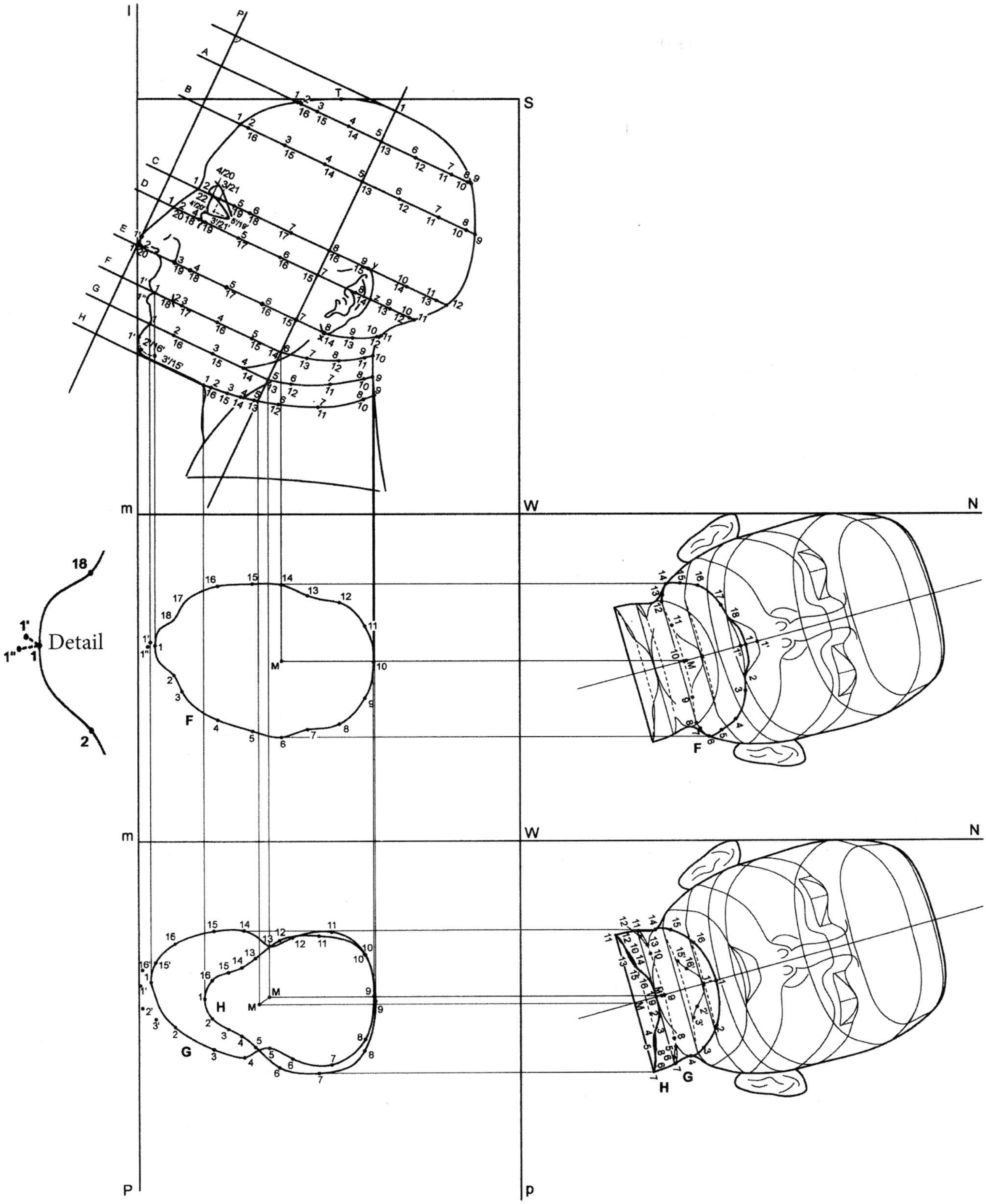

Figure B.149
Book 3, Proposition VIII.

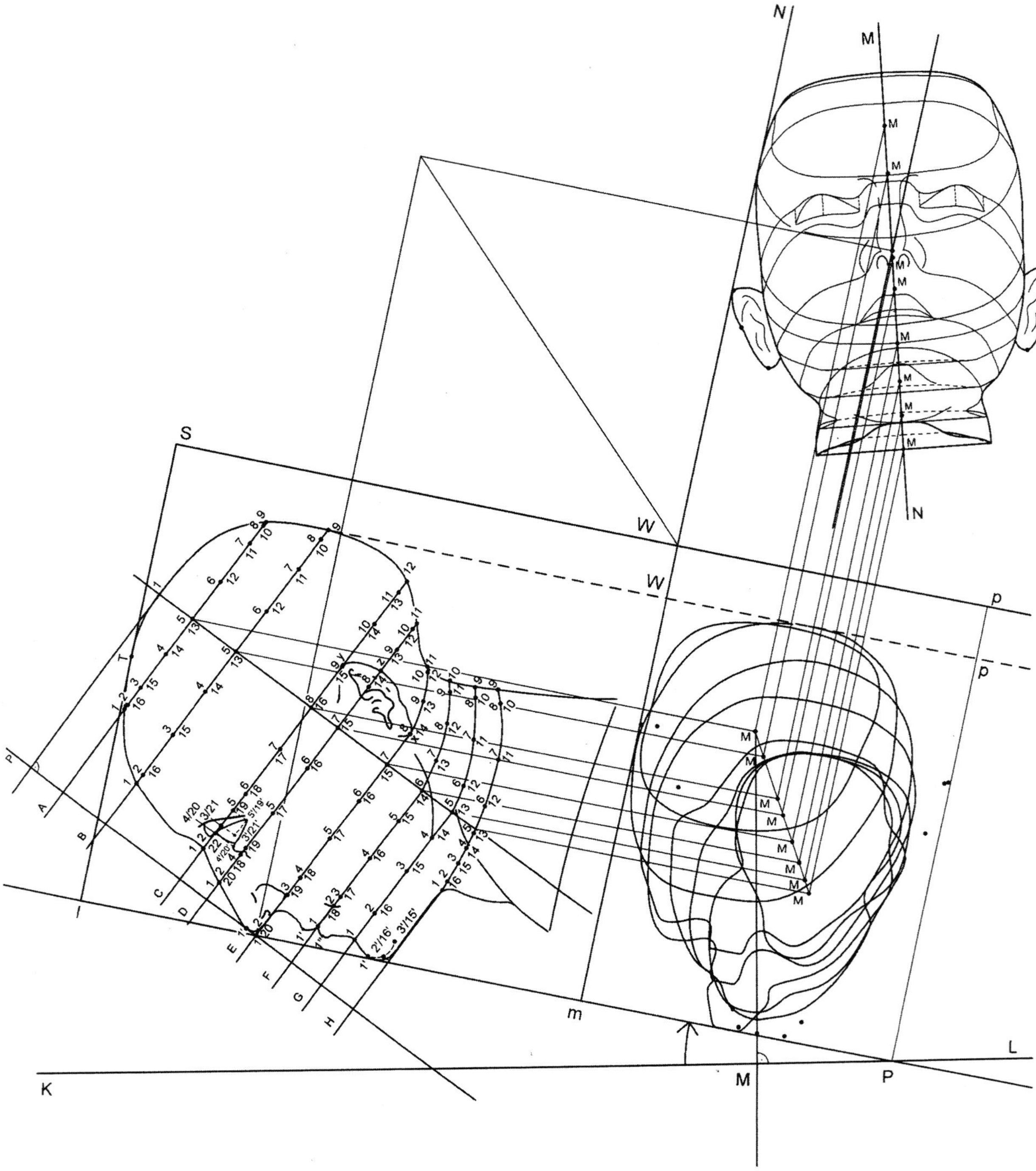

Figure B.150
Book 3, Proposition VIII.

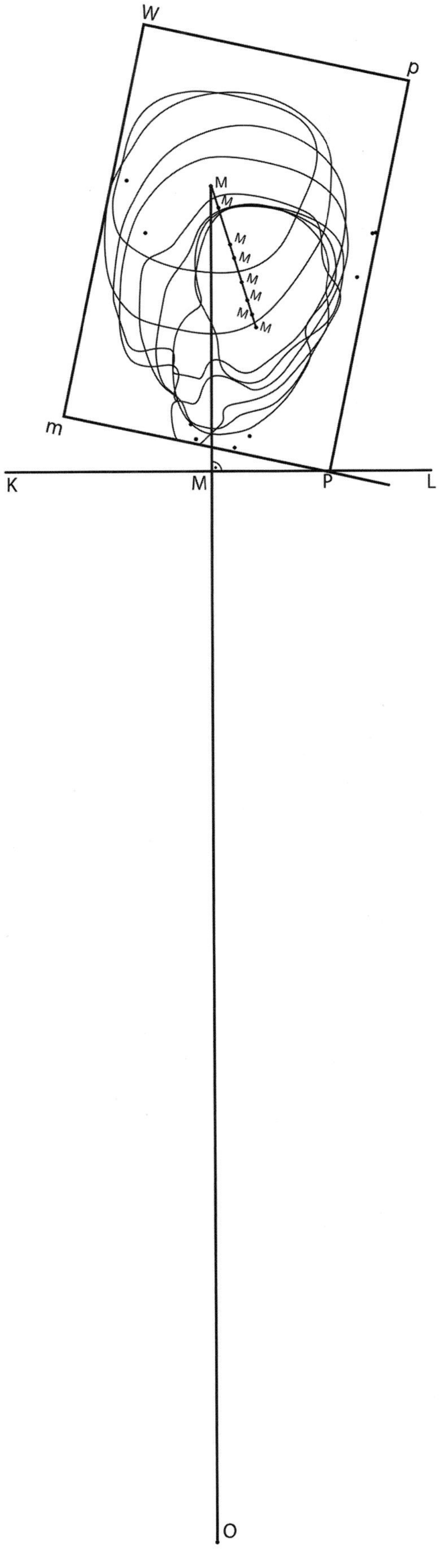

Figure B.151
Book 3, Proposition VIII.

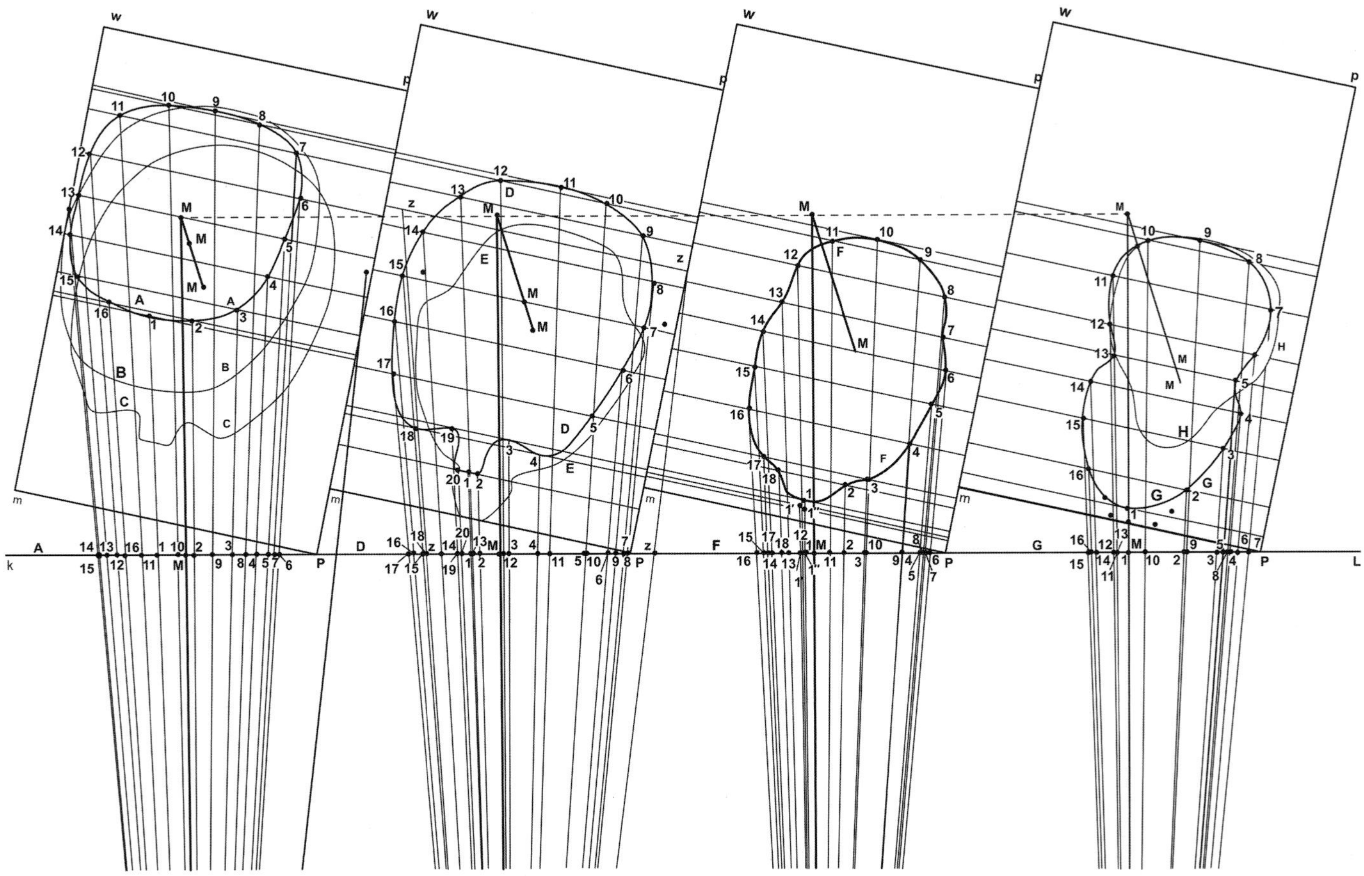

Figure B.152
Book 3, Proposition VIII.

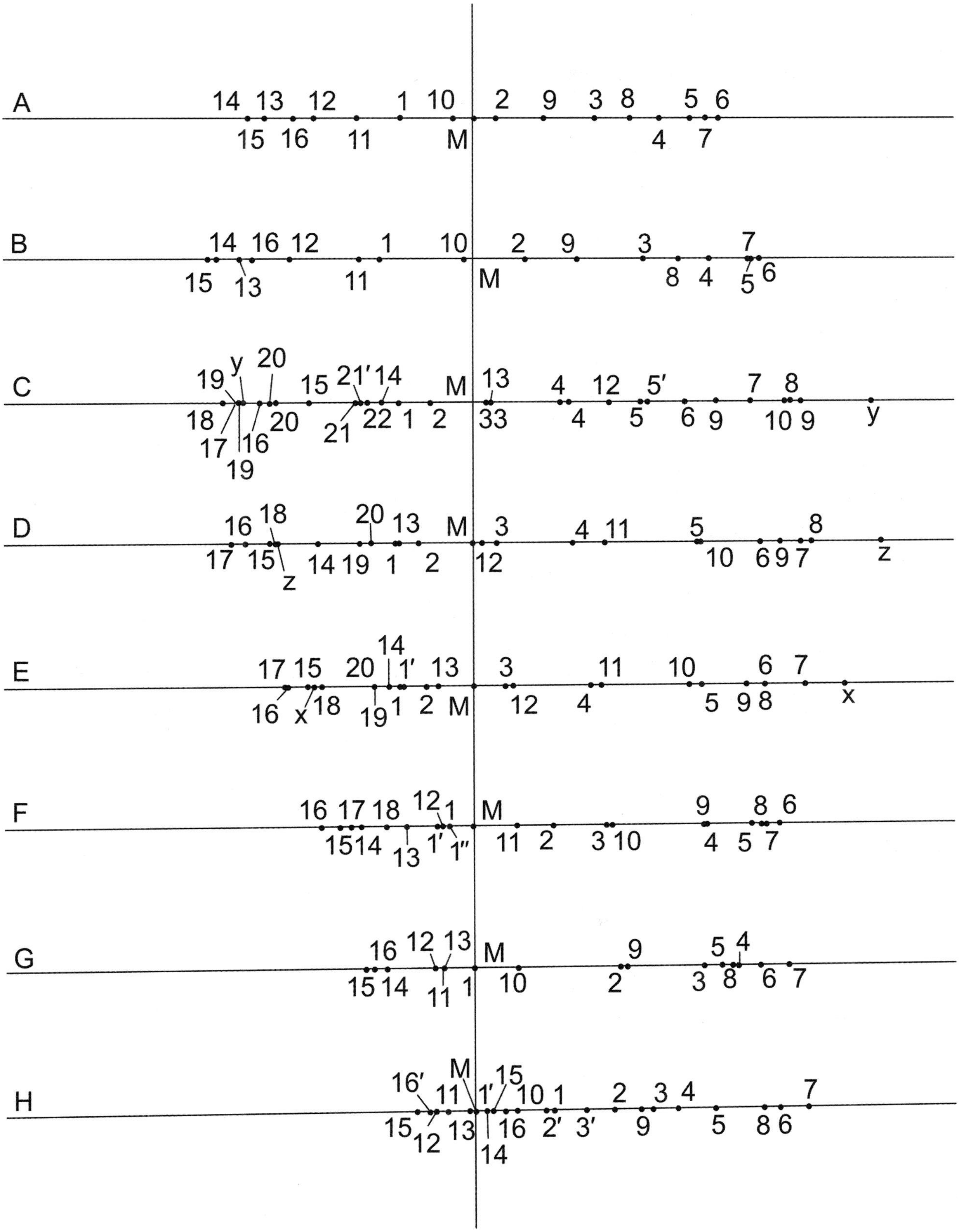

Figure B.153
Book 3, Proposition VIII.

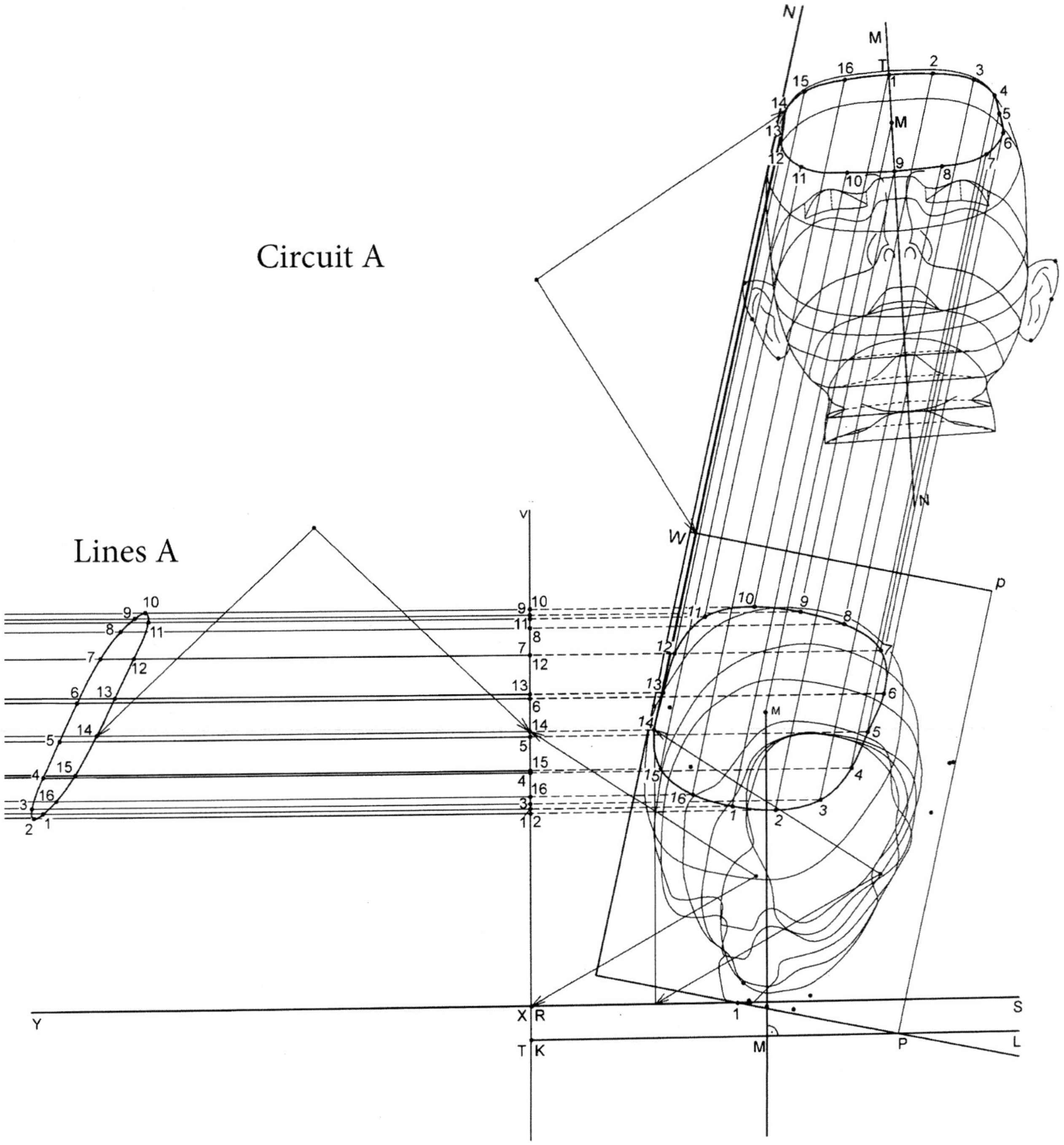

Figure B.154
Book 3, Proposition VIII.

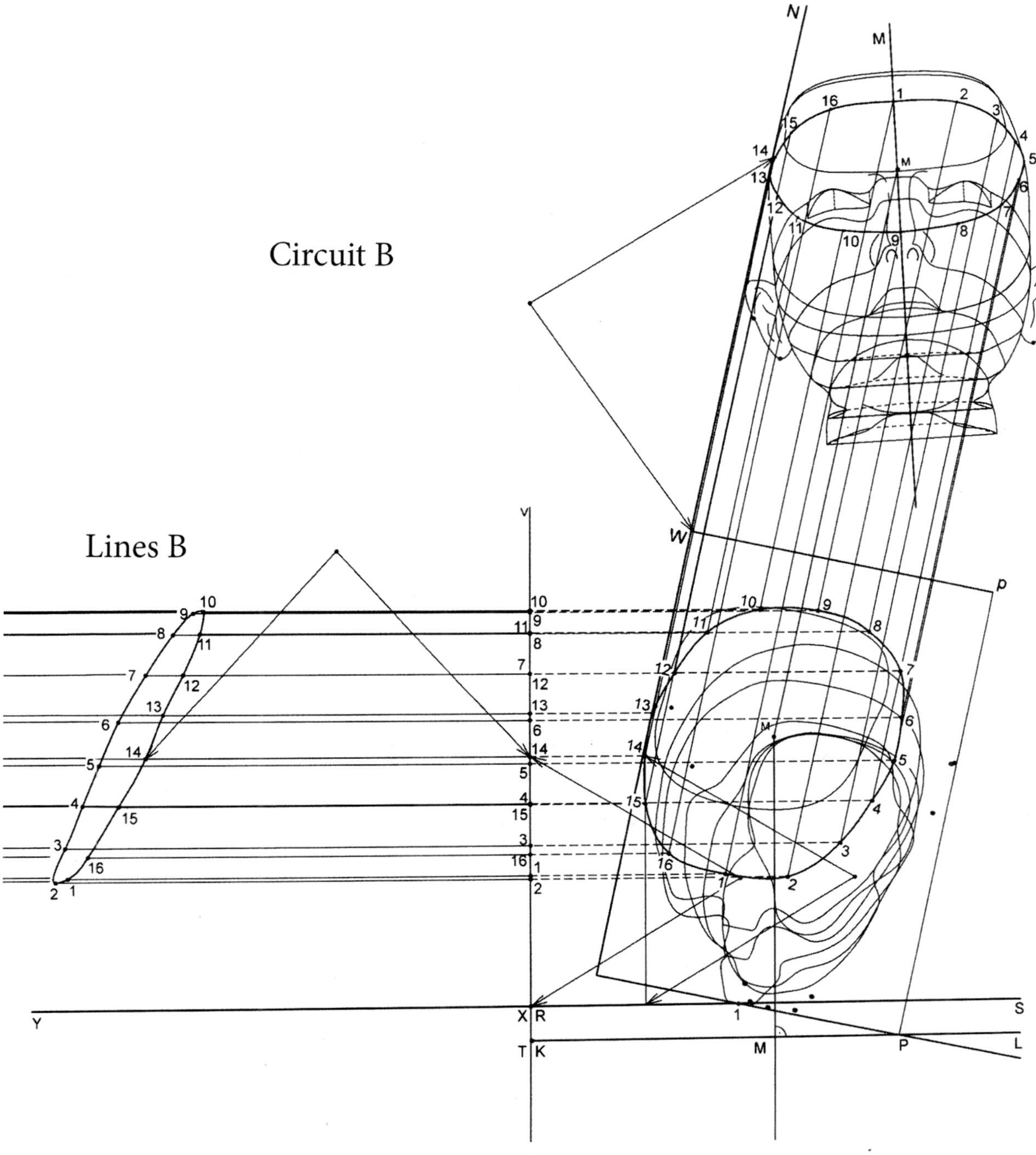

Figure B.155
Book 3, Proposition VIII.

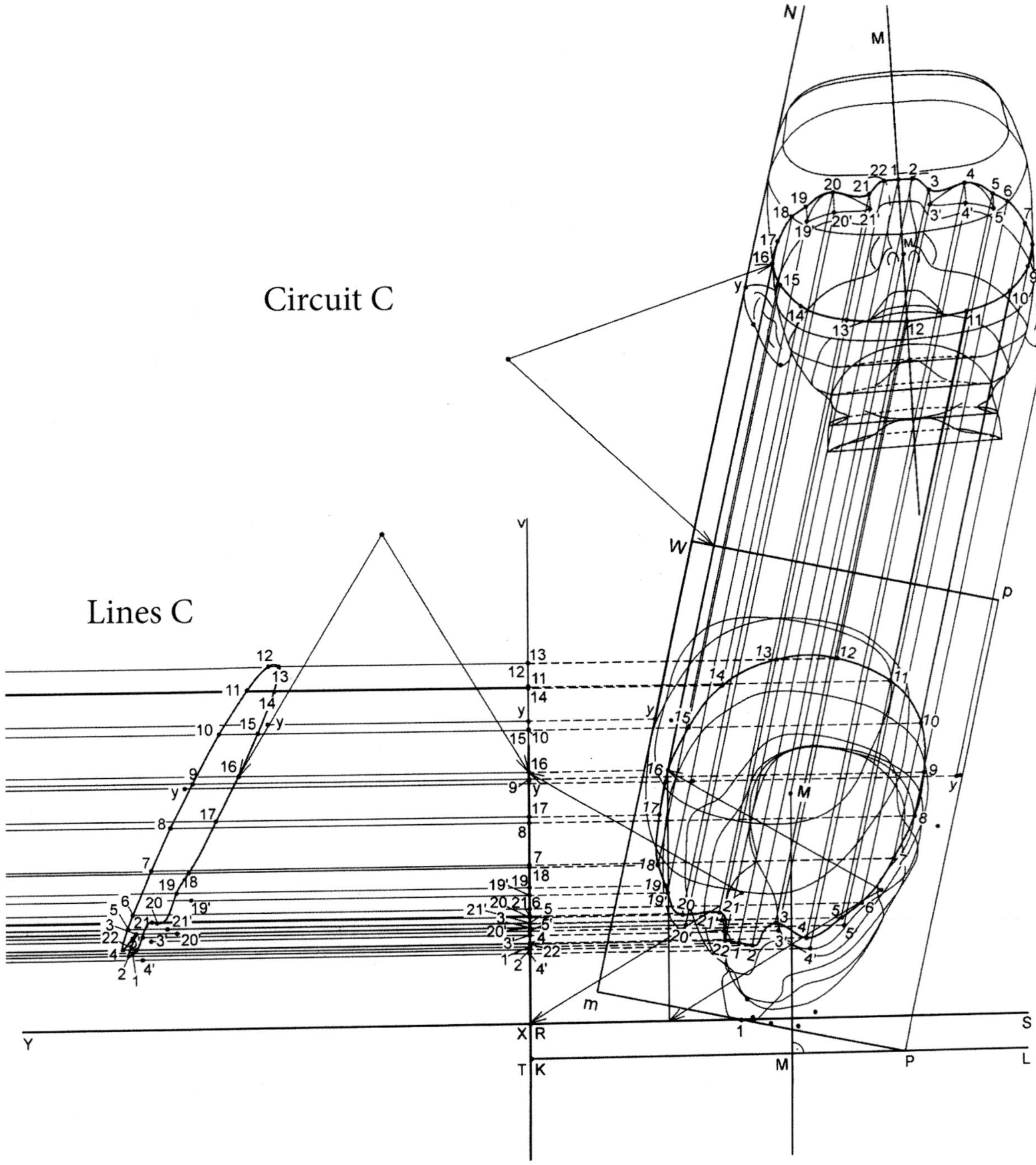

Figure B.156
Book 3, Proposition VIII.

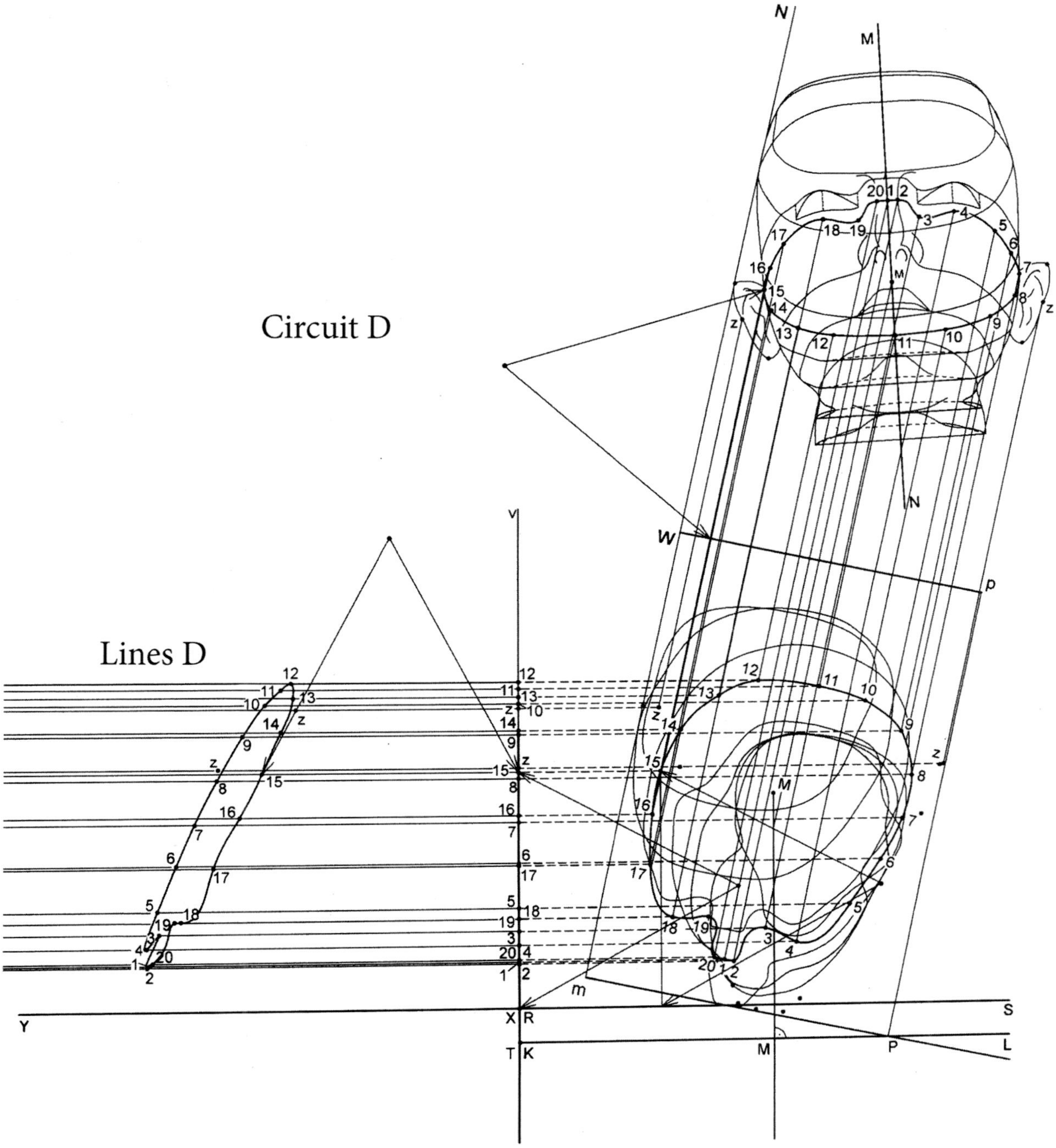

Figure B.157
Book 3, Proposition VIII.

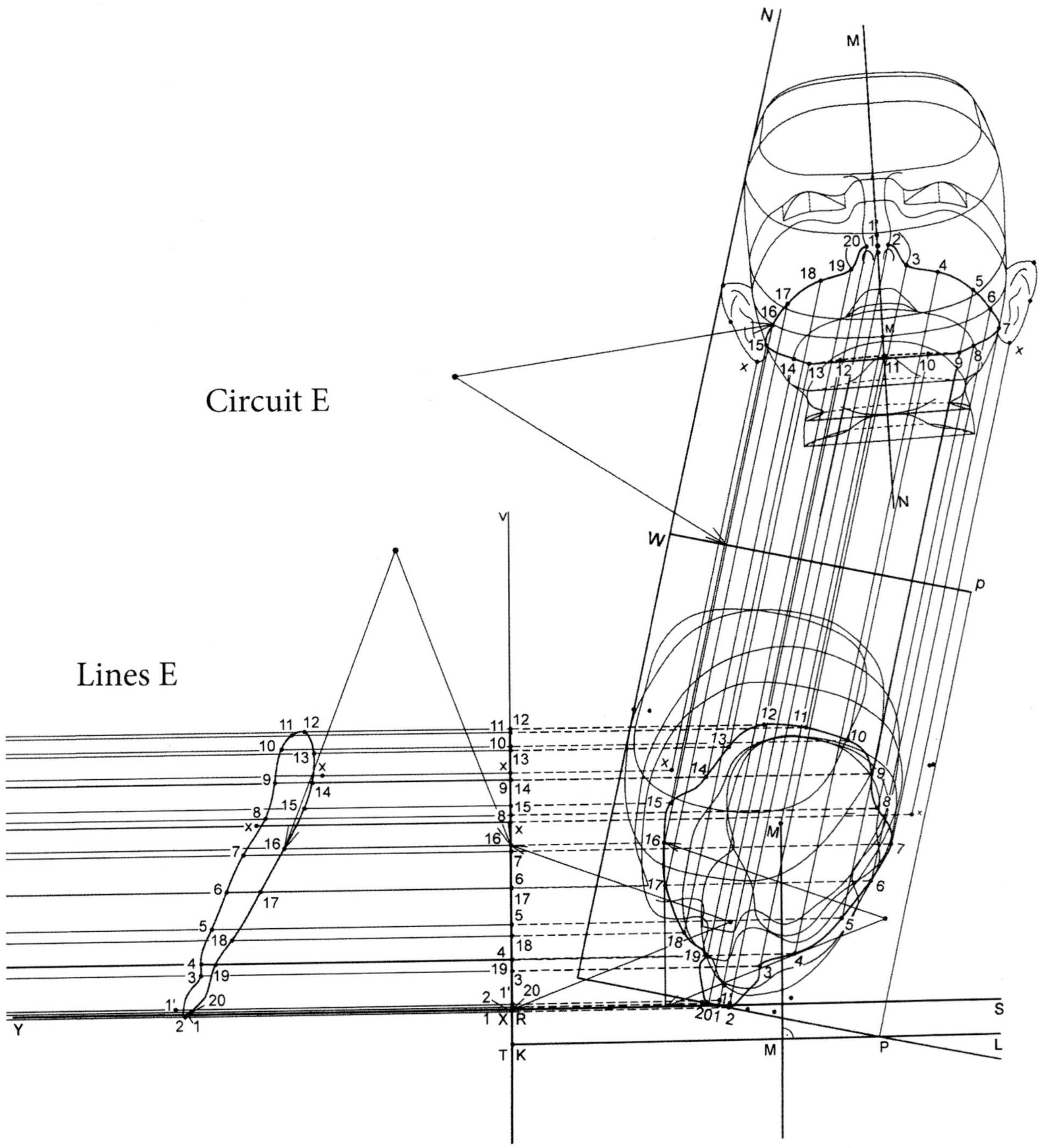

Figure B.158
Book 3, Proposition VIII.

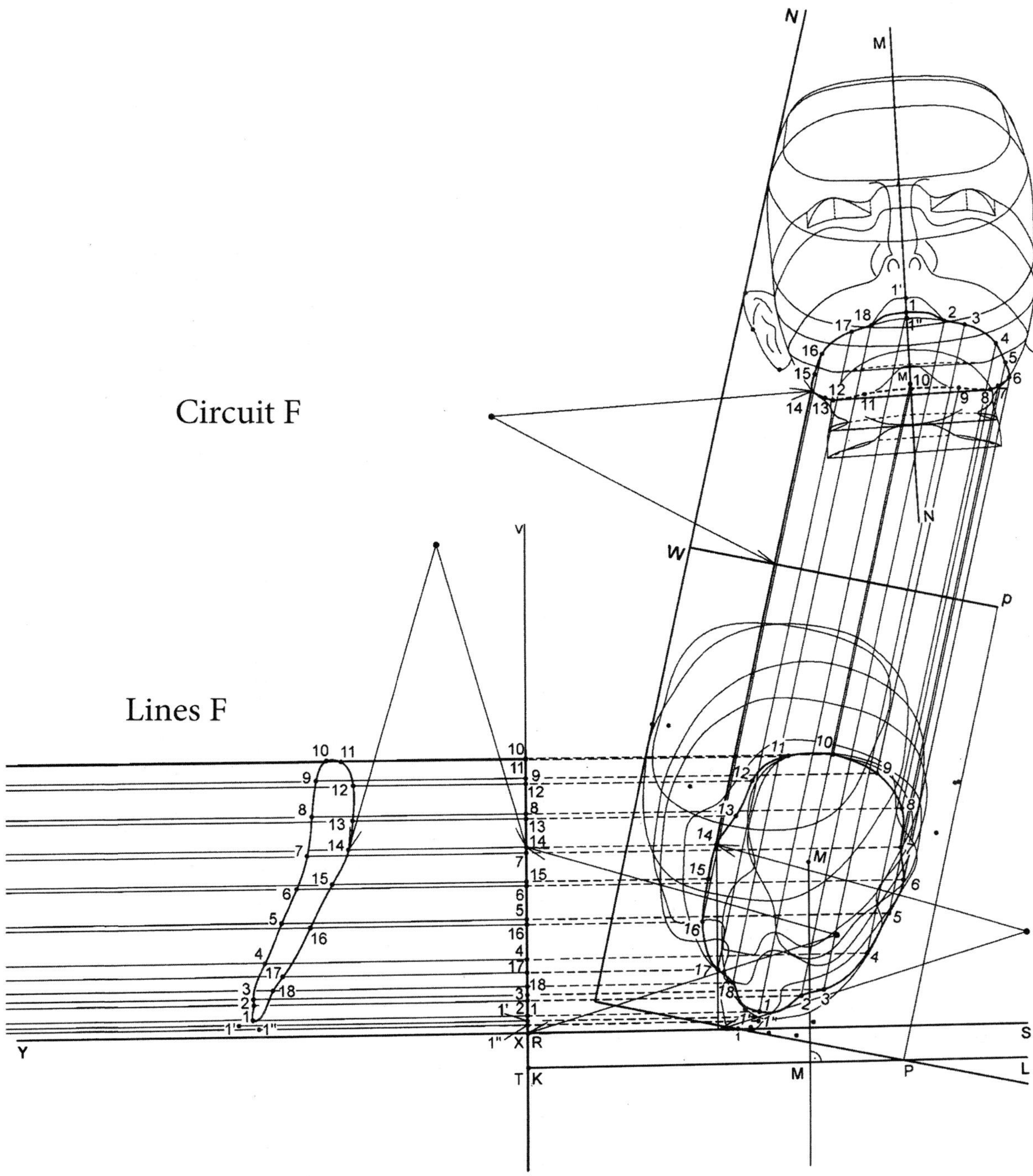

Figure B.159
Book 3, Proposition VIII.

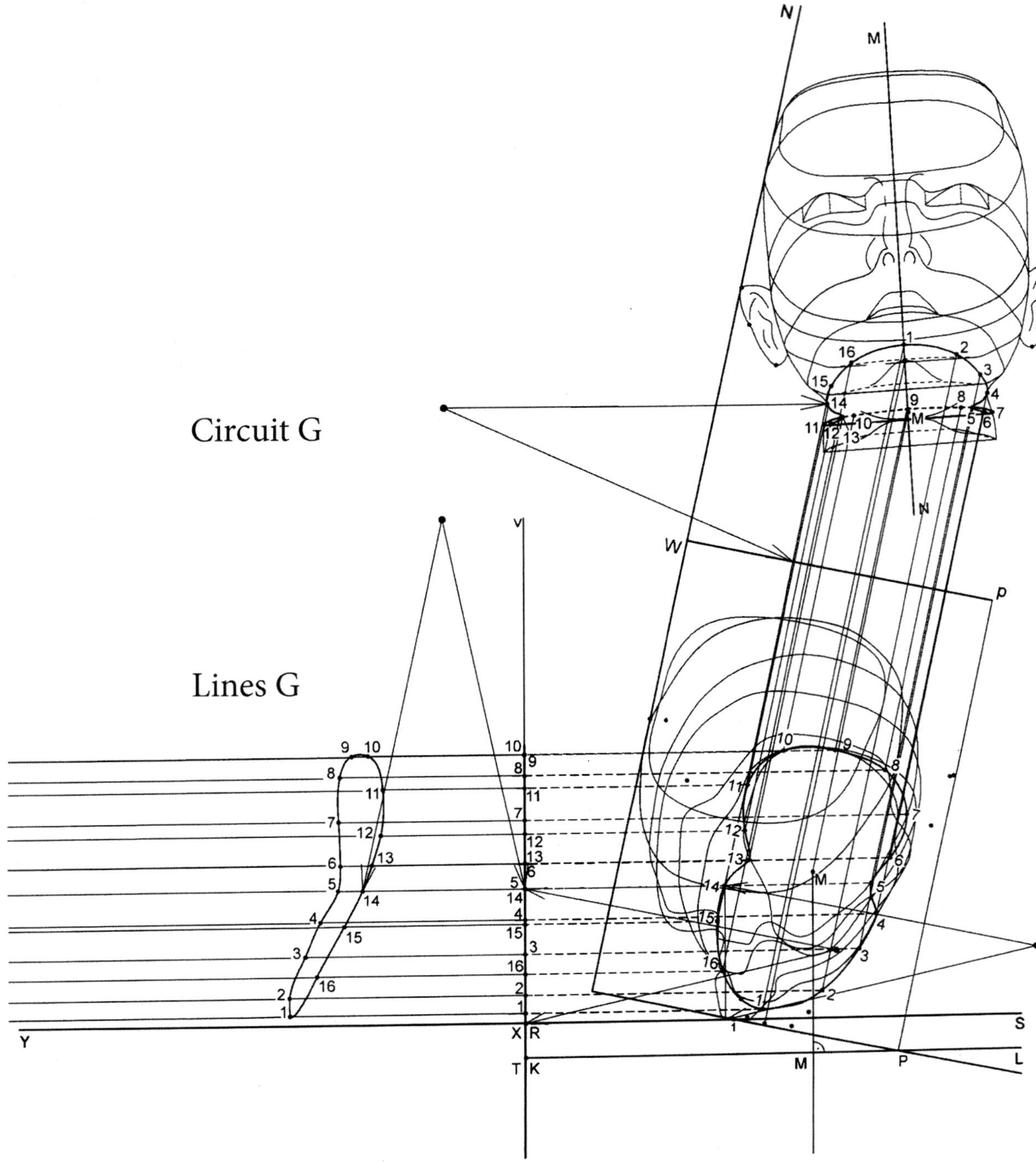

Figure B.160
Book 3, Proposition VIII.

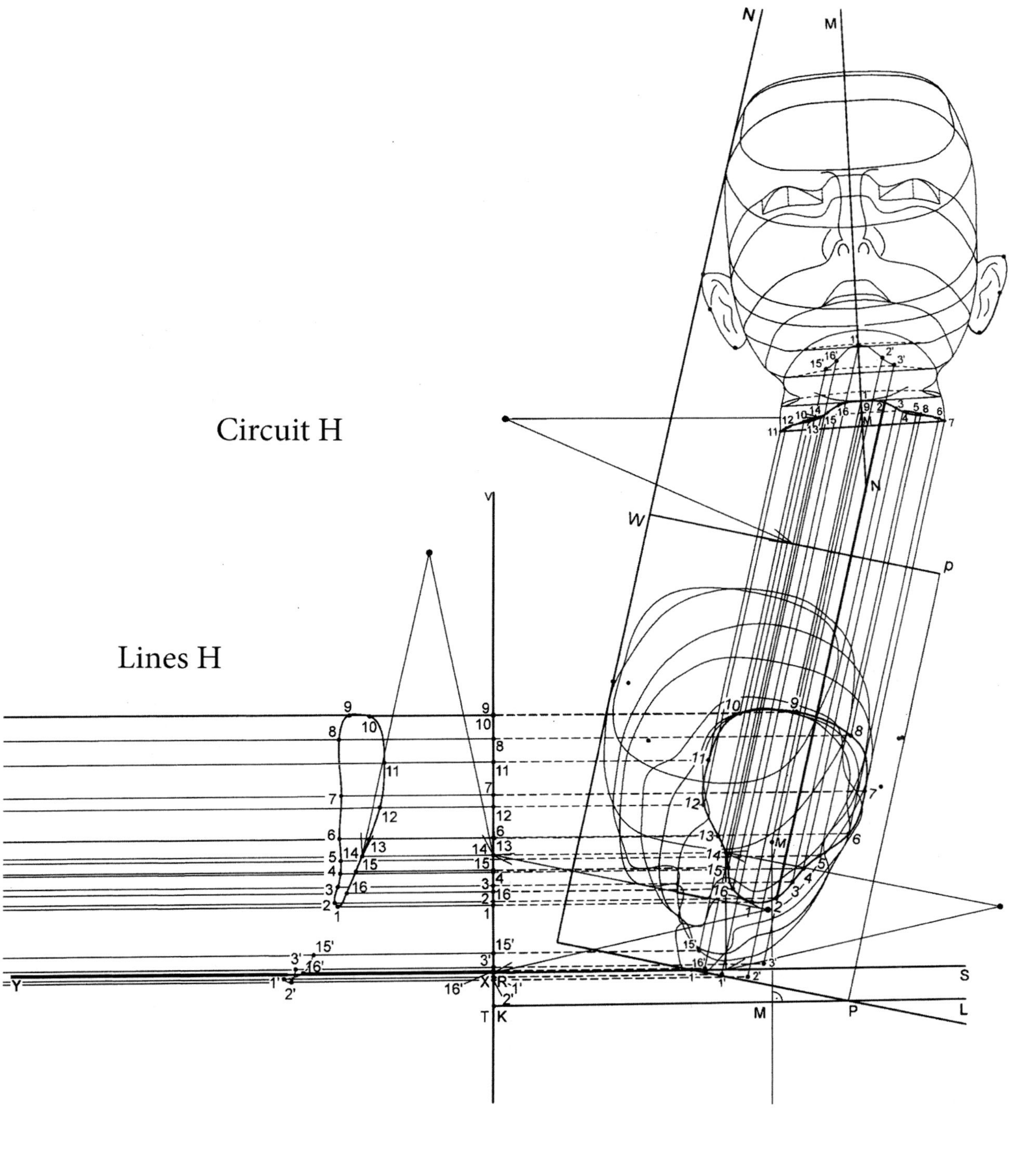

Figure B.161
Book 3, Proposition VIII.

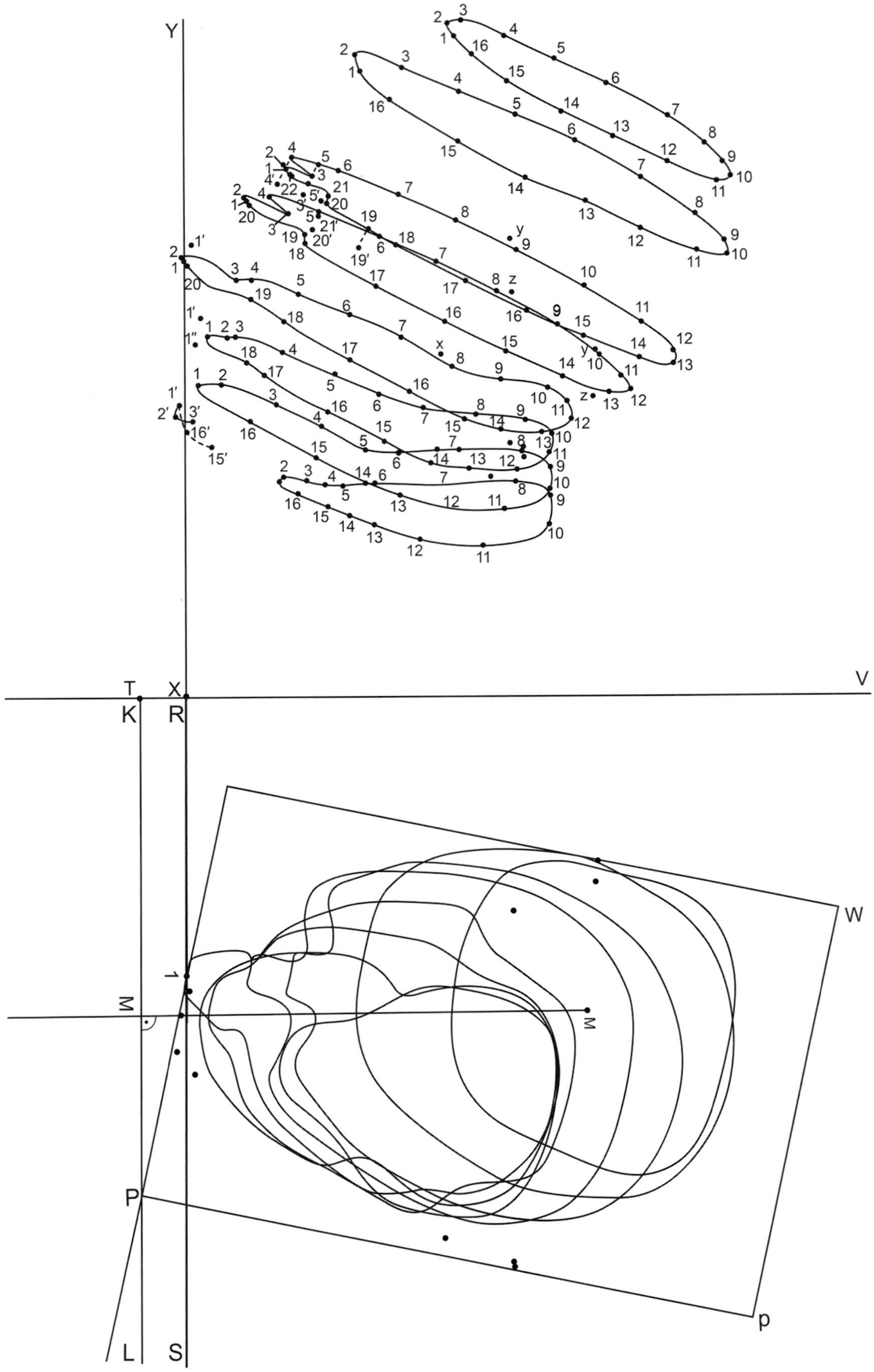

Figure B.162
Book 3, Proposition VIII.

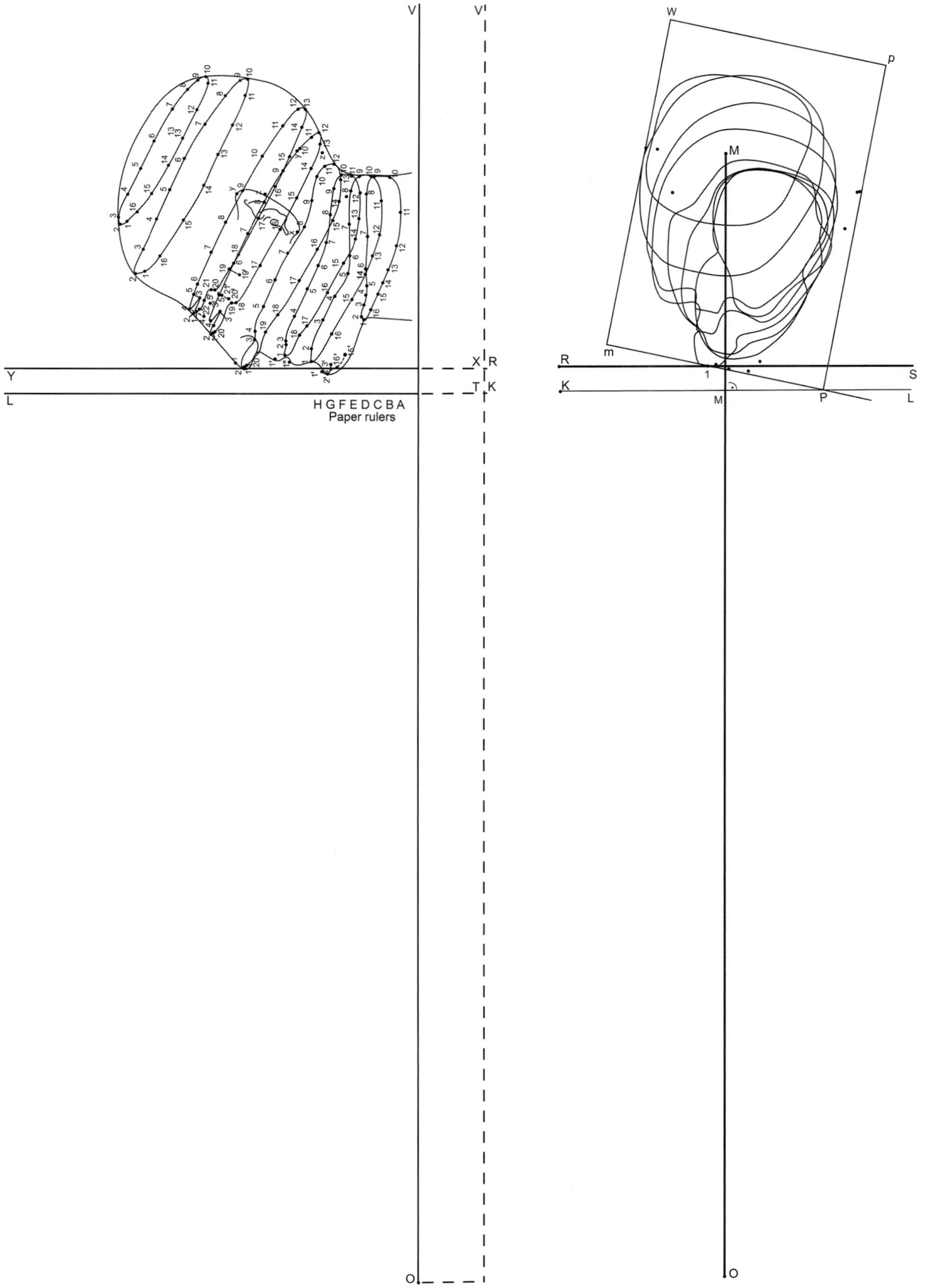

Figure B.163
Book 3, Proposition VIII.

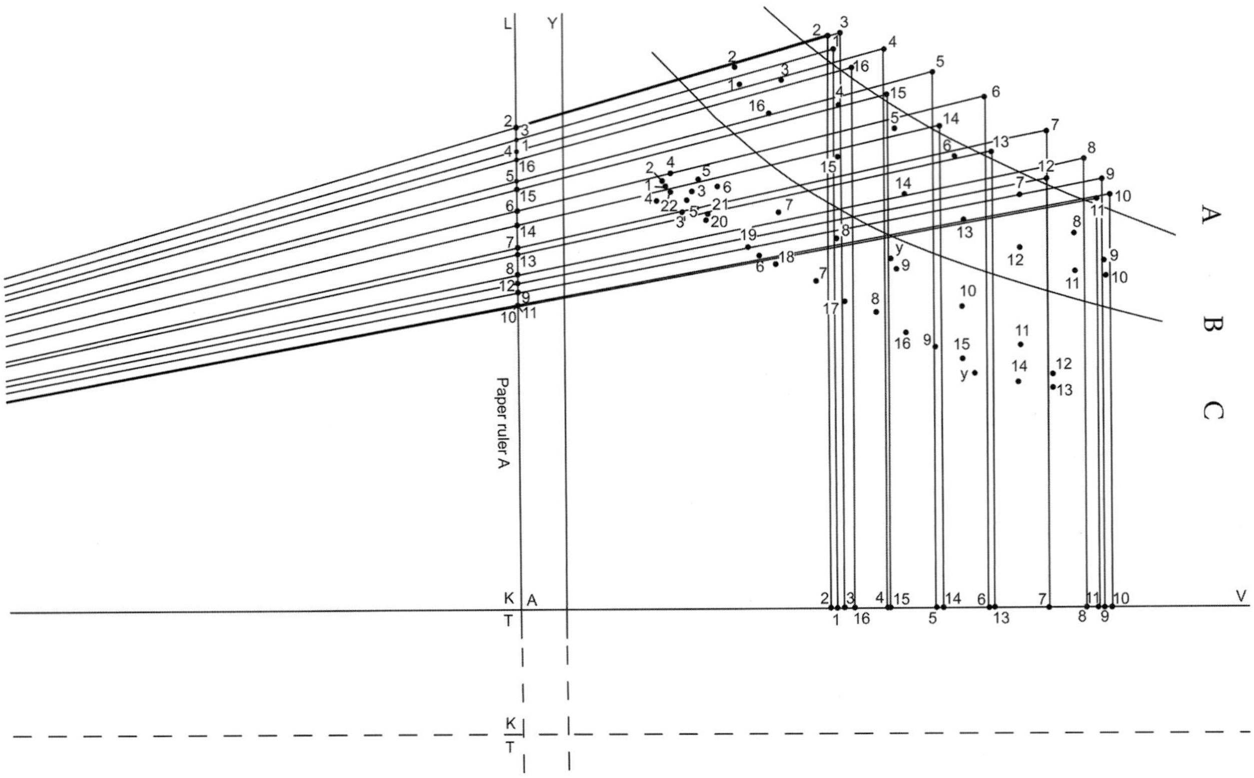

Figure B.164
Book 3, Proposition VIII.

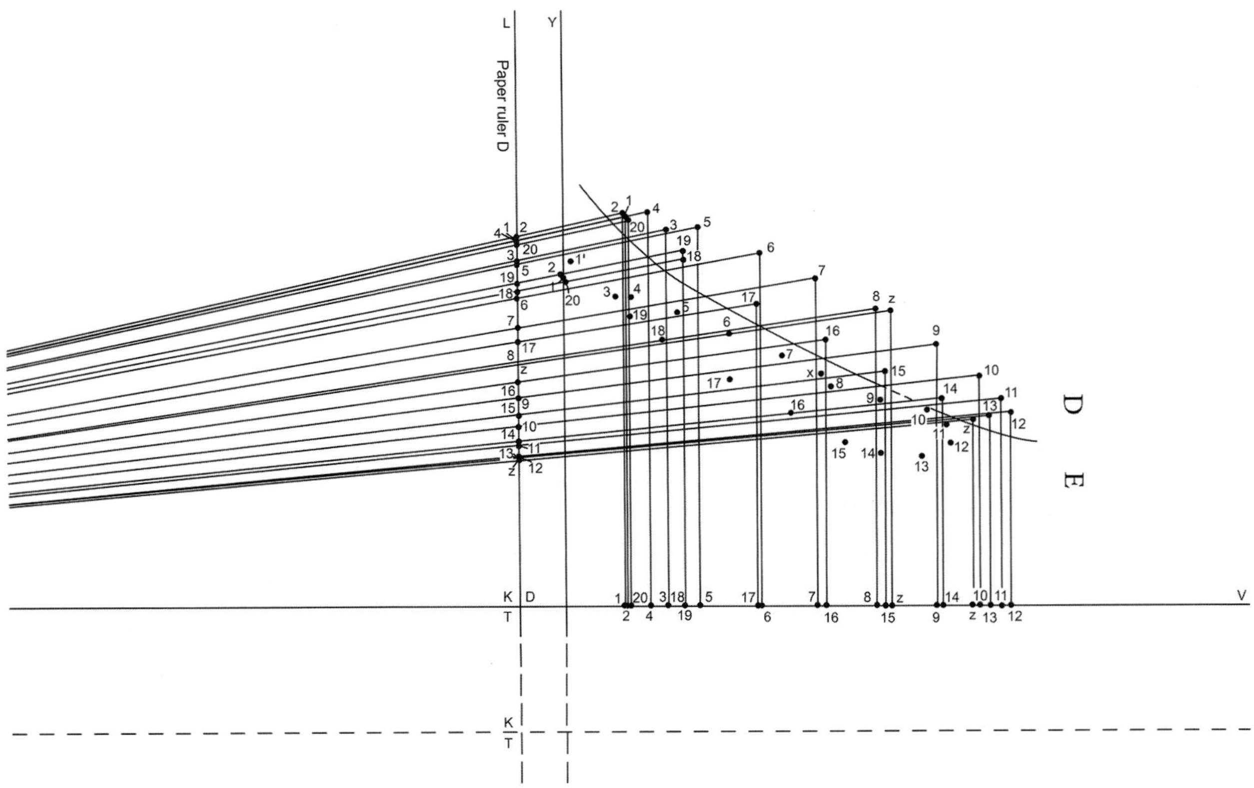

Figure B.165
Book 3, Proposition VIII.

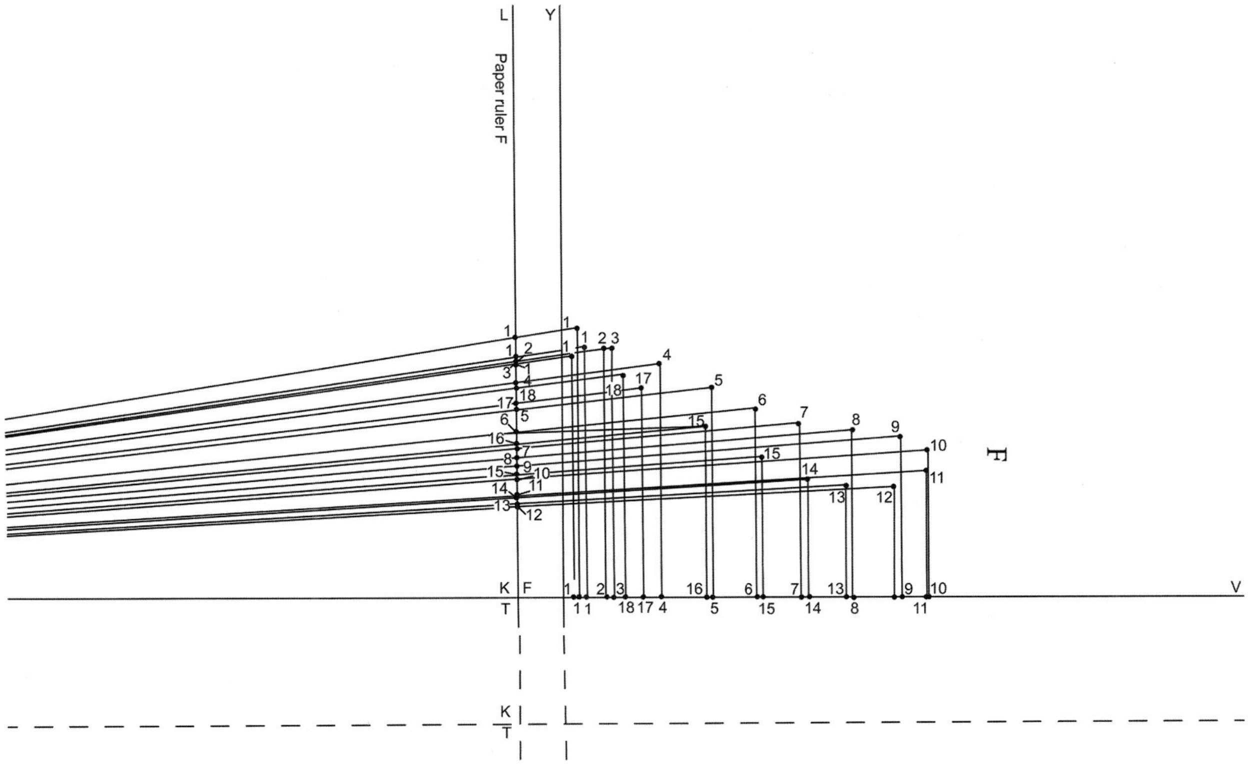

Figure B.166
Book 3, Proposition VIII.

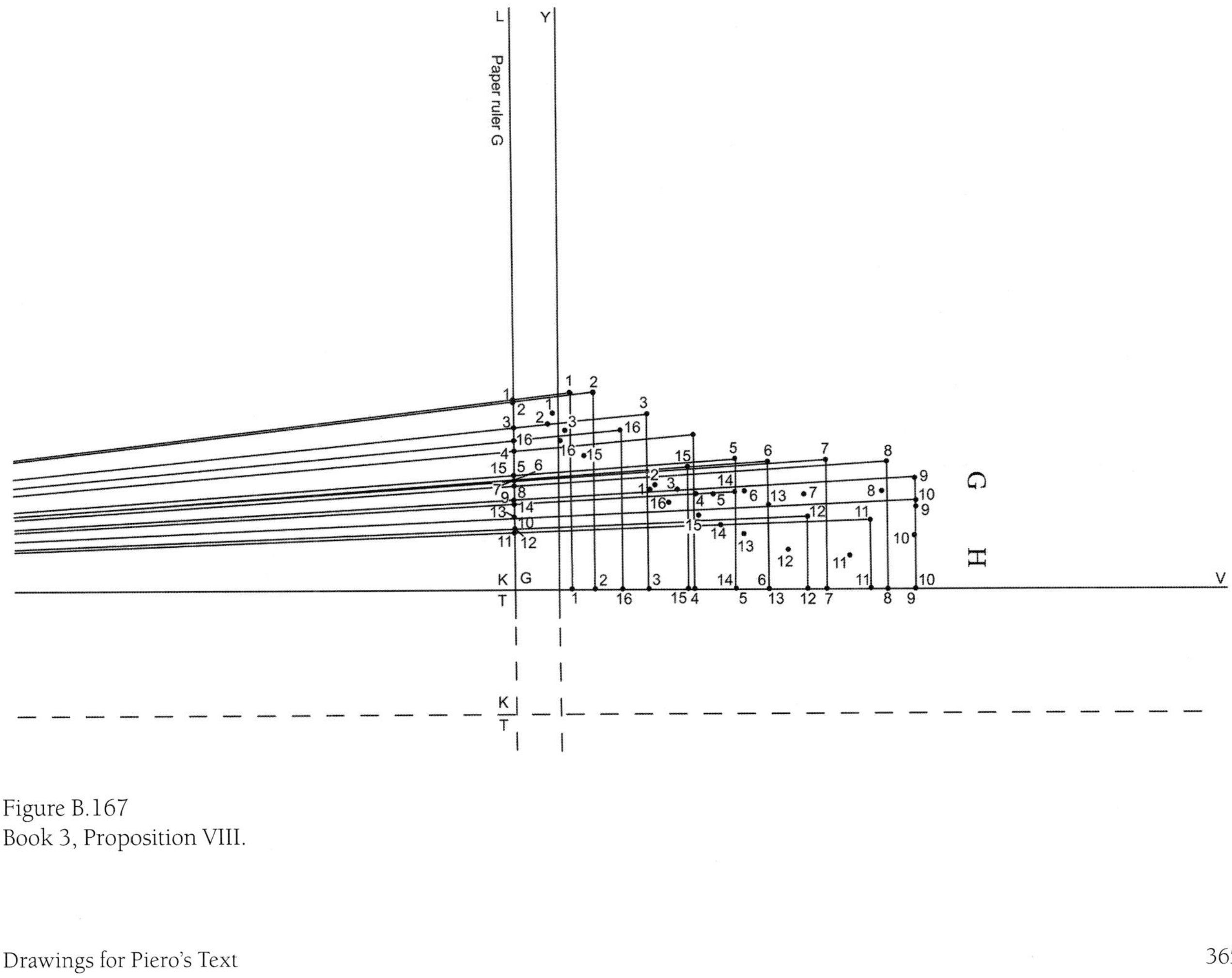

Figure B.167
Book 3, Proposition VIII.

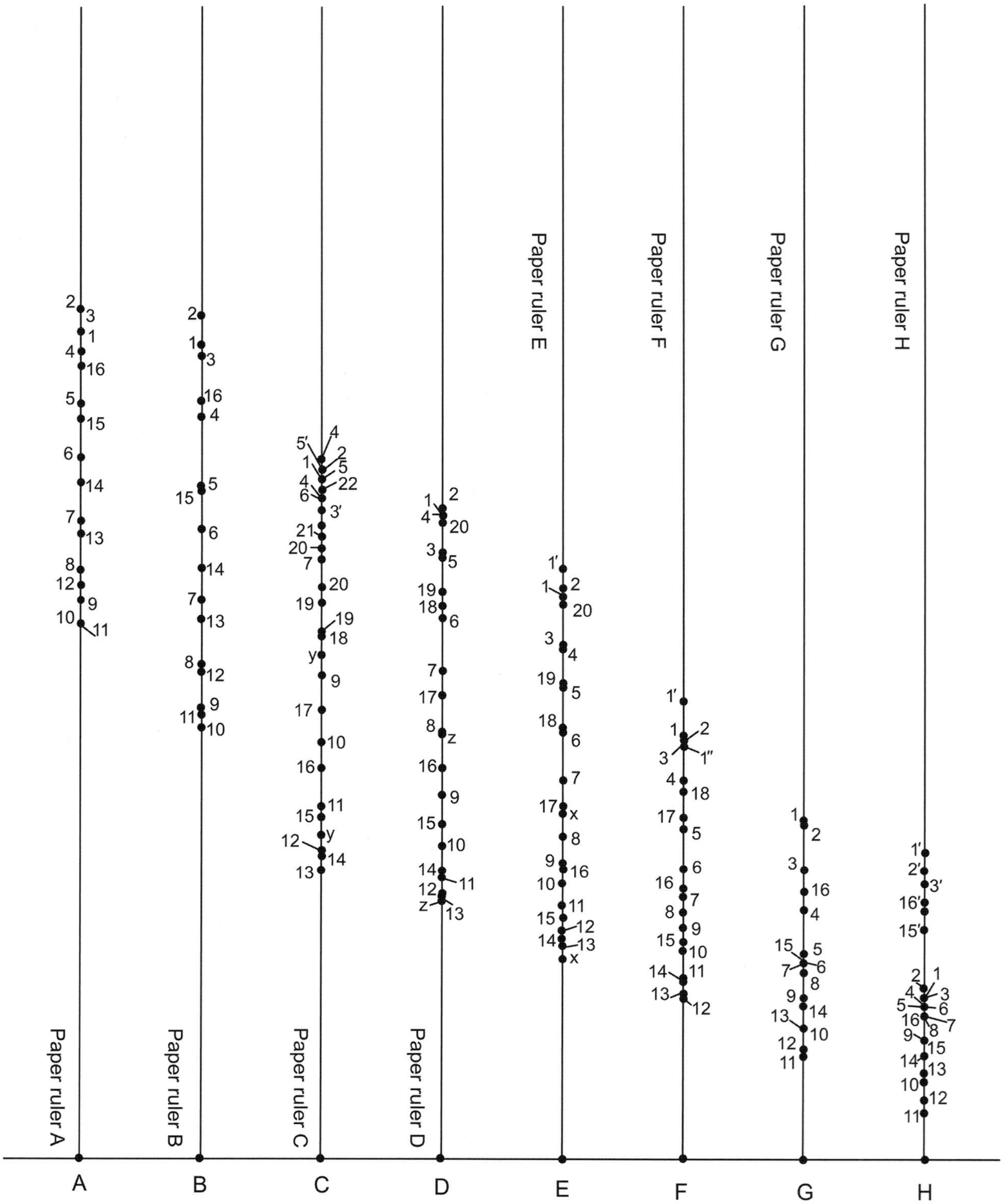

Figure B.168
Book 3, Proposition VIII.

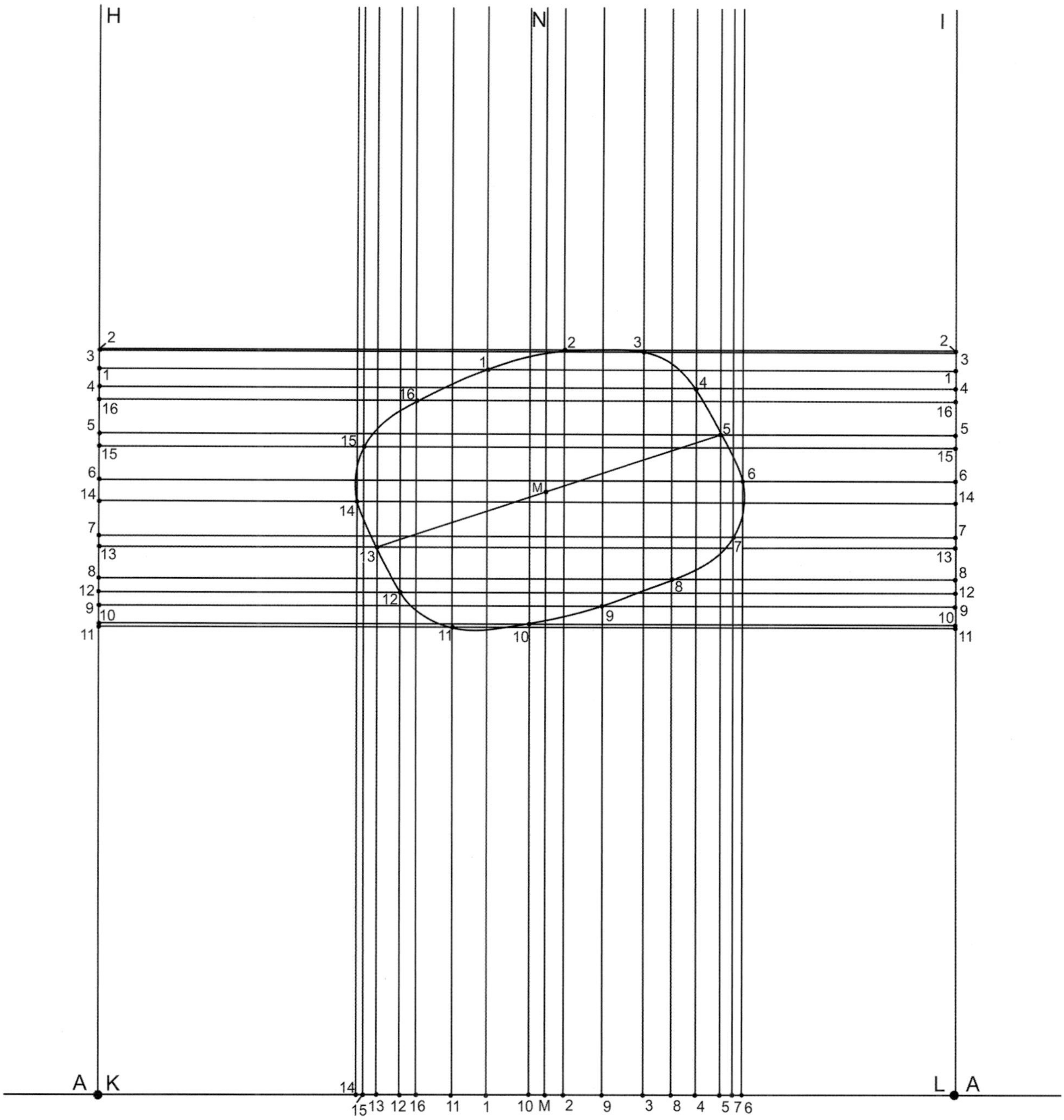

Figure B.169
Book 3, Proposition VIII.

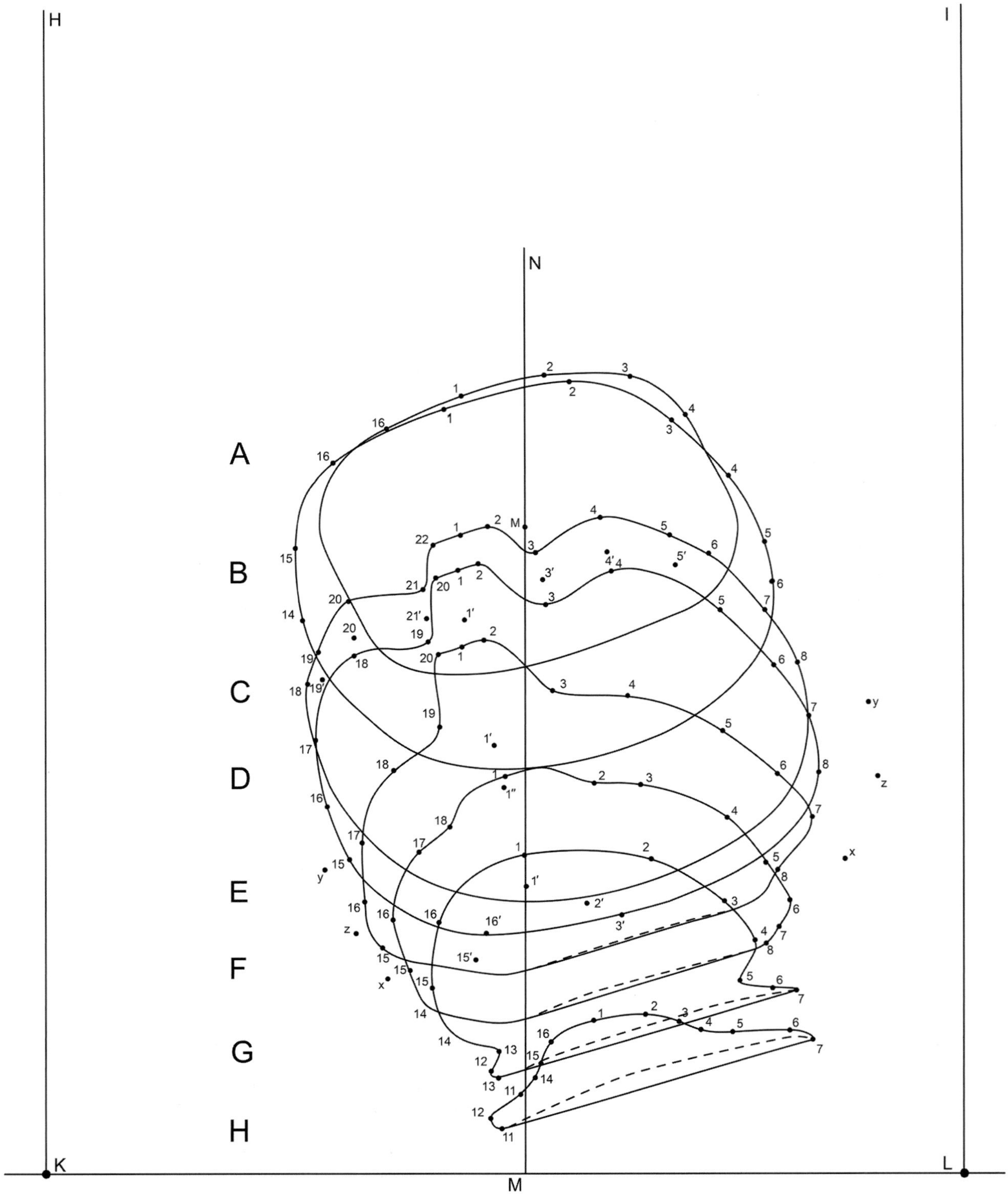

Figure B.170
Book 3, Proposition VIII.

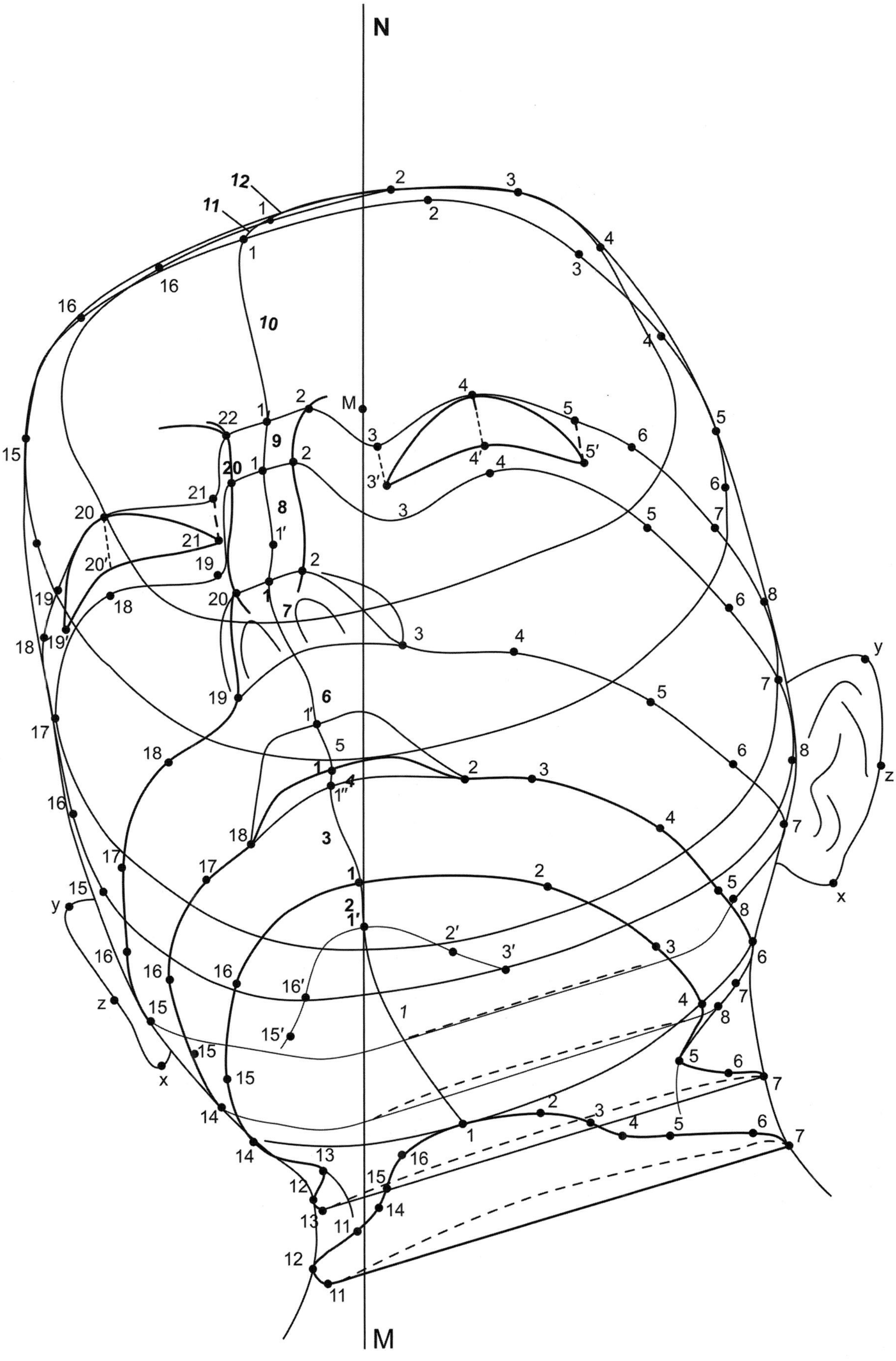

Figure B.171
Book 3, Proposition VIII.

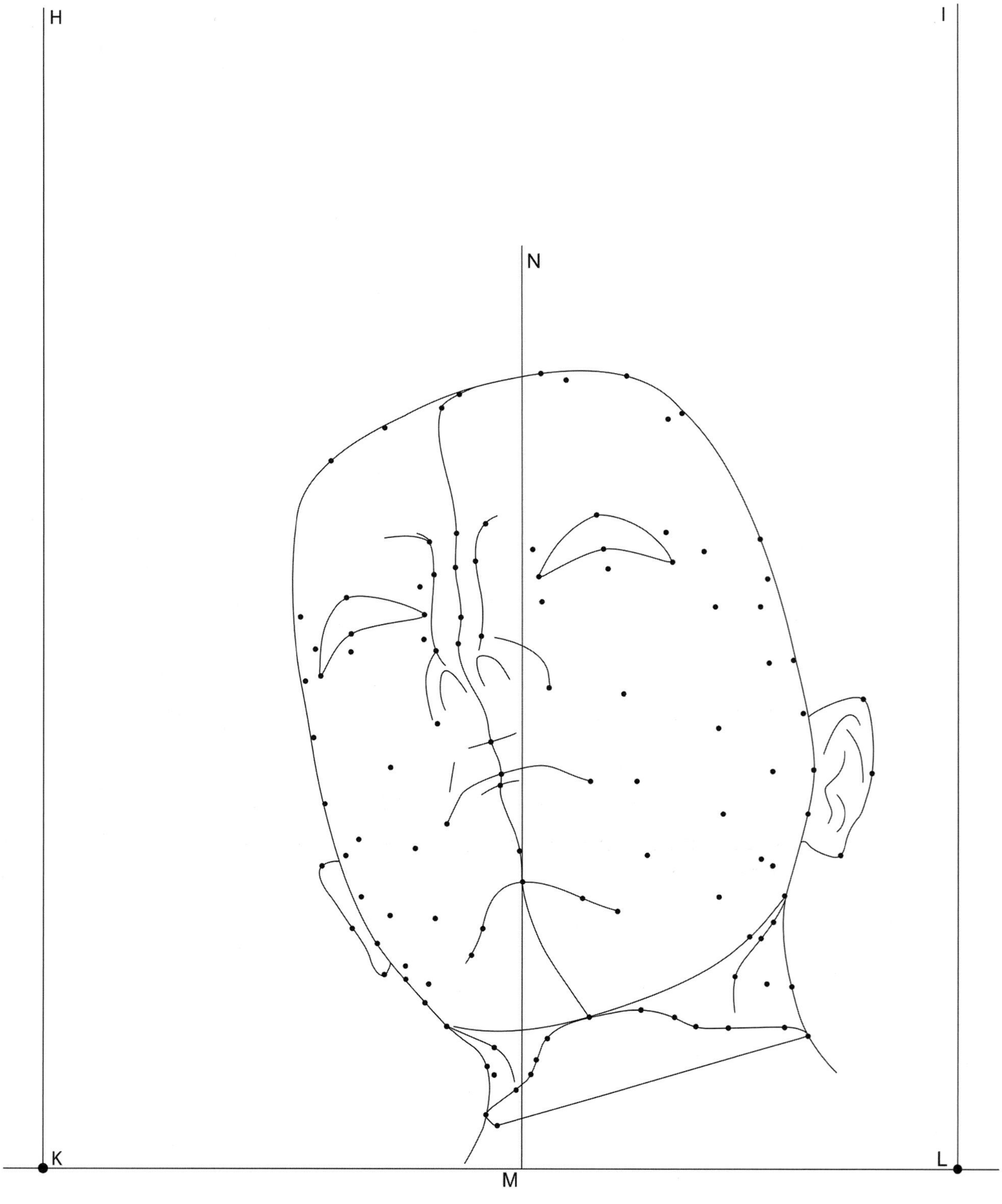

Figure B.172
Book 3, Proposition VIII.

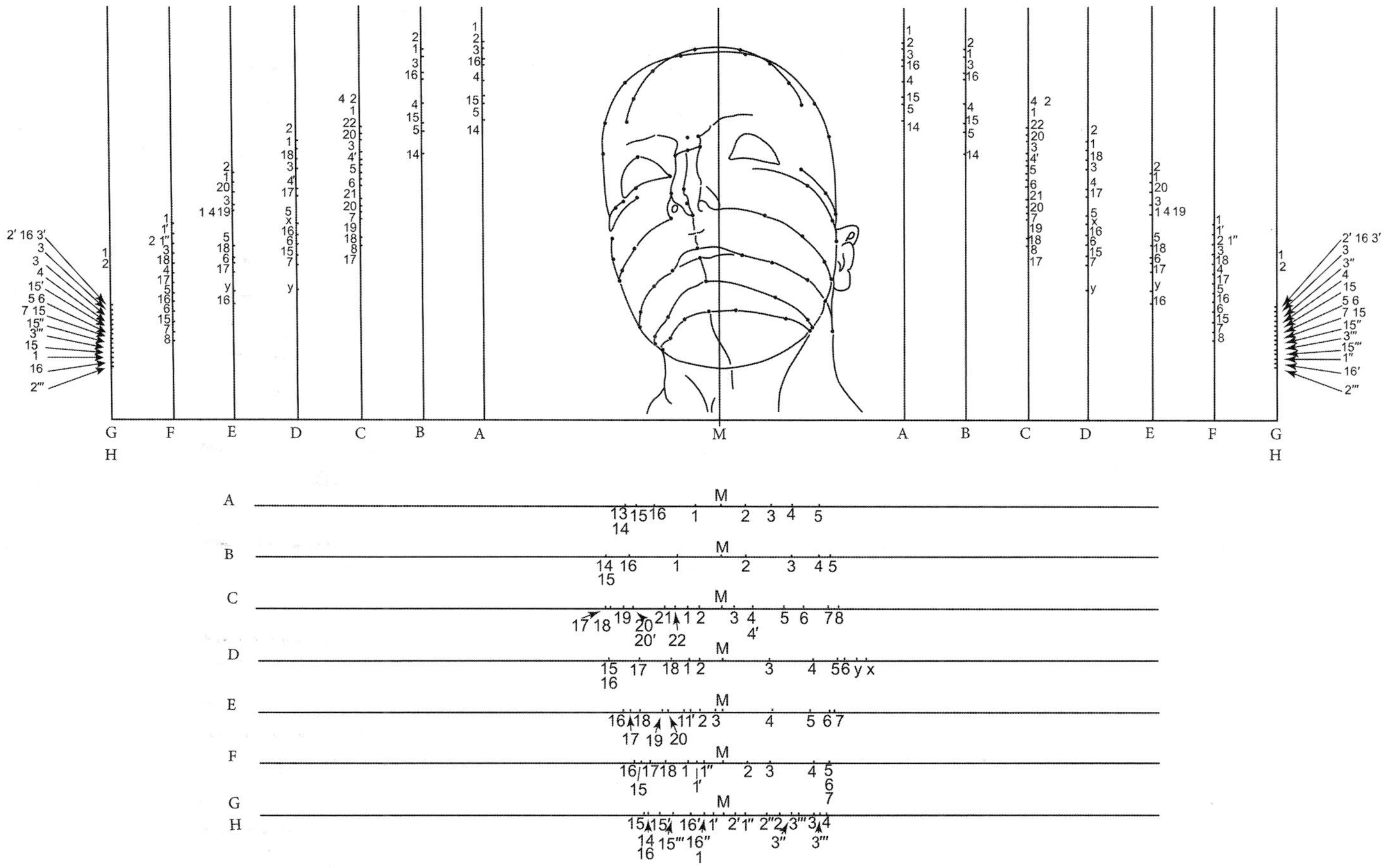

Figure B.173
Book 3, Proposition VIII.

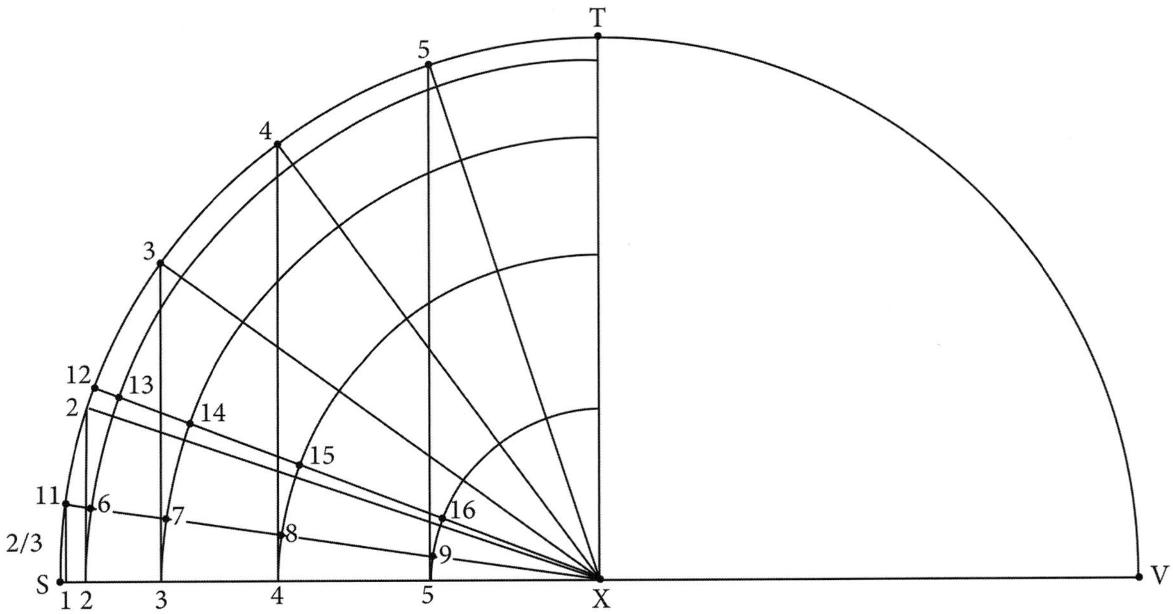

Figure B.174
Book 3, Proposition IX.

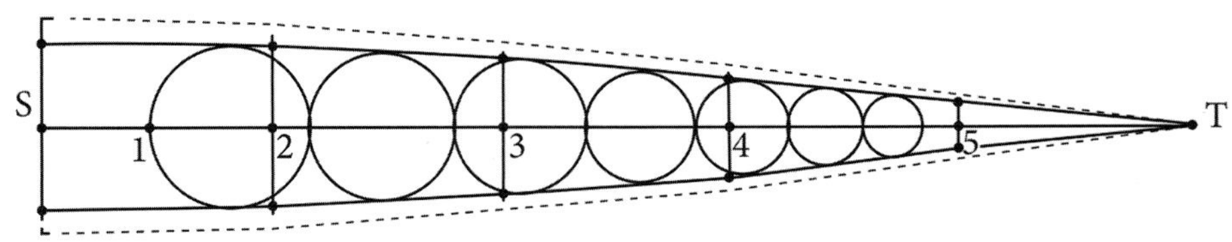

Figure B.175
Book 3, Proposition IX.

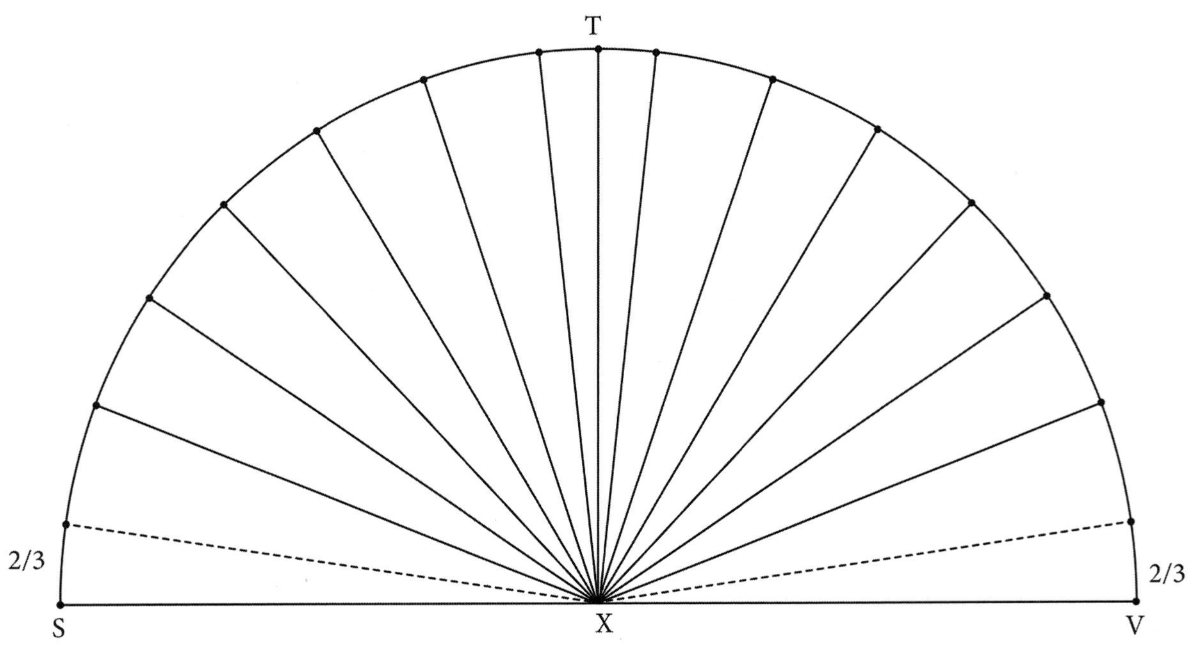

Figure B.176
Book 3, Proposition IX.

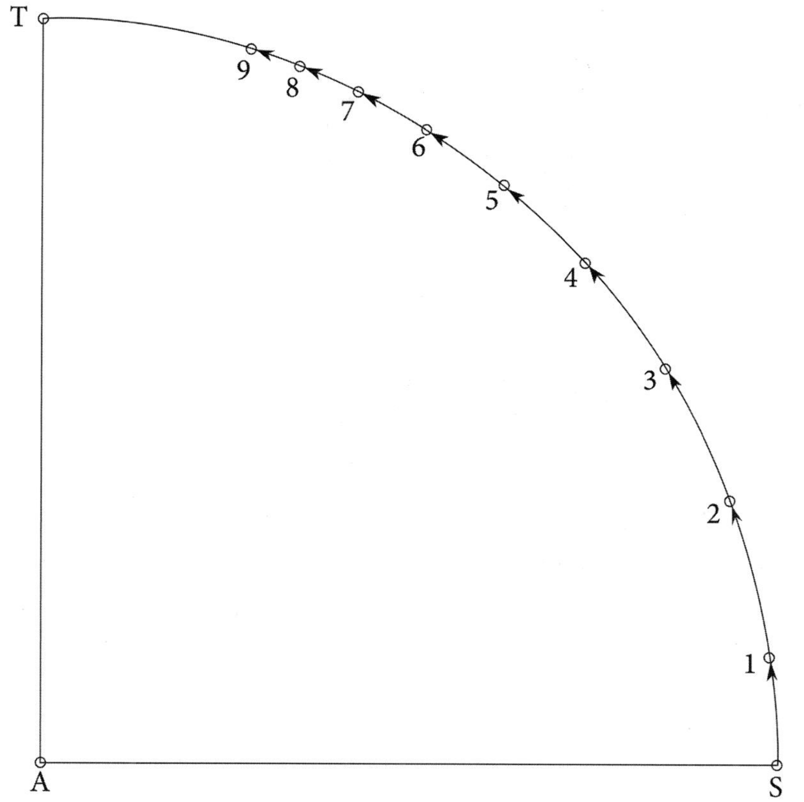

Figure B.177
Book 3, Proposition IX.

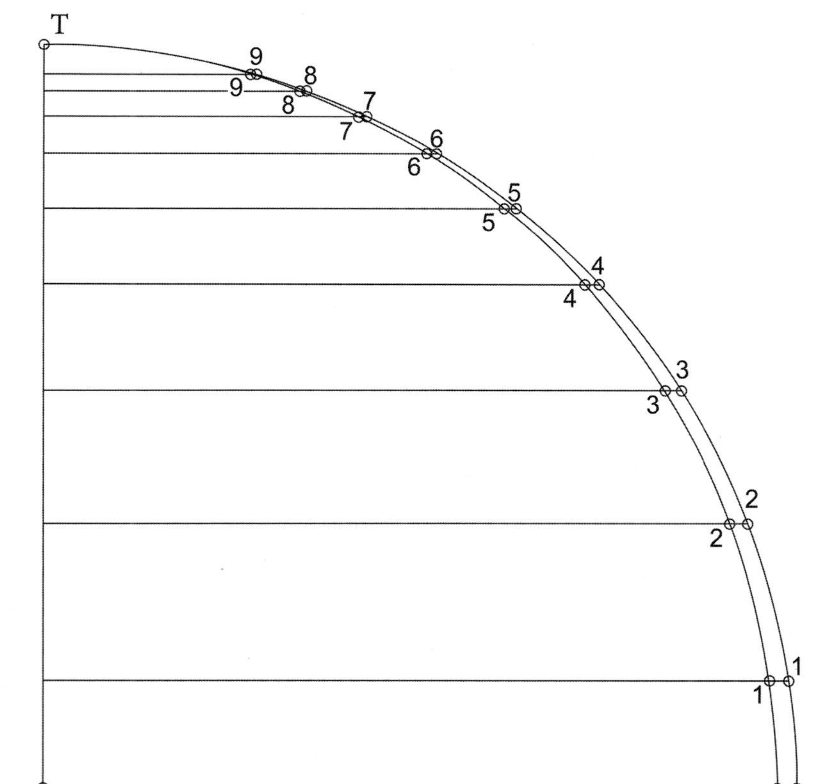

Figure B.178
Book 3, Proposition IX.

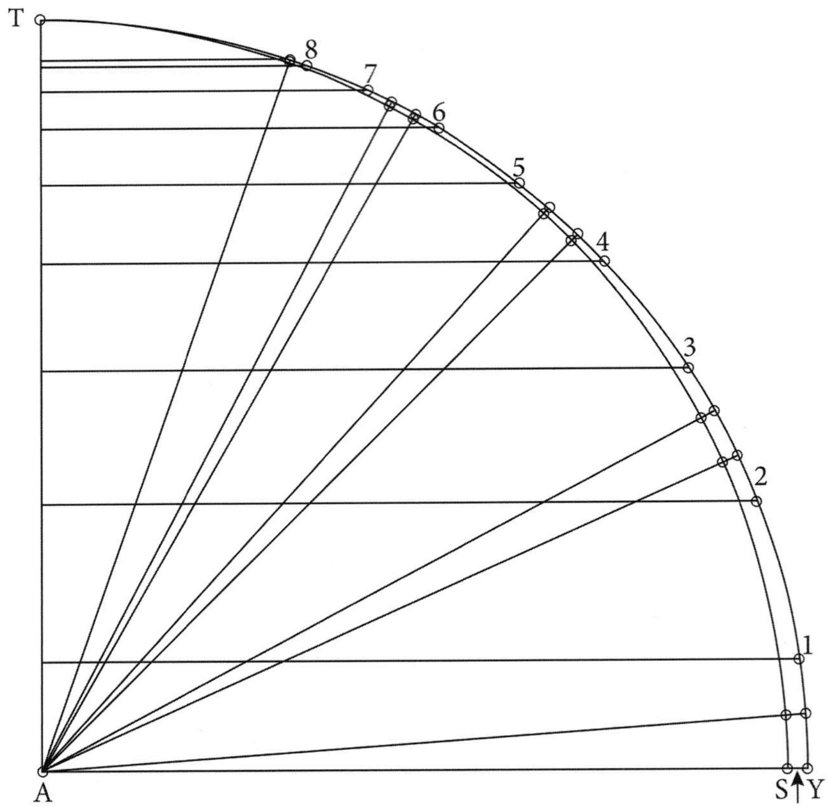

Figure B.179
Book 3, Proposition IX.

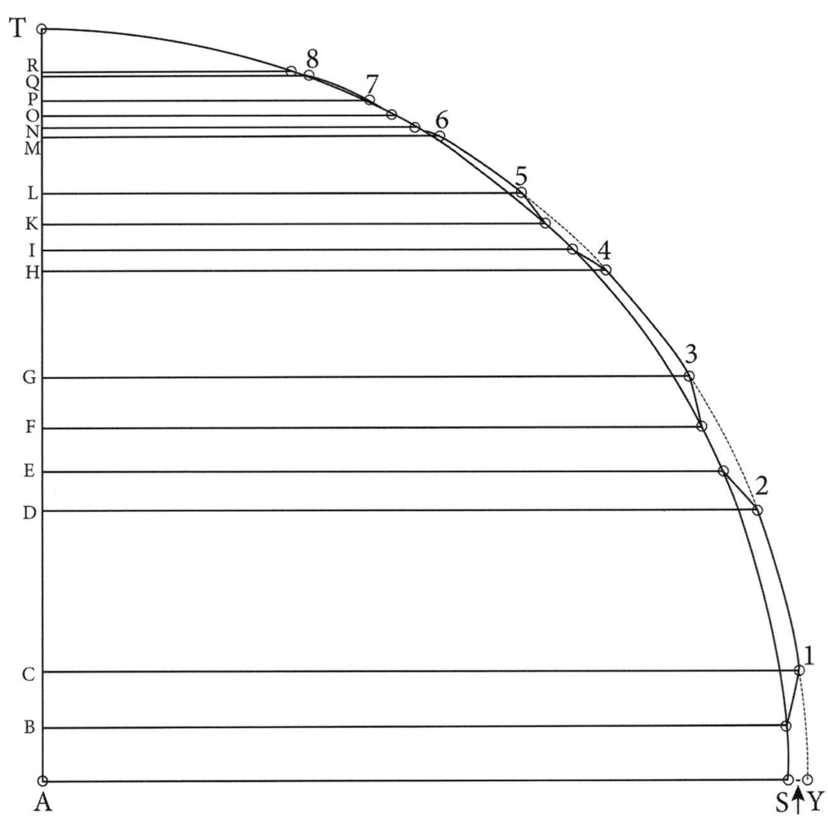

Figure B.180
Book 3, Proposition IX.

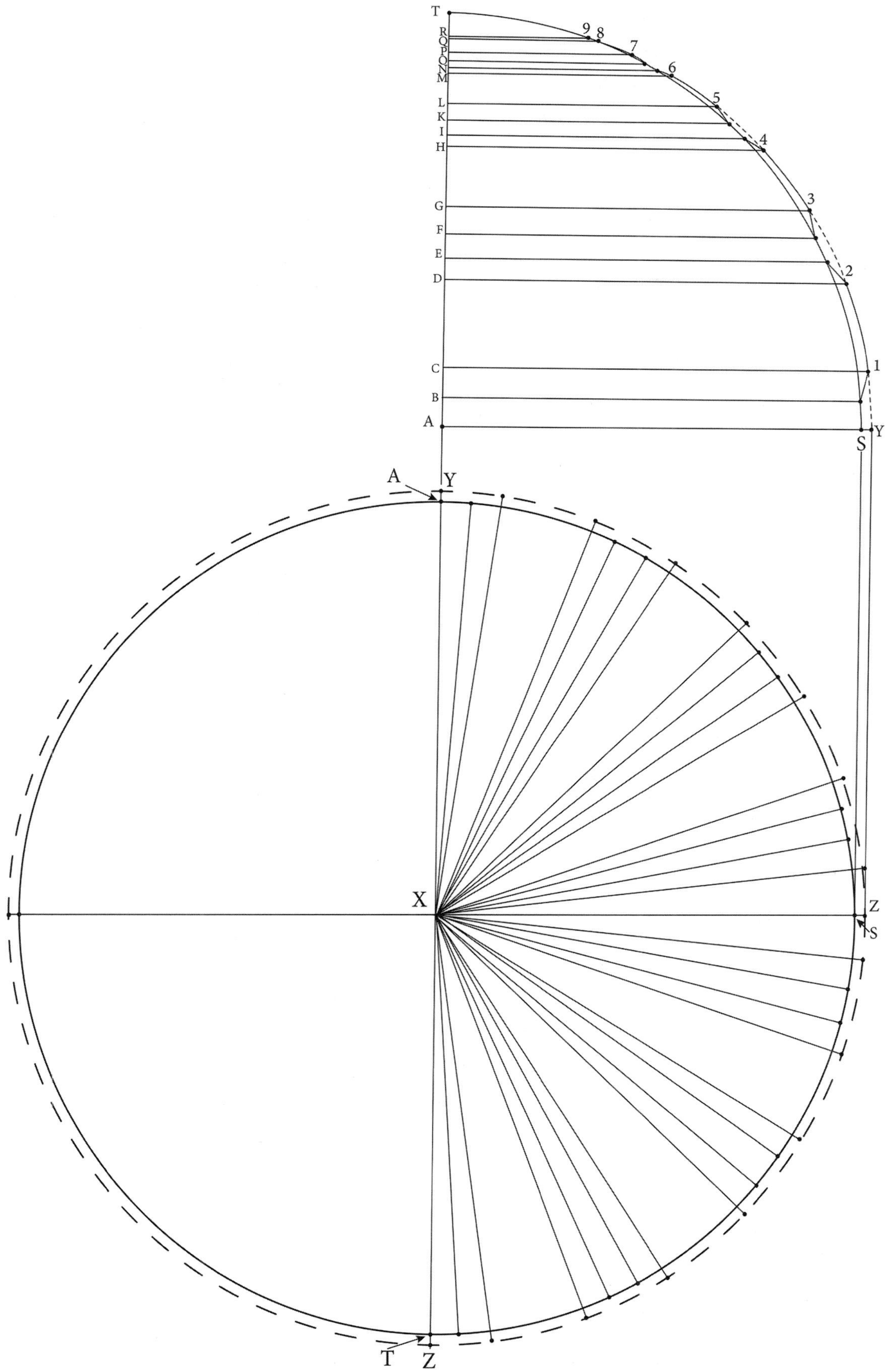

Figure B.181
Book 3, Proposition IX.

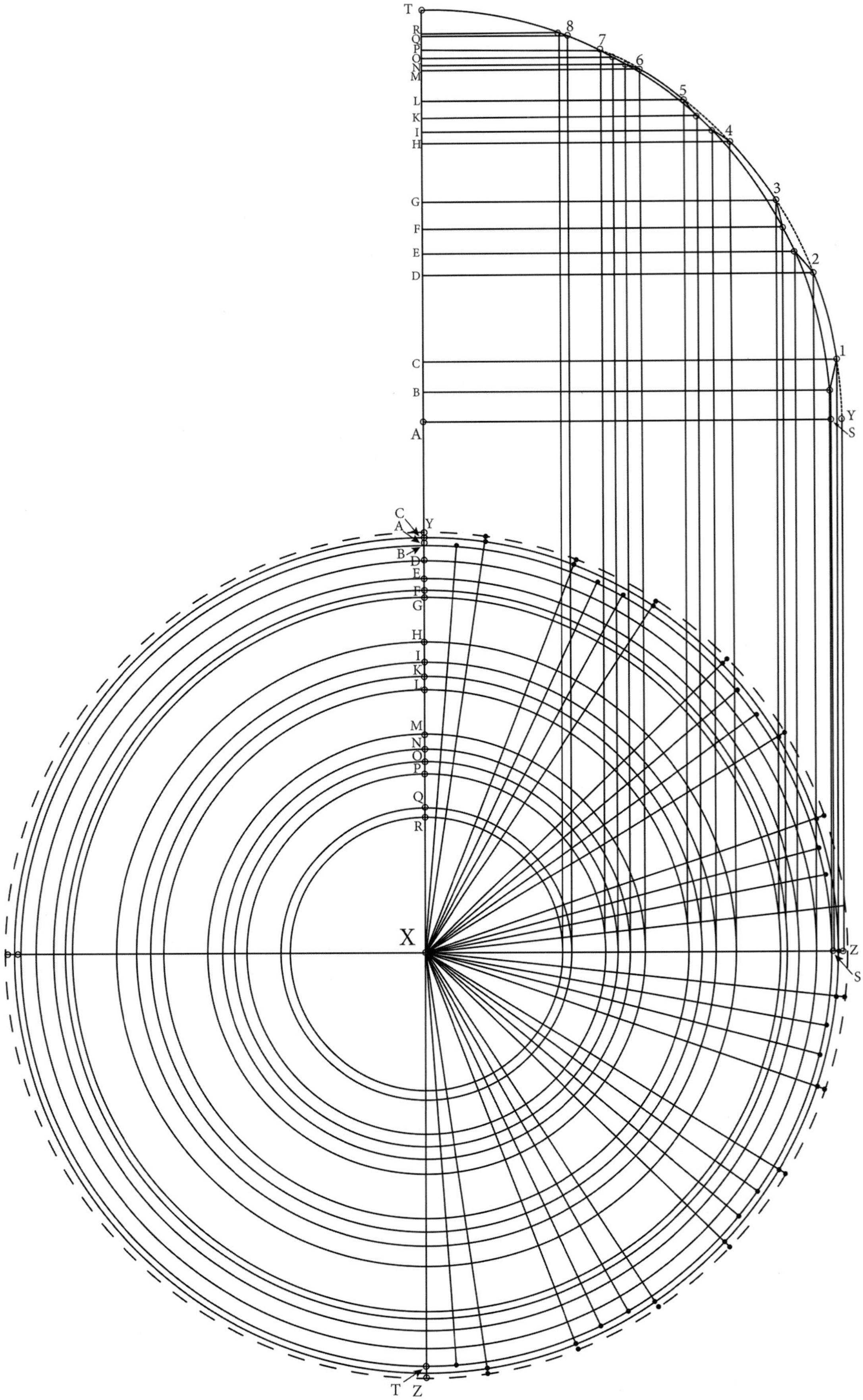

Figure B.182
Book 3, Proposition IX.

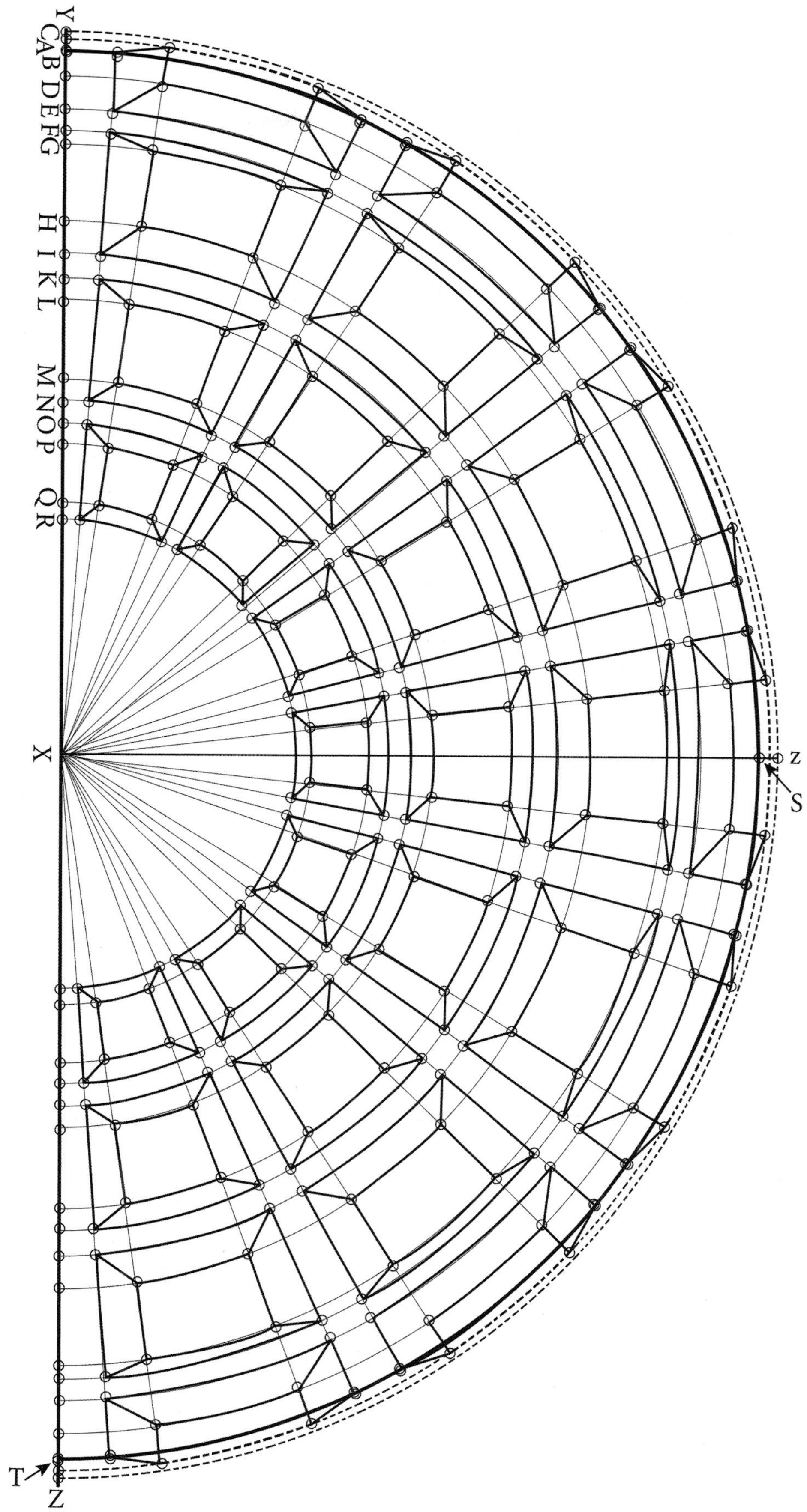

Figure B.183
Book 3, Proposition IX.

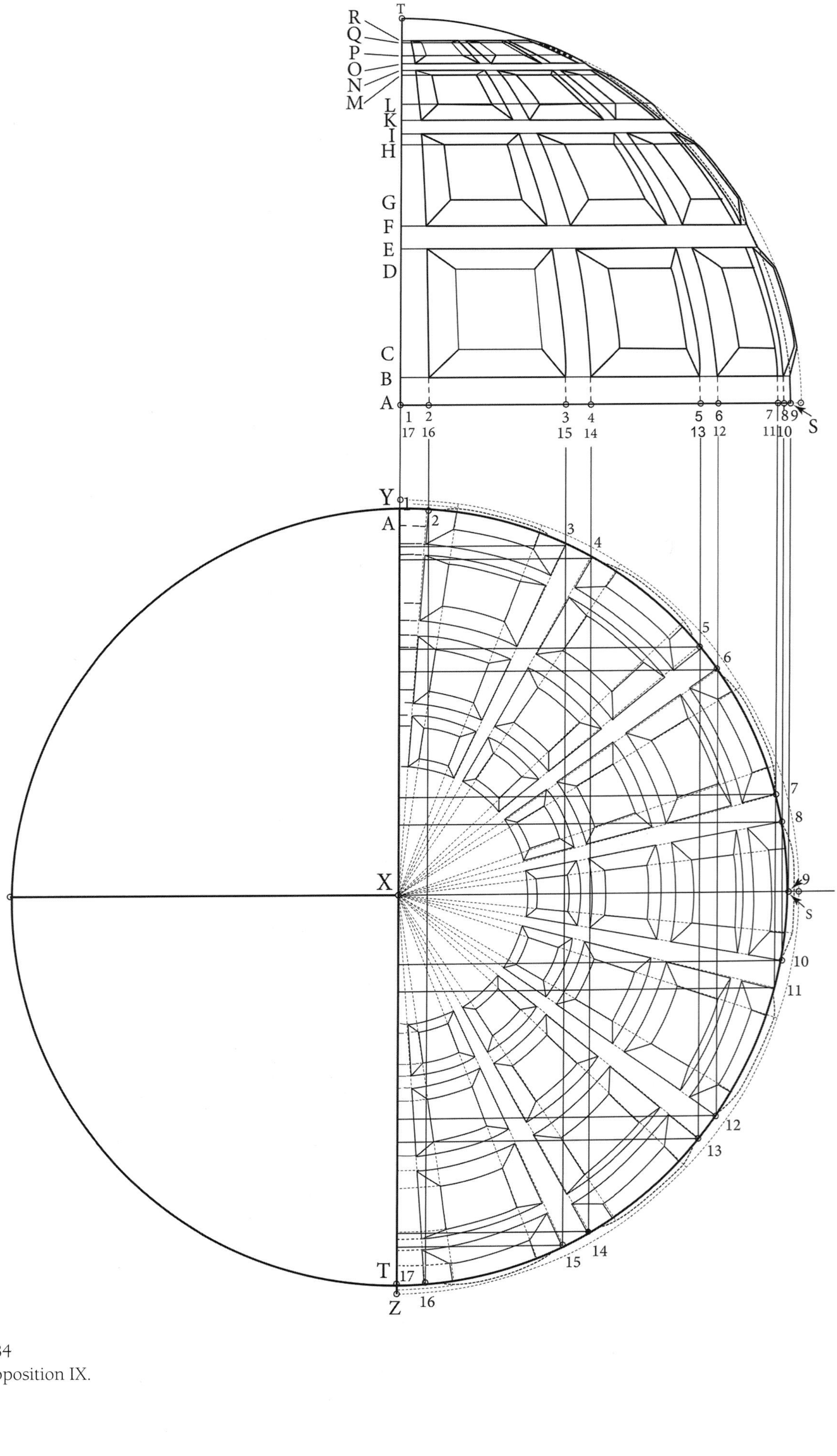

Figure B.184
Book 3, Proposition IX.

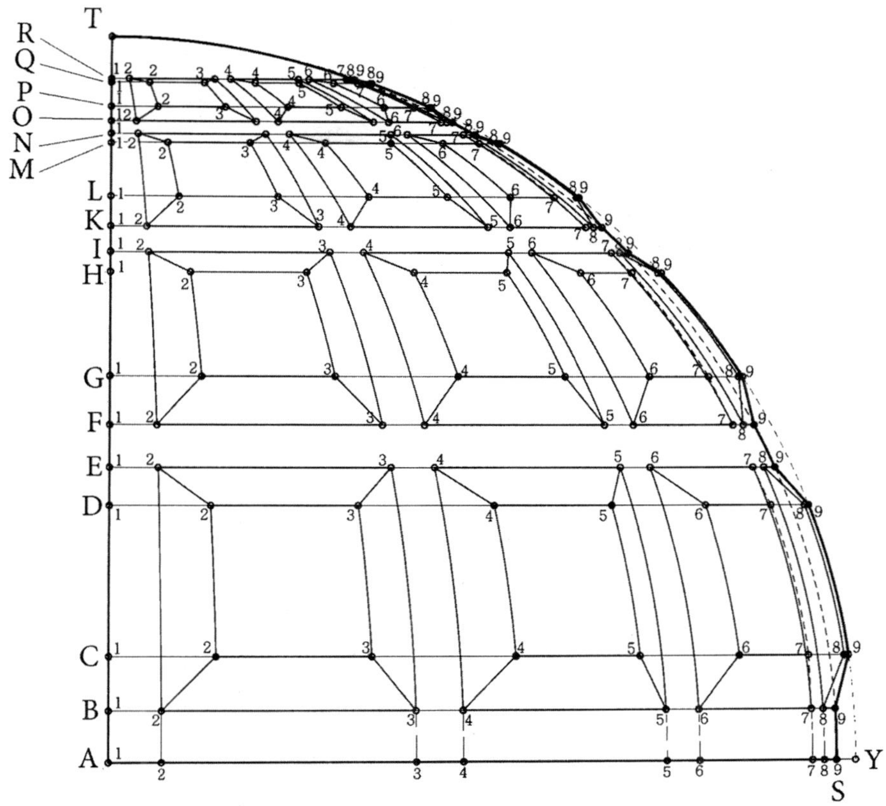

Figure B.185
Book 3, Proposition IX.

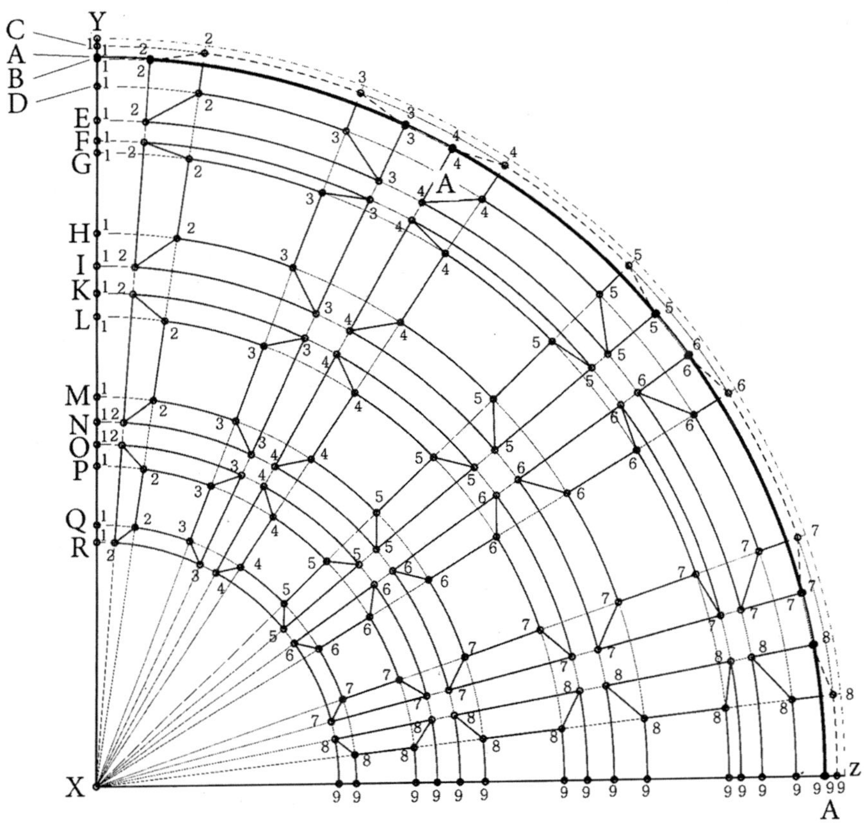

Figure B.186
Book 3, Proposition IX.

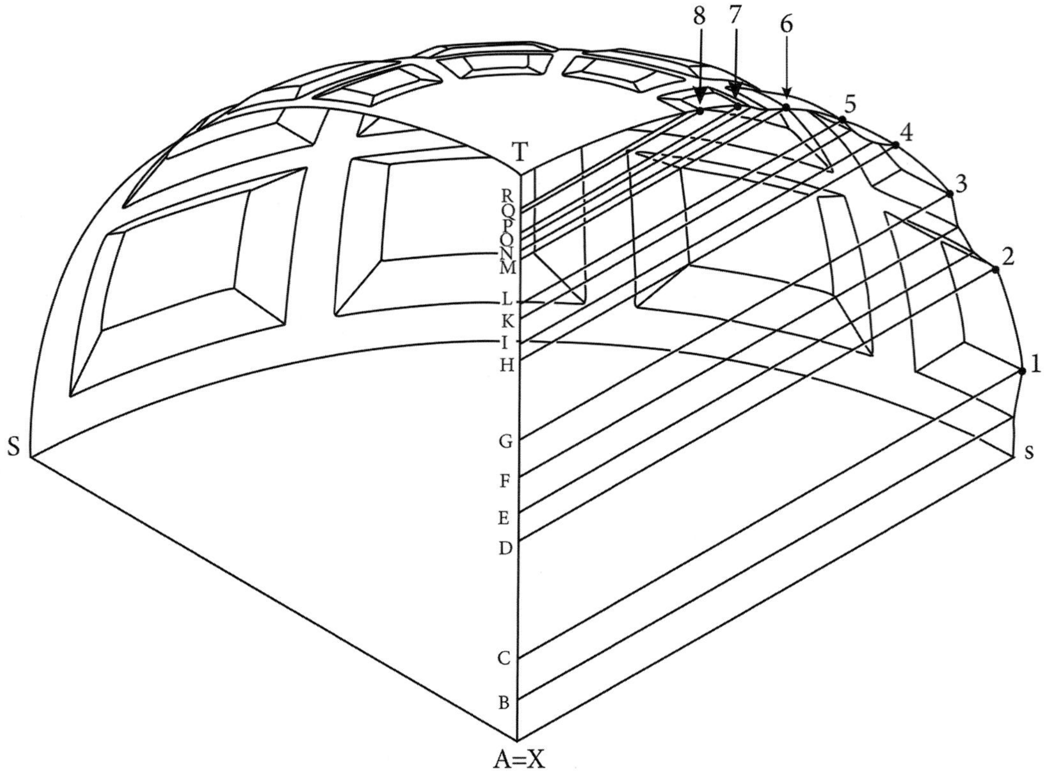

Figure B.187
Book 3, Proposition IX.

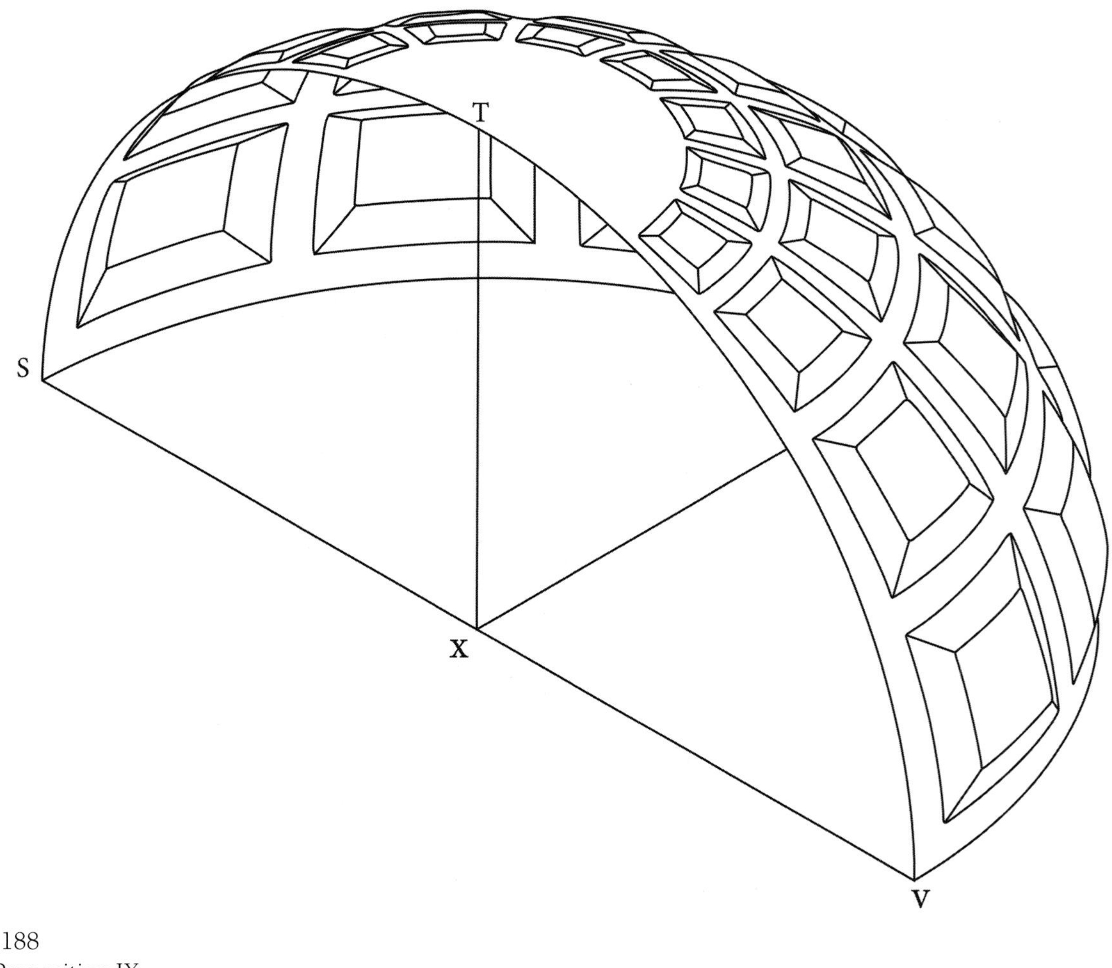

Figure B.188
Book 3, Proposition IX.

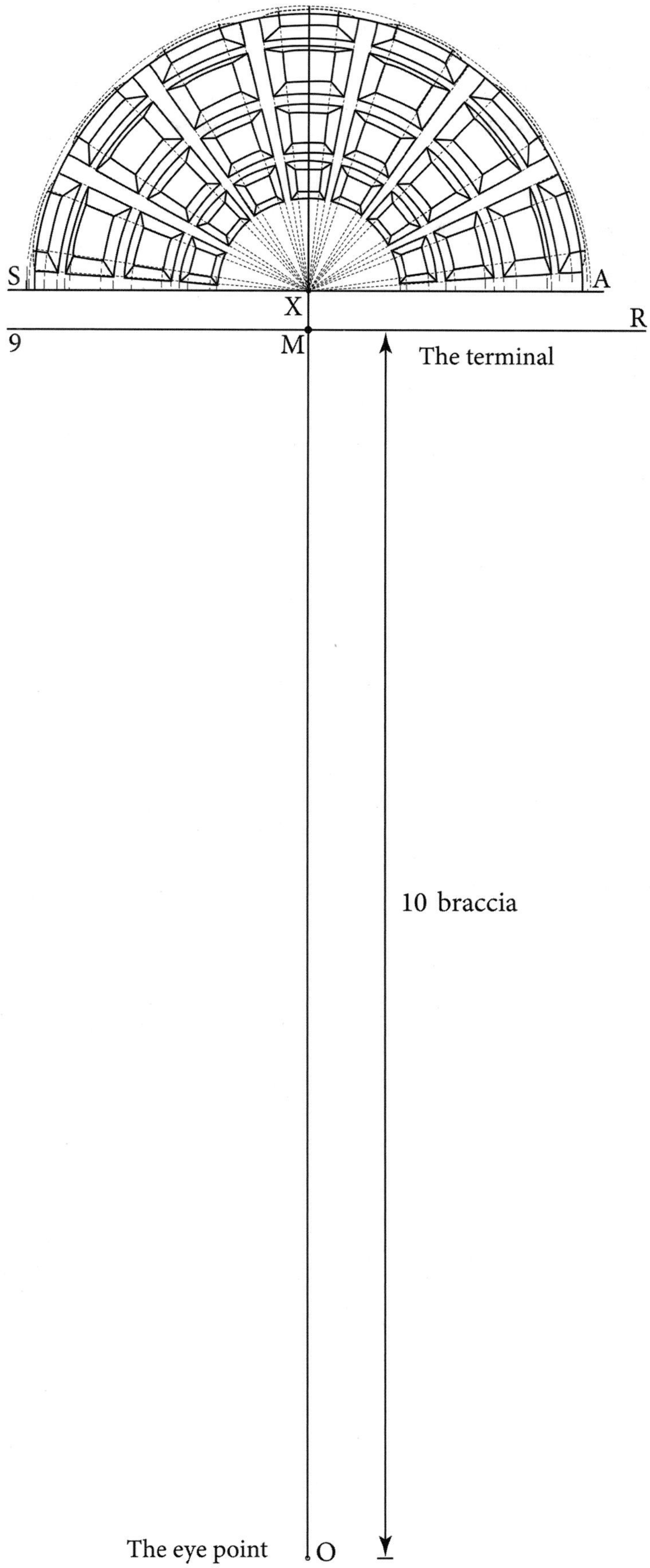

Figure B.189
Book 3, Proposition IX.

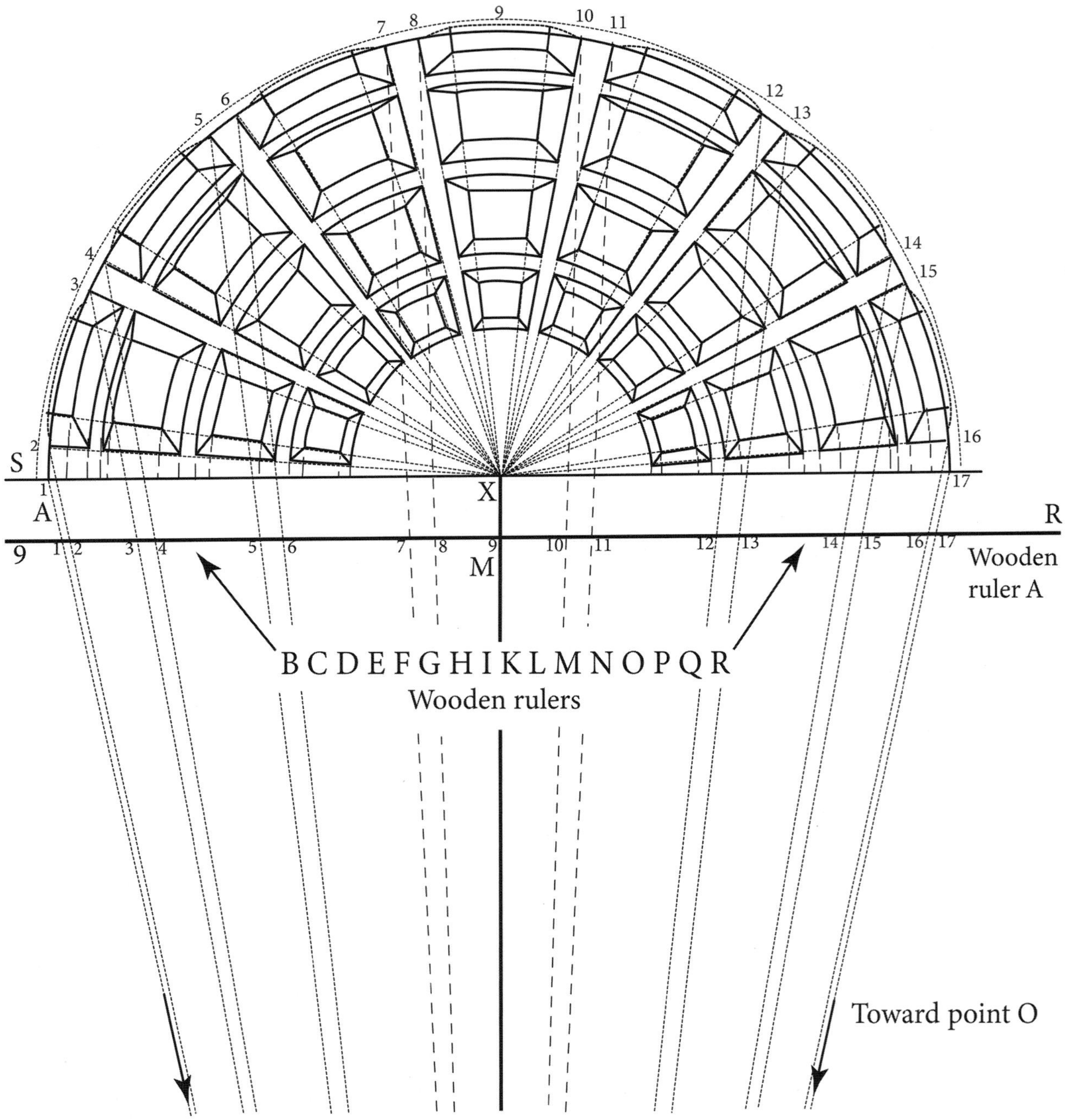

Figure B.190
Book 3, Proposition IX.

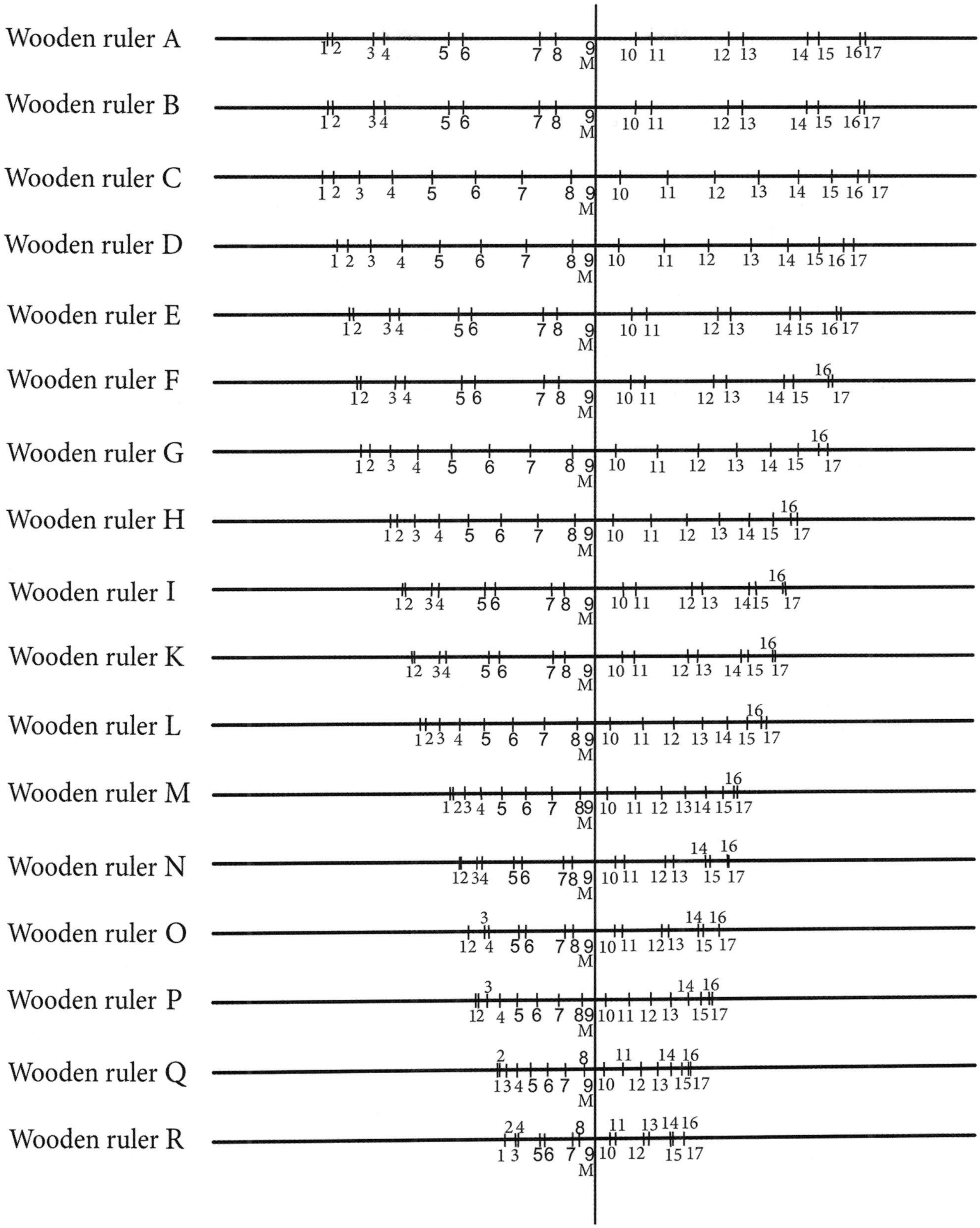

Figure B.191
Book 3, Proposition IX.

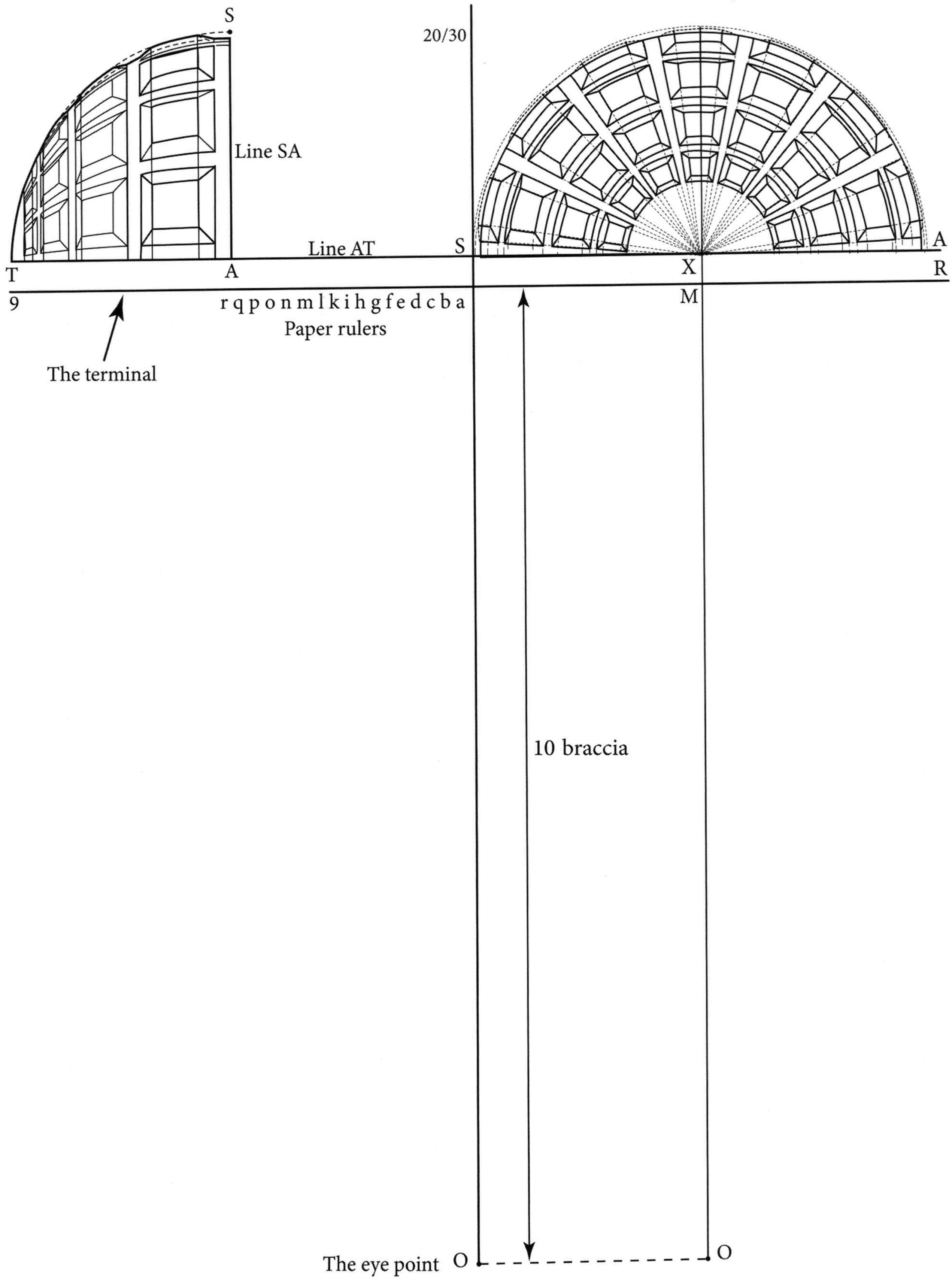

Figure B.192
Book 3, Proposition IX.

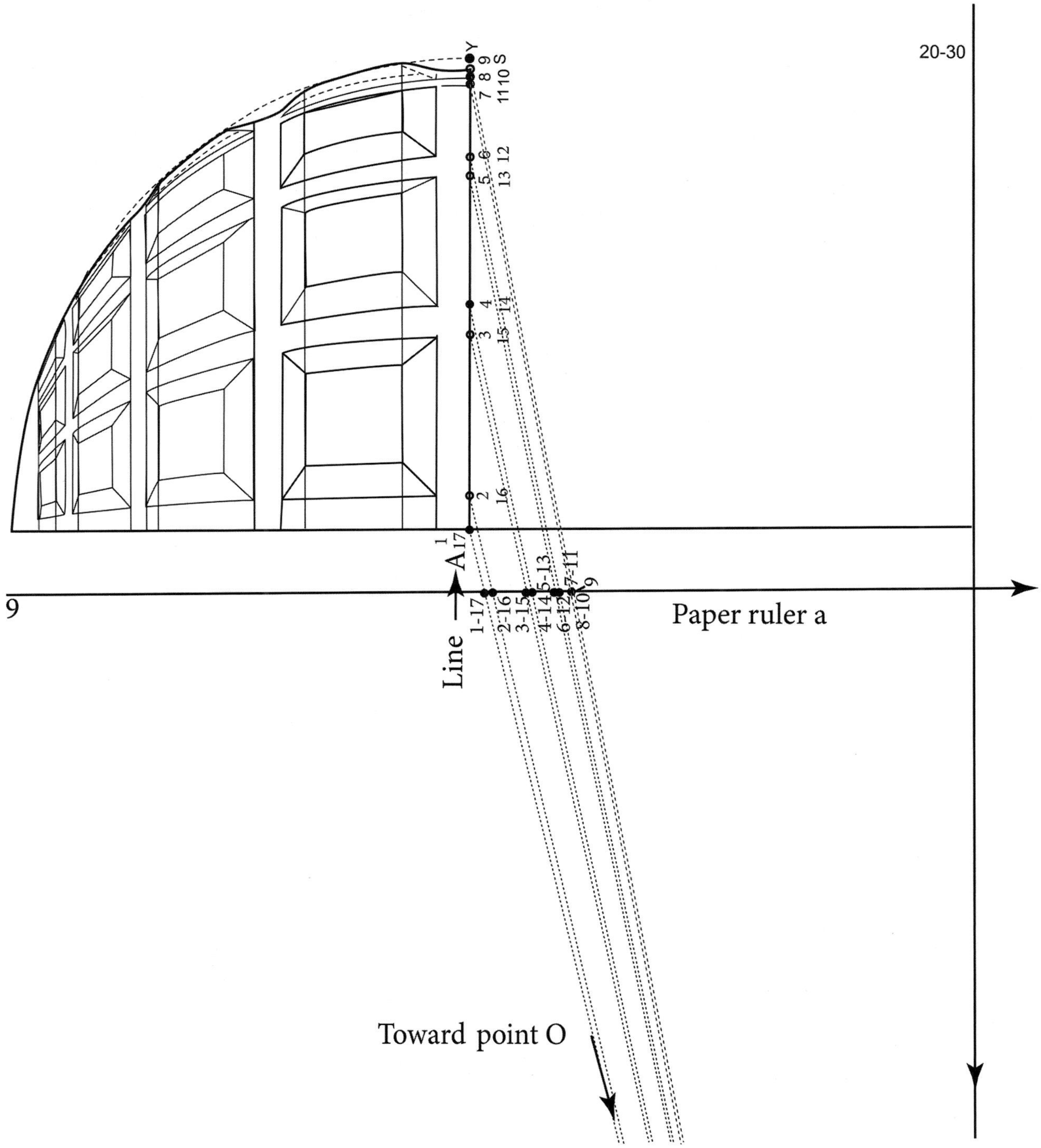

Figure B.193
Book 3, Proposition IX.

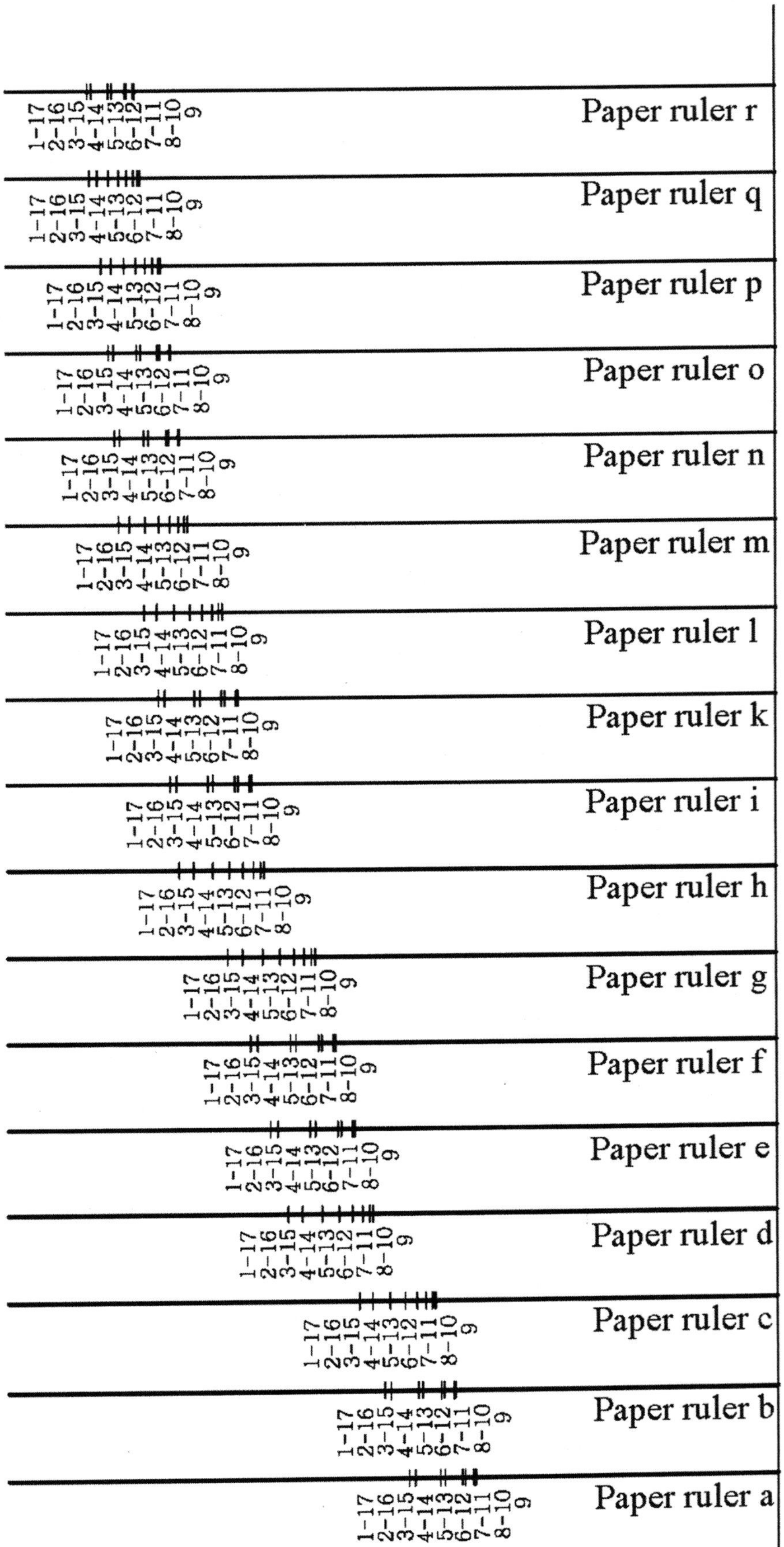

Figure B.194
Book 3, Proposition IX.

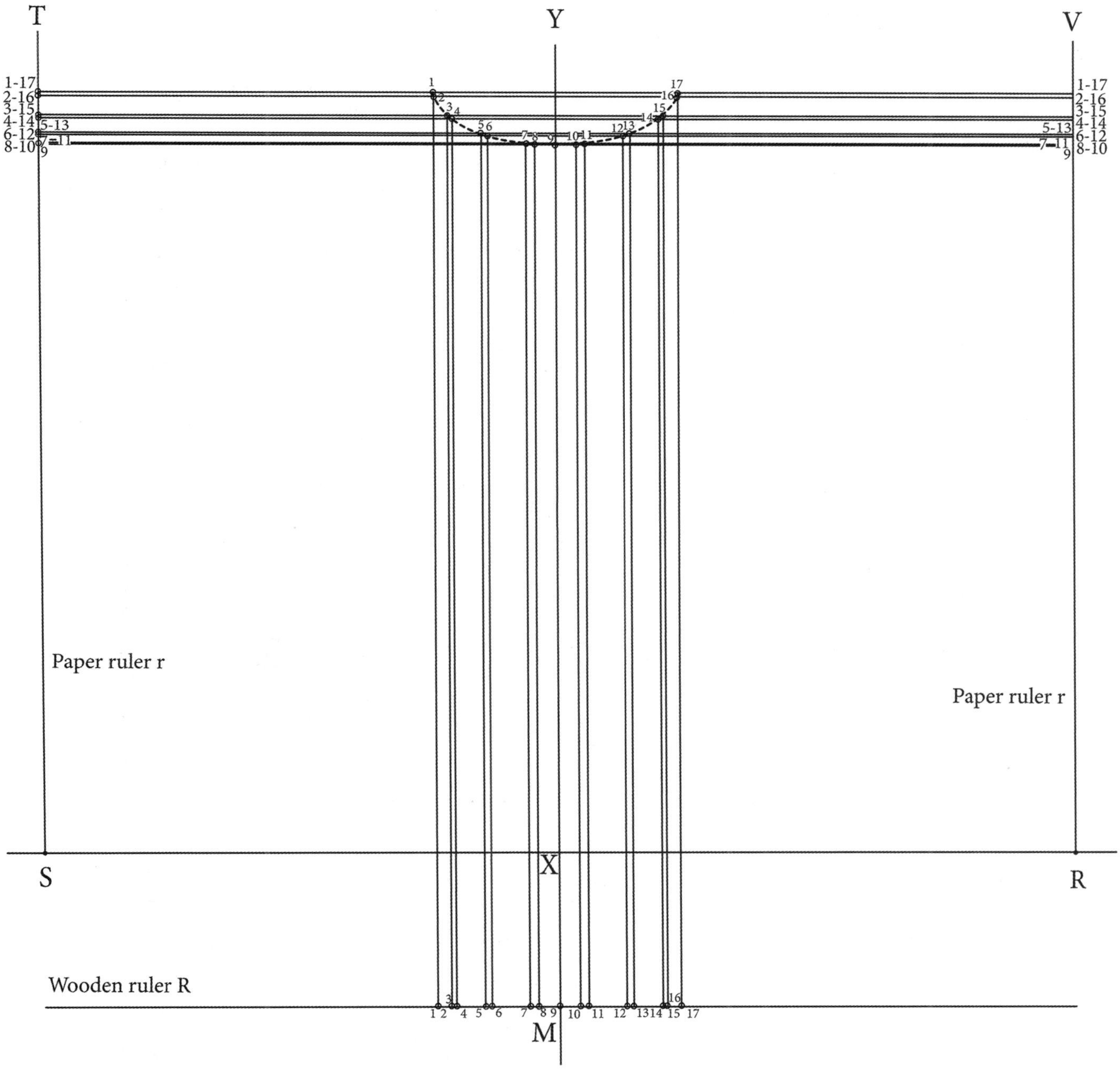

Figure B.195
Book 3, Proposition IX.

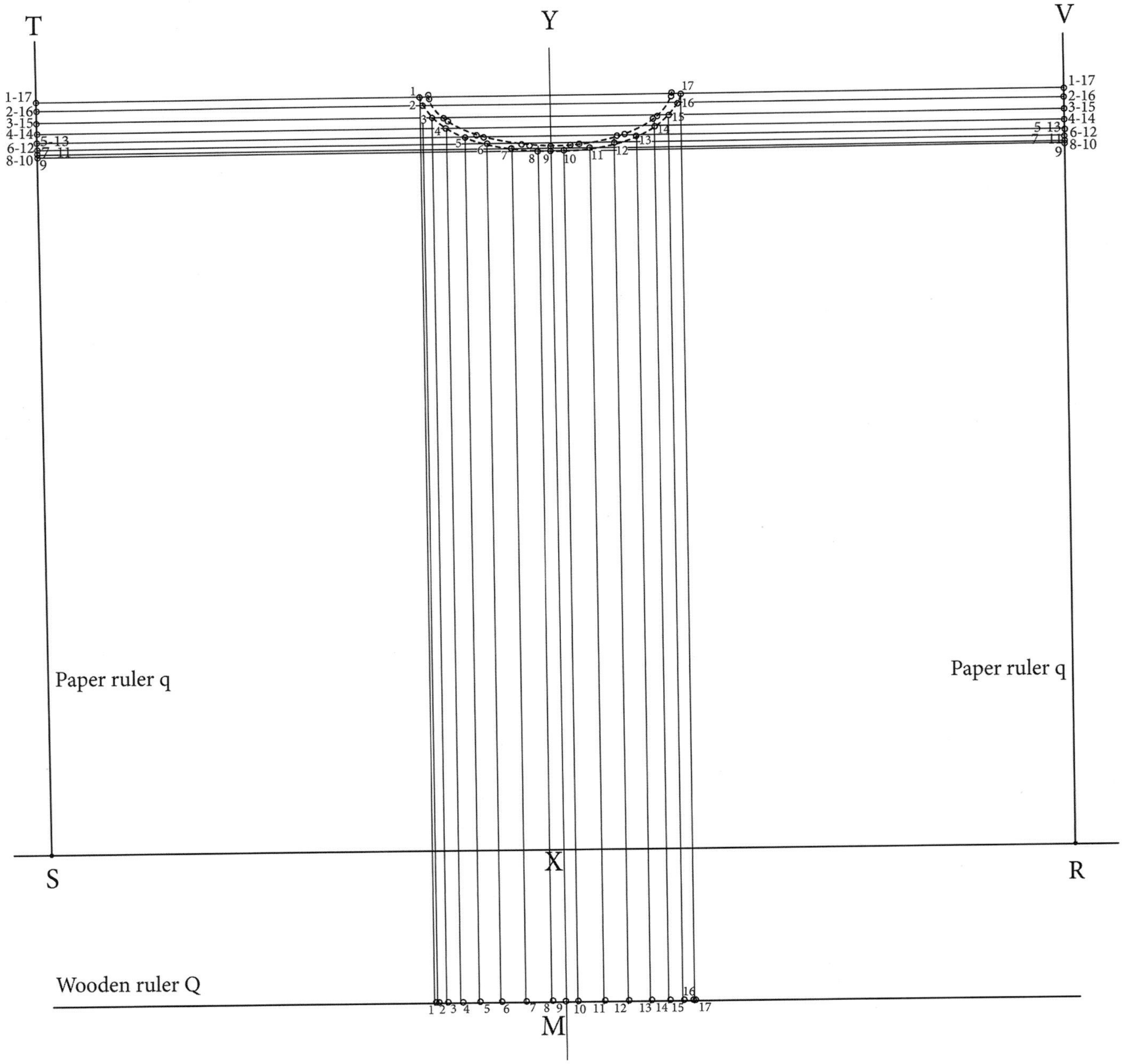

Figure B.196
Book 3, Proposition IX.

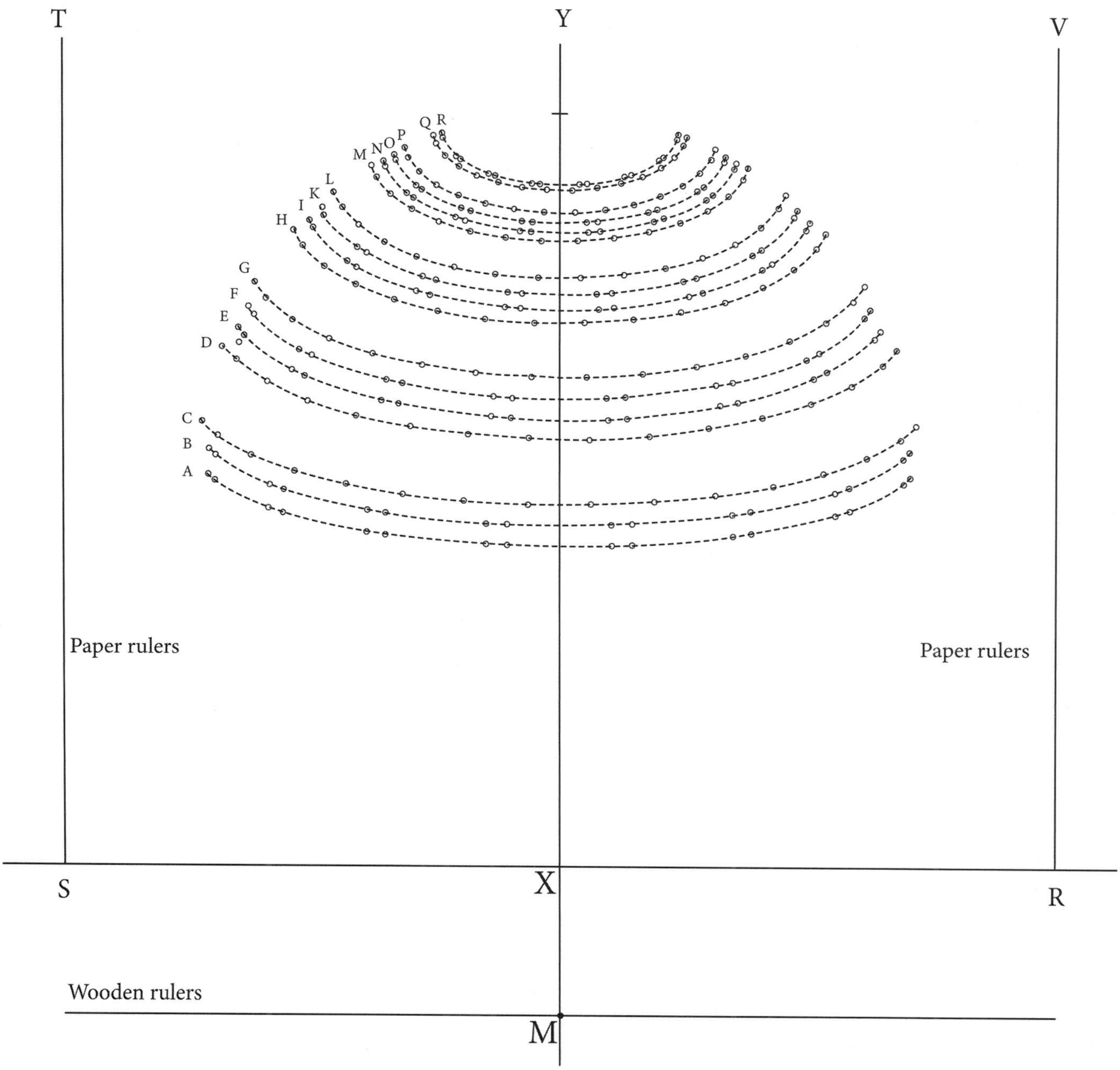

Figure B.197
Book 3, Proposition IX.

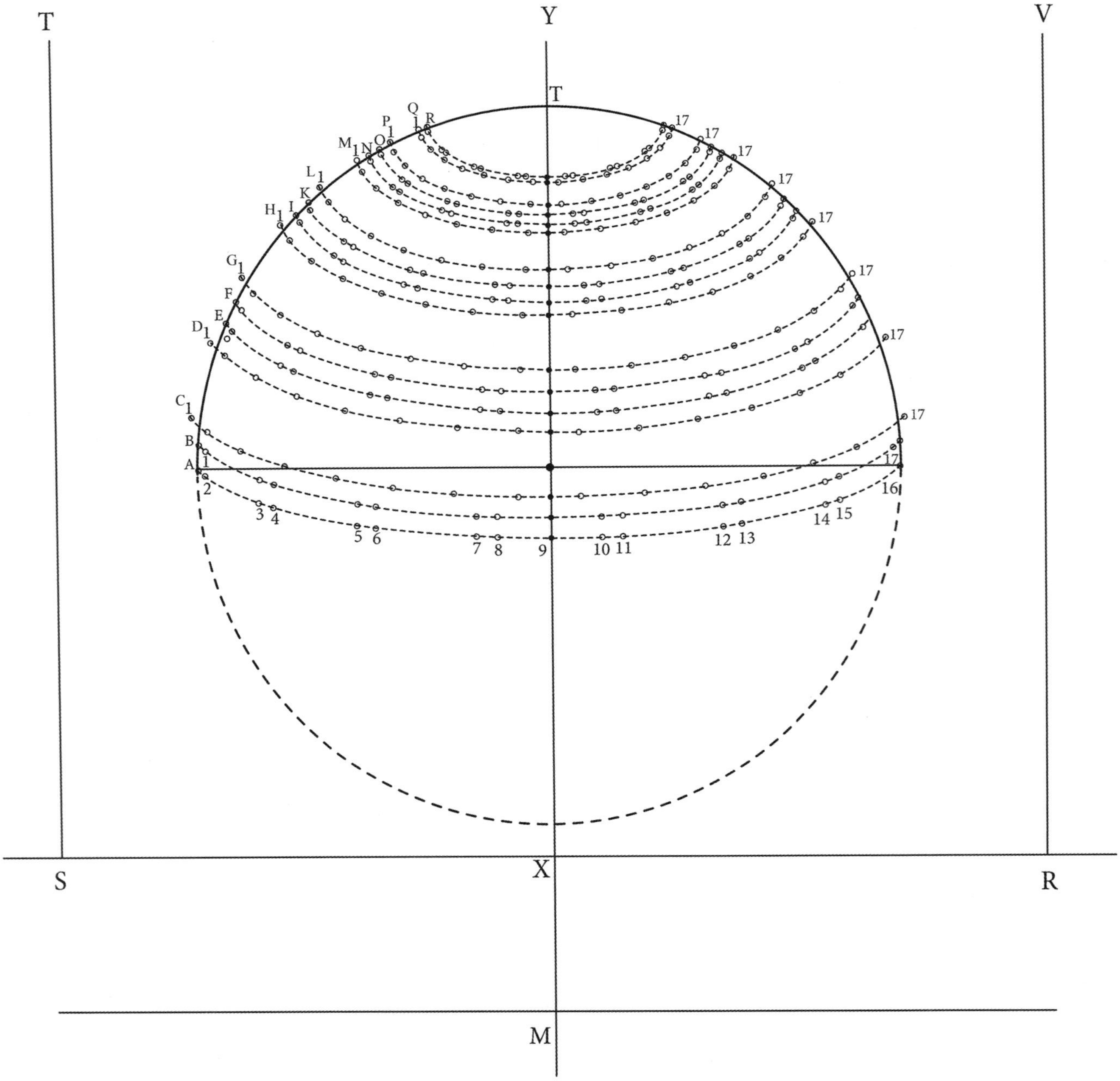

Figure B.198
Book 3, Proposition IX.

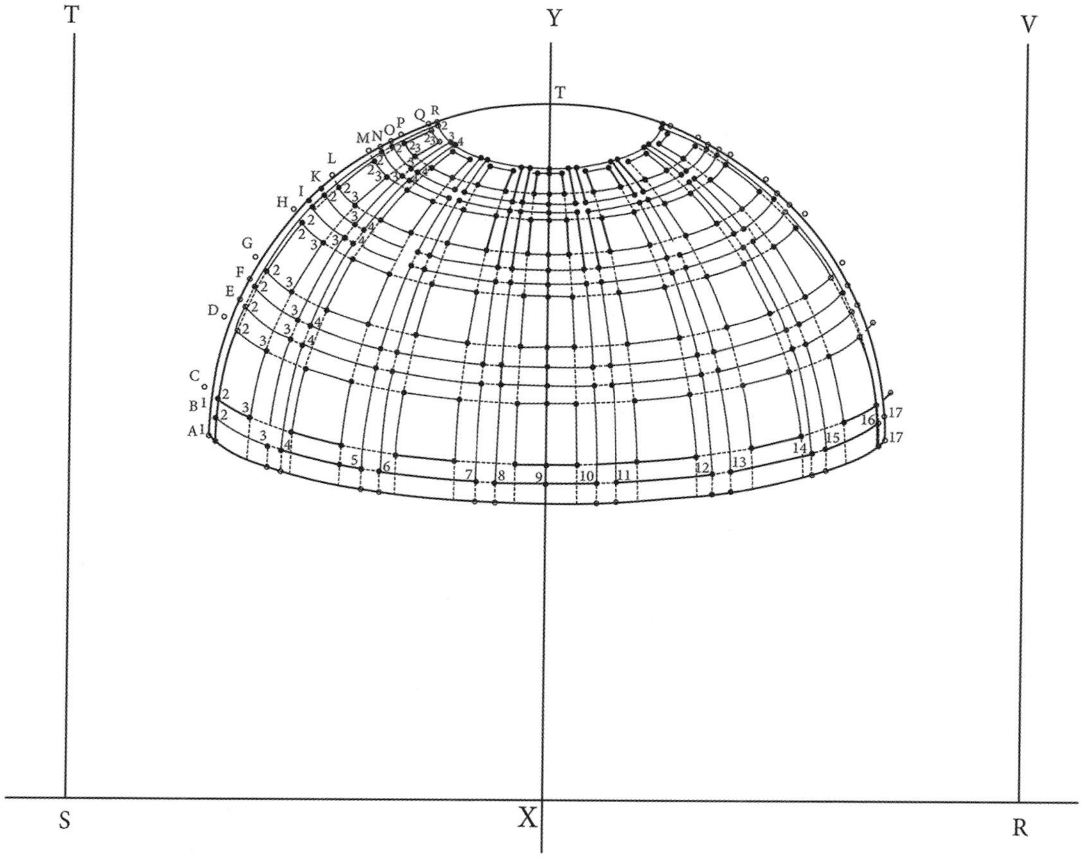

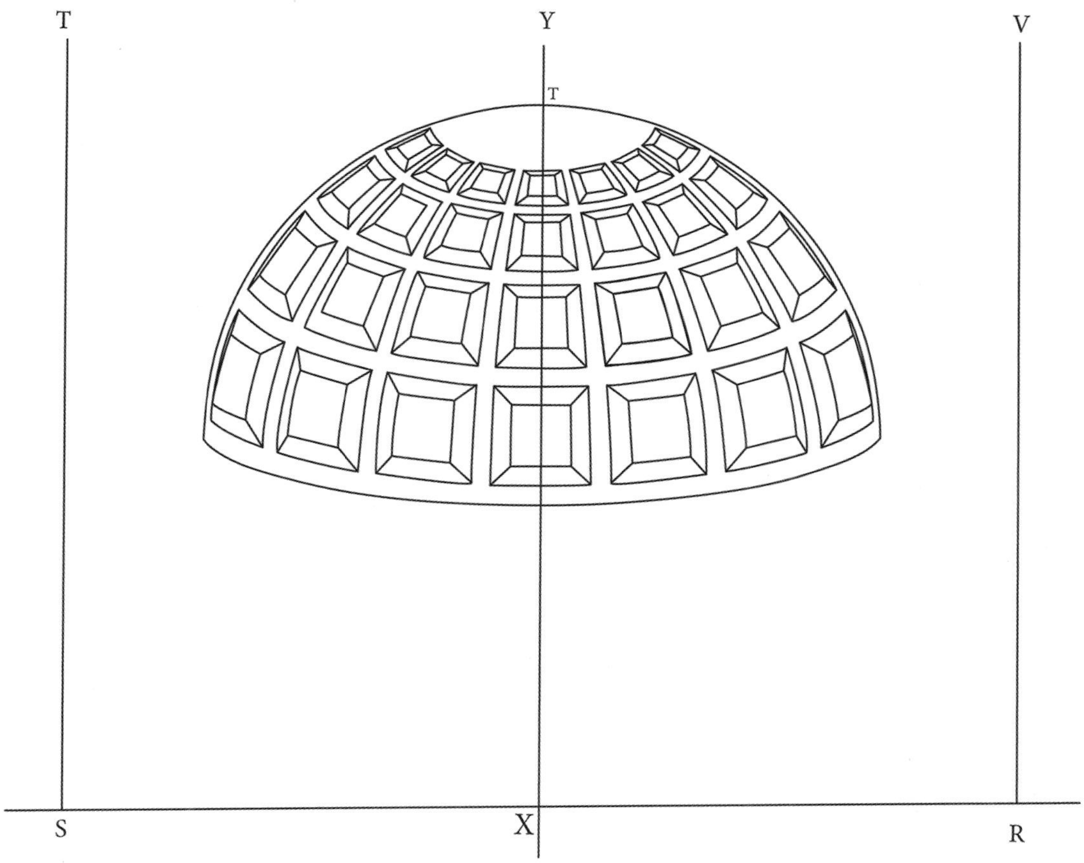

Figure B.199
Book 3, Proposition IX.

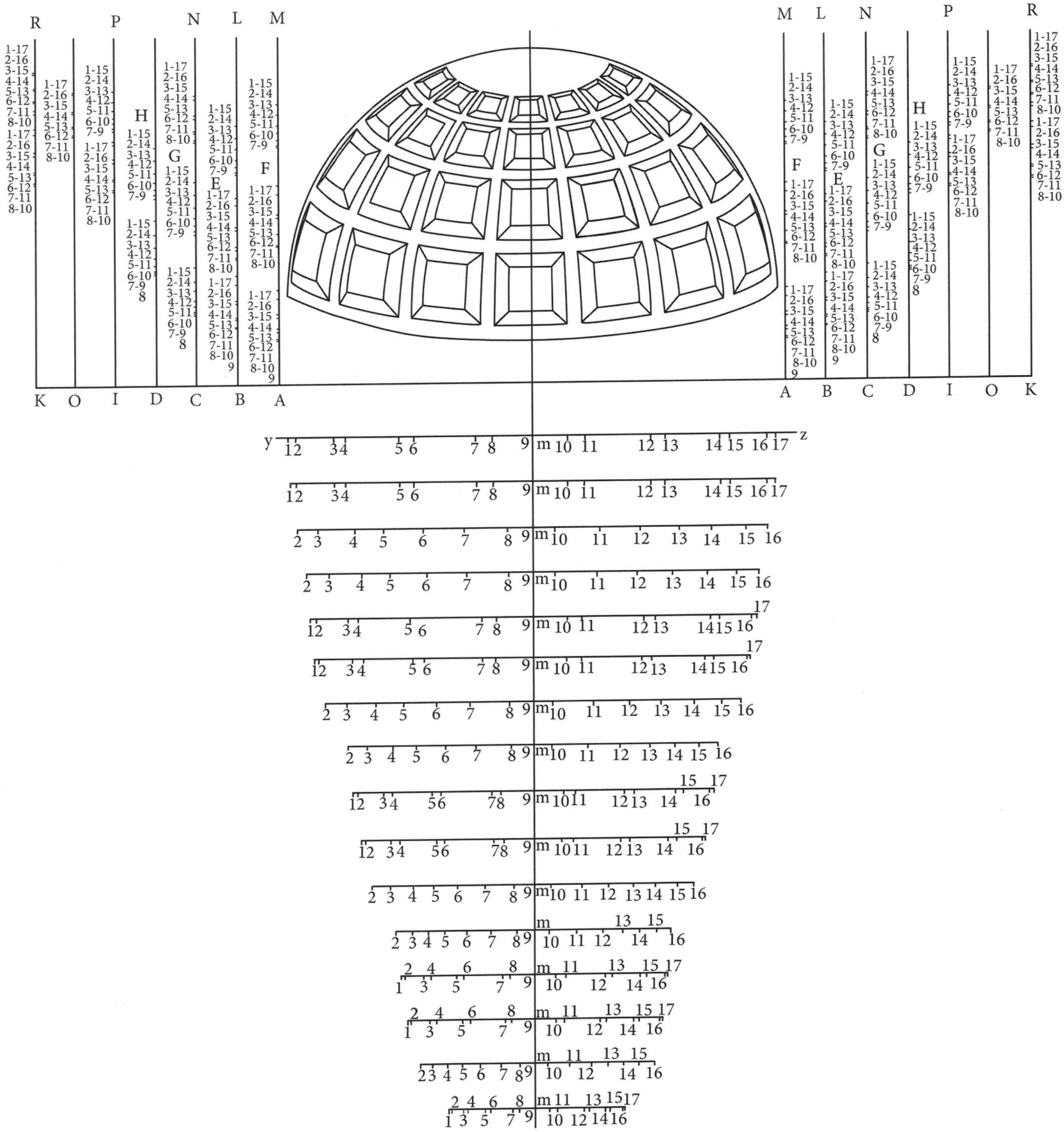

Figure B.200
Book 3, Proposition IX.

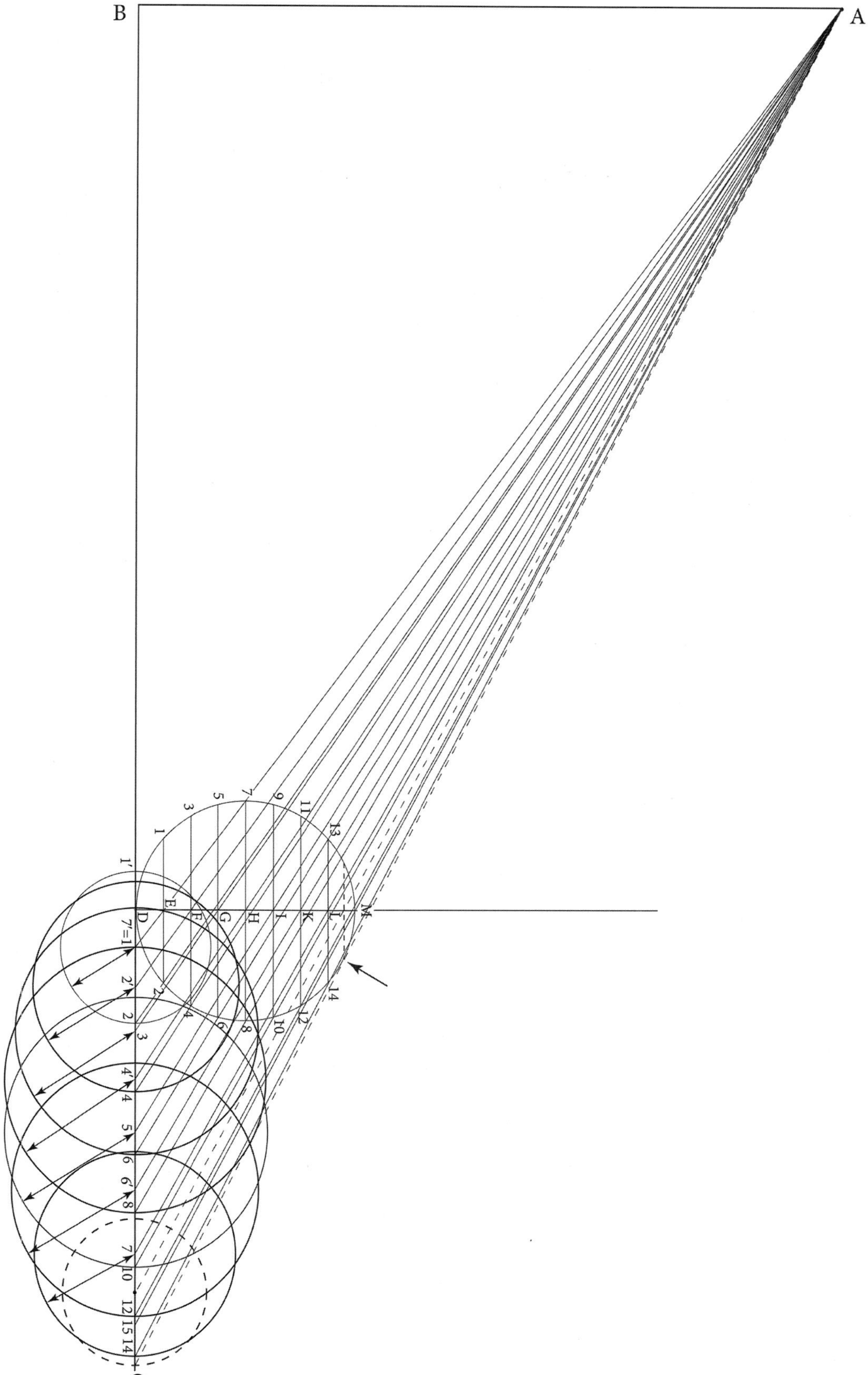

Figure B.201
Book 3, Proposition X.

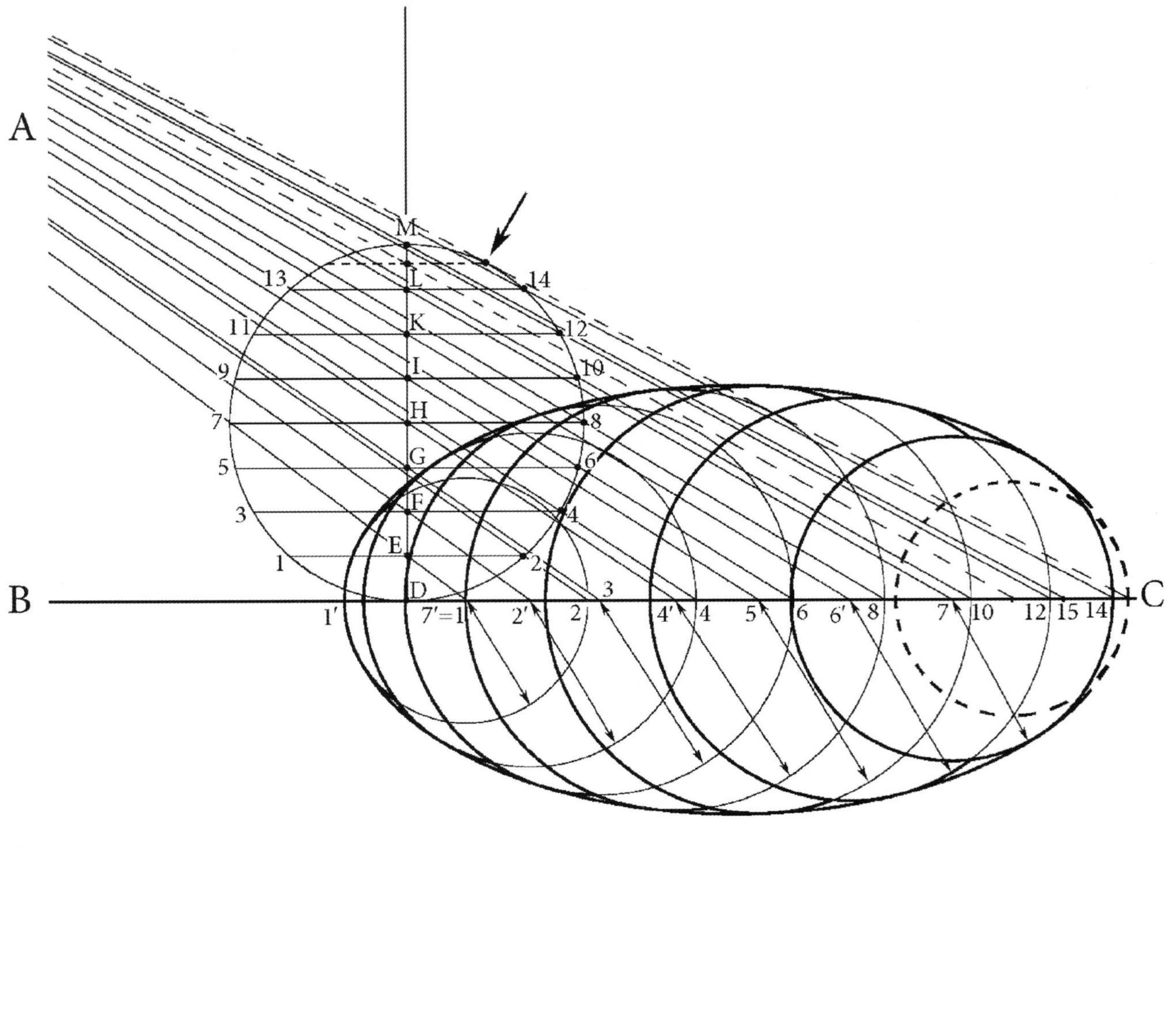

Figure B.202
Book 3, Proposition X.

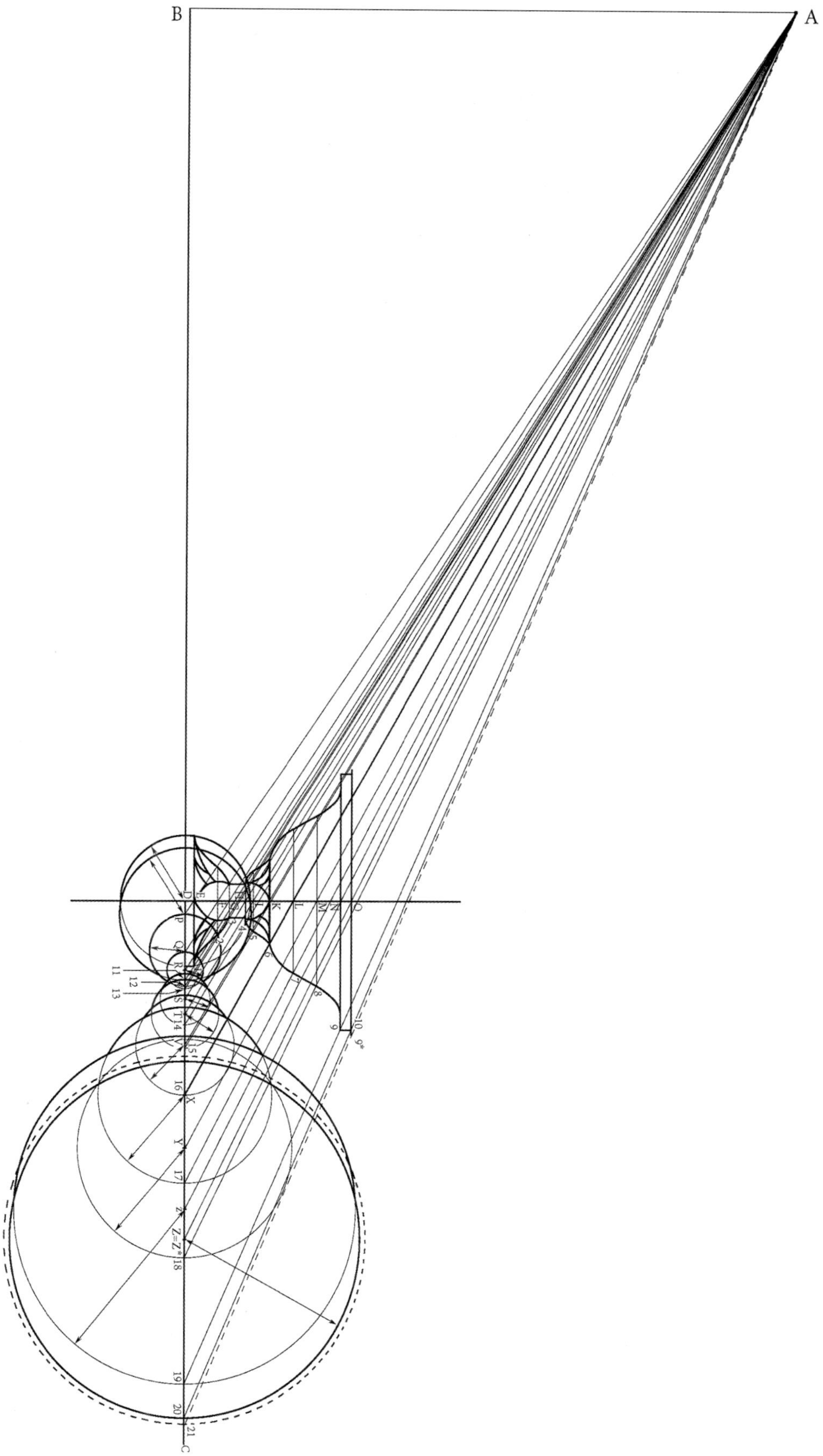

Figure B.203
Book 3, Proposition XI.

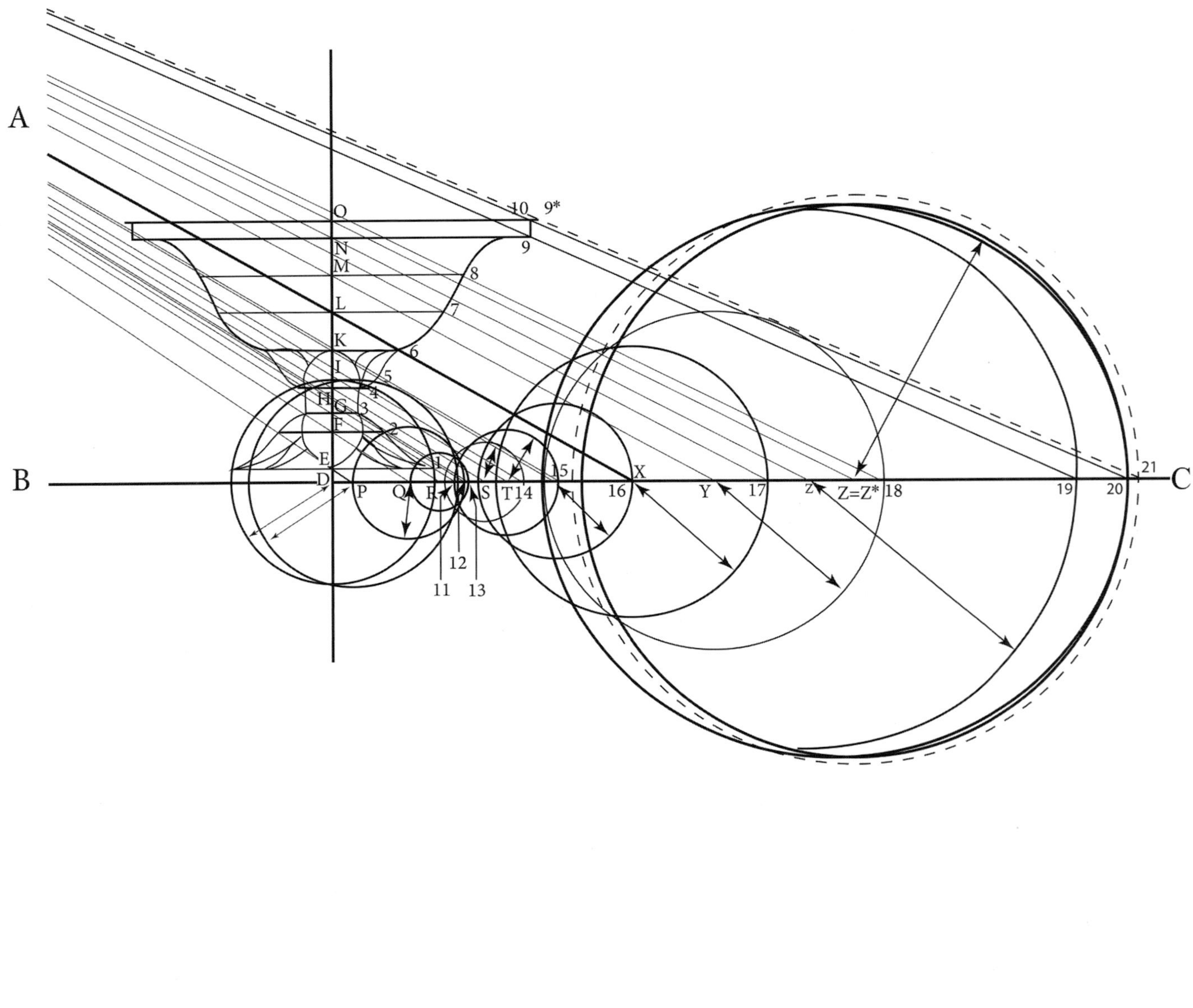

Figure B.204
Book 3, Proposition XI.

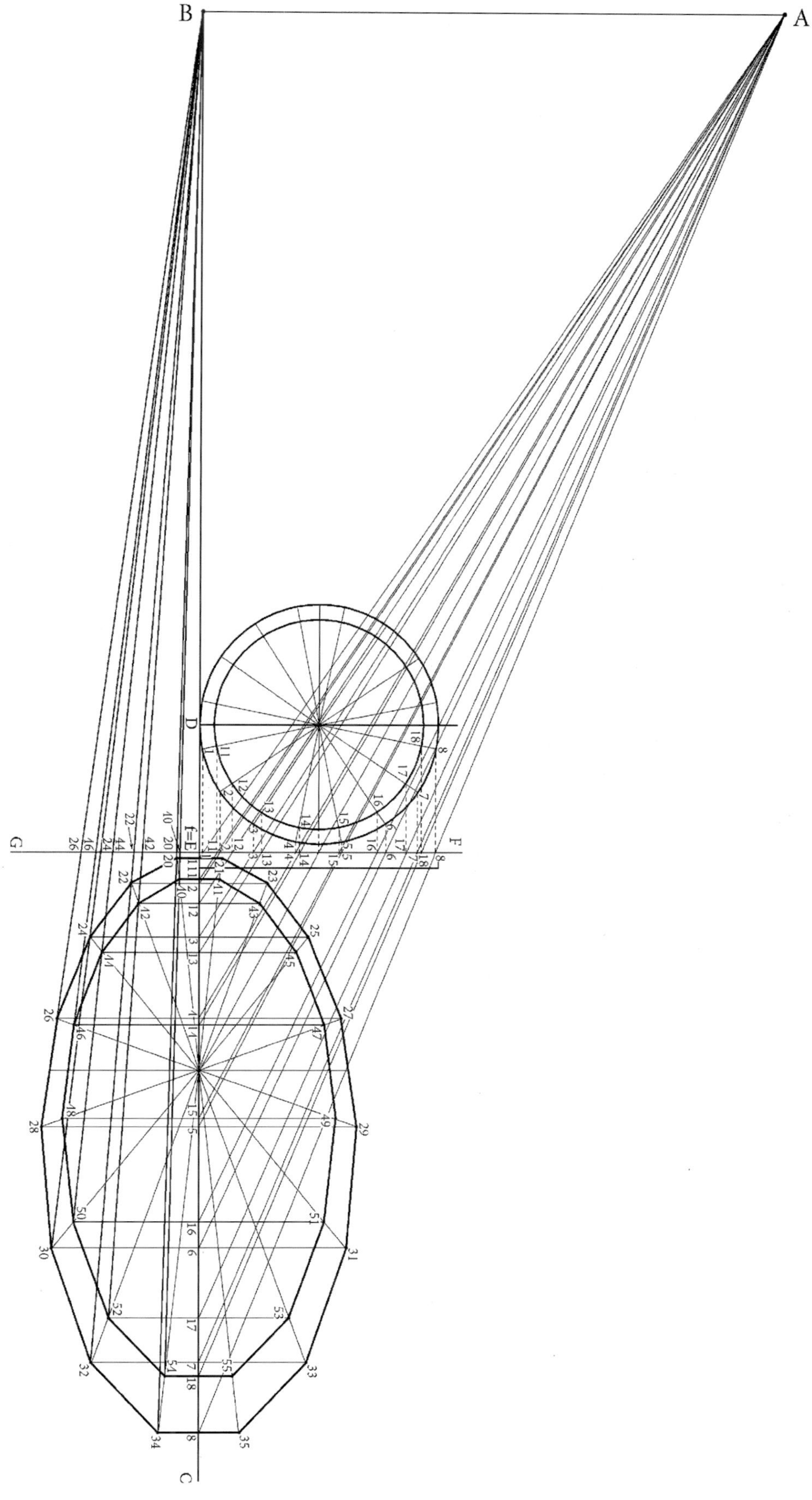

Figure B.205
Book 3, Proposition XII.

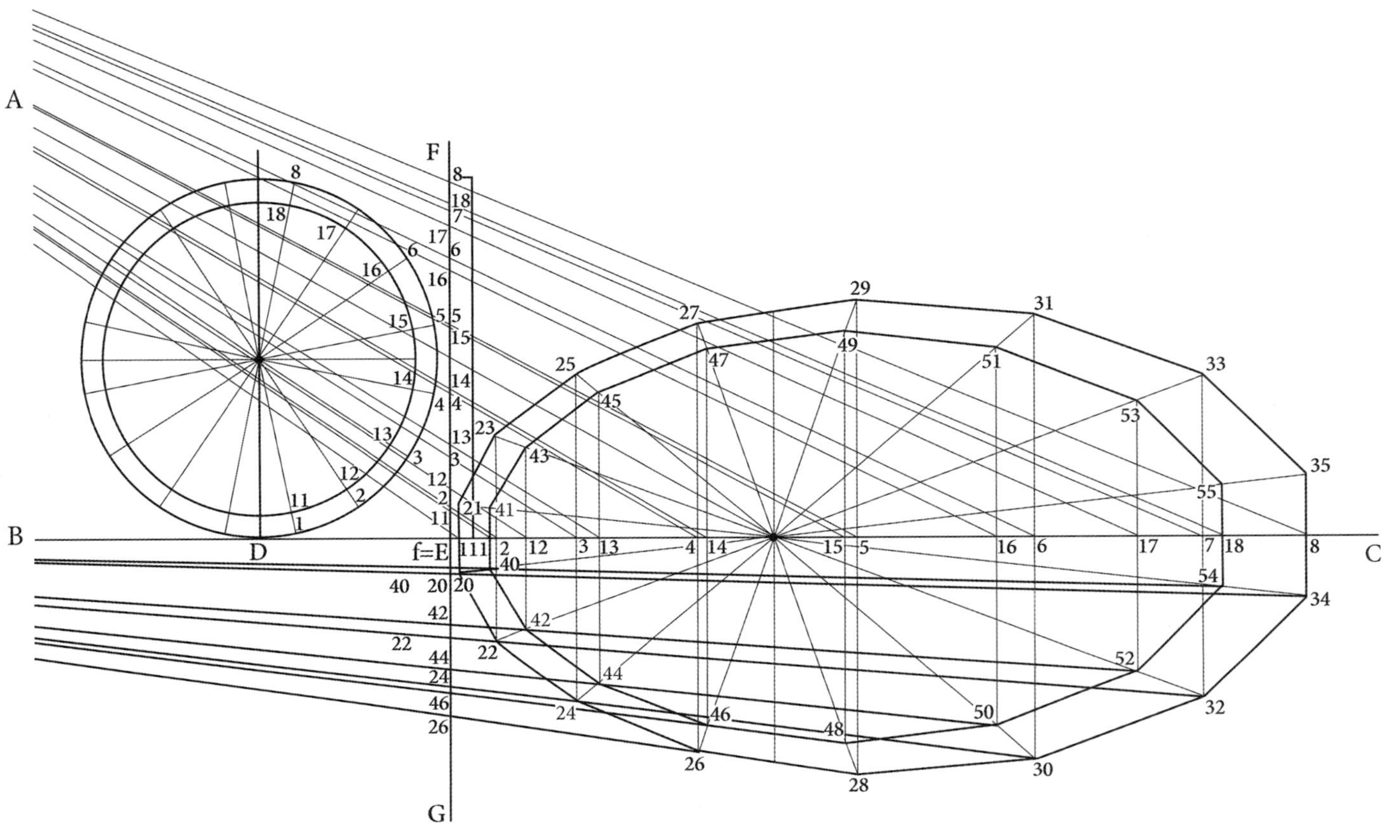

Figure B.206
Book 3, Proposition XII.

Notes

Prologue

1. See Alhazen (Ibn al-Haytham), *Book of Optics* (*Kitāb al-Manāẓir*), ca. 1021.

2. Roger Bacon, *The Mirror of Alchimy*, trans. E.W. (London: The Alchemical Press, 1582), chap. IV.

Introduction

1. "Life of Filippo Brunelleschi (Filippo di Ser Brunellesco) Sculptor and Architect," in Giorgio Vasari, *Lives of the Most Eminent Painters Sculptors and Architects*, trans. Gaston Du C. De Vere (London: Macmillan; Medici Society, 1912–1914).

2. Antonio di Tuccio Manetti, *The Life of Brunelleschi*, ed. Howard Saalman, trans. Catherine Enggass (University Park: Pennsylvania State University Press, 1970).

3. These figures have been taken from Rocco Sinisgalli, *Verso una storia organica della prospettiva* (Rome: Kappa, 2000).

4. Rocco Sinisgalli, *Perspective in the Visual Culture of Classical Antiquity* (New York: Cambridge University Press, 2012), 139–153.

5. Leon Battista Alberti, *On Painting: A New Translation and Critical Edition*, ed. and trans. Rocco Sinisgalli (New York: Cambridge University Press, 2011).

6. This figure has been taken from Alberti, *On Painting: A New Translation and Critical Edition*, Book I, paragraph 20, p. 173.

7. This figure has been taken from Alberti, *On Painting: A New Translation and Critical Edition*, Book I, paragraph 20, p. 174.

8. This figure has been taken from Alberti, *On Painting: A New Translation and Critical Edition*, Book II, paragraph 33, p. 180.

9. This figure has been taken from Alberti, *On Painting: A New Translation and Critical Edition*, Book II, paragraph 33, p. 181.

10. Alberti, *On Painting: A New Translation and Critical Edition*, Book II, paragraph 31, p. 51.

11. This figure has been taken from Alberti, *On Painting: A New Translation and Critical Edition*, Book II, paragraph 31, p. 176.

12. This figure has been taken from Alberti, *On Painting: A New Translation and Critical Edition*, Book II, paragraph 31, p. 177; for the extracts, see p. 51 of that edition.

13. As we were able to see in the paragraph about Brunelleschi, and as I showed in the book *Perspective in the Visual Culture of Classical Antiquity*, 139–153, the experiments on mirrors were at the base of the ancients' science of perspective, just as they were at the base of the moderns' rediscovery of perspective. Although Piero does not hint at the theory of mirrors, he does show a deep knowledge of the theory of mirrors with this proposition when he presents the procedure which, as anyone can verify, derives from the theory of mirrors.

14. Regarding this, compare *De prospectiva pingendi*, critical edition by Giusta Nicco Fasola (1942), republished with comments by Eugenio Battisti and Franco Ghione, 2 vols. (Florence: Le Lettere, 1984), 40. For the two terms abscissa and ordinate, which do not appear in the text of *De prospectiva pingendi*, it is good to specify that they have been added in the drawings of figures A.13, A.14, and A.15 to highlight how, with the second perspective method, Piero della Francesca was the forerunner of René Descartes (1596–1650) and the Cartesian axes, at least as regards the identification of a point in the plane.

15. The four vernacular manuscripts are: the Parma manuscript, Biblioteca Palatina, Ms 1576, dated mid-fifteenth century (considered to be entirely an autograph manuscript by Piero, both for text and drawings); the Reggio Emilia manuscript, Biblioteca Municipale A. Panizzi, Reggio Emilia codex A 41/2, dated mid-fifteenth century (the drawings in this manuscript are thought to have been executed by Piero himself); the manuscript of Milan, Biblioteca Ambrosiana, D 200 inf., dated late sixteenth century (this manuscript only contains text and no drawings); and the manuscript of the Archiginnasio di Bologna, ms. A 266, sixteenth century (partial text and content). The four Latin manuscripts are: the manuscript of Milan, Biblioteca Ambrosiana, ms. S.P. 6 bis, already C 307 inf. (the drawings in this book are thought to have been executed by Piero himself); the manuscript of Bordeaux, Ms. 616, Bibliothèque Municipale (the drawings in this book are thought to have been executed by Piero himself); the manuscript of London, Ms. Additional 10366, the British Library; and the manuscript of Paris, Ms. Lat. 9337, Bibliothèque Nationale de France.

16. Luca Pacioli, *Summa de aritmetica*, Dist. VI, tratt. I, art. II (Venice: Paganini, 1494).

17. James R. Banker, "Piero della Francesca's Friend and Translator: Maestro Matteo di ser Paolo d'Anghiari," *Rivista d'Arte. Studi documentari per la storia delle arti in Toscana* 4 (1992): 331–340. See also James R. Banker, *Piero della Francesca: Artist and Man* (Oxford: Oxford University Press, 2014).

18. *Petrus Pictor Burgensis De prospectiva pingendi: nach dem Codex der Königlichen Bibliothek zu Parma*, ed. Costantin Winterberg (Strassburg: Heitz & Mündel, 1899).

19. *De prospectiva pingendi*, ed. Nicco Fasola. In 2016 a digital edition of the Parma text was published, edited by Chiara Gizzi (Venice: Edizioni Ca' Foscari).

20. *De la perspective en peinture*, ed. Jean-Pierre Le Goff and Jean-Pierre Néraudau (Paris: Éditions in Medias Res, 1998).

21. The facsimile editions are: (1) *De prospectiva pingendi: A Facsimile of Parma, Biblioteca Palatina, MS 1576*, ed. Jane Andrews Aiken, Liana de Girolami Cheney, and Leonardo Farinelli (New York: Broude International Editions, 1992); (2) *De prospectiva pingendi*, ed. Massimo Mussini and Luigi Grasselli, facsimile of Reggio Emilia codex (Sansepolcro: Aboca Edizioni, 2008); (3) *De prospectiva pingendi. Edizione critica del testo volgare. Edizione critica dei disegni*, facsimile of Parma codex 1576, Biblioteca Palatina, Parma (Rome: Istituto Poligrafico e Zecca dello Stato, 2016); (4) *De prospectiva pingendi. Edizione critica del testo latino. Edizione critica dei disegni*, facsimile of codex 616, Bibliothèque Municipale, Bordeaux (Rome: Istituto Poligrafico e Zecca dello Stato, 2017).

22. *Scritti scelti di Leonardo da Vinci*, ed. Anna Maria Brizio (Turin: Unione Tipografica-Editrice Torinese, rpt. 1980), 664. The fact that a copy of *De prospectiva pingendi* ended up in Leonardo's hands makes one think that his drawing *Busto di un uomo visto di profilo* (Galleria dell'Accademia, Venice) is to be somewhat associated with Piero's head. Leonardo, in fact, cuts this head almost as Piero does, and just like Piero he goes on to number its relative sections.

23. Compare August Schmarsow, *Melozzo da Forlì* (Berlin and Stuttgart, 1886), 69ff, and Piero della Francesca, *De prospectiva pingendi*, ed. Nicco Fasola, 41.

24. Compare the Latin edition by Joachim Camerarius (Kammermeister) entitled *Albertus Durerus institutionum geometricarum libri quatuor, lineas, superficies et solida corpora tractavit, adhitis designationibus ad eam rem accomodatis* (Paris: Wechel, 1532).

25. *Alberti Dureri clarissimi pictoris et geometrae de sym[m]etria partium in rectis formis hu[ma]norum corporum libri in latinum conversi* [by Camerarius] (Nuremberg: in aedib. Viduae Durerianae, 1532). The drawings of figure A.16b have been taken from the edition published in Venice in 1594

entitled *Di Alberto Durero pittore e geometra chiarissimo. Della simmetria dei corpi umani. Libri quattro.*

26. Giorgio Vasari, "Vita di Piero della Francesca pittore," in *Le vite dei più eccellenti scultori, pittori e architetti* (Forlì: Orsa Maggiore Editrice, 1991). Compare also Vasari, "Life of Piero della Francesca," in *Lives of the Most Eminent Painters, Sculptors and Architects*, trans. De Vere, vol. 3: 18. We must specify that Vasari stated of Piero: "Now Maestro Luca [Pacioli] dal Borgo, a friar of St. Francis, who wrote about the regular geometrical bodies, was his pupil; and when Piero, having written many books, grew old and finally died, the said Maestro Luca, claiming the authorship of these books, had them printed as his own, for they had fallen into his hands after the death of Piero." He is talking about the *Libellus de quinque corporibus regolaribus*: a book of many chapters which Piero called "treatises," which Luca Pacioli translated from Latin and published as his own book in 1509. Regarding this, please compare G. Pittarelli, "Luca Pacioli usurpó per se stesso qualche libro di Piero deì Franceschi? [Did Luca Pacioli usurp some book by Piero deì Franceschi for himself?]," in *Atti del IV Congresso Internazionale dei Matematici, Roma 6–11 Aprile 1908*, vol. III (Rome: Tipografia della Regia Accademia dei Lincei, 1909), 436–440.

27. Compare Daniel Barbaro, *La pratica della perspettiva* (Venice: appresso Camillo e Rutilio Borgominieri, 1569).

28. Daniele Barbaro, "Proemio," in *La pratica della perspettiva*, 3. The phrase "Borgo Santo Stefano" is wrong; it should be "Borgo San Sepolcro."

29. Compare Gino Loria, *Storia della geometria descrittiva dalle origini sino ai giorni nostri* (Milan: Ulrico Hoepli, 1921), 13.

30. Compare the "Prefazione" to *Le due regole della prospettiva pratica*, by Jacopo Barozzi da Vignola. This book was published posthumously in Bologna, 1582, by Ignazio Danti.

31. Noël-Germinal Poudra, *Histoire de la perspective ancienne et moderne* (Paris: J. Corréard éditeur, 1864), 119ff.

32. Kirsti Andersen, in her *The Geometry of an Art: The History of the Mathematical Theory of Perspective from Alberti to Monge* (New York: Springer, 2007), illustrates Piero's work excellently; however, she does not mention the perspective of a human head randomly placed in space.

33. Rocco Sinisgalli, *Playing with Piero della Francesca: Geometry at the Service of the Arts* (Certaldo [FI]: Federighi Editori, 2012), 5.

34. Compare the section I have titled "The Subdivisions of the Circuits," Proposition VIIIa of Book 3.

35. Compare the section titled "The Points of Tangency," Proposition VIIIa of Book 3.

36. Compare for this circuit A, compliant with the original, figures A.35, A.36, and A.37, which follow, relative to the Parma manuscript, folio 59 verso, folio 61 recto, folio 63 verso: where circuit A, compliant with the original, is the first internal circuit.

37. Compare the section titled "The Twelve Units," Proposition VIIIa of Book 3.

38. Compare the section titled "The Protrusions of the Head," Proposition VIIIa of Book 3.

39. Compare the section titled "The Protrusions of the Head," Proposition VIIIa of Book 3.

40. The figure reproduced here is taken from the original of the Parma manuscript, folio 59 verso.

41. This is a serious deficiency. Without the plans of the eight circuits, figure A.18 (previously examined) would not make sense.

42. The figure reproduced here is taken from the original of the Parma manuscript, folio 61 recto.

43. This is a serious deficiency. Without the plans of the four lower circuits, figure A.19 (previously examined) would not make sense.

44. This is a serious deficiency. This folio, without the point of view in elevation and with the terminal (the picture plane) missing all the numbers, does not make sense.

45. The drawings on folio 63 verso of Parma, as well as the drawings on folio 71 verso of Reggio Emilia, where all circuits are reproduced in plan, are missing the corresponding drawings in elevation.

46. The figure reproduced here is taken from the original of the Bordeaux manuscript, folio 74 verso.

47. These letters do not appear in the corresponding folio 59 verso of Parma.

48. When observing the drawing in figure A.43, taken from folio 79 recto of Bordeaux, take into account that the distance between the drawings of the first four circuits on the left and the second four circuits on the right was slightly widened, to better identify the different positions of the points of view in plan.

49. Take into account that in the manuscript of Reggio Emilia, folio 71 verso, on the plans of the eight circuits of the head the points of view are missing as well as the relative radii.

50. The perspectives that appear in the Latin manuscripts of Milan, London, and Paris are very similar to the one of Bordeaux.

51. Compare figures A.16b and A.17.

52. Since the drawings in figure A.60 were hardly traceable on the copy of the original of folio 93 verso of Milan that I had in my possession, I extracted the tilted head with one eye only from folio 86 recto of Bordeaux and the tilted head with two eyes from folio 86 verso of Bordeaux, without any number specifications. Thus joined, the two drawings offer an appropriate idea of the figure present in the Milan manuscript, folio 93 verso.

53. Take into account that the Reggio Emilia perspective is similar to the one of Parma, and that the perspectives of the manuscripts of Milan, London, and Paris are similar to the one of Bordeaux.

54. These steps can be followed by reading Proposition VII of the third book.

55. In the manuscript of Milan, on folio 74 verso, two points of view instead of one have been wrongly reproduced in the drawing of the capital in elevation, with their respective overlapping visual rays that converge toward two separate points of view.

56. I would like to specify that this incongruence present in figure A.69 only appears in the Parma manuscript.

57. To find the wooden rulers, all Latin and vernacular manuscripts report a second drawing of the plan. These second drawings of the plans are no longer indicative of constructions and numbers with respect to the drawing of figure A.70.

58. Giorgio Vasari was right when he said: "Truly unhappy are those who, labouring at their studies in order to benefit others and make their own name famous, are hindered by infirmity and sometimes by death from carrying to perfection the works that they have begun" (Vasari, "Life of Piero della Francesca"). Vasari specifies that Piero become blind at sixty; but James Banker (*Piero della Francesca: Artist and Man*) believes that Piero become blind in the last months of his life. Here is a brief explanation of how the manuscript of Parma came to be considered "autograph": "In the flyleaf [of the Parma manuscript], a scholar, whom Mancini identifies as Michele Colombo, praises the Codex and hypothesizes that is an autograph" (*De prospectiva pingendi*, ed. Nicco Fasola, 46). In fact, Michele Colombo (1747–1838), who was the owner of this manuscript, wrote in the flyleaf: "This Codex is immensely commendable; according to all appearance, it is an autograph of the author himself. Certainly the pictures, and the heads in particular, have been drawn with a precision and exactness that are not achieved but by a very expert hand." Colombo believed that the precision of the pictures, in particular the heads, was such that only the author could have executed them. Girolamo Mancini (1832–1924) wrote in 1915 a "Memoria della Regia Accademia dei Lincei," entitled *L'opera De corporibus regularibus di Pietro Franceschi detto della Francesca usurpata da fra L. Pacioli*. In "Appendice II" of this "Memoria," after reporting the Colombo passage, Mancini wrote: "In May 1913 I went to Parma to reexamine the volume. I had to look at Colombo's conjecture as a divination [i.e., as an intuition]. In fact, I found the vowel 'e' of the special proper form [i.e., which belongs] to Franceschi; I verified that the entire text of the treatise [was written] by his hand, as well as the 12 corrections in the margins of the folios and the substitution of 35 numbers [relative] to as many [numbers] deleted at folio 63. The pictures made of straight lines, curves, or both straight and curved lines have been drawn with confident strokes and with that special precision [which belongs] to those masters who execute complex drawings; and these [pictures] of the precious Codex are so well drawn that they are judged to be by

Franceschi's hand, just like the text." In addition to concurring with Colombo regarding the exactness of the drawings which, for this very reason, had to be Piero's autograph drawings, Mancini claimed that the text was also written by Piero himself because the form of the letter *e* matched what he had seen in Piero's manuscript of *De corporibus regularibus*. Ever since then, the Parma manuscript has been thought to be Piero's autograph for both drawings and text. In folio 1 recto of this manuscript, in the title of the treatise, written in capital letters and in Latin, *Petrus pictor burgensis De prospectiva pingendi*, the word "Prospectiva" is missing the letter "c"; however this term is always written correctly in all the titles of all the other manuscripts.

On the Perspective of Painting

1. The specification "That part consists of five parts" only appears in the vernacular manuscripts; it is missing in the Latin manuscripts.

2. Here "terminal" means "the picture plane."

3. This sentence is one of the premises, or definitions, of Euclid's *Optics*, representing a postulate that is at the basis of vision. The Latin manuscripts report "oculo," whilst the vernacular manuscripts report "angolo." We prefer "angle" to "eye."

4. In the Latin manuscripts the drawings relative to these figures, which are copied here as they appear in the vernacular manuscripts, do not appear. The figure titled "Parallelogram" shows the diction "simile halmuachin" ("similar to the *halmuachin*," which is an Arabic term for parallelogram). This means that Piero has studied Euclid in an Arab-to-Latin translation, perhaps the one by Adelard of Bath (first half of the twelfth century). Piero confirms this diction at folio 87 verso of his *Trattato d'abaco* in the Codex Ashburnham at the Biblioteca Laurenziana of Florence.

5. There's no title of the proposition with the specification "Proposition" or "Theorem" in any of the manuscripts in the vernacular or in Latin. In the two vernacular manuscripts of Parma and Reggio Emilia, before the title of each proposition, most of the time there are cardinal numbers in a very small format to indicate the numerical progression of the propositions; the same cardinal numbers have been reported on the related drawings with the same aim. At the beginning of each proposition in the Milan manuscript in Latin, there is one ordinal number; and on each drawing relating to the same proposition, a cardinal number corresponding to the ordinal number. Always in the Milan manuscript, the first letter with which the first word of each title of a proposition starts is drawn in capitals; it is very big. In the same manner, the first letter of the first word of the text relative to the proposition is drawn in capitals. In the Latin manuscript of Bordeaux the ordinal numbers are missing at the beginning of each proposition, but, as in Milan, the first letter of the first word of each title of a proposition is big. In the same manner, the first letter of the first word of the proposition text is big. On each drawing, in the Bordeaux manuscript there is a progressive cardinal number that indicates the succession of the drawing with respect to the drawings that precede it. The London Latin manuscript presents bigger letters, as in Bordeaux, and on each drawing there is a progressive cardinal number as in Bordeaux. The Paris Latin manuscript does not present bigger letters: it has cardinal numbers on the drawings and it has the title of each proposition separated from the following text by an empty ruler. Considering these discrepancies, I have assigned to all theorems an ordinal number preceded by the specification "Proposition."

6. The Latin manuscripts report "attingentes," which should be "attingente" and corresponds to "that touches."

7. This sentence is repeated twice in the Latin manuscripts. In the Bordeaux manuscript there is an additional "vacat" which means "Be it without."

8. Euclid's *Elements*, Book I, Proposition XXIV: "If two triangles have two sides equal to two sides, respectively, but [one] has the angle encompassed by the equal straight-lines greater than the [corresponding] angle [in the other], then [the former triangle] will also have a base greater than the base [of the latter]." See Euclid's *Elements of Geometry*, the Greek text of J. L. Heiberg (Leipzig: B. G. Teubner, 1883–1885), ed. and trans. Richard Fitzpatrick, rev. ed. (2007; London, 2008). In all further references to Euclid's *Elements*, I will be citing from this text.

9. In the two Latin manuscripts of Milan and Bordeaux, but also in the two vernacular manuscripts of Parma and Reggio Emilia, there is no extension DF of line AF, which stops in D.

10. We prefer to draw the figure again with the numbers indicated.

11. The two Latin manuscripts of Milan and Bordeaux report, after "BD [equal to] 6," a second time "AC [equal to] 21." Observe that since ACB and AED are similar triangles, we have EC : AE = FE : DE; which means that AC : AE = CB : ED; this means that the product of the medians is equal to the product of the extremes.

12. The two Latin manuscripts of Milan and Bordeaux report BC instead of BD.

13. We must naturally imagine point A as if it were the eye that observes lines BD and CE not in plan but in elevation.

14. Both in vernacular and in Latin manuscripts, Piero reports the title *De aspectuum diversitate* by which Euclid's *Optics* was known during the Renaissance, which means "On the diversity of appearances." Euclid's Proposition X says: Sections of planes further away placed below the eye appear higher. See *Euclidis Optica, opticorum recensio theonis, catoptrica, cum scholiis antiquis*, ed. J. L. Heiberg, vol. VII of *Euclidis Opera Omnia*, ed. J. L. Heiberg and H. Menge (Leipzig: B. G. Teubner, 1895).

15. See note 8.

16. In the two Latin manuscripts of Milan and Bordeaux, this last term is reported as N instead of NI. The vernacular versions correctly report NI.

17. *Elements*, Book VI, Proposition XXI: Rectilinear figures similar to the same rectilinear figure are also similar to one another.

18. The two Latin manuscripts of Milan and Bordeaux have B instead of C.

19. *Elements*, Book I, Proposition XXXIV: In parallelogram figures, the opposite sides and angles are equal to one another, and a diagonal cuts them in half.

20. The two Latin manuscripts of Milan and Bordeaux report "per tertiam." The amanuensis wrote "tertiam" instead of "trigesimam." The vernacular manuscripts do not report the specification "first"; they simply report: "for the 30a of Euclid." *Elements*, Book I, Proposition XXX: [Straight lines] parallel to the same straight line are also parallel to one another.

21. The two Latin manuscripts of Milan and Bordeaux report "BC DG," where BC was added by mistake.

22. The vernacular manuscripts of Parma and Reggio Emilia report "for XXV," but this is wrong. In the same manner the Latin texts of Milan and Bordeaux err, because they report "vigesimam," which means "the twentieth." The correct proposition is XXIV (*Elements*, Book VI, Proposition XXIV): In any parallelogram, the parallelograms about the diagonal are similar to the whole, and to one another.

23. The Latin manuscripts of Milan and Bordeaux wrongly report "sexdecim," which means "sixteen," instead of "sex," which means "six."

24. The braccio (plural braccia), literally "arm," was a typical unit of measurement of Piero's time.

25. Ratio of "sexquiquarta" = $1+\frac{1}{4}$; that is, $\frac{5}{4}$. Ratio of "sexquiquinta" = $1+\frac{1}{5}$; that is, $\frac{6}{5}$. Ratio of "sexquisexta" = $1+\frac{1}{6}$; that is, $\frac{7}{6}$.

26. "Sexquiquarta": $105:5 \times 4 = 84$; "sexquiquinta": $84:6 \times 5 = 70$; "sexquisexta": $70:7 \times 6 = 60$.

27. Practically we have $84:7 = 12, \times 6 = 72$; $72:8 = 9, \times 7 = 63$; $63:9 = 7, \times 8 = 56$; where the numbers 7, 8, and 9 with which the numbers 84, 72, 63 are divided are immediately successive to 5, to 6, and to 7 of the "sexquiquarta," of the "sexquiquinta," and of the "sexquisexta."

28. In the Latin manuscripts of Milan and Bordeaux, only D is reported.

29. Piero specifies that since the perspective can be executed on a wall or on a picture plane, the distance between the eye of the observer and the wall can be considered as a gap (interspace); in the same way the distance between the eye of the observer and the picture plane can be considered as a gap (interspace).

30. See note 14.

31. The Latin version better specifies here; the vernacular limits itself to saying: Reduce the shortened plane to a square.

32. The vernacular manuscripts report "de equali lati," which means "with equal sides." The Latin manuscripts rightly report "parium angulorum," which means "with equal angles."

33. For the tenth proposition of *De aspectuum diversitate*, compare note 14. Proposition XI says: Sections of planes further away, placed above the eye, appear lower.

34. The Latin corresponding to "has been sectioned according to the same proportion as BC. In fact BF is to DK what BC is to DE" is missing in the Milan manuscript.

35. The Milan and Bordeaux Latin manuscripts wrongly report "per sextam." However, in the Bordeaux manuscript a corrective 5 has been marked on the word "sextam" which means "for the sixth proposition." The vernacular versions correctly report "per la 5a," which means "for the fifth proposition."

36. These two last sentences are completely translated well in Latin.

37. The manuscripts of Parma and Reggio Emilia report "puncto visibile"; the Latin manuscripts of Milan and Bordeaux report "punctum visibile." Both these definitions indicate the "point that has the faculty of seeing," which I translated as "the eye point."

38. The Latin manuscripts of Milan and Bordeaux report the "fourth" proposition of this book; the vernacular manuscripts of Parma and Reggio Emilia report the "third"; but of course it is referring to the thirteenth. Please note on the frontal elevation of the head in figure A.19: (1) that the tip of the nose is rounded and that the end of this tip passes through point 1 of circuit E; (2) that this rounding is in accordance with the drawings relating to the nose, present on folio 59 verso and on folio 61 recto of the Parma manuscript. Please note, on the lateral elevation of the head, that the outline of the nose is tangent, upward, to the vertical line P and, downward, to the horizontal line E, in accordance with how it appears in the drawings relating to the nose, also present on folio 59 verso and on folio 61 recto of the Parma manuscript. Please note, in the "Detail of the nose" in figure A.19, the alternative outline of the nose that has been marked with dotted lines in plan and elevation. Keep in mind that, in perspective, the apparent outline of the nose will be represented according to the two tracks of this alternative outline, as delineated by the dotted lines in plan and elevation, although these two tracks appear only in this "Detail of the nose."

39. Though "25a" is found in both the Latin and vernacular manuscripts, this is instead the twenty-fourth. Compare note 22.

40. The sentence "in the way we have demonstrated earlier" means "as we have demonstrated in the proposition that precedes."

41. The drawings in the Latin manuscripts of Bordeaux and Milan report the circumference circumscribed to the figure with sixteen sides; this circumference does not appear in the drawings of the vernacular manuscripts of Parma and Reggio Emilia.

42. "GH" is missing in the two Latin manuscripts of Milan and Bordeaux; it is also missing in the two vernacular manuscripts of Parma and Reggio Emilia. Keep in mind that side FG is considered in perspective: in effect it is found on the terminal, where it appears in real length.

43. The Latin manuscripts of Milan and Bordeaux report "DC at" repeated twice.

44. The Latin manuscripts of Milan and Bordeaux wrongly report "equidistantibus BC" instead of "equidistans BC," which has been translated with "a line parallel to BC."

45. In the Latin manuscript of Milan the following is missing: "BE and I will draw a line from point M to point A. Where [this line] will intersect the diagonal." The Latin of Bordeaux reports it in the sheet's margin.

46. This sentence is missing in the two Latin manuscripts of Milan and Bordeaux; it is present in the two vernacular manuscripts of Parma and Reggio Emilia.

47. The two Latin manuscripts of Milan and Bordeaux report the correction of "eptagonum" (heptagon) to "exagonum" (hexagon)

48. In the drawings of the two Latin manuscripts of Milan and Bordeaux there is the circle containing the hexagon, but this circle is missing in the two vernacular manuscripts of Parma and Reggio Emilia.

49. Although these extensions do not appear in the drawings of the Latin and vernacular manuscripts, they have been drawn here.

50. Although these lines do not appear in the Latin or the vernacular manuscripts, they have been drawn here. In the drawings of the Latin manuscripts of Bordeaux and Milan the pentagon appears inscribed in a circle; this circle does not appear in the drawings of the manuscripts of Parma and Reggio Emilia.

51. Letters O, Q, R, P of the shortened surface do not appear in the drawing of the manuscript of Parma.

52. This sentence is missing in the Latin manuscripts of Milan and Bordeaux: it was taken from the vernacular manuscripts of Parma and Reggio Emilia.

53. The vernacular manuscripts of Parma and Reggio Emilia report: "as it was demonstrated in Propositions XV and XVII"; however, it is correct to speak of Propositions XI and XV, as is done in the Latin manuscripts of Milan and Bordeaux.

54. The two Latin manuscripts of Milan and Bordeaux report DE instead of BD.

55. The text that follows till the end of the proposition does not appear in the texts of the two Latin manuscripts of Milan and Bordeaux.

56. In the two vernacular manuscripts of Parma and Reggio Emilia this reads "as demonstrated in Proposition XI," but it should be "as demonstrated in Proposition XII."

57. The passage "BD at L and CE at M. Then I shall extend DE to P so that" is missing in the texts of the two Latin manuscripts of Milan and Bordeaux. It was taken from the vernacular manuscripts of Parma and Reggio Emilia.

58. This sentence "First, draw a parallel to BC from F, which will intersect the diagonal BE" is missing in the texts of the two Latin manuscripts of Milan and Bordeaux. It was taken from the vernacular manuscripts of Parma and Reggio Emilia.

59. Figure B.32 should be considered superimposed on figure B.31.

60. The passage "where it will intersect the line [that goes through point] 39 there will be 31; where it will intersect the line [that goes through point] 40" is missing in the texts of the two Latin manuscripts of Milan and Bordeaux. It was taken from the vernacular versions of Parma and Reggio Emilia.

61. Both the two Latin manuscripts of Milan and Bordeaux as well as the vernacular manuscripts of Parma and Reggio Emilia report XXVIIa when it should be XXVIa.

62. This passage "of the shortened surface, which is divided into two equal parts" is not present in the vernacular manuscripts of Parma and Reggio Emilia.

63. These last four sentences are a very precise conclusion compared to the one present in the vernacular versions.

64. The issue of distortions is tackled here.

65. This part of the sentence "and a smaller angle than a right angle" is not present in the two vernacular manuscripts.

66. These last five sentences in Latin are written very clearly compared to the way they are written in the vernacular manuscripts.

67. In the Latin manuscript of Milan we have the word "latitudine" twice, translated here as "width."

68. The specification "one with the shape of a round pyramid" does not appear in the vernacular manuscripts of Parma and Reggio Emilia.

69. "And the proportion of HI and FG is the same as between DE and BC": this sentence is missing in the vernacular manuscripts of Parma and Reggio Emilia.

70. In the Latin manuscripts of Milan and Bordeaux we have "the twenty-seventh proposition," but it is the twenty-sixth proposition, as is correctly reported in the vernacular manuscripts of Parma and Reggio Emilia. Please keep in mind that figure B.37 appears inverted, meaning reflected (as if in a mirror), in all the drawings of the Latin manuscripts of Milan, Bordeaux, London, and Paris.

71. The Latin manuscripts of Milan and Bordeaux report "through the twenty-seventh proposition"; but it is the twenty-sixth proposition, as is correctly reported in the vernacular manuscripts of Parma and Reggio Emilia.

72. This sentence is not present in the vernacular manuscript of Parma.

73. Line DE, seen in figure B.40, is missing in the respective drawings of the two vernacular manuscripts of Parma and Reggio Emilia and in the drawing of the Latin manuscript of Bordeaux. There is a hint of this in the drawing of the Latin manuscript of Milan.

74. Line DE, seen in figure B.41, is missing in the respective drawings of all vernacular and Latin manuscripts.

75. The text of the Latin manuscripts of Milan and Bordeaux improves the text of the vernacular manuscripts of Parma and Reggio Emilia because they add the expression "ex ambitu alio," which has been translated with "considering the second perimeter." This specification can also be found later when discussing the third and fourth perimeter.

76. The two Latin manuscripts of Milan and Bordeaux report "eptagoni" which means "of seven sides" instead of "exagoni," which means "of six sides." The expression "on a surface of six sides" is missing in the vernacular manuscripts of Parma and Reggio Emilia in any case.

77. The specification with numbers 5, 8, 16 does not appear in the vernacular manuscripts of Parma and Reggio Emilia.

78. In the drawing of figure B.42, the plane BCDE does not appear in the drawings of the vernacular or Latin manuscripts.

79. In the vernacular manuscripts of Parma and Reggio Emilia we can find this addition: "in the way [we have seen] in the twenty-seventh proposition of the first book."

80. By observing the drawing of figure B.43 we understand that the height F-11 is equal, in reality, to base F-1. Therefore the height H-13 is equal to base H-3 in reality.

81. The Latin manuscripts of Milan and Bordeaux report number 18 instead of 17 which is present in the preceding sentence; the same manuscripts report 17 instead of 18 in this sentence.

82. The Latin manuscripts of Milan and Bordeaux better specify this. The vernacular manuscripts of Parma and Reggio Emilia simply say: "And the cube is given with the base and the cymatium, things I asked to make."

83. The vernacular manuscripts of Parma and Reggio Emilia report "the twenty-sixth proposition."

84. This specification "which will be as long as the thickness of the column" is missing in the Latin manuscripts of Milan and Bordeaux.

85. The vernacular manuscripts of Parma and Reggio Emilia report "the twenty-sixth."

86. Although the points on line LA of figure B.44 are not indicated with numbers, they are easily identifiable.

87. Although the points on line KA of figure B.44 are not indicated with numbers, they are easily identifiable.

88. Observe figures B.31 and B.32 relating to the twenty-eighth proposition of the first book. It is easy to understand that BCDE is the shortened plane of the plane in real shape BCDE.

89. The reference to the twenty-eighth proposition of the first book implies an invitation to consider shortened plane BCDE, which corresponds to the plane in real shape BCDE of the twenty-eighth proposition of the first book. Having made this premise, the author can start by assigning—in perspective (figure B.45)—the external perimeter FGHI of the assigned house and the internal perimeter KLMN of the same house.

90. The vernacular manuscripts of Parma and Reggio Emilia are missing "the eleventh one 40-6."

91. Compare figure B.46, the detail of the windows: this figure has been added; there is no trace of this figure in the Latin or vernacular manuscripts.

92. An asterisk has been added to this 10 to differentiate it from the 10 assigned to the beginning of the proposition, when the author indicated with F-9 and G-10 the size of the wall.

93. Both vernacular and Latin manuscripts are missing "to 17*." I preferred adding 17* with an asterisk. Distance 15-17* corresponds to the width of the wall. In fact, this distance corresponds to distance F-9*, which is successively confirmed with specification 16-20 which corresponds to G-30.

94. In the drawings of the Latin or vernacular manuscripts, Piero does not report the points and the lines discussed, considering the material difficulty of transferring them onto drawings that are that small, like the ones of the foreshortened windows.

95. The vernacular manuscripts of Parma and Reggio Emilia only say: "Now the doors are to be made."

96. Please note that the point of number 158 coincides with the point of number 157, both in the vernacular and Latin manuscripts.

97. Both the vernacular manuscripts of Parma and Reggio Emilia and the two Latin manuscripts of Milan and Bordeaux report "till it [meaning this parallel line] touches line 152-145." This specification does not make sense because the vertical line through 200 is not in the same plane as line 152-145. This is certainly an involuntary mistake.

98. Both the vernacular manuscripts of Parma and Reggio Emilia and the two Latin manuscripts of Milan and Bordeaux report "till it [meaning this parallel line] touches line 154–149." This specification does not make sense because the vertical line through 201 is not in the same plane as line 154–149. This is certainly an involuntary mistake.

99. This sentence is missing in the Latin manuscripts of Milan and Bordeaux.

100. Please note that number 203 coincides with the same point of number 199, both in the vernacular and Latin manuscripts.

101. The Latin version improves the text. The vernacular texts only say: "these lines complete the doors and the proposed house."

102. The vernacular manuscripts of Parma and Reggio Emilia report "Proposition XXVII," but it should be "Proposition XXVI," as the Latin manuscripts of Milan and Bordeaux rightly report. In the same Latin manuscripts "ut docemur" was added, and it has been translated "as . . . we have learnt."

103. In the Latin manuscript of Milan we see HM instead of GL. The sentence is missing in the Latin manuscript of Bordeaux.

104. Please note how these lines do not appear in any of the drawings of the Latin or vernacular manuscripts.

105. The specification "which offers us the first view" is missing in the vernacular manuscripts of Parma and Reggio Emilia.

106. This sentence is missing in the vernacular manuscripts of Parma and Reggio Emilia. These manuscripts show the sentence "ed è finito il primoocchio," which means "and so the first circular opening is finished."

107. This sentence is missing in the vernacular manuscripts of Parma and Reggio Emilia.

108. This sentence is missing in the vernacular manuscripts of Parma and Reggio Emilia.

109. The vernacular manuscripts of Parma and Reggio Emilia, together with Latin manuscripts of Bordeaux report "in the twenty-sixth proposition of the first book." It is obvious that this should be "in the sixteenth proposition of the first book," as it is in the context and in the Latin manuscript of Milan.

110. For what concerns this sentence, the vernacular manuscripts of Parma and Reggio Emilia only say: "And we have the eight faces."

111. This sentence is missing in the vernacular manuscripts of Parma and Reggio Emilia.

112. The Latin manuscripts of Milan and Bordeaux report "in degradata planitie"; the vernacular manuscripts of Parma and Reggio Emilia report "nel piano degradato": the translation of these two versions is "on a shortened plane." This sentence should be corrected to "on a plane to be shortened." In fact, the plane we are talking about is the ground floor. This plane is not shortened at all, but it will have to be shortened, once we realize the perspective on the terminal (i.e., on the picture plane).

113. The Latin manuscript of Milan reports: "Habemus degradatam planitiem BCDE"; the Latin manuscript of Bordeaux reports "Habemus degradatam planitiem bcde." The translation of these versions is "We have the shortened plane bcde," or "We have the shortened plane BCDE." It is proper to correct these sentences with "We have the plane to be shortened," because the plane discussed, as hinted in the previous note, is the ground floor.

114. This sentence is missing in the Latin manuscripts of Milan and Bordeaux.

115. The vernacular manuscripts report the adjective "shortened" as well as the addition "or augmented."

116. The vernacular manuscripts report "Zeuxis" too. Regarding this list of painters, compare Vitruvius, *Ten Books on Architecture*, trans. Ingrid D. Rowland, with commentary by others, Book III, introduction (New York: Cambridge University Press, 1999).

117. This sentence is missing in the vernacular manuscripts.

118. The vernacular manuscripts change this sentence to: "commo presuntuoso ho preso ardire scrivere questa particella de prospectiva apartinente alla pictura [As, presumptuous, I dared write this particle of the perspective belonging to painting]."

119. The Latin manuscripts report "nisi a primo ordine recederemus," which means "if we do not renounce"; it has to be "si a primo ordine recederemus" which means "if we renounce the first procedure," as it was translated.

120. This sentence is missing in the Latin manuscripts.

121. The Latin manuscript of Bordeaux reports "acum quondam," which means "a certain needle." I prefer the sentence "acum unam," which means "a needle" which is present in Milan's Latin manuscript.

122. In each picture of each proposition, Piero always established a point M which is needed to fix the vertical axis of reference MN (figure B.55). This is necessary to identify the points in perspective. On this axis MN, in fact, the heights of the points can be read. These heights are marked on the paper rulers, starting from point a which is reported on each paper ruler. Starting from this MN axis, you can read on the wooden rulers, as if they were abscissas, the width of the points to the left and right of the MN axis.

123. Piero reproduces square BCDE above CE in figure B.54, as if to remember that the square that needs to be represented is on CE.

124. See Proposition I of this book.

125. The specification "the ruler of width" appears in the vernacular manuscripts.

126. Piero reports square BCDE above CE of this figure B.57, with the drawing of the octagon, as if to remember that the square that needs to be represented is on CE.

127. Right after the title of Proposition III, in the vernacular manuscripts, this sentence appears: "We intend to shorten four circles that surround only one center and are subdivided into twelve equal parts, as has been said." This is a repetitive sentence which is not present in the Latin manuscripts.

128. Observe figure B.60: we are talking about distance MO. This distance is equal to distance PQ in figure B.61.

129. The "apex" is a term for a diacritical mark corresponding to the prime symbol (′); for clarity, the translation will employ the prime symbol where the manuscript text has "with an apex." However, in all the drawings of the Latin and vernacular manuscripts of this proposition, the numbers with apexes (prime symbols) are not reported. These apexes must be imagined (figure B.61) to be on the intersections of straight line AF with lines 6-8, 5-9, 4-10, 3-11, 2-12, and naturally on points 7 and 1 of the three internal circles. The numbers of the intersections of the circle going through B will have an apex (e.g., 1′). The numbers of the intersections of the circle going through C will have two apexes (a double prime symbol, e.g., 1″); the numbers of the intersections going through D will have three apexes (a triple prime symbol, e.g., 1‴). We will not report these apexes (prime symbols) on figure B.61 either.

130. This sentence is missing in the vernacular manuscripts.

131. This "with an apex" is missing in the vernacular and Latin manuscripts: I think it was appropriate to add it because all numbers from 2-12 till 7 report it.

132. This sentence is missing in the Latin manuscripts.

133. To indicate "where it coincides," the texts of the Latin manuscripts use the following expressions: "ubi convenit," "ubi assistit," "ubi congressum habet," "ubi occurrit," "ubi coniunctum habet," "ubi congreditur," "ubi . . . stet," "ubi coniungitur." These expressions have been translated with "it coincides." This phrase best simplifies the concept that in the vernacular texts is always rendered with "combacia."

134. This sentence is missing in the Latin manuscripts.

135. This sentence is missing in the Latin manuscripts.

136. This sentence is missing in the vernacular manuscripts.

137. This sentence is missing in the vernacular manuscripts.

138. The vernacular manuscripts repeat all the operations of taking the thread and marking the points in which the thread touches the wooden ruler till point 12.

139. The specification "of ruler A" is missing in the Latin manuscripts.

140. The vernacular manuscripts report, one after the other, all the numbers of the subdivisions, from 1 to 12, and M.

141. This first part of the sentence is missing in the vernacular manuscripts.

142. This sentence and the four following ones are missing in the vernacular manuscript of Parma, but they are present in the vernacular manuscript of Reggio Emilia.

143. The Latin manuscripts abbreviate, while the vernacular manuscript of Parma repeats the instruction to place the thread on points 2-12, 3-11, 4-10, 5-9, 6-8 and to mark one by one all the numbers where the thread hits the ruler.

144. This sentence is missing in the vernacular manuscript of Parma.

145. This sentence is missing in the vernacular manuscript of Parma.

146. This sentence is missing in the Latin manuscripts.

147. This sentence is missing in the Latin manuscripts.

148. This sentence is missing in the vernacular manuscript of Parma.

149. With this sentence, the Latin manuscript abbreviates, while the vernacular manuscript of Parma repeats the instruction to place the thread on points 6-8, 5-9, 4-10, 3-11, 2-12, and 1 and each and every point has to be marked wherever the thread hits the ruler.

150. This sentence is missing in the vernacular manuscript of Parma.

151. The vernacular manuscript of Parma concludes "and you finished the torus we decided to make."

152. Proposition V is completely missing in the Latin manuscript of Bordeaux; this proposition is present in all other Latin and vernacular manuscripts. I'm translating from the Latin manuscript of Milan.

153. Observe the plan of the cube. On this plan which is the lower base of the solid, project on point A also point B of the upper base. In the same manner, the remaining three points C, G, H of the upper base are respectively projected onto points D, F, I of the lower base. For this reason, the points in plan are indicated with two letters. These are AB, DC, FG, IH. Observe again that since the plan of the cube exactly corresponds to the shape of a square, it is better to translate the Latin sentence "propriam . . . formam" and the vernacular "in propria forma" as "in real shape."

154. This means the surface of the upper base of the cube that, as mentioned in the previous note, coincides with the lower base in plan.

155. In both the Latin manuscript of Milan and in the vernacular manuscript of Parma only B appears in the specification about the corner, but it is better to say "corner BG," which is inclined in plan if compared to straight line RS. The addition of G is a specification that helps to better understand. In fact, the cube lies on the horizontal plane of the plan and is positioned at an angle with respect to line RS. The first rotation of the cube has been executed with respect to straight line RS, as may be seen from the angle indicated by the arrow.

156. The representation of the elevation of the cube begins.

157. In the vernacular manuscripts of Parma and Reggio Emilia this line goes through D, F, and V but does not go through HI. In the Latin manuscripts of Milan and London line DFV goes through HI.

158. Take the BC side of the cube in plan and place it in elevation. The cube's edges are all the same, both horizontal and vertical.

159. The second necessary rotation is established to fix the cube in space in any random position. Piero sees this rotation in the corner on the right looking at point F, but, as may be noticed, it can be read also in the corner marked on the left of F. The two corners are in fact complementary to each other.

160. The Latin manuscript of Milan exactly reads: "cubus in planitie propriam formam habente positus." The translation of this sentence, which may be read in the text, makes us understand that we are facing a very special representation. In fact, this is an orthogonal axonometry of the cube. Consider that the Latin adjective "proprius" can be correctly

translated with the adjective "special." It is on the plan representation of this cube, however inclined in space, that Piero della Francesca chooses the point M: the visual axis will start from this point, in accordance with the choice of terminal. On this visual axis for M Piero positions eye O in plan, at the desired distance.

161. This specification "which is the third figure" does not appear in the Latin manuscript of Milan. In all manuscripts, both vernacular and Latin, figures B.66, B.67, and B.68 seem summarized in figure B.69.

162. This sentence is missing in the vernacular manuscript of Parma.

163. This sentence is missing in the Latin manuscript of Milan.

164. Piero makes us understand that this figure of the elevation is a very special representation. In fact, this is another orthogonal axonometry of the cube, with the projection plane placed vertically through PQ on the plane of the third figure. All the heights of the points that delimit the representation of the cube in elevation have been taken from the second figure, and they have been transferred in their real dimensions onto the plane of this second axonometry.

165. This last part of the sentence, starting from "which," does not appear in the vernacular manuscript of Parma.

166. In this proposition and in the next, Piero interprets the fundamental parts of the Corinthian Roman column following Vitruvius's and Alberti's rules. Compare *De prospectiva pingendi*, critical edition by Giusta Nicco Fasola (1942), republished with comments by Eugenio Battisti and Franco Ghione, 2 vols. (Florence: Le Lettere, 1984), 148.

167. In the Latin manuscript of Bordeaux, the text from here till note 187 is missing: the translation has been executed using the Latin of the Milan manuscript.

168. This point E″ identifies the upper plane which separates the base of the column from the shaft of the column.

169. This means in the middle of the torus.

170. The second part of this sentence, starting from "which," is missing in the vernacular manuscript of Parma.

171. The following four sentences, concerning points 2-10, 4-12, 6-14, 8-16, are missing in the Latin manuscript of Milan.

172. There are two lines GH in elevation: the one below is the one that is meant.

173. It is obvious that right and left have been swapped. This is due to the presence of point M and the reference axis of points 5-13. We should look at the drawing in figure B.73 from the opposite side of the sheet of paper on which the drawing is represented. The line of points 5-13 (which we could consider a real reference axis) has all points 2 and 16 on its right. The same line of points 5-13 has all points 8 and 10 on the left; and similarly for all other points.

174. This sentence and the following are missing in the Latin manuscript of Milan.

175. This sentence appears in the Latin manuscript of Milan after what follows.

176. This sentence and what follows are missing in the Latin manuscript of Milan.

177. This sentence appears in the Latin manuscript of Milan after what follows.

178. The four sentences about circle E are missing in the Latin manuscript of Milan.

179. The four sentences about circle F are missing in the Latin manuscript of Milan.

180. The vernacular manuscripts summarize the last two sentences in the following sentence: "Now one must execute the shortening on the rulers."

181. This sentence is missing in the vernacular manuscript of Parma.

182. This sentence is missing in the vernacular manuscript of Parma.

183. This sentence is missing in the Latin manuscript of Milan.

184. In figure B.76, you may observe all divisions present on the wooden and paper rulers. Rulers A, C, and E are indicated with the significant additions A/G/G′, C/C′, and E/E′/E″, as may be read in the text.

185. When citing the subdivisions of circles B, C, D, E, F, the Latin manuscript of Milan, as we have seen, summarizes and abbreviates. The vernacular manuscript of Parma specifies time after time: take ruler A and place it aside, place wooden ruler B so that it stays still, then take the thread and place it on all the points of circle B and above M; and wherever the thread hits ruler B, mark all points, including point M.

186. This line G is the wooden ruler relative to the base of the plinth, whose letters K, I, H, G coincide with the letters of the upper base of the plinth.

187. From here the text of the Latin manuscript of Bordeaux starts again: "charta 61recto," first line.

188. In figure B.75 paper rulers G and A are overlapping paper ruler G′.

189. The Latin manuscript of Milan reports 8 instead of 16.

190. The Latin manuscript of Milan reports 7 instead of 15.

191. This sentence is missing in the vernacular manuscript of Parma. The Latin manuscript of Milan reports 6 instead of 14.

192. In the Latin manuscript of Milan point 13 is missing.

193. In the Latin manuscript of Milan this sentence is missing; the three following sentences about points 7-11, 8-10, 9 are missing; and the following sentence is missing: "And you have reported the marks of line A on ruler A."

194. In figure B.76 all divisions present on paper and wooden rulers may be observed. Rulers A, C, and E are indicated with the significant additions A/G/G′, C/C′, and E/E′/E″, as may be read in the text.

195. This sentence is missing in the Latin manuscript of Milan.

196. The Latin manuscript of Milan reports 8 instead of 16.

197. The Latin manuscript of Milan reports 7 instead of 15.

198. The Latin manuscript of Milan reports 6 instead of 14.

199. In the Latin manuscript of Bordeaux point 13 is missing.

200. In the Latin manuscript of Milan this sentence is missing; the three following sentences relative to points 7-11, 8-10, 9 are missing; the following sentence is missing: "These are the points of line B on ruler B."

201. The words "through point 6, through point 7, through point 8, through point 9" are missing in the Latin manuscript of Milan.

202. The Latin manuscript of Milan shows 8 instead of 16.

203. The Latin manuscript of Milan shows 7 instead of 15.

204. The Latin manuscript of Milan shows 6 instead of 14.

205. In the Latin manuscript of Milan point 13 is missing.

206. In the Latin manuscript of Milan this sentence and the three that follow about points 7-11, 8-10, 9 are missing.

207. The words "through point 6, through point 7, through point 8, through point 9′" are missing in the Latin manuscript of Milan.

208. The words "through point 6, through point 7, through point 8, through point 9′" are missing in the Latin manuscript of Milan.

209. The words "through point 6, through point 7, through point 8, through point 9 of ruler E″" are missing in the Latin manuscript of Milan.

210. In the vernacular manuscript of Parma, "as you made in the first ruler and in the others" has been added.

211. This sentence and the following one, "Put both aside," do not appear in the Latin manuscript of Milan.

212. In the vernacular manuscript of Parma the following appears: "then place the thread above 1, above 2, above 3, above 4, above 5, above 6, above 7, above 8, above 9 of line E′ of the height."

213. This sentence does not appear in the vernacular manuscript of Parma. However, always in the vernacular manuscript of Parma, there is: "then place the thread above 1, above, 2, above 3, above 4, above 5, above 6, above 7, above 8, above 9 of line f of the height."

214. This sentence appears in the Latin manuscript of Milan. In the vernacular manuscript of Parma, there is the list of all points from 1 to 9 of line E, on which the thread must be placed. Then we are prompted to mark the corresponding numbers, that is, 1, 2-16, 3-15, 4-14, 5-13, 6-12, 7-11, 8-10, 9.

215. Both in the Latin manuscript of Milan and in the vernacular manuscript of Parma "in two equal parts" appears. However, in all the original drawings of both the other Latin manuscripts and of the vernacular manuscript of Reggio Emilia, line KL is not divided at M into two equal parts. This does not make the drawing wrong. It is surely a way to make the paper rulers fit with the drawing of the perspective on the page of the manuscript.

216. The Latin manuscript of Milan only shows these numbers for rulers E. The vernacular manuscript of Parma instead shows all points from 1 to 9 of the rulers E. Then, in the vernacular manuscript of Parma, we are prompted to mark the corresponding numbers; that is, 1, 2-16, 3-15, 4-14, 5-13, 6-12, 7-11, 8-10, 9.

217. As we were able to show and as we will see later, the Latin text often abbreviates rather than repeating the specifications "place," "put," "trace."

218. In the Latin manuscript of Milan, there are only these numbers for rulers F. The vernacular manuscript of Parma instead shows all points from 1 to 9 of the rulers F. Then, always in the vernacular manuscript of Parma, we are prompted to mark the corresponding numbers; that is, 1, 2-16, 3-15, 4-14, 5-13, 6-12, 7-11, 8-10, 9.

219. In the Latin manuscript of Milan, there are only these numbers for rulers E′. The vernacular manuscript of Parma instead shows all points from 1 to 9 of the rulers E′. Then, always in the vernacular manuscript of Parma, we are prompted to mark the corresponding numbers; that is, 1, 2-16, 3-15, 4-14, 5-13, 6-12, 7-11, 8-10, 9.

220. In the Latin manuscript of Milan as well as the vernacular manuscript of Parma the numbers for rulers E″ are missing. In the same manner, the numbers have not been written both in the Latin and vernacular versions for rulers D, C, C′, B, A that follow.

221. This sentence is missing in the vernacular manuscript of Parma.

222. Piero begins by drawing the quadrangle at the plinth: GH is the first side in depth, IK is the second side in depth.

223. This sentence is missing in the Latin manuscript of Milan. Piero has drawn the lower quadrangle of the plinth.

224. This sentence is missing in the vernacular manuscript of Parma. Piero has drawn the upper quadrangle of the plinth.

225. Piero prompts us to draw 9 circles through 16 parts of lines. The vernacular manuscript of Parma shows all the numbers of the 16 parts, the Latin manuscript of Milan only shows the first eight parts.

226. In all the original Latin and vernacular manuscripts, relative to the perspective of the base of the column there are no numbers or letters, with the exception of letters MN which identify the vertical axis MN.

227. The Latin manuscript of Milan only shows numbers 1 to 8, but in the vernacular manuscript of Parma "till 16" has been added.

228. The Latin manuscript of Milan only shows numbers 1 to 8, but in the vernacular manuscript of Parma "till 16" has been added.

229. It is good to specify that in the perspective of figure B.77 the face of the plinth is detached from line KL: only the drawing in the vernacular manuscript of Parma presents this detail. The vernacular of Reggio Emilia and all other Latin manuscripts show line IG, in the lower part of the frontal rectangle of the plinth, matching line KL. This specificity implies that line KL goes through the two points G and I on paper rulers A/G/G′. The dotted line, in fact, goes through the two points G and I and coincides with the line which I marked with letters G and I on the face of the plinth.

230. The term "puncto" which means "point" is missing in the Latin manuscript of Bordeaux.

231. We prefer to translate "propria in forma" as "in appropriate shape," which corresponds to the one in plan and in elevation.

232. It is clear that the measurement on which Piero bases the compositional characteristics of the capital is the unitary model, a standard quantity that serves as a basis for the measurement of other quantities of the same nature or even to carry out, in this case, the design of a geometrically "proportioned" capital.

233. Naturally the line which goes through B′ must equal the line that goes through B.

234. Naturally the line which goes through E′ must equal the line that goes through E.

235. The specification "either twigs or supports" is missing in the vernacular manuscript of Parma.

236. Both the vernacular manuscripts of Parma and Reggio Emilia claim that the circuit of the volute must go through the extremity of line D. Both Latin manuscripts of Milan and Bordeaux make clear that the curve of the volute does not go through the extremity of line D but is somewhat far from this extremity, as claimed in the text.

237. As may be observed in the elevation of figure B.79, the flower has been identified in the center by the four sides of a quadrangle; the dotted lines identify the elevations of the flowers on the sides. Naturally, the flowers are four, as may be observed in plan in the same figure.

238. We prefer to translate Latin "propriam formam" as "real shape," which shows the real shape of the square.

239. Circle E is needed for four circles, because lines F, E, E′, I all measure $4\frac{1}{2}$. In the same manner, the two lines C and G are $5\frac{4}{7}$ long and generate one circle only. Thus the circles that derive from the lines in elevation are eight, even if the circles to be drawn are nine, as can shown in figure B.79.

240. In the graphics produced by the computer, the big four arcs of circuit A, which are tangent to circuit D, are not exactly tangent to circle D; they present a calculated error of 0.0178.

241. The specification "circuit A at point 2" is missing in both the Latin manuscripts of Milan and Bordeaux and the vernacular manuscripts of Parma and Reggio Emilia.

242. The last two numbers are missing in both the Latin manuscripts of Milan and Bordeaux and the vernacular manuscripts of Parma and Reggio Emilia. We preferred adding them with the aim of doing a correct reading of the text.

243. The vernacular manuscripts of Parma and Reggio Emilia show "the width," which is clearly a contradiction.

244. Concerning left and right, Piero is looking at the capital from the back and not from the front: it would be to say from Y to X and not vice versa. Compare note 173.

245. The XY axis is indicated in Bordeaux and Reggio Emilia with letters TV. We prefer XY, which appears in Milan and Parma.

246. In the Latin manuscript of Milan one can see the correction of wrong numbers.

247. In the Latin manuscript of Milan one can see the correction of wrong numbers.

248. This sentence is missing in the vernacular manuscripts of Parma and Reggio Emilia; it is also missing in the Latin manuscripts of Milan and Bordeaux. We preferred to add it, with the objective of a better reading of the text. In the three following sentences, there are corrections made on several numbers, both in the vernacular manuscripts of Parma and Reggio Emilia and the Latin manuscripts of Milan and Bordeaux.

249. At this point, there is a sentence that does not comply with the numbers of the correct progression we examined both in the vernacular manuscripts of Parma and Reggio Emilia and the Latin manuscripts of Milan and Bordeaux

250. This sentence and the three following ones present mistakes and corrections of numbers in both the vernacular and the Latin texts.

251. Starting here till the end of this paragraph, in the Latin manuscript of Milan there appear corrections of wrong numbers.

252. A geometrical reconstruction of the overlapping circuits E-E′ and circuit D has been attempted. In all six manuscripts (the vernacular of Parma and Reggio Emilia, the Latin of Milan, Bordeaux, London, and Paris) whose drawings I was able to check, there is no trace of the real tracings in plan of these circuits.

253. At this point there are no numbers to report concerning circuit E′, in either the vernacular or Latin manuscripts, except for the manuscript of Reggio Emilia where the following numbers appear: 33-3, 27-9, 39-45, 21-15, 40-44, 16-20, 41-43, 19-17. We prefer to represent the entire circuit with all the numbers in figure B.81.

254. In the Latin manuscript of Milan we have "on the right" instead of "on the left." This sentence, which is in the right place in Reggio Emilia's vernacular, appears five sentences later in Parma's vernacular.

255. In the Latin manuscript of Milan one can see corrections of wrong numbers.

256. This sentence is missing in the Latin manuscripts of Milan and Bordeaux and in the vernacular manuscript of Reggio Emilia. It is present only in Parma. We preferred adding it with the purpose of a correct reading of the text.

257. In the Latin manuscript of Milan one can see corrections of wrong numbers.

258. Corrections of numbers in the manuscripts of Milan, Bordeaux, and Reggio Emilia; in Parma the whole sentence is missing.

259. Corrections of numbers in the manuscripts of Milan, Bordeaux, and Reggio Emilia, in which this sentence was rewritten and then deleted.

260. In the examined Latin manuscript of Milan, right and left have been swapped.

261. This sentence is missing in the two vernacular manuscripts of Parma and Reggio Emilia.

262. This sentence is missing in the two vernacular manuscripts of Parma and Reggio Emilia.

263. After this sentence, the Latin manuscripts of Milan and Bordeaux repeat the sentence with numbers 6-30.

264. This sentence and the five following ones are missing in the Latin manuscript of Milan.

265. There are corrections of numbers in the manuscripts of Milan, Reggio Emilia, and Bordeaux in this sentence and the two following ones.

266. At this point, we carried out a long graphic study of the volute separately, but it is not possible to report it here. Piero does not hint at the smoothed volute, neither is there a reference to it in any of the drawings of all the manuscripts. For this reason, we have limited ourselves to drawing the margins without chamfers, as they can be seen in the drawings.

267. The wording "on both lines B" means "on the right and on the left of b" on line B.

268. The vernacular manuscripts of Parma and Reggio Emilia do not contain "so that each [part] contains its signs."

269. At this point, there are no numbers to be reported in either the Latin or vernacular manuscripts. This is because in the paragraphs "Wooden Rulers or Rulers of the Width" and "Papers Rulers or Rulers of the Height" we find that there must be many more numbers, and since in the same paragraphs it is specified that circuit B and circuit B′ (which follows) must have the same numbers, we have preferred to report all the numbers as they are found in the drawings' reconstruction.

270. At this point, in the Latin manuscript of Milan there is a sentence where 11 is reported with the apex (11′), which is not on circuit B′ but on circuit A. Do not forget that points 3′-19′, 5′-17′, 26′-24′, 10′-12′ of the flowers belong to the plan of circuit B′, as may be observed in figure B.86.

271. This specification does not appear in the vernacular manuscripts.

272. This specification does not appear in the vernacular manuscripts.

273. This specification does not appear in the vernacular manuscripts. Do not forget that points 4′-18′, 11′, and 25′ of the flowers belong to the plan of circuit A, as may be observed in figure B.87.

274. Now it is possible to complete the cymatium and the flowers. Compare note 266.

275. Imagine the final representation of the capital in plan and in elevation, with all the points' numbers indicated. Imagine the confusion of numbers which this type of overlapping would have caused in a figure! This is why we preferred to reproduce figure B.88 with no numbers.

276. Naturally line KL must be perpendicular to lines PR, TV, QS.

277. The two vernacular manuscripts specify "The bristles of the pony-tail would be better." This sentence is not present in the Latin manuscripts of Milan and Bordeaux.

278. The two vernacular manuscripts of Parma and Reggio Emilia say: "this is to be always understood." This sentence does not appear in the Latin manuscripts of Milan and Bordeaux.

279. This sentence is missing in the Latin manuscripts of Milan and Bordeaux.

280. This sentence is missing in the Latin manuscripts of Milan and Bordeaux.

281. This sentence is missing in the Latin manuscripts of Milan and Bordeaux.

282. Please remember that wooden ruler B′, which does not appear in the Latin or vernacular texts, coincides with wooden ruler B in plan.

283. Useful specification: one must operate only on the points that are visible to the eye. Please remember that wooden ruler E′, which does not appear in the examined Latin or vernacular texts, coincides with wooden ruler E in plan.

284. Naturally line KL must be perpendicular to lines PR, TV, and QS.

285. This is the partially dotted line Kp.

286. Point a in figure B.91 represents the point from which the line of paper ruler a overlaps pL. In the same manner, starting at point a, on pL points b, b′, c, d, e, e′, f, g, h, i, k are overlapped. These are departure points of the remaining wooden rulers in elevation.

287. The vernacular manuscripts of Reggio Emilia and Parma only report 4. Even the Latin manuscript of Bordeaux only reports 4, while the Latin manuscript of Milan reports 4 and 18.

288. In Bordeaux, Parma, and Reggio Emilia this sentence is reported after number 11. Piero only reports the visible points. In the drawing of figure B.92 all points present on the entire circuit have been reported.

289. Remember that in the drawings' reconstruction all the points on every circuit have been reported.

290. With the exception of points 4′ and 11′.

291. In the Latin manuscript of Milan numbers 21, 22, 26 are missing.

292. This sentence is missing in the vernacular manuscript of Parma.

293. In the Latin manuscript of Milan there's one more 8 and the 5 is missing.

294. There are some discrepancies in the numbers for ruler K in all manuscripts.

295. In the vernacular manuscript of Parma this part of the sentence is missing: "two by letter e′."

296. As previously mentioned, keep in mind that in the drawings' reconstruction all the points present on every circuit have been reported, as can be seen in the previous figures.

297. In the two Latin manuscripts of Milan and Bordeaux and in the vernacular manuscript of Reggio Emilia all these numbers (taken from the Parma manuscript) are missing.

298. The numbers' lists reported till here and later are very lacking in the manuscripts, and they often show corrections. There are also many discrepancies between the numbers of the texts and the numbers of the rulers of the drawings. By comparing the vernacular and Latin texts and by reconstructing all the circuits, we wanted to offer the reader the easiest path for reading Piero correctly.

299. As can be seen in figure B.95, the entire circuit was shortened. In the Latin manuscripts of Milan and Bordeaux some specific points of circuit B are not mentioned; while in the vernacular manuscripts of Reggio Emilia and Parma we have the following points: 1, 2, 3, 4, 5, 6, 7, 8, 9 10, 11, 12, 13, 14, 15.

300. As can be seen in figure B.96, the entire circuit B′ was shortened. In the texts of the various manuscripts there is a big discrepancy between the numbers as well as many repetitions and corrections. The Latin manuscript of Milan shows only two number corrections.

301. The Latin manuscript of Milan reports 4′, but it should be 5′.

302. The Latin manuscript of Milan reports 4′, but it should be 5′.

303. The Latin manuscript of Milan reports 11′, but it should be 12′.

304. Starting from circuit C, no numbers have been reported in any of the other manuscripts. Only the vernacular manuscript of Parma reports the numbers of this circuit and all other circuits, with all the passages we may imagine. In figure B.97, all points of all circuits have been found in perspective.

305. E′ has been added. Please note that in the drawing of the Latin manuscript of Milan, wooden rulers B′ and E′ have not been reported.

306. These last four sentences, which are not present in the vernacular manuscripts, indicate the will to abbreviate: Piero actually does this by leaving out all other circuits and the series of numbers that only appear in the vernacular of Parma. To better understand how the volute must appear in perspective, observe in figure B.98 the strip (of the volute) between horizontal lines E and C. In figure B.99, we offer the definite perspective of the capital. Figure B.100 corresponds to the final table: I copied it, as far as possible, according to the figure present in the vernacular manuscript of Parma. With this figure, I wanted to offer an idea of the constructive complexity of Piero's procedure: this perspective includes all wooden

rulers, all paper rulers, and all numbers, which I have tried to report as clearly as possible.

307. The Latin language explains better than the vernacular how to make these lines go from one head to the other.

308. Parma also adds one line that goes through the beginning of the mouth: this line has been reported in the drawing.

309. Parma and Reggio Emilia also report a line at the beginning of the forehead: this line has been reported in the drawing.

310. This sentence is missing in Parma.

311. This line does not coincide with the line that goes through the beginning of the ear, but it lies a little after this beginning. Two more vertical lines are added: one that goes through the upper extremity of the ear and one that goes through the lower extremity of the ear. These lines are necessary to find points y and x which are placed on those extremities. The line "coming from the upper end of the ear" can be found in the paragraph "How to draw circuit C in real shape."

312. Reggio Emilia confirms the Latin. Parma talks about a line "uscente dal contatto del principio dell'orecchio," which means "which is originating from the contact of the beginning of the ear." This is an obvious discrepancy. It is good to specify that Piero is talking about the vertical line going through points 5, 1, 13 in plan and through point 1 of the top of the head with one eye only, as can be seen in figure B.104.

313. The Latin version reports "in summo frontis" which means "on the top of the forehead." However, "The top of the forehead" coincides with "The top of the head." Reggio Emilia confirms the Latin, while Parma talks about a line "uscente dal principio de l'urechia" which means "which is originating from the contact of the beginning of the ear." This is an obvious incongruence.

314. 20 appears in Milan and Reggio Emilia, but Bordeaux and Parma report 22.

315. 19 appears in Milan and Reggio Emilia, but Bordeaux and Parma report 21.

316. 18 appears in Milan and Reggio Emilia, but Bordeaux and Parma report 20.

317. In Parma it says "L'altro stendi al fine de l'urechia," which means "Extend the other end to the end of the ear"; but it should be "Extend the other end to the beginning of the ear," as Milan and Bordeaux show and as can be seen in the elevation of the head with two eyes.

318. The Latin says "ab auricule principio" which means "from the beginning of the ear"; but it should be "from the top of the head." Parma is missing this sentence.

319. 9 does not appear in Milan and Bordeaux.

320. Instead of 20, Parma, Reggio Emilia, and Milan report 18. Bordeaux shows that 18 has been deleted, and in figure B.112 it is reported as 20 in compliance with the subdivisions of circuit D.

321. The Latin reports "a principio auricule," which means "from the beginning of the ear"; but it should be "a summo frontis," which means "from the top of the head." In Parma too, it says "dal principio de l'urechia" which means "from the beginning of the ear."

322. Reggio Emilia, Bordeaux, and Milan report 14 and 6. Parma reports 13 and 7. In compliance with the subdivisions of circuit D, as may be seen in figure B.112, numbers 15 and 7 have been reported in figure B.107.

323. As can be seen in the drawings of figure B.108, the numbers 20, 1, 2, even though they belong to circuit E, do not lie on the contour of the tip of the nose, either in the profile head or in the frontal head. I have proportioned, in fact, the two contours of the tip of the nose, as they appear, in order to create in the endpoint 1 of line E the center of the two rotations of the second head.

324. The Latin reports "a principio auricule" which means "from the beginning of the ear," but it should be "a summo frontis" which means "from the top of the head." In Parma too, it says "dal principio de l'urechia"—"from the beginning of the ear."

325. Parma and Bordeaux report: "Place one of the two ends of the compass on point 1 of the line originating at the chin, and with the other end mark 16′ on the side of the head and 2′ below." (In these manuscripts, numbers 16 and 2 are marked "col titulo" and "cum titulo," which mean "with the apex.") However, we decided to show points 16 and 2 (without the apex) for the following reasons. The first is that, in compliance with all the other numbers of all circuits that have appeared without an apex until now, there is no reference to the construction of the arc of the chin in this context. On which—as you will see—numbers 16′ and 2′ have been placed. The second reason is that, between the vertical lines in the paragraph headed "Vertical Lines," Piero only hints at the line "that aligns with the chin," without any reference to another line on which (if elongated) we should find in plan points 16′ and 2′ of the chin. In the same manner, as can be seen, the line "that aligns with the chin" cannot possibly go through the other points of the arc of the chin which are 1′, 15′, and 3′. The third reason is that, in the paragraph headed "Vertical Lines," Piero directly says "trace . . . another [line] at the beginning of the throat." Consequently we decided to change "originating at the chin" to "originating at the throat."

326. Keep in mind that this line does not exactly go through the middle of the neck; therefore "from the middle of the neck" becomes a way of speaking, just as the previous sentence ("With the other end touch the end of the neck") is a way of speaking. In fact, points 13 and 5, as can be noticed, are not at the lateral extremity of the neck, but in the drawing it seems that the neck ends exactly on the line going through points 13 and 5, where the throat should begin. Piero was forced to utilize these ways of speaking because he was working with the human body, which he represents geometrically for the first time.

327. This sentence appears in the Latin manuscripts and in Reggio Emilia; it is missing in Parma.

328. The Latin says "pro arbitrio multitudo cyrculorum," which means "a multitude of circuits at one's own will." The use of the Latin term "multidudo" is significant to the aim of introducing the concept of construction of the apparent contour of the head. In effect, the apparent contour can be obtained through a "multitude" of circuits at one's own will, therefore sections. Reggio Emilia says "[you] will be able to do however many [circuits] you will want to do"; Parma says "[you] will be able to do however many [circuits] a man wants."

329. The Latin reports "in linea egrediente a principio auricule," which means "on the line starting from the beginning of the ear." Even in Parma and Reggio Emilia there is "sulla linea uscente dal principio de l'urecchia," which means "on the line starting from the beginning of the ear." It was appropriate to change the sentence to "on the line that originates at the top of the head."

330. The vernacular manuscripts report "in sedici parti uguali" and the Latin texts report "in partes 16 que sint aequales" which means the same thing. However, since the circuit cannot be divided into sixteen equal parts, we preferred translating the Latin term "aequales" as "proportionate," which is what all Latin dictionaries use.

331. Parma and Reggio Emilia talk about two lines and eighteen subdivisions. The Latin talks about four lines and twenty subdivisions. The circuits' reconstructions, as may be seen in figure B.112, prompt one to choose four lines and twenty subdivisions, which confirms the Latin version's exactness.

332. Figure B.113 corresponds to the first table of drawings which are present in the manuscript of Parma (folio 59 verso) except for the incongruences present in the drawings of this manuscript. The incongruences in this manuscript have been analyzed in the Introduction (under the heading "First Head"). With figure B.113, we want to show how Piero summarized with one drawing all the drawings present in all the figures that have been presented till now, from figure B.101 to figure B.112.

333. Lines I and P identify the same vertical line.

334. In Parma and Reggio Emilia the bigger numbers have been reported below line A, while the smaller ones have been reported above line A. It is different in Bordeaux, London, and Paris where the bigger numbers are above line A and the smaller ones are below. In the Milan manuscript they do not follow a precise order.

335. In Milan, Bordeaux, and Reggio Emilia there are deletions and corrections of numbers. Parma reports numbers 4′ and 20′; however, the numbers 3′ and 21′, 5′ and 19′ are missing; these are needed to identify

the points relative to the eyes. These are numbers that have to be added, starting now, in both the texts and the drawings of the figures where they need to appear

336. Remember that point 1′ is the point through which the tangent to the nose goes.

337. Remember that point 1′ corresponds to the upper lip and point 1″ corresponds to the lower lip.

338. The order in which the cited numbers have been reported does not correspond to the one in the texts of the manuscripts. In this particular case, the order present in the lines of the circuits was used, as can be seen in the reconstructed drawings. In Reggio Emilia there's number 16, which is missing in Milan. In Bordeaux and Parma number 15 is missing, while 16′ appears (with an apex). As may be seen, the missing numbers of the arc of the chin have been added, that is, 1′, 2′, 3′, and 15′. Considering that the subdivisions of circuit H, as we have seen, are and stay sixteen, we preferred adding numbers 1′, 2′, 3′, 16′, and 15′ for the arc of the chin, setting them onto the head with one eye only in its most appropriate shape and in compliance with the analyzed drawings examined in figures A.39 and A.46.

339. In the following, you will find the reversal of right with left with reference to axis MN. Basically, we are invited to observe the head without looking at it, in plan, from left to right, but also from right to left, which means from N toward the center M of the circuit.

340. This last sentence is missing in Milan, Bordeaux, and Reggio Emilia. It was taken from Parma.

341. This last sentence is missing in Milan, Bordeaux, and Reggio Emilia. It was taken from Parma.

342. Numbers 19′ and 21′ as well as numbers 3′ and 5′ are missing both in the vernacular and Latin manuscripts. All four of these numbers are naturally reported on the elevations of the two heads in the appropriate places.

343. This last sentence is missing in Milan, Bordeaux, and Reggio Emilia. It was taken from Parma.

344. Numbers 19 and 20 are missing both in the vernacular and Latin manuscripts.

345. Keep in mind that point 1′ is the tangent point of straight line P with the nose.

346. Point 1′ corresponds to the extremity of the upper lip; point 1″ corresponds to the extremity of the lower lip.

347. In the vernacular of Parma there are 4′, 14′, 15′, 16′. In the Latin of Bordeaux we have 1″, 4′, 14′. None of these numbers is on circuit G. Keep in mind that because of the arc of the chin, which is not mentioned nor described, there is a discrepancy between the numbers that belong to circuits G and H. The numbers with the apexes that belong to the arc of the chin should appear on circuit H. For the arc of the chin, also compare figure B.117 and figures A.39 and A.46.

348. These are the numbers as they appear in Parma: 1, 1′, 9, 9′, 2, 2′, 3, 3′, 4, 5, 6, 7, 8, 10, 11, 12, 13, 14, 15, 15′, 16, 16′. These are the numbers as they appear in Bordeaux: 1′, 1, 9, 2, 2′, 3, 3′, 4, 5, 6, 7, 8, 10, 11, 12, 13, 14, 15, 15′, 16, 16′. Compare the previous note.

349. The vernacular of Reggio Emilia is in agreement with the Latin. This sentence is missing in the vernacular of Parma. Figure B.122 corresponds to the second table of drawings in the manuscript of Parma (folio 61 recto), except for the incongruences in the drawings of this manuscript. These incongruences have been analyzed in the Introduction (under the heading "First Head"). With figure B.122 we want to demonstrate, in any case, how Piero wanted to summarize, in one figure, all the drawings presented from figure B.114 to figure B.121.

350. Reggio Emilia does not have a drawing in which the point of view is reported in plan or elevation.

351. Parma and Reggio Emilia report: "Make sure the ruler is really still and do the same for all other rulers." In Latin there is no such sentence.

352. Point 14* is between points 14 and 15. 14* is the contact point of the thread, that is, the tangent that goes from the eye in plan. This is the point of delimitation of the apparent contour of the circuit on the left. In the same manner, point 7* lies between points 7 and 6; point 7* is the contact point of the thread, that is, the tangent that goes from the eye in plan. This is the point of delimitation of the apparent contour of the circuit on the right. Piero speaks about this problem twice in detail: at the end of wooden ruler B which follows, and at the end of the text concerning this head. The search for the two points 14* and 7* has led us to expand the original circuits, which looked rather round and narrow. Regarding this, compare figures A.23 and A.24.

353. In Latin, this sentence and the two following are placed at the end of the paragraph. Keep in mind that Piero does not name the points of contact of the threads, that is, of the tangents led from the eye on the left and on the right of circuit B: the points are 14* and 6*.

354. At this point, in Parma there are these abnormal sentences: "Take ruler C and place it so that it touches line KL. Place the thread on point 16 of circuit C."

355. Piero likes to specify that these tangent numbers must be reported on the figure in elevation of the head with two eyes, as was done earlier. Regarding this, compare figures B.118–B.121. Successively Piero specifies that these same tangent numbers must be reported on the figure in elevation of the head with one eye only, on which the paper rulers will be placed. Therefore, all the tangent numbers that we have found, both on the head with two eyes and with one eye, have been reported.

356. We do not understand why these observations (concerning the points of contacts of the threads, that is, the tangents led from the eye on the right and on the left of the circuit) have been placed at the end of wooden ruler B and not at the end of wooden ruler A.

357. These last two sentences with 16 and 17 are not present in the Latin manuscripts; they were taken from Parma and Reggio Emilia.

358. Keep in mind that Piero does not mention the points of contact of the threads, that is, of the tangent lines led from the eye on the left and on the right of circuit C, which are points 17* and 9*. The following six numbers are also not mentioned, although they should be: 19′, 20′, 21′, 3′, 4′, 5′. Two more points y of the ears are not mentioned, although they belong to circuit D.

359. Points 19 and 20 (which do not appear in the Latin and vernacular manuscripts) have been added between points 18 and 1. Keep in mind that Piero does not mention the points of contact of the threads, that is, the tangent lines led from the eye on the left and on the right of circuit D: these points are 17* and 7*. Moreover, the two points z of the ears that also belong to circuit D are not mentioned.

360. Keep in mind that Piero does not mention the points of contact of the threads, that is, the tangent lines led from the eye on the left and on the right of circuit E. These two points in this case become only one point 16*, since the tangent on the right goes through point 7. Moreover, the two points x of the ears that also belong to circuit E are not mentioned.

361. Keep in mind that Piero does not mention the points of contact of the threads, that is, the tangent lines led from the eye on the left and on the right of circuit F. These two points in this case become only one point 16*, since the tangent on the right goes through point 6.

362. Keep in mind that Piero does not mention the points of contact of the threads, that is, the tangent lines led from the eye on the left and on the right of circuit G: these points are 15* and 12* on the left, while on the right there is only 4*, since the second tangent on the right goes through point 7. The numbers identified in the text are all the visible points, meaning the points that can be seen, as the graphic reconstructions demonstrate. Only point 5 on the right is an exception, and it must be reported, even if it is not visible, in order to indicate the continuity of the circuit between points 4* and 6. The points on the left, 15, 14, 13, 12, are an exception even if they are not visible; they identify the continuity of the circuit between numbers 15* and 12*.

363. Keep in mind that Piero does not mention the points of contact of the threads, that is, the tangent lines led from the eye on the left and on the right of circuit H. These points are 12* on the left and 7* on the right. Points 12 and 13 are missing on the left.

364. This sentence is missing in both the Latin and vernacular manuscripts. It was inserted because in Parma there are numbers 1′ and 2′, which certainly belong to the arc of the chin. Therefore it was necessary

to report all the numbers present on the arc of the chin; these are 1′, 2′, 3′, 15′, 16′.

365. The visible numbers mentioned in Parma and Bordeaux are: 14, 15, 16, 1, 2, 3, 4, 5, 6, 7. The numbers are the same in the Latin and vernacular texts; the only differences are that 14 becomes 14* and 7 becomes 7*. Naturally in figure B.124 and in figure B.132, all the numbers of circuit A (including the ones that are not visible) have been reported on wooden ruler A. The visible points on circuit A (compare figure B.139) are the following: 14*, 15, 16, 1, 2, 3, 4, 5, 6, 7*.

366. As can be seen in the reconstruction of circuit B (compare figure B.125), only the first eight numbers of circuit B coincide with the first eight numbers of circuit A (compare figure B.124). Because the point of contact of the thread, that is, the tangent led from the eye on the right of circuit B, lies between 5 and 6 (compare figure B.125), we are forced to take point 6* as the point of contact of the tangent. The visible points on circuit B (compare figure B.139) are the following: 14*, 15, 16, 1, 2, 3, 4, 5, 6*. This is because we have had to plan the path of circuit B again. Naturally in figure B.132, on wooden ruler B, there are all the numbers of circuit B, including the ones that are not visible.

367. The numbers of rulers C that are mentioned in Parma are: 18, 19, 20, 21, 22, 1, 2, 3, 4, 5, 6, 7, 8, 9, x. The numbers that are mentioned in Bordeaux are: 18, 19, 20, 20′, 21, 22, 1, 2, 3, 4′, 5, 6, 7, 8, 9. The numbers reported in the text are all visible, as is shown (compare figure B.139) in the examined graphic reconstructions. Please observe that although in the vernacular of Parma there's an x, we have preferred to use y because on the two paper rulers C, in the drawing about the perspective of the head of Parma, there are two y.

368. The numbers of ruler D mentioned in Parma are: 16, 17, 18, 1, 2, 3, 4, 5, 6, z. The numbers mentioned in Bordeaux are: 15, 16, 17, 18, 1, 2, 3, 4, 5, 6, 7, 8, x. The numbers reported in the text are all visible; this is demonstrated in the graphic reconstructions (compare figure B.139). Point z (naturally only the one on the right is visible) is called z because, apart from being called z in Parma, it is also called z in the drawing in perspective in Bordeaux both on wooden ruler D and on paper rulers D.

369. The numbers of ruler E which are named in Parma are: 17, 18, 19, 20, 1, 2, 3, 4, 5, 6, 7, 8, 9. The numbers mentioned in Bordeaux are: 16, 18, 20, 1′, 2, 3, 4, 5, 6, 7, 8, 9. The numbers reported in the text are all visible (i.e., they are visible to eyesight). This is demonstrated in the graphic reconstructions (compare figure B.139).

370. The numbers of ruler F which are named in Parma are: 16, 17, 18, 1, 1′, 1″, 2, 3, 4, 5, 6. The numbers mentioned in Bordeaux are: 15, 16, 17, 18, 1″, 2, 3, 4, 5, 6, 7, 8. The numbers reported in the text are all visible, as is demonstrated in the graphic reconstructions (compare figure B.139).

371. The numbers of ruler G which are named in Parma are: 15, 16, 18, 1, 1′, 1″, 2, 3, 4, 5, 6, 7. The numbers mentioned in Bordeaux are: 14, 15, 16, 1, 2, 3, 4, 5, 6, 7. The numbers reported in the text are all visible (i.e., they are visible to eyesight). This is demonstrated in the graphic reconstructions (compare figure B.139); there is only one exception, that is, point 5, which must be reported (even if it is not visible) in order to indicate the continuity of the circuit between points 4* and 6. On the left the exceptions are points 15, 14, 13, 12, and although they are not visible, they indicate the continuity between the numbers 15* and 12*.

372. The numbers of ruler H which are named in Parma are: 15, 16, 18, 1, 1′, 1″, 2, 3, 4, 5, 6, 7. They are the same numbers as in the previous note for ruler G. In reality, the vernacular of Parma directly unites the two circuits by saying: "On ruler G and on ruler H we have points: 15, 16, 18, 1, 1′, 1″, 2, 3, 4, 5, 6, 7." It is obvious that this is a serious anomaly. The numbers mentioned in Bordeaux are: 14, 15, 16, 1, 2, 3, 4, 5, 6, 7. The numbers reported in the text are all visible numbers (compare figure B.139), as is demonstrated in the graphic reconstructions. Naturally, the numbers with apexes refer to the numbers about the arc of the chin. Keep in mind that numbers 11, 10, 9, and 8 which lie on circuit H of figure B.139 have been made visible to delimit the apparent contour of figure B.140 as the route of the circuit seen from the bottom up.

373. All tangent points have been reported on the head with one eye only in figure B.135, in the way they have been identified in figures B.131 and B.132.

374. This passage justifies the principal optical axis, which ends above the chin in the drawings of the Latin manuscripts. Regarding this, compare figure A.45 and the related text in the Introduction.

375. This sentence is missing in the Latin and vernacular manuscripts. It has been inserted to add the tangent point 14*. Point 14* with an asterisk has been identified in plan in figure B.124.

376. This sentence is missing in the Latin and vernacular manuscripts. It has been inserted to add the tangent point 7*. Point 7* with has been identified in plan in figure B.124.

377. For all the following paper rulers, we are referring to figures B.135 and B.137.

378. This sentence is missing in the Latin and vernacular manuscripts. It has been inserted to add the tangent point 14*. Point 14* has been identified in plan in figure B.125.

379. This sentence is missing in the Latin and vernacular manuscripts. It has been inserted to add the tangent point 6*. Point 6* has been identified in plan in figure B.125.

380. For what concerns this paragraph, in Parma we have the following numbers: 4′ and 20′, 5 and 19, 6 and 18, 7 and 17, 8, 9, x, y; in Bordeaux we have 4′, 18′, 20′, 5, 17, 6, 18, 7, 8, 9. Concerning points x and y which appear in Parma, it is obvious one must take y (already chosen in the past). The points of contact of the tangent lines, led from the eye on the right and on the left of circuit C, have been added,: these are points 17* and 9*. The numbers reported in the text through the end of this paragraph are all visible, as is demonstrated in the graphic reconstructions; compare figure B.139.

381. The numbers mentioned in Parma and Bordeaux match, and they are the following: 1, 2, 18, 3, 17, 4, 16, 5, 15, 6, 7, 8. The numbers reported in the text are all visible, as is demonstrated in the graphic reconstructions; compare figure B.139.

382. The numbers mentioned in Parma are: 1, 1′, 2, 20, 3, 19, 4, 18, 5, 17, 6, 7, 8, 9. The numbers mentioned in Bordeaux are: 1′, 2, 20, 3, 19, 4, 18, 5, 17, 6, 7, 8, 9. The numbers reported in the text are all visible, as is demonstrated in the graphic reconstructions; compare figure B.139.

383. For what concerns Ruler F, the numbers mentioned in Parma are: 1, 1′, 1″, 2, 18, 3, 17, 4, 16, 5, 15, 6, 7, 8. The numbers mentioned in Bordeaux are: 1′, 1″, 2, 18, 3, 17, 4, 16, 5, 15, 6, 7, 8. The numbers reported in the text are all visible, as is demonstrated in the graphic reconstructions; compare figure B.139.

384. For what concerns Ruler G, the numbers mentioned in Parma are: 1, 2, 16, 3, 15, 3′, 15′, 4, 14, 5, 6, 7. The numbers mentioned in Bordeaux are: 1, 2, 16, 3, 15, 4, 14′, 14, 5, 6, 7. The numbers reported in the text are all visible, as is demonstrated in the graphic reconstructions; compare figure B.139. There is an exception on the right, that is, point 5, which we have to report (even it is not visible) to indicate the continuity of the circuit between points 4* and 6; the exceptions on the left are points 15, 14, 13, 12; even if they are not visible, they are reported in order to indicate the continuity of the circuit between numbers 15* and 12*.

385. For what concerns Ruler H, the numbers mentioned in Parma are: 1, 2, 16, 3, 15, 1′, 2, 16′, 3, 15′, 4, 14, 5, 6, 7. The numbers mentioned in Bordeaux are: 1, 1′, 2, 2′, 16, 16′, 3, 3′, 15, 15′, 14, 5, 6, 7. The numbers reported in the text are all visible, as is demonstrated in the graphic reconstructions. Keep in mind that the numbers 11, 10, 9, 8 present on circuit H of figure B.139 have been made visible to delimit the apparent contour of figure B.140 as the route of the circuit seen from the bottom up.

386. The vernacular manuscripts repeat that these last two paper rulers—like all the other paper rulers—must report the indications of all the marks, that is, of all the numbers and distances.

387. Remember that Piero only reports the visible points of the circuits, that is, all the points that are essential to the reconstruction of the visible part of the circuits. In figure B.138 we have the points of tangency 14* and 7*, as can be seen in the reconstruction of the circuit. For this reason, points 14 and 7 are not visible. However, these last two points have been

marked on figure B.138 just to appropriately be compared with points 14* and 7*.

388. For all rulers B/B/B, C/C/C, D/D/D, E/E/E, F/F/F, G/G/G, H/H/H which follow, we will refer to figure B.139 where all the visible points have been reported that these rulers identify on the terminal (i.e., the picture plane). Figure B.141 offers the whole head's representation with the totality of all the paper and wooden rulers. I have tried to copy this drawing, as far as was possible, in compliance with the drawing on folio 66 verso of the Parma manuscript.

389. Remember that Piero only reports the visible points of the circuits, that is, all the points that are essential to the reconstruction of the visible part of the circuits. In figure B.139, between the points identified on the terminal by rulers B/B/B, there are tangent points 14* and 6*, as they are shown in the reconstruction of the circuit. For this reason points 14, 7, and 6 are not visible.

390. The Latin version reports 20 as well as 18, which is anomalous.

391. The Latin version reports 4 and 18′; the vernacular reports 4 and 20′: the numbers should be 4′and 20′.

392. The vernacular correctly reports the numbers. The Latin version states: "where point 4 lies, mark point 4′; where point 18-20′ lies, mark point 18-20." These numbers are incorrect.

393. The vernacular correctly reports the numbers. The Latin adds 17 between 5 and 19; this is not right.

394. The vernacular correctly reports the numbers. The Latin reports 7-15 first and then 17-10; both these numbers are incorrect.

395. This sentence is missing in Bordeaux. It is taken from the vernacular of Parma.

396. When rereading this paragraph and carefully observing figure B.139, one notices that the numbers 19′, 21′, 3′, and 5′ appear on ruler C. There is no reference to these numbers in the Latin or vernacular versions. At the same time, one notices that there should be two y, one on the right and one on the left; the latter appears below the numbers of ruler D. These two y too are never mentioned in the Latin or vernacular versions.

397. For circuit H, we prefer to report points 11, 10, 9, 8 too, because this is a bottom-up perspective and they are visible as well, unless one wants to cover (as Piero did) by tracing an imaginary contour of the chest.

398. This sentence is missing in the Latin and in Reggio Emilia; it is taken from Parma.

399. Remember that Piero only reports the visible points of the circuits, those points that are essential to reconstruct the visible part of the circuits. The numbers that appear in Parma are correct with the only exceptions of points 14 and 7. We must add an asterisk to these points, making them 14* and 7*, because they are tangent points, as may be seen in the reconstruction of the circuit. The numbers that appear in Bordeaux are 14/15, 15/16, 1, 1/2, 2/3, 3/4, 4/5, 5/6, 6/6.

400. For what concerns the description of contour B, the numbers that appear in Parma are 4/15, 15/16, 16/1, 1/2, 2/3, 3/4, 4/5, 5/6, 6/7. The numbers that appear in Bordeaux are 14/15, 15/16, 1, 1/2, 2/3, 3/4, 4/5, 5/6, 6/7. The numbers reported are those that originate from the circuit's reconstruction.

401. The numbers that appear in Parma are: 17/18, 18/19, 19/20, 20/21, 21/22, 19/20′, 20′/21, 22/1, 1/2, 2/3, 3/4, 4/5, 3/4′, 4′/5, 5/6, 6/7, 7/8, 8/9. An asterisk was added to numbers 17 and 9, thus they have become 17* and 9*, since they are tangent points, as the reconstruction of the circuit demonstrates. Parma is missing the following numbers: 19′, 21′, 3′, and 5′. We preferred to join 19 and 20′, 20′ and 21, 3 and 4′, 4′ and 5 with some lines with the aim of indicating the positions of the eyes, which will be drawn later on. The numbers that appear in Bordeaux are: 16/17, 17/18, 20, 18, 20/19, 21, 18, 20′/22, 22/1, 1/2, 2/3, 3/4, 4′, 4/5, 5/6, 6/7, 7/8, 8/9; missing are 19′, 21′, 3′, and 5′.

402. The numbers that appear in Parma are: 15/16, 16/17, 17/18, 18/1, 1/2, 2/3, 3/4, 4/5, 5/6, 6/7, 7/8; numbers 18/1 are incorrect; the following are missing: 18/19, 19/20, 20/1. The numbers that appear in Bordeaux are 15/16, 16/17, 17/18, 18/1, 1/2, 2/3′, 3/4, 4/5, 5/6, 6/7, 7/8; numbers 18/1 are incorrect; missing are 18/19, 19/20, 20/1; 2/3 is reported as 2/3′. Remember that the numbers reported in the text are the ones that originate in the reconstruction of the circuit.

403. The numbers that appear in Parma are: 16/17, 17/18, 18/19, 19/20, 20/1, 1/1′, 1/2, 2/3, 3/4, 4/5, 5/6, 6/7, 7/8, 8/9. The numbers that appear in Bordeaux are: 16/17, 17/18, 18/19, 19/20, 20/1′, 1/2′, 2/3, 3/4, 4/5, 5/6, 6/7, 7/8, 8/9; numbers 20/1′, 1/2′ are incorrect. Remember that the numbers reported on the text are the ones that originate in the reconstruction of the circuit.

404. This sentence is missing in Parma; it is present in Reggio Emilia and in Bordeaux.

405. This statement is missing in Parma; it is present in Reggio Emilia and in Bordeaux. It implies the construction of the middle vertical section of the head. This section is visible in all perspectives of the original drawings. Concerning this, compare figure B.141 in compliance with the original. It was possible to identify twelve units in which this section is subdivided, thus finding the point of tangency S. This point has been identified in figure B.139 and figure B.140; it also appears in the figure compliant with the original, figure B.141. Point S can be found by drawing in elevation (figure B.135) the tangent from the eye to the elevation of the head with one eye only; then finding again, in elevation, the height QS on the paper ruler and finding again, in plan, the width MS on the wooden ruler; then reporting the position of point S in perspective (figure B.139). Only by finding this point S in perspective can we identify the twelve units and draw the middle vertical section of the head, as may be seen in figure B.139. Numbers from 1 to 12 that identify the twelve units in figure B.139 are shown in bold italic.

406. This paragraph is missing in Parma but is present in Reggio Emilia and Bordeaux.

407. Reggio Emilia reports 17, 20, 21 as the upper numbers (number 17 is anomalous), and 17, 20′, 21 are reported as the lower numbers (where 20′ is the only correct number). Bordeaux reports 17, 18, 19 as the upper numbers, where 17 and 18 are wrong; 17, 18′, 19 are reported as the lower numbers, and all three numbers are wrong.

408. Reggio Emilia correctly reports 3, 4, 5 as the upper numbers and 3′, 4′ as the lower numbers, but it adds a 5 without an apex. Bordeaux reports 3 and 5 as the upper numbers but it forgets number 4; as the lower numbers it reports 4′, but it adds 3 and 5 without the apexes.

409. In Reggio Emilia, all numbers are correct and they correspond to the ones reported in the text. In Bordeaux, we have 19, 20, 20′, 1, 2′, 3, where only 19, 20, 1, and 3 are right and 1′ and 2 are missing.

410. In Reggio Emilia, all numbers are correct and correspond to the ones reported in the text. In Bordeaux, number 1 is missing.

411. These are the outlines of the chin and the ears and of the outlines that delimit the head externally, as may be observed in figure B.140.

412. This is a very important observation; it was developed in the Introduction, in the discussion of the remade circuit A. We urge you to carefully read this paragraph in the Introduction and to observe the drawings in figures A.23 and A.24. This point of contact has been identified with 14*; an asterisk has been added so as not to confuse it with 14 which is right next to it. All other tangency points have been identified using an asterisk. The appropriate search for these tangent points has led us to plan circuit A again. Regarding this, also compare figure B.124.

413. The second specification "and the same for the head that has one eye only" does not appear in the Latin manuscripts.

414. Reggio Emilia correctly uses "between line I," while Parma and Bordeaux have "between 1."

415. It is natural that Piero reminds us to move this new point 14* on the face with one eye only (figure B.133), if we wish to find the image of this new point. The search for this tangent point 14* has led us to plan circuit A again, as has been emphasized in the Introduction, in figures A.23 and A.24. Also compare figure B.124.

416. Replicate what has been done for point 14* for point 7*: this is the point of contact generated by the second tangent to circuit A. Place this new point 7* on the face with one eye only (figure B.131) if you wish to find the image of this new point. The search for this tangent point has led

us to plan circuit A again, as has been emphasized in the Introduction, in figures A.23 and A.24. Also compare figure B.124.

417. In figure B.131, points 14* and 7* are visible and are correctly reported on line A of the entire face.

418. Let's take a look. Tangent points 14* and 7* of circuit A have been reported on the head with one eye only and on the head that shows the entire face, following the passages that have been shown and indicated. This is what has been realized till now by reconstructing all the circuits, through the careful search for all points of tangency, just as Piero suggests. However, for what concerns circuit A, we did not limit ourselves to rebuilding it: instead we provided all necessary comparisons, as may be read and observed in the Introduction. Figure A.37 corresponds to the third table of the drawings present in the Parma manuscript (folio 63 verso), except for the incongruences present in the drawings of the manuscript. Figure A.38 corresponds to the fourth table of the drawings present in the Parma manuscript (folio 65 recto), except for the incongruences present in the drawings of the manuscript. Figure B.141 corresponds to the fifth table of the drawings present in the Parma manuscript, folio 66 verso.

419. These are the three conditions that lie at the base of the representation of the second head.

420. As may be observed in figure B.142b, lines E, F, G, H have been appropriately bent in their rear part, in compliance with the drawings from the Latin and vernacular manuscripts. Line IP naturally goes through point 1, which borders the base of the nose and is tangent to the chin. Piero rightly bends the rear part of the neck according to its natural curvature which is generated by the rotation. The numbers on the curved parts of the sections have been reported in compliance with the numbers in the manuscripts of Parma and Bordeaux.

421. In the two vernacular manuscripts there is an added specification: "legermente collo piombo" which means the lines are drawn "lightly with lead graphite."

422. This specification does not appear in the Latin manuscripts; it is taken from Parma. The vernacular reports TV. We prefer to use PV to avoid confusing it with another line which will later be identified as line TV. With line PV traced, the construction of the rotated head has been completed.

423. For what concerns left and right, keep in mind that Piero is looking at the head from the back and not from the front.

424. Milan reports 1 instead of TV, as does Reggio Emilia. Bordeaux and Parma report TV.

425. Points 9 and 1 must be on vertical line MN.

426. The adjective "curved" has been added to better specify.

427. The vernacular of Parma adds: "from TV to 9"; however, the Latin, in which this specification does not appear, is correct, because 9 should be on vertical line MN where points 9 and 1 match.

428. Bordeaux reports 12 correctly changed to 16.

429. The Latin says "C has numbers from 1′ to 22." The Latin reports 1′ (with the apex), but this number should not have an apex. The Latin also fails to identify the two numbers y as well as 19′, 20′, 21′, 3′, 4′, 5′. The vernacular of Parma reports the following numbers: 1, 2, 3, 4, 4′, 5, 6, 7, 8, 9, x, 10, 11, 12, 13, 14, 15, y, 16, 17, 18, 19, 20, 20′, 21, 22. As may be observed, in Parma the numbers 19′, 21′, 3′, 5′ are missing, while there is an x which should be on circuit E.

430. Both Latin and vernacular texts report "from 1 to 18," while it is clear that it should be "from 1 to 20."

431. This sentence is taken from the vernacular of Parma. The Latin instead has: "The last three circuits have the same numbers. Any one of them ends with number 16 and opens with number 1." It is obvious that this sentence is anomalous and does not correspond to reality. Only circuits G and H end with number 16; circuit F continues with 17 and 18. Moreover, circuit F has two numbers with apexes and circuit H has five: the latter identify the chin.

432. This sentence is taken from the vernacular of Parma, which is missing 15 and has an extra 3′ and 15′.

433. This sentence is taken from the vernacular of Parma, which is missing 16, 3′, 15′.

434. In figure B.146 the example of number 2 is highlighted with the compass. The rotation of the head with a full face has been highlighted in this figure. This rotation happens around number 1 of line E, that is, around number 1 which limits the base of the nose.

435. The Latin and vernacular manuscripts only report the word "line." However, to be more specific the adjective "curved" should be added to the word "line," which would indicate circuit A on the elevation of the head with two eyes, that is, the elevation of the head "that shows the whole face."

436. The specification "that shows the whole face" is missing in the vernacular.

437. This sentence is missing in Milan.

438. This sentence is missing in Parma.

439. This sentence is missing in Milan; it is present in Bordeaux and Parma. Bordeaux and Parma report x, but we have chosen to identify this point with y, as established for the first head.

440. As may be observed in the reconstructed drawings, numbers 19′, 21′, 3′, 5′ are missing both in Parma and Bordeaux. These numbers delimit the lateral extremities of the eyes.

441. As specified in the Introduction (under the heading "Second Head"), the vernacular manuscripts of Parma and Reggio Emilia and the Latin manuscript of Milan are missing the new representations in plan of the eight circuits. These new representation in plan of the eight circuits are present in the Latin manuscripts of Bordeaux, London, and Paris, even if with all the incongruences described in the Introduction.

442. As may be observed in figure B.148, point 11 must also be present; this point is missing both in the Latin and in the vernacular manuscripts.

443. As may be observed in the reconstruction of the drawings, points 19 and 20 are missing here; the marking of points should continue until point 20, not point 18. Also keep in mind that neither the Latin nor the vernacular manuscripts contain any hint of the z points which should be found on the two eyes.

444. As mentioned in the previous note, points 19 and 20 are missing, therefore 18 with 1 should be eliminated and 18 with 19, 19 with 20, 20 with 1 should be added.

445. The Latin manuscripts wrongly report "2 with 2" instead of "1 with 2."

446. As may be observed in figure B.149, point 10 must also be present; this point is missing both in the Latin and in the vernacular manuscripts.

447. As may be observed in the detail of figure B.149, this sentence does not correspond to reality. Point 1″ cannot be joined with 2. Moreover, since circuit F does not end with 16 but with 18, 16 with 1 should be eliminated, and 16 with 17, 17 with 18, 18 with 1 should be added.

448. This is referring to a real sheet of paper that should be placed on top of the drawing of the other circuits. It is necessary to indicate this sheet of paper with the letter F. Compare figure B.149.

449. This is referring to a real sheet of paper that should be placed on top of the drawing of the other circuits. It is necessary to indicate this sheet of paper with the letter G. Compare figure B.149.

450. In the Latin manuscripts, 4′ appears after the number 4, and after 14 we have 14′. The vernacular says that the following points must be joined: 1 and 2, 2 and 3, 3′ and 4, 4 and 5, 5 and 6, 6 and 7, 7 and 8, 8 and 9, 9 and 10, 10 and 11, 11 and 12, 12 and 13, 13 and 14, 14 and 15, 15 and 16, 16 and 1. Because circuit G does not present numbers with apexes, as may be seen in the reconstructed drawings, it is anomalous to have points 4′ and 14′ in the Latin and point 3′ in the vernacular.

451. This sentence is missing in the vernacular manuscripts.

452. The vernacular text instructs to join these numbers: 1 and 2, 2 and 3, 3 and 3′, 1′ and 2′ and 3″, 3 and 4, 4 and 5, 5 and 6, 6 and 7, 7 and 8, 8 and 9, 9 and 10, 10 and 11, 11 and 12, 12 and 13, 13 and 14, 14 and 15, 15 and 16, 16 and 1. The Latin text instructs to join these numbers: 1 and 1′, 2 and 2′, 2 and 3′, 3 and 4, 4 and 5, 5 and 6, 6 and 7, 7 and 8, 8 and 9, 9 and 10, 10 and 11, 11 and 12, 12 and 13, 13 and 14, 14 and 15, 15 and 15′, 16 and 16′. According to the analysis of the remade drawings, we proceeded with the proposal as written in the text.

453. This is an invitation to start with the wooden rulers.

454. The following points are necessary to ensure a correct analysis of figure B.150: (1) an arrow shows the rotation that the new circuits' plans make with respect to the terminal, that is, with respect to the chosen picture plane; (2) straight line MM, which is orthogonal to KP in plan, is the principal optical axis on which the eye in plan is chosen in figure B.151; (3) the inclined straight line, identified by the eight lowercase letters m present in plan and elevation, represents the axis that goes through the centers of all the circuits.

455. Let's begin from the last circuit H on the right, looking at figure B.152. Remember that all the circuits of this figure should overlap one another, as can be seen in the plans of figures B.150 and B.151.

456. It is obvious that Piero reports only the numbers he deems visible, otherwise he would have had to list all of them.

457. The numbers listed are in the Latin of Bordeaux, where letters x and y are present: we use y, as it was established for the first head. The references to numbers 19′, 21′, 3′, 5′ are missing: these numbers delimit the lateral extremities of the eyes. Parma reports the following numbers: 17, 18, 19, 20, 20′, 21, 22, 1, 2, 3, 4, 4′, 5, 6, 7, 8, x.

458. Both Parma and Bordeaux have no hint of letters z which are on circuit D.

459. The listed numbers are the ones present in Parma, where the letter y appears; this corresponds to letter x, as it was established for the first head. Bordeaux reports the following numbers: 16, 17, 18, 19, 20, 1, 1′, 2, 3, 4, 5, 6, 7, 8, 8′, 9.

460. The listed numbers are present in Bordeaux. Parma reports the following numbers: 16, 17, 18, 1, 1′, 1″, 2, 4, 5, 6.

461. The listed numbers 14′ and 4′ are present in Bordeaux but have been taken out because they cannot belong to circuit G. Parma reports numbers 1, 2, 3, 3′, 4, 5, 6, 7.

462. The listed numbers are present in Bordeaux; 14′ and 4′ have been taken out because they are anomalous. Parma reports the following numbers: 1, 2, 1′, 2′, 3, 4, 5, 6, 7.

463. The heights of the circuits must be found, that is, the heights of the numbers of all the circuits on the vertical plane that goes through TV and which is orthogonal to straight line KL, as may be seen in figure B.154. Basically, the new elevation of all the circuits will have to be drawn again; this is the new auxiliary elevation of the head with one eye only. Successively, we will have to choose the eye and set the paper rulers on this elevation.

464. This is referring to the head with one eye only and the head represented in plan with all the circuits, as may be seen in figure B.150.

465. Bordeaux and Milan incorrectly report 15.

466. The first step to report the height of the points in the new elevation has been completed. Notice that Parma, Bordeaux, and Milan only report numbers 14, 15, 1, 6, 7, while in the reconstructions we have realized, all the numbers of the circuits have been reported. This will be done for all circuits.

467. In figure B.146, the rotation of the head with the whole face has been emphasized. This entailed the displacement of the head with two eyes (see figure B.150) from the position of the first Wp (drawn with a dotted line) to the second Wp, which was moved up. The first Wp, in fact, "detaches from the back of the head with one eye only," and when it touches the back of the head it "divides mN in right angles," as Piero specifies in the following statement.

468. At this point, the Latin text of Milan goes from folio 95 verso, line 19, to folio 96 verso, line 1, to then end with folio 97 recto, line 16. In the same manner, the Latin text of Bordeaux goes from folio 88 recto, line 6, to folio 89 recto, line 1, to end with folio 89 verso, line 4.

469. In the following sentences, we have preferred to merge the meanings of these three Latin verbs: *videbis-mensurabis-capies*, which are in the future tense and respectively mean "you will see," "you will measure," and "you will take." They have been simply translated "take the distance" with the aim of letting the grammar flow.

470. The Latin of this sentence is missing in Milan, but it is present in Bordeaux.

471. The Latin of this sentence is missing in Milan, but it is present in Bordeaux.

472. The Latin of this sentence is missing in Milan, but it is present in Bordeaux.

473. This Latin sentence "Take the distance from RS to point 8′ of the ear: place it on line XV and mark 8″" is missing in Parma. Because 8′ is a point that does not appear on circuit C, this sentence is incorrect.

474. This sentence is missing in Milan and Bordeaux; it is taken from Parma. In Parma we have x instead of y; this is the point on the ear, which was indicated with y in the first head.

475. This sentence is missing in Parma.

476. The Latin of this sentence is missing in Milan, but it is present in Bordeaux.

477. The Latin of this sentence is missing in Milan, but it is present in Bordeaux.

478. This sentence and the following one are missing in Milan but are present in Bordeaux.

479. The sentence "Take the distance between line Wp and point 8′ of circuit C and place this distance on the line originating from point 8′ of line XV: mark 8″" appears here in the Latin manuscripts but is missing in Parma. Because 8′ is a point that does not appear on circuit C, this sentence is incorrect.

480. This sentence is missing in Milan and Bordeaux; it is taken from Parma. In Parma we have x instead of y; this is the point on the ear, which was indicated with y in the first head. As may be observed, the numbers 5′ and 19′ are missing. In order to correctly reconstruct the drawings, all the numbers present on circuit C have been reported in figure B.156, including numbers 9, 10, 11, 12, 13, 14, 15.

481. At this point, the Latin text of Milan goes from folio 97 recto, line 16, to folio 98 recto, line 1, to end with folio 98 recto, line 28. In the same manner, the Latin text of Bordeaux goes from folio 89 verso, line 4, to folio 90 verso, line 1, to end with folio 90 verso, line 22.

482. "Now take another paper on which you have executed the representation of three circuits: D, E, F. This paper must be placed so that it touches mN and in a way so that all the marks you made coincide. Having done this, a certain paper must be brought onto line XV so that it touches X and V. Moreover, line TV must be extended and perpendicular Y must be drawn on TV which intersects TV on point X: you will have straight line XY." These four sentences appear in Latin; with these Piero prompts us to separate the circuits to avoid mixing up the lines.

483. The corresponding Latin text of this sentence is missing in Milan but is present in Bordeaux.

484. This sentence is missing in Parma.

485. This sentence is missing in Milan and Bordeaux; it is taken from Parma. Keep in mind that in Parma, we have x instead of z; this is the point on the ear, which in the first head was indicated with z.

486. As may be observed, points 16 and 17 are missing; in the same manner, points 19 and 20, which lie between 18 and 1, are missing, as are points 3, 4, and 5. After a correct reconstruction of the drawings, all the numbers present on circuit D have been reported, including numbers 8, 9, 10, 11, 12 13, 14.

487. At this point this sentence appears in Latin: "Go back to circuit E of the width." It is not useful to include it in the text as a matter of consistency. At this point, the Latin of Milan goes from folio 98 recto, line 28, to folio 99 recto, line 1, to end with folio 99 recto, line 33. The Latin text of Bordeaux goes from folio 90 verso, line 22, to folio 91 verso, line 1, to end with folio 91 verso, line 27.

488. The Latin corresponding text of this sentence is missing in Milan but is present in Bordeaux.

489. This sentence is missing in Milan and Bordeaux; it is taken from Parma. Keep in mind that in Parma, we have y instead of x; this is the point on the ear, which in the first head has been indicated with x.

490. As may be observed, numbers 17 and 18 are missing, as are 4, 5 and 6. According to the correct reconstruction of the drawings, all the numbers on circuit E have been reported, including numbers 10, 11, 12 13, 14, 15.

491. This sentence is missing in Milan and Bordeaux; it is taken from Parma. Keep in mind that in Parma, we have y instead of x; this is the point on the ear, which in the first head has been indicated with x.

492. As may be observed, numbers 17 and 18 are missing and so are 4, 5, and 6. According to the correct reconstruction of the drawings, all numbers present on circuit E have been reported, including numbers 10, 11, 12 13, 14, 15. At this point, this sentence appears in Latin: "Go back to circuits of the width." It is not appropriate to include it in the text for reasons of consistency. At this point, the Latin text of Milan goes from folio 99 recto, line 33, to folio 100 recto, line 1, to end with folio 100 recto, line 32. The Latin text of Bordeaux goes from folio 91 verso, line 27, to folio 92 verso, line 1, to end with folio 92 verso, line 25.

493. This sentence is missing in Bordeaux.

494. This sentence is missing in Milan and Parma but is present in Bordeaux.

495. As may be observed, numbers 17 and 18, which lie between 16 and 1, are missing; so are numbers 3 and 4 which lie between 2 and 5. According to the correct reconstruction of the drawings, all numbers present on circuit F have been reported, including numbers 8, 9, 10, 11, 12, 13. At this point, the Latin text of Milan goes from folio 100 recto, line 32, to folio 101 recto, line 1, to end with folio 101 recto, line 24. The Latin text of Bordeaux goes from folio 92 verso, line 25, to folio 93 verso, line 1, to end with folio 93 verso, line 20.

496. As may be observed, numbers 15, 17, 18, 3, and 4 are missing. According to the correct reconstruction of the drawings, all numbers present on circuit F have been reported, including numbers 8, 9, 10, 11, 12, 13. At this point, the Latin text of Milan goes from folio 100 recto, line 32, to folio 101 recto, line 1, to end with folio 101 recto, line 24. The Latin text of Bordeaux goes from folio 92 verso, line 25, to folio 93 verso, line 1, to end with folio 93 verso, line 20.

497. At this point, we have the two following sentences in the Latin manuscripts: "After this, take that paper that contains the representation of the two circuits G, H and place it so that it touches mN and that q touches line IP [at P]. For the above-mentioned paper extend line RS." We have preferred to translate from the vernacular of Parma, which states: "Then for what concerns circuit G . . .".

498. The corresponding Latin text of this sentence is missing in Milan, but it is present in Bordeaux. Keep in mind that in Parma, we have this sentence: "Take the distance from RS to 15 and 15′ of circuit G; place it on XV and mark 15 and 15′." These two specifications are incorrect, because there's no number 15′ on circuit G.

499. The corresponding Latin text of this sentence is missing in Milan, but it is present in Bordeaux. Soon after this sentence in Parma, we have the following statement: "Take the distance from RS to 3′ of circuit G; place it on XV and mark 3′." This sentence is incorrect, because there's no number 3′ on circuit G.

500. At this point in both the Latin manuscripts of Milan and Bordeaux as well as the vernacular of Parma, we read the following: "Take the distance from RS to 4′ of circuit G; place it on XV and mark 4′." This sentence is incorrect, because there's no number 4′ on circuit G.

501. The corresponding Latin text of this sentence is missing in Milan, but it is present in Bordeaux.

502. Keeping in mind the considerations in the previous five notes, we can state that no other numbers are missing. According to the correct reconstruction of the drawings, all numbers on circuit G have been reported.

503. At this point in Parma, we have the following sentence: "Take the distance between line Wp and point 14′ of circuit G and place this distance on the line originating at point 14′ of line XV: mark 14′." This sentence is incorrect, because there's no number 14′ on circuit G.

504. The corresponding Latin text of this sentence is missing in Milan, but it is present in Bordeaux. At this point in Parma, we have the following sentence: "Take the distance between line Wp and point 15′ of circuit G and place this distance on the line originating from point 15′ of line XV: mark 15′." This sentence is incorrect, because there's no number 15′ on circuit G.

505. The corresponding Latin text of this and the following sentence is missing in Milan, but it is present in Bordeaux.

506. At this point in Parma, we have the following sentence: "Take the distance between line Wp and point 3′ of circuit G and place this distance on the line originating from point 3′ of line XV: mark 3′." This sentence is incorrect, because there's no number 3′ on circuit G.

507. At this point in Parma, Milan, and Bordeaux, we have the following sentence: "Take the distance between line Wp and point 4′ of circuit G and place this distance on the line originating from point 4′ of line XV: mark 4′." This sentence is incorrect, because there's no number 4′ on circuit G.

508. This sentence is missing in Parma, Milan, and Bordeaux; it was taken from Parma.

509. According to the correct reconstruction of the drawings, all numbers present on circuit G have been reported.

510. At this point in the Latin, we have the following anomalous sentence: "Go back to the circuits of the width." At this point, the Latin text of Milan goes from folio 101 recto, line 24, to folio 102 recto, line 1, to end with folio 102 verso, line 14. The Latin text of Bordeaux goes from folio 93 verso, line 20, to folio 94 verso, line 1, to end with folio 95 recto, line 4.

511. This sentence is missing in Parma.

512. This sentence is missing in Milan and Bordeaux; it has been taken from Parma.

513. Keeping in mind the considerations in the previous two notes, we can state that no other numbers are missing. According to the correct reconstruction of the drawings, all numbers present on circuit H have been reported.

514. This sentence is missing in Parma.

515. All the mentioned numbers 15, 15′, 16, 16′, 1, 1′, 2, 2′, 3, 3′ do not follow the order in the vernacular or the Latin manuscripts, where they appear in a disorderly fashion.

516. According to the correct reconstruction of the drawings, all numbers on circuit H have been reported. In figure B.162, all points have been reported for all circuits. Keep in mind that Piero does not draw the route of the circuits, he only marks the points. I have tried to rebuild the apparent contour of the new elevation in figure B.163.

517. At this point in Parma, we find a duplicate of the preceding paragraph (the text for the complete "Second Step" of the "Heights of Points of Circuit H"). At this point, the Latin text of Milan goes from folio 102 verso, line 14, to folio 95 verso, line 19, to end with folio 96 recto, line 8. The Latin text of Bordeaux goes from folio 95 recto, line 4, to folio 88 recto, line 6, to end with folio 88 recto, line 25.

518. This sentence is missing in Milan and Bordeaux; it is taken from Parma.

519. This sentence and the following three are missing in Milan and Bordeaux; they are taken from Parma.

520. All numbers of circuit A have been reported on paper ruler A in figures B.164 and B.168.

521. This sentence is missing in Milan and Bordeaux; it is taken from Parma.

522. This sentence and the following three are missing in Milan and Bordeaux; they are taken from Parma.

523. All numbers of circuit B have been reported on paper ruler B in figure B.168.

524. At this point, the Latin text of Milan goes from folio 96 recto, line 8, to folio 97 recto, line 16, to end with folio 97 recto, line 36. The Latin text of Bordeaux goes from folio 88 recto, line 25, to folio 89 verso, line 4, to end with folio 89 verso, line 19.

525. At this point, in the Latin texts of Milan and Bordeaux, we have the following sentence: "Place the thread on point 8′ of line C: mark 8′ where it intersects the ruler." This sentence is incorrect, because there is no number 8′ on circuit C.

526. This sentence is missing in Milan, but it is present in Bordeaux.

527. The reference regarding at least one number y of the ears is missing in both the vernacular of Parma and the Latin texts of Milan and Bordeaux. All numbers regarding circuit C on paper ruler C have been reported in figure B.168.

528. At this point, the Latin text of Milan goes from folio 97 recto, line 36, to folio 98 recto, line 28, to end with folio 98 verso, line 3. The Latin text of Bordeaux goes from folio 89 verso, line 19, to folio 90 verso, line 22, to end with folio 90 verso, line 31.

529. This sentence is missing in Parma.

530. This sentence is missing in Parma.

531. The reference regarding at least one number z of the ears is missing in both the vernacular of Parma and the Latin texts of Milan and Bordeaux. All numbers regarding circuit D on paper ruler D have been reported in figures B.165 and B.168.

532. At this point, the Latin text of Milan goes from folio 98 verso, line 3, to folio 99 recto, line 33, to end with folio 99 verso, line 8. The Latin text of Bordeaux goes from folio 90 verso, line 31, to folio 91 verso, line 27, to end with folio 91 verso, line 36.

533. This sentence, with the exception of the phrase "This being the situation," is missing in Milan and Bordeaux; it was taken from Parma.

534. From this point to the end of the paragraph, Parma instructs to place the thread on numbers 16, 17, 18, 19, 20, 1, 1′, 2, 3, 4, 5, 6, 7, y, 8, and to simply mark ruler E with the points where the thread intersects the ruler: 16, 17, 18, 19, 20, 1, 1′, 2, 3, 6, 7, y, 8. Notice that 4, 5, and 9 are missing in the second list of numbers (but they do appear in the Latin manuscripts), and that y appears instead of x, which is not compliant with what was established for the first head.

535. The corresponding Latin text for this and the following sentence is missing in Milan, but it is present in Bordeaux.

536. This sentence is missing in Milan and Bordeaux: it was reconstructed utilizing number 1′ present in Parma.

537. The corresponding Latin text of this sentence is missing in Milan, but it is present in Bordeaux.

538. The corresponding Latin text of this sentence and the two following is missing in Milan, but it is present in Bordeaux.

539. This sentence is missing in Milan and Bordeaux: it was reconstructed utilizing the original letter y, present in Parma, which is indicated with x, as it was established for the first head.

540. This sentence is missing in Parma.

541. In figures B.165 and B.168 all the numbers related to circuit E on paper ruler E have been reported. At this point, the Latin text of Milan goes from folio 99 verso, line 8, to folio 100 recto, line 32, to end with folio 100 verso, line 10. The Latin text of Bordeaux goes from folio 91 verso, line 36, to folio 92 verso, line 25, to end with folio 92 verso, line 35. Also note: in Parma, this sentence is missing, as well as the following text regarding paper rulers F, G, H; in Reggio Emilia, this sentence is missing, as well as the following text regarding paper rulers F and G.

542. The corresponding Latin text of this sentence is missing in Milan, but it is present in Bordeaux.

543. The corresponding Latin text of this sentence is missing in Milan, but it is present in Bordeaux.

544. The corresponding Latin text of this and the following sentence is missing in Milan, but it is present in Bordeaux.

545. In figures B.166 and B.168 all the numbers related to circuit F on paper ruler F have been reported.

546. At this point, the Latin text of Milan goes from folio 100 verso, line 10, to folio 101 recto, line 24, to end with folio 101 recto, line 36. The Latin text of Bordeaux goes from folio 92 verso, line 35, to folio 93 verso, line 20, to end with folio 93 verso, line 31.

547. The corresponding Latin text of this and the following sentence is missing in Milan, but it is present in Bordeaux.

548. The corresponding Latin text of this sentence is missing in Milan, but it is present in Bordeaux.

549. At this point in Milan and Bordeaux, we have this sentence: "Place the thread on point 4′ of line G: mark 4′ where it intersects the ruler." This sentence is incorrect because there is no number 4′ on circuit G.

550. In figures B.167 and B.168 all the numbers related to circuit G on paper ruler G have been reported.

551. At this point, the Latin text of Milan goes from folio 101 recto, line 36, to folio 102 verso, line 14, to end with folio 102 verso, line 34. The Latin text of Bordeaux goes from folio 93 verso, line 31, to folio 95 recto, line 4, to end with folio 95 recto, line 18. From this point onward, the vernacular of Reggio Emilia (folio 90 verso, line 12) matches the Latin texts.

552. Latin text of Milan: we are on folio 102 verso, line 34; Latin text of Bordeaux: we are on folio 95 recto, line 18. From this point onward, the text of Parma (folio 75 recto, line 2) matches the Latin texts.

553. In figures B.167 and B.168 all the numbers related to circuit H on paper ruler H have been reported.

554. The first part of this sentence is missing in Bordeaux.

555. Differing from what Piero affirms, we should remember that we have preferred to report all the numbers present on all the circuits and on all the rulers of the width and the heights.

556. A vertical median section of the head was realized in figure B.171 by using the twelve parts into which the head was subdivided, even if Piero does not hint at this section nor the twelve parts in the text. By observing the apparent contours in figures B.171 and B.172, we will notice the presence of the left ear, which is missing in figure B.173. In this figure, in fact, I reproduced, as far as was possible, the drawing on folio 76 verso of the manuscript of Parma in compliance with the original. To replace the missing left ear, the apparent alternative contours were realized: compare figures A.61, A.62, A.63, A.64 and the text in the Introduction under the heading "Second Head."

557. It is useful to point out that before this sentence, in the Parma manuscript, the following sentence appears: "[Bisogna] che tu facci prima la linea recta che sia SV, che sia di quantità che tu voi che sia larga la cupola, e sopra quella volgi il semicirculo che sia STV." In translation: "[You need] to draw first the straight line SV, which is as long as you want the dome to be; on the upper part of that line describe the semicircle STV."

558. In the drawings of the Latin and vernacular manuscripts, there are two lateral stripes that have not been mentioned in any part of the texts; these stripes appear on the outside of the dome's slice (figure B.175). The two stripes are dotted in this figure because they probably indicate the width of the vertical stripes of the coffers.

559. Naturally, the same process that was just applied on the left may be applied to the right, that is, on the opposite side.

560. The seven circles Piero mentions and invites us to draw are not reported in any of the Latin or vernacular manuscripts. It is possible to realize the construction of these circles through the following elementary theorem: Given two lines and a point on the bisecting line of the angle between the two lines, trace a circle that goes through that point and which is tangent to the two straight lines.

561. Successively Piero will introduce straight line R: for the identification of this line we will need to draw the eighth circle and point 9.

562. This second circuit is a quarter of an ellipse. Piero does not mention the construction of an ellipse. We know that the construction of an ellipse was realized by Federico Commandino (1509–1575), who scientifically explained it in his *De horologiorum descriptione*, published in Rome in 1552; compare Rocco Sinisgalli and Salvatore Vastola, *La rappresentazione degli orologi solari* (Florence: Cadmo Editrice, 1994). However, Piero could have known the scientific transformation of a circle into an ellipse, possibly from the tradition of antiquity, or he could have personally invented it.

563. Although the texts of the Latin and vernacular manuscript report "the first," we will say "the second," following the enumeration of the lines drawn again in the figures. In fact, in figure B.180, the first line corresponds to the horizontal line going through point A, while the second line corresponds to the horizontal line that goes through point B.

564. In all the drawings in the vernacular and Latin manuscripts, there are no lines or radii which touch point A.

565. Keep in mind that numerous sections of the coffers are neither drawn correctly nor present in the drawings in the vernacular and Latin manuscripts.

566. This is straight line R, mentioned in note 561.

567. On the circle of radius XA in plan or of radius AS in elevation rests the inner spherical cap of the dome. On the circle of radius XY in plan or of radius AY in elevation rests the outer (elliptical) cap of the dome. There are no points to be represented on this second circle.

568. Here, by "second figure" is meant the elevation of the dome. In all vernacular and Latin manuscripts, the elevation and the plan appear to be horizontally aligned without any vertical correspondence.

569. In addition to circle A, on which the inner spherical cap of the dome rests, there are the remaining sixteen circles which have lines B, C, D, E, F, G, H, I, K, L, M, N, O, P, Q, R as radii, until these lines intersect the relative arcs ST and YZ. Because circles A and B almost overlap, keep in mind that circle A is external with respect to circle B.

570. The Latin manuscripts of Milan and Bordeaux and the vernacular manuscript of Reggio Emilia introduce the concept of drawing the semicircles around X for the lines that follow. This specification is missing for circles L, N, O, Q. The manuscript of Parma continues to call these semicircles "circles."

571. The examination of the letters that has been carried out in the various manuscripts (both in the text and in the drawings) has brought us to indicate this semicircle with letters AST.

572. These seventeen points include point A, S, and T in circle AST.

573. These points of the coffers lie on the external vault of the dome. These points include point z but not points Y and Z on circle YzZ. To see these points on the vault, compare the images of circuits C, D, G, H, L, M, P, Q (figure B.198), in which all points 1 and 17 are external to the frontal circle of the vault. For this reason, they do not appear in perspective and they have been excluded from the count.

574. We preferred the wording "we need point 1 to serve for 1 and 17," present in the Latin manuscripts of Bordeaux, Paris, and London, rather than "we need point A to serve for 1 and for 17," present in all other manuscripts.

575. In figures B.185 and B.186, we have limited ourselves to representing a detail of the dome in plan and in elevation. Naturally, only numbers from 1 to 9 have been reported. We will imagine the remaining numbers 10, 11, 12, 13, 14, 15, 16, 17 in their appropriate location, by observing figure B.184. In figures B.187 and B.188, we offer an isometric view of the dome in its entirety and a cross section of the same dome with all the lines from A to R.

576. We prefer "9-R" in Bordeaux, London, and Paris. The vernacular of Parma shows number 9 followed by an unreadable letter.

577. The Latin manuscripts report "Sed ponamus ut unius ulne spatio contineatur" which means "But let's suppose [that the distance of the terminal from SA] is contained in the size of a braccio" (a typical unit of measurement at the time; see note 24). Even in Reggio we have "Let's suppose that it is a braccio." However, in Parma we only have "quanto a te piace" (which means "as much as you would like it to be").

578. The distance between line 9-R (the terminal) and point O (the eye in plan) should be equal to 10 braccia, which is also reported in the two vernacular manuscripts. As may also be seen in the drawings of all the vernacular and Latin manuscripts, distance OM is more than 30 times the distance between the terminal 9-R and line SA. For this reason, the distance between terminal 9-R and line SA cannot be one braccio, but it should be "as much as you would like it to be."

579. Although all the texts of the Latin and vernacular manuscripts report X instead of M, because this point is marked with the letter M on all the drawings of all the Latin and vernacular manuscripts, we prefer to indicate it with an M, and this shall be done all other times that X appears.

580. As may be seen in figure B.183, circle B is internal with respect to circle A. All the numbers present on wooden ruler B may be observed in figure B.191.

581. All the circles in plan may be observed in figure B.183. All the numbers on all the wooden rulers may be observed in figure B.191.

582. At this point, the Latin manuscripts have "ulne spatio," which means "at a braccio's distance"; the same sentence appears in the vernacular of Reggio Emilia. This is an incongruence that could make us think about a mistake in the drawings of the manuscripts, because the distance between AT and 9-R cannot be equal to the tenth part between point 9 (i.e., terminal 9-R goes through this point) and eye O. The Parma manuscript becomes useful here because we do not have the sentence "at a braccio's distance."

583. This is point O of the eye in elevation.

584. Make sure that on the lines of figure B.185 there are only numbers 1, 2, 3, 4, 5, 6, 7, 8, 9, which we should imagine to be 1-17, 2-16, 3-15, 4-14, 5-13, 6-12, 7-11, 8-10, 9.

585. This means the dome should be represented on the dome.

586. Bordeaux states: "Divide SR into two equal parts at point M and trace a perpendicular line through Y above M: you will have line MY′." We preferred the letters X, Y, and XY present in Milan, London, Paris, Parma.

587. The manuscripts of Parma, Milan, London, and Paris report X instead of M. Bordeaux and Reggio Emilia report M. Because we see M and not X on all the wooden rulers in the drawings of all manuscripts, we prefer M.

588. Here and below, when we see "the ruler," what is intended is "the wooden ruler"; when we see "the two rulers," what is intended is "the paper rulers."

589. On the terminal, that is, the picture plane. In all Latin and vernacular manuscripts, on the folios where the perspective of the dome is reproduced, the paper rulers q and the wooden ruler Q do not appear. In the Latin manuscripts of Bordeaux, London, and Paris, under the wooden ruler R at the bottom, the words "Q non abbisogna" appear. These words mean "the Q is not necessary." This is undoubtedly due to the fact that the copyists believed that the penultimate curve Q at the top did not appear because, in perspective, it was covered or coincided with the last curve R at the top, and that therefore it was not necessary to draw it. And indeed, in the dome perspective present in all the manuscripts, no graphic indication of this penultimate curve Q appears.

590. The frontal circle of the vault, in perspective, has as a diameter the distance from 1 to 17 of wooden ruler A (figure B.190); therefore we only need to join points 1 and 17 (figure B.198) of circuit A to find the center of the frontal circle. Points 1 and 17 of circuits C, D, G, H, L, M, P, Q are external to the frontal circle and do not appear in perspective.

591. In Milan we have a repetition of points 2 which are related to C and D, G and H, L and M, P and Q. This repetition is not present in Bordeaux, London, and Paris.

592. Milan, Bordeaux, London, Paris report M. Reggio Emilia and Parma report N, which is the correct letter.

593. The Latin manuscripts of Milan and Bordeaux and the vernacular manuscript of Reggio Emilia stop at 8-9. Parma also reports numbers 10, 11, 12, 13, 14, 15, which have been added to the text, but incorrectly prompts us to draw the lines from 10 to 11, from 12 to 13, from 14 to 15. We prefer to add the specification "this is circuit B" to the text; this sentence is present in Parma.

594. In figure B.200, we reproduced the last drawing in compliance with the original drawing of Parma folio 82 recto, as far as possible. Please notice how in the representation of the coffers, the lines that join all vertices of the single squares are missing. Concerning this, compare figure B.199 (bottom) where these single segments have been drawn. This absence (compare figure B.200) occurs in all the perspectives of all the vernacular and Latin manuscripts.

595. Piero encourages us to draw "the cornice." These are suggestions to keep in mind, starting from the semicircle STV in figures B.174 and B.176, for the best accomplishment of the entire construction process.

596. The dotted circle is missing in the drawings of the Latin and vernacular manuscripts. To complete the good shape of the apparent contour of the sphere that Piero invites us to pursue, it is also necessary to project the circle that goes through the tip of the arrow: this tip is identified by the tangent that goes from A to the vertical circle DM.

597. As we have already specified, the *De aspectuum diversitate* corresponds to the *Optics* of Euclid.

598. Points 9* and 21 do not appear in the drawings of all the Latin and vernacular manuscripts.

599. This circle does not appear in the drawings of all the Latin and vernacular manuscripts.

600. This second vertical line does not appear in the drawings of all the Latin and vernacular manuscripts.

601. The ring lies on point E, which divides the ring in half.

602. The Latin manuscripts of Bordeaux, Paris, and London have the term "Finis" as the last word of the text, which is missing in the Latin manuscript of Milan and in the vernacular manuscripts of both Reggio Emilia and Parma.

603. This dedication and what follows are missing in Reggio Emilia. The texts appear in Milan, but the titles "To the Author" and "To the Reader" are not present.

604. The name Borgo is the first part of Piero della Francesca's native town. The town was called Borgo Sansepolcro; today it is simply called Sansepolcro.

Selected Bibliography

Editions of *De prospectiva pingendi*

Petrus Pictor Burgensis De prospectiva pingendi: nach dem Codex der Königlichen Bibliothek zu Parma. Edited by Costantin Winterberg. Strassburg: Heitz & Mündel, 1899.

De prospectiva pingendi. Edited by Giusta Nicco Fasola. Florence: Le Lettere, 1942, republished in 1984 with comments of Eugenio Battisti and Franco Ghione.

De prospectiva pingendi: A Facsimile of Parma, Biblioteca Palatina, MS 1576. Edited by Jane Andrews Aiken, Liana De Girolami Cheney, and Leonardo Farinelli. New York: Broude International Editions, 1992.

De prospectiva pingendi. De la perspective en peinture. Edited by Jean-Pierre Le Goff and Jean-Pierre Néraudau. Paris: Éditions in Medias Res, 1998.

De prospectiva pingendi. Photographic facsimile of the Reggio Emilia Codex, edited by Massimo Mussini and Luigi Grasselli. Sansepolcro: Aboca Edizioni, 2008.

De prospectiva pingendi. Edited by Chiara Gizzi. Venice: Edizioni Ca' Foscari, 2016.

De prospectiva pingendi, Tomo I: Testo volgare, Edizione critica del testo volgare; Tomo II: Disegni, Edizione critica dei disegni; Tomo III: Stampa anastatica del codice di Parma 1576, Biblioteca Palatina di Parma. Edizione Nazionale degli Scritti di Piero della Francesca III A. Rome: Istituto Poligrafico e Zecca dello Stato, 2016.

De prospectiva pingendi, Tomo I: Testo latino, Edizione critica del testo latino; Tomo II: Disegni, Edizione critica dei disegni; Tomo III: Stampa anastatica del codice 616, Bibliothèque Municipale de Bordeaux. Edizione Nazionale degli Scritti di Piero della Francesca III B. Rome: Istituto Poligrafico e Zecca dello Stato, 2017.

Selected Books and Articles

Alberti, Leon Battista. *On Painting*. Trans. Cecil Grayson, intro. Martin Kemp. London: Penguin, 1991.

Alberti, Leon Battista. *On Painting: A New Translation and Critical Edition*. Ed. and trans. Rocco Sinisgalli. New York: Cambridge University Press, 2011.

Andersen, Kirsti. *The Geometry of an Art: The History of the Mathematical Theory of Perspective from Alberti to Monge*. New York: Springer, 2007.

Banker, James R. *Piero della Francesca: Artist and Man*. Oxford: Oxford University Press, 2014.

Banker, James R., "Piero della Francesca's Friend and Translator: Maestro Matteo di ser Paolo d'Anghiari." *Rivista d'Arte. Studi documentari per la storia delle arti in Toscana* (1992).

Barbaro, Daniele. *La pratica della perspettiva*. Venice: appresso Camillo e Rutilio Borgominieri, 1569.

Battisti, Eugenio. *Piero della Francesca*. Milan: Istituto Editoriale Italiano, 1971–1972.

Bouleau, Charles. *Charpentes. La géométrie secrète des peintres*. Paris: Éditions du Seuil, 1963.

Calvesi, Maurizio. *Piero della Francesca nel XV e nel XX secolo*. Rome: Lithos, 1994.

Campisi, Michele. *L'impresentabile storia. Misfatti, delitti e cronache su monumenti, collezioni e antichità*. Rome: Gangemi Editore International, 2018.

Clark, Kenneth. *Piero della Francesca*. London: Phaidon, 1969.

Dalai Emiliani, Marisa, and Valter Curzi, eds. *Piero della Francesca tra arte e scienza*. Venice: Marsilio, 1996.

Davis, Margaret Daly. *Piero della Francesca's Mathematical Treatises: The "Trattato d'abaco" and "Libellus de quinque corporibus regularibus."* Ravenna: Longo, 1977.

Derenzini, Giovanna. "Note autografe di Piero della Francesca nel Codice 616 della Bibliothèque Municipale di Bordeaux. Per la storia testuale del '*De prospectiva pingendi*.'" *Filologia Antica e Moderna* 9 (1995), 29–55.

Dürer, Albrecht. *Alberti Dureri clarissimi pictoris et geometrae De simmetria partium in rectis formis humanorum corporum libri in latinum conversi (a Cristophoro Colero)*. Nuremburg: in aedib. Viduae Durerianae, 1532.

Dürer, Albrecht. *Albertus Durerus institutionum geometricarum libri quatuor, lineas, superficies et solida corpora tractavit, adhitis designationibus ad eam rem accomodatis*. Paris: Wechel, 1532.

Euclid. *Euclidis Optica, Opticorum recensio theonis, Catoptrica, cum scholiis antiquis*. Vol. VII of *Euclidis opera omnia*, ed. J. L. Heiberg and H. Menge. Leipzig: B. G. Teubner, 1895.

Euclid. *Euclid's Elements of Geometry*. The Greek text of J. L. Heiberg (Leipzig: B. G. Teubner, 1883–1885), ed. and trans. Richard Fitzpatrick. London, 2008.

Federici-Vescovini, Graziella. "Contributo per la storia di Alhazen in Italia: il volgarizzamento del Ms. Vaticano 4595 ed il Commentario Terzo del Ghiberti." *Rinascimento* 5 (1965), 17–49.

Field, Judith Veronica. *Piero della Francesca: A Mathematician's Art*. New Haven: Yale University Press, 2005.

Gilson, Simon A. *Medieval Optics and Theories of Light in the Works of Dante*. Lewiston, NY: E. Mellen Press, 2000.

Kemp, Martin. *The Science of Art: Optical Themes in Western Art from Brunelleschi to Seurat*. New Haven: Yale University Press, 1990.

Lavin, Marilyn Aronberg. *Piero della Francesca*. New York: Abrams, 1992.

Lavin, Marilyn Aronberg. *Piero della Francesca and His Legacy*. Washington, DC: National Gallery of Art; Hanover, NH: University Press of New England, 1995.

Leonardo da Vinci. *Scritti scelti di Leonardo da Vinci*. Ed. Anna Maria Brizio. Turin: Unione Tipografico-Editrice Torinese, 1980.

Lévy, Bernard-Henri. *Piero della Francesca*. Paris: Editions de la Différence, 1992.

Lightbown, Ronald. *Piero della Francesca*. New York: Abbeville Press, 1992.

Lindberg, David C. *Theories of Vision from Al-Kindi to Kepler*. Chicago: University of Chicago Press, 1976.

Loria, Gino. *Storia della geometria descrittiva dalle origini sino ai giorni nostril*. Milan: Ulrico Hoepli, 1921.

Loria, Gino. *Storia delle matematiche*. Milan: Ulrico Hoepli, 1929.

Manetti, Antonio di Tuccio. *The Life of Brunelleschi*. Ed. Howard Saalman, trans. Catherine Enggass. University Park: Pennsylvania State University Press, 1970.

Manetti, Antonio di Tuccio. *Vita di Filippo Brunelleschi, preceduta da la novella del Grasso*. Ed. Domenico De Robertis, intro. Giuliano Tanturli. Milan: Il Polifilo, 1976.

Pacioli, Luca. *Summa de arithmetica*. Venice: Paganini, 1494.

Panti, Cecilia. "The Commentario Terzo by Lorenzo Ghiberti and His volgarizzamenti of Vitruvius, Alhazen and the Medieval Perspectivists." *Micrologus* 29 (2021), 121–156.

Pittarelli, G[iulio]. "Intorno al libro '*De prospectiva pingendi*' di Piero deì Franceschi." In *Atti del Congresso Internazionale di Scienze Storiche* 12, section 8a (Rome, 1904).

Pittarelli, G[iulio]. "Luca Pacioli usurpò per se stesso qualche libro di Piero deì Franceschi?" In *Atti del IV Congresso Internazionale dei Matematici, Roma 6–11 Aprile 1908*, vol. 3 (Rome: Tipografia della Regia Accademia dei Lincei, 1909).

Poudra, Noël-Germinal. *Histoire de la perspective ancienne et moderne*. Paris: J. Corréard, 1864.

Schmarsow, August. *Melozzo da Forlì*. Berlin and Stuttgart, 1886.

Sinisgalli, Rocco. *Il contributo di Simon Stevin allo sviluppo scientifico della prospettiva artificiale*. Rome: L'Erma di Bretschneider, 1978.

Sinisgalli, Rocco. *Per la storia della prospettiva (1405–1605)*. Rome: L'Erma di Bretschneider, 1978.

Sinisgalli, Rocco. *Perspective in the Visual Culture of Classical Antiquity*. New York: Cambridge University Press, 2012.

Sinisgalli, Rocco. *Piero della Francesca: De prospectiva pingendi*. Milan: Silvana Editoriale, 2021.

Sinisgalli, Rocco. *Playing with Piero della Francesca: Geometry at the Service of the Arts*. Certaldo (FI): Federighi Editori, 2012.

Sinisgalli, Rocco. *I sei libri della Prospettiva di Guidobaldo dei marchesi Del Monte dal latino tradotti interpretati e commentati*. Rome: L'Erma di Bretschneider, 1984.

Sinisgalli, Rocco. *Verso una storia organica della prospettiva*. Rome: Kappa, 2000.

Sinisgalli, Rocco, and Lasse Hodne. *Il Battistero di Firenze*. Florence: Cadmo Editrice, 2000.

Sinisgalli, Rocco, and Salvatore Vastola. *La rappresentazione degli orologi solari*. Florence: Cadmo Editrice, 1994.

Smith, A. Mark. *From Sight to Light: The Passage from Ancient to Modern Optics*. Chicago: University of Chicago Press, 2014.

Vagnetti, Luigi. *De naturali et artificiali perspectiva*. Florence: Libreria Editrice Fiorentina, 1979.

Vagnetti, Luigi. "Riflessioni sul '*De prospectiva pingendi*.'" *Commentari* (Florence) 26, no. 1–2 (1975).

Vasari, Giorgio. *Lives of the Most Eminent Painters, Sculptors and Architects*. Trans. Gaston Du C. De Vere: London: Macmillan; Medici Society, 1912–1914.

Vasari, Giorgio. *Le vite dei più eccellenti scultori, pittori e architetti*. Forlì: Orsa Maggiore Editrice, 1991.

Veltman, Kim H. *Studies on Leonardo da Vinci I. Linear Perspective and the Visual Dimensions of Science and Art*. In collaboration with K. D. Keele. Munich: Deutscher Kunstverlag, 1986.

Veltman, Kim H. "Visualisation and Perspective." In *Leonardo e l'età della Ragione*, ed. E. Bellone and P. Rossi, 185–223. Milan, 1982.

Venturi, Adolfo. *Piero della Francesca*. Florence: G. & P. Alinari, 1922.

Vignola, Jacopo Barozzi da. *Le regole della prospettiva pratica*. Bologna, 1582.

Vitruvius. *De architectura*. Ed. Pierre Gros, trans. and commentary by Antonio Corso and Elisa Romano. Turin: Einaudi, 1997.

Vitruvius. *Ten Books on Architecture*, Book III. Trans. Ingrid D. Rowland, with commentary by others. New York: Cambridge University Press, 1999.

Witham, Larry. *Piero's Light*. New York: Pegasus Books, 2014.